Mathematical Statistics
Basic Ideas and Selected Topics
Volume I

Second Edition

Second Edition

Mathematical Statistics
Basic Ideas and Selected Topics
Volume I

Peter J. Bickel
University of California

Kjell A. Doksum
University of Wisconsin

PEARSON
Prentice
Hall

Upper Saddle River, New Jersey 07458

Library of Congress Cataloging-in-Publication Data

Bickel, Peter J.
    Mathematical statistics: basic ideas and selected topics / Peter J. Bickel, Kjell A.
    Doksum—2nd ed.
        p. cm.
    Updated printing.
    Includes bibliographical references and index.
    ISBN 0-13-230637-9(v. 1)
    1. Mathematical statistics. I. Doksum, Kjell A. II. Title.
    QA276.B47 2007
    519.5--dc22   2006044755

Executive Acquisition Editor: *Petra Recter*
Editor in Chief: *Sally Yagan*
Project Manager: *Michael Bell*
Production Editor: *Lynn Savino Wendel*
Executive Managing Editor: *Kathleen Schiaparelli*
Senior Managing Editor: *Linda Mihatov Behrens*
Manufacturing Buyer: *Maura Zaldivar*
Manufacturing Manager: *Alexis Heydt-Long*
Marketing Manager: *Wayne Parkins*
Marketing Assistant: *Jennifer de Leeuwerk*
Director of Marketing: *Patrice Jones*
Editorial Assistant: *Joanne Wendelken*
Art Director: *Jayne Conte*
Cover Design: *Karen Salzbach*

©2007, 2001, 1977 Pearson Education, Inc.
Pearson Prentice Hall
Pearson Education, Inc.
Upper Saddle River, New Jersey 07458

Pearson Prentice Hall™ is a trademark of Pearson Education, Inc.

Printed in the United States of America

10 9 8 7 6 5 4 3 2

ISBN: 0-13-230637-9

Pearson Education LTD., *London*
Pearson Education Australia PTY, Limited, *Sydney*
Pearson Education Singapore, Pte. Ltd
Pearson Education North Asia Ltd, *Hong Kong*
Pearson Education Canada, Ltd., *Toronto*
Pearson Educación de Mexico, S.A. de C.V.
Pearson Education—Japan, *Tokyo*
Pearson Education Malaysia, Pte. Ltd

To Erich L. Lehmann

CONTENTS

PREFACE TO THE SECOND
EDITION: VOLUME I

In the twenty-nine years that have passed since the first edition of our book appeared, statistics has changed enormously under the impact of several forces:

(1) The generation of what were once unusual types of data such as images, trees (phylogenetic and other), and other types of combinatorial objects.

(2) The generation of enormous amounts of data—terabytes (the equivalent of 10^{12} characters) for an astronomical survey over three years.

(3) The possibility of implementing computations of a magnitude that would have once been unthinkable.

The underlying sources of these changes have been the exponential change in computing speed (Moore's "law") and the development of devices (computer controlled) using novel instruments and scientific techniques (e.g., NMR tomography, gene sequencing). These techniques often have a strong intrinsic computational component. Tomographic data are the result of mathematically based processing. Sequencing is done by applying computational algorithms to raw gel electrophoresis data.

As a consequence the emphasis of statistical theory has shifted away from the small sample optimality results that were a major theme of our book in a number of directions:

(1) Methods for inference based on larger numbers of observations and minimal assumptions—asymptotic methods in non- and semiparametric models, models with "infinite" number of parameters.

(2) The construction of models for time series, temporal spatial series, and other complex data structures using sophisticated probability modeling but again relying for analytical results on asymptotic approximation. Multiparameter models are the rule.

(3) The use of methods of inference involving simulation as a key element such as the bootstrap and Markov Chain Monte Carlo.

xiii

(4) The development of techniques not describable in "closed mathematical form" but rather through elaborate algorithms for which problems of existence of solutions are important and far from obvious.

(5) The study of the interplay between numerical and statistical considerations. Despite advances in computing speed, some methods run quickly in real time. Others do not and some though theoretically attractive cannot be implemented in a human lifetime.

(6) The study of the interplay between the number of observations and the number of parameters of a model and the beginnings of appropriate asymptotic theories.

There have, of course, been other important consequences such as the extensive development of graphical and other exploratory methods for which theoretical development and connection with mathematics have been minimal. These will not be dealt with in our work.

As a consequence our second edition, reflecting what we now teach our graduate students, is much changed from the first. Our one long book has grown to two volumes, each to be only a little shorter than the first edition.

Volume I, which we present in 2006, covers material we now view as important for all beginning graduate students in statistics and science and engineering graduate students whose research will involve statistics intrinsically rather than as an aid in drawing conclusions.

In this edition we pursue our philosophy of describing the basic concepts of mathematical statistics relating theory to practice. However, our focus and order of presentation have changed.

Volume I covers the material of Chapters 1–6 and Chapter 10 of the first edition with pieces of Chapters 7–9 and includes Appendix A on basic probability theory. However, Chapter 1 now has become part of a larger Appendix B, which includes more advanced topics from probability theory such as the multivariate Gaussian distribution, weak convergence in Euclidean spaces, and probability inequalities as well as more advanced topics in matrix theory and analysis. The latter include the principal axis and spectral theorems for Euclidean space and the elementary theory of convex functions on R^d as well as an elementary introduction to Hilbert space theory. As in the first edition, we do not require measure theory but assume from the start that our models are what we call "regular." That is, we assume either a discrete probability whose support does not depend on the parameter set, or the absolutely continuous case with a density. Hilbert space theory is not needed, but for those who know this topic Appendix B points out interesting connections to prediction and linear regression analysis.

Appendix B is as self-contained as possible with proofs of most statements, problems, and references to the literature for proofs of the deepest results such as the spectral theorem. The reason for these additions are the changes in subject matter necessitated by the current areas of importance in the field.

Specifically, instead of beginning with parametrized models we include from the start non- and semiparametric models, then go to parameters and parametric models stressing the role of identifiability. From the beginning we stress function-valued parameters, such as the density, and function-valued statistics, such as the empirical distribution function. We also, from the start, include examples that are important in applications, such as regression experiments. There is more material on Bayesian models and analysis. Save for these changes of emphasis the other major new elements of Chapter 1, which parallels Chapter 2 of the first edition, are an extended discussion of prediction and an expanded introduction to k-parameter exponential families. These objects that are the building blocks of most modern models require concepts involving moments of random vectors and convexity that are given in Appendix B.

Chapter 2 of this edition parallels Chapter 3 of the first and deals with estimation. Major differences here are a greatly expanded treatment of maximum likelihood estimates (MLEs), including a complete study of MLEs in canonical k-parameter exponential families. Other novel features of this chapter include a detailed analysis including proofs of convergence of a standard but slow algorithm for computing MLEs in multiparameter exponential families and an introduction to the EM algorithm, one of the main ingredients of most modern algorithms for inference. Chapters 3 and 4 parallel the treatment of Chapters 4 and 5 of the first edition on the theory of testing and confidence regions, including some optimality theory for estimation as well and elementary robustness considerations. The main difference in our new treatment is the downplaying of unbiasedness both in estimation and testing and the presentation of the decision theory of Chapter 10 of the first edition at this stage.

Chapter 5 of the new edition is devoted to asymptotic approximations. It includes the initial theory presented in the first edition but goes much further with proofs of consistency and asymptotic normality and optimality of maximum likelihood procedures in inference. Also new is a section relating Bayesian and frequentist inference via the Bernstein–von Mises theorem.

Finally, Chapter 6 is devoted to inference in multivariate (multiparameter) models. Included are asymptotic normality of maximum likelihood estimates, inference in the general linear model, Wilks theorem on the asymptotic distribution of the likelihood ratio test, the Wald and Rao statistics and associated confidence regions, and some parallels to the optimality theory and comparisons of Bayes and frequentist procedures given in the univariate case in Chapter 5. Generalized linear models are introduced as examples. Robustness from an asymptotic theory point of view appears also. This chapter uses multivariate calculus in an intrinsic way and can be viewed as an essential prerequisite for the more advanced topics of Volume II.

As in the first edition, problems play a critical role by elucidating and often substantially expanding the text. Almost all the previous ones have been kept with an approximately equal number of new ones added—to correspond to our new topics and point of view. The conventions established on footnotes and notation in the first edition remain, if somewhat augmented.

Chapters 1–4 develop the basic principles and examples of statistics. Nevertheless, we star sections that could be omitted by instructors with a classical bent and others that could be omitted by instructors with more computational emphasis. Although we believe the material of Chapters 5 and 6 has now become fundamental, there is clearly much that could be omitted at a first reading that we also star. There are clear dependencies between starred sections that follow.

$$5.4.2 \ \rightarrow \ 5.4.3 \ \rightarrow \ 6.2 \ \rightarrow \ 6.3 \ \rightarrow \ 6.4 \ \rightarrow \ 6.5$$
$$\llcorner\rightarrow \ 6.6$$

Volume II is expected to be forthcoming in 2010. Topics to be covered include permutation and rank tests and their basis in completeness and equivariance. Examples of application such as the Cox model in survival analysis, other transformation models, and the classical nonparametric k sample and independence problems will be included. Semiparametric estimation and testing will be considered more generally, greatly extending the material in Chapter 8 of the first edition. The topic presently in Chapter 8, density estimation, will be studied in the context of nonparametric function estimation. We also expect to discuss classification and model selection using the elementary theory of empirical processes. The basic asymptotic tools that will be developed or presented, in part in the text and, in part in appendices, are weak convergence for random processes, elementary empirical process theory, and the functional delta method.

A final major topic in Volume II will be Monte Carlo methods such as the bootstrap and Markov Chain Monte Carlo.

With the tools and concepts developed in this second volume students will be ready for advanced research in modern statistics.

For the first volume of the second edition we would like to add thanks to new colleagues, particularly Jianging Fan, Michael Jordan, Jianhua Huang, Ying Qing Chen, and Carl Spruill and the many students who were guinea pigs in the basic theory course at Berkeley. We also thank Faye Yeager for typing, Michael Ostland and Simon Cawley for producing the graphs, Yoram Gat for proofreading that found not only typos but serious errors, and Prentice Hall for generous production support.

Last and most important we would like to thank our wives, Nancy Kramer Bickel and Joan H. Fujimura, and our families for support, encouragement, and active participation in an enterprise that at times seemed endless, appeared gratifyingly ended in 1976 but has, with the field, taken on a new life.

Peter J. Bickel
bickel@stat.berkeley.edu
Kjell Doksum
doksum@stat.wisc.edu

PREFACE TO THE FIRST EDITION

This book presents our view of what an introduction to mathematical statistics for students with a good mathematics background should be. By a good mathematics background we mean linear algebra and matrix theory and advanced calculus (but no measure theory). Because the book is an introduction to statistics, we need probability theory and expect readers to have had a course at the level of, for instance, Hoel, Port, and Stone's *Introduction to Probability Theory*. Our appendix does give all the probability that is needed. However, the treatment is abridged with few proofs and no examples or problems.

We feel such an introduction should at least do the following:

(1) Describe the basic concepts of mathematical statistics indicating the relation of theory to practice.

(2) Give careful proofs of the major "elementary" results such as the Neyman–Pearson lemma, the Lehmann–Scheffé theorem, the information inequality, and the Gauss–Markoff theorem.

(3) Give heuristic discussions of more advanced results such as the large sample theory of maximum likelihood estimates, and the structure of both Bayes and admissible solutions in decision theory. The extent to which holes in the discussion can be patched and where patches can be found should be clearly indicated.

(4) Show how the ideas and results apply in a variety of important subfields such as Gaussian linear models, multinomial models, and nonparametric models.

Although there are several good books available for this purpose, we feel that none has quite the mix of coverage and depth desirable at this level. The work of Rao, *Linear Statistical Inference and Its Applications*, 2nd ed., covers most of the material we do and much more but at a more abstract level employing measure theory. At the other end of the scale of difficulty for books at this level is the work of Hogg and Craig, *Introduction to Mathematical Statistics*, 3rd ed. These authors also discuss most of the topics we deal with but in many instances do not include detailed discussion of topics we consider essential such as existence and computation of procedures and large sample behavior.

Our book contains more material than can be covered in two quarters. In the two-quarter courses for graduate students in mathematics, statistics, the physical sciences, and engineering that we have taught we cover the core Chapters 2 to 7, which go from modeling through estimation and testing to linear models. In addition we feel Chapter 10 on decision theory is essential and cover at least the first two sections. Finally, we select topics from

Chapter 8 on discrete data and Chapter 9 on nonparametric models.

Chapter 1 covers probability theory rather than statistics. Much of this material unfortunately does not appear in basic probability texts but we need to draw on it for the rest of the book. It may be integrated with the material of Chapters 2–7 as the course proceeds rather than being given at the start; or it may be included at the end of an introductory probability course that precedes the statistics course.

A special feature of the book is its many problems. They range from trivial numerical exercises and elementary problems intended to familiarize the students with the concepts to material more difficult than that worked out in the text. They are included both as a check on the student's mastery of the material and as pointers to the wealth of ideas and results that for obvious reasons of space could not be put into the body of the text.

Conventions: (i) In order to minimize the number of footnotes we have added a section of comments at the end of each chapter preceding the problem section. These comments are ordered by the section to which they pertain. Within each section of the text the presence of comments at the end of the chapter is signaled by one or more numbers, 1 for the first, 2 for the second, and so on. The comments contain digressions, reservations, and additional references. They need to be read only as the reader's curiosity is piqued.

(i) Various notational conventions and abbreviations are used in the text. A list of the most frequently occurring ones indicating where they are introduced is given at the end of the text.

(iii) Basic notation for probabilistic objects such as random variables and vectors, densities, distribution functions, and moments is established in the appendix.

We would like to acknowledge our indebtedness to colleagues, students, and friends who helped us during the various stages (notes, preliminary edition, final draft) through which this book passed. E. L. Lehmann's wise advice has played a decisive role at many points. R. Pyke's careful reading of a next-to-final version caught a number of infelicities of style and content. Many careless mistakes and typographical errors in an earlier version were caught by D. Minassian who sent us an exhaustive and helpful listing. W. Carmichael, in proofreading the final version, caught more mistakes than both authors together. A serious error in Problem 2.2.5 was discovered by F. Scholz. Among many others who helped in the same way we would like to mention C. Chen, S. J. Chou, G. Drew, C. Gray, U. Gupta, P. X. Quang, and A. Samulon. Without Winston Chow's lovely plots Section 9.6 would probably not have been written and without Julia Rubalcava's impeccable typing and tolerance this text would never have seen the light of day.

We would also like to thank the colleagues and friends who inspired and helped us to enter the field of statistics. The foundation of our statistical knowledge was obtained in the lucid, enthusiastic, and stimulating lectures of Joe Hodges and Chuck Bell, respectively. Later we were both very much influenced by Erich Lehmann whose ideas are strongly reflected in this book.

Peter J. Bickel
Kjell Doksum

Berkeley
1976

Mathematical Statistics
Basic Ideas and Selected Topics
Volume I

Second Edition

Chapter 1

STATISTICAL MODELS, GOALS, AND PERFORMANCE CRITERIA

1.1 DATA, MODELS, PARAMETERS AND STATISTICS

1.1.1 Data and Models

Most studies and experiments, scientific or industrial, large scale or small, produce data whose analysis is the ultimate object of the endeavor.

Data can consist of:

(1) Vectors of scalars, measurements, and/or characters, for example, a single time series of measurements.

(2) Matrices of scalars and/or characters, for example, digitized pictures or more routinely measurements of covariates and response on a set of n individuals—see Example 1.1.4 and Sections 2.2.1 and 6.1.

(3) Arrays of scalars and/or characters as in contingency tables—see Chapter 6—or more generally multifactor multiresponse data on a number of individuals.

(4) All of the above and more, in particular, functions as in signal processing, trees as in evolutionary phylogenies, and so on.

The goals of science and society, which statisticians share, are to draw useful information from data using everything that we know. The particular angle of mathematical statistics is to view data as the outcome of a random experiment that we model mathematically.

A detailed discussion of the appropriateness of the models we shall discuss in particular situations is beyond the scope of this book, but we will introduce general model diagnostic tools in Volume 2, Chapter 1. Moreover, we shall parenthetically discuss features of the sources of data that can make apparently suitable models grossly misleading. A generic source of trouble often called *gross errors* is discussed in greater detail in the section on robustness (Section 3.5.3). In any case all our models are generic and, as usual, "The Devil is in the details!" All the principles we discuss and calculations we perform should only be suggestive guides in successful applications of statistical analysis in science and policy. Subject matter specialists usually have to be principal guides in model formulation. A

1

priori, in the words of George Box (1979), "Models of course, are never true but fortunately it is only necessary that they be useful."

In this book we will study how, starting with tentative models:

(1) We can conceptualize the data structure and our goals more precisely. We begin this in the simple examples that follow and continue in Sections 1.2–1.5 and throughout the book.

(2) We can derive methods of extracting useful information from data and, in particular, give methods that assess the generalizability of experimental results. For instance, if we observe an effect in our data, to what extent can we expect the same effect more generally? Estimation, testing, confidence regions, and more general procedures will be discussed in Chapters 2–4.

(3) We can assess the effectiveness of the methods we propose. We begin this discussion with decision theory in Section 1.3 and continue with optimality principles in Chapters 3 and 4.

(4) We can decide if the models we propose are approximations to the mechanism generating the data adequate for our purposes. Goodness of fit tests, robustness, and diagnostics are discussed in Volume 2, Chapter 1.

(5) We can be guided to alternative or more general descriptions that might fit better. Hierarchies of models are discussed throughout.

Here are some examples:

(a) We are faced with a population of N elements, for instance, a shipment of manufactured items. An unknown number $N\theta$ of these elements are defective. It is too expensive to examine all of the items. So to get information about θ, a sample of n is drawn without replacement and inspected. The data gathered are the number of defectives found in the sample.

(b) We want to study how a physical or economic feature, for example, height or income, is distributed in a large population. An exhaustive census is impossible so the study is based on measurements and a sample of n individuals drawn at random from the population. The population is so large that, for modeling purposes, we approximate the actual process of sampling without replacement by sampling with replacement.

(c) An experimenter makes n independent determinations of the value of a physical constant μ. His or her measurements are subject to random fluctuations (error) and the data can be thought of as μ plus some random errors.

(d) We want to compare the efficacy of two ways of doing something under similar conditions such as brewing coffee, reducing pollution, treating a disease, producing energy, learning a maze, and so on. This can be thought of as a problem of comparing the efficacy of two methods applied to the members of a certain population. We run $m + n$ independent experiments as follows: $m + n$ members of the population are picked at random and m of these are assigned to the first method and the remaining n are assigned to the second method. In this manner, we obtain one or more quantitative or qualitative measures of efficacy from each experiment. For instance, we can assign two drugs, A to m, and B to n, randomly selected patients and then measure temperature and blood pressure, have the patients rated qualitatively for improvement by physicians, and so on. Random variability here would come primarily from differing responses among patients to the same drug but also from error in the measurements and variation in the purity of the drugs.

We shall use these examples to arrive at our formulation of statistical models and to indicate some of the difficulties of constructing such models. First consider situation (a), which we refer to as:

Example 1.1.1. *Sampling Inspection.* The mathematical model suggested by the description is well defined. A random experiment has been performed. The sample space consists of the numbers $0, 1, \ldots, n$ corresponding to the number of defective items found. On this space we can define a random variable X given by $X(k) = k, k = 0, 1, \ldots, n$. If $N\theta$ is the number of defective items in the population sampled, then by (A.13.6)

$$P[X = k] = \frac{\binom{N\theta}{k} \binom{N - N\theta}{n - k}}{\binom{N}{n}} \tag{1.1.1}$$

$$\text{if } \max(n - N(1 - \theta), 0) \leq k \leq \min(N\theta, n).$$

Thus, X has an hypergeometric, $\mathcal{H}(N\theta, N, n)$ distribution.

The main difference that our model exhibits from the usual probability model is that $N\theta$ is *unknown* and, in principle, can take on any value between 0 and N. So, although the sample space is well defined, we cannot specify the probability structure completely but rather only give a family $\{\mathcal{H}(N\theta, N, n)\}$ of probability distributions for X, any one of which could have generated the data actually observed. □

Example 1.1.2. *Sample from a Population. One-Sample Models.* Situation (b) can be thought of as a generalization of (a) in that a quantitative measure is taken rather than simply recording "defective" or not. It can also be thought of as a limiting case in which $N = \infty$, so that sampling with replacement replaces sampling without. Formally, if the measurements are scalar, we observe $x_1, \ldots, x_n$, which are modeled as realizations of $X_1, \ldots, X_n$ independent, identically distributed (i.i.d.) random variables with common unknown distribution function F. We often refer to such $X_1, \ldots, X_n$ as a *random sample* from F, and also write that $X_1, \ldots, X_n$ are i.i.d. as X with $X \sim F$, where "$\sim$" stands for "is distributed as." The model is fully described by the set $\mathcal{F}$ of distributions that we specify. The same model also arises naturally in situation (c). Here we can write the n determinations of μ as

$$X_i = \mu + \epsilon_i, \ 1 \leq i \leq n \tag{1.1.2}$$

where $\epsilon = (\epsilon_1, \ldots, \epsilon_n)^T$ is the vector of random errors. What should we assume about the distribution of ϵ, which together with μ completely specifies the joint distribution of $X_1, \ldots, X_n$? Of course, that depends on how the experiment is carried out. Given the description in (c), we postulate

(1) The value of the error committed on one determination does not affect the value of the error at other times. That is, $\epsilon_1, \ldots, \epsilon_n$ are independent.

(2) The distribution of the error at one determination is the same as that at another. Thus, $\epsilon_1, \ldots, \epsilon_n$ are identically distributed.

(3) The distribution of ϵ is independent of μ.

Equivalently $X_1, \ldots, X_n$ are a random sample and, if we let G be the distribution function of ϵ_1 and F that of X_1, then

$$F(x) = G(x - \mu) \tag{1.1.3}$$

and the model is alternatively specified by $\mathcal{F}$, the set of F's we postulate, or by $\{(\mu, G) : \mu \in R,\ G \in \mathcal{G}\}$ where $\mathcal{G}$ is the set of all allowable error distributions that we postulate. Commonly considered $\mathcal{G}$'s are all distributions with center of symmetry 0, or alternatively all distributions with expectation 0. The classical default model is:

(4) The common distribution of the errors is $\mathcal{N}(0, \sigma^2)$, where σ^2 is unknown. That is, the X_i are a sample from a $\mathcal{N}(\mu, \sigma^2)$ population or equivalently $\mathcal{F} = \{\Phi\left(\frac{\cdot - \mu}{\sigma}\right) : \mu \in R,\ \sigma > 0\}$ where Φ is the standard normal distribution. □

This default model is also frequently postulated for measurements taken on units obtained by random sampling from populations, for instance, heights of individuals or log incomes. It is important to remember that these are assumptions at best only approximately valid. All actual measurements are discrete rather than continuous. There are absolute bounds on most quantities—100 ft high men are impossible. Heights are always nonnegative. The Gaussian distribution, whatever be μ and σ, will have none of this.

Now consider situation (d).

Example 1.1.3. *Two-Sample Models.* Let $x_1, \ldots, x_m$; $y_1, \ldots, y_n$, respectively, be the responses of m subjects having a given disease given drug A and n other similarly diseased subjects given drug B. By convention, if drug A is a standard or placebo, we refer to the x's as *control observations*. A placebo is a substance such as water that is expected to have no effect on the disease and is used to correct for the well-documented placebo effect, that is, patients improve even if they only think they are being treated. We let the y's denote the responses of subjects given a new drug or treatment that is being evaluated by comparing its effect with that of the placebo. We call the y's *treatment observations*.

Natural initial assumptions here are:

(1) The x's and y's are realizations of $X_1, \ldots, X_m$ a sample from F, and $Y_1, \ldots, Y_n$ a sample from G, so that the model is specified by the set of possible (F, G) pairs.

To specify this set more closely the critical *constant treatment* effect assumption is often made.

(2) Suppose that if treatment A had been administered to a subject response x would have been obtained. Then if treatment B had been administered to the same subject instead of treatment A, response $y = x + \Delta$ would be obtained *where Δ does not depend on x*. This implies that if F is the distribution of a control, then $G(\cdot) = F(\cdot - \Delta)$. We call this the *shift model* with parameter Δ.

Often the final simplification is made.

(3) The control responses are normally distributed. Then if F is the $\mathcal{N}(\mu, \sigma^2)$ distribution and G is the $\mathcal{N}(\mu + \Delta, \sigma^2)$ distribution, we have specified the Gaussian two sample model with equal variances. □

How do we settle on a set of assumptions? Evidently by a mixture of experience and physical considerations. The advantage of piling on assumptions such as (1)–(4) of Example 1.1.2 is that, if they are true, we know how to combine our measurements to estimate μ in a highly efficient way and also assess the accuracy of our estimation procedure (Example 4.4.1). The danger is that, if they are false, our analyses, though correct for the model written down, may be quite irrelevant to the experiment that was actually performed. As our examples suggest, there is tremendous variation in the degree of knowledge and control we have concerning experiments.

In some applications we often have a tested theoretical model and the danger is small. The number of defectives in the first example clearly has a hypergeometric distribution; the number of α particles emitted by a radioactive substance in a small length of time is well known to be approximately Poisson distributed.

In others, we can be reasonably secure about some aspects, but not others. For instance, in Example 1.1.2, we can ensure independence and identical distribution of the observations by using different, equally trained observers with no knowledge of each other's findings. However, we have little control over what kind of distribution of errors we get and will need to investigate the properties of methods derived from specific error distribution assumptions when these assumptions are violated. This will be done in Sections 3.5.3 and 6.6.

Experiments in medicine and the social sciences often pose particular difficulties. For instance, in comparative experiments such as those of Example 1.1.3 the group of patients to whom drugs A and B are to be administered may be haphazard rather than a random sample from the population of sufferers from a disease. In this situation (and generally) it is important to *randomize*. That is, we use a random number table or other random mechanism so that the m patients administered drug A are a sample without replacement from the set of $m + n$ available patients. Without this device we could not know whether observed differences in drug performance might not (possibly) be due to unconscious bias on the part of the experimenter. All the severely ill patients might, for instance, have been assigned to B. The study of the model based on the minimal assumption of randomization is complicated and further conceptual issues arise. Fortunately, the methods needed for its analysis are much the same as those appropriate for the situation of Example 1.1.3 when F, G are assumed arbitrary. Statistical methods for models of this kind are given in Volume 2.

Using our first three examples for illustrative purposes, we now define the elements of a statistical model. A review of necessary concepts and notation from probability theory are given in the appendices.

We are given a random experiment with sample space Ω. On this sample space we have defined a random vector $\mathbf{X} = (X_1, \ldots, X_n)$. When ω is the outcome of the experiment, $\mathbf{X}(\omega)$ is referred to as the *observations* or *data*. It is often convenient to identify the random vector $\mathbf{X}$ with its realization, the data $\mathbf{X}(\omega)$. Since it is only $\mathbf{X}$ that we observe, we need only consider its probability distribution. This distribution is assumed to be a member of a family $\mathcal{P}$ of probability distributions on R^n. $\mathcal{P}$ is referred to as the *model*. For instance, in Example 1.1.1, we observe X and the family $\mathcal{P}$ is that of all hypergeometric distributions with sample size n and population size N. In Example 1.1.2, if (1)–(4) hold, $\mathcal{P}$ is the family of all distributions according to which $X_1, \ldots, X_n$ are independent and identically distributed with a common $\mathcal{N}(\mu, \sigma^2)$ distribution.

1.1.2   Parametrizations and Parameters

To describe $\mathcal{P}$ we use a *parametrization*, that is, a map, $\theta \to P_\theta$ from a space of labels, the parameter space Θ, to $\mathcal{P}$; or equivalently write $\mathcal{P} = \{P_\theta : \theta \in \Theta\}$. Thus, in Example 1.1.1 we take θ to be the fraction of defectives in the shipment, $\Theta = \{0, \frac{1}{N}, \ldots, 1\}$ and P_θ the $\mathcal{H}(N\theta, N, n)$ distribution. In Example 1.1.2 with assumptions (1)–(4) we have implicitly taken $\Theta = R \times R^+$ and, if $\theta = (\mu, \sigma^2)$, P_θ the distribution on R^n with density $\prod_{i=1}^{n} \frac{1}{\sigma} \varphi\left(\frac{x_i - \mu}{\sigma}\right)$ where φ is the standard normal density. If, still in this example, we know we are measuring a positive quantity in this model, we have $\Theta = R^+ \times R^+$. If, on the other hand, we only wish to make assumptions (1)–(3) with ϵ having expectation 0, we can take $\Theta = \{(\mu, G) : \mu \in R, \ G \text{ with density } g \text{ such that } \int xg(x)dx = 0\}$ and $P_{(\mu, G)}$ has density $\prod_{i=1}^{n} g(x_i - \mu)$.

When we can take Θ to be a nice subset of Euclidean space and the maps $\theta \to P_\theta$ are smooth, in senses to be made precise later, models $\mathcal{P}$ are called *parametric*. Models such as that of Example 1.1.2 with assumptions (1)–(3) are called *semiparametric*. Finally, models such as that of Example 1.1.3 with only (1) holding and F, G taken to be arbitrary are called *nonparametric*. It's important to note that even nonparametric models make substantial assumptions—in Example 1.1.3 that $X_1, \ldots, X_m$ are independent of each other and $Y_1, \ldots, Y_n$; moreover, $X_1, \ldots, X_m$ are identically distributed as are $Y_1, \ldots, Y_n$. The only truly nonparametric but useless model for $\mathbf{X} \in R^n$ is to assume that its (joint) distribution can be anything.

Note that there are many ways of choosing a parametrization in these and all other problems. We may take any one-to-one function of θ as a new parameter. For instance, in Example 1.1.1 we can use the number of defectives in the population, $N\theta$, as a parameter and in Example 1.1.2, under assumptions (1)–(4), we may parametrize the model by the first and second moments of the normal distribution of the observations (i.e., by $(\mu, \mu^2 + \sigma^2)$).

What parametrization we choose is usually suggested by the phenomenon we are modeling; θ is the fraction of defectives, μ is the unknown constant being measured. However, as we shall see later, the first parametrization we arrive at is not necessarily the one leading to the simplest analysis. Of even greater concern is the possibility that the parametrization is not one-to-one, that is, such that we can have $\theta_1 \neq \theta_2$ and yet $P_{\theta_1} = P_{\theta_2}$. Such parametrizations are called *unidentifiable*. For instance, in (1.1.2) suppose that we permit G to be arbitrary. Then the map sending $\theta = (\mu, G)$ into the distribution of $(X_1, \ldots, X_n)$ remains the same but $\Theta = \{(\mu, G) : \mu \in R, \ G \text{ has (arbitrary) density } g\}$. Now the parametrization is unidentifiable because, for example, $\mu = 0$ and $\mathcal{N}(0, 1)$ errors lead to the same distribution of the observations as $\mu = 1$ and $\mathcal{N}(-1, 1)$ errors. The critical problem with such parametrizations is that even with "infinite amounts of data," that is, knowledge of the true P_θ, parts of θ remain unknowable. Thus, we will need to ensure that our parametrizations are *identifiable*, that is, $\theta_1 \neq \theta_2 \Rightarrow P_{\theta_1} \neq P_{\theta_2}$.

Dual to the notion of a parametrization, a map from some Θ to $\mathcal{P}$, is that of a *parameter*, formally a map, ν, from $\mathcal{P}$ to another space $\mathcal{N}$. A parameter is a feature $\nu(P)$ of the distribution of X. For instance, in Example 1.1.1, the fraction of defectives θ can be thought of as the mean of X/n. In Example 1.1.3 with assumptions (1)–(2) we are interested in Δ, which can be thought of as the difference in the means of the two populations of responses. In addition to the parameters of interest, there are also usually *nuisance parameters*, which correspond to other unknown features of the distribution of $\mathbf{X}$. For instance, in Example 1.1.2, if the errors are normally distributed with unknown variance σ^2, then σ^2 is a nuisance parameter. We usually try to combine parameters of interest and nuisance parameters into a single grand parameter θ, which indexes the family $\mathcal{P}$, that is, make $\theta \rightarrow P_\theta$ into a parametrization of $\mathcal{P}$. Implicit in this description is the assumption that θ is a parameter in the sense we have just defined. But given a parametrization $\theta \rightarrow P_\theta$, θ is a parameter if and only if the parametrization is identifiable. Formally, we can define the (well defined) parameter $\theta : \mathcal{P} \rightarrow \Theta$ as the inverse of the map $\theta \rightarrow P_\theta$, from Θ to its range $\mathcal{P}$ iff the latter map is 1-1, that is, if $P_{\theta_1} = P_{\theta_2}$ implies $\theta_1 = \theta_2$. Note that $\theta(P_\theta) = \theta$.

More generally, a function $q : \Theta \rightarrow \mathcal{N}$ can be identified with a parameter $\nu(P)$ iff $P_{\theta_1} = P_{\theta_2}$ implies $q(\theta_1) = q(\theta_2)$ and then $\nu(P_\theta) \equiv q(\theta)$.

Here are two points to note:

(1) A parameter can have many representations. For instance, in Example 1.1.2 with assumptions (1)–(4) the parameter of interest $\mu \equiv \mu(P)$ can be characterized as the mean of P, or the median of P, or the midpoint of the interquantile range of P, or more generally as the center of symmetry of P, as long as $\mathcal{P}$ is the set of all Gaussian distributions.

(2) A vector parametrization that is unidentifiable may still have components that are parameters (identifiable). For instance, consider Example 1.1.2 again in which we assume the error ϵ to be Gaussian but with arbitrary mean Δ. Then P is parametrized by $\theta = (\mu, \Delta, \sigma^2)$, where σ^2 is the variance of ϵ. As we have seen this parametrization is unidentifiable and neither μ nor Δ are parameters in the sense we've defined. But $\sigma^2 = \text{Var}(X_1)$ evidently is and so is $\mu + \Delta$.

Sometimes the choice of $\mathcal{P}$ starts by the consideration of a particular parameter. For instance, our interest in studying a population of incomes may precisely be in the mean income. When we sample, say with replacement, and observe $X_1, \ldots, X_n$ independent with common distribution, it is natural to write

$$X_i = \mu + \epsilon_i, \ 1 \le i \le n$$

where μ denotes the mean income and, thus, $E(\epsilon_i) = 0$. The (μ, G) parametrization of Example 1.1.2 is now well defined and identifiable by (1.1.3) and $\mathcal{G} = \{G : \int x \, dG(x) = 0\}$.

Similarly, in Example 1.1.3, instead of postulating a constant treatment effect Δ, we can start by making the difference of the means, $\delta = \mu_Y - \mu_X$, the focus of the study. Then δ is identifiable whenever μ_X and μ_Y exist.

1.1.3   Statistics as Functions on the Sample Space

Models and parametrizations are creations of the statistician, but the true values of parameters are secrets of nature. Our aim is to use the data inductively, to narrow down in useful
ways our ideas of what the "true" P is. The link for us are things we can compute, statistics.
Formally, a *statistic* T is a map from the sample space $\mathcal{X}$ to some space of values $\mathcal{T}$, usually
a Euclidean space. Informally, $T(x)$ is what we can compute if we observe $X = x$. Thus,
in Example 1.1.1, the fraction defective in the sample, $T(x) = x/n$. In Example 1.1.2
a common estimate of μ is the statistic $T(X_1, \ldots, X_n) = \bar{X} \equiv \frac{1}{n} \sum_{i=1}^{n} X_i$, a common
estimate of σ^2 is the statistic

$$s^2 \equiv \frac{1}{n-1} \sum_{i=1}^{n} (X_i - \bar{X})^2.$$

$\bar{X}$ and s^2 are called the *sample mean* and *sample variance*. How we use statistics in estimation and other decision procedures is the subject of the next section.

For future reference we note that a statistic just as a parameter need not be real or
Euclidean valued. For instance, a statistic we shall study extensively in Chapter 2 is the
function valued statistic $\widehat{F}$, called the *empirical distribution function*, which evaluated at
$x \in R$ is

$$\widehat{F}(X_1, \ldots, X_n)(x) = \frac{1}{n} \sum_{i=1}^{n} 1(X_i \leq x)$$

where $(X_1, \ldots, X_n)$ are a sample from a probability P on R and $1(A)$ is the indicator
of the event A. This statistic takes values in the set of all distribution functions on R. It
estimates the function valued parameter F defined by its evaluation at $x \in R$,

$$F(P)(x) = P[X_1 \leq x].$$

Deciding which statistics are important is closely connected to deciding which parameters are important and, hence, can be related to model formulation as we saw earlier. For
instance, consider situation (d) listed at the beginning of this section. If we suppose there is
a single numerical measure of performance of the drugs and the difference in performance
of the drugs for any given patient is a constant irrespective of the patient, then our attention
naturally focuses on estimating this constant. If, however, this difference depends on the
patient in a complex manner (the effect of each drug is complex), we have to formulate a
relevant measure of the difference in performance of the drugs and decide how to estimate
this measure.

Often the outcome of the experiment is used to decide on the model and the appropriate measure of difference. Next this model, which now depends on the data, is used to
decide what estimate of the measure of difference should be employed (cf., for example,
Mandel, 1964). Data-based model selection can make it difficult to ascertain or even assign
a meaning to the accuracy of estimates or the probability of reaching correct conclusions.
Nevertheless, we can draw guidelines from our numbers and cautiously proceed. These
issues will be discussed further in Volume 2. In this volume we assume that the model has

been selected prior to the current experiment. This selection is based on experience with previous similar experiments (cf. Lehmann, 1990).

There are also situations in which selection of what data will be observed depends on the experimenter and on his or her methods of reaching a conclusion. For instance, in situation (d) again, patients may be considered one at a time, sequentially, and the decision of which drug to administer for a given patient may be made using the knowledge of what happened to the previous patients. The experimenter may, for example, assign the drugs alternatively to every other patient in the beginning and then, after a while, assign the drug that seems to be working better to a higher proportion of patients. Moreover, the statistical procedure can be designed so that the experimenter stops experimenting as soon as he or she has significant evidence to the effect that one drug is better than the other. Thus, the number of patients in the study (the sample size) is random. Problems such as these lie in the fields of *sequential analysis* and *experimental design.* They are not covered under our general model and will not be treated in this book. We refer the reader to Wetherill and Glazebrook (1986) and Kendall and Stuart (1966) for more information.

Notation. Regular models. When dependence on θ has to be observed, we shall denote the distribution corresponding to any particular parameter value θ by P_θ. Expectations calculated under the assumption that $\mathbf{X} \sim P_\theta$ will be written E_θ. Distribution functions will be denoted by $F(\cdot, \theta)$, density and frequency functions by $p(\cdot, \theta)$. However, these and other subscripts and arguments will be omitted where no confusion can arise.

It will be convenient to assume[1] from now on that in any parametric model we consider either:

(1) All of the P_θ are continuous with densities $p(\mathbf{x}, \theta)$;
(2) All of the P_θ are discrete with frequency functions $p(\mathbf{x}, \theta)$, and the set $\{\mathbf{x}_1, \mathbf{x}_2, \dots \}$ $\equiv \{\mathbf{x} \colon p(\mathbf{x}, \theta) > 0\}$ is the same set for all $\theta \in \Theta$.

Such models will be called *regular parametric models.* In the discrete case we will use both the terms *frequency function* and *density* for $p(\mathbf{x}, \theta)$. See A.10.

1.1.4   Examples, Regression Models

We end this section with two further important examples indicating the wide scope of the notions we have introduced.

In most studies we are interested in studying relations between responses and several other variables not just treatment or control as in Example 1.1.3. This is the stage for the following.

Example 1.1.4.   *Regression Models.*   We observe $(\mathbf{z}_1, Y_1), \dots, (\mathbf{z}_n, Y_n)$ where $Y_1, \dots, Y_n$ are independent. The distribution of the response Y_i for the ith subject or case in the study is postulated to depend on certain characteristics $\mathbf{z}_i$ of the ith subject. Thus, $\mathbf{z}_i$ is a d dimensional vector that gives characteristics such as sex, age, height, weight, and so on of the ith subject in a study. For instance, in Example 1.1.3 we could take z to be the treatment label and write our observations as $(A, X_1), (A, X_m), (B, Y_1), \dots, (B, Y_n)$. This is obviously overkill but suppose that, in the study, drugs A and B are given at several dose levels. Then, $d = 2$ and $\mathbf{z}_i^T$ can denote the pair (Treatment Label, Treatment Dose Level) for patient i.

In general, $\mathbf{z}_i$ is a nonrandom vector of values called a *covariate* vector or a vector of *explanatory variables* whereas Y_i is random and referred to as the *response variable* or *dependent* variable in the sense that its distribution depends on $\mathbf{z}_i$. If we let $f(y_i \mid \mathbf{z}_i)$ denote the density of Y_i for a subject with covariate vector $\mathbf{z}_i$, then the model is

(a) $$p(y_1, \ldots, y_n) = \prod_{i=1}^{n} f(y_i \mid \mathbf{z}_i).$$

If we let $\mu(\mathbf{z})$ denote the expected value of a response with given covariate vector $\mathbf{z}$, then we can write,

(b) $$Y_i = \mu(\mathbf{z}_i) + \epsilon_i, \ i = 1, \ldots, n$$

where $\epsilon_i = Y_i - E(Y_i)$, $i = 1, \ldots, n$. Here $\mu(\mathbf{z})$ is an unknown function from R^d to R that we are interested in. For instance, in Example 1.1.3 with the Gaussian two-sample model $\mu(A) = \mu$, $\mu(B) = \mu + \Delta$. We usually need to postulate more. A common (but often violated assumption) is

(1) The ϵ_i are identically distributed with distribution F. That is, the effect of $\mathbf{z}$ on Y is through $\mu(\mathbf{z})$ only. In the two sample models this is implied by the constant treatment effect assumption. See Problem 1.1.8.

On the basis of subject matter knowledge and/or convenience it is usually postulated that

(2) $\mu(\mathbf{z}) = g(\boldsymbol{\beta}, \mathbf{z})$ where g is known except for a vector $\boldsymbol{\beta} = (\beta_1, \ldots, \beta_d)^T$ of unknowns. The most common choice of g is the linear form,

(3) $g(\boldsymbol{\beta}, \mathbf{z}) = \sum_{j=1}^{d} \beta_j z_j = \mathbf{z}^T \boldsymbol{\beta}$ so that (b) becomes

(b') $$Y_i = \mathbf{z}_i^T \boldsymbol{\beta} + \epsilon_i, \ 1 \leq i \leq n.$$

This is the *linear model*. Often the following final assumption is made:

(4) The distribution F of (1) is $\mathcal{N}(0, \sigma^2)$ with σ^2 unknown. Then we have the classical *Gaussian linear model*, which we can write in vector matrix form,

(c) $$\mathbf{Y} \sim \mathcal{N}_n(\mathbf{Z}\boldsymbol{\beta}, \sigma^2 J)$$

where $\mathbf{Z}_{n \times d} = (\mathbf{z}_1^T, \ldots, \mathbf{z}_n^T)^T$ and J is the $n \times n$ identity.

Clearly, Example 1.1.3(3) is a special case of this model. So is Example 1.1.2 with assumptions (1)–(4). In fact by varying our assumptions this class of models includes any situation in which we have independent but not necessarily identically distributed observations. By varying the assumptions we obtain parametric models as with (1), (3) and (4) above, semiparametric as with (1) and (2) with F arbitrary, and nonparametric if we drop (1) and simply treat the $\mathbf{z}_i$ as a label of the completely unknown distributions of Y_i. Identifiability of these parametrizations and the status of their components as parameters are discussed in the problems.   □

Finally, we give an example in which the responses are dependent.

Example 1.1.5.   *Measurement Model with Autoregressive Errors.*   Let $X_1, \ldots, X_n$ be the n determinations of a physical constant μ. Consider the model where

$$X_i = \mu + e_i, \ i = 1, \ldots, n$$

and assume

$$e_i = \beta e_{i-1} + \epsilon_i, \ i = 1, \ldots, n, \ e_0 = 0$$

where ϵ_i are independent identically distributed with density f. Here the errors $e_1, \ldots, e_n$ are dependent as are the X's. In fact we can write

$$X_i = \mu(1 - \beta) + \beta X_{i-1} + \epsilon_i, \ i = 2, \ldots, n, \ X_1 = \mu + \epsilon_1. \tag{1.1.4}$$

An example would be, say, the elapsed times $X_1, \ldots, X_n$ spent above a fixed high level for a series of n consecutive wave records at a point on the seashore. Let $\mu = E(X_i)$ be the average time for an infinite series of records. It is plausible that e_i depends on e_{i-1} because long waves tend to be followed by long waves. A second example is consecutive measurements X_i of a constant μ made by the same observer who seeks to compensate for apparent errors. Of course, model (1.1.4) assumes much more but it may be a reasonable first approximation in these situations.

To find the density $p(x_1, \ldots, x_n)$, we start by finding the density of $e_1, \ldots, e_n$. Using conditional probability theory and $e_i = \beta e_{i-1} + \epsilon_i$, we have

$$
\begin{aligned}
p(e_1, \ldots, e_n) &= p(e_1)p(e_2 \mid e_1)p(e_3 \mid e_1, e_2) \ldots p(e_n \mid e_1, \ldots, e_{n-1}) \\
&= p(e_1)p(e_2 \mid e_1)p(e_3 \mid e_2) \ldots p(e_n \mid e_{n-1}) \\
&= f(e_1)f(e_2 - \beta e_1) \ldots f(e_n - \beta e_{n-1}).
\end{aligned}
$$

Because $e_i = X_i - \mu$, the model for $X_1, \ldots, X_n$ is

$$p(x_1, \ldots, x_n) = f(x_1 - \mu) \prod_{j=2}^{n} f(x_j - \beta x_{j-1} - (1 - \beta)\mu).$$

The default assumption, at best an approximation for the wave example, is that f is the $N(0, \sigma^2)$ density. Then we have what is called the AR(1) *Gaussian model*

$$p(x_1, \ldots, x_n) =$$

$$(2\pi)^{-\frac{1}{2}n} \sigma^{-n} \exp\left\{ -\frac{1}{2\sigma^2} \left[(x_1 - \mu)^2 + \sum_{i=2}^{n} (x_i - \beta x_{i-1} - (1 - \beta)\mu)^2 \right] \right\}.$$

We include this example to illustrate that we need not be limited by independence. However, save for a brief discussion in Volume 2, the conceptual issues of stationarity, ergodicity, and the associated probability theory models and inference for dependent data are beyond the scope of this book.   □

Summary. In this section we introduced the first basic notions and formalism of mathematical statistics, vector *observations* $\mathbf{X}$ with unknown probability distributions P ranging over *models* $\mathcal{P}$. The notions of *parametrization* and *identifiability* are introduced. The general definition of *parameters* and *statistics* is given and the connection between parameters and parametrizations elucidated. This is done in the context of a number of classical examples, the most important of which is the workhorse of statistics, the *regression model*. We view statistical models as useful tools for learning from the outcomes of experiments and studies. They are useful in understanding how the outcomes can be used to draw inferences that go beyond the particular experiment. Models are approximations to the mechanisms generating the observations. How useful a particular model is is a complex mix of how good the approximation is and how much insight it gives into drawing inferences.

1.2 BAYESIAN MODELS

Throughout our discussion so far we have assumed that there is no information available about the true value of the parameter beyond that provided by the data. There are situations in which most statisticians would agree that more can be said. For instance, in the inspection Example 1.1.1, it is possible that, in the past, we have had many shipments of size N that have subsequently been distributed. If the customers have provided accurate records of the number of defective items that they have found, we can construct a frequency distribution $\{\pi_0, \ldots, \pi_N\}$ for the proportion θ of defectives in past shipments. That is, π_i is the frequency of shipments with i defective items, $i = 0, \ldots, N$. Now it is reasonable to suppose that the value of θ in the present shipment is the realization of a random variable $\boldsymbol{\theta}$ with distribution given by

$$P[\boldsymbol{\theta} = \frac{i}{N}] = \pi_i, \ i = 0, \ldots, N. \tag{1.2.1}$$

Our model is then specified by the joint distribution of the observed number X of defectives in the sample and the random variable $\boldsymbol{\theta}$. We know that, given $\boldsymbol{\theta} = i/N$, X has the hypergeometric distribution $\mathcal{H}(i, N, n)$. Thus,

$$\begin{aligned}
P[X = k, \boldsymbol{\theta} = \frac{i}{N}] &= P[\boldsymbol{\theta} = \frac{i}{N}] P[X = k \mid \boldsymbol{\theta} = \frac{i}{N}] \\
&= \pi_i \frac{\dbinom{i}{k} \dbinom{N-i}{n-k}}{\dbinom{N}{n}}.
\end{aligned} \tag{1.2.2}$$

This is an example of a *Bayesian* model.

There is a substantial number of statisticians who feel that it is always reasonable, and indeed necessary, to think of the true value of the parameter θ as being the realization of a random variable $\boldsymbol{\theta}$ with a known distribution. This distribution does not always correspond to an experiment that is physically realizable but rather is thought of as a measure of the beliefs of the experimenter concerning the true value of θ before he or she takes any data.

Thus, the resulting statistical inference becomes subjective. The theory of this school is expounded by L. J. Savage (1954), Raiffa and Schlaiffer (1961), Lindley (1965), De Groot (1969), and Berger (1985). An interesting discussion of a variety of points of view on these questions may be found in Savage et al. (1962). There is an even greater range of viewpoints in the statistical community from people who consider all statistical statements as purely subjective to ones who restrict the use of such models to situations such as that of the inspection example in which the distribution of $\boldsymbol{\theta}$ has an objective interpretation in terms of frequencies. Our own point of view is that subjective elements including the views of subject matter experts are an essential element in all model building. However, insofar as possible we prefer to take the frequentist point of view in validating statistical statements and avoid making final claims in terms of subjective posterior probabilities (see later). However, by giving θ a distribution purely as a theoretical tool to which no subjective significance is attached, we can obtain important and useful results and insights. We shall return to the Bayesian framework repeatedly in our discussion.

In this section we shall define and discuss the basic elements of Bayesian models. Suppose that we have a regular parametric model $\{P_\theta : \theta \in \Theta\}$. To get a Bayesian model we introduce a random vector $\boldsymbol{\theta}$, whose range is contained in Θ, with density or frequency function π. The function π represents our belief or information about the parameter θ before the experiment and is called the *prior density* or *frequency function*. We now think of P_θ as the conditional distribution of $\mathbf{X}$ given $\boldsymbol{\theta} = \theta$. The joint distribution of $(\boldsymbol{\theta}, \mathbf{X})$ is that of the outcome of a random experiment in which we first select $\boldsymbol{\theta} = \theta$ according to π and then, given $\boldsymbol{\theta} = \theta$, select $\mathbf{X}$ according to P_θ. If both $\mathbf{X}$ and $\boldsymbol{\theta}$ are continuous or both are discrete, then by (B.1.3), $(\boldsymbol{\theta}, \mathbf{X})$ is appropriately continuous or discrete with density or frequency function,

$$f(\theta, \mathbf{x}) = \pi(\theta)p(\mathbf{x}, \theta). \tag{1.2.3}$$

Because we now think of $p(\mathbf{x}, \theta)$ as a conditional density or frequency function given $\boldsymbol{\theta} = \theta$, we will denote it by $p(\mathbf{x} \mid \theta)$ for the remainder of this section.

Equation (1.2.2) is an example of (1.2.3). In the "mixed" cases such as $\boldsymbol{\theta}$ continuous $\mathbf{X}$ discrete, the joint distribution is neither continuous nor discrete.

The most important feature of a Bayesian model is the conditional distribution of $\boldsymbol{\theta}$ given $\mathbf{X} = \mathbf{x}$, which is called the *posterior* distribution of $\boldsymbol{\theta}$. Before the experiment is performed, the information or belief about the true value of the parameter is described by the prior distribution. After the value $\mathbf{x}$ has been obtained for $\mathbf{X}$, the information about θ is described by the posterior distribution.

For a concrete illustration, let us turn again to Example 1.1.1. For instance, suppose that $N = 100$ and that from past experience we believe that each item has probability .1 of being defective independently of the other members of the shipment. This would lead to the prior distribution

$$\pi_i = \binom{100}{i} (0.1)^i (0.9)^{100-i}, \tag{1.2.4}$$

for $i = 0, 1, \ldots, 100$. *Before* sampling any items the chance that a given shipment contains

20 or more bad items is by the normal approximation with continuity correction, (A.15.10),

$$P[100\boldsymbol{\theta} \geq 20] = P\left[\frac{100\boldsymbol{\theta} - 10}{\sqrt{100(0.1)(0.9)}} \geq \frac{10}{\sqrt{100(0.1)(0.9)}}\right]$$

$$\approx 1 - \Phi\left(\frac{9.5}{3}\right) = 0.001.$$

$(1.2.5)$

Now suppose that a sample of 19 has been drawn in which 10 defective items are found. This leads to

$$P[100\boldsymbol{\theta} \geq 20 \mid X = 10] \approx 0.30. \tag{1.2.6}$$

To calculate the posterior probability given in (1.2.6) we argue loosely as follows: If before the drawing each item was defective with probability .1 and good with probability .9 independently of the other items, this will continue to be the case for the items left in the lot after the 19 sample items have been drawn. Therefore, $100\boldsymbol{\theta} - X$, the number of defectives left after the drawing, is independent of X and has a $\mathcal{B}(81, 0.1)$ distribution. Thus,

$$P[100\boldsymbol{\theta} \geq 20 \mid X = 10] = P[100\boldsymbol{\theta} - X \geq 10 \mid X = 10]$$

$$= P\left[\frac{(100\boldsymbol{\theta} - X) - 8.1}{\sqrt{81(0.9)(0.1)}} \geq \frac{1.9}{\sqrt{81(0.9)(0.1)}}\right] \tag{1.2.7}$$

$$\approx 1 - \Phi(0.52)$$

$$= 0.30.$$

In general, to calculate the posterior, some variant of Bayes' Theorem (B.1.4) can be used. Specifically,

(i) The posterior distribution is discrete or continuous according as the prior distribution is discrete or continuous.

(ii) If we denote the corresponding (posterior) frequency function or density by $\pi(\theta \mid \mathbf{x})$, then

$$\pi(\theta \mid \mathbf{x}) = \frac{\pi(\theta)p(\mathbf{x} \mid \theta)}{\sum_t \pi(t)p(\mathbf{x} \mid t)} \qquad \text{if } \boldsymbol{\theta} \text{ is discrete,}$$

$$= \frac{\pi(\theta)p(\mathbf{x} \mid \theta)}{\int_{-\infty}^{\infty} \pi(t)p(\mathbf{x} \mid t)dt} \qquad \text{if } \boldsymbol{\theta} \text{ is continuous.}$$

$(1.2.8)$

In the cases where $\boldsymbol{\theta}$ and $\mathbf{X}$ are both continuous or both discrete this is precisely Bayes' rule applied to the joint distribution of $(\boldsymbol{\theta}, \mathbf{X})$ given by (1.2.3). Here is an example.

Example 1.2.1. *Bernoulli Trials.* Suppose that $X_1, \ldots, X_n$ are indicators of n Bernoulli trials with probability of success θ where $0 < \theta < 1$. If we assume that θ has a priori distribution with density π, we obtain by (1.2.8) as posterior density of θ,

$$\pi(\theta \mid x_1, \ldots, x_n) = \frac{\pi(\theta)\theta^k(1-\theta)^{n-k}}{\int_0^1 \pi(t)t^k(1-t)^{n-k}dt} \tag{1.2.9}$$

for $0 < \theta < 1$, $x_i = 0$ or 1, $i = 1, \ldots, n$, $k = \sum_{i=1}^{n} x_i$.

Note that the posterior density depends on the data only through the total number of successes, $\sum_{i=1}^{n} X_i$. We also obtain the same posterior density if θ has prior density π and we only observe $\sum_{i=1}^{n} X_i$, which has a $\mathcal{B}(n, \theta)$ distribution given $\theta = \theta$ (Problem 1.2.9). We can thus write $\pi(\theta \mid k)$ for $\pi(\theta \mid x_1, \ldots, x_n)$, where $k = \sum_{i=1}^{n} x_i$.

To choose a prior π, we need a class of distributions that concentrate on the interval $(0, 1)$. One such class is the two-parameter beta family. This class of distributions has the remarkable property that the resulting posterior distributions are again beta distributions. Specifically, upon substituting the $\beta(r, s)$ density (B.2.11) in (1.2.9) we obtain

$$\pi(\theta \mid k) = \frac{\theta^{r-1}(1-\theta)^{s-1}\theta^k(1-\theta)^{n-k}}{c} = \frac{\theta^{k+r-1}(1-\theta)^{n-k+s-1}}{c}. \qquad (1.2.10)$$

The proportionality constant c, which depends on k, r, and s only, must (see (B.2.11)) be $B(k + r, n - k + s)$ where $B(\cdot, \cdot)$ is the beta function, and the posterior distribution of θ given $\sum X_i = k$ is $\beta(k + r, n - k + s)$.

As Figure B.2.2 indicates, the beta family provides a wide variety of shapes that can approximate many reasonable prior distributions though by no means all. For instance, non-U-shaped bimodal distributions are not permitted.

Suppose, for instance, we are interested in the proportion θ of "geniuses" ($IQ \geq 160$) in a particular city. To get information we take a sample of n individuals from the city. If n is small compared to the size of the city, (A.15.13) leads us to assume that the number X of geniuses observed has approximately a $\mathcal{B}(n, \theta)$ distribution. Now we may either have some information about the proportion of geniuses in similar cities of the country or we may merely have prejudices that we are willing to express in the form of a prior distribution on θ. We may want to assume that θ has a density with maximum value at 0 such as that drawn with a dotted line in Figure B.2.2. Or else we may think that $\pi(\theta)$ concentrates its mass near a small number, say 0.05. Then we can choose r and s in the $\beta(r, s)$ distribution, so that the mean is $r/(r + s) = 0.05$ and its variance is very small. The result might be a density such as the one marked with a solid line in Figure B.2.2.

If we were interested in some proportion about which we have no information or belief, we might take θ to be uniformly distributed on $(0, 1)$, which corresponds to using the beta distribution with $r = s = 1$. □

A feature of Bayesian models exhibited by this example is that there are natural parametric families of priors such that the posterior distributions also belong to this family. Such families are called *conjugate*. Evidently the beta family is conjugate to the binomial. Another bigger conjugate family is that of finite mixtures of beta distributions—see Problem 1.2.16. We return to conjugate families in Section 1.6.

Summary. We present an elementary discussion of *Bayesian models*, introduce the notions of *prior* and *posterior* distributions and give *Bayes rule*. We also by example introduce the notion of a *conjugate family* of distributions.

1.3   THE DECISION THEORETIC FRAMEWORK

Given a statistical model, the information we want to draw from data can be put in various forms depending on the purposes of our analysis. We may wish to produce "best guesses" of the values of important parameters, for instance, the fraction defective θ in Example 1.1.1 or the physical constant μ in Example 1.1.2. These are *estimation* problems. In other situations certain P are "special" and we may primarily wish to know whether the data support "specialness" or not. For instance, in Example 1.1.3, P's that correspond to no treatment effect (i.e., placebo and treatment are equally effective) are special because the FDA (Food and Drug Administration) does not wish to permit the marketing of drugs that do no good. If μ_0 is the critical matter density in the universe so that $\mu < \mu_0$ means the universe is expanding forever and $\mu \geq \mu_0$ correspond to an eternal alternation of Big Bangs and expansions, then depending on one's philosophy one could take either P's corresponding to $\mu < \mu_0$ or those corresponding to $\mu \geq \mu_0$ as special. Making determinations of "specialness" corresponds to *testing significance*. As the second example suggests, there are many problems of this type in which it's unclear which of two disjoint sets of P's; $\mathcal{P}_0$ or $\mathcal{P}_0^c$ is special and the general *testing problem* is really one of discriminating between $\mathcal{P}_0$ and $\mathcal{P}_0^c$. For instance, in Example 1.1.1 contractual agreement between shipper and receiver may penalize the return of "good" shipments, say, with $\theta < \theta_0$, whereas the receiver does not wish to keep "bad," $\theta \geq \theta_0$, shipments. Thus, the receiver wants to discriminate and may be able to attach monetary costs to making a mistake of either type: "keeping the bad shipment" or "returning a good shipment." In testing problems we, at a first cut, state which is supported by the data: "specialness" or, as it's usually called, "hypothesis" or "nonspecialness" (or alternative).

   We may have other goals as illustrated by the next two examples.

Example 1.3.1. *Ranking.* A consumer organization preparing (say) a report on air conditioners tests samples of several brands. On the basis of the sample outcomes the organization wants to give a ranking from best to worst of the brands (ties not permitted). Thus, if there are k different brands, there are $k!$ possible rankings or actions, one of which will be announced as more consistent with the data than others.   □

Example 1.3.2. *Prediction.* A very important class of situations arises when, as in Example 1.1.4, we have a vector $\mathbf{z}$, such as, say, (age, sex, drug dose)^T that can be used for prediction of a variable of interest Y, say a 50-year-old male patient's response to the level of a drug. Intuitively, and as we shall see formally later, a reasonable prediction rule for an unseen Y (response of a new patient) is the function $\mu(\mathbf{z})$, the expected value of Y given $\mathbf{z}$. Unfortunately $\mu(\mathbf{z})$ is unknown. However, if we have observations $(\mathbf{z}_i, Y_i)$, $1 \leq i \leq n$, we can try to estimate the function $\mu(\cdot)$. For instance, if we believe $\mu(\mathbf{z}) = g(\boldsymbol{\beta}, \mathbf{z})$ we can estimate $\boldsymbol{\beta}$ from our observations Y_i of $g(\boldsymbol{\beta}, \mathbf{z}_i)$ and then plug our estimate of $\boldsymbol{\beta}$ into g. Note that we really want to estimate the function $\mu(\cdot)$; our results will guide the selection of doses of drug for future patients.   □

   In all of the situations we have discussed it is clear that the analysis does not stop by specifying an estimate or a test or a ranking or a prediction function. There are many possible choices of estimates. In Example 1.1.1 do we use the observed fraction of defectives

X/n as our estimate or ignore the data and use historical information on past shipments, or combine them in some way? In Example 1.1.2 to estimate μ do we use the mean of the measurements, $\bar{X} = \frac{1}{n} \sum_{i=1}^{n} X_i$, or the median, defined as any value such that half the X_i are at least as large and half no bigger? The same type of question arises in all examples. The answer will depend on the model and, most significantly, on what criteria of performance we use. Intuitively, in estimation we care how far off we are, in testing whether we are right or wrong, in ranking what mistakes we've made, and so on. In any case, whatever our choice of procedure we need either a priori (before we have looked at the data) and/or a posteriori estimates of how well we're doing. In designing a study to compare treatments A and B we need to determine sample sizes that will be large enough to enable us to detect differences that matter. That is, we need a priori estimates of how well even the best procedure can do. For instance, in Example 1.1.3 even with the simplest Gaussian model it is intuitively clear and will be made precise later that, even if Δ is large, a large σ^2 will force a large m, n to give us a good chance of correctly deciding that the treatment effect is there. On the other hand, once a study is carried out we would probably want not only to estimate Δ but also know how reliable our estimate is. Thus, we would want a posteriori estimates of performance.

These examples motivate the decision theoretic framework: We need to
(1) clarify the objectives of a study,
(2) point to what the different possible actions are,
(3) provide assessments of risk, accuracy, and reliability of statistical procedures,
(4) provide guidance in the choice of procedures for analyzing outcomes of experiments.

1.3.1   Components of the Decision Theory Framework

As in Section 1.1, we begin with a statistical model with an observation vector $\mathbf{X}$ whose distribution P ranges over a set $\mathcal{P}$. We usually take $\mathcal{P}$ to be parametrized, $\mathcal{P} = \{P_\theta : \theta \in \Theta\}$.

Action space. A new component is an *action space* $\mathcal{A}$ of actions or decisions or claims that we can contemplate making. Here are action spaces for our examples.

Estimation. If we are estimating a real parameter such as the fraction θ of defectives, in Example 1.1.1, or μ in Example 1.1.2, it is natural to take $\mathcal{A} = R$ though smaller spaces may serve equally well, for instance, $\mathcal{A} = \{0, \frac{1}{N}, \ldots, 1\}$ in Example 1.1.1.

Testing. Here only two actions are contemplated: accepting or rejecting the "specialness" of P (or in more usual language the hypothesis $H : P \in \mathcal{P}_0$ in which we identify $\mathcal{P}_0$ with the set of "special" P's). By convention, $\mathcal{A} = \{0, 1\}$ with 1 corresponding to rejection of H. Thus, in Example 1.1.3, taking action 1 would mean deciding that $\Delta \neq 0$.

Ranking. Here quite naturally $\mathcal{A} = \{\text{Permutations } (i_1, \ldots, i_k) \text{ of } \{1, \ldots, k\}\}$. Thus, if we have three air conditioners, there are $3! = 6$ possible rankings,

$$\mathcal{A} = \{(1, 2, 3), (1, 3, 2), (2, 1, 3), (2, 3, 1), (3, 1, 2), (3, 2, 1)\}.$$

Prediction. Here $\mathcal{A}$ is much larger. If Y is real, and $\mathbf{z} \in Z$, $\mathcal{A} = \{a : a$ is a function from Z to $R\}$ with $a(\mathbf{z})$ representing the prediction we would make if the new unobserved Y had covariate value $\mathbf{z}$. Evidently Y could itself range over an arbitrary space $\mathcal{Y}$ and then R would be replaced by $\mathcal{Y}$ in the definition of $a(\cdot)$. For instance, if $Y = 0$ or 1 corresponds to, say, "does not respond" and "responds," respectively, and $\mathbf{z} = (\text{Treatment}, \text{Sex})^T$, then $a(B, M)$ would be our prediction of response or no response for a male given treatment B.

Loss function. Far more important than the choice of action space is the choice of *loss function* defined as a function $l : \mathcal{P} \times \mathcal{A} \to R^+$. The interpretation of $l(P, a)$, or $l(\theta, a)$ if $\mathcal{P}$ is parametrized, is the nonnegative loss incurred by the statistician if he or she takes action a and the true "state of Nature," that is, the probability distribution producing the data, is P. As we shall see, although loss functions, as the name suggests, sometimes can genuinely be quantified in economic terms, they usually are chosen to qualitatively reflect what we are trying to do and to be mathematically convenient.

Estimation. In estimating a real valued parameter $\nu(P)$ or $q(\theta)$ if $\mathcal{P}$ is parametrized the most commonly used loss function is,

Quadratic Loss: $l(P, a) = (\nu(P) - a)^2$ (or $l(\theta, a) = (q(\theta) - a)^2$).

Other choices that are, as we shall see (Section 5.1), less computationally convenient but perhaps more realistically penalize large errors less are *Absolute Value Loss:* $l(P; a) = |\nu(P) - a|$, and *truncated quadratic* loss: $l(P, a) = \min\{(\nu(P) - a)^2, d^2\}$. Closely related to the latter is what we shall call *confidence interval loss*, $l(P, a) = 0$, $|\nu(P) - a| \leq d$, $l(P, a) = 1$ otherwise. This loss expresses the notion that all errors within the limits $\pm d$ are tolerable and outside these limits equally intolerable. Although estimation loss functions are typically symmetric in ν and a, asymmetric loss functions can also be of importance. For instance, $l(P, a) = 1(\nu < a)$, which penalizes only overestimation and by the same amount arises naturally with lower confidence bounds as discussed in Example 1.3.3.

If $\nu = (\nu_1, \ldots, \nu_d) = (q_1(\boldsymbol{\theta}), \ldots, q_d(\boldsymbol{\theta}))$ and $\mathbf{a} = (a_1, \ldots, a_d)$ are vectors, examples of loss functions are

$$l(\theta, a) = \frac{1}{d}\sum(a_j - \nu_j)^2 = \text{ squared Euclidean distance}/d$$

$$l(\theta, a) = \frac{1}{d}\sum|a_j - \nu_j| = \text{ absolute distance}/d$$

$$l(\theta, a) = \max\{|a_j - \nu_j|, j = 1, \ldots, d\} = \text{ supremum distance.}$$

We can also consider function valued parameters. For instance, in the *prediction example* 1.3.2, $\mu(\cdot)$ is the parameter of interest. If we use $a(\cdot)$ as a predictor and the new $\mathbf{z}$ has marginal distribution Q then it is natural to consider,

$$l(P, a) = \int (\mu(\mathbf{z}) - a(\mathbf{z}))^2 dQ(\mathbf{z}),$$

the expected squared error if a is used. If, say, Q is the empirical distribution of the $\mathbf{z}_j$ in the training set $(\mathbf{z}_1, Y), \ldots, (\mathbf{z}_n, \mathbf{Y}_n)$, this leads to the commonly considered

$$l(P, a) = \frac{1}{n} \sum_{j=1}^{n} (\mu(\mathbf{z}_j) - a(\mathbf{z}_j))^2,$$

which is just n^{-1} times the squared Euclidean distance between the prediction vector $(a(\mathbf{z}_1), \ldots, a(\mathbf{z}_n))^T$ and the vector parameter $(\mu(\mathbf{z}_1), \ldots, \mu(\mathbf{z}_n))^T$.

Testing. We ask whether the parameter θ is in the subset Θ_0 or subset Θ_1 of Θ, where $\{\Theta_0, \Theta_1\}$, is a partition of Θ (or equivalently if $P \in \mathcal{P}_0$ or $P \in \mathcal{P}_1$). If we take action a when the parameter is in Θ_a, we have made the correct decision and the loss is zero. Otherwise, the decision is wrong and the loss is taken to equal one. This $0-1$ loss function can be written as

$$0-1 \text{ loss:}\quad l(\theta, a) = 0 \text{ if } \theta \in \Theta_a \text{ (The decision is correct)}$$

$$l(\theta, a) = 1 \text{ otherwise (The decision is wrong)}.$$

Of course, other economic loss functions may be appropriate. For instance, in Example 1.1.1 suppose returning a shipment with $\theta < \theta_0$ defectives results in a penalty of s dollars whereas every defective item sold results in an r dollar replacement cost. Then the appropriate loss function is

$$
\begin{aligned}
l(\theta, 1) &= s \text{ if } \theta < \theta_0 \\
l(\theta, 1) &= 0 \text{ if } \theta \geq \theta_0 \\
l(\theta, 0) &= rN\theta.
\end{aligned}
\tag{1.3.1}
$$

Decision procedures. We next give a representation of the process whereby the statistician uses the data to arrive at a decision. The data is a point $\mathbf{X} = \mathbf{x}$ in the outcome or sample space $\mathcal{X}$. We define a *decision rule* or *procedure* δ to be any function from the sample space taking its values in A. Using δ means that if $\mathbf{X} = \mathbf{x}$ is observed, the statistician takes action $\delta(\mathbf{x})$.

Estimation. For the problem of estimating the constant μ in the measurement model, we implicitly discussed two estimates or decision rules: $\delta_1(\mathbf{x}) = $ sample mean $\bar{x}$ and $\delta_2(\mathbf{x}) = \hat{x} = $ sample median.

Testing. In Example 1.1.3 with X and Y distributed as $\mathcal{N}(\mu, \sigma^2)$ and $\mathcal{N}(\mu + \Delta, \sigma^2)$, respectively, if we are asking whether the treatment effect parameter Δ is 0 or not, then a reasonable rule is to decide $\Delta = 0$ if our estimate $\bar{x} - \bar{y}$ is close to zero, and to decide $\Delta \neq 0$ if our estimate is not close to zero. Here we mean close to zero relative to the variability in the experiment, that is, relative to the standard deviation σ. In Section 4.9.3 we will show how to obtain an estimate $\hat{\sigma}$ of σ from the data. The decision rule can now be written

$$
\delta(\mathbf{x}, \mathbf{y}) = \quad 0 \text{ if } \frac{|\bar{x} - \bar{y}|}{\hat{\sigma}} < c
$$

$$
\qquad\qquad 1 \text{ if } \frac{|\bar{x} - \bar{y}|}{\hat{\sigma}} \geq c
\tag{1.3.2}
$$

where c is a positive constant called the *critical value*. How do we choose c? We need the next concept of the decision theoretic framework, the *risk* or *risk function*:

The risk function. If δ is the procedure used, l is the loss function, θ is the true value of the parameter, and $\mathbf{X} = \mathbf{x}$ is the outcome of the experiment, then the loss is $l(P, \delta(\mathbf{x}))$. We do not know the value of the loss because P is unknown. Moreover, we typically want procedures to have good properties not at just one particular $\mathbf{x}$, but for a range of plausible $\mathbf{x}$'s. Thus, we turn to the average or mean loss over the sample space. That is, we regard $l(P, \delta(\mathbf{X}))$ as a random variable and introduce the *risk function*

$$R(P, \delta) = E_P[l(P, \delta(\mathbf{X}))]$$

as the measure of the performance of the decision rule $\delta(\mathbf{X})$. Thus, for each δ, R maps $\mathcal{P}$ or Θ to R^+. $R(\cdot, \delta)$ is our a priori measure of the performance of δ. We illustrate computation of R and its a priori use in some examples.

Estimation. Suppose $\nu \equiv \nu(P)$ is the real parameter we wish to estimate and $\widehat{\nu} \equiv \widehat{\nu}(X)$ is our estimator (our decision rule). If we use quadratic loss, our risk function is called the *mean squared error* (MSE) of $\widehat{\nu}$ and is given by

$$MSE(\widehat{\nu}) = R(P, \widehat{\nu}) = E_P(\widehat{\nu}(X) - \nu(P))^2 \tag{1.3.3}$$

where for simplicity dependence on P is suppressed in MSE.

The MSE depends on the variance of $\widehat{\nu}$ and on what is called the *bias* of $\widehat{\nu}$ where

$$\text{Bias}(\widehat{\nu}) = E(\widehat{\nu}) - \nu$$

can be thought of as the "long-run average error" of $\widehat{\nu}$. A useful result is

Proposition 1.3.1.

$$MSE(\widehat{\nu}) = (Bias\ \widehat{\nu})^2 + Var(\widehat{\nu}).$$

Proof. Write the error as

$$(\widehat{\nu} - \nu) = [\widehat{\nu} - E(\widehat{\nu})] + [E(\widehat{\nu}) - \nu].$$

If we expand the square of the right-hand side keeping the brackets intact and take the expected value, the cross term will be zero because $E[\widehat{\nu} - E(\widehat{\nu})] = 0$. The other two terms are $(\text{Bias } \widehat{\nu})^2$ and $Var(\widehat{\nu})$. (If one side is infinite, so is the other and the result is trivially true.)   □

If $\text{Bias}(\widehat{\nu}) = 0$, $\widehat{\nu}$ is called *unbiased*. We next illustrate the computation and the a priori and a posteriori use of the risk function.

Example 1.3.3. *Estimation of μ (Continued).* Suppose $X_1, \ldots, X_n$ are i.i.d. measurements of μ with $\mathcal{N}(0, \sigma^2)$ errors. If we use the mean $\bar{X}$ as our estimate of μ and assume quadratic loss, then

$$\text{Bias}(\bar{X}) \quad = \quad E(\bar{X}) - \mu = 0$$

$$\text{Var}(\bar{X}) \quad = \quad \frac{1}{n^2} \sum_{i=1}^{n} \text{Var}(X_i) = \frac{\sigma^2}{n}$$

and, by Proposition 1.3.1

$$MSE(\bar{X}) = R(\mu, \sigma^2, \bar{X}) = \frac{\sigma^2}{n}, \qquad (1.3.4)$$

which doesn't depend on μ.

Suppose that the precision of the measuring instrument σ^2 is known and equal to σ_0^2 or where realistically it is known to be $\leq \sigma_0^2$. Then (1.3.4) can be used for an a priori estimate of the risk of $\bar{X}$. If we want to be guaranteed $MSE(\bar{X}) \leq \varepsilon^2$ we can do it by taking at least $n_0 = (\sigma_0/\varepsilon)^2$ measurements.

If we have no idea of the value of σ^2, planning is not possible but having taken n measurements we can then estimate σ^2, for instance by $\hat{\sigma}^2 = \frac{1}{n}\sum_{i=1}^{n}(X_i - \bar{X})^2$, or $n\hat{\sigma}^2/(n-1)$, an estimate we can justify later. The *a posteriori* estimate of risk $\hat{\sigma}^2/n$ is, of course, itself subject to random error.

Suppose that instead of quadratic loss we used the more natural[1] absolute value loss. Then

$$R(\mu, \sigma^2, \bar{X}) = E|\bar{X} - \mu| = E|\bar{\varepsilon}|$$

where $\varepsilon_i = X_i - \mu$. If, as we assumed, the ε_i are $\mathcal{N}(0, \sigma^2)$ then by (A.13.23), $(\sqrt{n}/\sigma)\bar{\varepsilon} \sim \mathcal{N}(0, 1)$ and

$$R(\mu, \sigma^2, \bar{X}) = \frac{\sigma}{\sqrt{n}} \int_{-\infty}^{\infty} |t|\varphi(t)dt = \frac{\sigma}{\sqrt{n}}\sqrt{\frac{2}{\pi}}. \qquad (1.3.5)$$

This harder calculation already suggests why quadratic loss is really favored. If we only assume, as we discussed in Example 1.1.2, that the ε_i are i.i.d. with mean 0 and variance $\sigma^2(P)$, then for quadratic loss, $R(P, \bar{X}) = \sigma^2(P)/n$ still, but for absolute value loss only approximate, analytic, or numerical and/or Monte Carlo computation, is possible. In fact, computational difficulties arise even with quadratic loss as soon as we think of estimates other than $\bar{X}$. For instance, define the *sample median* $\widehat{X}$ of the sample $X_1, \ldots, X_n$ as follows: When n is odd, $\widehat{X} = X_{(k)}$ where $k = \frac{1}{2}(n+1)$ and $X_{(1)}, \ldots, X_{(n)}$ denotes $X_1, \ldots, X_n$ ordered from smallest to largest. When n is even, $\widehat{X} = \frac{1}{2}[X_{(r)} + X_{(r+1)}]$, where $r = \frac{1}{2}n$. Now $E(\widehat{X} - \mu)^2 = E(\hat{\varepsilon}^2)$ can only be evaluated numerically (see Problem 1.3.6), or approximated asymptotically. $\quad\square$

We next give an example in which quadratic loss and the breakup of MSE given in Proposition 1.3.1 is useful for evaluating the performance of competing estimators.

Example 1.3.4. Let μ_0 denote the mean of a certain measurement included in the U.S. census, say, age or income. Next suppose we are interested in the mean μ of the same measurement for a certain area of the United States. If we have no data for area A, a natural guess for μ would be μ_0, whereas if we have a random sample of measurements $X_1, X_2, \ldots, X_n$ from area A, we may want to combine μ_0 and $\bar{X} = n^{-1}\sum_{i=1}^{n}X_i$ into an estimator, for instance,

$$\widehat{\mu} = (0.2)\mu_0 + (0.8)\bar{X}.$$

The choice of the weights 0.2 and 0.8 can only be made on the basis of additional knowledge about demography or the economy. We shall derive them in Section 1.6 through a formal Bayesian analysis using a normal prior to illustrate a way of bringing in additional knowledge. Here we compare the performances of $\widehat{\mu}$ and $\bar{X}$ as estimators of μ using MSE. We easily find

$$\begin{aligned}
\text{Bias}(\widehat{\mu}) &= 0.2\mu_0 + 0.8\mu - \mu = 0.2(\mu_0 - \mu) \\
\text{Var}(\widehat{\mu}) &= (0.8)^2 \text{Var}(\bar{X}) = (.64)\sigma^2/n \\
R(\mu, \widehat{\mu}) &= MSE(\widehat{\mu}) = .04(\mu_0 - \mu)^2 + (.64)\sigma^2/n.
\end{aligned}$$

If μ is close to μ_0, the risk $R(\mu, \widehat{\mu})$ of $\widehat{\mu}$ is smaller than the risk $R(\mu, \bar{X}) = \sigma^2/n$ of $\bar{X}$ with the minimum relative risk $\inf\{MSE(\widehat{\mu})/MSE(\bar{X}); \mu \in R\}$ being 0.64 when $\mu = \mu_0$. Figure 1.3.1 gives the graphs of $MSE(\widehat{\mu})$ and $MSE(\bar{X})$ as functions of μ. Because we do not know the value of μ, using MSE, neither estimator can be proclaimed as being better than the other. However, if we use as our criteria the maximum (over μ) of the MSE (called the *minimax* criteria), then $\bar{X}$ is optimal (Example 3.3.4).

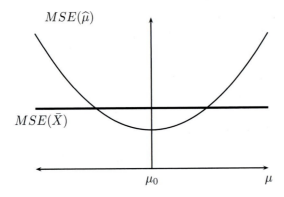

Figure 1.3.1. The mean squared errors of $\bar{X}$ and $\widehat{\mu}$. The two MSE curves cross at
$$\mu = \mu_0 \pm 3\sigma/\sqrt{n}.$$

□

Testing. The test rule (1.3.2) for deciding between $\Delta = 0$ and $\Delta \neq 0$ can only take on the two values 0 and 1; thus, the risk is

$$R(\Delta, \delta) = l(\Delta, 0)P[\delta(\mathbf{X}, \mathbf{Y}) = 0] + l(\Delta, 1)P[\delta(\mathbf{X}, \mathbf{Y}) = 1],$$

which in the case of $0 - 1$ loss is

$$\begin{aligned}
R(\Delta, \delta) &= P[\delta(\mathbf{X}, \mathbf{Y}) = 1] \text{ if } \Delta = 0 \\
&= P[\delta(\mathbf{X}, \mathbf{Y}) = 0] \text{ if } \Delta \neq 0.
\end{aligned}$$

In the general case $\mathcal{X}$ and Θ denote the outcome and parameter space, respectively, and we are to decide whether $\theta \in \Theta_0$ or $\theta \in \Theta_1$, where $\Theta = \Theta_0 \cup \Theta_1$, $\Theta_0 \cap \Theta_1 = \emptyset$. A *test function* is a decision rule $\delta(\mathbf{X})$ that equals 1 on a set $C \subset \mathcal{X}$ called the *critical region* and equals 0 on the complement of C; that is, $\delta(\mathbf{X}) = 1[\mathbf{X} \in C]$, where 1 denotes the indicator function. If $\delta(\mathbf{X}) = 1$ and we decide $\theta \in \Theta_1$ when in fact $\theta \in \Theta_0$, we call the error committed a *Type I error*, whereas if $\delta(\mathbf{X}) = 0$ and we decide $\theta \in \Theta_0$ when in fact $\theta \in \Theta_1$, we call the error a *Type II error*. Thus, the risk of $\delta(\mathbf{X})$ is

$$
\begin{aligned}
R(\theta, \delta) &= E(\delta(\mathbf{X})) = P(\delta(\mathbf{X}) = 1) \text{ if } \theta \in \Theta_0 \\
&= \text{Probability of Type I error} \\
R(\theta, \delta) &= P(\delta(\mathbf{X}) = 0) \text{ if } \theta \in \Theta_1 \\
&= \text{Probability of Type II error.}
\end{aligned} \tag{1.3.6}
$$

Finding good test functions corresponds to finding critical regions with small probabilities of error. In the *Neyman–Pearson* framework of statistical hypothesis testing, the focus is on first providing a small bound, say .05, on the probability of Type I error, and then trying to minimize the probability of a Type II error. The bound on the probability of a Type I error is called the *level of significance* and deciding Θ_1 is referred to as "Rejecting the hypothesis $H : \theta \in \Theta_0$ at level of significance α". For instance, in the treatments A and B example, we want to start by limiting the probability of falsely proclaiming one treatment superior to the other (deciding $\Delta \neq 0$ when $\Delta = 0$), and then next look for a procedure with low probability of proclaiming no difference if in fact one treatment is superior to the other (deciding $\Delta = 0$ when $\Delta \neq 0$).

This is not the only approach to testing. For instance, the loss function (1.3.1) and tests δ_k of the form, "Reject the shipment if and only if $X \geq k$," in Example 1.1.1 lead to (Problem 1.3.18).

$$
\begin{aligned}
R(\theta, \delta) &= sP_\theta[X \geq k] + rN\theta P_\theta[X < k], \ \theta < \theta_0 \\
&= rN\theta P_\theta[X < k], \ \theta \geq \theta_0.
\end{aligned} \tag{1.3.7}
$$

Confidence Bounds and Intervals

Decision theory enables us to think clearly about an important hybrid of testing and estimation, confidence bounds and intervals (and more generally regions). Suppose our primary interest in an estimation type of problem is to give an upper bound for the parameter ν. For instance, an accounting firm examining accounts receivable for a firm on the basis of a random sample of accounts would be primarily interested in an upper bound on the total amount owed. If (say) X represents the amount owed in the sample and ν is the unknown total amount owed, it is natural to seek $\bar{\nu}(X)$ such that

$$
P[\bar{\nu}(X) \geq \nu] \geq 1 - \alpha \tag{1.3.8}
$$

for all possible distributions P of X. Such a $\bar{\nu}$ is called a $(1 - \alpha)$ upper confidence bound on ν. Here α is small, usually .05 or .01 or less. This corresponds to an a priori bound on the risk of α on $\bar{\nu}(X)$ viewed as a decision procedure with action space R and loss function,

$$
\begin{aligned}
l(P, a) &= 0, \quad a \geq \nu(P) \\
&= 1, \quad a < \nu(P)
\end{aligned}
$$

an asymmetric estimation type loss function. The $0 - 1$ nature makes it resemble a testing loss function and, as we shall see in Chapter 4, the connection is close. It is clear, though, that this formulation is inadequate because by taking $\bar{\nu} \equiv \infty$ we can achieve risk $\equiv 0$. What is missing is the fact that, though upper bounding is the primary goal, in fact it is important to get close to the truth—knowing that at most ∞ dollars are owed is of no use. The decision theoretic framework accommodates by adding a component reflecting this. For instance

$$
\begin{aligned}
l(P, a) &= a - \nu(P) \quad , a \geq \nu(P) \\
&= c \qquad\quad , a < \nu(P),
\end{aligned}
$$

for some constant $c > 0$. Typically, rather than this Lagrangian form, it is customary to first fix α in (1.3.8) and then see what one can do to control (say) $R(P, \bar{\nu}) = E(\bar{\nu}(X) - \nu(P))_+$, where $x_+ = x1(x \geq 0)$.

The same issue arises when we are interested in a *confidence interval* $[\underline{\nu}(X), \bar{\nu}(X)]$ for ν defined by the requirement that

$$
P[\underline{\nu}(X) \leq \nu(P) \leq \bar{\nu}(X)] \geq 1 - \alpha
$$

for all $P \in \mathcal{P}$. We shall go into this further in Chapter 4.

We next turn to the final topic of this section, general criteria for selecting "optimal" procedures.

1.3.2   Comparison of Decision Procedures

In this section we introduce a variety of concepts used in the comparison of decision procedures. We shall illustrate some of the relationships between these ideas using the following simple example in which Θ has two members, A has three points, and the risk of all possible decision procedures can be computed and plotted. We conclude by indicating to what extent the relationships suggested by this picture carry over to the general decision theoretic model.

Example 1.3.5. Suppose we have two possible states of nature, which we represent by θ_1 and θ_2. For instance, a component in a piece of equipment either works or does not work; a certain location either contains oil or does not; a patient either has a certain disease or does not, and so on. Suppose that three possible actions, a_1, a_2, and a_3, are available. In the context of the foregoing examples, we could leave the component in, replace it, or repair it; we could drill for oil, sell the location, or sell partial rights; we could operate, administer drugs, or wait and see. Suppose the following loss function is decided on

TABLE 1.3.1. The loss function $l(\theta, a)$

		(Drill) a_1	(Sell) a_2	(Partial rights) a_3
(Oil)	θ_1	0	10	5
(No oil)	θ_2	12	1	6

Thus, if there is oil and we drill, the loss is zero, whereas if there is no oil and we drill, the loss is 12, and so on. Next, an experiment is conducted to obtain information about θ resulting in the random variable X with possible values coded as 0, 1, and frequency function $p(x, \theta)$ given by the following table

TABLE 1.3.2. The frequency function $p(x, \theta_i)$; $i = 1, 2$

		Rock formation x	
		0	1
(Oil)	θ_1	0.3	0.7
(No oil)	θ_2	0.6	0.4

Thus, X may represent a certain geological formation, and when there is oil, it is known that formation 0 occurs with frequency 0.3 and formation 1 with frequency 0.7, whereas if there is no oil, formations 0 and 1 occur with frequencies 0.6 and 0.4. We list all possible decision rules in the following table.

TABLE 1.3.3. Possible decision rules $\delta_i(x)$

					i				
	1	2	3	4	5	6	7	8	9
$x = 0$	a_1	a_1	a_1	a_2	a_2	a_2	a_3	a_3	a_3
$x = 1$	a_1	a_2	a_3	a_1	a_2	a_3	a_1	a_2	a_3

Here, δ_1 represents "Take action a_1 regardless of the value of X," δ_2 corresponds to "Take action a_1, if $X = 0$; take action a_2, if $X = 1$," and so on.

The risk of δ at θ is

$$
\begin{aligned}
R(\theta, \delta) &= E[l(\theta, \delta(X))] = l(\theta, a_1)P[\delta(X) = a_1] \\
&\quad + l(\theta, a_2)P[\delta(X) = a_2] + l(\theta, a_3)P[\delta(X) = a_3].
\end{aligned}
$$

For instance,

$$
\begin{aligned}
R(\theta_1, \delta_2) &= 0(0.3) + 10(0.7) = 7 \\
R(\theta_2, \delta_2) &= 12(0.6) + 1(0.4) = 7.6.
\end{aligned}
$$

If Θ is finite and has k members, we can represent the whole risk function of a procedure δ by a point in k-dimensional Euclidean space, $(R(\theta_1, \delta), \ldots, R(\theta_k, \delta))$ and if $k = 2$ we can plot the set of all such points obtained by varying δ. The *risk points* $(R(\theta_1, \delta_i), R(\theta_2, \delta_i))$ are given in Table 1.3.4 and graphed in Figure 1.3.2 for $i = 1, \ldots, 9$.

TABLE 1.3.4. Risk points $(R(\theta_1, \delta_1), R(\theta_2, \delta_1))$

i	1	2	3	4	5	6	7	8	9
$R(\theta_1, \delta_i)$	0	7	3.5	3	10	6.5	1.5	8.5	5
$R(\theta_2, \delta_i)$	12	7.6	9.6	5.4	1	3	8.4	4.0	6

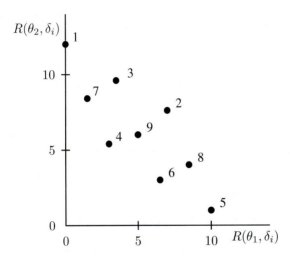

Figure 1.3.2. The risk points $(R(\theta_1, \delta_i), R(\theta_2, \delta_i))$, $i = 1, \ldots, 9$.

It remains to pick out the rules that are "good" or "best." Criteria for doing this will be introduced in the next subsection. □

1.3.3 Bayes and Minimax Criteria

The difficulties of comparing decision procedures have already been discussed in the special contexts of estimation and testing. We say that a procedure δ *improves* a procedure δ' if, and only if,

$$R(\theta, \delta) \leq R(\theta, \delta')$$

for all θ with strict inequality for some θ. It is easy to see that there is typically no rule δ that improves all others. For instance, in estimating $\theta \in R$ when $X \sim N(\theta, \sigma_0^2)$, if we ignore the data and use the estimate $\widehat{\theta} = 0$, we obtain $MSE(\widehat{\theta}) = \theta^2$. The absurd rule "$\delta^*(X) = 0$" cannot be improved on at the value $\theta = 0$ because $E_0(\delta^2(X)) = 0$ if and only if $\delta(X) = 0$. Usually, if δ and δ' are two rules, neither improves the other. Consider, for instance, δ_4 and δ_6 in our example. Here $R(\theta_1, \delta_4) < R(\theta_1, \delta_6)$ but $R(\theta_2, \delta_4) > R(\theta_2, \delta_6)$.

The problem of selecting good decision procedures has been attacked in a variety of ways.

(1) Narrow classes of procedures have been proposed using criteria such as considerations of symmetry, unbiasedness (for estimates and tests), or level of significance (for tests). Researchers have then sought procedures that improve all others within the class. We shall pursue this approach further in Chapter 3. Extensions of unbiasedness ideas may be found in Lehmann (1997, Section 1.5). Symmetry (or invariance) restrictions are discussed in Ferguson (1967).

(2) A second major approach has been to compare risk functions by global criteria rather than on a pointwise basis. We shall discuss the Bayes and minimax criteria.

Bayes: The Bayesian point of view leads to a natural global criterion. Recall that in the Bayesian model θ is the realization of a random variable or vector $\boldsymbol{\theta}$ and that P_θ is the conditional distribution of $\mathbf{X}$ given $\boldsymbol{\theta} = \theta$. In this framework $R(\theta, \delta)$ is just $E[l(\boldsymbol{\theta}, \delta(\mathbf{X})) \mid \boldsymbol{\theta} = \theta]$, the expected loss, if we use δ and $\boldsymbol{\theta} = \theta$. If we adopt the Bayesian point of view, we need not stop at this point, but can proceed to calculate what we expect to lose on the average as $\boldsymbol{\theta}$ varies. This quantity which we shall call the *Bayes risk of δ* and denote $r(\delta)$ is then, given by

$$r(\delta) = E[R(\boldsymbol{\theta}, \delta)] = E[l(\boldsymbol{\theta}, \delta(\mathbf{X}))]. \tag{1.3.9}$$

The second preceding identity is a consequence of the double expectation theorem (B.1.20) in Appendix B.

To illustrate, suppose that in the oil drilling example an expert thinks the chance of finding oil is .2. Then we treat the parameter as a random variable $\boldsymbol{\theta}$ with possible values θ_1, θ_2 and frequency function

$$\pi(\theta_1) = 0.2, \ \pi(\theta_2) = 0.8.$$

The Bayes risk of δ is, therefore,

$$r(\delta) = 0.2R(\theta_1, \delta) + 0.8R(\theta_2, \delta). \tag{1.3.10}$$

Table 1.3.5 gives $r(\delta_1), \ldots, r(\delta_9)$ specified by (1.3.9).

TABLE 1.3.5. Bayes and maximum risks of the procedures of Table 1.3.3.

i	1	2	3	4	5	6	7	8	9
$r(\delta_i)$	9.6	7.48	8.38	4.92	2.8	3.7	7.02	4.9	5.8
$\max\{R(\theta_1, \delta_i), R(\theta_2, \delta_i)\}$	12	7.6	9.6	5.4	10	6.5	8.4	8.5	6

In the Bayesian framework δ is preferable to δ' if, and only if, it has smaller Bayes risk. If there is a rule δ^*, which attains the minimum Bayes risk, that is, such that

$$r(\delta^*) = \min_\delta r(\delta)$$

then it is called a *Bayes rule*. From Table 1.3.5 we see that rule δ_5 is the unique Bayes rule for our prior.

The method of computing Bayes procedures by listing all available δ and their Bayes risk is impracticable in general. We postpone the consideration of posterior analysis, the only reasonable computational method, to Section 3.2.

Note that the Bayes approach leads us to compare procedures on the basis of,

$$r(\delta) = \Sigma_\theta R(\theta, \delta)\pi(\theta),$$

if θ is discrete with frequency function $\pi(\theta)$, and

$$r(\delta) = \int R(\theta, \delta)\pi(\theta)d\theta,$$

if θ is continuous with density $\pi(\theta)$. Such comparisons make sense even if we do not interpret π as a prior density or frequency, but only as a weight function for averaging the values of the function $R(\theta, \delta)$. For instance, in Example 1.3.5 we might feel that both values of the risk were equally important. It is then natural to compare procedures using the simple average $\frac{1}{2}[R(\theta_1, \delta) + R(\theta_2, \delta)]$. But this is just Bayes comparison where π places equal probability on θ_1 and θ_2.

Minimax: Instead of averaging the risk as the Bayesian does we can look at the worst possible risk. This is, we prefer δ to δ', if and only if,

$$\sup_\theta R(\theta, \delta) < \sup_\theta R(\theta, \delta').$$

A procedure δ^*, which has

$$\sup_\theta R(\theta, \delta^*) = \inf_\delta \sup_\theta R(\theta, \delta),$$

is called *minimax* (*mini*mizes the *max*imum risk).

The criterion comes from the general theory of two-person zero sum games of von Neumann.[2] We briefly indicate "the game of decision theory." Nature (Player I) picks a point $\theta \in \Theta$ independently of the statistician (Player II), who picks a decision procedure δ from $\mathcal{D}$, the set of all decision procedures. Player II then pays Player I, $R(\theta, \delta)$. The maximum risk of δ^* is the upper pure value of the game.

This criterion of optimality is very conservative. It aims to give maximum protection against the worst that can happen, Nature's choosing a θ, which makes the risk as large as possible. The principle would be compelling, if the statistician believed that the parameter value is being chosen by a malevolent opponent who knows what decision procedure will be used. Of course, Nature's intentions and degree of foreknowledge are not that clear and most statisticians find the minimax principle too conservative to employ as a general rule. Nevertheless, in many cases the principle can lead to very reasonable procedures.

To illustrate computation of the minimax rule we turn to Table 1.3.4. From the listing of $\max(R(\theta_1, \delta), R(\theta_2, \delta))$ we see that δ_4 is minimax with a maximum risk of 5.4.

Students of game theory will realize at this point that the statistician may be able to lower the maximum risk without requiring any further information by using a random mechanism to determine which rule to employ. For instance, suppose that, in Example 1.3.5, we toss a fair coin and use δ_4 if the coin lands heads and δ_6 otherwise. Our expected risk would be,

$$\frac{1}{2}R(\theta, \delta_4) + \frac{1}{2}R(\theta, \delta_6) \quad = \quad 4.75 \text{ if } \theta = \theta_1$$

$$= \quad 4.20 \text{ if } \theta = \theta_2.$$

The maximum risk 4.75 is strictly less than that of δ_4.

Randomized decision rules: In general, if $\mathcal{D}$ is the class of all decision procedures (nonrandomized), a *randomized decision procedure* can be thought of as a random experiment whose outcomes are members of $\mathcal{D}$. For simplicity we shall discuss only randomized procedures that select among a finite set $\delta_1, \ldots, \delta_q$ of nonrandomized procedures. If the randomized procedure δ selects δ_i with probability λ_i, $i = 1, \ldots, q$, $\sum_{i=1}^q \lambda_i = 1$, we then define

$$R(\theta, \delta) = \sum_{i=1}^q \lambda_i R(\theta, \delta_i). \tag{1.3.11}$$

Similarly we can define, given a prior π on Θ, the *Bayes risk* of δ

$$r(\delta) = \sum_{i=1}^q \lambda_i E[R(\boldsymbol{\theta}, \delta_i)]. \tag{1.3.12}$$

A randomized Bayes procedure δ^* minimizes $r(\delta)$ *among all randomized procedures.* A randomized minimax procedure minimizes $\max_\theta R(\theta, \delta)$ *among all randomized procedures.*

We now want to study the relations between randomized and nonrandomized Bayes and minimax procedures in the context of Example 1.3.5. We will then indicate how much of what we learn carries over to the general case. As in Example 1.3.5, we represent the risk of any procedure δ by the vector $(R(\theta_1, \delta), R(\theta_2, \delta))$ and consider the *risk set*

$$S = \{(R(\theta_1, \delta), R(\theta_2, \delta)) : \delta \in \mathcal{D}^*\}$$

where $\mathcal{D}^*$ is the set of all procedures, including randomized ones.

By (1.3.11),

$$S = \left\{ (r_1, r_2) : r_1 = \sum_{i=1}^9 \lambda_i R(\theta_1, \delta_i), \ r_2 = \sum_{i=1}^9 \lambda_i R(\theta_2, \delta_i), \ \lambda_i \geq 0, \ \sum_{i=1}^9 \lambda_i = 1 \right\}.$$

That is, S is the convex hull of the risk points $(R(\theta_1, \delta_i), R(\theta_2, \delta_i))$, $i = 1, \ldots, 9$ (Figure 1.3.3).

If $\pi(\theta_1) = \gamma = 1 - \pi(\theta_2)$, $0 \leq \gamma \leq 1$, then all rules having Bayes risk c correspond to points in S that lie on the line

$$\gamma r_1 + (1 - \gamma)r_2 = c. \tag{1.3.13}$$

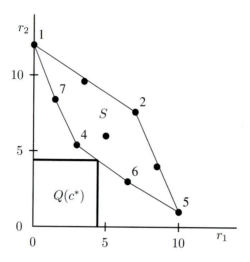

Figure 1.3.3. The convex hull S of the risk points $(R(\theta_1, \delta_i), R(\theta_2, \delta_i))$, $i = 1, \ldots, 9$. The point where the square $Q(c^*)$ defined by (1.3.16) touches S is the risk point of the minimax rule.

As c varies, (1.3.13) defines a family of parallel lines with slope $-\gamma/(1 - \gamma)$. Finding the Bayes rule corresponds to finding the smallest c for which the line (1.3.13) intersects S. This is that line with slope $-\gamma/(1 - \gamma)$ that is tangent to S at the lower boundary of S. All points of S that are on the tangent are Bayes. Two cases arise:

(1) The tangent has a unique point of contact with a risk point corresponding to a nonrandomized rule. For instance, when $\gamma = 0.2$, this point is $(10, 1)$, which is the risk point of the Bayes rule δ_5 (see Figure 1.3.3).

(2) The tangent is the line connecting two "nonrandomized" risk points δ_i, δ_j. A point (r_1, r_2) on this line can be written

$$
\begin{aligned}
r_1 &= \lambda R(\theta_1, \delta_i) + (1 - \lambda) R(\theta_1, \delta_j), \\
r_2 &= \lambda R(\theta_2, \delta_i) + (1 - \lambda) R(\theta_2, \delta_j),
\end{aligned}
\tag{1.3.14}
$$

where $0 \le \lambda \le 1$, and, thus, by (1.3.11) corresponds to the values

$$
\begin{aligned}
\delta &= \delta_i \text{ with probability } \lambda \\
&= \delta_j \text{ with probability } (1 - \lambda), 0 \le \lambda \le 1.
\end{aligned}
\tag{1.3.15}
$$

Each one of these rules, as λ ranges from 0 to 1, is Bayes against π. We can choose two nonrandomized Bayes rules from this class, namely δ_i (take $\lambda = 1$) and δ_j (take $\lambda = 0$).

Because changing the prior π corresponds to changing the slope $-\gamma/(1-\gamma)$ of the line given by (1.3.13), the set B of all risk points corresponding to procedures Bayes with respect to some prior is just the lower left boundary of S (i.e., all points on the lower boundary of S that have as tangents the y axis or lines with nonpositive slopes).

To locate the risk point of the minimax rule consider the family of squares,

$$Q(c) = \{(r_1, r_2) : 0 \le r_1 \le c, \ 0 \le r_2 \le c\} \qquad (1.3.16)$$

whose diagonal is the line $r_1 = r_2$. Let c^* be the smallest c for which $Q(c) \cap S \ne \emptyset$ (i.e., the first square that touches S). Then $Q(c^*) \cap S$ is either a point or a horizontal or vertical line segment. See Figure 1.3.3. It is the set of risk points of minimax rules because any point with smaller maximum risk would belong to $Q(c) \cap S$ with $c < c^*$ contradicting the choice of c^*. In our example, the first point of contact between the squares and S is the intersection between $r_1 = r_2$ and the line connecting the two points corresponding to δ_4 and δ_6. Thus, the minimax rule is given by (1.3.14) with $i = 4$, $j = 6$ and λ the solution of

$$r_1 = \lambda R(\theta_1, \delta_4) + (1 - \lambda)R(\theta_1, \delta_6) = \lambda R(\theta_2, \delta_4) + (1 - \lambda)R(\theta_2, \delta_6) = r_2.$$

From Table 1.3.4, this equation becomes

$$3\lambda + 6.5(1 - \lambda) = 5.4\lambda + 3(1 - \lambda),$$

which yields $\lambda \cong 0.59$.

There is another important concept that we want to discuss in the context of the risk set. A decision rule δ is said to be *inadmissible* if there exists another rule δ' such that δ' improves δ. Naturally, all rules that are not inadmissible are called *admissible*. Using Table 1.3.4 we can see, for instance, that δ_2 is inadmissible because δ_4 improves it (i.e., $R(\theta_1, \delta_4) = 3 < 7 = R(\theta_1, \delta_2)$ and $R(\theta_2, \delta_4) = 5.4 < 7.6 = R(\theta_2, \delta_2)$).

To gain some insight into the class of all admissible procedures (randomized and non-randomized) we again use the risk set. A rule δ with risk point (r_1, r_2) is admissible, if and only if, there is no (x, y) in S such that $x \le r_1$ and $y \le r_2$, or equivalently, if and only if, $\{(x, y) : x \le r_1, \ y \le r_2\}$ has only (r_1, r_2) in common with S. From the figure it is clear that such points must be on the lower left boundary. In fact, the set of all lower left boundary points of S corresponds to the class of admissible rules and, thus, agrees with the set of risk points of Bayes procedures.

If Θ is finite, $\Theta = \{\theta_1, \ldots, \theta_k\}$, we can define the risk set in general as

$$S = \{(R(\theta_1, \delta), \ldots, R(\theta_k, \delta)) : \delta \in \mathcal{D}^*\}$$

where $\mathcal{D}^*$ is the set of all randomized decision procedures. The following features exhibited by the risk set by Example 1.3.5 can be shown to hold generally (see Ferguson, 1967, for instance).

(a) For any prior there is always a nonrandomized Bayes procedure, if there is a randomized one. Randomized Bayes procedures are mixtures of nonrandomized ones in the sense of (1.3.14).

(b) The set B of risk points of Bayes procedures consists of risk points on the lower boundary of S whose tangent hyperplanes have normals pointing into the positive quadrant.

(c) If Θ is finite and minimax procedures exist, they are Bayes procedures.

(d) All admissible procedures are Bayes procedures.

(e) If a Bayes prior has $\pi(\theta_i) > 0$ for all i, then any Bayes procedure corresponding to π is admissible.

If Θ is not finite there are typically admissible procedures that are not Bayes. However, under some conditions, all admissible procedures are either Bayes procedures or limits of Bayes procedures (in various senses). These remarkable results, at least in their original form, are due essentially to Wald. They are useful because the property of being Bayes is easier to analyze than admissibility.

Other theorems are available characterizing larger but more manageable classes of procedures, which include the admissible rules, at least when procedures with the same risk function are identified. An important example is the class of procedures that depend only on knowledge of a sufficient statistic (see Ferguson, 1967; Section 3.4). We stress that looking at randomized procedures is essential for these conclusions, although it usually turns out that all admissible procedures of interest are indeed nonrandomized. For more information on these topics, we refer to Blackwell and Girshick (1954) and Ferguson (1967).

Summary. We introduce the decision theoretic foundation of statistics including the notions of *action space, decision rule, loss function,* and *risk* through various examples including *estimation, testing, confidence bounds, ranking,* and *prediction.* The basic *bias-variance* decomposition of *mean square error* is presented. The basic global comparison criteria *Bayes* and *minimax* are presented as well as a discussion of optimality by restriction and notions of *admissibility.*

1.4 PREDICTION

The prediction Example 1.3.2 presented important situations in which a vector $\mathbf{z}$ of covariates can be used to predict an unseen response Y. Here are some further examples of the kind of situation that prompts our study in this section. A college admissions officer has available the College Board scores at entrance and first-year grade point averages of freshman classes for a period of several years. Using this information, he wants to predict the first-year grade point averages of entering freshmen on the basis of their College Board scores. A stockholder wants to predict the value of his holdings at some time in the future on the basis of his past experience with the market and his portfolio. A meteorologist wants to estimate the amount of rainfall in the coming spring. A government expert wants to predict the amount of heating oil needed next winter. Similar problems abound in every field. The frame we shall fit them into is the following.

We assume that we know the joint probability distribution of a random vector (or variable) $\mathbf{Z}$ and a random variable Y. We want to find a function g defined on the range of

$\mathbf{Z}$ such that $g(\mathbf{Z})$ (the *predictor*) is "close" to Y. In terms of our preceding discussion, $\mathbf{Z}$ is the information that we have and Y the quantity to be predicted. For example, in the college admissions situation, $\mathbf{Z}$ would be the College Board score of an entering freshman and Y his or her first-year grade point average. The joint distribution of $\mathbf{Z}$ and Y can be calculated (or rather well estimated) from the records of previous years that the admissions officer has at his disposal. Next we must specify what *close* means. One reasonable measure of "distance" is $(g(\mathbf{Z}) - Y)^2$, which is the *squared prediction error* when $g(\mathbf{Z})$ is used to predict Y. Since Y is not known, we turn to the *mean squared prediction error* (MSPE)

$$\Delta^2(Y, g(\mathbf{Z})) = E[g(\mathbf{Z}) - Y]^2$$

or its square root $\sqrt{E(g(\mathbf{Z}) - Y)^2}$. The MSPE is the measure traditionally used in the mathematical theory of prediction whose deeper results (see, for example, Grenander and Rosenblatt, 1957) presuppose it. The method that we employ to prove our elementary theorems *does* generalize to other measures of distance than $\Delta(Y, g(\mathbf{Z}))$ such as the mean absolute error $E(|g(\mathbf{Z}) - Y|)$ (Problems 1.4.7–11). Just how widely applicable the notions of this section are will become apparent in Remark 1.4.5 and Section 3.2 where the problem of MSPE prediction is identified with the optimal decision problem of Bayesian statistics with squared error loss.

The class $\mathcal{G}$ of possible predictors g may be the nonparametric class $\mathcal{G}_{NP}$ of all g : $R^d \to R$ or it may be to some subset of this class. See Remark 1.4.6. In this section we consider $\mathcal{G}_{NP}$ and the class $\mathcal{G}_L$ of linear predictors of the form $a + \sum_{j=1}^d b_j Z_j$.

We begin the search for the best predictor in the sense of minimizing MSPE by considering the case in which there is no covariate information, or equivalently, in which $\mathbf{Z}$ is a constant; see Example 1.3.4. In this situation all predictors are constant and the best one is that number c_0 that minimizes $E(Y - c)^2$ as a function of c.

Lemma 1.4.1. *$E(Y - c)^2$ is either ∞ for all c or is minimized uniquely by $c = \mu = E(Y)$. In fact, when $EY^2 < \infty$,*

$$E(Y - c)^2 = \text{Var } Y + (c - \mu)^2. \qquad (1.4.1)$$

Proof. $EY^2 < \infty$ if and only if $E(Y - c)^2 < \infty$ for all c; see Problem 1.4.25. $EY^2 < \infty$ implies that μ exists, and by expanding

$$Y - c = (Y - \mu) + (\mu - c)$$

(1.4.1) follows because $E(Y - \mu) = 0$ makes the cross product term vanish. We see that $E(Y - c)^2$ has a unique minimum at $c = \mu$ and the lemma follows. $\qquad \square$

Now we can solve the problem of finding the best MSPE predictor of Y, given a vector $\mathbf{Z}$; that is, we can find the g that minimizes $E(Y - g(\mathbf{Z}))^2$. By the substitution theorem for conditional expectations (B.1.16), we have

$$E[(Y - g(\mathbf{Z}))^2 \mid \mathbf{Z} = \mathbf{z}] = E[(Y - g(\mathbf{z}))^2 \mid \mathbf{Z} = \mathbf{z}]. \qquad (1.4.2)$$

Let

$$\mu(\mathbf{z}) = E(Y \mid \mathbf{Z} = \mathbf{z}).$$

Because $g(\mathbf{z})$ is a constant, Lemma 1.4.1 assures us that

$$E[(Y - g(\mathbf{z}))^2 \mid \mathbf{Z} = \mathbf{z}] = E[(Y - \mu(\mathbf{z}))^2 \mid \mathbf{Z} = \mathbf{z}] + [g(\mathbf{z}) - \mu(\mathbf{z})]^2. \qquad (1.4.3)$$

If we now take expectations of both sides and employ the double expectation theorem (B.1.20), we can conclude that

Theorem 1.4.1. *If $\mathbf{Z}$ is any random vector and Y any random variable, then either $E(Y - g(\mathbf{Z}))^2 = \infty$ for every function g or*

$$E(Y - \mu(\mathbf{Z}))^2 \leq E(Y - g(\mathbf{Z}))^2 \qquad (1.4.4)$$

for every g with strict inequality holding unless $g(\mathbf{Z}) = \mu(\mathbf{Z})$. That is, $\mu(\mathbf{Z})$ is the unique best MSPE predictor. In fact, when $E(Y^2) < \infty$,

$$E(Y - g(\mathbf{Z}))^2 = E(Y - \mu(\mathbf{Z}))^2 + E(g(\mathbf{Z}) - \mu(\mathbf{Z}))^2. \qquad (1.4.5)$$

An important special case of (1.4.5) is obtained by taking $g(\mathbf{z}) = E(Y)$ for all $\mathbf{z}$. Write $\mathrm{Var}(Y \mid \mathbf{z})$ for the variance of the condition distribution of Y given $\mathbf{Z} = \mathbf{z}$, that is, $\mathrm{Var}(Y \mid \mathbf{z}) = E([Y - E(Y \mid \mathbf{z})]^2 \mid \mathbf{z})$, and recall (B.1.20), then (1.4.5) becomes,

$$\mathrm{Var}\, Y = E(\mathrm{Var}(Y \mid \mathbf{Z})) + \mathrm{Var}(E(Y \mid \mathbf{Z})), \qquad (1.4.6)$$

which is generally valid because if one side is infinite, so is the other.

Property (1.4.6) is linked to a notion that we now define: Two random variables U and V with $E|UV| < \infty$ are said to be *uncorrelated* if

$$E[V - E(V)][U - E(U)] = 0.$$

Equivalently U and V are uncorrelated if either $EV[U - E(U)] = 0$ or $EU[V - E(V)] = 0$. Let $\epsilon = Y - \mu(\mathbf{Z})$ denote the random prediction error, then we can write

$$Y = \mu(\mathbf{Z}) + \epsilon.$$

Proposition 1.4.1. *Suppose that $\mathrm{Var}\, Y < \infty$, then*
(a) *ϵ is uncorrelated with every function of $\mathbf{Z}$*
(b) *$\mu(\mathbf{Z})$ and ϵ are uncorrelated*
(c) *$\mathrm{Var}(Y) = \mathrm{Var}\, \mu(\mathbf{Z}) + \mathrm{Var}\, \epsilon$.*

Proof. To show (a), let $h(\mathbf{Z})$ be any function of $\mathbf{Z}$, then by the iterated expectation theorem,

$$
\begin{aligned}
E\{h(\mathbf{Z})\epsilon\} &= E\{E[h(\mathbf{Z})\epsilon \mid \mathbf{Z}]\} \\
&= E\{h(\mathbf{Z})E[Y - \mu(\mathbf{Z}) \mid \mathbf{Z}]\} = 0
\end{aligned}
$$

because $E[Y - \mu(\mathbf{Z}) \mid \mathbf{Z}] = \mu(\mathbf{Z}) - \mu(\mathbf{Z}) = 0$. Properties (b) and (c) follow from (a). □

Note that Proposition 1.4.1(c) is equivalent to (1.4.6) and that (1.4.5) follows from (a) because (a) implies that the cross product term in the expansion of $E\{[Y - \mu(\mathbf{z})] + [\mu(\mathbf{z}) - g(\mathbf{z})]\}^2$ vanishes.

As a consequence of (1.4.6), we can derive the following theorem, which will prove of importance in estimation theory.

Theorem 1.4.2. *If $E(|Y|) < \infty$ but $\mathbf{Z}$ and Y are otherwise arbitrary, then*

$$\mathrm{Var}(E(Y \mid \mathbf{Z})) \leq \mathrm{Var}\, Y. \tag{1.4.7}$$

If $\mathrm{Var}\, Y < \infty$ *strict inequality holds unless*

$$Y = E(Y \mid \mathbf{Z}) \tag{1.4.8}$$

or equivalently unless Y is a function of $\mathbf{Z}$.

Proof. The assertion (1.4.7) follows immediately from (1.4.6). Equality in (1.4.7) can hold if, and only if,

$$E(\mathrm{Var}(Y \mid \mathbf{Z})) = E(Y - E(Y \mid \mathbf{Z}))^2 = 0.$$

By (A.11.9) this can hold if, and only if, (1.4.8) is true.   □

Example 1.4.1. An assembly line operates either at full, half, or quarter capacity. Within any given month the capacity status does not change. Each day there can be 0, 1, 2, or 3 shutdowns due to mechanical failure. The following table gives the frequency function $p(z, y) = P(Z = z, Y = y)$ of the number of shutdowns Y and the capacity state Z of the line for a randomly chosen day. The row sums of the entries $p_Z(z)$ (given at the end of each row) represent the frequency with which the assembly line is in the appropriate capacity state, whereas the column sums $p_Y(y)$ yield the frequency of 0, 1, 2, or 3 failures among all days. We want to predict the number of failures for a given day knowing the state of the assembly line for the month. We find

$$E(Y \mid Z = 1) = \sum_{i=1}^{3} iP[Y = i \mid Z = 1] = 2.45,$$

$$E\left(Y \mid Z = \tfrac{1}{2}\right) = 2.10, \; E\left(Y \mid Z = \tfrac{1}{4}\right) = 1.20.$$

These fractional figures are not too meaningful as predictors of the natural number values of Y. But this predictor is also the right one, if we are trying to guess, as we reasonably might, the average number of failures per day in a given month. In this case if Y_i represents the number of failures on day i and Z the state of the assembly line, the best predictor is $E\left(30^{-1}\sum_{i=1}^{30} Y_i \mid Z\right) = E(Y \mid Z)$, also.

$z\backslash y$	0	1	2	3	$p_Z(z)$
	$p(z,y)$				
$\frac{1}{4}$	0.10	0.05	0.05	0.05	0.25
$\frac{1}{2}$	0.025	0.025	0.10	0.10	0.25
1	0.025	0.025	0.15	0.30	0.50
$p_Y(y)$	0.15	0.10	0.30	0.45	1

The MSPE of the best predictor can be calculated in two ways. The first is direct.

$$E(Y - E(Y \mid Z))^2 = \sum_z \sum_{y=0}^{3} (y - E(Y \mid Z = z))^2 p(z, y) = 0.885.$$

The second way is to use (1.4.6) writing,

$$\begin{aligned} E(Y - E(Y \mid Z))^2 &= \text{Var}\, Y - \text{Var}(E(Y \mid Z)) \\ &= E(Y^2) - E[(E(Y \mid Z))^2] \\ &= \sum_y y^2 p_Y(y) - \sum_z [E(Y \mid Z = z)]^2 p_Z(z) \\ &= 0.885 \end{aligned}$$

as before. □

Example 1.4.2. *The Bivariate Normal Distribution. Regression toward the mean.* If (Z, Y) has a $\mathcal{N}(\mu_Z, \mu_Y, \sigma_Z^2, \sigma_Y^2, \rho)$ distribution, Theorem B.4.2 tells us that the conditional distribution of Y given $Z = z$ is $\mathcal{N}(\mu_Y + \rho(\sigma_Y/\sigma_Z)(z - \mu_Z), \sigma_Y^2(1 - \rho^2))$. Therefore, the best predictor of Y using Z is the linear function

$$\mu_0(Z) = \mu_Y + \rho(\sigma_Y/\sigma_Z)(Z - \mu_Z).$$

Because

$$E((Y - E(Y \mid Z = z))^2 \mid Z = z) = \sigma_Y^2(1 - \rho^2) \tag{1.4.9}$$

is independent of z, the MSPE of our predictor is given by,

$$E(Y - E(Y \mid Z))^2 = \sigma_Y^2(1 - \rho^2). \tag{1.4.10}$$

The qualitative behavior of this predictor and of its MSPE gives some insight into the structure of the bivariate normal distribution. If $\rho > 0$, the predictor is a monotone increasing function of Z indicating that large (small) values of Y tend to be associated with large (small) values of Z. Similarly, $\rho < 0$ indicates that large values of Z tend to go with small values of Y and we have negative dependence. If $\rho = 0$, the best predictor is just the constant μ_Y as we would expect in the case of independence. One minus the ratio of the MSPE of the best predictor of Y given Z to Var Y, which is the MSPE of the best constant predictor, can reasonably be thought of as a measure of dependence. The larger this quantity the more dependent Z and Y are. In the bivariate normal case, this quantity is just ρ^2. Thus, for this family of distributions the sign of the correlation coefficient gives the type of dependence between Z and Y, whereas its magnitude measures the degree of such dependence.

Because of (1.4.6) we can also write

$$\rho^2 = \frac{\text{Var } \mu_0(Z)}{\text{Var } Y}. \tag{1.4.11}$$

The line $y = \mu_Y + \rho(\sigma_Y/\sigma_Z)(z - \mu_Z)$, which corresponds to the best predictor of Y given Z in the bivariate normal model, is usually called the *regression* (line) of Y on Z. The term *regression* was coined by Francis Galton and is based on the following observation. Suppose Y and Z are bivariate normal random variables with the same mean μ, variance σ^2, and positive correlation ρ. In Galton's case, these were the heights of a randomly selected father (Z) and his son (Y) from a large human population. Then the predicted height of the son, or the average height of sons whose fathers are the height Z, $(1 - \rho)\mu + \rho Z$, is closer to the population mean of heights μ than is the height of the father. Thus, tall fathers tend to have shorter sons; there is "regression toward the mean." This is compensated for by "progression" toward the mean among the sons of shorter fathers and there is no paradox. The variability of the predicted value about μ should, consequently, be less than that of the actual heights and indeed $\text{Var}((1 - \rho)\mu + \rho Z) = \rho^2\sigma^2$. Note that in practice, in particular in Galton's studies, the distribution of (Z, Y) is unavailable and the regression line is estimated on the basis of a sample $(Z_1, Y_1), \ldots, (Z_n, Y_n)$ from the population. We shall see how to do this in Chapter 2.   □

Example 1.4.3. *The Multivariate Normal Distribution.* Let $\mathbf{Z} = (Z_1, \ldots, Z_d)^T$ be a $d \times 1$ covariate vector with mean $\boldsymbol{\mu_Z} = (\mu_1, \ldots, \mu_d)^T$ and suppose that $(\mathbf{Z}^T, Y)^T$ has a $(d + 1)$ multivariate normal, $\mathcal{N}_{d+1}(\boldsymbol{\mu}, \Sigma)$, distribution (Section B.6) in which $\boldsymbol{\mu} = (\mu_{\mathbf{Z}}^T, \mu_Y)^T, \mu_Y = E(Y)$,

$$\Sigma = \begin{pmatrix} \Sigma_{\mathbf{ZZ}} & \Sigma_{\mathbf{Z}Y} \\ \Sigma_{Y\mathbf{Z}} & \sigma_{YY} \end{pmatrix},$$

$\Sigma_{\mathbf{ZZ}}$ is the $d \times d$ variance-covariance matrix $\text{Var}(\mathbf{Z})$,

$$\Sigma_{\mathbf{Z}Y} = (\text{Cov}(Z_1, Y), \ldots, \text{Cov}(Z_d, Y))^T = \Sigma_{Y\mathbf{Z}}^T$$

and $\sigma_{YY} = \text{Var}(Y)$. Theorem B.6.5 states that the conditional distribution of Y given $\mathbf{Z} = \mathbf{z}$ is $\mathcal{N}(\mu_Y + (\mathbf{z} - \boldsymbol{\mu_z})^T\boldsymbol{\beta}, \sigma_{YY|\mathbf{z}})$ where $\boldsymbol{\beta} = \Sigma_{\mathbf{ZZ}}^{-1}\Sigma_{\mathbf{Z}Y}$ and $\sigma_{YY|\mathbf{z}} = \sigma_{YY} - \Sigma_{Y\mathbf{Z}}\Sigma_{\mathbf{ZZ}}^{-1}\Sigma_{\mathbf{Z}Y}$. Thus, the best predictor $E(Y \mid \mathbf{Z})$ of Y is the linear function

$$\mu_0(\mathbf{Z}) = \mu_Y + (\mathbf{Z} - \boldsymbol{\mu_Z})^T\boldsymbol{\beta} \tag{1.4.12}$$

with MSPE

$$E[Y - \mu_0(\mathbf{Z})]^2 = E\{E[Y - \mu_0(\mathbf{Z})]^2 \mid \mathbf{Z}\} = E(\sigma_{YY|\mathbf{z}}) = \sigma_{YY} - \Sigma_{Y\mathbf{Z}}\Sigma_{\mathbf{ZZ}}^{-1}\Sigma_{\mathbf{Z}Y}.$$

The quadratic form $\Sigma_{YZ}\Sigma_{ZZ}^{-1}\Sigma_{ZY}$ is positive except when the joint normal distribution is degenerate, so the MSPE of $\mu_0(\mathbf{Z})$ is smaller than the MSPE of the constant predictor μ_Y. One minus the ratio of these MSPEs is a measure of how strongly the covariates are associated with Y. This quantity is called the *multiple correlation coefficient* (MCC), *coefficient of determination* or *population R-squared*. We write

$$MCC = \rho_{\mathbf{Z}Y}^2 = 1 - \frac{E[Y - \mu_0(\mathbf{Z})]^2}{\text{Var } Y} = \frac{\text{Var } \mu_0(\mathbf{Z})}{\text{Var } Y}$$

where the last identity follows from (1.4.6). By (1.4.11), the MCC equals the square of the usual correlation coefficient $\rho = \sigma_{ZY}/\sigma_{YY}^{\frac{1}{2}}\sigma_{ZZ}^{\frac{1}{2}}$ when $d = 1$.

For example, let Y and $\mathbf{Z} = (Z_1, Z_2)^T$ be the heights in inches of a 10-year-old girl and her parents ($Z_1 = $ mother's height, $Z_2 = $ father's height). Suppose[1] that $(\mathbf{Z}^T, Y)^T$ is trivariate normal with $\text{Var}(Y) = 6.39$

$$\Sigma_{\mathbf{ZZ}} = \begin{pmatrix} 7.74 & 2.92 \\ 2.92 & 6.67 \end{pmatrix}, \quad \Sigma_{\mathbf{Z}Y} = (4.07, 2.98)^T.$$

Then the strength of association between a girl's height and those of her mother and father, respectively; and parents, are

$$\rho_{Z_1,Y}^2 = .335, \quad \rho_{Z_2,Y}^2 = .209, \quad \rho_{\mathbf{Z}Y}^2 = .393.$$

In words, knowing the mother's height reduces the mean squared prediction error over the constant predictor by 33.5%. The percentage reductions knowing the father's and both parent's heights are 20.9% and 39.3%, respectively. In practice, when the distribution of $(\mathbf{Z}^T, Y)^T$ is unknown, the linear predictor $\mu_0(\mathbf{Z})$ and its MSPE will be estimated using a sample $(\mathbf{Z}_1^T, Y_1)^T, \ldots, (\mathbf{Z}_n^T, Y_n)^T$. See Sections 2.1 and 2.2.   □

The best linear predictor. The problem of finding the best MSPE predictor is solved by Theorem 1.4.1. Two difficulties of the solution are that we need fairly precise knowledge of the joint distribution of $\mathbf{Z}$ and Y in order to calculate $E(Y \mid \mathbf{Z})$ and that the best predictor may be a complicated function of $\mathbf{Z}$. If we are willing to sacrifice absolute excellence, we can avoid both objections by looking for a predictor that is best within a class of simple predictors. The natural class to begin with is that of linear combinations of components of $\mathbf{Z}$. We first do the one-dimensional case.

Let us call any random variable of the form $a + bZ$ a *linear predictor* and any such variable with $a = 0$ a *zero intercept linear predictor*. What is the best (zero intercept) linear predictor of Y in the sense of minimizing MSPE? The answer is given by:

Theorem 1.4.3. *Suppose that $E(Z^2)$ and $E(Y^2)$ are finite and Z and Y are not constant. Then the unique best zero intercept linear predictor is obtained by taking*

$$b = b_0 = \frac{E(ZY)}{E(Z^2)},$$

whereas the unique best linear predictor is $\mu_L(Z) = a_1 + b_1 Z$ where

$$b_1 = \frac{\text{Cov}(Z, Y)}{\text{Var } Z}, \quad a_1 = E(Y) - b_1 E(Z).$$

Proof. We expand $\{Y - bZ\}^2 = \{Y - [Z(b - b_0) + Zb_0]\}^2$ to get

$$E(Y - bZ)^2 = E(Y^2) + E(Z^2)(b - b_0)^2 - E(Z^2)b_0^2.$$

Therefore, $E(Y - bZ)^2$ is uniquely minimized by $b = b_0$, and

$$E(Y - b_0 Z)^2 = E(Y^2) - \frac{[E(ZY)]^2}{E(Z^2)}. \qquad (1.4.13)$$

To prove the second assertion of the theorem note that by (1.4.1),

$$E(Y - a - bZ)^2 = \text{Var}(Y - bZ) + (E(Y) - bE(Z) - a)^2.$$

Therefore, whatever be b, $E(Y - a - bZ)^2$ is uniquely minimized by taking

$$a = E(Y) - bE(Z).$$

Substituting this value of a in $E(Y - a - bZ)^2$ we see that the b we seek minimizes $E[(Y - E(Y)) - b(Z - E(Z))]^2$. We can now apply the result on zero intercept linear predictors to the variables $Z - E(Z)$ and $Y - E(Y)$ to conclude that b_1 is the unique minimizing value. $\qquad\square$

Remark 1.4.1. From (1.4.13) we obtain the proof of the Cauchy–Schwarz inequality (A.11.17) in the appendix. This is because $E(Y - b_0 Z)^2 \geq 0$ is equivalent to the Cauchy–Schwarz inequality with equality holding if, and only if, $E(Y - b_0 Z)^2 = 0$, which corresponds to $Y = b_0 Z$. We could similarly obtain (A.11.16) directly by calculating $E(Y - a_1 - b_1 Z)^2$. $\qquad\square$

Note that if $E(Y \mid Z)$ is of the form $a + bZ$, then $a = a_1$ and $b = b_1$, because, by (1.4.5), if the best predictor is linear, it must coincide with the best linear predictor. This is in accordance with our evaluation of $E(Y \mid Z)$ in Example 1.4.2. In that example nothing is lost by using linear prediction. On the other hand, in Example 1.4.1 the best linear predictor and best predictor differ (see Figure 1.4.1). A loss of about 5% is incurred by using the best linear predictor. That is,

$$\frac{E[Y - \mu_L(Z)]^2}{E[Y - \mu(Z)]^2} = 1.05.$$

Best Multivariate Linear Predictor. Our linear predictor is of the form

$$\mu_l(\mathbf{Z}) = a + \sum_{j=1}^{d} b_j Z_j = a + \mathbf{Z}^T \mathbf{b}$$

where $\mathbf{Z} = (Z_1, \ldots, Z_d)^T$ and $\mathbf{b} = (b_1, \ldots, b_d)^T$. Let

$$\boldsymbol{\beta} = (E([\mathbf{Z} - E(\mathbf{Z})][\mathbf{Z} - E(\mathbf{Z})]^T))^{-1} E([\mathbf{Z} - E(\mathbf{Z})][Y - E(Y)]) = \Sigma_{\mathbf{ZZ}}^{-1} \Sigma_{\mathbf{Z}Y}.$$

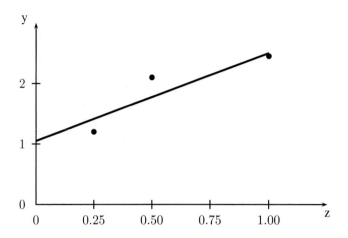

Figure 1.4.1. The three dots give the best predictor. The line represents the best linear predictor $y = 1.05 + 1.45z$.

Theorem 1.4.4. *If EY^2 and $(E([\mathbf{Z} - E(\mathbf{Z})]^T[\mathbf{Z} - E(\mathbf{Z})]))^{-1}$ exist, then the unique best linear* MSPE *predictor is*

$$\mu_L(\mathbf{Z}) = \mu_Y + (\mathbf{Z} - \boldsymbol{\mu}_\mathbf{Z})^T\boldsymbol{\beta}. \tag{1.4.14}$$

Proof. Note that $R(a, \mathbf{b}) \equiv E_p[Y - \mu_l(\mathbf{Z})]^2$ depends on the joint distribution P of $\mathbf{X} = (\mathbf{Z}^T, Y)^T$ only through the expectation $\boldsymbol{\mu}$ and covariance Σ of $\mathbf{X}$. Let P_0 denote the multivariate normal, $\mathcal{N}(\boldsymbol{\mu}, \Sigma)$, distribution and let $R_0(a, \mathbf{b}) = E_{P_0}[Y - \mu_l(\mathbf{Z})]^2$. By Example 1.4.3, $R_0(a, \mathbf{b})$ is minimized by (1.4.14). Because P and P_0 have the same $\boldsymbol{\mu}$ and Σ, $R(a, \mathbf{b}) = R_0(a, \mathbf{b})$, and $R(a, \mathbf{b})$ is minimized by (1.4.14).   □

Remark 1.4.2. We could also have established Theorem 1.4.4 by extending the proof of Theorem 1.4.3 to $d > 1$. However, our new proof shows how second-moment results sometimes can be established by "connecting" them to the normal distribution. A third approach using calculus is given in Problem 1.4.19.   □

Remark 1.4.3. In the general, not necessarily normal, case the *multiple correlation coefficient* (MCC) or *coefficient of determination* is defined as the correlation between Y and the best linear predictor of Y; that is,

$$\rho_{\mathbf{Z}Y}^2 = \text{Corr}^2(Y, \mu_L(\mathbf{Z})).$$

Thus, the MCC gives the strength of the *linear* relationship between $\mathbf{Z}$ and Y. See Problem 1.4.17 for an overall measure of the strength of this relationship.   □

Remark 1.4.4. Suppose the model for $\mu(\mathbf{Z})$ is linear; that is,

$$\mu(\mathbf{Z}) = E(Y \mid \mathbf{Z}) = \alpha + \mathbf{Z}^T \boldsymbol{\beta}$$

for unknown $\alpha \in R$ and $\boldsymbol{\beta} \in R^d$. We want to express α and $\boldsymbol{\beta}$ in terms of moments of $(\mathbf{Z}, Y)$. Set $Z_0 = 1$. By Proposition 1.4.1(a), $\epsilon = Y - \mu(\mathbf{Z})$ and each of $Z_0, \ldots, Z_d$ are uncorrelated; thus,

$$E(Z_j[Y - (\alpha + \mathbf{Z}^T \boldsymbol{\beta})]) = 0, \ j = 0, \ldots, d. \qquad (1.4.15)$$

Solving (1.4.15) for α and $\boldsymbol{\beta}$ gives (1.4.14) (Problem 1.4.23). Because the multivariate normal model is a linear model, this gives a new derivation of (1.4.12).

Remark 1.4.5. Consider the Bayesian model of Section 1.2 and the Bayes risk (1.3.8) defined by $r(\delta) = E[l(\boldsymbol{\theta}, \delta(\mathbf{X}))]$. If we identify $\boldsymbol{\theta}$ with Y and $\mathbf{X}$ with $\mathbf{Z}$, we see that $r(\delta) = \text{MSPE}$ for squared error loss $l(\theta, \delta) = (\theta - \delta)^2$. Thus, the optimal MSPE predictor $E(\boldsymbol{\theta} \mid \mathbf{X})$ is the Bayes procedure for squared error loss. We return to this in Section 3.2. $\qquad\qquad\qquad\qquad\qquad\qquad\qquad\qquad\qquad\qquad\qquad\qquad\qquad\qquad\qquad\quad\square$

Remark 1.4.6. When the class $\mathcal{G}$ of possible predictors g with $E|g(\mathbf{Z})| < \infty$ form a Hilbert space as defined in Section B.10 and there is a $g_0 \in \mathcal{G}$ such that

$$g_0 = \arg \inf\{\Delta(Y, g(\mathbf{Z})) : g \in \mathcal{G}\},$$

then $g_0(\mathbf{Z})$ is called the *projection* of Y on the space $\mathcal{G}$ of functions of $\mathbf{Z}$ and we write $g_0(\mathbf{Z}) = \pi(Y \mid \mathcal{G})$. Moreover, $g(\mathbf{Z})$ and $h(\mathbf{Z})$ are said to be *orthogonal* if at least one has expected value zero and $E[g(\mathbf{Z})h(\mathbf{Z})] = 0$. With these concepts the results of this section are linked to the general Hilbert space results of Section B.10. Using the distance Δ and projection π notation, we can conclude that

$$\mu(\mathbf{Z}) = \pi(Y \mid \mathcal{G}_{NP}), \ \mu_L(\mathbf{Z}) = \pi(Y \mid \mathcal{G}_L) = \pi(\mu(\mathbf{Z}) \mid \mathcal{G}_L)$$
$$\Delta^2(Y, \mu_L(\mathbf{Z})) = \Delta^2(\mu_L(\mathbf{Z}), \mu(\mathbf{Z})) + \Delta^2(Y, \mu(\mathbf{Z})) \qquad (1.4.16)$$
$$Y - \mu(\mathbf{Z}) \text{ is orthogonal to } \mu(\mathbf{Z}) \text{ and to } \mu_L(\mathbf{Z}). \qquad (1.4.17)$$

Note that (1.4.16) is the Pythagorean identity. $\qquad\qquad\qquad\qquad\qquad\qquad\qquad\qquad\qquad\quad\square$

Summary. We consider situations in which the goal is to predict the (perhaps in the future) value of a random variable Y. The notion of *mean squared prediction error* (MSPE) is introduced, and it is shown that if we want to predict Y on the basis of information contained in a random vector $\mathbf{Z}$, the optimal MSPE predictor is the conditional expected value of Y given $\mathbf{Z}$. The optimal MSPE predictor in the multivariate normal distribution is presented. It is shown to coincide with the optimal MSPE predictor when the model is left general but the class of possible predictors is restricted to be linear.

1.5   SUFFICIENCY

Once we have postulated a statistical model, we would clearly like to separate out any aspects of the data that are irrelevant in the context of the model and that may obscure our understanding of the situation.

We begin by formalizing what we mean by "a reduction of the data" $\mathbf{X} \in \mathcal{X}$. Recall that a *statistic* is any function of the observations generically denoted by $T(\mathbf{X})$ or T. The range of T is any space of objects $\mathcal{T}$, usually R or R^k, but as we have seen in Section 1.1.3, can also be a set of functions. If T assigns the same value to different sample points, then by recording or taking into account only the value of $T(\mathbf{X})$ we have a reduction of the data. Thus, $T(\mathbf{X}) = \bar{X}$ loses information about the X_i as soon as $n > 1$. Even $T(X_1, \dots, X_n) = (X_{(1)}, \dots, X_{(n)})$, loses information about the labels of the X_i. The idea of sufficiency is to reduce the data with statistics whose use involves no loss of information, in the context of a model $\mathcal{P} = \{P_\theta : \theta \in \Theta\}$.

For instance, suppose that in Example 1.1.1 we had sampled the manufactured items in order, recording at each stage whether the examined item was defective or not. We could then represent the data by a vector $\mathbf{X} = (X_1, \dots, X_n)$ where $X_i = 1$ if the ith item sampled is defective and $X_i = 0$ otherwise. The total number of defective items observed, $T = \sum_{i=1}^{n} X_i$, is a statistic that maps many different values of $(X_1, \dots, X_n)$ into the same number. However, it is intuitively clear that if we are interested in the proportion θ of defective items nothing is lost in this situation by recording and using only T.

One way of making the notion "a statistic whose use involves no loss of information" precise is the following. A statistic $T(\mathbf{X})$ is called *sufficient* for $P \in \mathcal{P}$ or the parameter θ if the conditional distribution of $\mathbf{X}$ given $T(\mathbf{X}) = t$ does not involve θ. Thus, once the value of a sufficient statistic T is known, the sample $\mathbf{X} = (X_1, \dots, X_n)$ does not contain any further information about θ or equivalently P, *given that $\mathcal{P}$ is valid*. We give a decision theory interpretation that follows. The most trivial example of a sufficient statistic is $T(\mathbf{X}) = \mathbf{X}$ because by any interpretation the conditional distribution of $\mathbf{X}$ given $T(\mathbf{X}) = \mathbf{X}$ is point mass at $\mathbf{X}$.

Example 1.5.1. A machine produces n items in succession. Each item produced is good with probability θ and defective with probability $1 - \theta$, where θ is unknown. Suppose there is no dependence between the quality of the items produced and let $X_i = 1$ if the ith item is good and 0 otherwise. Then $\mathbf{X} = (X_1, \dots, X_n)$ is the record of n Bernoulli trials with probability θ. By (A.9.5),

$$P[X_1 = x_1, \dots, X_n = x_n] = \theta^t (1 - \theta)^{n-t} \qquad (1.5.1)$$

where x_i is 0 or 1 and $t = \sum_{i=1}^{n} x_i$. By Example B.1.1, the conditional distribution of $\mathbf{X}$ given $T = \sum_{i=1}^{n} X_i = t$ does not involve θ. Thus, T is a sufficient statistic for θ. □

Example 1.5.2. Suppose that arrival of customers at a service counter follows a Poisson process with arrival rate (parameter) θ. Let X_1 be the time of arrival of the first customer, X_2 the time between the arrival of the first and second customers. By (A.16.4), X_1 and X_2 are independent and identically distributed exponential random variables with parameter θ. We prove that $T = X_1 + X_2$ is sufficient for θ. Begin by noting that according to Theorem B.2.3, whatever be θ, $X_1/(X_1 + X_2)$ and $X_1 + X_2$ are independent and the first of these statistics has a uniform distribution on $(0, 1)$. Therefore, the conditional distribution of $X_1/(X_1 + X_2)$ given $X_1 + X_2 = t$ is $\mathcal{U}(0, 1)$ whatever be t. Using our discussion in Section B.1.1 we see that given $X_1 + X_2 = t$, the conditional distribution of $X_1 = [X_1/(X_1 + X_2)](X_1 + X_2)$ and that of $X_1 t/(X_1 + X_2)$ are the same and we can conclude that given $X_1 + X_2 = t$, X_1 has a $\mathcal{U}(0, t)$ distribution. It follows that, when $X_1 + X_2 = t$, whatever be θ, (X_1, X_2) is conditionally distributed as (X, Y) where X is uniform on $(0, t)$ and $Y = t - X$. Thus, $X_1 + X_2$ is sufficient. $\qquad\qquad\qquad\qquad\qquad\qquad\qquad\qquad\square$

In both of the foregoing examples considerable reduction has been achieved. Instead of keeping track of several numbers, we need only record one. Although the sufficient statistics we have obtained are "natural," it is important to notice that there are many others that will do the same job. Being told that the numbers of successes in five trials is three is the same as knowing that the difference between the numbers of successes and the number of failures is one. More generally, if T_1 and T_2 are any two statistics such that $T_1(\mathbf{x}) = T_1(\mathbf{y})$ if and only if $T_2(\mathbf{x}) = T_2(\mathbf{y})$, then T_1 and T_2 provide the same information and achieve the same reduction of the data. Such statistics are called *equivalent*.

In general, checking sufficiency directly is difficult because we need to compute the conditional distribution. Fortunately, a simple necessary and sufficient criterion for a statistic to be sufficient is available. This result was proved in various forms by Fisher, Neyman, and Halmos and Savage. It is often referred to as the *factorization theorem* for sufficient statistics.

Theorem 1.5.1. *In a regular model, a statistic $T(\mathbf{X})$ with range $\mathcal{T}$ is sufficient for θ if, and only if, there exists a function $g(t, \theta)$ defined for t in $\mathcal{T}$ and θ in Θ and a function h defined on $\mathcal{X}$ such that*

$$p(\mathbf{x}, \theta) = g(T(\mathbf{x}), \theta)h(\mathbf{x}) \qquad\qquad\qquad (1.5.2)$$

for all $\mathbf{x} \in \mathcal{X}$, $\theta \in \Theta$.

We shall give the proof in the discrete case. The complete result is established for instance by Lehmann (1997, Section 2.6).

Proof. Let $(\mathbf{x}_1, \mathbf{x}_2, \dots)$ be the set of possible realizations of $\mathbf{X}$ and let $t_i = T(\mathbf{x}_i)$. Then T is discrete and $\sum_{i=1}^{\infty} P_\theta[T = t_i] = 1$ for every θ. To prove the sufficiency of (1.5.2), we need only show that $P_\theta[\mathbf{X} = \mathbf{x}_j | T = t_i]$ is independent of θ for every i and j. By our definition of conditional probability in the discrete case, it is enough to show that $P_\theta[\mathbf{X} = \mathbf{x}_j | T = t_i]$ is independent of θ on each of the sets $S_i = \{\theta : P_\theta[T = t_i] > 0\}$, $i = 1, 2, \dots$. Now, if (1.5.2) holds,

$$P_\theta[T = t_i] = \sum_{\{\mathbf{x}:T(\mathbf{x})=t_i\}} p(\mathbf{x}, \theta) = g(t_i, \theta) \sum_{\{\mathbf{x}:T(\mathbf{x})=t_i\}} h(\mathbf{x}). \qquad (1.5.3)$$

By (B.1.1) and (1.5.2), for $\theta \in S_i$,

$$
\begin{aligned}
P_\theta[\mathbf{X} = \mathbf{x}_j | T = t_i] &= P_\theta[\mathbf{X} = \mathbf{x}_j, T = t_i]/P_\theta[T = t_i] \\[2mm]
&= \frac{p(\mathbf{x}_j, \theta)}{P_\theta[T = t_i]} \\[2mm]
&= \frac{g(t_i, \theta)h(\mathbf{x}_j)}{P_\theta[T = t_i]} \quad \text{if } T(\mathbf{x}_j) = t_i \\[2mm]
&= 0 \text{ if } T(\mathbf{x}_j) \neq t_i.
\end{aligned}
\qquad (1.5.4)
$$

Applying (1.5.3) we arrive at,

$$P_\theta[\mathbf{X} = \mathbf{x}_j | T = t_i] \quad = \quad 0 \text{ if } T(\mathbf{x}_j) \neq t_i$$

$$= \quad \frac{h(\mathbf{x}_j)}{\sum_{\{\mathbf{x}_k : T(\mathbf{x}_k) = t_i\}} h(\mathbf{x}_k)} \text{ if } T(\mathbf{x}_j) = t_i. \quad (1.5.5)$$

Therefore, T is sufficient. Conversely, if T is sufficient, let

$$g(t_i, \theta) = P_\theta[T = t_i], h(\mathbf{x}) = P_\theta[\mathbf{X} = \mathbf{x} | T(\mathbf{X}) = t_i] \quad (1.5.6)$$

Then

$$p(\mathbf{x}, \theta) = P_\theta[\mathbf{X} = \mathbf{x}, T = T(\mathbf{x})] = g(T(\mathbf{x}), \theta) h(\mathbf{x}) \quad (1.5.7)$$

by (B.1.3). $\square$

Example 1.5.2 (continued). If $X_1, \ldots, X_n$ are the interarrival times for n customers, then the joint density of $(X_1, \ldots, X_n)$ is given by (see (A.16.4)),

$$p(x_1, \ldots, x_n, \theta) = \theta^n \exp[-\theta \sum_{i=1}^{n} x_i] \quad (1.5.8)$$

if all the x_i are > 0, and $p(x_1, \ldots, x_n, \theta) = 0$ otherwise. We may apply Theorem 1.5.1 to conclude that $T(X_1, \ldots, X_n) = \sum_{i=1}^{n} X_i$ is sufficient. Take $g(t, \theta) = \theta^n e^{-\theta t}$ if $t > 0$, $\theta > 0$, and $h(x_1, \ldots, x_n) = 1$ if all the x_i are > 0, and both functions $= 0$ otherwise. A whole class of distributions, which admits simple sufficient statistics and to which this example belongs, are introduced in the next section. $\square$

Example 1.5.3. *Estimating the Size of a Population.* Consider a population with θ members labeled consecutively from 1 to θ. The population is sampled with replacement and n members of the population are observed and their labels $X_1, \ldots, X_n$ are recorded. Common sense indicates that to get information about θ, we need only keep track of $X_{(n)} = \max(X_1, \ldots X_n)$. In fact, we can show that $X_{(n)}$ is sufficient. The probability distribution of $\mathbf{X}$ is given by

$$p(x_1, \ldots, x_n, \theta) = \theta^{-n} \quad (1.5.9)$$

if every x_i is an integer between 1 and θ and $p(x_1, \ldots, x_n, \theta) = 0$ otherwise. Expression (1.5.9) can be rewritten as

$$p(x_1, \ldots, x_n, \theta) = \theta^{-n} 1\{x_{(n)} \leq \theta\}, \quad (1.5.10)$$

where $x_{(n)} = \max(x_1, \ldots, x_n)$. By Theorem 1.5.1, $X_{(n)}$ is a sufficient statistic for θ. $\square$

Example 1.5.4. Let $X_1, \ldots, X_n$ be independent and identically distributed random variables each having a normal distribution with mean μ and variance σ^2, both of which are unknown. Let $\theta = (\mu, \sigma^2)$. Then the density of $(X_1, \ldots, X_n)$ is given by

$$p(x_1, \ldots, x_n, \theta) \;=\; [2\pi\sigma^2]^{-n/2} \exp\{-\frac{1}{2\sigma^2} \sum_{i=1}^{n}(x_i - \mu)^2\}$$

$$= \; [2\pi\sigma^2]^{-n/2}[\exp\{-\frac{n\mu^2}{2\sigma^2}\}][\exp\{-\frac{1}{2\sigma^2}(\sum_{i=1}^{n} x_i^2 - 2\mu \sum_{i=1}^{n} x_i)\}].$$

$$(1.5.11)$$

Evidently $p(x_1, \ldots, x_n, \theta)$ is itself a function of $(\sum_{i=1}^{n} x_i, \sum_{i=1}^{n} x_i^2)$ and θ only and upon applying Theorem 1.5.1 we can conclude that

$$T(X_1, \ldots, X_n) = (\sum_{i=1}^{n} X_i, \sum_{i=1}^{n} X_i^2)$$

is sufficient for θ. An equivalent sufficient statistic in this situation that is frequently used is

$$S(X_1, \ldots, X_n) = [(1/n) \sum_{i=1}^{n} X_i, [1/(n-1)] \sum_{i=1}^{n}(X_i - \bar{X})^2],$$

where $\bar{X} = (1/n) \sum_{i=1}^{n} X_i$. The first and second components of this vector are called the *sample mean* and the *sample variance*, respectively. □

Example 1.5.5. Suppose, as in Example 1.1.4 with $d = 2$, that $Y_1, \ldots, Y_n$ are independent, $Y_i \sim \mathcal{N}(\mu_i, \sigma^2)$, with μ_i following the linear regresssion model

$$\mu_i = \beta_1 + \beta_2 z_i, \; i = 1, \ldots, n,$$

where we assume that the given constants $\{z_i\}$ are not all identical. Then $\boldsymbol{\theta} = (\beta_1, \beta_2, \sigma^2)^T$ is identifiable (Problem 1.1.9) and

$$p(\mathbf{y}, \boldsymbol{\theta}) = (2\pi\sigma^2)^{-\frac{n}{2}} \exp\left\{\frac{-\Sigma(\beta_1 + \beta_2 z_i)^2}{2\sigma^2}\right\} \exp\left\{\frac{-\Sigma Y_i^2 + 2\beta_1 \Sigma Y_i + 2\beta_2 \Sigma z_i Y_i}{2\sigma^2}\right\}.$$

Thus, $\mathbf{T} = (\Sigma Y_i, \Sigma Y_i^2, \Sigma z_i Y_i)$ is sufficient for $\boldsymbol{\theta}$. □

Sufficiency and decision theory

Sufficiency can be given a clear operational interpretation in the decision theoretic setting. Specifically, if $T(\mathbf{X})$ is sufficient, we can, for any decision procedure $\delta(\mathbf{x})$, find a randomized decision rule $\delta^*(T(\mathbf{X}))$ depending only on $T(\mathbf{X})$ that does as well as $\delta(\mathbf{X})$ in the sense of having the same risk function; that is,

$$R(\theta, \delta) = R(\theta, \delta^*) \text{ for all } \theta. \qquad (1.5.12)$$

By *randomized* we mean that $\delta^*(T(\mathbf{X}))$ can be generated from the value t of $T(\mathbf{X})$ and a random mechanism not depending on θ.

Here is an example.

Example 1.5.6. Suppose $X_1, \ldots, X_n$ are independent identically $\mathcal{N}(\theta, 1)$ distributed. Then

$$p(\mathbf{x}, \theta) = \exp\{n\theta(\bar{x} - \tfrac{1}{2}\theta)\}(2\pi)^{-\frac{1}{2}n} \exp\{-\tfrac{1}{2}\sum x_i^2\}$$

By the factorization theorem, $\bar{X}$ is sufficient. Let $\delta(\mathbf{X}) = X_1$. Using only $\bar{X}$, we construct a rule $\delta^*(\mathbf{X})$ with the same risk = mean squared error as $\delta(\mathbf{X})$ as follows: Conditionally, given $\bar{X} = t$, choose $T^* = \delta^*(\mathbf{X})$ from the normal $\mathcal{N}(t, \frac{n-1}{n})$ distribution. Using Section B.1 and (1.4.6), we find

$$E(T^*) = E[E(T^*|\bar{X})] = E(\bar{X}) = \theta = E(X_1)$$

$$\text{Var}(T^*) = E[\text{Var}(T^*|\bar{X})] + \text{Var}[E(T^*|\bar{X})] = \frac{n-1}{n} + \frac{1}{n} = 1 = \text{Var}(X_1).$$

Thus, $\delta^*(\mathbf{X})$ and $\delta(\mathbf{X})$ have the same mean squared error.   □

The proof of (1.5.12) follows along the lines of the preceding example: Given $T(\mathbf{X}) = t$, the distribution of $\delta(\mathbf{X})$ does not depend on θ. Now draw δ^* randomly from this conditional distribution. This $\delta^*(T(\mathbf{X}))$ will have the same risk as $\delta(\mathbf{X})$ because, by the double expectation theorem,

$$R(\theta, \delta^*) = E\{E[\ell(\theta, \delta^*(T))|T]\} = E\{E[\ell(\theta, \delta(\mathbf{X}))|T]\} = R(\theta, \delta).$$

□

Sufficiency and Bayes models

There is a natural notion of sufficiency of a statistic T in the Bayesian context where in addition to the model $\mathcal{P} = \{P_\theta : \theta \in \Theta\}$ we postulate a prior distribution Π for Θ.

In Example 1.2.1 (Bernoulli trials) we saw that the posterior distribution given $\mathbf{X} = \mathbf{x}$ is the same as the posterior distribution given $T(\mathbf{X}) = \sum_{i=1}^n X_i = k$, where $k = \sum_{i=1}^n X_i$. In this situation we call T Bayes sufficient.

Definition. $T(\mathbf{X})$ is *Bayes sufficient* for Π if the posterior distribution of θ given $\mathbf{X} = \mathbf{x}$ is the same as the posterior (conditional) distribution of θ given $T(\mathbf{X}) = T(\mathbf{x})$ for all $\mathbf{x}$.

Equivalently, θ and $\mathbf{X}$ are independent given $T(\mathbf{X})$.

Theorem 1.5.2. (Kolmogorov). *If $T(\mathbf{X})$ is sufficient for θ, it is Bayes sufficient for every Π.*

This result and a partial converse is the subject of Problem 1.5.14.

Minimal sufficiency

For any model there are many sufficient statistics: Thus, if $X_1, \ldots, X_n$ is a $\mathcal{N}(\mu, \sigma^2)$ sample $n \geq 2$, then $T(\mathbf{X}) = (\sum_{i=1}^n X_i, \sum_{i=1}^n X_i^2)$ and $S(\mathbf{X}) = (X_1, \ldots, X_n)$ are both sufficient. But $T(\mathbf{X})$ provides a greater reduction of the data. We define the statistic $T(\mathbf{X})$ to be *minimally sufficient* if it is sufficient and provides a greater reduction of the data than any other sufficient statistic $S(\mathbf{X})$, in that, we can find a transformation r such that $T(\mathbf{X}) = r(S(\mathbf{X}))$.

Example 1.5.1 (continued). In this Bernoulli trials case, $T = \sum_{i=1}^{n} X_i$ was shown to be sufficient. Let $S(\mathbf{X})$ be any other sufficient statistic. Then by the factorization theorem we can write $p(\mathbf{x}, \theta)$ as

$$p(\mathbf{x}, \theta) = g(S(\mathbf{x}), \theta)h(\mathbf{x})$$

Combining this with (1.5.1), we find

$$\theta^T (1 - \theta)^{n-T} = g(S(\mathbf{x}), \theta)h(\mathbf{x}) \text{ for all } \theta.$$

For any two fixed θ_1 and θ_2, the ratio of both sides of the foregoing gives

$$(\theta_1/\theta_2)^T [(1 - \theta_1)/(1 - \theta_2)]^{n-T} = g(S(\mathbf{x}), \theta_1)/g(S(\mathbf{x}), \theta_2).$$

In particular, if we set $\theta_1 = 2/3$ and $\theta_2 = 1/3$, take the log of both sides of this equation and solve for T, we find

$$T = r(S(\mathbf{x})) = \{\log[2^n g(S(\mathbf{x}), 2/3)/g(S(\mathbf{x}), 1/3)]\}/2 \log 2.$$

Thus, T is minimally sufficient. □

The likelihood function

The preceding example shows how we can use $p(\mathbf{x}, \theta)$ for different values of θ and the factorization theorem to establish that a sufficient statistic is minimally sufficient. We define the likelihood function L for a given observed data vector $\mathbf{x}$ as

$$L_{\mathbf{x}}(\theta) = p(\mathbf{x}, \theta), \theta \in \Theta.$$

Thus, $L_{\mathbf{x}}$ is a map from the sample space $\mathcal{X}$ to the class $\mathcal{T}$ of functions $\{\theta \to p(\mathbf{x}, \theta) : \mathbf{x} \in \mathcal{X}\}$. It is a statistic whose values are functions; if $\mathbf{X} = \mathbf{x}$, the statistic L takes on the value $L_{\mathbf{x}}$. In the discrete case, for a given θ, $L_{\mathbf{x}}(\theta)$ gives the probability of observing the point $\mathbf{x}$. In the continuous case it is approximately proportional to the probability of observing a point in a small rectangle around $\mathbf{x}$. However, when we think of $L_{\mathbf{x}}(\theta)$ as a function of θ, it gives, for a given observed $\mathbf{x}$, the "likelihood" or "plausibility" of various θ. The formula (1.2.8) for the posterior distribution can then be remembered as Posterior $\propto$ (Prior) $\times$ (Likelihood) where the sign $\propto$ denotes proportionality as functions of θ.

Example 1.5.4 (continued). In this $\mathcal{N}(\mu, \sigma^2)$ example, the likelihood function (1.5.11) is determined by the two-dimensional sufficient statistic

$$T = (T_1, T_2) = (\sum_{i=1}^{n} X_i, \sum_{i=1}^{n} X_i^2).$$

Set $\theta = (\theta_1, \theta_2) = (\mu, \sigma^2)$, then

$$L_{\mathbf{x}}(\theta) = (2\pi\theta_2)^{-n/2} \exp\{-\frac{n\theta_1^2}{2\theta_2}\} \exp\{-\frac{1}{2\theta_2}(t_2 - 2\theta_1 t_1)\}.$$

Now, as a function of θ, $L_{\mathbf{x}}(\cdot)$ determines (t_1, t_2) because, for example,

$$t_2 = -2\log L_{\mathbf{x}}(0, 1) - n\log 2\pi$$

with a similar expression for t_1 in terms of $L_{\mathbf{x}}(0, 1)$ and $L_{\mathbf{x}}(1, 1)$ (Problem 1.5.17). Thus, L is a statistic that is equivalent to (t_1, t_2) and, hence, itself sufficient. By arguing as in Example 1.5.1 (continued) we can show that T and, hence, L is minimal sufficient. □

In fact, a statistic closely related to L solves the minimal sufficiency problem in general. Suppose there exists θ_0 such that

$$\{\mathbf{x} : p(\mathbf{x}, \theta) > 0\} \subset \{\mathbf{x} : p(\mathbf{x}, \theta_0) > 0\}$$

for all θ. Let $\Lambda_{\mathbf{x}} = \frac{L_{\mathbf{x}}}{L_{\mathbf{x}}(\theta_0)}$. Thus, $\Lambda_{\mathbf{x}}$ is the function valued statistic that at θ takes on the value $\frac{p(\mathbf{x}, \theta)}{p(\mathbf{x}, \theta_0)}$, the *likelihood ratio* of θ to θ_0. Then $\Lambda_{\mathbf{x}}$ is minimal sufficient. See Problem 1.5.12 for a proof of this theorem of Dynkin, Lehmann, and Scheffé.

The "irrelevant" part of the data

We can always rewrite the original $\mathbf{X}$ as $(T(\mathbf{X}), S(\mathbf{X}))$ where $S(\mathbf{X})$ is a statistic needed to uniquely determine $\mathbf{x}$ once we know the sufficient statistic $T(\mathbf{x})$. For instance, if $T(\mathbf{X}) = \bar{X}$ we can take $S(\mathbf{X}) = (X_1 - \bar{X}, \dots, X_n - \bar{X})$, the *residuals*; or if $T(\mathbf{X}) = (X_{(1)}, \dots, X_{(n)})$, the order statistics, $S(\mathbf{X}) = (R_1, \dots, R_n)$, the *ranks*, where $R_i = \sum_{j=1}^{n} 1(X_j \leq X_i)$. $S(\mathbf{X})$ becomes irrelevant (*ancillary*) for inference if $T(\mathbf{X})$ is known *but only if $\mathcal{P}$ is valid*. Thus, in Example 1.5.5, if $\sigma^2 = 1$ is postulated, $\bar{X}$ is sufficient, but if in fact $\sigma^2 \neq 1$ all information about σ^2 is contained in the residuals. If, as in the Example 1.5.4, σ^2 is assumed unknown, $(\bar{X}, \sum_{i=1}^{n}(X_i - \bar{X})^2)$ is sufficient, but if in fact the common distribution of the observations is not Gaussian all the information needed to estimate this distribution is contained in the corresponding $S(\mathbf{X})$—see Problem 1.5.13. If $\mathcal{P}$ specifies that $X_1, \dots, X_n$ are a random sample, $(X_{(1)}, \dots, X_{(n)})$ is sufficient. But the ranks are needed if we want to look for possible dependencies in the observations as in Example 1.1.5.

Summary. Consider an experiment with observation vector $\mathbf{X} = (X_1, \dots, X_n)$. Suppose that $\mathbf{X}$ has distribution in the class $\mathcal{P} = \{P_\theta : \theta \in \Theta\}$. We say that a statistic $T(\mathbf{X})$ is *sufficient* for $P \in \mathcal{P}$, or for the parameter θ, if the conditional distribution of $\mathbf{X}$ given $T(\mathbf{X}) = t$ does not involve θ. Let $p(\mathbf{X}, \theta)$ denote the frequency function or density of $\mathbf{X}$. The *factorization theorem* states that $T(\mathbf{X})$ is sufficient for θ if and only if there exist functions $g(t, \theta)$ and $h(\mathbf{X})$ such that

$$p(\mathbf{X}, \theta) = g(T(\mathbf{X}), \theta)h(\mathbf{X}).$$

We show the following result: If $T(\mathbf{X})$ is sufficient for θ, then for any decision procedure $\delta(\mathbf{X})$, we can find a randomized decision rule $\delta^*(T(\mathbf{X}))$ depending only on the value of $t = T(\mathbf{X})$ and not on θ such that δ and δ^* have identical risk functions. We define a statistic $T(\mathbf{X})$ to be *Bayes sufficient* for a prior π if the posterior distribution of θ given $\mathbf{X} = \mathbf{x}$ is the same as the posterior distribution of θ given $T(\mathbf{X}) = T(\mathbf{x})$ for all $\mathbf{X}$. If

$T(\mathbf{X})$ is sufficient for θ, it is Bayes sufficient for θ. A sufficient statistic $T(\mathbf{X})$ is *minimally sufficient* for θ if for any other sufficient statistic $S(\mathbf{X})$ we can find a transformation r such that $T(\mathbf{X}) = r(S(\mathbf{X}))$. The likelihood function is defined for a given data vector of observations $\mathbf{X}$ to be the function of θ defined by $L_{\mathbf{X}}(\theta) = p(\mathbf{X}, \theta)$, $\theta \in \Theta$. If $T(\mathbf{X})$ is sufficient for θ, and if there is a value $\theta_0 \in \Theta$ such that

$$\{\mathbf{x} : p(\mathbf{x}, \theta) > 0\} \subset \{\mathbf{x} : p(\mathbf{x}, \theta_0) > 0\}, \ \theta \in \Theta,$$

then, by the factorization theorem, the *likelihood ratio*

$$\Lambda_{\mathbf{X}}(\theta) = \frac{L_{\mathbf{X}}(\theta)}{L_{\mathbf{X}}(\theta_0)}$$

depends on $\mathbf{X}$ through $T(\mathbf{X})$ only. $\Lambda_{\mathbf{X}}(\theta)$ is a minimally sufficient statistic.

1.6 EXPONENTIAL FAMILIES

The binomial and normal models considered in the last section exhibit the interesting feature that there is a natural sufficient statistic whose dimension as a random vector is independent of the sample size. The class of families of distributions that we introduce in this section was first discovered in statistics independently by Koopman, Pitman, and Darmois through investigations of this property[1]. Subsequently, many other common features of these families were discovered and they have become important in much of the modern theory of statistics.

Probability models with these common features include normal, binomial, Poisson, gamma, beta, and multinomial regression models used to relate a response variable Y to a set of predictor variables. More generally, these families form the basis for an important class of models called generalized linear models. We return to these models in Chapter 2. They will reappear in several connections in this book.

1.6.1 The One-Parameter Case

The family of distributions of a model $\{P_\theta : \theta \in \Theta\}$, is said to be a *one-parameter exponential family*, if there exist real-valued functions $\eta(\theta)$, $B(\theta)$ on Θ, real-valued functions T and h on R^q, such that the density (frequency) functions $p(x, \theta)$ of the P_θ may be written

$$p(x, \theta) = h(x) \exp\{\eta(\theta)T(x) - B(\theta)\} \tag{1.6.1}$$

where $x \in \mathcal{X} \subset R^q$. Note that the functions η, B, and T are not unique.

In a one-parameter exponential family the random variable $T(X)$ is sufficient for θ. This is clear because we need only identify $\exp\{\eta(\theta)T(x) - B(\theta)\}$ with $g(T(x), \theta)$ and $h(x)$ with itself in the factorization theorem. We shall refer to T as a *natural sufficient statistic* of the family.

Here are some examples.

Example 1.6.1. *The Poisson Distribution.* Let P_θ be the Poisson distribution with unknown mean θ. Then, for $x \in \{0, 1, 2, \cdots\}$,

$$p(x, \theta) = \frac{\theta^x e^{-\theta}}{x!} = \frac{1}{x!} \exp\{x \log \theta - \theta\}, \ \theta > 0. \tag{1.6.2}$$

Therefore, the P_θ form a one-parameter exponential family with

$$q = 1, \eta(\theta) = \log \theta, B(\theta) = \theta, T(x) = x, h(x) = \frac{1}{x!}. \tag{1.6.3}$$

Example 1.6.2. *The Binomial Family.* Suppose X has a $\mathcal{B}(n, \theta)$ distribution, $0 < \theta < 1$. Then, for $x \in \{0, 1, \ldots, n\}$

$$
\begin{aligned}
p(x, \theta) &= \begin{pmatrix} n \\ x \end{pmatrix} \theta^x (1 - \theta)^{n-x} \\
&= \begin{pmatrix} n \\ x \end{pmatrix} \exp[x \log(\frac{\theta}{1 - \theta}) + n \log(1 - \theta)].
\end{aligned} \tag{1.6.4}
$$

Therefore, the family of distributions of X is a one-parameter exponential family with

$$q = 1, \eta(\theta) = \log(\frac{\theta}{1 - \theta}), B(\theta) = -n \log(1 - \theta), T(x) = x, h(x) = \begin{pmatrix} n \\ x \end{pmatrix}. \tag{1.6.5}$$

□

Here is an example where $q = 2$.

Example 1.6.3. Suppose $X = (Z, Y)^T$ where $Y = Z + \theta W$, $\theta > 0$, Z and W are independent $\mathcal{N}(0, 1)$. Then

$$
\begin{aligned}
f(x, \theta) &= f(z, y, \theta) = f(z) f_\theta(y \mid z) = \varphi(z) \theta^{-1} \varphi((y - z)\theta^{-1}) \\
&= (2\pi\theta)^{-1} \exp\left\{ -\frac{1}{2}[z^2 + (y - z)^2 \theta^{-2}] \right\} \\
&= (2\pi)^{-1} \exp\left\{ -\frac{1}{2}z^2 \right\} \exp\left\{ -\frac{1}{2}\theta^{-2}(y - z)^2 - \log \theta \right\}.
\end{aligned}
$$

This is a one-parameter exponential family distribution with

$$q = 2, \eta(\theta) = -\frac{1}{2}\theta^{-2}, B(\theta) = \log \theta, T(x) = (y - z)^2, h(x) = (2\pi)^{-1} \exp\left\{ -\frac{1}{2}z^2 \right\}.$$

□

The families of distributions obtained by sampling from one-parameter exponential families are themselves one-parameter exponential families. Specifically, suppose $X_1, \ldots, X_m$ are independent and identically distributed with common distribution P_θ, where the P_θ form a one-parameter exponential family as in (1.6.1). If $\{P_\theta^{(m)}\}$, $\theta \in \Theta$, is the family of distributions of $\mathbf{X} = (X_1, \ldots, X_m)$ considered as a random vector in R^{mq} and $p(\mathbf{x}, \theta)$ are the corresponding density (frequency) functions, we have

$$
\begin{aligned}
p(\mathbf{x}, \theta) &= \prod_{i=1}^{m} h(x_i) \exp[\eta(\theta)T(x_i) - B(\theta)] \\
&= \left[\prod_{i=1}^{m} h(x_i)\right] \exp\left[\eta(\theta)\sum_{i=1}^{m} T(x_i) - mB(\theta)\right]
\end{aligned}
\tag{1.6.6}
$$

where $\mathbf{x} = (x_1, \ldots, x_m)$. Therefore, the $P_\theta^{(m)}$ form a one-parameter exponential family. If we use the superscript m to denote the corresponding T, η, B, and h, then $q^{(m)} = mq$, and

$$
\eta^{(m)}(\theta) = \eta(\theta),
$$

$$
T^{(m)}(\mathbf{x}) = \sum_{i=1}^{m} T(x_i), B^{(m)}(\theta) = mB(\theta), h^{(m)}(\mathbf{x}) = \prod_{i=1}^{m} h(x_i).
\tag{1.6.7}
$$

Note that the natural sufficient statistic $T^{(m)}$ is one-dimensional whatever be m. For example, if $\mathbf{X} = (X_1, \ldots, X_m)$ is a vector of independent and identically distributed $\mathcal{P}(\theta)$ random variables and $P_\theta^{(m)}$ is the family of distributions of $\mathbf{x}$, then the $P_\theta^{(m)}$ form a one-parameter exponential family with natural sufficient statistic $T^{(m)}(\mathbf{X}) = \sum_{i=1}^{m} T(X_i)$.

Some other important examples are summarized in the following table. We leave the proof of these assertions to the reader.

TABLE 1.6.1

Family of distributions		$\eta(\theta)$	$T(x)$
$\mathcal{N}(\mu, \sigma^2)$	σ^2 fixed	μ/σ^2	x
	μ fixed	$-1/2\sigma^2$	$(x - \mu)^2$
$\Gamma(p, \lambda)$	p fixed	$-\lambda$	x
	λ fixed	$(p - 1)$	$\log x$
$\beta(r, s)$	r fixed	$(s - 1)$	$\log(1 - x)$
	s fixed	$(r - 1)$	$\log x$

The statistic $T^{(m)}(X_1, \ldots, X_m)$ corresponding to the one-parameter exponential family of distributions of a sample from any of the foregoing is just $\sum_{i=1}^{m} T(X_i)$.

In our first Example 1.6.1 the sufficient statistic $T^{(m)}(X_1, \ldots, X_m) = \sum_{i=1}^{m} X_i$ is distributed as $\mathcal{P}(m\theta)$. This family of Poisson distributions is one-parameter exponential whatever be m. In the discrete case we can establish the following general result.

Theorem 1.6.1. Let $\{P_\theta\}$ be a one-parameter exponential family of discrete distributions with corresponding functions T, η, B, and h, then the family of distributions of the statistic $T(X)$ is a one-parameter exponential family of discrete distributions whose frequency functions may be written

$$h^*(t) \exp\{\eta(\theta)t - B(\theta)\}$$

for suitable h^*.

Proof. By definition,

$$
\begin{aligned}
P_\theta[T(x) = t] &= \sum_{\{x:T(x)=t\}} p(x, \theta) \\
&= \sum_{\{x:T(x)=t\}} h(x) \exp[\eta(\theta)T(x) - B(\theta)] \qquad (1.6.8) \\
&= \exp[\eta(\theta)t - B(\theta)]\{ \sum_{\{x:T(x)=t\}} h(x)\}.
\end{aligned}
$$

If we let $h^*(t) = \sum_{\{x:T(x)=t\}} h(x)$, the result follows. □

A similar theorem holds in the continuous case if the distributions of $T(X)$ are themselves continuous.

Canonical exponential families. We obtain an important and useful reparametrization of the exponential family (1.6.1) by letting the model be indexed by η rather than θ. The exponential family then has the form

$$q(x, \eta) = h(x) \exp[\eta T(x) - A(\eta)], \quad x \in \mathcal{X} \subset R^q \qquad (1.6.9)$$

where $A(\eta) = \log \int \cdots \int h(x) \exp[\eta T(x)] dx$ in the continuous case and the integral is replaced by a sum in the discrete case. If $\theta \in \Theta$, then $A(\eta)$ must be finite, if q is definable. Let $\mathcal{E}$ be the collection of all η such that $A(\eta)$ is finite. Then as we show in Theorem 1.6.3, $\mathcal{E}$ is either an interval or all of R and the class of models (1.6.9) with $\eta \in \mathcal{E}$ contains the class of models with $\theta \in \Theta$. The model given by (1.6.9) with η ranging over $\mathcal{E}$ is called the *canonical one-parameter exponential family generated by T and h. $\mathcal{E}$ is called the natural parameter space* and T is called the *natural sufficient statistic*.

Example 1.6.1. *(continued).* The Poisson family in canonical form is

$$q(x, \eta) = (1/x!) \exp\{\eta x - \exp[\eta]\}, \quad x \in \{0, 1, 2, \dots\},$$

where $\eta = \log \theta$,

$$\exp\{A(\eta)\} = \sum_{x=0}^{\infty} (e^{\eta x}/x!) = \sum_{x=0}^{\infty} (e^\eta)^x/x! = \exp(e^\eta),$$

and $\mathcal{E} = R$. □

Here is a useful result.

Theorem 1.6.2. *If X is distributed according to (1.6.9) and η is an interior point of $\mathcal{E}$, the moment-generating function of $T(X)$ exists and is given by*

$$M(s) = \exp[A(s + \eta) - A(\eta)]$$

for s in some neighborhood of 0.

Moreover,

$$E(T(X)) = A'(\eta), \quad \text{Var}(T(X)) = A''(\eta).$$

Proof. We give the proof in the continuous case. We compute

$$M(s) = E(\exp(sT(X))) = \int \cdots \int h(x) exp[(s+\eta)T(x) - A(\eta)]dx$$

$$= \{exp[A(s+\eta) - A(\eta)]\} \int \cdots \int h(x) exp[(s+\eta)T(x) - A(s+\eta)]dx$$

$$= exp[A(s+\eta) - A(\eta)]$$

because the last factor, being the integral of a density, is one. The rest of the theorem follows from the moment generating property of $M(s)$ (see Section A.12).   □

Here is a typical application of this result.

Example 1.6.4 Suppose $X_1, \ldots, X_n$ is a sample from a population with density

$$p(x, \theta) = (x/\theta^2) \exp(-x^2/2\theta^2), \quad x > 0, \theta > 0.$$

This is known as the *Rayleigh* distribution. It is used to model the density of "time until failure" for certain types of equipment. Now

$$p(\mathbf{x}, \theta) = (\prod_{i=1}^{n}(x_i/\theta^2)) \exp(-\sum_{i=1}^{n} x_i^2/2\theta^2)$$

$$= (\prod_{i=1}^{n} x_i) \exp[\frac{-1}{2\theta^2} \sum_{i=1}^{n} x_i^2 - n\log\theta^2].$$

Here $\eta = -1/2\theta^2$, $\theta^2 = -1/2\eta$, $B(\theta) = n\log\theta^2$ and $A(\eta) = -n\log(-2\eta)$. Therefore, the natural sufficient statistic $\sum_{i=1}^{n} X_i^2$ has mean $-n/\eta = 2n\theta^2$ and variance $n/\eta^2 = 4n\theta^4$. Direct computation of these moments is more complicated.   □

1.6.2   The Multiparameter Case

Our discussion of the "natural form" suggests that one-parameter exponential families are naturally indexed by a one-dimensional real parameter η and admit a one-dimensional sufficient statistic $T(x)$. More generally, Koopman, Pitman, and Darmois were led in their investigations to the following family of distributions, which is naturally indexed by a k-dimensional parameter and admit a k-dimensional sufficient statistic.

A family of distributions $\{P_{\boldsymbol{\theta}} : \boldsymbol{\theta} \in \Theta\}$, $\Theta \subset R^k$, is said to be a *k-parameter exponential family*, if there exist real-valued functions $\eta_1, \ldots, \eta_k$ and B of $\boldsymbol{\theta}$, and real-valued functions $T_1, \ldots, T_k$, h on R^q such that the density (frequency) functions of the P_θ may be written as,

$$p(x, \boldsymbol{\theta}) = h(x) \exp[\sum_{j=1}^{k} \eta_j(\boldsymbol{\theta})T_j(x) - B(\boldsymbol{\theta})], \quad x \in \mathcal{X} \subset R^q. \tag{1.6.10}$$

By Theorem 1.5.1, the vector $\mathbf{T}(X) = (T_1(X), \dots, T_k(X))^T$ is sufficient. It will be referred to as a *natural sufficient statistic* of the family.

Again, suppose $\mathbf{X} = (X_1, \dots, X_m)^T$ where the X_i are independent and identically distributed and their common distribution ranges over a k-parameter exponential family given by (1.6.10). Then the distributions of $\mathbf{X}$ form a k-parameter exponential family with natural sufficient statistic

$$\mathbf{T}^{(m)}(\mathbf{X}) = (\sum_{i=1}^{m} T_1(X_i), \dots, \sum_{i=1}^{m} T_k(X_i))^T.$$

Example 1.6.5. *The Normal Family.* Suppose that $P_\theta = \mathcal{N}(\mu, \sigma^2)$, $\Theta = \{(\mu, \sigma^2) : -\infty < \mu < \infty, \sigma^2 > 0\}$. The density of P_θ may be written as

$$p(x, \boldsymbol{\theta}) = \exp[\frac{\mu}{\sigma^2}x - \frac{x^2}{2\sigma^2} - \frac{1}{2}(\frac{\mu^2}{\sigma^2} + \log(2\pi\sigma^2))], \qquad (1.6.11)$$

which corresponds to a two-parameter exponential family with $q = 1$, $\theta_1 = \mu$, $\theta_2 = \sigma^2$, and

$$\eta_1(\boldsymbol{\theta}) \;=\; \frac{\mu}{\sigma^2}, \;\; T_1(x) = x, \;\; \eta_2(\boldsymbol{\theta}) = -\frac{1}{2\sigma^2}, \;\; T_2(x) = x^2,$$

$$B(\boldsymbol{\theta}) \;=\; \frac{1}{2}(\frac{\mu^2}{\sigma^2} + \log(2\pi\sigma^2)), \;\; h(x) = 1.$$

If we observe a sample $\mathbf{X} = (X_1, \dots, X_m)$ from a $\mathcal{N}(\mu, \sigma^2)$ population, then the preceding discussion leads us to the natural sufficient statistic

$$(\sum_{i=1}^{m} X_i, \sum_{i=1}^{m} X_i^2)^T,$$

which we obtained in the previous section (Example 1.5.4). $\square$

Again it will be convenient to consider the "biggest" families, letting the model be indexed by $\boldsymbol{\eta} = (\eta_1, \dots, \eta_k)^T$ rather than $\boldsymbol{\theta}$. Thus, *the canonical k-parameter exponential family generated by $\mathbf{T}$ and h is*

$$q(x, \boldsymbol{\eta}) = h(x) \exp\{\mathbf{T}^T(x)\boldsymbol{\eta} - A(\boldsymbol{\eta})\}, x \in \mathcal{X} \subset R^q$$

where $\mathbf{T}(x) = (T_1(x), \dots, T_k(x))^T$ and, in the continuous case,

$$A(\boldsymbol{\eta}) = \log \int_{-\infty}^{\infty} \cdots \int_{-\infty}^{\infty} h(x) \exp\{\mathbf{T}^T(x)\boldsymbol{\eta}\}dx.$$

In the discrete case, $A(\boldsymbol{\eta})$ is defined in the same way except integrals over R^q are replaced by sums. In either case, we define the natural parameter space as

$$\mathcal{E} = \{\boldsymbol{\eta} \in R^k : -\infty < A(\boldsymbol{\eta}) < \infty\}.$$

Example 1.6.5. ($\mathcal{N}(\mu, \sigma^2)$ *continued*). In this example, $k = 2$, $\mathbf{T}^T(x) = (x, x^2) = (T_1(x), T_2(x))$, $\eta_1 = \mu/\sigma^2$, $\eta_2 = -1/2\sigma^2$, $A(\boldsymbol{\eta}) = \frac{1}{2}[(-\eta_1^2/2\eta_2) + \log(\pi/(-\eta_2))]$, $h(x) = 1$ and $\mathcal{E} = R \times R^- = \{(\eta_1, \eta_2) : \eta_1 \in R, \eta_2 < 0\}$.

Example 1.6.6. *Linear Regression.* Suppose as in Examples 1.1.4 and 1.5.5 that $Y_1, \ldots, Y_n$ are independent, $Y_i \sim \mathcal{N}(\mu_i, \sigma^2)$, with $\mu_i = \beta_1 + \beta_2 z_i$, $i = 1, \ldots, n$. From Example 1.5.5, the density of $\mathbf{Y} = (Y_1, \ldots, Y_n)^T$ can be put in canonical form with $k = 3$, $\mathbf{T}(\mathbf{Y}) = (\Sigma Y_i, \Sigma z_i Y_i, \Sigma Y_i^2,)^T$, $\eta_1 = \beta_1/\sigma^2$, $\eta_2 = \beta_2/\sigma^2$, $\eta_3 = -1/2\sigma^2$,

$$A(\boldsymbol{\eta}) = \frac{-n}{4\eta_3}[\eta_1^2 + \widehat{m}_2\eta_2^2 + \bar{z}\eta_1\eta_2 + 2\log(\pi/-\eta_3)],$$

and $\mathcal{E} = \{(\eta_1, \eta_2, \eta_3) : \eta_1 \in R, \eta_2 \in R, \eta_3 < 0\}$, where $\widehat{m}_2 = n^{-1}\Sigma z_i^2$.

Example 1.6.7. *Multinomial Trials.* We observe the outcomes of n independent trials where each trial can end up in one of k possible categories. We write the outcome vector as $\mathbf{X} = (X_1, \ldots, X_n)^T$ where the X_i are i.i.d. as X and the sample space of each X_i is the k categories $\{1, 2, \ldots, k\}$. Let $T_j(\mathbf{x}) = \sum_{i=1}^{n} 1[X_i = j]$, and $\lambda_j = P(X_i = j)$. Then $p(\mathbf{x}, \boldsymbol{\lambda}) = \prod_{j=1}^{k} \lambda_j^{T_j(\mathbf{x})}$, $\boldsymbol{\lambda} \in \Lambda$, where Λ is the simplex $\{\boldsymbol{\lambda} \in R^k : 0 < \lambda_j < 1, j = 1, \ldots, k, \sum_{j=1}^{k} \lambda_j = 1\}$. It will often be more convenient to work with unrestricted parameters. In this example, we can achieve this by the reparametrization

$$\lambda_j = e^{\alpha_j} / \sum_{j=1}^{k} e^{\alpha_j}, j = 1, \ldots, k, \boldsymbol{\alpha} \in R^k.$$

Now we can write the likelihood as

$$q_0(\mathbf{x}, \boldsymbol{\alpha}) = \exp\{\sum_{j=1}^{k} \alpha_j T_j(\mathbf{x}) - n \log \sum_{j=1}^{k} \exp(\alpha_j)\}.$$

This is a k-parameter canonical exponential family generated by $T_1, \ldots, T_k$ and $h(\mathbf{x}) = \prod_{i=1}^{n} 1[x_i \in \{1, \ldots, k\}]$ with canonical parameter $\boldsymbol{\alpha}$ and $\mathcal{E} = R^k$. However $\boldsymbol{\alpha}$ is not identifiable because $q_0(\mathbf{x}, \boldsymbol{\alpha} + c\mathbf{1}) = q_0(\mathbf{x}, \boldsymbol{\alpha})$ for $\mathbf{1} = (1, \ldots, 1)^T$ and all c. This can be remedied by considering

$$\mathbf{T}_{(k-1)}(\mathbf{x}) \equiv (T_1(\mathbf{x}), \ldots, T_{k-1}(\mathbf{x}))^T,$$

$\eta_j = \log(\lambda_j/\lambda_k) = \alpha_j - \alpha_k$, $1 \leq j \leq k - 1$, and rewriting

$$q(\mathbf{x}, \boldsymbol{\eta}) = \exp\{\mathbf{T}_{(k-1)}^T(\mathbf{x})\boldsymbol{\eta} - n \log(1 + \sum_{j=1}^{k-1} e^{\eta_j})\}$$

where

$$\lambda_j = \frac{e^{\eta_j}}{1 + \sum_{j=1}^{k-1} e^{\eta_j}} = \frac{e^{\alpha_j}}{\sum_{j=1}^{k} e^{\alpha_j}}, \quad j = 1, \ldots, k - 1.$$

Note that $q(\mathbf{x}, \boldsymbol{\eta})$ is a $k-1$ parameter canonical exponential family generated by $\mathbf{T}_{(k-1)}$ and $h(\mathbf{x}) = \prod_{i=1}^{n} 1[x_i \in \{1, \ldots, k\}]$ with canonical parameter $\boldsymbol{\eta}$ and $\mathcal{E} = R^{k-1}$. Moreover, the parameters $\eta_j = \log(P_{\boldsymbol{\eta}}[X = j]/P_{\boldsymbol{\eta}}[X = k])$, $1 \leq j \leq k-1$, are identifiable. Note that the model for $\mathbf{X}$ is unchanged.   □

1.6.3 Building Exponential Families

Submodels

A *submodel* of a k-parameter canonical exponential family $\{q(\mathbf{x}, \eta); \eta \in \mathcal{E} \subset R^k\}$ is an exponential family defined by

$$p(x, \boldsymbol{\theta}) = q(x, \boldsymbol{\eta}(\boldsymbol{\theta})) \tag{1.6.12}$$

where $\boldsymbol{\theta} \in \Theta \subset R^l, l \leq k$, and $\boldsymbol{\eta}$ is a map from Θ to a subset of R^k. Thus, if X is discrete taking on k values as in Example 1.6.7 and $\mathbf{X} = (X_1, \ldots, X_n)^T$ where the X_i are i.i.d. as X, then *all* models for $\mathbf{X}$ are exponential families because they are submodels of the multinomial trials model.

Affine transformations

If $\mathcal{P}$ is the canonical family generated by $\mathbf{T}_{k \times 1}$ and h and $\mathbf{M}$ is the affine transformation from R^k to R^l defined by

$$\mathbf{M}(\mathbf{T}) = M_{\ell \times k} \mathbf{T} + \mathbf{b}_{\ell \times 1},$$

it is easy to see that the family generated by $\mathbf{M}(\mathbf{T}(X))$ and h is the subfamily of $\mathcal{P}$ corresponding to

$$\Theta = [\boldsymbol{\eta}^{-1}](\mathcal{E}) \subset R^\ell$$

and

$$\boldsymbol{\eta}(\boldsymbol{\theta}) = M^T \boldsymbol{\theta}.$$

Similarly, if $\Theta \subset R^\ell$ and $\boldsymbol{\eta}(\boldsymbol{\theta}) = B_{k \times \ell} \boldsymbol{\theta} \subset R^k$, then the resulting submodel of $\mathcal{P}$ above is a submodel of the exponential family generated by $B^T \mathbf{T}(X)$ and h. See Problem 1.6.17 for details. Here is an example of affine transformations of $\boldsymbol{\theta}$ and $\mathbf{T}$.

Example 1.6.8. *Logistic Regression.* Let Y_i be independent binomial, $\mathcal{B}(n_i, \lambda_i)$, $1 \leq i \leq n$. If the λ_i are unrestricted, $0 < \lambda_i < 1$, $1 \leq i \leq n$, this, from Example 1.6.2, is an n-parameter canonical exponential family with $\mathcal{Y}_i \equiv$ integers from 0 to n_i generated by $\mathbf{T}(Y_1, \ldots, Y_n) = \mathbf{Y}, h(\mathbf{y}) = \prod_{i=1}^{n} \binom{n_i}{y_i} 1(0 \leq y_i \leq n_i)$. Here $\eta_i = \log \frac{\lambda_i}{1-\lambda_i}$, $A(\boldsymbol{\eta}) = \sum_{i=1}^{n} n_i \log(1 + e^{\eta_i})$. However, let $x_1 < \ldots < x_n$ be specified levels and

$$\eta_i(\boldsymbol{\theta}) = \theta_1 + \theta_2 x_i, \quad 1 \leq i \leq n, \quad \boldsymbol{\theta} = (\theta_1, \theta_2)^T \in R^2. \tag{1.6.13}$$

This is a linear transformation $\boldsymbol{\eta}(\boldsymbol{\theta}) = B_{n \times 2}\boldsymbol{\theta}$ corresponding to $B_{n \times 2} = (\mathbf{1}, \mathbf{x})$, where $\mathbf{1}$ is $(1, \dots, 1)^T$, $\mathbf{x} = (x_1, \dots, x_n)^T$. Set $M = B^T$, then this is the two-parameter canonical exponential family generated by $M\mathbf{Y} = (\sum_{i=1}^{n} Y_i, \sum_{i=1}^{n} x_i Y_i)^T$ and h with

$$A(\theta_1, \theta_2) = \sum_{i=1}^{n} n_i \log(1 + \exp(\theta_1 + \theta_2 x_i)).$$

This model is sometimes applied in experiments to determine the toxicity of a substance. The Y_i represent the number of animals dying out of n_i when exposed to level x_i of the substance. It is assumed that each animal has a random toxicity threshold X such that death results if and only if a substance level on or above X is applied. Assume also:

(a) No interaction between animals (independence) in relation to drug effects

(b) The distribution of X in the animal population is *logistic*; that is,

$$P[X \leq x] = [1 + \exp\{-(\theta_1 + \theta_2 x)\}]^{-1}, \tag{1.6.14}$$

$\theta_1 \in R$, $\theta_2 > 0$. Then (and only then),

$$\log(P[X \leq x]/(1 - P[X \leq x])) = \theta_1 + \theta_2 x$$

and (1.6.13) holds. □

Curved exponential families

Exponential families (1.6.12) with the range of $\boldsymbol{\eta}(\boldsymbol{\theta})$ restricted to a subset of dimension l with $l \leq k - 1$, are called *curved exponential families* provided they do not form a canonical exponential family in the $\boldsymbol{\theta}$ parametrization.

Example 1.6.9. *Gaussian with Fixed Signal-to-Noise Ratio.* In the normal case with $X_1, \dots, X_n$ i.i.d. $\mathcal{N}(\mu, \sigma^2)$, suppose the ratio $|\mu|/\sigma$, which is called the *coefficient of variation* or *signal-to-noise ratio*, is a known constant $\lambda_0 > 0$. Then, with $\theta = \mu$, we can write

$$p(\mathbf{x}, \theta) = \exp\left\{\lambda_0^2 \theta^{-1} T_1 - \frac{1}{2}\lambda_0^2 \theta^{-2} T_2 - \frac{1}{2}n[\lambda_0^2 + \log(2\pi\lambda_0^{-2}\theta^2)]\right\}$$

where $T_1 = \sum_{i=1}^{n} x_i$, $T_2 = \sum_{i=1}^{n} x_i^2$, $\eta_1(\theta) = \lambda_0^2 \theta^{-1}$ and $\eta_2(\theta) = -\frac{1}{2}\lambda_0^2 \theta^{-2}$. This is a curved exponential family with $l = 1$. □

In Example 1.6.8, the $\boldsymbol{\theta}$ parametrization has dimension 2, which is less than $k = n$ when $n > 3$. However, $p(x, \boldsymbol{\theta})$ in the $\boldsymbol{\theta}$ parametrization is a canonical exponential family, so it is not a curved family.

Example 1.6.10. *Location-Scale Regression.* Suppose that $Y_1, \dots, Y_n$ are independent, $Y_i \sim \mathcal{N}(\mu_i, \sigma_i^2)$. If each μ_i ranges over R and each σ_i^2 ranges over $(0, \infty)$, this is by Example 1.6.5 a $2n$-parameter canonical exponential family model with $\eta_i = \mu_i/\sigma_i^2$, and $\eta_{n+i} = -1/2\sigma_i^2$, $i = 1, \dots, n$, generated by

$$T(\mathbf{Y}) = (Y_1, \dots, Y_n, Y_1^2, \dots, Y_n^2)^T$$

and $h(\mathbf{Y}) = 1$. Next suppose that (μ_i, σ_i^2) depend on the value z_i of some covariate, say,

$$\mu_i = \theta_1 + \theta_2 z_i, \ \sigma_i^2 = \theta_3(\theta_1 + \theta_2 z_i)^2, \ z_1 < \cdots < z_n$$

for unknown parameters $\theta_1 \in R$, $\theta_2 \in R$, $\theta_3 > 0$ (e.g., Bickel, 1978; Carroll and Ruppert, 1988, Sections 2.1–2.5; and Snedecor and Cochran, 1989, Section 15.10). For $\boldsymbol{\theta} = (\theta_1, \theta_2, \theta_2)$, the map $\boldsymbol{\eta}(\boldsymbol{\theta})$ is

$$\eta_i(\boldsymbol{\theta}) = \theta_3^{-1}(\theta_1 + \theta_2 z_i)^{-1}, \ \eta_{n+i}(\boldsymbol{\theta}) = \frac{1}{2}\theta_3^{-1}(\theta_1 + \theta_2 z_i)^{-2}, \ i = 1, \ldots, n.$$

Because $\sum_{i=1}^n \eta_i(\boldsymbol{\theta})Y_i + \sum_{i=1}^n \eta_{n+i}(\boldsymbol{\theta})Y_i^2$ cannot be written in the form $\sum_{j=1}^3 \eta_j^*(\boldsymbol{\theta})T_j^*(\mathbf{Y})$ for some $\eta_j^*(\boldsymbol{\theta})$, $T_j^*(\mathbf{Y})$, then $p(\mathbf{y}, \boldsymbol{\theta}) = q(\mathbf{y}, \boldsymbol{\eta}(\boldsymbol{\theta}))$ as defined in (6.1.12) is not an exponential family model, but a curved exponential family model with $l = 3$.   □

Models in which the variance $\mathrm{Var}(Y_i)$ depends on i are called *heteroscedastic* whereas models in which $\mathrm{Var}(Y_i)$ does not depend on i are called *homoscedastic*. Thus, Examples 1.6.10 and 1.6.6 are heteroscedastic and homoscedastic models, respectively.

We return to curved exponential family models in Section 2.3.

Supermodels

We have already noted that the exponential family structure is preserved under i.i.d. sampling. Even more is true. Let Y_j, $1 \le j \le n$, be independent, $Y_j \in \mathcal{Y}_j \subset R^q$, with an exponential family density

$$q_j(y_j, \boldsymbol{\theta}) = \exp\{\mathbf{T}_j^T(y_j)\boldsymbol{\eta}(\boldsymbol{\theta}) - B_j(\boldsymbol{\theta})\}h_j(y_j), \ \boldsymbol{\theta} \in \Theta \subset R^k.$$

Then $\mathbf{Y} \equiv (Y_1, \ldots, Y_n)^T$ is modeled by the exponential family generated by $\mathbf{T}(\mathbf{Y}) = \sum_{j=1}^n \mathbf{T}_j(Y_j)$ and $\prod_{j=1}^n h_j(y_j)$, with parameter $\boldsymbol{\eta}(\boldsymbol{\theta})$, and $B(\boldsymbol{\theta}) = \sum_{j=1}^n B_j(\boldsymbol{\theta})$.

In Example 1.6.8 note that (1.6.13) exhibits Y_j as being distributed according to a two-parameter family generated by $T_j(Y_j) = (Y_j, x_j Y_j)$ and we can apply the supermodel approach to reach the same conclusion as before.

1.6.4   Properties of Exponential Families

Theorem 1.6.1 generalizes directly to k-parameter families as does its continuous analogue. We extend the statement of Theorem 1.6.2.

Recall from Section B.5 that for any random vector $\mathbf{T}_{k \times 1}$, we define

$$M(\mathbf{s}) \equiv E e^{\mathbf{s}^T \mathbf{T}}$$

as the moment-generating function, and

$$E(\mathbf{T}) \equiv (E(T_1), \ldots, E(T_k))^T$$

$$\mathrm{Var}(\mathbf{T}) = \|\mathrm{Cov}(T_a, T_b)\|_{k \times k}.$$

Theorem 1.6.3. *Let $\mathcal{P}$ be a canonical k-parameter exponential family generated by $(\mathbf{T}, h)$ with corresponding natural parameter space $\mathcal{E}$ and function $A(\boldsymbol{\eta})$. Then*

(a) $\mathcal{E}$ *is convex*

(b) $A : \mathcal{E} \to R$ *is convex*

(c) *If $\mathcal{E}$ has nonempty interior $\mathcal{E}^0$ in R^k and $\boldsymbol{\eta}_0 \in \mathcal{E}^0$, then $\mathbf{T}(X)$ has under $\boldsymbol{\eta}_0$ a moment-generating function M given by*

$$M(\mathbf{s}) = \exp\{A(\boldsymbol{\eta}_0 + \mathbf{s}) - A(\boldsymbol{\eta}_0)\}$$

valid for all $\mathbf{s}$ such that $\boldsymbol{\eta}_0 + \mathbf{s} \in \mathcal{E}$. Since $\boldsymbol{\eta}_0$ is an interior point this set of $\mathbf{s}$ includes a ball about $\mathbf{0}$.

Corollary 1.6.1. *Under the conditions of Theorem 1.6.3*

$$E_{\boldsymbol{\eta}_0} \mathbf{T}(X) = \dot{A}(\boldsymbol{\eta}_0)$$

$$\mathrm{Var}_{\boldsymbol{\eta}_0} \mathbf{T}(X) = \ddot{A}(\boldsymbol{\eta}_0)$$

where $\dot{A}(\boldsymbol{\eta}_0) = (\frac{\partial A}{\partial \eta_1}(\boldsymbol{\eta}_0), \dots, \frac{\partial A}{\partial \eta_k}(\boldsymbol{\eta}_0))^T$, $\ddot{A}(\boldsymbol{\eta}_0) = \|\frac{\partial^2 A}{\partial \eta_a \partial \eta_b}(\boldsymbol{\eta}_0)\|$.

The corollary follows immediately from Theorem B.5.1 and Theorem 1.6.3(c).

Proof of Theorem 1.6.3. We prove (b) first. Suppose $\boldsymbol{\eta}_1, \boldsymbol{\eta}_2 \in \mathcal{E}$ and $0 \leq \alpha \leq 1$. By the Hölder inequality (B.9.4), for any $u(x), v(x), h(x) \geq 0$, $r, s > 0$ with $\frac{1}{r} + \frac{1}{s} = 1$,

$$\int u(x)v(x)h(x)dx \leq \left(\int u^r(x)h(x)dx\right)^{\frac{1}{r}} \left(\int v^s(x)h(x)dx\right)^{\frac{1}{s}}.$$

Substitute $\frac{1}{r} = \alpha$, $\frac{1}{s} = 1 - \alpha$, $u(x) = \exp(\alpha \boldsymbol{\eta}_1^T \mathbf{T}(x))$, $v(x) = \exp((1-\alpha)\boldsymbol{\eta}_2^T \mathbf{T}(x))$ and take logs of both sides to obtain, (with ∞ permitted on either side),

$$A(\alpha\boldsymbol{\eta}_1 + (1-\alpha)\boldsymbol{\eta}_2) \leq \alpha A(\boldsymbol{\eta}_1) + (1-\alpha)A(\boldsymbol{\eta}_2) \tag{1.6.15}$$

which is (b). If $\boldsymbol{\eta}_1, \boldsymbol{\eta}_2 \in \mathcal{E}$ the right-hand side of (1.6.15) is finite. Because

$$\int \exp(\boldsymbol{\eta}^T \mathbf{T}(x))h(x)dx > 0$$

for all $\boldsymbol{\eta}$ we conclude from (1.6.15) that $\alpha\boldsymbol{\eta}_1 + (1-\alpha)\boldsymbol{\eta}_2 \in \mathcal{E}$ and (a) follows. Finally (c) is proved in exactly the same way as Theorem 1.6.2.   □

The formulae of Corollary 1.6.1 give a classical result in Example 1.6.6.

Example 1.6.7. *(continued).* Here, using the α parametrization,

$$A(\alpha) = n\log(\sum_{j=1}^{k} e^{\alpha_j})$$

and

$$E_{\lambda}(T_j(\mathbf{X})) = nP_{\lambda}[X = j] \equiv n\lambda_j = ne^{\alpha_j}/\sum_{\ell=1}^{k} e^{\alpha_\ell}$$

$$\text{Cov}_{\lambda}(T_i, T_j) = \frac{\partial^2 A}{\partial\alpha_i\partial\alpha_j}(\alpha) = -n\frac{e^{\alpha_j}e^{\alpha_i}}{(\sum_{\ell=1}^{k} e^{\alpha_\ell})^2} = -n\lambda_i\lambda_j, \ \ i \neq j$$

$$\text{Var}_{\lambda}(T_i) = \frac{\partial^2 A}{\partial\alpha_i^2}(\alpha) = n\lambda_i(1 - \lambda_i).$$

$\square$

The rank of an exponential family

Evidently every k-parameter exponential family is also k'-dimensional with $k' > k$. However, there is a minimal dimension.

An exponential family is of *rank* k iff the generating statistic $\mathbf{T}$ is k-dimensional and 1, $T_1(X), \ldots, T_k(X)$ are linearly independent with positive probability. Formally, $P_{\eta}[\sum_{j=1}^{k} a_j T_j(X) = a_{k+1}] < 1$ unless all a_j are 0.

Note that $P_{\theta}(A) = 0$ or $P_{\theta}(A) < 1$ for some θ iff the corresponding statement holds for all θ because $0 < \frac{p(x,\theta_1)}{p(x,\theta_2)} < \infty$ for all x, θ_1, θ_2 such that $h(x) > 0$.

Going back to Example 1.6.7 we can see that the multinomial family is of rank at most $k - 1$. It is intuitively clear that $k - 1$ is in fact its rank and this is seen in Theorem 1.6.4 that follows. Similarly, in Example 1.6.8, if $n = 1$, and $\eta_1(\theta) = \theta_1 + \theta_2 x_1$ we are writing the one-parameter binomial family corresponding to Y_1 as a two-parameter family with generating statistic $(Y_1, x_1 Y_1)$. But the rank of the family is 1 and θ_1 and θ_2 are not identifiable. However, if we consider $\mathbf{Y}$ with $n \geq 2$ and $x_1 < x_n$ the family as we have seen remains of rank ≤ 2 and is in fact of rank 2. Our discussion suggests a link between rank and identifiability of the η parameterization. We establish the connection and other fundamental relationships in Theorem 1.6.4.

Theorem 1.6.4. *Suppose* $\mathcal{P} = \{q(x, \eta); \ \eta \in \mathcal{E}\}$ *is a canonical exponential family generated by* $(\mathbf{T}_{k\times 1}, h)$ *with natural parameter space* $\mathcal{E}$ *such that* $\mathcal{E}$ *is open. Then the following are equivalent.*

(i) $\mathcal{P}$ *is of rank* k.

(ii) η *is a parameter (identifiable).*

(iii) $\text{Var}_{\eta}(\mathbf{T})$ *is positive definite.*

(iv) $\eta \to \dot{A}(\eta)$ *is* 1-1 *on* $\mathcal{E}$.

(v) *A is strictly convex on* $\mathcal{E}$.

Note that, by Theorem 1.6.3, because $\mathcal{E}$ is open, $\dot{A}$ is defined on all of $\mathcal{E}$.

Proof. We give a detailed proof for $k = 1$. The proof for $k > 1$ is then sketched with details left to a problem. Let $\sim (\cdot)$ denote "$(\cdot)$ is false." Then

$\sim$(i) $\Leftrightarrow P_\eta[a_1 T = a_2] = 1$ for $a_1 \neq 0$. This is equivalent to $\mathrm{Var}_\eta(T) = 0 \Leftrightarrow \sim$ (iii)

$\sim$(ii) $\Leftrightarrow$ There exist $\eta_1 \neq \eta_2$ such that $P_{\eta_1} = P_{\eta_2}$.

Equivalently

$$\exp\{\eta_1 T(x) - A(\eta_1)\}h(x) = \exp\{\eta_2 T(x) - A(\eta_2)\}h(x).$$

Taking logs we obtain $(\eta_1 - \eta_2)T(X) = A(\eta_2) - A(\eta_1)$ with probability $1 \equiv \sim$(i). We, thus, have (i) $\equiv$ (ii) $\equiv$ (iii). Now (iii) $\Rightarrow A''(\eta) > 0$ by Theorem 1.6.2 and, hence, $A'(\eta)$ is strictly monotone increasing and 1-1. Conversely, $A''(\eta_0) = 0$ for some η_0 implies that $T \equiv c$, with probability 1, for all η, by our remarks in the discussion of rank, which implies that $A''(\eta) = 0$ for all η and, hence, A' is constant. Thus, (iii) $\equiv$ (iv) and the same discussion shows that (iii) $\equiv$ (v).

Proof of the general case sketched

I. $\sim$ (i) $\equiv \sim$ (iii)

$\sim$ (i) $\equiv P_\eta[\mathbf{a}^T \mathbf{T} = c] = 1$ for some $\mathbf{a} \neq 0$, all η
$\sim$ (iii) $\equiv \mathbf{a}^T \mathrm{Var}_\eta(\mathbf{T})\mathbf{a} = \mathrm{Var}_\eta(\mathbf{a}^T \mathbf{T}) = 0$ for some $\mathbf{a} \neq 0$, all $\eta \equiv (\sim i)$

II. $\sim$ (ii) $\equiv \sim$ (i)

$\sim$ (ii) $\equiv P_{\eta_1} = P_{\eta_0}$ some $\eta_1 \neq \eta_0$. Let

$$\mathcal{Q} = \{P_{\eta_0 + c(\eta_1 - \eta_0)} : \eta_0 + c(\eta_1 - \eta_0) \in \mathcal{E}\}.$$

$\mathcal{Q}$ is the exponential family (one-parameter) generated by $(\eta_1 - \eta_0)^T \mathbf{T}$. Apply the case $k = 1$ to $\mathcal{Q}$ to get $\sim$ (ii) $\equiv \sim$ (i).

III. (iv) $\equiv$ (v) $\equiv$ (iii)

Properties (iv) and (v) are equivalent to the statements holding for every $\mathcal{Q}$ defined as previously for arbitrary η_0, η_1. $\square$

Corollary 1.6.2. *Suppose that the conditions of Theorem* 1.6.4 *hold and* $\mathcal{P}$ *is of rank* k. *Then*

(a) $\mathcal{P}$ *may be uniquely parametrized by* $\mu(\eta) \equiv E_\eta \mathbf{T}(X)$ *where* μ *ranges over* $\dot{A}(\mathcal{E})$,
(b) $\log q(x, \eta)$ *is a strictly concave function of* η *on* $\mathcal{E}$.

Proof. This is just a restatement of (iv) and (v) of the theorem. $\square$

The relation in (a) is sometimes evident and the μ parametrization is close to the initial parametrization of classical $\mathcal{P}$. Thus, the $\mathcal{B}(n, \boldsymbol{\theta})$ family is parametrized by $E(X)$, where X is the Bernoulli trial, the $\mathcal{N}(\mu, \sigma_0^2)$ family by $E(X)$. For $\{\mathcal{N}(\mu, \sigma^2)\}$, $E(X, X^2) = (\mu, \sigma^2 + \mu^2)$, which is obviously a 1-1 function of (μ, σ^2). However, the relation in (a) may be far from obvious (see Problem 1.6.21). The corollary will prove very important in estimation theory. See Section 2.3. We close the present discussion of exponential families with the following example.

Example 1.6.11. *The p Variate Gaussian Family.* An important exponential family is based on the multivariate Gaussian distributions of Section B.6. Recall that $\mathbf{Y}_{p \times 1}$ has a p variate Gaussian distribution, $\mathcal{N}_p(\boldsymbol{\mu}, \Sigma)$, with mean $\boldsymbol{\mu}_{p \times 1}$ and positive definite variance covariance matrix $\Sigma_{p \times p}$, iff its density is

$$f(\mathbf{Y}, \boldsymbol{\mu}, \Sigma) = |\det(\Sigma)|^{-p/2} (2\pi)^{-p/2} \exp\{-\frac{1}{2}(\mathbf{Y} - \boldsymbol{\mu})^T \Sigma^{-1} (\mathbf{Y} - \boldsymbol{\mu})\}. \quad (1.6.16)$$

Rewriting the exponent we obtain

$$\begin{aligned}
\log f(\mathbf{Y}, \boldsymbol{\mu}, \Sigma) &= -\frac{1}{2} \mathbf{Y}^T \Sigma^{-1} \mathbf{Y} + (\Sigma^{-1} \boldsymbol{\mu})^T \mathbf{Y} \\
&\quad - \frac{1}{2}(\log |\det(\Sigma)| + \boldsymbol{\mu}^T \Sigma^{-1} \boldsymbol{\mu}) - \frac{p}{2} \log \pi.
\end{aligned} \quad (1.6.17)$$

The first two terms on the right in (1.6.17) can be rewritten

$$-\left(\sum_{1 \le i < j \le p} \sigma^{ij} Y_i Y_j + \frac{1}{2} \sum_{i=1}^{p} \sigma^{ii} Y_i^2 \right) + \sum_{i=1}^{p} \left(\sum_{j=1}^{p} \sigma^{ij} \mu_j \right) Y_i$$

where $\Sigma^{-1} \equiv \|\sigma^{ij}\|$, revealing that this is a $k = p(p+3)/2$ parameter exponential family with statistics $(Y_1, \ldots, Y_p, \{Y_i Y_j\}_{1 \le i \le j \le p})$, $h(\mathbf{Y}) \equiv 1$, $\boldsymbol{\theta} = (\boldsymbol{\mu}, \Sigma)$, $B(\boldsymbol{\theta}) = \frac{1}{2}(\log |\det(\Sigma)| + \boldsymbol{\mu}^T \Sigma^{-1} \boldsymbol{\mu})$. By our supermodel discussion, if $\mathbf{Y}_1, \ldots, \mathbf{Y}_n$ are iid $\mathcal{N}_p(\boldsymbol{\mu}, \Sigma)$, then $\mathbf{X} \equiv (\mathbf{Y}_1, \ldots, \mathbf{Y}_n)^T$ follows the $k = p(p+3)/2$ parameter exponential family with $\mathbf{T} = (\Sigma_i \mathbf{Y}_i, \Sigma_i \mathbf{Y}_i \mathbf{Y}_i^T)$, where we identify the second element of $\mathbf{T}$, which is a $p \times p$ symmetric matrix, with its distinct $p(p+1)/2$ entries. It may be shown (Problem 1.6.29) that $\mathbf{T}$ (and $h \equiv 1$) generate this family and that the rank of the family is indeed $p(p+3)/2$, generalizing Example 1.6.5, and that $\mathcal{E}$ is open, so that Theorem 1.6.4 applies.   □

1.6.5   Conjugate Families of Prior Distributions

In Section 1.2 we considered beta prior distributions for the probability of success in n Bernoulli trials. This is a special case of *conjugate families* of priors, families to which the posterior after sampling also belongs.

Suppose $X_1, \ldots, X_n$ is a sample from the k-parameter exponential family (1.6.10), and, as we always do in the Bayesian context, write $p(\mathbf{x} \mid \boldsymbol{\theta})$ for $p(\mathbf{x}, \boldsymbol{\theta})$. Then

$$p(\mathbf{x}|\boldsymbol{\theta}) = \left[\prod_{i=1}^{n} h(x_i) \right] \exp\left\{ \sum_{j=1}^{k} \eta_j(\boldsymbol{\theta}) \sum_{i=1}^{n} T_j(x_i) - nB(\boldsymbol{\theta}) \right\}. \quad (1.6.18)$$

where $\boldsymbol{\theta} \in \Theta$, which is k-dimensional. A conjugate exponential family is obtained from (1.6.18) by letting n and $t_j = \sum_{i=1}^{n} T_j(x_i)$, $j = 1, \dots, k$, be "parameters" and treating $\boldsymbol{\theta}$ as the variable of interest. That is, let $\mathbf{t} = (t_1, \dots, t_{k+1})^T$ and

$$
\omega(\mathbf{t}) = \int_{-\infty}^{\infty} \cdots \int_{-\infty}^{\infty} \exp\{\sum_{j=1}^{k} t_j \eta_j(\boldsymbol{\theta}) - t_{k+1} B(\boldsymbol{\theta})\} d\theta_1 \cdots d\theta_k
\tag{1.6.19}
$$

$$
\Omega = \{(t_1, \dots, t_{k+1}) : 0 < \omega(t_1, \dots, t_{k+1}) < \infty\}
$$

with integrals replaced by sums in the discrete case. We assume that Ω is nonempty (see Problem 1.6.36), then

Proposition 1.6.1. *The $(k+1)$-parameter exponential family given by*

$$
\pi_{\mathbf{t}}(\boldsymbol{\theta}) = \exp\{\sum_{j=1}^{k} \eta_j(\boldsymbol{\theta}) t_j - t_{k+1} B(\boldsymbol{\theta}) - \log \omega(\mathbf{t})\}
\tag{1.6.20}
$$

where $\mathbf{t} = (t_1, \dots, t_{k+1}) \in \Omega$, *is a conjugate prior to* $p(\mathbf{x}|\boldsymbol{\theta})$ *given by (1.6.18).*

Proof. If $p(\mathbf{x}|\boldsymbol{\theta})$ is given by (1.6.18) and π by (1.6.20), then

$$
\pi(\boldsymbol{\theta}|\mathbf{x}) \propto p(\mathbf{x}|\boldsymbol{\theta})\pi_{\mathbf{t}}(\boldsymbol{\theta}) \propto \exp\{\sum_{j=1}^{k} \eta_j(\boldsymbol{\theta})(\sum_{i=1}^{n} T_j(x_i) + t_j) - (t_{k+1} + n)B(\boldsymbol{\theta})\}
$$

$$
\propto \pi_{\mathbf{s}}(\boldsymbol{\theta}),
\tag{1.6.21}
$$

where

$$
\mathbf{s} = (s_1, \dots, s_{k+1})^T = \left(t_1 + \sum_{i=1}^{n} T_1(x_i), \dots, t_k + \sum_{i=1}^{n} T_k(x_i), t_{k+1} + n\right)^T
$$

and $\propto$ indicates that the two sides are proportional functions of $\boldsymbol{\theta}$. Because two probability densities that are proportional must be equal, $\pi(\boldsymbol{\theta}|\mathbf{x})$ is the member of the exponential family (1.6.20) given by the last expression in (1.6.21) and our assertion follows. $\square$

Remark 1.6.1. Note that (1.6.21) is an updating formula in the sense that as data $x_1, \dots, x_n$ become available, the parameter $\mathbf{t}$ of the prior distribution is updated to $\mathbf{s} = (\mathbf{t} + \mathbf{a})$, where $\mathbf{a} = (\sum_{i=1}^{n} T_1(x_i), \dots, \sum_{i=1}^{n} T_k(x_i), n)^T$. $\square$

It is easy to check that the beta distributions are obtained as conjugate to the binomial in this way.

Example 1.6.12. Suppose $X_1, \dots, X_n$ is a $\mathcal{N}(\theta, \sigma_0^2)$ sample, where σ_0^2 is known and θ is unknown. To choose a prior distribution for θ, we consider the conjugate family of the model defined by (1.6.20). For $n = 1$

$$
p(x|\theta) \propto \exp\{\frac{\theta x}{\sigma_0^2} - \frac{\theta^2}{2\sigma_0^2}\}.
\tag{1.6.22}
$$

This is a one-parameter exponential family with

$$T_1(x) = x, \ \eta_1(\theta) = \frac{\theta}{\sigma_0^2}, \ B(\theta) = \frac{\theta^2}{2\sigma_0^2}.$$

The conjugate two-parameter exponential family given by (1.6.20) has density

$$\pi_{\mathbf{t}}(\theta) = \exp\{\frac{\theta}{\sigma_0^2}t_1 - \frac{\theta^2}{2\sigma_0^2}t_2 - \log \omega(t_1, t_2)\}. \tag{1.6.23}$$

Upon completing the square, we obtain

$$\pi_{\mathbf{t}}(\theta) \propto \exp\{-\frac{t_2}{2\sigma_0^2}(\theta - \frac{t_1}{t_2})^2\}. \tag{1.6.24}$$

Thus, $\pi_{\mathbf{t}}(\theta)$ is defined only for $t_2 > 0$ and all t_1 and is the $\mathcal{N}(t_1/t_2, \sigma_0^2/t_2)$ density. Our conjugate family, therefore, consists of all $\mathcal{N}(\eta_0, \tau_0^2)$ distributions where η_0 varies freely and τ_0^2 is positive.

If we start with a $\mathcal{N}(\eta_0, \tau_0^2)$ prior density, we must have in the (t_1, t_2) parametrization

$$t_2 = \frac{\sigma_0^2}{\tau_0^2}, \ t_1 = \frac{\eta_0 \sigma_0^2}{\tau_0^2}. \tag{1.6.25}$$

By (1.6.21), if we observe $\Sigma X_i = s$, the posterior has a density (1.6.23) with

$$t_2(n) = \frac{\sigma_0^2}{\tau_0^2} + n, \ t_1(s) = \frac{\eta_0 \sigma_0^2}{\tau_0^2} + s.$$

Using (1.6.24), we find that $\pi(\theta|\mathbf{x})$ is a normal density with mean

$$\mu(s, n) = \frac{t_1(s)}{t_2(n)} = (\frac{\sigma_0^2}{\tau_0^2} + n)^{-1}[s + \frac{\eta_0 \sigma_0^2}{\tau_0^2}] \tag{1.6.26}$$

and variance

$$\tau_0^2(n) = \frac{\sigma_0^2}{t_2(n)} = (\frac{1}{\tau_0^2} + \frac{n}{\sigma_0^2})^{-1}. \tag{1.6.27}$$

Note that we can rewrite (1.6.26) intuitively as

$$\mu(s, n) = w_1 \bar{x} + w_2 \eta_0 \tag{1.6.28}$$

where $w_1 = n\tau_0^2(n)/\sigma_0^2$, $w_2 = \tau_0^2(n)/\tau_0^2$ so that $w_2 = 1 - w_1$. □

These formulae can be generalized to the case $\mathbf{X}_i$ i.i.d. $\mathcal{N}_p(\boldsymbol{\theta}, \Sigma_0)$, $1 \leq i \leq n$, Σ_0 known, $\boldsymbol{\theta} \sim \mathcal{N}_p(\boldsymbol{\eta}_0, \tau_0^2 \mathbf{I})$ where $\boldsymbol{\eta}_0$ varies over R^p, τ_0^2 is scalar with $\tau_0 > 0$ and $\mathbf{I}$ is the $p \times p$ identity matrix (Problem 1.6.37). Moreover, it can be shown (Problem 1.6.30) that the $\mathcal{N}_p(\boldsymbol{\lambda}, \Gamma)$ family with $\boldsymbol{\lambda} \in R^p$ and Γ symmetric positive definite is a conjugate family to $\mathcal{N}_p(\boldsymbol{\theta}, \Sigma_0)$, but a richer one than we've defined in (1.6.20) except for $p = 1$ because $\mathcal{N}_p(\boldsymbol{\lambda}, \Gamma)$ is a $p(p+3)/2$ rather than a $p+1$ parameter family. In fact, the conditions of Proposition 1.6.1 are often too restrictive. In the one-dimensional Gaussian case the members of the Gaussian conjugate family are unimodal and symmetric and have the same shape. It is easy to see that one can construct conjugate priors for which one gets reasonable formulae for the parameters indexing the model and yet have as great a richness of the shape variable as one wishes by considering finite mixtures of members of the family defined in (1.6.20). See Problems 1.6.31 and 1.6.32.

Discussion

Note that the uniform $\mathcal{U}(\{1, 2, \dots, \theta\})$ model of Example 1.5.3 is *not* covered by this theory. The natural sufficient statistic $\max(X_1, \dots, X_n)$, which is one-dimensional whatever be the sample size, is not of the form $\sum_{i=1}^{n} T(X_i)$. In fact, the family of distributions in this example and the family $\mathcal{U}(0, \theta)$ are not exponential. Despite the existence of classes of examples such as these, starting with Koopman, Pitman, and Darmois, a theory has been built up that indicates that under suitable regularity conditions families of distributions, which admit k-dimensional sufficient statistics for all sample sizes, must be k-parameter exponential families. Some interesting results and a survey of the literature may be found in Brown (1986). Problem 1.6.10 is a special result of this type.

Summary. $\{P_{\boldsymbol{\theta}} : \boldsymbol{\theta} \in \Theta\}$, $\Theta \subset R^k$, is a *k-parameter exponential family* of distributions if there are real-valued functions $\eta_1, \dots, \eta_k$ and B on Θ, and real-valued functions $T_1, \dots, T_k$, h on R^q such that the density (frequency) function of $P_{\boldsymbol{\theta}}$ can be written as

$$p(x, \boldsymbol{\theta}) = h(x) \exp[\sum_{j=1}^{k} \eta_j(\boldsymbol{\theta}) T_i(x) - B(\boldsymbol{\theta})], x \in \mathcal{X} \subset R^q. \tag{1.6.29}$$

$\mathbf{T}(X) = (T_1(X), \dots, T_k(X))^T$ is called the *natural sufficient statistic* of the family. The *canonical k-parameter exponential family generated by $\mathbf{T}$ and h* is

$$q(x, \boldsymbol{\eta}) = h(x) \exp\{\mathbf{T}^T(x)\boldsymbol{\eta} - A(\boldsymbol{\eta})\}$$

where

$$A(\boldsymbol{\eta}) = \log \int_{-\infty}^{\infty} \cdots \int_{-\infty}^{\infty} h(x) \exp\{\mathbf{T}^T(x)\boldsymbol{\eta}\} dx$$

in the continuous case, with integrals replaced by sums in the discrete case. The set

$$\mathcal{E} = \{\boldsymbol{\eta} \in R^k : -\infty < A(\boldsymbol{\eta}) < \infty\}$$

is called the *natural parameter space*. The set $\mathcal{E}$ is convex, the map $A : \mathcal{E} \to R$ is convex. If $\mathcal{E}$ has a nonempty interior in R^k and $\boldsymbol{\eta}_0 \in \mathcal{E}$, then $\mathbf{T}(X)$ has for $X \sim P_{\boldsymbol{\eta}_0}$ the moment-generating function

$$\psi(\mathbf{s}) = \exp\{A(\boldsymbol{\eta}_0 + \mathbf{s}) - A(\boldsymbol{\eta}_0)\}$$

for all s such that $\boldsymbol{\eta}_0 + \mathbf{s}$ is in $\mathcal{E}$. Moreover $E_{\boldsymbol{\eta}_0}[\mathbf{T}(X)] = \dot{A}(\boldsymbol{\eta}_0)$ and $Var_{\boldsymbol{\eta}_0}[\mathbf{T}(X)] = \ddot{A}(\boldsymbol{\eta}_0)$ where $\dot{A}$ and $\ddot{A}$ denote the gradient and Hessian of A.

An exponential family is said to be of *rank* k if $\mathbf{T}$ is k-dimensional and $1, T_1, \ldots, T_k$ are linearly independent with positive $P_{\boldsymbol{\theta}}$ probability for some $\boldsymbol{\theta} \in \Theta$. If $\mathcal{P}$ is a canonical exponential family with $\mathcal{E}$ open, then the following are equivalent:

(i) $\mathcal{P}$ is of rank k,

(ii) $\boldsymbol{\eta}$ is identifiable,

(iii) $\mathrm{Var}_{\boldsymbol{\eta}}(\mathbf{T})$ is positive definite,

(iv) the map $\boldsymbol{\eta} \to \dot{A}(\boldsymbol{\eta})$ is 1 - 1 on $\mathcal{E}$,

(v) A is strictly convex on $\mathcal{E}$.

A family $\mathcal{F}$ of prior distributions for a parameter vector $\boldsymbol{\theta}$ is called a *conjugate family* of priors to $p(x \mid \boldsymbol{\theta})$ if the posterior distribution of $\boldsymbol{\theta}$ given $\mathbf{x}$ is a member of $\mathcal{F}$. The $(k+1)$-parameter exponential family

$$\pi_{\mathbf{t}}(\boldsymbol{\theta}) = \exp\{\sum_{j=1}^{k} \eta_j(\boldsymbol{\theta})t_j - B(\boldsymbol{\theta})t_{k+1} - \log \omega\}$$

where

$$\omega = \int_{-\infty}^{\infty} \cdots \int_{-\infty}^{\infty} \exp\{\Sigma \eta_j(\boldsymbol{\theta})t_j - B(\boldsymbol{\theta})t_{k+1}\}d\boldsymbol{\theta},$$

and

$$\mathbf{t} = (t_1, \ldots, t_{k+1}) \in \Omega = \{(t_1, \ldots, t_{k+1}) \in R^{k+1} : 0 < \omega < \infty\},$$

is conjugate to the exponential family $p(x|\boldsymbol{\theta})$ defined in (1.6.29).

1.7   PROBLEMS AND COMPLEMENTS

Problems for Section 1.1

1. Give a formal statement of the following models identifying the probability laws of the data and the parameter space. State whether the model in question is parametric or nonparametric.

(a) A geologist measures the diameters of a large number n of pebbles in an old stream bed. Theoretical considerations lead him to believe that the logarithm of pebble diameter is normally distributed with mean μ and variance σ^2. He wishes to use his observations to obtain some information about μ and σ^2 but has in advance no knowledge of the magnitudes of the two parameters.

(b) A measuring instrument is being used to obtain n independent determinations of a physical constant μ. Suppose that the measuring instrument is known to be biased to the positive side by 0.1 units. Assume that the errors are otherwise identically distributed normal random variables with known variance.

(c) In part (b) suppose that the amount of bias is positive but unknown. Can you perceive any difficulties in making statements about μ for this model?

(d) The number of eggs laid by an insect follows a Poisson distribution with unknown mean λ. Once laid, each egg has an unknown chance p of hatching and the hatching of one egg is independent of the hatching of the others. An entomologist studies a set of n such insects observing both the number of eggs laid and the number of eggs hatching for each nest.

2. Are the following parametrizations identifiable? (Prove or disprove.)

(a) The parametrization of Problem 1.1.1(c).

(b) The parametrization of Problem 1.1.1(d).

(c) The parametrization of Problem 1.1.1(d) if the entomologist observes *only* the number of eggs hatching but not the number of eggs laid in each case.

3. Which of the following parametrizations are identifiable? (Prove or disprove.)

(a) $X_1, \ldots, X_p$ are independent with $X_i \sim \mathcal{N}(\alpha_i + \nu, \sigma^2)$.

$$\theta = (\alpha_1, \alpha_2, \ldots, \alpha_p, \nu, \sigma^2)$$

and P_θ is the distribution of $\mathbf{X} = (X_1, \ldots, X_p)$.

(b) Same as (a) with $\alpha = (\alpha_1, \ldots, \alpha_p)$ restricted to

$$\{(a_1, \ldots, a_p) : \sum_{i=1}^{p} a_i = 0\}.$$

(c) X and Y are independent $\mathcal{N}(\mu_1, \sigma^2)$ and $\mathcal{N}(\mu_2, \sigma^2)$, $\theta = (\mu_1, \mu_2)$ and we observe $Y - X$.

(d) X_{ij}, $i = 1, \ldots, p$; $j = 1, \ldots, b$ are independent with $X_{ij} \sim \mathcal{N}(\mu_{ij}, \sigma^2)$ where $\mu_{ij} = \nu + \alpha_i + \lambda_j$, $\theta = (\alpha_1, \ldots, \alpha_p, \lambda_1, \ldots, \lambda_b, \nu, \sigma^2)$ and P_θ is the distribution of $X_{11}, \ldots, X_{pb}$.

(e) Same as (d) with $(\alpha_1, \ldots, \alpha_p)$ and $(\lambda_1, \ldots, \lambda_b)$ restricted to the sets where $\sum_{i=1}^{p} \alpha_i = 0$ and $\sum_{j=1}^{b} \lambda_j = 0$.

4. (a) Let U be any random variable and V be any other nonnegative random variable. Show that

$$F_{U+V}(t) \le F_U(t) \text{ for every } t.$$

(If F_X and F_Y are distribution functions such that $F_X(t) \le F_Y(t)$ for every t, then X is said to be *stochastically larger* than Y.)

(b) As in Problem 1.1.1 describe formally the following model. Two groups of n_1 and n_2 individuals, respectively, are sampled at random from a very large population. Each member of the second (treatment) group is administered the same dose of a certain drug believed to lower blood pressure and the blood pressure is measured after 1 hour. Each member of the first (control) group is administered an equal dose of a placebo and then has the blood pressure measured after 1 hour. It is known that the drug either has no effect or lowers blood pressure, but the distribution of blood pressure in the population sampled before and after administration of the drug is quite unknown.

5. The number n of graduate students entering a certain department is recorded. In each of k subsequent years the number of students graduating and of students dropping out is recorded. Let N_i be the number dropping out and M_i the number graduating during year i, $i = 1, \ldots, k$. The following model is proposed.

$$P_\theta[N_1 = n_1, M_1 = m_1, \ldots, N_k = n_k, M_k = m_k]$$

$$= \frac{n!}{n_1! \ldots n_k! m_1! \ldots m_k! r!} \mu_1^{n_1} \cdots \mu_k^{n_k} \nu_1^{m_1} \cdots \nu_k^{m_k} \rho^r$$

where

$$\mu_1 + \cdots + \mu_k + \nu_1 + \cdots + \nu_k + \rho = 1, \ 0 < \mu_i < 1, \ 0 < \nu_i < 1, \ 1 \le i \le k$$

$$n_1 + \cdots + n_k + m_1 + \cdots + m_k + r = n$$

and $\theta = (\mu_1, \ldots, \mu_k, \nu_1, \ldots, \nu_k)$ is unknown.

(a) What are the assumptions underlying this model?

(b) θ is very difficult to estimate here if k is large. The simplification $\mu_i = \pi(1 - \mu)^{i-1}\mu$, $\nu_i = (1 - \pi)(1 - \nu)^{i-1}\nu$ for $i = 1, \ldots, k$ is proposed where $0 < \pi < 1$, $0 < \mu < 1, 0 < \nu < 1$ are unknown. What assumptions underlie the simplification?

6. Which of the following models are regular? (Prove or disprove.)

(a) P_θ is the distribution of X when X is uniform on $(0, \theta)$, $\Theta = (0, \infty)$.

(b) P_θ is the distribution of X when X is uniform on $\{0, 1, 2, \ldots, \theta\}$, $\Theta = \{1, 2, \ldots\}$.

(c) Suppose $X \sim \mathcal{N}(\mu, \sigma^2)$. Let $Y = 1$ if $X \le 1$ and $Y = X$ if $X > 1$. $\theta = (\mu, \sigma^2)$ and P_θ is the distribution of Y.

(d) Suppose the possible control responses in an experiment are $0.1, 0.2, \ldots, 0.9$ and they occur with frequencies $p(0.1), p(0.2), \ldots, p(0.9)$. Suppose the effect of a treatment is to increase the control response by a fixed amount θ. Let P_θ be the distribution of a treatment response.

7. Show that $Y - c$ has the same distribution as $-Y + c$, if and only if, the density or frequency function p of Y satisfies $p(c + t) = p(c - t)$ for all t. Both Y and p are said to be *symmetric* about c.

Hint: If $Y - c$ has the same distribution as $-Y + c$, then $P(Y \le t + c) = P(-Y \le t - c) = P(Y \ge c - t) = 1 - P(Y < c - t)$.

8. Consider the two sample models of Examples 1.1.3(2) and 1.1.4(1).

(a) Show that if $Y \sim X + \delta(X)$, $\delta(x) = 2\mu + \Delta - 2x$ and $X \sim \mathcal{N}(\mu, \sigma^2)$, then $G(\cdot) = F(\cdot - \Delta)$. That is, the two cases $\delta(x) \equiv \Delta$ and $\delta(x) = 2\mu + \Delta - 2x$ yield the same distribution for the data $(X_1, \ldots, X_n)$, $(Y_1, \ldots, Y_n)$. Therefore, $G(\cdot) = F(\cdot - \Delta)$ does not imply the constant treatment effect assumption.

(b) In part (a), suppose X has a distribution F that is not necessarily normal. For what type of F is it possible to have $G(\cdot) = F(\cdot - \Delta)$ for both $\delta(x) \equiv \Delta$ and $\delta(x) = 2\mu + \Delta - 2x$?

(c) Suppose that $Y \sim X + \delta(X)$ where $X \sim \mathcal{N}(\mu, \sigma^2)$ and $\delta(x)$ is continuous. Show that if we assume that $\delta(x) + x$ is strictly increasing, then $G(\cdot) = F(\cdot - \Delta)$ implies that $\delta(x) \equiv \Delta$.

9. *Collinearity:* Suppose $Y_i = \sum_{j=1}^{p} z_{ij}\beta_j + \epsilon_i$, $\epsilon_i \sim \mathcal{N}(0, \sigma^2)$ independent, $1 \le i \le n$. Let $\mathbf{z}_j \equiv (z_{1j}, \ldots, z_{nj})^T$.

(a) Show that $(\beta_1, \ldots, \beta_p)$ are identifiable iff $\mathbf{z}_1, \ldots, \mathbf{z}_p$, are not collinear (linearly independent).

(b) Deduce that $(\beta_1, \ldots, \beta_p)$ are not identifiable if $n < p$, that is, if the number of parameters is larger than the number of observations.

10. Let $X = (\min(T, C), I(T \le C))$ where T, C are independent,

$$
\begin{aligned}
P[T = j] &= p(j), \ j = 0, \ldots, N, \\
P[C = j] &= r(j), \ i = 0, \ldots, N
\end{aligned}
$$

and (p, r) vary freely over $\mathcal{F} = \{(p, r) : p(j) > 0, r(j) > 0, 0 \le j \le N, \sum_{j=0}^{N} p(j) = 1, \sum_{j=0}^{N} r(j) = 1\}$ and N is known. Suppose $X_1, \ldots, X_n$ are observed i.i.d. according to the distribution of X.

Show that $\{p(j) : j = 0, \ldots, N\}$, $\{r(j) : j = 0, \ldots, N\}$ are identifiable.

Hint: Consider "hazard rates" for $Y \equiv \min(T, C)$,

$$
P[Y = j, Y = T \mid Y \ge j].
$$

11. *The Scale Model.* Positive random variables X and Y satisfy a *scale model* with parameter $\delta > 0$ if $P(Y \le t) = P(\delta X \le t)$ for all $t > 0$, or equivalently, $G(t) = F(t/\delta)$, $\delta > 0, t > 0$.

(a) Show that in this case, $\log X$ and $\log Y$ satisfy a shift model with parameter $\log \delta$.

(b) Show that if X and Y satisfy a shift model with parameter Δ, then e^X and e^Y satisfy a scale model with parameter e^Δ.

(c) Suppose a scale model holds for X, Y. Let $c > 0$ be a constant. Does $X' = X^c$, $Y' = Y^c$ satisfy a scale model? Does $\log X'$, $\log Y'$ satisfy a shift model?

12. *The Lehmann Two-Sample Model.* In Example 1.1.3 let $X_1, \ldots, X_m$ and $Y_1, \ldots, Y_n$ denote the survival times of two groups of patients receiving treatments A and B. $S_X(t) =$

$P(X > t) = 1 - F(t)$ and $S_Y(t) = P(Y > t) = 1 - G(t)$, $t > 0$, are called the *survival functions*. For the A group, survival beyond time t is modeled to occur if the events $T_1 > t, \ldots, T_a > t$ all occur, where $T_1, \ldots, T_a$ are unobservable and i.i.d. as T with survival function S_0. Similarly, for the B group, $Y > t$ occurs iff $T_1' > t, \ldots, T_b' > t$ where $T_1', \ldots, T_b'$ are i.i.d. as T.

   (a) Show that $S_Y(t) = S_X^{b/a}(t)$.

   (b) By extending (b/a) from the rationals to $\delta \in (0, \infty)$, we have the *Lehmann model*

$$S_Y(t) = S_X^{\delta}(t), \ t > 0. \tag{1.7.1}$$

Equivalently, $S_Y(t) = S_0^{\Delta}(t)$ with $\Delta = a\delta$, $t > 0$. Show that if S_0 is continuous, then $X' = -\log S_0(X)$ and $Y' = -\log S_0(Y)$ follow an exponential scale model (see Problem 1.1.11) with scale parameter δ^{-1}.

   Hint: By Problem B.2.12, $S_0(T)$ has a $\mathcal{U}(0, 1)$ distribution; thus, $-\log S_0(T)$ has an exponential distribution. Also note that $P(X > t) = S_0^a(t)$.

   (c) Suppose that T and Y have densities $f_0(t)$ and $g(t)$. Then $h_0(t) = f_0(t)/S_0(t)$ and $h_Y(t) = g(t)/S_Y(t)$ are called the *hazard rates* of T and Y. Moreover, $h_Y(t) = \Delta h_0(t)$ is called the *Cox proportional hazard* model. Show that $h_Y(t) = \Delta h_0(t)$ if and only if $S_Y(t) = S_0^{\Delta}(t)$.

13. *A proportional hazard model.* Let $f(t \mid \mathbf{z}_i)$ denote the density of the survival time Y_i of a patient with covariate vector $\mathbf{z}_i$ and define the *regression survival* and *hazard* functions of Y_i as

$$S_Y(t \mid \mathbf{z}_i) = \int_t^{\infty} f(y \mid \mathbf{z}_i)dy, \ h(t \mid \mathbf{z}_i) = f(t \mid \mathbf{z}_i)/S_Y(t \mid \mathbf{z}_i).$$

Let T denote a survival time with density $f_0(t)$ and hazard rate $h_0(t) = f_0(t)/P(T > t)$. The *Cox proportional hazard model* is defined as

$$h(t \mid \mathbf{z}) = h_0(t) \exp\{g(\boldsymbol{\beta}, \mathbf{z})\} \tag{1.7.2}$$

where $h_0(t)$ is called the *baseline* hazard function and g is known except for a vector $\boldsymbol{\beta} = (\beta_1, \ldots, \beta_p)^T$ of unknowns. The most common choice of g is the linear form $g(\boldsymbol{\beta}, \mathbf{z}) = \mathbf{z}^T\boldsymbol{\beta}$. Set $\Delta = \exp\{g(\boldsymbol{\beta}, \mathbf{z})\}$.

   (a) Show that $(1.7.2)$ is equivalent to $S_Y(t \mid \mathbf{z}) = S_T^{\Delta}(t)$.

   (b) Assume $(1.7.2)$ and that $F_0(t) = P(T \le t)$ is known and strictly increasing. Find an increasing function $Q(t)$ such that the regression survival function of $Y' = Q(Y)$ does not depend on $h_0(t)$.

   Hint: See Problem 1.1.12.

   (c) Under the assumptions of (b) above, show that there is an increasing function $Q^*(t)$ such that if $Y_i^* = Q^*(Y_i)$, then

$$Y_i^* = g(\boldsymbol{\beta}, \mathbf{z}_i) + \epsilon_i$$

for some appropriate ϵ_i. Specify the distribution of ϵ_i.

Hint: See Problems 1.1.11 and 1.1.12.

14. In Example 1.1.2 with assumptions (1)–(4), the parameter of interest can be characterized as the median $\nu = F^{-1}(0.5)$ or mean $\mu = \int_{-\infty}^{\infty} x dF(x) = \int_0^1 F^{-1}(u) du$. Generally, μ and ν are regarded as *centers* of the distribution F. When F is not symmetric, μ may be very much pulled in the direction of the longer tail of the density, and for this reason, the median is preferred in this case. Examples are the distribution of income and the distribution of wealth. Here is an example in which the mean is extreme and the median is not. Suppose the monthly salaries of state workers in a certain state are modeled by the *Pareto* distribution with distribution function

$$
\begin{aligned}
F(x, \theta) &= 1 - (x/c)^{-\theta}, \quad x \geq c \\
&= 0, \qquad\qquad\quad x < c
\end{aligned}
$$

where $\theta > 0$ and $c = 2,000$ is the minimum monthly salary for state workers. Find the median ν and the mean μ for the values of θ where the mean exists. Show how to choose θ to make $\mu - \nu$ arbitrarily large.

15. Let $X_1, \ldots, X_m$ be i.i.d. F, $Y_1, \ldots, Y_n$ be i.i.d. G, where the model $\{(F, G)\}$ is described by

$$
\psi(X_1) = Z_1, \quad \psi(Y_1) = Z_1' + \Delta,
$$

where ψ is an unknown strictly increasing differentiable map from R to R, $\psi' > 0$, $\psi(\pm\infty) = \pm\infty$, and Z_1 and Z_1' are independent random variables.

(a) Suppose Z_1, Z_1' have a $\mathcal{N}(0, 1)$ distribution. Show that both ψ and Δ are identifiable.

(b) Suppose Z_1 and Z_1' have a $\mathcal{N}(0, \sigma^2)$ distribution with σ^2 unknown. Are ψ and Δ still identifiable? If not, what parameters are?
Hint: (a) $P[X_1 \leq t] = \Phi(\psi(t))$.

Problems for Section 1.2

1. *Merging Opinions.* Consider a parameter space consisting of two points θ_1 and θ_2, and suppose that for given θ, an experiment leads to a random variable X whose frequency function $p(x \mid \theta)$ is given by

$\theta \backslash x$	0	1
θ_1	0.8	0.2
θ_2	0.4	0.6

Let π be the prior frequency function of θ defined by $\pi(\theta_1) = \frac{1}{2}$, $\pi(\theta_2) = \frac{1}{2}$.

(a) Find the posterior frequency function $\pi(\theta \mid x)$.

(b) Suppose $X_1, \ldots, X_n$ are independent with frequency function $p(x \mid \theta)$. Find the posterior $\pi(\theta \mid x_1, \ldots, x_n)$. Observe that it depends only on $\sum_{i=1}^n x_i$.

(c) Same as (b) except use the prior $\pi_1(\theta_1) = .25$, $\pi_1(\theta_2) = .75$.

(d) Give the values of $P(\theta = \theta_1 \mid \sum_{i=1}^{n} X_i = .5n)$ for the two priors π and π_1 when $n = 2$ and 100.

(e) Give the most probable values $\hat{\theta} = \arg\max_\theta \pi(\theta \mid \sum_{i=1}^{n} X_i = k)$ for the two priors π and π_1. Compare these $\hat{\theta}$'s for $n = 2$ and 100.

(f) Give the set on which the two $\hat{\theta}$'s disagree. Show that the probability of this set tends to zero as $n \to \infty$. Assume $X \sim p(x) = \sum_{i=1}^{2} \pi(\theta_i)p(x \mid \theta_i)$. For this convergence, does it matter which prior, π or π_1, is used in the formula for $p(x)$?

2. Consider an experiment in which, for given $\theta = \theta$, the outcome X has density $p(x \mid \theta) = (2x/\theta^2)$, $0 < x < \theta$. Let π denote a prior density for θ.

(a) Find the posterior density of θ when $\pi(\theta) = 1$, $0 \leq \theta \leq 1$.

(b) Find the posterior density of θ when $\pi(\theta) = 3\theta^2$, $0 \leq \theta \leq 1$.

(c) Find $E(\theta \mid x)$ for the two priors in (a) and (b).

(d) Suppose $X_1, \ldots, X_n$ are independent with the same distribution as X. Find the posterior density of θ given $X_1 = x_1, \ldots, X_n = x_n$ when $\pi(\theta) = 1$, $0 \leq \theta \leq 1$.

3. Let X be the number of failures before the first success in a sequence of Bernoulli trials with probability of success θ. Then $P_\theta[X = k] = (1-\theta)^k\theta$, $k = 0, 1, 2, \ldots$. This is called the *geometric distribution* $(\mathcal{G}(\theta))$. Suppose that for given $\theta = \theta$, X has the geometric distribution

(a) Find the posterior distribution of θ given $X = 2$ when the prior distribution of θ is uniform on $\{\frac{1}{4}, \frac{1}{2}, \frac{3}{4}\}$.

(b) Relative to (a), what is the most probable value of θ given $X = 2$? Given $X = k$?

(c) Find the posterior distribution of θ given $X = k$ when the prior distribution is beta, $\beta(r, s)$.

4. Let $X_1, \ldots, X_n$ be distributed as

$$p(x_1, \ldots, x_n \mid \theta) = \frac{1}{\theta^n}$$

where $x_1, \ldots, x_n$ are natural numbers between 1 and θ and $\Theta = \{1, 2, 3, \ldots\}$.

(a) Suppose θ has prior frequency,

$$\pi(j) = \frac{c(a)}{j^a}, \ j = 1, 2, \ldots,$$

where $a > 1$ and $c(a) = [\sum_{j=1}^{\infty} j^{-a}]^{-1}$. Show that

$$\pi(j \mid x_1, \ldots, x_n) = \frac{c(n + a, m)}{j^{n+a}}, \ j = m, m+1, \ldots,$$

where $m = \max(x_1, \ldots, x_n)$, $c(b, t) = [\sum_{j=t}^{\infty} j^{-b}]^{-1}$, $b > 1$.

(b) Suppose that $\max(x_1, \ldots, x_n) = x_1 = m$ for all n. Show that $\pi(m \mid x_1, \ldots, x_n) \to 1$ as $n \to \infty$ whatever be a. Interpret this result.

5. In Example 1.2.1 suppose n is large and $(1/n)\sum_{i=1}^{n} x_i = \bar{x}$ is not close to 0 or 1 and the prior distribution is beta, $\beta(r, s)$. Justify the following approximation to the posterior distribution

$$P[\theta \le t \mid X_1 = x_1, \ldots, X_n = x_n] \approx \Phi\left(\frac{t - \tilde{\mu}}{\tilde{\sigma}}\right)$$

where Φ is the standard normal distribution function and

$$\tilde{\mu} = \frac{n}{n+r+s}\bar{x} + \frac{r}{n+r+s}, \quad \tilde{\sigma}^2 = \frac{\tilde{\mu}(1 - \tilde{\mu})}{n+r+s}.$$

Hint: Let $\beta(a, b)$ denote the posterior distribution. If a and b are integers, then $\beta(a, b)$ is the distribution of $(a\bar{V}/b\bar{W})[1 + (a\bar{V}/b\bar{W})]^{-1}$, where $V_1, \ldots, V_a, W_1, \ldots, W_b$ are independent standard exponential. Next use the central limit theorem and Slutsky's theorem.

6. Show that a conjugate family of distributions for the Poisson family is the gamma family.

7. Show rigorously using (1.2.8) that if in Example 1.1.1, $D = N\theta$ has a $\mathcal{B}(N, \pi_0)$ distribution, then the posterior distribution of D given $X = k$ is that of $k + Z$ where Z has a $\mathcal{B}(N - n, \pi_0)$ distribution.

8. Let $(X_1, \ldots, X_{n+k})$ be a sample from a population with density $f(x \mid \theta)$, $\theta \in \Theta$. Let θ have prior density π. Show that the conditional distribution of $(\theta, X_{n+1}, \ldots, X_{n+k})$ given $X_1 = x_1, \ldots, X_n = x_n$ is that of $(Y, Z_1, \ldots, Z_k)$ where the marginal distribution of Y equals the posterior distribution of θ given $X_1 = x_1, \ldots, X_n = x_n$, and the conditional distribution of the Z_i's given $Y = t$ is that of sample from the population with density $f(x \mid t)$.

9. Show in Example 1.2.1 that the conditional distribution of θ given $\sum_{i=1}^{n} X_i = k$ agrees with the posterior distribution of θ given $X_1 = x_1, \ldots, X_n = x_n$, where $\sum_{i=1}^{n} x_i = k$.

10. Suppose $X_1, \ldots, X_n$ is a sample with $X_i \sim p(x \mid \theta)$, a regular model and integrable as a function of θ. Assume that $A = \{x : p(x \mid \theta) > 0\}$ does not involve θ.

(a) Show that the family of priors

$$\pi(\theta) = \prod_{i=1}^{N} p(\xi_i \mid \theta) \bigg/ \int_{\Theta} \prod_{i=1}^{N} p(\xi_i \mid \theta) d\theta$$

where $\xi_i \in A$ and $N \in \{1, 2, \ldots\}$ is a conjugate family of prior distributions for $p(\mathbf{x} \mid \theta)$ and that the posterior distribution of θ given $\mathbf{X} = \mathbf{x}$ is

$$\pi(\theta \mid \mathbf{x}) = \prod_{i=1}^{N'} p(\xi_i' \mid \theta) \bigg/ \int_{\Theta} \prod_{i=1}^{N'} p(\xi_i' \mid \theta) d\theta$$

where $N' = N + n$ and $(\xi'_i, \ldots, \xi'_{N'}) = (\xi_1, \ldots, \xi_N, x_1, \ldots, x_n)$.

(b) Use the result (a) to give $\pi(\theta)$ and $\pi(\theta \mid \mathbf{x})$ when

$$
\begin{aligned}
p(x \mid \theta) &= \theta \exp\{-\theta x\}, \ x > 0, \ \theta > 0 \\
&= 0 \text{ otherwise.}
\end{aligned}
$$

11. Let $p(x \mid \theta) = \exp\{-(x - \theta)\}, 0 < \theta < x$ and let $\pi(\theta) = 2\exp\{-2\theta\}, \theta > 0$. Find the posterior density $\pi(\theta \mid x)$.

12. Suppose $p(\mathbf{x} \mid \theta)$ is the density of i.i.d. $X_1, \ldots, X_n$, where $X_i \sim \mathcal{N}\left(\mu_0, \frac{1}{\theta}\right)$, μ_0 is known, and $\theta = \sigma^{-2}$ is (called) the *precision* of the distribution of X_i.

(a) Show that $p(\mathbf{x} \mid \theta) \propto \theta^{\frac{1}{2}n} \exp\left(-\frac{1}{2}t\theta\right)$ where $t = \sum_{i=1}^{n}(X_i - \mu_0)^2$ and $\propto$ denotes "proportional to" as a function of θ.

(b) Let $\pi(\theta) \propto \theta^{\frac{1}{2}(\lambda-2)} \exp\left\{-\frac{1}{2}\nu\theta\right\}, \nu > 0, \lambda > 0; \theta > 0$. Find the posterior distribution $\pi(\theta \mid \mathbf{x})$ and show that if λ is an integer, given $\mathbf{x}$, $\theta(t + \nu)$ has a $\chi^2_{\lambda+n}$ distribution. Note that, unconditionally, $\nu\theta$ has a χ^2_λ distribution.

(c) Find the posterior distribution of σ.

13. Show that if $X_1, \ldots, X_n$ are i.i.d. $\mathcal{N}(\mu, \sigma^2)$ and we formally put $\pi(\mu, \sigma) = \frac{1}{\sigma}$, then the posterior density $\pi(\mu \mid \bar{x}, s^2)$ of μ given $(\bar{x}, s^2)$ is such that $\sqrt{n}\frac{(\mu-\bar{X})}{s} \sim t_{n-1}$. Here $s^2 = \frac{1}{n-1}\sum(X_i - \bar{X})^2$.

Hint: Given μ and σ, $\bar{X}$ and s^2 are independent with $\bar{X} \sim \mathcal{N}(\mu, \sigma^2/n)$ and $(n-1)s^2/\sigma^2 \sim \chi^2_{n-1}$. This leads to $p(\bar{x}, s^2 \mid \mu, \sigma^2)$. Next use Bayes rule.

14. In a Bayesian model where $X_1, \ldots, X_n, X_{n+1}$ are i.i.d. $f(x \mid \theta), \theta \sim \pi$, the *predictive distribution* is the marginal distribution of X_{n+1}. The *posterior predictive distribution* is the conditional distribution of X_{n+1} given $X_1, \ldots, X_n$.

(a) If f and π are the $\mathcal{N}(\theta, \sigma_0^2)$ and $\mathcal{N}(\theta_0, \tau_0^2)$ densities, compute the predictive and posterior predictive distribution.

(b) Discuss the behavior of the two predictive distributions as $n \to \infty$.

15. *The Dirichlet distribution is a conjugate prior for the multinomial.* The Dirichlet distribution, $\mathcal{D}(\alpha), \alpha = (\alpha_1, \ldots, \alpha_r)^T, \alpha_j > 0, 1 \le j \le r$, has density

$$
f_\alpha(\mathbf{u}) = \frac{\Gamma\left(\sum_{j=1}^{r}\alpha_j\right)}{\prod_{j=1}^{r}\Gamma(\alpha_j)} \prod_{j=1}^{r} u_j^{\alpha_j-1}, \ 0 < u_j < 1, \ \sum_{j=1}^{r} u_j = 1.
$$

Let $\mathbf{N} = (N_1, \ldots, N_r)$ be multinomial

$$
\mathcal{M}(n, \boldsymbol{\theta}), \ \boldsymbol{\theta} = (\theta_1, \ldots, \theta_r)^T, \ 0 < \theta_j < 1, \ \sum_{j=1}^{r} \theta_j = 1.
$$

Show that if the prior $\pi(\boldsymbol{\theta})$ for $\boldsymbol{\theta}$ is $\mathcal{D}(\boldsymbol{\alpha})$, then the posterior $\pi(\boldsymbol{\theta} \mid \mathbf{N} = \mathbf{n})$ is $\mathcal{D}(\boldsymbol{\alpha} + \mathbf{n})$, where $\mathbf{n} = (n_1, \ldots, n_r)$.

Problems for Section 1.3

1. Suppose the possible states of nature are θ_1, θ_2, the possible actions are a_1, a_2, a_3, and the loss function $l(\theta, a)$ is given by

$\theta \backslash a$	a_1	a_2	a_3
θ_1	0	1	2
θ_2	2	0	1

Let X be a random variable with frequency function $p(x, \theta)$ given by

$\theta \backslash x$	0	1
θ_1	p	$(1 - p)$
θ_2	q	$(1 - q)$

and let $\delta_1, \ldots, \delta_9$ be the decision rules of Table 1.3.3. Compute and plot the risk points when

(a) $p = q = .1$,

(b) $p = 1 - q = .1$.

(c) Find the minimax rule among $\delta_1, \ldots, \delta_9$ for the preceding case (a).

(d) Suppose that θ has prior $\pi(\theta_1) = 0.5$, $\pi(\theta_2) = 0.5$. Find the Bayes rule for case (a).

2. Suppose that in Example 1.3.5, a new buyer makes a bid and the loss function is changed to

$\theta \backslash a$	a_1	a_2	a_3
θ_1	0	7	4
θ_2	12	1	6

(a) Compute and plot the risk points in this case for each rule $\delta_1, \ldots, \delta_9$ of Table 1.3.3.

(b) Find the minimax rule among $\{\delta_1, \ldots, \delta_9\}$.

(c) Find the minimax rule among the randomized rules.

(d) Suppose θ has prior $\pi(\theta_1) = \gamma$, $\pi(\theta_2) = 1 - \gamma$. Find the Bayes rule when (i) $\gamma = 0.5$ and (ii) $\gamma = 0.1$.

3. The problem of selecting the better of two treatments or of deciding whether the effect of one treatment is beneficial or not often reduces to the problem of deciding whether $\theta < 0$, $\theta = 0$ or $\theta > 0$ for some parameter θ. See Example 1.1.3. Let the actions corresponding to deciding whether $\theta < 0$, $\theta = 0$ or $\theta > 0$ be denoted by $-1, 0, 1$, respectively and suppose the loss function is given by (from Lehmann, 1957)

$\theta \backslash a$	-1	0	1
< 0	0	c	$b+c$
$= 0$	b	0	b
> 0	$b+c$	c	0

where b and c are positive. Suppose $\mathbf{X}$ is a $\mathcal{N}(\theta, 1)$ sample and consider the decision rule

$$\delta_{r,s}(\mathbf{X}) = \quad -1 \quad \text{if } \bar{X} < r$$
$$0 \quad \text{if } r \leq \bar{X} \leq s$$
$$1 \quad \text{if } \bar{X} > s.$$

(a) Show that the risk function is given by

$$
\begin{aligned}
R(\theta, \delta_{r,s}) \quad &= \quad c\bar{\Phi}(\sqrt{n}(r-\theta)) + b\bar{\Phi}(\sqrt{n}(s-\theta)), \quad \theta < 0 \\
&= \quad b\bar{\Phi}(\sqrt{n}s) + b\Phi(\sqrt{n}r), \quad \theta = 0 \\
&= \quad c\Phi(\sqrt{n}(s-\theta)) + b\Phi(\sqrt{n}(r-\theta)), \quad \theta > 0
\end{aligned}
$$

where $\bar{\Phi} = 1 - \Phi$, and Φ is the $\mathcal{N}(0, 1)$ distribution function.

(b) Plot the risk function when $b = c = 1$, $n = 1$ and

$$(\text{i}) \ r = -s = -1, \ (\text{ii}) \ r = -\frac{1}{2}s = -1.$$

For what values of θ does the procedure with $r = -s = -1$ have smaller risk than the procedure with $r = -\frac{1}{2}s = -1$?

4. *Stratified sampling.* We want to estimate the mean $\mu = E(X)$ of a population that has been divided (stratified) into s mutually exclusive parts (strata) (e.g., geographic locations or age groups). Within the jth stratum we have a sample of i.i.d. random variables $X_{1j}, \ldots, X_{n_j j}$; $j = 1, \ldots, s$, and a stratum sample mean $\bar{X}_j$; $j = 1, \ldots, s$. We assume that the s samples from different strata are independent. Suppose that the jth stratum has $100p_j\%$ of the population and that the jth stratum population mean and variances are μ_j and σ_j^2. Let $N = \sum_{j=1}^{s} n_j$ and consider the two estimators

$$\hat{\mu}_1 = N^{-1} \sum_{j=1}^{s} \sum_{i=1}^{n_j} X_{ij}, \ \hat{\mu}_2 = \sum_{j=1}^{s} p_j \bar{X}_j$$

where we assume that p_j, $1 \leq j \leq s$, are known.

(a) Compute the biases, variances, and MSEs of $\hat{\mu}_1$ and $\hat{\mu}_2$. How should n_j, $0 \leq j \leq s$, be chosen to make $\hat{\mu}_1$ unbiased?

(b) *Neyman allocation.* Assume that $0 < \sigma_j^2 < \infty$, $1 \leq j \leq s$, are known (estimates will be used in a later chapter). Show that the strata sample sizes that minimize $MSE(\hat{\mu}_2)$ are given by

$$n_k = N \frac{p_k \sigma_k}{\sum_{j=1}^{s} p_j \sigma_j}, \ k = 1, \ldots, s. \qquad (1.7.3)$$

Hint: You may use a Lagrange multiplier.

(c) Show that $MSE(\widehat{\mu}_1)$ with $n_k = p_k N$ minus $MSE(\widehat{\mu}_2)$ with n_k given by (1.7.3) is $N^{-1} \sum_{j=1}^{s} p_j (\sigma_j - \bar{\sigma})^2$, where $\bar{\sigma} = \sum_{j=1}^{s} p_j \sigma_j$.

5. Let $\bar{X}_b$ and $\widehat{X}_b$ denote the sample mean and the sample median of the sample $X_1 - b, \ldots, X_n - b$. If the parameters of interest are the population mean and median of $X_i - b$, respectively, show that $MSE(\bar{X}_b)$ and $MSE(\widehat{X}_b)$ are the same for all values of b (the MSEs of the sample mean and sample median are *invariant* with respect to shift).

6. Suppose that $X_1, \ldots, X_n$ are i.i.d. as $X \sim F$, that $\widehat{X}$ is the median of the sample, and that n is odd. We want to estimate "the" median ν of F, where ν is defined as a value satisfying $P(X \le \nu) \ge \frac{1}{2}$ and $P(X \ge \nu) \ge \frac{1}{2}$.

(a) Find the MSE of $\widehat{X}$ when

(i) F is discrete with $P(X = a) = P(X = c) = p$, $P(X = b) = 1 - 2p$, $0 < p < 1$, $a < b < c$.

Hint: Use Problem 1.3.5. The answer is $MSE(\widehat{X}) = [(a-b)^2 + (c-b)^2] P(S \ge k)$ where $k = .5(n+1)$ and $S \sim \mathcal{B}(n, p)$.

(ii) F is uniform, $\mathcal{U}(0, 1)$.

Hint: See Problem B.2.9.

(iii) F is normal, $\mathcal{N}(0, 1)$, $n = 1, 5, 25, 75$.

Hint: See Problem B.2.13. Use a numerical integration package.

(b) Compute the relative risk $RR = MSE(\widehat{X})/MSE(\bar{X})$ in question (i) when $b = 0$, $a = -\Delta$, $b = \Delta$, $p = .20, .40$, and $n = 1, 5, 15$.

(c) Same as (b) except when $n = 15$, plot RR for $p = .1, .2, .3, .4, .45$.

(d) Find $E|\widehat{X} - b|$ for the situation in (i). Also find $E|\bar{X} - b|$ when $n = 1$, and 2 and compare it to $E|\widehat{X} - b|$.

(e) Compute the relative risks $MSE(\widehat{X})/MSE(\bar{X})$ in questions (ii) and (iii).

7. Let $X_1, \ldots, X_n$ be a sample from a population with values

$$\theta - 2\Delta, \theta - \Delta, \theta, \theta + \Delta, \theta + 2\Delta; \ \Delta > 0.$$

Each value has probability .2. Let $\bar{X}$ and $\widehat{X}$ denote the sample mean and median. Suppose that n is odd.

(a) Find $MSE(\widehat{X})$ and the relative risk $RR = MSE(\widehat{X})/MSE(\bar{X})$.

(b) Evaluate RR when $n = 1, 3, 5$.

Hint: By Problem 1.3.5, set $\theta = 0$ without loss of generality. Next note that the distribution of $\widehat{X}$ involves Bernoulli and multinomial trials.

8. Let $X_1, \ldots, X_n$ be a sample from a population with variance $\sigma^2, 0 < \sigma^2 < \infty$.

(a) Show that $s^2 = (n-1)^{-1} \sum_{i=1}^n (X_i - \bar{X})^2$ is an unbiased estimator of σ^2.
Hint: Write $(X_i - \bar{X})^2 = ([X_i - \mu] - [\bar{X} - \mu])^2$, then expand $(X_i - \bar{X})^2$ keeping the square brackets intact.

(b) Suppose $X_i \sim \mathcal{N}(\mu, \sigma^2)$.

(i) Show that $MSE(s^2) = 2(n-1)^{-1}\sigma^4$.

(ii) Let $\hat{\sigma}_c^2 = c \sum_{i=1}^n (X_i - \bar{X})^2$. Show that the value of c that minimizes $MSE(\hat{\sigma}_c^2)$ is $c = (n+1)^{-1}$.

Hint for question (b): Recall (Theorem B.3.3) that $\sigma^{-2} \sum_{i=1}^n (X_i - \bar{X})^2$ has a χ_{n-1}^2 distribution. You may use the fact that $E(X_i - \mu)^4 = 3\sigma^4$.

9. Let θ denote the proportion of people working in a company who have a certain characteristic (e.g., being left-handed). It is known that in the state where the company is located, 10% have the characteristic. A person in charge of ordering equipment needs to estimate θ and uses

$$\hat{\theta} = (.2)(.10) + (.8)\hat{p}$$

where $\hat{p} = X/n$ is the proportion with the characteristic in a sample of size n from the company. Find $MSE(\hat{\theta})$ and $MSE(\hat{p})$. If the true θ is θ_0, for what θ_0 is

$$MSE(\hat{\theta})/MSE(\hat{p}) < 1?$$

Give the answer for $n = 25$ and $n = 100$.

10. In Problem 1.3.3(a) with $b = c = 1$ and $n = 1$, suppose θ is discrete with frequency function $\pi(0) = \pi\left(-\frac{1}{2}\right) = \pi\left(\frac{1}{2}\right) = \frac{1}{3}$. Compute the Bayes risk of $\delta_{r,s}$ when

(a) $r = -s = -1$

(b) $r = -\frac{1}{2}s = -1$.

Which one of the rules is the better one from the Bayes point of view?

11. A decision rule δ is said to be *unbiased* if

$$E_\theta(l(\theta, \delta(\mathbf{X}))) \leq E_\theta(l(\theta', \delta(\mathbf{X})))$$

for all $\theta, \theta' \in \Theta$.

(a) Show that if θ is real and $l(\theta, a) = (\theta - a)^2$, then this definition coincides with the definition of an unbiased estimate of θ.

(b) Show that if we use the $0-1$ loss function in testing, then a test function is unbiased in this sense if, and only if, the *power function*, defined by $\beta(\theta, \delta) = E_\theta(\delta(\mathbf{X}))$, satisfies

$$\beta(\theta', \delta) \geq \sup\{\beta(\theta, \delta) : \theta \in \Theta_0\},$$

for all $\theta' \in \Theta_1$.

12. In Problem 1.3.3, show that if $c \leq b$, $z > 0$, and

$$r = -s = -z \left(\frac{b}{b+c} \right) \bigg/ \sqrt{n},$$

then $\delta_{r,s}$ is unbiased

13. A (behavioral) *randomized test* of a hypothesis H is defined as any statistic $\varphi(\mathbf{X})$ such that $0 \leq \varphi(\mathbf{X}) \leq 1$. The interpretation of φ is the following. If $\mathbf{X} = \mathbf{x}$ and $\varphi(\mathbf{x}) = 0$ we decide Θ_0, if $\varphi(\mathbf{x}) = 1$, we decide Θ_1; but if $0 < \varphi(\mathbf{x}) < 1$, we perform a Bernoulli trial with probability $\varphi(\mathbf{x})$ of success and decide Θ_1 if we obtain a success and decide Θ_0 otherwise.

Define the nonrandomized test δ_u, $0 < u < 1$, by

$$
\begin{aligned}
\delta_u(\mathbf{X}) &= 1 \quad \text{if } \varphi(\mathbf{X}) \geq u \\
&= 0 \quad \text{if } \varphi(\mathbf{X}) < u.
\end{aligned}
$$

Suppose that $U \sim \mathcal{U}(0,1)$ and is independent of $\mathbf{X}$. Consider the following randomized test δ: Observe U. If $U = u$, use the test δ_u. Show that δ agrees with φ in the sense that,

$$P_\theta[\delta(\mathbf{X}) = 1] = 1 - P_\theta[\delta(\mathbf{X}) = 0] = E_\theta(\varphi(\mathbf{X})).$$

14. *Convexity of the risk set.* Suppose that the set of decision procedures is finite. Show that if δ_1 and δ_2 are two randomized procedures, then, given $0 < \alpha < 1$, there is a randomized procedure δ_3 such that $R(\theta, \delta_3) = \alpha R(\theta, \delta_1) + (1 - \alpha)R(\theta, \delta_2)$ for all θ.

15. Suppose that $P_{\theta_0}(B) = 0$ for some event B implies that $P_\theta(B) = 0$ for all $\theta \in \Theta$. Further suppose that $l(\theta_0, a_0) = 0$. Show that the procedure $\delta(X) \equiv a_0$ is admissible.

16. In Example 1.3.4, find the set of μ where $MSE(\hat{\mu}) \leq MSE(\bar{X})$. Your answer should depend on n, σ^2 and $\delta = |\mu - \mu_0|$.

17. In Example 1.3.4, consider the estimator

$$\hat{\mu}_w = w\mu_0 + (1 - w)\bar{X}.$$

If n, σ^2 and $\delta = |\mu - \mu_0|$ are known,

(a) find the value of w_0 that minimizes $MSE(\hat{\mu}_w)$,

(b) find the minimum relative risk of $\hat{\mu}_{w_0}$ to $\bar{X}$.

18. For Example 1.1.1, consider the loss function (1.3.1) and let δ_k be the decision rule "reject the shipment iff $X \geq k$."

(a) Show that the risk is given by (1.3.7).

(b) If $N = 10$, $s = r = 1$, $\theta_0 = .1$, and $k = 3$, plot $R(\theta, \delta_k)$ as a function of θ.

(c) Same as (b) except $k = 2$. Compare δ_2 and δ_3.

19. Consider a decision problem with the possible states of nature θ_1 and θ_2, and possible actions a_1 and a_2. Suppose the loss function $\ell(\theta, a)$ is

$\theta\backslash a$	a_1	a_2
θ_1	0	2
θ_2	3	1

Let X be a random variable with probability function $p(x \mid \theta)$

$\theta\backslash x$	0	1
θ_1	0.2	0.8
θ_2	0.4	0.6

(a) Compute and plot the risk points of the nonrandomized decision rules. Give the minimax rule among the nonrandomized decision rules.

(b) Give and plot the risk set S. Give the minimax rule among the randomized decision rules.

(c) Suppose θ has the prior distribution defined by $\pi(\theta_1) = 0.1$, $\pi(\theta_2) = 0.9$. What is the Bayes decision rule?

Problems for Section 1.4

1. An urn contains four red and four black balls. Four balls are drawn at random without replacement. Let Z be the number of red balls obtained in the first two draws and Y the total number of red balls drawn.

(a) Find the best predictor of Y given Z, the best linear predictor, and the best zero intercept linear predictor.

(b) Compute the MSPEs of the predictors in (a).

2. In Example 1.4.1 calculate explicitly the best zero intercept linear predictor, its MSPE, and the ratio of its MSPE to that of the best and best linear predictors.

3. In Problem B.1.7 find the best predictors of Y given X and of X given Y and calculate their MSPEs.

4. Let U_1, U_2 be independent standard normal random variables and set $Z = U_1^2 + U_2^2$, $Y = U_1$. Is Z of any value in predicting Y?

5. Give an example in which the best linear predictor of Y given Z is a constant (has no predictive value) whereas the best predictor of Y given Z predicts Y perfectly.

6. Give an example in which Z can be used to predict Y perfectly, but Y is of no value in predicting Z in the sense that $\mathrm{Var}(Z \mid Y) = \mathrm{Var}(Z)$.

7. Let Y be any random variable and let $R(c) = E(|Y-c|)$ be the *mean absolute prediction error*. Show that either $R(c) = \infty$ for all c or $R(c)$ is minimized by taking c to be any number such that $P[Y \geq c] \geq \frac{1}{2}$, $P[Y \leq c] \geq \frac{1}{2}$. A number satisfying these restrictions is called a *median* of (the distribution of) Y. The midpoint of the interval of such c is called the conventionally defined median or simply just *the median*.

Hint: If $c < c_0$,

$$E|Y - c_0| = E|Y - c| + (c - c_0)\{P[Y \geq c_0] - P[Y < c_0]\} + 2E[(c - Y)1[c < Y < c_0]].$$

8. Let Y have a $\mathcal{N}(\mu, \sigma^2)$ distribution.

   (a) Show that $E(|Y - c|) = \sigma \mathcal{Q}[|c - \mu|/\sigma]$ where $\mathcal{Q}(t) = 2[\varphi(t) + t\Phi(t)] - t$.

   (b) Show directly that μ minimizes $E(|Y - c|)$ as a function of c.

9. If Y and Z are any two random variables, exhibit a best predictor of Y given Z for mean absolute prediction error.

10. Suppose that Z has a density p, which is symmetric about c, $p(c + z) = p(c - z)$ for all z. Show that c is a median of Z.

11. Show that if (Z, Y) has a bivariate normal distribution the best predictor of Y given Z in the sense of MSPE coincides with the best predictor for mean absolute error.

12. Many observed biological variables such as height and weight can be thought of as the sum of unobservable genetic and environmental variables. Suppose that Z, Y are measurements on such a variable for a randomly selected father and son. Let Z', Z'', Y', Y'' be the corresponding genetic and environmental components $Z = Z' + Z''$, $Y = Y' + Y''$, where (Z', Y') have a $\mathcal{N}(\mu, \mu, \sigma^2, \sigma^2, \rho)$ distribution and Z'', Y'' are $\mathcal{N}(\nu, \tau^2)$ variables independent of each other and of (Z', Y').

   (a) Show that the relation between Z and Y is weaker than that between Z' and Y'; that is, $|\mathrm{Corr}(Z, Y)| < |\rho|$.

   (b) Show that the error of prediction (for the best predictor) incurred in using Z to predict Y is greater than that incurred in using Z' to predict Y'.

13. Suppose that Z has a density p, which is symmetric about c and which is *unimodal*; that is, $p(z)$ is nonincreasing for $z \geq c$.

   (a) Show that $P[|Z - t| \leq s]$ is maximized as a function of t for each $s > 0$ by $t = c$.

   (b) Suppose (Z, Y) has a bivariate normal distribution. Suppose that if we observe $Z = z$ and predict $\mu(z)$ for Y our loss is 1 unit if $|\mu(z) - Y| > s$, and 0 otherwise. Show that the predictor that minimizes our expected loss is again the best MSPE predictor.

14. Let Z_1 and Z_2 be independent and have exponential distributions with density $\lambda e^{-\lambda z}$, $z > 0$. Define $Z = Z_2$ and $Y = Z_1 + Z_1 Z_2$. Find

   (a) The best MSPE predictor $E(Y \mid Z = z)$ of Y given $Z = z$

   (b) $E(E(Y \mid Z))$

   (c) $\mathrm{Var}(E(Y \mid Z))$

   (d) $\mathrm{Var}(Y \mid Z = z)$

   (e) $E(\mathrm{Var}(Y \mid Z))$

(f) The best linear MSPE predictor of Y based on $Z = z$.
Hint: Recall that $E(Z_1) = E(Z_2) = 1/\lambda$ and $\text{Var}(Z_1) = \text{Var}(Z_2) = 1/\lambda^2$.

15. Let $\mu(\mathbf{z}) = E(Y \mid \mathbf{Z} = \mathbf{z})$. Show that

$$\text{Var}(\mu(\mathbf{Z}))/\text{Var}(Y) = \text{Corr}^2(Y, \mu(\mathbf{Z})) = \max_g \text{Corr}^2(Y, g(\mathbf{Z}))$$

where $g(\mathbf{Z})$ stands for any predictor.

16. Show that $\rho_{\mathbf{Z}Y}^2 = \text{Corr}^2(Y, \mu_L(\mathbf{Z})) = \max_{g \in \mathcal{L}} \text{Corr}^2(Y, g(\mathbf{Z}))$ where $\mathcal{L}$ is the set of linear predictors.

17. One minus the ratio of the smallest possible MSPE to the MSPE of the constant predictor is called *Pearson's correlation ratio* $\eta_{\mathbf{Z}Y}^2$; that is,

$$\eta_{\mathbf{Z}Y}^2 = 1 - E[Y - \mu(\mathbf{Z})]^2/\text{Var}(Y) = \text{Var}(\mu(\mathbf{Z}))/\text{Var}(Y).$$

(See Pearson, 1905, and Doksum and Samarov, 1995, on estimation of $\eta_{\mathbf{Z}Y}^2$.)

(a) Show that $\eta_{\mathbf{Z}Y}^2 \geq \rho_{\mathbf{Z}Y}^2$, where $\rho_{\mathbf{Z}Y}^2$ is the population multiple correlation coefficient of Remark 1.4.3.
Hint: See Problem 1.4.15.

(b) Show that if Z is one-dimensional and h is a 1-1 increasing transformation of Z, then $\eta_{h(Z)Y}^2 = \eta_{\mathbf{Z}Y}^2$. That is, η^2 is invariant under such h.

(c) Let $\epsilon_L = Y - \mu_L(\mathbf{Z})$ be the linear prediction error. Show that, in the linear model of Remark 1.4.4, ϵ_L is uncorrelated with $\mu_L(\mathbf{Z})$ and $\eta_{\mathbf{Z}Y}^2 = \rho_{\mathbf{Z}Y}^2$.

18. *Predicting the past from the present.* Consider a subject who walks into a clinic today, at time t, and is diagnosed with a certain disease. At the same time t a diagnostic indicator Z_0 of the severity of the disease (e.g., a blood cell or viral load measurement) is obtained. Let S be the unknown date in the past when the subject was infected. We are interested in the time $Y_0 = t - S$ from infection until detection. Assume that the conditional density of Z_0 (the present) given $Y_0 = y_0$ (the past) is

$$\mathcal{N}(\mu + \beta y_0, \sigma^2),$$

where μ and σ^2 are the mean and variance of the severity indicator Z_0 in the population of people without the disease. Here βy_0 gives the mean increase of Z_0 for infected subjects over the time period y_0; $\beta > 0$, $y_0 > 0$. It will be convenient to rescale the problem by introducing $Z = (Z_0 - \mu)/\sigma$ and $Y = \beta Y_0/\sigma$.

(a) Show that the conditional density $f(z \mid y)$ of Z given $Y = y$ is $\mathcal{N}(y, 1)$.

(b) Suppose that Y has the exponential density

$$\pi(y) = \lambda \exp\{-\lambda y\}, \ \lambda > 0, \ y > 0.$$

Show that the conditional distribution of Y (the past) given $Z = z$ (the present) has density

$$\pi(y \mid z) = (2\pi)^{-\frac{1}{2}} c^{-1} \exp\left\{-\frac{1}{2}[y - (z - \lambda)]^2\right\}, \ y > 0$$

where $c = \Phi(z - \lambda)$. This density is called the *truncated* (at zero) *normal*, $\mathcal{N}(z - \lambda, 1)$, *density.*

  Hint: Use Bayes' Theorem.

  (c) Find the conditional density $\pi_0(y_0 \mid z_0)$ of Y_0 given $Z_0 = z_0$.

  (d) Find the best predictor of Y_0 given $Z_0 = z_0$ using mean absolute prediction error $E|Y_0 - g(Z_0)|$.

  Hint: See Problems 1.4.7 and 1.4.9.

  (e) Show that the best MSPE predictor of Y given $Z = z$ is

$$E(Y \mid Z = z) = c^{-1}\varphi(\lambda - z) - (\lambda - z).$$

(In practice, all the unknowns, including the "prior" π, need to be estimated from cohort studies; see Berman, 1990, and Normand and Doksum, 2001).

19. Establish 1.4.14 by setting the derivatives of $R(a, \mathbf{b})$ equal to zero, solving for $(a, \mathbf{b})$, and checking convexity.

20. Let Y be the number of heads showing when X fair coins are tossed, where X is the number of spots showing when a fair die is rolled. Find

  (a) The mean and variance of Y.

  (b) The MSPE of the optimal predictor of Y based on X.

  (c) The optimal predictor of Y given $X = x$, $x = 1, \ldots, 6$.

21. Let $\mathbf{Y}$ be a vector and let $r(\mathbf{Y})$ and $s(\mathbf{Y})$ be real valued. Write $\mathrm{Cov}[r(\mathbf{Y}), s(\mathbf{Y}) \mid \mathbf{z}]$ for the covariance between $r(\mathbf{Y})$ and $s(\mathbf{Y})$ in the conditional distribution of $(r(\mathbf{Y}), s(\mathbf{Y}))$ given $\mathbf{Z} = \mathbf{z}$.

  (a) Show that if $\mathrm{Cov}[r(\mathbf{Y}), s(\mathbf{Y})] < \infty$, then

$$\mathrm{Cov}[r(\mathbf{Y}), s(\mathbf{Y})] = E\{\mathrm{Cov}[r(\mathbf{Y}), s(\mathbf{Y}) \mid \mathbf{Z}]\} + \mathrm{Cov}\{E[r(\mathbf{Y}) \mid \mathbf{Z}], E[s(\mathbf{Y}) \mid \mathbf{Z}]\}.$$

  (b) Show that (a) is equivalent to (1.4.6) when $r = s$.

  (c) Show that if Z is real, $\mathrm{Cov}[r(\mathbf{Y}), Z] = \mathrm{Cov}\{E[r(\mathbf{Y}) \mid Z], Z\}$.

  (d) Suppose $Y_1 = a_1 + b_1 Z_1 + W$ and $Y_2 = a_2 + b_2 Z_2 + W$, where Y_1 and Y_2 are responses of subjects 1 and 2 with common influence W and separate influences Z_1 and Z_2, where Z_1, Z_2 and W are independent with finite variances. Find $\mathrm{Corr}(Y_1, Y_2)$ using (a).

(e) In the preceding model (d), if $b_1 = b_2$ and Z_1, Z_2 and W have the same variance σ^2, we say that there is a 50% overlap between Y_1 and Y_2. In this case what is $\text{Corr}(Y_1, Y_2)$?

(f) In model (d), suppose that Z_1 and Z_2 are $\mathcal{N}(\mu, \sigma^2)$ and $W \sim \mathcal{N}(\mu_0, \sigma_0^2)$. Find the optimal predictor of Y_2 given (Y_1, Z_1, Z_2).

22. In Example 1.4.3, show that the MSPE of the optimal predictor is $\sigma_Y^2(1 - \rho_{ZY}^2)$.

23. Verify that solving (1.4.15) yields (1.4.14).

24. (a) Let $w(y, \mathbf{z})$ be a positive real-valued function. Then $[y - g(\mathbf{z})]^2/w(y, \mathbf{z}) = \delta_w(y, g(\mathbf{z}))$ is called *weighted squared prediction error*. Show that the mean weighted squared prediction error is minimized by $\mu_0(\mathbf{Z}) = E_0(Y \mid \mathbf{Z})$, where

$$p_0(y, \mathbf{z}) = cp(y, \mathbf{z})/w(y, \mathbf{z})$$

and c is the constant that makes p_0 a density. Assume that

$$E\delta_w(Y, g(\mathbf{Z})) < \infty$$

for some g and that p_0 is a density.

(b) Suppose that given $Y = y$, $Z \sim \mathcal{B}(n, y)$, $n \geq 2$, and suppose that Y has the beta, $\beta(r, s)$, density. Find $\mu_0(Z)$ when (i) $w(y, z) = 1$, and (ii) $w(y, z) = y(1 - y), 0 < y < 1$.
Hint: See Example 1.2.9.

25. Show that $EY^2 < \infty$ if and only if $E(Y - c)^2 < \infty$ for all c.
Hint: Whatever be Y and c,

$$\frac{1}{2}Y^2 - c^2 \leq (Y - c)^2 = Y^2 - 2cY + c^2 \leq 2(Y^2 + c^2).$$

Problems for Section 1.5

1. Let $X_1, \ldots, X_n$ be a sample from a Poisson, $\mathcal{P}(\theta)$, population where $\theta > 0$.

(a) Show directly that $\sum_{i=1}^{n} X_i$ is sufficient for θ.

(b) Establish the same result using the factorization theorem.

2. Let n items be drawn in order without replacement from a shipment of N items of which $N\theta$ are bad. Let $X_i = 1$ if the ith item drawn is bad, and $= 0$ otherwise. Show that $\sum_{i=1}^{n} X_i$ is sufficient for θ directly and by the factorization theorem.

3. Suppose $X_1, \ldots, X_n$ is a sample from a population with one of the following densities.

(a) $p(x, \theta) = \theta x^{\theta-1}, 0 < x < 1, \theta > 0$. This is the beta, $\beta(\theta, 1)$, density.

(b) $p(x, \theta) = \theta a x^{a-1} \exp(-\theta x^a), x > 0, \theta > 0, a > 0$.

This is known as the *Weibull* density.

(c) $p(x, \theta) = \theta a^\theta/x^{(\theta+1)}, x > a, \theta > 0, a > 0$.

This is known as the *Pareto* density.

In each case, find a real-valued sufficient statistic for θ, a fixed.

4. (a) Show that T_1 and T_2 are equivalent statistics if, and only if, we can write $T_2 = H(T_1)$ for some 1-1 transformation H of the range of T_1 into the range of T_2. Which of the following statistics are equivalent? (Prove or disprove.)

(b) $\prod_{i=1}^{n} x_i$ and $\sum_{i=1}^{n} \log x_i$, $x_i > 0$

(c) $\sum_{i=1}^{n} x_i$ and $\sum_{i=1}^{n} \log x_i$, $x_i > 0$

(d) $\left(\sum_{i=1}^{n} x_i, \sum_{i=1}^{n} x_i^2\right)$ and $\left(\sum_{i=1}^{n} x_i, \sum_{i=1}^{n} (x_i - \bar{x})^2\right)$

(e) $\left(\sum_{i=1}^{n} x_i, \sum_{i=1}^{n} x_i^3\right)$ and $\left(\sum_{i=1}^{n} x_i, \sum_{i=1}^{n} (x_i - \bar{x})^3\right)$.

5. Let $\theta = (\theta_1, \theta_2)$ be a bivariate parameter. Suppose that $T_1(\mathbf{X})$ is sufficient for θ_1 whenever θ_2 is fixed and known, whereas $T_2(\mathbf{X})$ is sufficient for θ_2 whenever θ_1 is fixed and known. Assume that θ_1, θ_2 vary independently, $\theta_1 \in \Theta_1$, $\theta_2 \in \Theta_2$ and that the set $S = \{\mathbf{x} : p(\mathbf{x}, \theta) > 0\}$ does not depend on θ.

(a) Show that if T_1 and T_2 do not depend on θ_2 and θ_1 respectively, then $(T_1(\mathbf{X}), T_2(\mathbf{X}))$ is sufficient for θ.

(b) Exhibit an example in which $(T_1(\mathbf{X}), T_2(\mathbf{X}))$ is sufficient for θ, $T_1(\mathbf{X})$ is sufficient for θ_1 whenever θ_2 is fixed and known, but $T_2(\mathbf{X})$ is not sufficient for θ_2, when θ_1 is fixed and known.

6. Let X take on the specified values $v_1, \ldots, v_k$ with probabilities $\theta_1, \ldots, \theta_k$, respectively. Suppose that $X_1, \ldots, X_n$ are independently and identically distributed as X. Suppose that $\theta = (\theta_1, \ldots, \theta_k)$ is unknown and may range over the set $\Theta = \{(\theta_1, \ldots, \theta_k) : \theta_i \geq 0, 1 \leq i \leq k, \sum_{i=1}^{k} \theta_i = 1\}$. Let N_j be the number of X_i which equal v_j.

(a) What is the distribution of $(N_1, \ldots, N_k)$?

(b) Show that $\mathbf{N} = (N_1, \ldots, N_{k-1})$ is sufficient for θ.

7. Let $X_1, \ldots, X_n$ be a sample from a population with density $p(x, \theta)$ given by

$$p(x, \theta) \quad = \quad \frac{1}{\sigma} \exp\left\{-\left(\frac{x - \mu}{\sigma}\right)\right\} \text{ if } x \geq \mu$$

$$= \quad 0 \text{ otherwise.}$$

Here $\theta = (\mu, \sigma)$ with $-\infty < \mu < \infty$, $\sigma > 0$.

(a) Show that $\min(X_1, \ldots, X_n)$ is sufficient for μ when σ is fixed.

(b) Find a one-dimensional sufficient statistic for σ when μ is fixed.

(c) Exhibit a two-dimensional sufficient statistic for θ.

8. Let $X_1, \ldots, X_n$ be a sample from some continuous distribution F with density f, which is unknown. Treating f as a parameter, show that the order statistics $X_{(1)}, \ldots, X_{(n)}$ (cf. Problem B.2.8) are sufficient for f.

9. Let $X_1, \ldots, X_n$ be a sample from a population with density

$$f_\theta(x) = a(\theta)h(x) \text{ if } \theta_1 \leq x \leq \theta_2$$
$$= 0 \text{ otherwise}$$

where $h(x) \geq 0$, $\theta = (\theta_1, \theta_2)$ with $-\infty < \theta_1 \leq \theta_2 < \infty$, and $a(\theta) = \left[\int_{\theta_1}^{\theta_2} h(x)dx\right]^{-1}$ is assumed to exist. Find a two-dimensional sufficient statistic for this problem and apply your result to the $\mathcal{U}[\theta_1, \theta_2]$ family of distributions.

10. Suppose $X_1, \ldots, X_n$ are i.i.d. with density $f(x, \theta) = \frac{1}{2}e^{-|x-\theta|}$. Show that $(X_{(1)}, \ldots, X_{(n)})$, the order statistics, are minimal sufficient.

  Hint: $\frac{\partial}{\partial\theta} \log L_\mathbf{X}(\theta) = -\sum_{i=1}^{n} \text{sgn}(X_i - \theta)$, $\theta \notin \{X_1, \ldots, X_n\}$, which determines $X_{(1)}, \ldots, X_{(n)}$.

11. Let $X_1, X_2, \ldots, X_n$ be a sample from the uniform, $\mathcal{U}(0, \theta)$, distribution. Show that $X_{(n)} = \max\{X_i; 1 \leq i \leq n\}$ is minimal sufficient for θ.

12. *Dynkin, Lehmann, Scheffé's Theorem.* Let $\mathcal{P} = \{P_\theta : \theta \in \Theta\}$ where P_θ is discrete concentrated on $\mathcal{X} = \{x_1, x_2, \ldots\}$. Let $p(x, \theta) \equiv P_\theta[X = x] \equiv L_x(\theta) > 0$ on $\mathcal{X}$. Show that $\frac{L_X(\cdot)}{L_X(\theta_0)}$ is minimial sufficient.

  Hint: Apply the factorization theorem.

13. Suppose that $\mathbf{X} = (X_1, \ldots, X_n)$ is a sample from a population with continuous distribution function $F(x)$. If $F(x)$ is $N(\mu, \sigma^2)$, $T(\mathbf{X}) = (\bar{X}, \hat{\sigma}^2)$, where $\hat{\sigma}^2 = n^{-1}\sum(X_i - \bar{X})^2$, is sufficient, and $S(\mathbf{X}) = (X'_{(1)}, \ldots, X'_{(n)})$, where $X'_{(i)} = (X_{(i)} - \bar{X})/\hat{\sigma}$, is "irrelevant" (ancillary) for (μ, σ^2). However, $S(\mathbf{X})$ is exactly what is needed to estimate the "shape" of $F(x)$ when $F(x)$ is unknown. The shape of F is represented by the equivalence class $\mathcal{F} = \{F((\cdot - a)/b) : b > 0, a \in R\}$. Thus a distribution G has the same shape as F iff $G \in \mathcal{F}$. For instance, one "estimator" of this shape is the scaled empirical distribution function

$$\hat{F}_s(x) = j/n, \ x'_{(j)} \leq x < x'_{(j+1)}, \ j = 1, \ldots, n-1$$
$$= 0, \ x < x'_{(1)}$$
$$= 1, \ x \geq x'_{(n)}.$$

Show that for fixed x, $\hat{F}_s((x - \bar{x})/\hat{\sigma})$ converges in probability to $F(x)$. Here we are using F to represent $\mathcal{F}$ because every member of $\mathcal{F}$ can be obtained from F.

14. *Kolmogorov's Theorem.* We are given a regular model with Θ finite.

  (a) Suppose that a statistic $T(\mathbf{X})$ has the property that for any prior distribution on θ, the posterior distribution of θ depends on $\mathbf{x}$ only through $T(\mathbf{x})$. Show that $T(\mathbf{X})$ is sufficient.

  (b) Conversely show that if $T(\mathbf{X})$ is sufficient, then, for any prior distribution, the posterior distribution depends on $\mathbf{x}$ only through $T(\mathbf{x})$.

Hint: Apply the factorization theorem.

15. Let $X_1, \ldots, X_n$ be a sample from $f(x - \theta)$, $\theta \in R$. Show that the order statistics are minimal sufficient when f is the density *Cauchy* $f(t) = 1/\pi(1 + t^2)$.

16. Let $X_1, \ldots, X_m$; $Y_1, \ldots, Y_n$ be independently distributed according to $\mathcal{N}(\mu, \sigma^2)$ and $\mathcal{N}(\eta, \tau^2)$, respectively. Find minimal sufficient statistics for the following three cases:

    (i) μ, η, σ, τ are arbitrary: $-\infty < \mu, \eta < \infty, 0 < \sigma, \tau$.

    (ii) $\sigma = \tau$ and μ, η, σ are arbitrary.

    (iii) $\mu = \eta$ and μ, σ, τ are arbitrary.

17. In Example 1.5.4, express t_1 as a function of $L_x(0, 1)$ and $L_x(1, 1)$.

Problems to Section 1.6

1. Prove the assertions of Table 1.6.1.

2. Suppose $X_1, \ldots, X_n$ is as in Problem 1.5.3. In each of the cases (a), (b) and (c), show that the distribution of $\mathbf{X}$ forms a one-parameter exponential family. Identify η, B, T, and h.

3. Let X be the number of failures before the first success in a sequence of Bernoulli trials with probability of success θ. Then $P_\theta[X = k] = (1 - \theta)^k \theta$, $k = 0, 1, 2, \ldots$ This is called the *geometric distribution* $(\mathcal{G}(\theta))$.

    (a) Show that the family of geometric distributions is a one-parameter exponential family with $T(x) = x$.

    (b) Deduce from Theorem 1.6.1 that if $X_1, \ldots, X_n$ is a sample from $\mathcal{G}(\theta)$, then the distributions of $\sum_{i=1}^{n} X_i$ form a one-parameter exponential family.

    (c) Show that $\sum_{i=1}^{n} X_i$ in part (b) has a *negative binomial* distribution with parameters (n, θ) defined by $P_\theta[\sum_{i=1}^{n} X_i = k] = \binom{n + k - 1}{k}(1 - \theta)^k \theta^n$, $k = 0, 1, 2, \ldots$ (The negative binomial distribution is that of the number of failures before the nth success in a sequence of Bernoulli trials with probability of success θ.)
    Hint: By Theorem 1.6.1, $P_\theta[\sum_{i=1}^{n} X_i = k] = c_k(1 - \theta)^k \theta^n$, $0 < \theta < 1$. If

$$\sum_{k=0}^{\infty} c_k \omega^k = \frac{1}{(1 - \omega)^n}, \quad 0 < \omega < 1, \text{ then } c_k = \frac{1}{k!} \frac{d^k}{d\omega^k}(1 - \omega)^{-n} \Big|_{\omega=0}.$$

4. Which of the following families of distributions are exponential families? (Prove or disprove.)

    (a) The $\mathcal{U}(0, \theta)$ family

(b) $p(x, \theta) = \{\exp[-2 \log \theta + \log(2x)]\} 1[x \in (0, \theta)]$

(c) $p(x, \theta) = \frac{1}{9}, x \in \{0.1 + \theta, \dots, 0.9 + \theta\}$

(d) The $\mathcal{N}(\theta, \theta^2)$ family, $\theta > 0$

(e) $p(x, \theta) = 2(x + \theta)/(1 + 2\theta), 0 < x < 1, \theta > 0$

(f) $p(x, \theta)$ is the conditional frequency function of a binomial, $\mathcal{B}(n, \theta)$, variable X, given that $X > 0$.

5. Show that the following families of distributions are two-parameter exponential families and identify the functions η, B, T, and h.

(a) The beta family.

(b) The gamma family.

6. Let X have the Dirichlet distribution, $\mathcal{D}(\alpha)$, of Problem 1.2.15.

Show the distribution of X form an r-parameter exponential family and identify η, B, T, and h.

7. Let $\mathbf{X} = ((X_1, Y_1), \dots, (X_n, Y_n))$ be a sample from a bivariate normal population. Show that the distributions of $\mathbf{X}$ form a five-parameter exponential family and identify η, B, T, and h.

8. Show that the family of distributions of Example 1.5.3 is not a one parameter exponential family.

Hint: If it were, there would be a set A such that $p(x, \theta) > 0$ on A for all θ.

9. Prove the analogue of Theorem 1.6.1 for discrete k-parameter exponential families.

10. Suppose that $f(x, \theta)$ is a positive density on the real line, which is continuous in x for each θ and such that if (X_1, X_2) is a sample of size 2 from $f(\cdot, \theta)$, then $X_1 + X_2$ is sufficient for θ. Show that $f(\cdot, \theta)$ corresponds to a one-parameter exponential family of distributions with $T(x) = x$.

Hint: There exist functions $g(t, \theta)$, $h(x_1, x_2)$ such that $\log f(x_1, \theta) + \log f(x_2, \theta) = g(x_1 + x_2, \theta) + h(x_1, x_2)$. Fix θ_0 and let $r(x, \theta) = \log f(x, \theta) - \log f(x, \theta_0)$, $q(x, \theta) = g(x, \theta) - g(x, \theta_0)$. Then, $q(x_1 + x_2, \theta) = r(x_1, \theta) + r(x_2, \theta)$, and hence, $[r(x_1, \theta) - r(0, \theta)] + [r(x_2, \theta) - r(0, \theta)] = r(x_1 + x_2, \theta) - r(0, \theta)$.

11. Use Theorems 1.6.2 and 1.6.3 to obtain moment-generating functions for the sufficient statistics when sampling from the following distributions.

(a) normal, $\boldsymbol{\theta} = (\mu, \sigma^2)$

(b) gamma, $\Gamma(p, \lambda)$, $\theta = \lambda$, p fixed

(c) binomial

(d) Poisson

(e) negative binomial (see Problem 1.6.3)

(f) gamma, $\Gamma(p, \lambda)$, $\boldsymbol{\theta} = (p, \lambda)$.

12. Show directly using the definition of the rank of an exponential family that the multinomial distribution, $\mathcal{M}(n; \theta_1, \dots, \theta_k)$, $0 < \theta_j < 1$, $1 \le j \le k$, $\sum_{j=1}^{k} \theta_j = 1$, is of rank $k - 1$.

13. Show that in Theorem 1.6.3, the condition that $\mathcal{E}$ has nonempty interior is equivalent to the condition that $\mathcal{E}$ is not contained in any $(k - 1)$-dimensional hyperplane.

14. Construct an exponential family of rank k for which $\mathcal{E}$ is not open and $\dot{A}$ is not defined on all of $\mathcal{E}$. Show that if $k = 1$ and $\mathcal{E}^0 \ne \emptyset$ and $\dot{A}, \ddot{A}$ are defined on all of $\mathcal{E}$, then Theorem 1.6.3 continues to hold.

15. Let $\mathcal{P} = \{P_\theta : \theta \in \Theta\}$ where P_θ is discrete and concentrated on $\mathcal{X} = \{x_1, x_2, \dots\}$, and let $p(x, \theta) = P_\theta[X = x]$. Show that if $\mathcal{P}$ is a (discrete) canonical exponential family generated by $(\mathbf{T}, h)$ and $\mathcal{E}^0 \ne \emptyset$, then $\mathbf{T}$ is minimal sufficient.
  Hint: $\frac{\partial \log}{\partial \eta_j} L_X(\boldsymbol{\eta}) = T_j(X) - E_{\boldsymbol{\eta}} T_j(X)$. Use Problem 1.5.12.

16. *Life testing.* Let $X_1, \dots, X_n$ be independently distributed with exponential density $(2\theta)^{-1} e^{-x/2\theta}$ for $x \ge 0$, and let the ordered X's be denoted by $Y_1 \le Y_2 \le \cdots \le Y_n$. It is assumed that Y_1 becomes available first, then Y_2, and so on, and that observation is continued until Y_r has been observed. This might arise, for example, in life testing where each X measures the length of life of, say, an electron tube, and n tubes are being tested simultaneously. Another application is to the disintegration of radioactive material, where n is the number of atoms, and observation is continued until r α-particles have been emitted. Show that

(i) The joint distribution of $Y_1, \dots, Y_r$ is an exponential family with density

$$\frac{1}{(2\theta)^r}\frac{n!}{(n-r)!} \exp\left[-\frac{\sum_{i=1}^{r} y_i + (n-r)y_r}{2\theta}\right], \quad 0 \le y_1 \le \cdots \le y_r.$$

(ii) The distribution of $[\sum_{i=1}^{r} Y_i + (n-r)Y_r]/\theta$ is χ^2 with $2r$ degrees of freedom.

(iii) Let $Y_1, Y_2, \dots$ denote the time required until the first, second,... event occurs in a Poisson process with parameter $1/2\theta'$ (see A.16). Then $Z_1 = Y_1/\theta'$, $Z_2 = (Y_2 - Y_1)/\theta'$, $Z_3 = (Y_3 - Y_2)/\theta', \dots$ are independently distributed as χ^2 with 2 degrees of freedom, and the joint density of $Y_1, \dots, Y_r$ is an exponential family with density

$$\frac{1}{(2\theta')^r} \exp\left(-\frac{y_r}{2\theta'}\right), \quad 0 \le y_1 \le \cdots \le y_r.$$

The distribution of Y_r/θ' is again χ^2 with $2r$ degrees of freedom.

(iv) The same model arises in the application to life testing if the number n of tubes is held constant by replacing each burned-out tube with a new one, and if Y_1 denotes the time at which the first tube burns out, Y_2 the time at which the second tube burns out, and so on, measured from some fixed time. The lifetimes are assumed to be exponentially distributed.

Hint (ii): The random variables $Z_i = (n - i + 1)(Y_i - Y_{i-1})/\theta$ $(i = 1, \ldots, r)$ are independently distributed as χ_2^2, $Y_0 = 0$, and $[\sum_{i=1}^r Y_i + (n-r)Y_r]/\theta = \sum_{i=1}^r Z_i$.

17. Suppose that $(\mathbf{T}_{k \times 1}, h)$ generate a canonical exponential family $\mathcal{P}$ with parameter $\eta_{k \times 1}$ and $\mathcal{E} = R^k$. Let

$$\mathcal{Q} = \{\mathcal{Q}_\theta : \mathcal{Q}_\theta = P_\eta \text{ with } \eta = B_{k \times l}\theta_{l \times 1} + \mathbf{c}_{k \times 1}\}, \quad l \le k.$$

(a) Show that $\mathcal{Q}$ is the exponential family generated by $\Pi_L \mathbf{T}$ and $h \exp\{\mathbf{c}^T \mathbf{T}\}$, where Π_L is the projection matrix onto $\mathcal{L} = \{\eta : \eta = B\theta, \theta \in R^l\}$.

(b) Show that if $\mathcal{P}$ has full rank k and B is of rank l, then $\mathcal{Q}$ has full rank l.
Hint: If B is of rank l, you may assume

$$\Pi_L = B[B^T B]^{-1} B^T.$$

18. Suppose $Y_1, \ldots, Y_n$ are independent with $Y_i \sim \mathcal{N}(\beta_1 + \beta_2 z_i, \sigma^2)$, where $z_1, \ldots, z_n$ are covariate values not all equal. (See Example 1.6.6.) Show that the family has rank 3. Give the mean vector and the variance matrix of $\mathbf{T}$.

19. *Logistic Regression.* We observe $(\mathbf{z}_1, Y_1), \ldots, (\mathbf{z}_n, Y_n)$ where the $Y_1, \ldots, Y_n$ are independent, $Y_i \sim \mathcal{B}(n_i, \lambda_i)$. The success probability λ_i depends on the characteristics $\mathbf{z}_i$ of the ith subject, for example, on the covariate vector $\mathbf{z}_i = $ (age, height, blood pressure)^T. The function $l(u) = \log[u/(1-u)]$ is called the *logit* function. In the logistic linear regression model it is assumed that $l(\lambda_i) = \mathbf{z}_i^T \boldsymbol{\beta}$ where $\boldsymbol{\beta} = (\beta_1, \ldots, \beta_d)^T$ and $\mathbf{z}_i$ is $d \times 1$. Show that $\mathbf{Y} = (Y_1, \ldots, Y_n)^T$ follow an exponential model with rank d iff $\mathbf{z}_1, \ldots, \mathbf{z}_d$ are not collinear (linearly independent) (cf. Examples 1.1.4, 1.6.8 and Problem 1.1.9).

20. (a) In part II of the proof of Theorem 1.6.4, fill in the details of the arguments that $\mathcal{Q}$ is generated by $(\eta_1 - \eta_0)^T T$ and that $\sim$(ii) $\equiv \sim$(i).

(b) Fill in the details of part III of the proof of Theorem 1.6.4.

21. Find $\mu(\eta) = E_\eta \mathbf{T}(X)$ for the gamma, $\Gamma(\alpha, \lambda)$, distribution, where $\theta = (\alpha, \lambda)$.

22. Let $X_1, \ldots, X_n$ be a sample from the k-parameter exponential family distribution (1.6.10). Let $\mathbf{T} = (\sum_{i=1}^n T_1(X_i), \ldots, \sum_{i=1}^n T_k(X_i))$ and let

$$\mathcal{S} = \{(\eta_1(\boldsymbol{\theta}), \ldots, \eta_k(\boldsymbol{\theta})) : \theta \in \Theta\}.$$

Show that if $\mathcal{S}$ contains a subset of $k + 1$ vectors $\mathbf{v}_0, \ldots, \mathbf{v}_{k+1}$ so that $\mathbf{v}_i - \mathbf{v}_0, 1 \le i \le k$, are not collinear (linearly independent), then $\mathbf{T}$ is minimally sufficient for θ.

23. Using (1.6.20), find a conjugate family of distributions for the gamma and beta families.

(a) With one parameter fixed.

(b) With both parameters free.

24. Using (1.6.20), find a conjugate family of distributions for the normal family using as parameter $\theta = (\theta_1, \theta_2)$ where $\theta_1 = E_\theta(X)$, $\theta_2 = 1/(\mathrm{Var}_\theta X)$ (cf. Problem 1.2.12).

25. Consider the linear Gaussian regression model of Examples 1.5.5 and 1.6.6 except with σ^2 known. Find a conjugate family of prior distributions for $(\beta_1, \beta_2)^T$.

26. Using (1.6.20), find a conjugate family of distributions for the multinomial distribution. See Problem 1.2.15.

27. Let $\mathcal{P}$ denote the canonical exponential family generated by $\mathbf{T}$ and h. For any $\eta_0 \in \mathcal{E}$, set $h_0(x) = q(x, \eta_0)$ where q is given by (1.6.9). Show that $\mathcal{P}$ is also the canonical exponential family generated by $\mathbf{T}$ and h_0.

28. *Exponential families are maximum entropy distributions.* The *entropy* $h(f)$ of a random variable X with density f is defined by

$$h(f) = E(-\log f(X)) = -\int_{-\infty}^{\infty} [\log f(x)] f(x) dx.$$

This quantity arises naturally in information theory; see Section 2.2.2 and Cover and Thomas (1991). Let $S = \{x : f(x) > 0\}$.

(a) Show that the canonical k-parameter exponential family density

$$f(x, \boldsymbol{\eta}) = \exp \left\{ \eta_0 + \sum_{j=1}^{k} \eta_j r_j(x) - A(\boldsymbol{\eta}) \right\}, \quad x \in S$$

maximizes $h(f)$ subject to the constraints

$$f(x) \geq 0, \quad \int_S f(x) dx = 1, \quad \int_S f(x) r_j(x) dx = \alpha_j, \ 1 \leq j \leq k,$$

where $\eta_0, \ldots, \eta_k$ are chosen so that f satisfies the constraints.
   Hint: You may use Lagrange multipliers. Maximize the integrand.

(b) Find the maximum entropy densities when $r_j(x) = x^j$ and (i) $S = (0, \infty)$, $k = 1$, $\alpha_1 > 0$; (ii) $S = R$, $k = 2$, $\alpha_1 \in R$, $\alpha_2 > 0$; (iii) $S = R$, $k = 3$, $\alpha_1 \in R$, $\alpha_2 > 0$, $\alpha_3 \in R$.

29. As in Example 1.6.11, suppose that $\mathbf{Y}_1, \ldots, \mathbf{Y}_n$ are i.i.d. $\mathcal{N}_p(\boldsymbol{\mu}, \Sigma)$ where $\boldsymbol{\mu}$ varies freely in R^p and Σ ranges freely over the class of all $p \times p$ symmetric positive definite matrices. Show that the distribution of $\mathbf{Y} = (Y_1, \ldots, Y_n)$ is the $p(p+3)/2$ canonical exponential family generated by $h = 1$ and the $p(p+3)/2$ statistics

$$T_j = \sum_{i=1}^{n} Y_{ij}, \ 1 \leq j \leq p; \ T_{jl} = \sum_{i=1}^{n} Y_{ij} Y_{il}, \ 1 \leq j \leq l \leq p$$

where $\mathbf{Y}_i = (Y_{i1}, \ldots, Y_{ip})$. Show that $\mathcal{E}$ is open and that this family is of rank $p(p+3)/2$.
   Hint: Without loss of generality, take $n = 1$. We want to show that $h = 1$ and the $m = p(p+3)/2$ statistics $T_j(\mathbf{Y}) = Y_j$, $1 \leq j \leq p$, and $T_{jl}(\mathbf{Y}) = Y_j Y_l$, $1 \leq j \leq l \leq p$,

generate $\mathcal{N}_p(\boldsymbol{\mu}, \Sigma)$. As Σ ranges over all $p \times p$ symmetric positive definite matrices, so does Σ^{-1}. Next establish that for symmetric matrices M,

$$\int \exp\{-\mathbf{u}^T M \mathbf{u}\} d\mathbf{u} < \infty \text{ iff } M \text{ is positive definite}$$

by using the spectral decomposition (see B.10.1.2)

$$M = \sum_{j=1}^{p} \lambda_j \mathbf{e}_j \mathbf{e}_j^T \text{ for } \mathbf{e}_1, \ldots, \mathbf{e}_p \text{ orthogonal}, \lambda_j \in R.$$

To show that the family has full rank m, use induction on p to show that if $Z_1, \ldots, Z_p$ are i.i.d. $\mathcal{N}(0, 1)$ and if $B_{p \times p} = (b_{jl})$ is symmetric, then

$$P\left(\sum_{j=1}^{p} a_j Z_j + \sum_{j,l} b_{jl} Z_j Z_l = c\right) = P(\mathbf{a}^T \mathbf{Z} + \mathbf{Z}^T B \mathbf{Z} = c) = 0$$

unless $\mathbf{a} = \mathbf{0}$, $B = 0$, $c = 0$. Next recall (Appendix B.6) that since $\mathbf{Y} \sim \mathcal{N}_p(\boldsymbol{\mu}, \Sigma)$, then $\mathbf{Y} = S\mathbf{Z}$ for some nonsingular $p \times p$ matrix S.

30. Show that if $\mathbf{X}_1, \ldots, \mathbf{X}_n$ are i.i.d. $\mathcal{N}_p(\boldsymbol{\theta}, \Sigma_0)$ given $\boldsymbol{\theta}$ where Σ_0 is known, then the $\mathcal{N}_p(\boldsymbol{\lambda}, \Gamma)$ family is conjugate to $\mathcal{N}_p(\boldsymbol{\theta}, \Sigma_0)$, where $\boldsymbol{\lambda}$ varies freely in R^p and Γ ranges over all $p \times p$ symmetric positive definite matrices.

31. *Conjugate Normal Mixture Distributions. A Hierarchical Bayesian Normal Model.* Let $\{(\mu_j, \tau_j) : 1 \leq j \leq k\}$ be a given collection of pairs with $\mu_j \in R$, $\tau_j > 0$. Let (μ, σ) be a random pair with $\lambda_j = P((\mu, \sigma) = (\mu_j, \tau_j))$, $0 < \lambda_j < 1$, $\sum_{j=1}^{k} \lambda_j = 1$. Let θ be a random variable whose conditional distribution given $(\mu, \sigma) = (\mu_j, \tau_j)$ is normal, $\mathcal{N}(\mu_j, \tau_j^2)$. Consider the model $X = \theta + \epsilon$, where θ and ϵ are independent and $\epsilon \sim \mathcal{N}(0, \sigma_0^2)$, σ_0^2 known. Note that θ has the prior density

$$\pi(\theta) = \sum_{j=1}^{k} \lambda_j \varphi_{\tau_j}(\theta - \mu_j) \tag{1.7.4}$$

where φ_τ denotes the $\mathcal{N}(0, \tau^2)$ density. Also note that $(X \mid \theta)$ has the $\mathcal{N}(\theta, \sigma_0^2)$ distribution.

(a) Find the posterior

$$\pi(\theta \mid x) = \sum_{j=1}^{k} P((\mu, \sigma) = (\mu_j, \tau_j) \mid x) \pi(\theta \mid (\mu_j, \tau_j), x)$$

and write it in the form

$$\sum_{j=1}^{k} \lambda_j(x) \varphi_{\tau_j(x)}(\theta - \mu_j(x))$$

for appropriate $\lambda_j(x)$, $\tau_j(x)$ and $\mu_j(x)$. This shows that (1.7.4) defines a conjugate prior for the $\mathcal{N}(\theta, \sigma_0^2)$, distribution.

(b) Let $X_i = \theta + \epsilon_i$, $1 \le i \le n$, where θ is as previously and $\epsilon_1, \ldots, \epsilon_n$ are i.i.d. $\mathcal{N}(0, \sigma_0^2)$. Find the posterior $\pi(\theta \mid x_1, \ldots, x_n)$, and show that it belongs to class (1.7.4).
   Hint: Consider the sufficient statistic for $p(\mathbf{x} \mid \theta)$.

32. *A Hierarchical Binomial–Beta Model.* Let $\{(r_j, s_j) : 1 \le j \le k\}$ be a given collection of pairs with $r_j > 0$, $s_j > 0$, let (R, S) be a random pair with $P(R = r_j, S = s_j) = \lambda_j$, $0 < \lambda_j < 1$, $\sum_{j=1}^{k} \lambda_j = 1$, and let θ be a random variable whose conditional density $\pi(\theta, r, s)$ given $R = r$, $S = s$ is beta, $\beta(r, s)$. Consider the model in which $(X \mid \theta)$ has the binomial, $\mathcal{B}(n, \theta)$, distribution. Note that θ has the prior density

$$\pi(\theta) = \sum_{j=1}^{k} \lambda_j \pi(\theta, r_j, s_j). \tag{1.7.5}$$

Find the posterior

$$\pi(\theta \mid x) = \sum_{j=1}^{k} P(R = r_j, S = s_j \mid x)\pi(\theta \mid (r_j, s_j), x)$$

and show that it can be written in the form $\sum \lambda_j(x)\pi(\theta, r_j(\mathbf{x}), s_j(\mathbf{x}))$ for appropriate $\lambda_j(x)$, $r_j(x)$ and $s_j(x)$. This shows that (1.7.5) defines a class of conjugate priors for the $\mathcal{B}(n, \theta)$ distribution.

33. Let $p(x, \eta)$ be a one parameter canonical exponential family generated by $T(x) = x$ and $h(x)$, $x \in \mathcal{X} \subset R$, and let $\psi(x)$ be a nonconstant, nondecreasing function. Show that $E_\eta \psi(X)$ is strictly increasing in η.
   Hint:

$$\frac{\partial}{\partial \eta} E_\eta \psi(X) = \operatorname{Cov}_\eta(\psi(X), X)$$

$$= \frac{1}{2} E\{(X - X')[\psi(X) - \psi(X')]\}$$

where X and X' are independent identically distributed as X (see A.11.12).

34. Let $(X_1, \ldots, X_n)$ be a stationary Markov chain with two states 0 and 1. That is,

$$P[X_i = \epsilon_i \mid X_1 = \epsilon_1, \ldots, X_{i-1} = \epsilon_{i-1}] = P[X_i = \epsilon_i \mid X_{i-1} = \epsilon_{i-1}] = p_{\epsilon_{i-1}\epsilon_i}$$

where $\begin{pmatrix} p_{00} & p_{01} \\ p_{10} & p_{11} \end{pmatrix}$ is the matrix of transition probabilities. Suppose further that

(i) $p_{00} = p_{11} = p$, so that, $p_{10} = p_{01} = 1 - p$.

(ii) $P[X_1 = 0] = P[X_1 = 1] = \frac{1}{2}$.

(a) Show that if $0 < p < 1$ is unknown this is a full rank, one-parameter exponential family with $T = N_{00} + N_{11}$ where $N_{ij} \equiv$ the number of transitions from i to j. For example, 01011 has $N_{01} = 2$, $N_{11} = 1$, $N_{00} = 0$, $N_{10} = 1$.

(b) Show that $E(T) = (n-1)p$ (by the method of indicators or otherwise).

35. *A Conjugate Prior for the Two-Sample Problem.* Suppose that $X_1, \ldots, X_n$ and $Y_1, \ldots, Y_n$ are independent $\mathcal{N}(\mu_1, \sigma^2)$ and $\mathcal{N}(\mu_2, \sigma^2)$ samples, respectively. Consider the prior π for which for some $r > 0$, $k > 0$, $r\sigma^{-2}$ has a χ_k^2 distribution and given σ^2, μ_1 and μ_2 are independent with $\mathcal{N}(\xi_1, \sigma^2/k_1)$ and $\mathcal{N}(\xi_2, \sigma^2/k_2)$ distributions, respectively, where $\xi_j \in R$, $k_j > 0$, $j = 1, 2$. Show that π is a conjugate prior.

36. The *inverse Gaussian* density, $IG(\mu, \lambda)$, is

$$f(x, \mu, \lambda) = [\lambda/2\pi]^{1/2} x^{-3/2} \exp\{-\lambda(x-\mu)^2/2\mu^2 x\}, \quad x > 0, \quad \mu > 0, \quad \lambda > 0.$$

(a) Show that this is an exponential family generated by $\mathbf{T}(X) = -\frac{1}{2}(X, X^{-1})^T$ and $h(x) = (2\pi)^{-1/2} x^{-3/2}$.

(b) Show that the canonical parameters η_1, η_2 are given by $\eta_1 = \mu^{-2}\lambda$, $\eta_2 = \lambda$, and that $A(\eta_1, \eta_2) = -\left[\frac{1}{2}\log(\eta_2) + \sqrt{\eta_1\eta_2}\right]$, $\mathcal{E} = [0, \infty) \times (0, \infty)$.

(c) Find the moment-generating function of $\mathbf{T}$ and show that $E(X) = \mu$, $\operatorname{Var}(X) = \mu^{-3}\lambda$, $E(X^{-1}) = \mu^{-1} + \lambda^{-1}$, $\operatorname{Var}(X^{-1}) = (\lambda\mu)^{-1} + 2\lambda^{-2}$.

(d) Suppose $\mu = \mu_0$ is known. Show that the gamma family, $\Gamma(\alpha, \beta)$, is a conjugate prior.

(e) Suppose that $\lambda = \lambda_0$ is known. Show that the conjugate prior formula (1.6.20) produces a function that is not integrable with respect to μ. That is, Ω defined in (1.6.19) is empty.

(f) Suppose that μ and λ are both unknown. Show that (1.6.20) produces a function that is not integrable; that is, Ω defined in (1.6.19) is empty.

37. Let $X_1, \ldots, X_n$ be i.i.d. as $X \sim \mathcal{N}_p(\boldsymbol{\theta}, \Sigma_0)$ where Σ_0 is known. Show that the conjugate prior generated by (1.6.20) is the $\mathcal{N}_p(\boldsymbol{\eta}_0, \tau_0^2 \mathbf{I})$ family, where $\boldsymbol{\eta}_0$ varies freely in R^p, $\tau_0^2 > 0$ and $\mathbf{I}$ is the $p \times p$ identity matrix.

38. Let $X_i = (Z_i, Y_i)^T$ be i.i.d. as $X = (Z, Y)^T$, $1 \leq i \leq n$, where X has the density of Example 1.6.3. Write the density of $X_1, \ldots, X_n$ as a canonical exponential family and identify T, h, A, and $\mathcal{E}$. Find the expected value and variance of the sufficient statistic.

39. Suppose that $Y_1, \ldots, Y_n$ are independent, $Y_i \sim \mathcal{N}(\mu_i, \sigma^2)$, $n \geq 4$.

(a) Write the distribution of $Y_1, \ldots, Y_n$ in canonical exponential family form. Identify T, h, $\boldsymbol{\eta}$, A, and $\mathcal{E}$.

(b) Next suppose that μ_i depends on the value z_i of some covariate and consider the submodel defined by the map $\boldsymbol{\eta} : (\theta_1, \theta_2, \theta_3)^T \to (\boldsymbol{\mu}^T, \sigma^2)^T$ where $\boldsymbol{\eta}$ is determined by

$$\mu_i = \exp\{\theta_1 + \theta_2 z_i\}, \ z_1 < z_2 < \cdots < z_n; \ \sigma^2 = \theta_3$$

where $\theta_1 \in R$, $\theta_2 \in R$, $\theta_3 > 0$. This model is sometimes used when μ_i is restricted to be positive. Show that $p(\mathbf{y}, \boldsymbol{\theta})$ as given by (1.6.12) is a curved exponential family model with $l = 3$.

40. Suppose $Y_1, \ldots, Y_n$ are independent exponentially, $\mathcal{E}(\lambda_i)$, distributed survival times, $n \geq 3$.

   (a) Write the distribution of $Y_1, \ldots, Y_n$ in canonical exponential family form. Identify $\mathbf{T}$, h, $\boldsymbol{\eta}$, A, and $\mathcal{E}$.

   (b) Recall that $\mu_i = E(Y_i) = \lambda_i^{-1}$. Suppose μ_i depends on the value z_i of a covariate. Because $\mu_i > 0$, μ_i is sometimes modeled as

$$\mu_i = \exp\{\theta_1 + \theta_2 z_i\}, \ i = 1, \ldots, n$$

where not all the z's are equal. Show that $p(\mathbf{y}, \boldsymbol{\theta})$ as given by (1.6.12) is a curved exponential family model with $l = 2$.

1.8 NOTES

Note for Section 1.1

(1) For the measure theoretically minded we can assume more generally that the P_θ are all dominated by a σ finite measure μ and that $p(x, \theta)$ denotes $\frac{dP_\theta}{d\mu}$, the Radon Nikodym derivative.

Notes for Section 1.3

(1) More natural in the sense of measuring the Euclidean distance between the estimate $\widehat{\theta}$ and the "truth" θ. Squared error gives much more weight to those $\widehat{\theta}$ that are far away from θ than those close to θ.

(2) We define the lower boundary of a convex set simply to be the set of all boundary points r such that the set lies completely on or above any tangent to the set at r.

Note for Section 1.4

(1) Source: Hodges, Jr., J. L., D. Kretch, and R. S. Crutchfield. (1975)

Notes for Section 1.6

(1) Exponential families arose much earlier in the work of Boltzmann in statistical mechanics as laws for the distribution of the states of systems of particles—see Feynman (1963), for instance. The connection is through the concept of entropy, which also plays a key role in information theory—see Cover and Thomas (1991).

(2) The restriction that's $x \in R^q$ and that these families be discrete or continuous is artificial. In general if μ is a σ finite measure on the sample space $\mathcal{X}$, $p(x, \theta)$ as given by (1.6.1) can be taken to be the density of X with respect to μ—see Lehmann (1997), for instance.

This permits consideration of data such as images, positions, and spheres (e.g., the Earth), and so on.

Note for Section 1.7

(1) $\mathbf{u}^T M \mathbf{u} > 0$ for all $p \times 1$ vectors $\mathbf{u} \neq 0$.

1.9 REFERENCES

BERGER, J. O., *Statistical Decision Theory and Bayesian Analysis* New York: Springer, 1985.

BERMAN, S. M., "A Stochastic Model for the Distribution of HIV Latency Time Based on T4 Counts," *Biometika, 77,* 733–741 (1990).

BICKEL, P. J., "Using Residuals Robustly I: Tests for Heteroscedasticity, Nonlinearity," *Ann. Statist.* 6, 266–291 (1978).

BLACKWELL, D. AND M. A. GIRSHICK, *Theory of Games and Statistical Decisions* New York: Wiley, 1954.

BOX, G. E. P., "Sampling and Bayes Inference in Scientific Modelling and Robustness (with Discussion)," *J. Royal Statist. Soc. A 143*, 383–430 (1979).

BROWN, L., *Fundamentals of Statistical Exponential Families with Applications in Statistical Decision Theory*, IMS Lecture Notes—Monograph Series, Hayward, 1986.

CARROLL, R. J. AND D. RUPPERT, *Transformation and Weighting in Regression* New York: Chapman and Hall, 1988.

COVER, T. M. AND J. A. THOMAS, *Elements of Information Theory* New York: Wiley, 1991.

DE GROOT, M. H., *Optimal Statistical Decisions* New York: McGraw–Hill, 1969.

DOKSUM, K. A. AND A. SAMAROV, "Nonparametric Estimation of Global Functionals and a Measure of the Explanatory Power of Covariates in Regression," *Ann. Statist. 23*, 1443–1473 (1995).

FERGUSON, T. S., *Mathematical Statistics* New York: Academic Press, 1967.

FEYNMAN, R. P., *The Feynman Lectures on Physics*, v. 1, R. P. Feynman, R. B. Leighton, and M. Sands, Eds., Ch. 40 *Statistical Mechanics of Physics* Reading, MA: Addison-Wesley, 1963.

GRENANDER, U. AND M. ROSENBLATT, *Statistical Analysis of Stationary Time Series* New York: Wiley, 1957.

HODGES, JR., J. L., D. KRETCH AND R. S. CRUTCHFIELD, *Statlab: An Empirical Introduction to Statistics* New York: McGraw–Hill, 1975.

KENDALL, M. G. AND A. STUART, *The Advanced Theory of Statistics*, Vols. II, III New York: Hafner Publishing Co., 1961, 1966.

LEHMANN, E. L., "A Theory of Some Multiple Decision Problems, I and II," *Ann. Math. Statist. 22*, 1–25, 547–572 (1957).

LEHMANN, E. L., "Model Specification: The Views of Fisher and Neyman, and Later Developments," *Statist. Science 5*, 160–168 (1990).

LEHMANN, E. L., *Testing Statistical Hypotheses*, 2nd ed. New York: Springer, 1997.

LINDLEY, D. V., *Introduction to Probability and Statistics from a Bayesian Point of View*, Part I: *Probability*; Part II: *Inference* London: Cambridge University Press, 1965.

MANDEL, J., *The Statistical Analysis of Experimental Data* New York: J. Wiley & Sons, 1964.

NORMAND, S-L. AND K. A. DOKSUM, "Empirical Bayes Procedures for a Change Point Problem with Application to HIV/AIDS Data," *Empirical Bayes and Likelihood Inference*, 67–79, Editors: S. E. Ahmed and N. Reid. New York: Springer, Lecture Notes in Statistics, 2001.

PEARSON, K., "On the General Theory of Skew Correlation and Nonlinear Regression," *Proc. Roy. Soc. London 71*, 303 (1905). (Draper's Research Memoirs, Dulan & Co, Biometrics Series II.)

RAIFFA, H. AND R. SCHLAIFFER, *Applied Statistical Decision Theory*, Division of Research, Graduate School of Business Administration, Harvard University, Boston, 1961.

SAVAGE, L. J., *The Foundations of Statistics*, J. Wiley & Sons, New York, 1954.

SAVAGE, L. J. ET AL., *The Foundation of Statistical Inference* London: Methuen & Co., 1962.

SNEDECOR, G. W. AND W. G. COCHRAN, *Statistical Methods*, 8th Ed. Ames, IA: Iowa State University Press, 1989.

WETHERILL, G. B. AND K. D. GLAZEBROOK, *Sequential Methods in Statistics* New York: Chapman and Hall, 1986.

Chapter 2

METHODS OF ESTIMATION

2.1 BASIC HEURISTICS OF ESTIMATION

2.1.1 Minimum Contrast Estimates; Estimating Equations

Our basic framework is as before, that is, $X \in \mathcal{X}$, $X \sim P \in \mathcal{P}$, usually parametrized as $\mathcal{P} = \{P_{\boldsymbol{\theta}} : \boldsymbol{\theta} \in \Theta\}$. See Section 1.1.2. In this parametric case, how do we select reasonable estimates for $\boldsymbol{\theta}$ itself? That is, how do we find a function $\widehat{\boldsymbol{\theta}}(X)$ of the vector observation X that in some sense "is close" to the unknown $\boldsymbol{\theta}$? The fundamental heuristic is typically the following. Consider a function

$$\rho : \mathcal{X} \times \Theta \to R$$

and define

$$D(\boldsymbol{\theta}_0, \boldsymbol{\theta}) \equiv E_{\boldsymbol{\theta}_0} \rho(X, \boldsymbol{\theta}).$$

Suppose that as a function of $\boldsymbol{\theta}$, $D(\boldsymbol{\theta}_0, \boldsymbol{\theta})$ measures the (population) *discrepancy* between $\boldsymbol{\theta}$ and the true value $\boldsymbol{\theta}_0$ of the parameter in the sense that $D(\boldsymbol{\theta}_0, \boldsymbol{\theta})$ is uniquely minimized for $\boldsymbol{\theta} = \boldsymbol{\theta}_0$. That is, if $P_{\boldsymbol{\theta}_0}$ were true and we knew $D(\boldsymbol{\theta}_0, \boldsymbol{\theta})$ as a function of $\boldsymbol{\theta}$, we could obtain $\boldsymbol{\theta}_0$ as the minimizer. Of course, we don't know the truth so this is inoperable, but $\rho(X, \boldsymbol{\theta})$ is the optimal MSPE predictor of $D(\boldsymbol{\theta}_0, \boldsymbol{\theta})$ (Lemma 1.4.1). So it is natural to consider $\widehat{\boldsymbol{\theta}}(X)$ minimizing $\rho(X, \boldsymbol{\theta})$ as an estimate of $\boldsymbol{\theta}_0$. Under these assumptions we call $\rho(\cdot, \cdot)$ a *contrast function* and $\widehat{\boldsymbol{\theta}}(X)$ a *minimum contrast estimate*.

Now suppose Θ is Euclidean $\subset R^d$, the true $\boldsymbol{\theta}_0$ is an interior point of Θ, and $\boldsymbol{\theta} \to D(\boldsymbol{\theta}_0, \boldsymbol{\theta})$ is smooth. Then we expect

$$\nabla_{\boldsymbol{\theta}} D(\boldsymbol{\theta}_0, \boldsymbol{\theta})\big|_{\boldsymbol{\theta}=\boldsymbol{\theta}_0} = 0 \qquad (2.1.1)$$

where ∇ denotes the *gradient*,

$$\nabla_{\boldsymbol{\theta}} = \left(\frac{\partial}{\partial \theta_1}, \ldots, \frac{\partial}{\partial \theta_d} \right)^T.$$

Arguing heuristically again we are led to estimates $\widehat{\boldsymbol{\theta}}$ that solve

$$\nabla_{\boldsymbol{\theta}} \rho(X, \boldsymbol{\theta}) = \mathbf{0}. \qquad (2.1.2)$$

The equations (2.1.2) define a special form of *estimating equations*.

More generally, suppose we are given a function $\mathbf{\Psi} : \mathcal{X} \times R^d \to R^d$, $\mathbf{\Psi} \equiv (\psi_1, \dots, \psi_d)^T$ and define

$$\mathbf{V}(\boldsymbol{\theta}_0, \boldsymbol{\theta}) = E_{\boldsymbol{\theta}_0} \mathbf{\Psi}(X, \boldsymbol{\theta}). \tag{2.1.3}$$

Suppose $\mathbf{V}(\boldsymbol{\theta}_0, \boldsymbol{\theta}) = \mathbf{0}$ has $\boldsymbol{\theta}_0$ as its unique solution for all $\boldsymbol{\theta}_0 \in \Theta$. Then we say $\widehat{\boldsymbol{\theta}}$ solving

$$\mathbf{\Psi}(X, \widehat{\boldsymbol{\theta}}) = \mathbf{0} \tag{2.1.4}$$

is an *estimating equation estimate*. Evidently, there is a substantial overlap between the two classes of estimates. Here is an example to be pursued later.

Example 2.1.1. *Least Squares.* Consider the parametric version of the regression model of Example 1.1.4 with $\mu(\mathbf{z}) = g(\boldsymbol{\beta}, \mathbf{z})$, $\boldsymbol{\beta} \in R^d$, where the function g is known. Here the data are $X = \{(\mathbf{z}_i, Y_i) : 1 \le i \le n\}$ where $Y_1, \dots, Y_n$ are independent. A natural[1] function $\rho(X, \boldsymbol{\beta})$ to consider is the squared Euclidean distance between the vector $\mathbf{Y}$ of observed Y_i and the vector expectation of $\mathbf{Y}$, $\boldsymbol{\mu}(\mathbf{z}) \equiv (g(\boldsymbol{\beta}, \mathbf{z}_1), \dots, g(\boldsymbol{\beta}, \mathbf{z}_n))^T$. That is, we take

$$\rho(X, \boldsymbol{\beta}) = |\mathbf{Y} - \boldsymbol{\mu}|^2 = \sum_{i=1}^n [Y_i - g(\boldsymbol{\beta}, \mathbf{z}_i)]^2. \tag{2.1.5}$$

Strictly speaking $\mathcal{P}$ is not fully defined here and this is a point we shall explore later. But, for convenience, suppose we postulate that the ϵ_i of Example 1.1.4 are i.i.d. $\mathcal{N}(0, \sigma_0^2)$. Then $\boldsymbol{\beta}$ parametrizes the model and we can compute (see Problem 2.1.16),

$$
\begin{aligned}
D(\boldsymbol{\beta}_0, \boldsymbol{\beta}) &= E_{\boldsymbol{\beta}_0} \rho(X, \boldsymbol{\beta}) \\
&= n\sigma_0^2 + \sum_{i=1}^n [g(\boldsymbol{\beta}_0, \mathbf{z}_i) - g(\boldsymbol{\beta}, \mathbf{z}_i)]^2,
\end{aligned}
\tag{2.1.6}
$$

which is indeed minimized at $\boldsymbol{\beta} = \boldsymbol{\beta}_0$ and uniquely so if and only if the parametrization is identifiable. An estimate $\widehat{\boldsymbol{\beta}}$ that minimizes $\rho(X, \boldsymbol{\beta})$ exists if $g(\boldsymbol{\beta}, \mathbf{z})$ is continuous in $\boldsymbol{\beta}$ and

$$\lim\{|g(\boldsymbol{\beta}, \mathbf{z})| : |\boldsymbol{\beta}| \to \infty\} = \infty$$

(Problem 2.1.10). The estimate $\widehat{\boldsymbol{\beta}}$ is called *the least squares estimate*.

If, further, $g(\boldsymbol{\beta}, \mathbf{z})$ is differentiable in $\boldsymbol{\beta}$, then $\widehat{\boldsymbol{\beta}}$ satisfies the equation (2.1.2) or equivalently the system of estimating equations,

$$\sum_{i=1}^n \frac{\partial g}{\partial \beta_j}(\widehat{\boldsymbol{\beta}}, \mathbf{z}_i) Y_i = \sum_{i=1}^n \frac{\partial g}{\partial \beta_j}(\widehat{\boldsymbol{\beta}}, \mathbf{z}_i) g(\widehat{\boldsymbol{\beta}}, \mathbf{z}_i), \ 1 \le j \le d. \tag{2.1.7}$$

In the important linear case,

$$g(\boldsymbol{\beta}, \mathbf{z}_i) = \sum_{j=1}^d z_{ij} \beta_j \text{ and } \mathbf{z}_i = (z_{i1}, \dots, z_{id})^T$$

the system becomes

$$\sum_{i=1}^{n} z_{ij} Y_i = \sum_{k=1}^{d} \left(\sum_{i=1}^{n} z_{ij} z_{ik} \right) \widehat{\beta}_k, \tag{2.1.8}$$

the *normal equations.* These equations are commonly written in matrix form

$$\mathbf{Z}_D^T \mathbf{Y} = \mathbf{Z}_D^T \mathbf{Z}_D \beta \tag{2.1.9}$$

where $\mathbf{Z}_D \equiv \|z_{ij}\|_{n \times d}$ is the *design matrix.* Least squares, thus, provides a first example of both minimum contrast and estimating equation methods.

We return to the remark that this estimating method is well defined even if the ϵ_i are not i.i.d. $\mathcal{N}(0, \sigma_0^2)$. In fact, once defined we have a method of computing a statistic $\widehat{\beta}$ from the data $X = \{(\mathbf{z}_i, Y_i),\ 1 \leq i \leq n\}$, which can be judged on its merits whatever the true P governing X is. This very important example is pursued further in Section 2.2 and Chapter 6. □

Here is another basic estimating equation example.

Example 2.1.2. *Method of Moments (MOM).* Suppose $X_1, \ldots, X_n$ are i.i.d. as $X \sim P_{\boldsymbol{\theta}}$, $\boldsymbol{\theta} \in R^d$ and $\boldsymbol{\theta}$ is identifiable. Suppose that $\mu_1(\boldsymbol{\theta}), \ldots, \mu_d(\boldsymbol{\theta})$ are the first d moments of the population we are sampling from. Thus, we assume the existence of

$$\mu_j(\boldsymbol{\theta}) = \mu_j = E_{\boldsymbol{\theta}}(X^j),\ 1 \leq j \leq d.$$

Define the jth *sample moment* $\widehat{\mu}_j$ by,

$$\widehat{\mu}_j = \frac{1}{n} \sum_{i=1}^{n} X_i^j,\ 1 \leq j \leq d.$$

To apply the method of moments to the problem of estimating $\boldsymbol{\theta}$, we need to be able to express $\boldsymbol{\theta}$ as a continuous function g of the first d moments. Thus, suppose

$$\boldsymbol{\theta} \rightarrow (\mu_1(\boldsymbol{\theta}), \ldots, \mu_d(\boldsymbol{\theta}))$$

is $1 - 1$ from R^d to R^d. The *method of moments* prescribes that we estimate $\boldsymbol{\theta}$ by the solution of

$$\widehat{\mu}_j = \mu_j(\widehat{\boldsymbol{\theta}}),\ 1 \leq j \leq d$$

if it exists. The motivation of this simplest estimating equation example is the law of large numbers: For $X \sim P_{\boldsymbol{\theta}}$, $\widehat{\mu}_j$ converges in probability to $\mu_j(\theta)$.

More generally, if we want to estimate a R^k-valued function $q(\boldsymbol{\theta})$ of $\boldsymbol{\theta}$, we obtain a MOM estimate of $q(\boldsymbol{\theta})$ by expressing $q(\boldsymbol{\theta})$ as a function of any of the first d moments $\mu_1, \ldots, \mu_d$ of X, say $q(\boldsymbol{\theta}) = h(\mu_1, \ldots, \mu_d)$, $d \geq k$, and then using $h(\widehat{\mu}_1, \ldots, \widehat{\mu}_d)$ as the estimate of $q(\boldsymbol{\theta})$.

For instance, consider a study in which the survival time X is modeled to have a gamma distribution, $\Gamma(\alpha, \lambda)$, with density

$$[\lambda^\alpha/\Gamma(\alpha)]x^{\alpha-1}\exp\{-\lambda x\},\ x > 0;\ \alpha > 0,\ \lambda > 0.$$

In this case $\boldsymbol{\theta} = (\alpha, \lambda)$, $\mu_1 = E(X) = \alpha/\lambda$, and $\mu_2 = E(X^2) = \alpha(1+\alpha)/\lambda^2$. Solving for $\boldsymbol{\theta}$ gives

$$\alpha = (\mu_1/\sigma)^2,\quad \widehat{\alpha} = (\bar{X}/\widehat{\sigma})^2;$$
$$\lambda = \mu_1/\sigma^2,\quad \widehat{\lambda} = \bar{X}/\widehat{\sigma}^2$$

where $\sigma^2 = \mu_2 - \mu_1^2$ and $\widehat{\sigma}^2 = n^{-1}\Sigma X_i^2 - \bar{X}^2$. In this example, the method of moment estimator is not unique. We can, for instance, express θ as a function of μ_1 and $\mu_3 = E(X^3)$ and obtain a method of moment estimator based on $\widehat{\mu}_1$ and $\widehat{\mu}_3$ (Problem 2.1.11).   □

Algorithmic issues

We note that, in general, neither minimum contrast estimates nor estimating equation solutions can be obtained in closed form. There are many algorithms for optimization and root finding that can be employed. An algorithm for estimating equations frequently used when computation of $M(X, \cdot) \equiv D\boldsymbol{\Psi}(X, \cdot) \equiv \left\|\frac{\partial\psi_i}{\partial\theta_j}(X, \cdot)\right\|_{d\times d}$ is quick and M is nonsingular with high probability is the *Newton–Raphson algorithm*. It is defined by initializing with $\boldsymbol{\theta}_0$, then setting

$$\widehat{\boldsymbol{\theta}}_{j+1} = \widehat{\boldsymbol{\theta}}_j - [M(X, \widehat{\boldsymbol{\theta}}_j)]^{-1}\boldsymbol{\Psi}(X, \widehat{\boldsymbol{\theta}}_j). \tag{2.1.10}$$

This algorithm and others will be discussed more extensively in Section 2.4 and in Chapter 6, in particular Problem 6.6.10.

2.1.2   The Plug-In and Extension Principles

We can view the method of moments as an example of what we call the *plug-in* (or *substitution*) and *extension* principles, two other basic heuristics particularly applicable in the i.i.d. case. We introduce these principles in the context of multinomial trials and then abstract them and relate them to the method of moments.

Example 2.1.3. *Frequency Plug-in*[2] *and Extension.* Suppose we observe multinomial trials in which the values $v_1, \ldots, v_k$ of the population being sampled are known, but their respective probabilities $p_1, \ldots, p_k$ are completely unknown. If we let $X_1, \ldots, X_n$ be i.i.d. as X and

$$N_i \equiv \text{number of indices } j \text{ such that } X_j = v_i,$$

then the natural estimate of $p_i = P[X = v_i]$ suggested by the law of large numbers is N_i/n, the proportion of sample values equal to v_i. As an illustration consider a population of men whose occupations fall in one of five different job categories, 1, 2, 3, 4, or 5. Here $k = 5$, $v_i = i$, $i = 1, \ldots, 5$, p_i is the proportion of men in the population in the ith job category and N_i/n is the sample proportion in this category. Here is some job category data (Mosteller, 1968).

Job Category

i	1	2	3	4	5	
N_i	23	84	289	217	95	$n = \sum_{i=1}^{5} N_i = 708$
$\widehat{p}_i$	0.03	0.12	0.41	0.31	0.13	$\sum_{i=1}^{5} \widehat{p}_i = 1$

for Danish men whose fathers were in category 3, together with the estimates $\widehat{p}_i = N_i/n$.

Next consider the more general problem of estimating a continuous function $q(p_1, \dots, p_k)$ of the population proportions. The *frequency plug-in principle* simply proposes to re-place the unknown population frequencies $p_1, \dots, p_k$ by the observable sample frequencies $N_1/n, \dots, N_k/n$. That is, use

$$T(X_1, \dots, X_n) = q\left(\frac{N_1}{n}, \dots, \frac{N_k}{n}\right) \tag{2.1.11}$$

to estimate $q(p_1, \dots, p_k)$. For instance, suppose that in the previous job category table, categories 4 and 5 correspond to blue-collar jobs, whereas categories 2 and 3 correspond to white-collar jobs. We would be interested in estimating

$$q(p_1, \dots, p_5) = (p_4 + p_5) - (p_2 + p_3),$$

the difference in the proportions of blue-collar and white-collar workers. If we use the frequency substitution principle, the estimate is

$$T(X_1, \dots, X_n) = \left(\frac{N_4}{n} + \frac{N_5}{n}\right) - \left(\frac{N_2}{n} + \frac{N_3}{n}\right),$$

which in our case is $0.44 - 0.53 = -0.09$.

Equivalently, let P denote $\mathbf{p} = (p_1, \dots, p_k)$ with $p_i = P[X = v_i]$, $1 \leq i \leq k$, and think of this model as $\mathcal{P} = \{$all probability distributions P on $\{v_1, \dots, v_k\}\}$. Then $q(\mathbf{p})$ can be identified with a parameter $\nu : \mathcal{P} \to R$, that is, $\nu(P) = (p_4 + p_5) - (p_2 + p_3)$, and the frequency plug-in principle simply says to replace $P = (p_1, \dots, p_k)$ in $\nu(P)$ by $\widehat{P} = \left(\frac{N_1}{n}, \dots, \frac{N_k}{n}\right)$, the multinomial empirical distribution of $X_1, \dots, X_n$. ☐

Now suppose that the proportions $p_1, \dots, p_k$ do not vary freely but are continuous functions of some d-dimensional parameter $\boldsymbol{\theta} = (\theta_1, \dots, \theta_d)$ and that we want to estimate a component of $\boldsymbol{\theta}$ or more generally a function $q(\boldsymbol{\theta})$. Many of the models arising in the analysis of discrete data discussed in Chapter 6 are of this type.

Example 2.1.4. *Hardy–Weinberg Equilibrium.* Consider a sample from a population in genetic equilibrium with respect to a single gene with two alleles. If we assume the three different genotypes are identifiable, we are led to suppose that there are three types of individuals whose frequencies are given by the so-called *Hardy–Weinberg proportions*

$$p_1 = \theta^2, \; p_2 = 2\theta(1 - \theta), \; p_3 = (1 - \theta)^2, \; 0 < \theta < 1. \tag{2.1.12}$$

If N_i is the number of individuals of type i in the sample of size n, then (N_1, N_2, N_3) has a multinomial distribution with parameters (n, p_1, p_2, p_3) given by (2.1.12). Suppose we want to estimate θ, the frequency of one of the alleles. Because $\theta = \sqrt{p_1}$, we can use the principle we have introduced and estimate by $\sqrt{N_1/n}$. Note, however, that we can also write $\theta = 1 - \sqrt{p_3}$ and, thus, $1 - \sqrt{N_3/n}$ is also a plausible estimate of θ.   □

In general, suppose that we want to estimate a continuous R^l-valued function q of θ. If $p_1, \ldots, p_k$ are continuous functions of θ, we can usually express $q(\theta)$ as a continuous function of $p_1, \ldots, p_k$, that is,

$$q(\theta) = h(p_1(\theta), \ldots, p_k(\theta)), \tag{2.1.13}$$

with h defined and continuous on

$$\mathcal{P} = \left\{ (p_1, \ldots, p_k) : p_i \geq 0, \sum_{i=1}^{k} p_i = 1 \right\}.$$

Given h we can apply the *extension principle* to estimate $q(\theta)$ as,

$$T(X_1, \ldots, X_n) = h\left(\frac{N_1}{n}, \ldots, \frac{N_k}{n} \right). \tag{2.1.14}$$

As we saw in the Hardy–Weinberg case, the representation (2.1.13) and estimate (2.1.14) are not unique. We shall consider in Chapters 3 (Example 3.4.4) and 5 how to choose among such estimates.

We can think of the extension principle alternatively as follows. Let

$$\mathcal{P}_0 = \{P_\theta = (p_1(\theta), \ldots, p_k(\theta)) : \theta \in \Theta\}$$

be a submodel of $\mathcal{P}$. Now $q(\theta)$ can be identified if θ is identifiable by a parameter ν : $\mathcal{P}_0 \to R$ given by $\nu(P_\theta) = q(\theta)$. Then (2.1.13) defines an extension of ν from $\mathcal{P}_0$ to $\mathcal{P}$ via $\bar{\nu} : \mathcal{P} \to R$ where $\bar{\nu}(P) \equiv h(\mathbf{p})$ and $\bar{\nu}(P) = \nu(P)$ for $P \in \mathcal{P}_0$.

The plug-in and extension principles can be abstractly stated as follows:

Plug-in principle. If we have an estimate $\widetilde{P}$ of $P \in \mathcal{P}$ such that $\widetilde{P} \in \mathcal{P}$ and $\nu : \mathcal{P} \to \mathcal{T}$ is a parameter, then $\nu(\widetilde{P})$ is the plug-in estimate of ν. In particular, in the i.i.d. case if $\mathcal{P}$ is the space of all distributions of X and $X_1, \ldots, X_n$ are i.i.d. as $X \sim P$, the *empirical distribution* $\widehat{P}$ of X given by

$$\widehat{P}[X \in A] = \frac{1}{n} \sum_{i=1}^{n} 1(X_i \in A) \tag{2.1.15}$$

is, by the law of large numbers, a natural estimate of P and $\nu(\widehat{P})$ is a plug-in estimate of $\nu(P)$ in this nonparametric context. For instance, if X is real and $F(x) = P(X \leq x)$ is the distribution function (d.f.), consider $\nu_\alpha(P) = \frac{1}{2}[F^{-1}(\alpha) + F_U^{-1}(\alpha)]$, where $\alpha \in (0, 1)$ and

$$F^{-1}(\alpha) = \inf\{x : F(x) \geq \alpha\}, \ F_U^{-1}(\alpha) = \sup\{x : F(x) \leq \alpha\}, \tag{2.1.16}$$

then $\nu_\alpha(P)$ is the αth *population quantile* x_α. Here $x_{\frac{1}{2}} = \nu_{\frac{1}{2}}(P)$ is called the *population median*. A natural estimate is the αth *sample quantile*

$$\widehat{x}_\alpha = \frac{1}{2}[\widehat{F}^{-1}(\alpha) + \widehat{F}_U^{-1}(\alpha)], \tag{2.1.17}$$

where $\widehat{F}$ is the empirical d.f. Here $\widehat{x}_{\frac{1}{2}}$ is called the *sample median*.

For a second example, if X is real and $\mathcal{P}$ is the class of distributions with $E|X|^j < \infty$, then the plug-in estimate of the jth moment $\nu(P) = \mu_j = E(X^j)$ in this nonparametric context is the jth sample moment $\nu(\widehat{P}) = \int x^j d\widehat{F}(x) = n^{-1} \sum_{i=1}^n X_i^j$.

Extension principle. Suppose $\mathcal{P}_0$ is a submodel of $\mathcal{P}$ and $\widehat{P}$ is an element of $\mathcal{P}$ but not necessarily $\mathcal{P}_0$ and suppose $\nu : \mathcal{P}_0 \to \mathcal{T}$ is a parameter. If $\bar{\nu} : \mathcal{P} \to \mathcal{T}$ is an extension of ν in the sense that $\bar{\nu}(P) = \nu(P)$ on $\mathcal{P}_0$, then $\bar{\nu}(\widehat{P})$ is an extension (and plug-in) estimate of $\nu(P)$.

With this general statement we can see precisely how method of moment estimates can be obtained as extension and frequency plug-in estimates for multinomial trials because

$$\mu_j(\boldsymbol{\theta}) = \sum_{i=1}^k v_i^j p_i(\boldsymbol{\theta}) = h(\mathbf{p}(\boldsymbol{\theta})) = \nu(P_{\boldsymbol{\theta}})$$

where

$$h(\mathbf{p}) = \sum_{i=1}^k v_i^j p_i = \bar{\nu}(P),$$

$$\widehat{\mu}_j = \frac{1}{n} \sum_{l=1}^n X_l^j = \sum_{i=1}^k v_i^j \frac{N_i}{n} = h\left(\frac{\mathbf{N}}{n}\right) = \bar{\nu}(\widehat{P})$$

and $\widehat{P}$ is the empirical distribution. This reasoning extends to the general i.i.d. case (Problem 2.1.12) and to more general method of moment estimates (Problem 2.1.13). As stated, these principles are general. However, they are mainly applied in the i.i.d. case—but see Problem 2.1.14.

Remark 2.1.1. The plug-in and extension principles are used when P_θ, ν, and $\bar{\nu}$ are continuous. For instance, in the multinomial examples 2.1.3 and 2.1.4, P_θ as given by the Hardy–Weinberg $\mathbf{p}(\theta)$, is a continuous map from $\Theta = [0,1]$ to $\mathcal{P}$, $\nu(P_{\boldsymbol{\theta}}) = q(\boldsymbol{\theta}) = h(\mathbf{p}(\boldsymbol{\theta}))$ is a continuous map from Θ to R and $\bar{\nu}(P) = h(\mathbf{p})$ is a continuous map from $\mathcal{P}$ to R.

Remark 2.1.2. The plug-in and extension principles must be calibrated with the target parameter. For instance, let $\mathcal{P}_0$ be the class of distributions of $X = \theta + \epsilon$ where $\theta \in R$ and the distribution of ϵ ranges over the class of symmetric distributions with mean zero. Let $\nu(P)$ be the mean of X and let $\mathcal{P}$ be the class of distributions of $X = \theta + \epsilon$ where $\theta \in R$ and the distribution of ϵ ranges over the class of distributions with mean zero. In this case both $\bar{\nu}_1(P) = E_P(X)$ and $\bar{\nu}_2(P) = $ "median of P" satisfy $\bar{\nu}(P) = \nu(P)$, $P \in \mathcal{P}_0$, but only $\bar{\nu}_1(\widehat{P}) = \bar{X}$ is a sensible estimate of $\nu(P)$, $P \notin \mathcal{P}_0$, because when P is not symmetric, the sample median $\bar{\nu}_2(\widehat{P})$ does not converge in probability to $E_P(X)$.

Here are three further simple examples illustrating reasonable and unreasonable MOM estimates.

Example 2.1.5. Suppose that $X_1, \ldots, X_n$ is a $\mathcal{N}(\mu, \sigma^2)$ sample as in Example 1.1.2 with assumptions (1)–(4) holding. The method of moments estimates of μ and σ^2 are $\bar{X}$ and $\widehat{\sigma}^2$. $\qquad\qquad\qquad\square$

Example 2.1.6. Suppose $X_1, \ldots, X_n$ are the indicators of a set of Bernoulli trials with probability of success θ. Because $\mu_1(\theta) = \theta$ the method of moments leads to the natural estimate of θ, $\bar{X}$, the frequency of successes. To estimate the population variance $\theta(1 - \theta)$ we are led by the first moment to the estimate, $\bar{X}(1 - \bar{X})$. Because we are dealing with (unrestricted) Bernoulli trials, these are *the* frequency plug-in (substitution) estimates (see Problem 2.1.1). $\qquad\qquad\qquad\square$

Example 2.1.7. *Estimating the Size of a Population (continued).* In Example 1.5.3 where $X_1, \ldots, X_n$ are i.i.d. $\mathcal{U}\{1, 2, \ldots, \theta\}$, we find $\mu = E_\theta(X_i) = \frac{1}{2}(\theta + 1)$. Thus, $\theta = 2\mu - 1$ and $2\bar{X} - 1$ is a method of moments estimate of θ. This is clearly a foolish estimate if $X_{(n)} = \max X_i > 2\bar{X} - 1$ because in this model θ is always at least as large as $X_{(n)}$. $\quad\square$

As we have seen, there are often several method of moments estimates for the same $q(\boldsymbol{\theta})$. For example, if we are sampling from a Poisson population with parameter θ, then θ is both the population mean and the population variance. The method of moments can lead to either the sample mean or the sample variance. Moreover, because $p_0 = P(X = 0) = \exp\{-\theta\}$, a frequency plug-in estimate of θ is $-\log \widehat{p}_0$, where $\widehat{p}_0$ is $n^{-1}[\#X_i = 0]$. We will make a selection among such procedures in Chapter 3.

Remark 2.1.3. What are the good points of the method of moments and frequency plug-in?

(a) They generally lead to procedures that are easy to compute and are, therefore, valuable as preliminary estimates in algorithms that search for more efficient estimates. See Section 2.4.

(b) If the sample size is large, these estimates are likely to be close to the value estimated (consistency). This minimal property is discussed in Section 5.2.

It does turn out that there are "best" frequency plug-in estimates, those obtained by the method of maximum likelihood, a special type of minimum contrast and estimating equation method. Unfortunately, as we shall see in Section 2.4, they are often difficult to compute. Algorithms for their computation will be introduced in Section 2.4.

Discussion. When we consider optimality principles, we may arrive at different types of estimates than those discussed in this section. For instance, as we shall see in Chapter 3, estimation of θ real with quadratic loss and Bayes priors lead to procedures that are data weighted averages of θ values rather than minimizers of functions $\rho(\theta, X)$. Plug-in is not the optimal way to go for the Bayes, minimax, or uniformly minimum variance unbiased (UMVU) principles we discuss briefly in Chapter 3. However, a saving grace becomes apparent in Chapters 5 and 6. If the model fits, for large amounts of data, optimality principle solutions agree to first order with the best minimum contrast and estimating equation solutions, the plug-in principle is justified, and there are best extensions.

Summary. We consider principles that suggest how we can use the outcome X of an experiment to estimate unknown parameters.

For the model $\{P_\theta : \theta \in \Theta\}$ a *contrast* ρ is a function from $\mathcal{X} \times \Theta$ to R such that the *discrepancy*

$$D(\boldsymbol{\theta}_0, \boldsymbol{\theta}) = E_{\boldsymbol{\theta}_0}\rho(X, \boldsymbol{\theta}), \ \theta \in \Theta \subset R^d$$

is uniquely minimized at the true value $\boldsymbol{\theta} = \boldsymbol{\theta}_0$ of the parameter. A *minimum contrast estimator* is a minimizer of $\rho(\mathbf{X}, \boldsymbol{\theta})$, and the *contrast estimating equations* are $\nabla_{\boldsymbol{\theta}} \rho(X, \boldsymbol{\theta}) = 0$.

For data $\{(\mathbf{z}_i, Y_i) : 1 \leq i \leq n\}$ with Y_i independent and $E(Y_i) = g(\boldsymbol{\beta}, \mathbf{z}_i), 1 \leq i \leq n$, where g is a known function and $\boldsymbol{\beta} \in R^d$ is a vector of unknown regression coefficients, a *least squares estimate* of $\boldsymbol{\beta}$ is a minimizer of

$$\rho(X, \boldsymbol{\beta}) = \sum [Y_i - g(\boldsymbol{\beta}, \mathbf{z}_i)]^2.$$

For this contrast, when $g(\boldsymbol{\beta}, \mathbf{z}) = \mathbf{z}^T \boldsymbol{\beta}$, the associated estimating equations are called the *normal equations* and are given by $\mathbf{Z}_D^T \mathbf{Y} = \mathbf{Z}_D^T \mathbf{Z}_D \boldsymbol{\beta}$, where $\mathbf{Z}_D = \|z_{ij}\|_{n \times d}$ is called the *design matrix*.

Suppose $X \sim P$. The *plug-in estimate* (PIE) for a vector parameter $\nu = \nu(P)$ is obtained by setting $\widehat{\nu} = \nu(\widehat{P})$ where $\widehat{P}$ is an estimate of P. When $\widehat{P}$ is the *empirical probability distribution* $\widehat{P}_E$ defined by $\widehat{P}_E(A) = n^{-1} \sum_{i=1}^{n} 1[X_i \in A]$, then $\widehat{\nu}$ is called the *empirical* PIE. If $P = P_{\boldsymbol{\theta}}, \theta \in \Theta$, is parametric and a vector $q(\boldsymbol{\theta})$ is to be estimated, we find a parameter ν such that $\nu(P_\theta) = q(\boldsymbol{\theta})$ and call $\nu(\widehat{P})$ a plug-in estimator of $q(\boldsymbol{\theta})$. Method of moment estimates are empirical PIEs based on $\nu(P) = (\mu_1, \ldots, \mu_d)^T$ where $\mu_j = E(X^j), 1 \leq j \leq d$. In the multinomial case the *frequency plug-in* estimators are empirical PIEs based on $\nu(P) = (p_1, \ldots, p_k)$, where p_j is the probability of the jth category, $1 \leq j \leq k$.

Let $\mathcal{P}_0$ and $\mathcal{P}$ be two statistical models for X with $\mathcal{P}_0 \subset \mathcal{P}$. An *extension* $\bar{\nu}$ of ν from $\mathcal{P}_0$ to $\mathcal{P}$ is a parameter satisfying $\bar{\nu}(P) = \nu(P), P \in \mathcal{P}_0$. If $\widehat{P}$ is an estimate of P with $\widehat{P} \in \mathcal{P}, \bar{\nu}(\widehat{P})$ is called the *extension plug-in estimate* of $\nu(P)$. The general principles are shown to be related to each other.

2.2   MINIMUM CONTRAST ESTIMATES AND ESTIMATING EQUATIONS

2.2.1   Least Squares and Weighted Least Squares

Least squares[1] was advanced early in the nineteenth century by Gauss and Legendre for estimation in problems of astronomical measurement. It is of great importance in many areas of statistics such as the analysis of variance and regression theory. In this section we shall introduce the approach and give a few examples leaving detailed development to Chapter 6.

In Example 2.1.1 we considered the nonlinear (and linear) Gaussian model $\mathcal{P}_0$ given by

$$Y_i = g(\boldsymbol{\beta}, \mathbf{z}_i) + \epsilon_i, \ 1 \le i \le n \tag{2.2.1}$$

where ϵ_i are i.i.d. $\mathcal{N}(0, \sigma_0^2)$ and $\boldsymbol{\beta}$ ranges over R^d or an open subset. The contrast

$$\rho(X, \boldsymbol{\beta}) = \sum_{i=1}^{n} [Y_i - g(\boldsymbol{\beta}, \mathbf{z}_i)]^2$$

led to the least squares estimates (LSEs) $\widehat{\boldsymbol{\beta}}$ of $\boldsymbol{\beta}$. Suppose that we enlarge $\mathcal{P}_0$ to $\mathcal{P}$ where we retain the independence of the Y_i but only require

$$\mu(\mathbf{z}_i) = E(Y_i) = g(\boldsymbol{\beta}, \mathbf{z}_i), \ E(\epsilon_i) = 0, \ \mathrm{Var}(\epsilon_i) < \infty. \tag{2.2.2}$$

Are least squares estimates still reasonable? For P in the semiparametric model $\mathcal{P}$, which satisfies (but is not fully specified by) (2.2.2), we can compute still

$$
\begin{aligned}
D_P(\boldsymbol{\beta}_0, \boldsymbol{\beta}) &= E_P \rho(X, \boldsymbol{\beta}) \\
&= \sum_{i=1}^{n} \mathrm{Var}_P(\epsilon_i) + \sum_{i=1}^{n} [g(\boldsymbol{\beta}_0, \mathbf{z}_i) - g(\boldsymbol{\beta}, \mathbf{z}_i)]^2,
\end{aligned}
\tag{2.2.3}
$$

which is again minimized as a function of $\boldsymbol{\beta}$ by $\boldsymbol{\beta} = \boldsymbol{\beta}_0$ and uniquely so if the map $\boldsymbol{\beta} \to (g(\boldsymbol{\beta}, \mathbf{z}_1), \dots, g(\boldsymbol{\beta}, \mathbf{z}_n))^T$ is $1 - 1$.

The estimates continue to be reasonable under the *Gauss–Markov* assumptions,

$$E(\epsilon_i) = 0, \ 1 \le i \le n, \tag{2.2.4}$$

$$\mathrm{Var}(\epsilon_i) = \sigma^2 > 0, \ 1 \le i \le n, \tag{2.2.5}$$

$$\mathrm{Cov}(\epsilon_i, \epsilon_j) = 0, \ 1 \le i < j \le n \tag{2.2.6}$$

because (2.2.3) continues to be valid.

Note that the joint distribution H of $(\epsilon_1, \dots, \epsilon_n)$ is any distribution satisfying the Gauss–Markov assumptions. That is, the model is semiparametric with $\boldsymbol{\beta}, \sigma^2$ and H unknown. The least squares method of estimation applies only to the parameters $\beta_1, \dots, \beta_d$ and is often applied in situations in which specification of the model beyond (2.2.4)–(2.2.6) is difficult.

Sometimes $\mathbf{z}$ can be viewed as the realization of a population variable $\mathbf{Z}$, that is, $(\mathbf{Z}_i, Y_i)$, $1 \le i \le n$, is modeled as a sample from a joint distribution. This is frequently the case for studies in the social and biological sciences. For instance, Z could be educational level and Y income, or Z could be height and Y log weight. Then we can write the conditional model of Y given $\mathbf{Z}_j = \mathbf{z}_j$, $1 \le j \le n$, as in (a) of Example 1.1.4 with ϵ_j simply defined as $Y_j - E(Y_j \mid \mathbf{Z}_j = \mathbf{z}_j)$. If we consider this model, $(\mathbf{Z}_i, Y_i)$ i.i.d. as $(\mathbf{Z}, Y) \sim P \in \mathcal{P} = \{$All joint distributions of $(\mathbf{Z}, Y)$ such that $E(Y \mid \mathbf{Z} = \mathbf{z}) = g(\boldsymbol{\beta}, \mathbf{z})$, $\boldsymbol{\beta} \in R^d\}$, and $\boldsymbol{\beta} \to (g(\boldsymbol{\beta}, \mathbf{z}_1), \dots, g(\boldsymbol{\beta}, \mathbf{z}_n))^T$ is $1 - 1$, then $\boldsymbol{\beta}$ has an interpretation as a parameter on $\mathcal{P}$, that is, $\boldsymbol{\beta} = \boldsymbol{\beta}(P)$ is the minimizer of $E(Y - g(\boldsymbol{\beta}, \mathbf{Z}))^2$. This follows from Theorem 1.4.1. In this case we recognize the LSE $\widehat{\beta}$ as simply being the usual plug-in estimate $\beta(\widehat{P})$, where $\widehat{P}$ is the empirical distribution assigning mass n^{-1} to each of the n pairs $(\mathbf{Z}_i, Y_i)$.

As we noted in Example 2.1.1 the most commonly used g in these models is $g(\boldsymbol{\beta}, \mathbf{z}) = \mathbf{z}^T \boldsymbol{\beta}$, which, in conjunction with (2.2.1), (2.2.4), (2.2.5) and (2.2.6), is called the *linear* (multiple) *regression model.* For the data $\{(\mathbf{z}_i, Y_i); \; i = 1, \ldots, n\}$ we write this model in matrix form as

$$\mathbf{Y} = \mathbf{Z}_D \boldsymbol{\beta} + \boldsymbol{\epsilon}$$

where $\mathbf{Z}_D = \|\mathbf{z}_{ij}\|$ is the design matrix. We continue our discussion for this important special case for which explicit formulae and theory have been derived. For nonlinear cases we can use numerical methods to solve the estimating equations (2.1.7). See also Problem 2.2.41, Sections 6.4.3 and 6.5, and Seber and Wild (1989). The linear model is often the default model for a number of reasons:

(1) If the range of the $\mathbf{z}$'s is relatively small and $\mu(\mathbf{z})$ is smooth, we can approximate $\mu(\mathbf{z})$ by

$$\mu(\mathbf{z}) = \mu(\mathbf{z}_0) + \sum_{j=1}^{d} \frac{\partial \mu}{\partial z_j}(\mathbf{z}_0)(z_j - z_{0j}),$$

for $\mathbf{z}_0 = (z_{01}, \ldots, z_{0d})^T$ an interior point of the domain. We can then treat $\mu(\mathbf{z}_0) - \sum_{j=1}^{d} \frac{\partial \mu}{\partial z_j}(\mathbf{z}_0) z_{0j}$ as an unknown β_0 and identify $\frac{\partial \mu}{\partial z_j}(\mathbf{z}_0)$ with β_j to give an approximate $(d+1)$-dimensional linear model with $z_0 \equiv 1$ and z_j as before and

$$\mu(\mathbf{z}) = \sum_{j=0}^{d} \beta_j z_j.$$

This type of approximation is the basis for nonlinear regression analysis based on local polynomials, see Ruppert and Wand (1994), Fan and Gijbels (1996), and Volume II.

(2) If as we discussed earlier, we are in a situation in which it is plausible to assume that $(\mathbf{Z}_i, Y_i)$ are a sample from a $(d+1)$-dimensional distribution and the covariates that are the coordinates of $\mathbf{Z}$ are continuous, a further modeling step is often taken and it is assumed that $(Z_1, \ldots, Z_d, Y)^T$ has a nondegenerate multivariate Gaussian distribution $\mathcal{N}_{d+1}(\boldsymbol{\mu}, \Sigma)$. In that case, as we have seen in Section 1.4, $E(Y \mid \mathbf{Z} = \mathbf{z})$ can be written as

$$\mu(\mathbf{z}) = \beta_0 + \sum_{j=1}^{d} \beta_j z_j \tag{2.2.7}$$

where

$$(\beta_1, \ldots, \beta_d)^T = \Sigma_{\mathbf{ZZ}}^{-1} \Sigma_{\mathbf{Z}Y} \tag{2.2.8}$$

$$\beta_0 = (-(\beta_1, \ldots, \beta_d), 1)\boldsymbol{\mu} = \mu_Y - \sum_{j=1}^{d} \beta_j \mu_j. \tag{2.2.9}$$

Furthermore,

$$\epsilon \equiv Y - \mu(\mathbf{Z})$$

is independent of $\mathbf{Z}$ and has a $\mathcal{N}(0, \sigma^2)$ distribution where

$$\sigma^2 = \sigma_{YY} - \Sigma_{Y\mathbf{Z}}\Sigma_{\mathbf{Z}\mathbf{Z}}^{-1}\Sigma_{\mathbf{Z}Y}.$$

Therefore, given $\mathbf{Z}_i = \mathbf{z}_i$, $1 \le i \le n$, we have a Gaussian linear regression model for Y_i, $1 \le i \le n$.

Estimation of β in the linear regression model. We have already argued in Example 2.1.1 that, if the parametrization $\beta \to \mathbf{Z}_D\beta$ is identifiable, the least squares estimate, $\widehat{\beta}$, exists and is unique and satisfies the normal equations (2.1.8). The parametrization is identifiable if and only if $\mathbf{Z}_D$ is of rank d or equivalently if $\mathbf{Z}_D^T\mathbf{Z}_D$ is of full rank d; see Problem 2.2.25. In that case, necessarily, the solution of the normal equations can be given "explicitly" by

$$\widehat{\beta} = [\mathbf{Z}_D^T\mathbf{Z}_D]^{-1}\mathbf{Z}_D^T\mathbf{Y}. \qquad (2.2.10)$$

Here are some examples.

Example 2.2.1. In the measurement model in which Y_i is the determination of a constant β_1, $d = 1$, $g(z, \beta_1) = \beta_1$, $\frac{\partial}{\partial \beta_1}g(z, \beta_1) = 1$ and the normal equation is $\sum_{i=1}^{n}(y_i - \beta_1) = 0$, whose solution is $\widehat{\beta}_1 = (1/n)\sum_{i=1}^{n} y_i = \bar{y}$. $\qquad \square$

Example 2.2.2. We want to find out how increasing the amount z of a certain chemical or fertilizer in the soil increases the amount y of that chemical in the plants grown in that soil. For certain chemicals and plants, the relationship between z and y can be approximated well by a linear equation $y = \beta_1 + \beta_2 z$ provided z is restricted to a reasonably small interval. If we run several experiments with the same z using plants and soils that are as nearly identical as possible, we will find that the values of y will not be the same. For this reason, we assume that for a given z, Y is random with a distribution $P(y \mid z)$.

Following are the results of an experiment to which a regression model can be applied (Snedecor and Cochran, 1967, p. 139). Nine samples of soil were treated with different amounts z of phosphorus. Y is the amount of phosphorus found in corn plants grown for 38 days in the different samples of soil.

z_i	1	4	5	9	11	13	23	23	28
Y_i	64	71	54	81	76	93	77	95	109

The points (z_i, Y_i) and an estimate of the line $\beta_1 + \beta_2 z$ are plotted in Figure 2.2.1.

We want to estimate β_1 and β_2. The normal equations are

$$\sum_{i=1}^{n}(y_i - \beta_1 - \beta_2 z_i) = 0, \quad \sum_{i=1}^{n} z_i(y_i - \beta_1 - \beta_2 z_i) = 0. \qquad (2.2.11)$$

When the z_i's are not all equal, we get the solutions

$$\widehat{\beta}_2 = \frac{\sum_{i=1}^{n}(z_i - \bar{z})y_i}{\sum_{i=1}^{n}(z_i - \bar{z})^2} = \frac{\sum_{i=1}^{n}(z_i - \bar{z})(y_i - \bar{y})}{\sum_{i=1}^{n}(z_i - \bar{z})^2} \qquad (2.2.12)$$

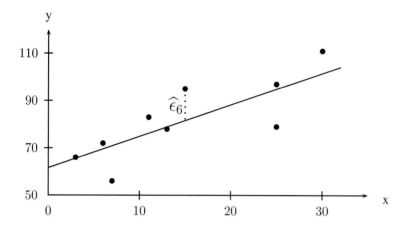

Figure 2.2.1. Scatter plot $\{(z_i, y_i);\ i = 1, \ldots, n\}$ and sample regression line for the phosphorus data. $\widehat{\epsilon}_6$ is the residual for (z_6, y_6).

and

$$\widehat{\beta}_1 = \bar{y} - \widehat{\beta}_2 \bar{z} \qquad\qquad (2.2.13)$$

where $\bar{z} = (1/n) \sum_{i=1}^{n} z_i$, and $\bar{y} = (1/n) \sum_{i=1}^{n} y_i$.

The line $y = \widehat{\beta}_1 + \widehat{\beta}_2 z$ is known as the *sample regression line* or *line of best fit* of $y_1, \ldots, y_n$ on $z_1, \ldots, z_n$. Geometrically, if we measure the distance between a point (z_i, y_i) and a line $y = a + bz$ vertically by $d_i = |y_i - (a + bz_i)|$, then the regression line minimizes the sum of the squared distances to the n points $(z_1, y_1), \ldots, (z_n, y_n)$. The vertical distances $\widehat{\epsilon}_i = [y_i - (\widehat{\beta}_1 + \widehat{\beta}_2 z_i)]$ are called the *residuals* of the fit, $i = 1, \ldots, n$. The line $y = \widehat{\beta}_1 + \widehat{\beta}_2 z$ is an estimate of the best linear MSPE predictor $a_1 + b_1 z$ of Theorem 1.4.3. This connection to prediction explains the use of vertical distances in regression. The regression line for the phosphorus data is given in Figure 2.2.1. Here $\widehat{\beta}_1 = 61.58$ and $\widehat{\beta}_2 = 1.42$. $\qquad\qquad\qquad\qquad\qquad\qquad\qquad\qquad\qquad\qquad\qquad\qquad\quad \square$

Remark 2.2.1. The linear regression model is considerably more general than appears at first sight. For instance, suppose we select p real-valued functions of $\mathbf{z}$, $g_1, \ldots, g_p, p \geq d$ and postulate that $\mu(\mathbf{z})$ is a linear combination of $g_1(\mathbf{z}), \ldots, g_p(\mathbf{z})$; that is

$$\mu(\mathbf{z}) = \sum_{j=1}^{p} \theta_j g_j(\mathbf{z}).$$

Then we are still dealing with a linear regression model because we can define $\mathbf{w}_{p \times 1} =$

$(g_1(\mathbf{z}), \ldots, g_p(\mathbf{z}))^T$ as our covariate and consider the linear model

$$Y_i = \sum_{j=1}^{p} \theta_j w_{ij} + \epsilon_i, \ 1 \leq i \leq n$$

where

$$w_{ij} \equiv g_j(\mathbf{z}_i).$$

For instance, if $d = 1$ and we take $g_j(z) = z^j, 0 \leq j \leq 2$, we arrive at quadratic regression, $Y_i = \theta_0 + \theta_1 z_i + \theta_2 z_i^2 + \epsilon_i$—see Problem 2.2.24 for more on *polynomial regression*.

Whether any linear model is appropriate in particular situations is a delicate matter partially explorable through further analysis of the data and knowledge of the subject matter. We return to this in Volume II.

Weighted least squares. In Example 2.2.2 and many similar situations it may not be reasonable to assume that the variances of the errors ϵ_i are the same for all levels z_i of the covariate variable. However, we may be able to characterize the dependence of $\mathrm{Var}(\epsilon_i)$ on z_i at least up to a multiplicative constant. That is, we can write

$$\mathrm{Var}(\epsilon_i) = w_i \sigma^2 \tag{2.2.14}$$

where σ^2 is unknown as before, but the w_i are known weights. Such models are called *heteroscedastic* (as opposed to the equal variance models that are *homoscedastic*). The method of least squares may not be appropriate because (2.2.5) fails. Note that the variables

$$\widetilde{Y}_i \equiv \frac{Y_i}{\sqrt{w_i}} = \frac{g(\boldsymbol{\beta}, \mathbf{z}_i)}{\sqrt{w_i}} + \frac{\epsilon_i}{\sqrt{w_i}}, \ 1 \leq i \leq n,$$

are sufficient for the Y_i and that $\mathrm{Var}(\epsilon_i/\sqrt{w_i}) = w_i \sigma^2/w_i = \sigma^2$. Thus, if we set $\widetilde{g}(\boldsymbol{\beta}, \mathbf{z}_i) = g(\boldsymbol{\beta}, \mathbf{z}_i)/\sqrt{w_i}, \widetilde{\epsilon}_i = \epsilon_i/\sqrt{w_i}$, then

$$\widetilde{Y}_i = \widetilde{g}(\boldsymbol{\beta}, \mathbf{z}_i) + \widetilde{\epsilon}_i, \ 1 \leq i \leq n \tag{2.2.15}$$

and the $\widetilde{Y}_i$ satisfy the assumption (2.2.5). The *weighted least squares estimate* of $\boldsymbol{\beta}$ is now the value $\widehat{\boldsymbol{\beta}}$, which for given $\widetilde{y}_i = y_i/\sqrt{w_i}$ minimizes

$$\sum_{i=1}^{n} [\widetilde{y}_i - \widetilde{g}(\boldsymbol{\beta}, \mathbf{z}_i)]^2 = \sum_{i=1}^{n} \frac{1}{w_i} [y_i - g(\boldsymbol{\beta}, \mathbf{z}_i)]^2 \tag{2.2.16}$$

as a function of $\boldsymbol{\beta}$.

Example 2.2.3. *Weighted Linear Regression.* Consider the case in which $d = 2$, $z_{i1} = 1$, $z_{i2} = z_i$, and $g(\boldsymbol{\beta}, \mathbf{z}_i) = \beta_1 + \beta_2 z_i$, $i = 1, \ldots, n$. We need to find the values $\widehat{\beta}_1$ and $\widehat{\beta}_2$ of β_1 and β_2 that minimize

$$\sum_{i=1}^{n} v_i [y_i - (\beta_1 + \beta_2 z_i)]^2 \tag{2.2.17}$$

where $v_i = 1/w_i$. This problem may be solved by setting up analogues to the normal equations (2.1.8). We can also use the results on prediction in Section 1.4 as follows. Let (Z^*, Y^*) denote a pair of discrete random variables with possible values $(z_1, y_1), \ldots, (z_n, y_n)$ and probability distribution given by

$$P[(Z^*, Y^*) = (z_i, y_i)] = u_i, \ i = 1, \ldots, n$$

where

$$u_i = v_i / \sum_{i=1}^{n} v_i, \ i = 1, \ldots, n.$$

If $\mu_l(Z^*) = \beta_1 + \beta_2 Z^*$ denotes a linear predictor of Y^* based on Z^*, then its MSPE is given by

$$E[Y^* - \mu_l(Z^*)]^2 = \sum_{i=1}^{n} u_i[y_i - (\beta_1 + \beta_2 z_i)]^2.$$

It follows that the problem of minimizing (2.2.17) is equivalent to finding the best linear MSPE predictor of Y^*. Thus, using Theorem 1.4.3,

$$\widehat{\beta}_2 = \frac{\text{Cov}(Z^*, Y^*)}{\text{Var}(Z^*)} = \frac{\sum_{i=1}^{n} u_i z_i y_i - (\sum_{i=1}^{n} u_i y_i)(\sum_{i=1}^{n} u_i z_i)}{\sum_{i=1}^{n} u_i z_i^2 - (\sum_{i=1}^{n} u_i z_i)^2} \quad (2.2.18)$$

and

$$\widehat{\beta}_1 = E(Y^*) - \widehat{\beta}_2 E(Z^*) = \sum_{i=1}^{n} u_i y_i - \widehat{\beta}_2 \sum_{i=1}^{n} u_i z_i.$$

This computation suggests, as we make precise in Problem 2.2.26, that weighted least squares estimates are also plug-in estimates. □

Next consider finding the $\widehat{\beta}$ that minimizes (2.2.16) for $g(\beta, z_i) = z_i^T \beta$ and for general d. By following the steps of Example 2.1.1 leading to (2.1.7), we find (Problem 2.2.27) that $\widehat{\beta}$ satisfy the *weighted least squares normal equations*

$$Z_D^T W^{-1} Y = (Z_D^T W^{-1} Z_D)\widehat{\beta} \quad (2.2.19)$$

where $W = \text{diag}(w_1, \ldots, w_n)$ and $Z_D = \|z_{ij}\|_{n \times d}$ is the design matrix. When Z_D has rank d and $w_i > 0$, $1 \leq i \leq n$, we can write

$$\widehat{\beta} = (Z_D^T W^{-1} Z_D)^{-1} Z_D^T W^{-1} Y. \quad (2.2.20)$$

Remark 2.2.2. More generally, we may allow for correlation between the errors $\{\epsilon_i\}$. That is, suppose $\text{Var}(\epsilon) = \sigma^2 W$ for some invertible matrix $W_{n \times n}$. Then it can be shown (Problem 2.2.28) that the model $Y = Z_D \beta + \epsilon$ can be transformed to one satisfying (2.2.1) and (2.2.4)–(2.2.6). Moreover, when $g(\beta, z) = z^T \beta$, the $\widehat{\beta}$ minimizing the least squares contrast in this transformed model is given by (2.2.19) and (2.2.20).

Remark 2.2.3. Here are some applications of weighted least squares: When the ith response Y_i is an average of n_i equally variable observations, then $\text{Var}(Y_i) = \sigma^2/n_i$, and $w_i = n_i^{-1}$. If Y_i is the sum of n_i equally variable observations, then $w_i = n_i$. If the variance of Y_i is proportional to some covariate, say z_1, then $\text{Var}(Y_i) = z_{i1}\sigma^2$ and $w_i = z_{i1}$. In time series and repeated measures, a covariance structure is often specified for ϵ (see Problems 2.2.29 and 2.2.42).

2.2.2 Maximum Likelihood

The method of maximum likelihood was first proposed by the German mathematician C. F. Gauss in 1821. However, the approach is usually credited to the English statistician R. A. Fisher (1922) who rediscovered the idea and first investigated the properties of the method. In the form we shall give, this approach *makes sense only in regular parametric models.* Suppose that $p(\mathbf{x}, \theta)$ is the frequency or density function of $\mathbf{X}$ if θ is true and that Θ is a subset of d-dimensional space.

Recall $L_\mathbf{x}(\theta)$, the likelihood function of θ, defined in Section 1.5, which is just $p(\mathbf{x}, \theta)$ considered as a function of θ for fixed $\mathbf{x}$. Thus, if $\mathbf{X}$ is discrete, then for each θ, $L_\mathbf{x}(\theta)$ gives the probability of observing $\mathbf{x}$. If Θ is finite and π is the uniform prior distribution on Θ, then the posterior probability that $\boldsymbol{\theta} = \theta$ given $\mathbf{X} = \mathbf{x}$ satisfies $\pi(\theta \mid \mathbf{x}) \propto L_\mathbf{x}(\theta)$, where the proportionality is up to a function of $\mathbf{x}$. Thus, we can think of $L_\mathbf{x}(\theta)$ as a measure of how "likely" θ is to have produced the observed $\mathbf{x}$. A similar interpretation applies to the continuous case (see A.7.10).

The *method* of *maximum likelihood* consists of finding that value $\widehat{\theta}(\mathbf{x})$ of the parameter that is "most likely" to have produced the data. That is, if $\mathbf{X} = \mathbf{x}$, we seek $\widehat{\theta}(\mathbf{x})$ that satisfies

$$L_\mathbf{x}(\widehat{\theta}(\mathbf{x})) = p(\mathbf{x}, \widehat{\theta}(\mathbf{x})) = \max\{p(\mathbf{x}, \theta) : \theta \in \Theta\} = \max\{L_\mathbf{x}(\theta) : \theta \in \Theta\}.$$

By our previous remarks, if Θ is finite and π is uniform, or, more generally, the prior density π on Θ is constant, such a $\widehat{\theta}(\mathbf{x})$ is a mode of the posterior distribution. If such a $\widehat{\theta}$ exists, we estimate any function $q(\theta)$ by $q(\widehat{\theta}(\mathbf{x}))$. The estimate $q(\widehat{\theta}(\mathbf{x}))$ is called the *maximum likelihood estimate* (MLE) of $q(\theta)$. This definition of $q(\widehat{\theta})$ is consistent. That is, suppose $\mathbf{q}$ is 1-1 from Θ to Ω; set $\omega = q(\theta)$ and write the density of $\mathbf{X}$ as $p_0(\mathbf{x}, \omega) = p(\mathbf{x}, q^{-1}(\omega))$. If $\widehat{\omega}$ maximizes $p_0(\mathbf{x}, \omega)$ then $\widehat{\omega} = q(\widehat{\theta})$ (Problem 2.2.16(a)). If q is not 1-1, the MLE of $\omega = q(\theta)$ is still $q(\widehat{\theta})$ (Problem 2.2.16(b)).

Here is a simple numerical example.

Suppose $\theta = 0$ or $\frac{1}{2}$ and $p(x, \theta)$ is given by the following table.

$x \backslash \theta$	0	$\frac{1}{2}$
1	0	0.10
2	1	0.90

Then $\widehat{\theta}(1) = \frac{1}{2}$, $\widehat{\theta}(2) = 0$. If $X = 1$ the only reasonable estimate of θ is $\frac{1}{2}$ because the value 1 could not have been produced when $\theta = 0$.

Maximum likelihood estimates need neither exist nor be unique (Problems 2.2.14 and 2.2.13). In the rest of this section we identify them as of minimum contrast and estimating equation type, relate them to the plug-in and extension principles and some notions in information theory, and compute them in some important special cases in which they exist, are unique, and expressible in closed form. In the rest of this chapter we study more detailed conditions for existence and uniqueness and algorithms for calculation of MLEs when closed forms are not available.

When θ is real, MLEs can often be obtained by inspection as we see in a pair of important examples.

Example 2.2.4. *The Normal Distribution with Known Variance.* Suppose $X \sim \mathcal{N}(\theta, \sigma^2)$, where σ^2 is known, and let φ denote the standard normal density. Then the likelihood function

$$L_x(\theta) = \frac{1}{\sigma} \varphi \left(\frac{\theta - x}{\sigma} \right)$$

is a normal density with mean x and variance σ^2. The maximum is, therefore, achieved uniquely for

$$\widehat{\theta}(x) = x.$$

Suppose more generally that $X_1, \ldots, X_n$ is a sample from a $\mathcal{N}(\theta, \sigma^2)$ population. It is a consequence of Problem 2.2.15 that the MLE of θ based on $X_1, \ldots, X_n$ is the same as that based on the sufficient statistic $\bar{X}$, which has a $\mathcal{N}(\theta, \sigma^2/n)$ distribution. In view of our result for $n = 1$ we can conclude that

$$\widehat{\theta}(X_1, \ldots, X_n) = \bar{X}$$

is the MLE of θ.   □

Example 2.2.5.   *Estimating the Size of a Population (continued).*   Suppose $X_1, \ldots, X_n$ are i.i.d. $\mathcal{U}\{1, 2, \ldots, \theta\}$ with θ an integer ≥ 1. We have seen in Example 2.1.7 that the method of moments leads to the unreasonable estimate $2\bar{X} - 1$ for the size θ of the population. What is the maximum likelihood estimate of θ? From (1.5.10) we see that $L_{\mathbf{x}}(\theta)$ is 0 for $\theta = 1, \ldots, \max(x_1, \ldots, x_n) - 1$, then jumps to $[\max(x_1, \ldots, x_n)]^{-n}$ and equals the monotone decreasing function θ^{-n} from then on. Figure 2.2.2 illustrates the situation. Clearly, $\max(X_1, \ldots, X_n)$ is the MLE of θ.   □

Maximum likelihood as a minimum contrast and estimating equation method. Define

$$l_x(\theta) = \log L_x(\theta) = \log p(x, \theta).$$

By definition the MLE $\widehat{\theta}(X)$ if it exists minimizes $-\log p$ because $-l_x(\theta)$ is a strictly decreasing function of p. Log p turns out to be the best monotone function of p to consider for many reasons. A prototypical one is that if the X_i are independent with densities or frequency function $f(x, \theta)$ for $i = 1, \ldots, n$, then, with $\mathbf{X} = (X_1, \ldots, X_n)$

$$l_{\mathbf{X}}(\theta) = \log p(\mathbf{X}, \theta) = \log \prod_{i=1}^{n} f(X_i, \theta) = \sum_{i=1}^{n} \log f(X_i, \theta) \qquad (2.2.21)$$

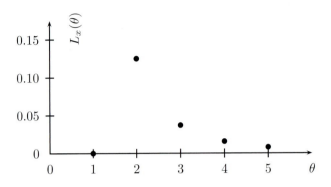

Figure 2.2.2. The likelihood function for Example 2.2.5. Here $n = 3$ and
$$\mathbf{x} = (x_1, x_2, x_3) = (1, 2, 1).$$

which when considered as a random quantity is a sum of independent variables, the most
studied object of probability theory.

Another reason for turning to $\log p$ is that we can, by making a connection to informa-
tion theory, identify $\rho(x, \theta) \equiv -l_x(\theta)$ as a contrast and $\widehat{\theta}$ as a minimum contrast estimate.
In this case,

$$D(\theta_0, \theta) = -E_{\theta_0} \log p(X, \theta) \tag{2.2.22}$$

and to say that $D(\theta_0, \theta)$ is minimized uniquely when $P_\theta = P_{\theta_0}$ is equivalent to

$$
\begin{aligned}
D(\theta_0, \theta) - D(\theta_0, \theta_0) &= -(E_{\theta_0} \log p(X, \theta) - E_{\theta_0} \log p(X, \theta_0)) \\
&= -E_{\theta_0} \log \frac{p(X, \theta)}{p(X, \theta_0)} > 0
\end{aligned}
\tag{2.2.23}
$$

unless $\theta = \theta_0$. Here $D(\theta_0, \theta_0)$ is called the *entropy* of X. See Problem 1.6.28.

Let X have frequency or density function p_0 or p_1 and let E_0 denote expectation
with respect to p_0. Define the *mutual entropy* or *Kullback–Leibler information divergence*
$K(p_0, p_1)$ between p_0 and p_1 by

$$K(p_0, p_1) = -E_0 \log \frac{p_1}{p_0}(X) = -\sum_x p_0(x) \log \frac{p_1}{p_0}(x) \tag{2.2.24}$$

for X discrete and replacing sums by integrals for X continuous. By convention $\frac{0}{0} = 0$
and $0 \times \infty = 0$ so that if $p_0(x) = 0$, $p_0(x) \log(p_1(x)/p_0(x)) = 0$ and if $p_1(x) = 0$ also,
then $p_1(x)/p_0(x) = 0$. Then (2.2.23) is equivalent to

Lemma 2.2.1 (Shannon, 1948)[2] *The mutual entropy $K(p_0, p_1)$ is always well defined and
$K(p_0, p_1) \geq 0$ with equality if and only if $\{x : p_0(x) = p_1(x)\}$ has probability 1 under
both P_0 and P_1.*

Proof. We give the proof for the discrete case. Kolmogorov (1956) gave the proof in the general case. Recall Jensen's inequality (B.9.3): If $g : R \to R$ is strictly convex and Z is any random variable such that $E(Z)$ is well defined, then $Eg(Z)$ is well defined and $Eg(Z) \ge g(E(Z))$ with = iff $P[Z = c_0] = 1$ where c_0 is a constant. Given X distributed according to P_0, let $Z = p_1(X)/p_0(X)$ and $g(z) = -\log z$. Because $g''(z) = z^{-2} > 0$, g is strictly convex. Thus, with our convention about 0 and ∞,

$$
\begin{aligned}
Eg(Z) &= \sum_x p_0(x) \left(-\log \frac{p_1}{p_0}(x) \right) \ge g(E(Z)) \\
&= -\log \left(\sum_x \frac{p_1}{p_0}(x) p_0(x) \right).
\end{aligned}
\tag{2.2.25}
$$

Because $\frac{p_1}{p_0}(x) p_0(x) \le p_1(x)$ we must have $g(E(Z)) \ge -\log \sum_x p_1(x) = 0$. Equality holds iff $p_1(x) > 0$ implies $p_0(x) > 0$ and $p_1(x)/p_0(x) = c$ if $p_0(x) > 0$. Then $1 = \sum_x p_1(x) = c \sum_x p_0(x) = c$, and we conclude that $p_0(x) = p_1(x)$ for all x.   □

Lemma 2.2.1 shows that in the case of $X_1, \ldots, X_n$ i.i.d.

$$
\rho(X, \theta) = -\frac{1}{n} \sum \log p(X_i, \theta)
$$

satisfies the condition of being a contrast function, and we have shown that the MLE is a minimum contrast estimate.

Next let $\mathcal{P}_0 = \{P_\theta : \theta \in \Theta\}$ and define $\nu : \mathcal{P}_0 \to \Theta$ by

$$
\nu(P_{\theta_0}) = \arg \min\{-E_{\theta_0} \log p(X, \theta) : \theta \in \Theta\}.
$$

The extension $\bar{\nu}$ of ν to $\mathcal{P} =$ all probabilities on $\mathcal{X}$ is

$$
\bar{\nu}(P) = \arg \min\{-E_P \log p(X, \theta) : \theta \in \Theta\}.
$$

Now the MLE is $\bar{\nu}(\widehat{P})$. That is, the MLE is the value of θ that minimizes the Kullback–Leibler divergence between the empirical probability $\widehat{P}$ and P_θ.

Likelihood equations

If Θ is open, $l_X(\theta)$ is differentiable in θ and $\widehat{\theta}$ exists then $\widehat{\theta}$ must satisfy the estimating equation

$$
\nabla_\theta l_X(\theta) = 0.
\tag{2.2.26}
$$

This is known as the *likelihood equation.* If the X_i are independent with densities $f_i(x, \theta)$ the likelihood equation simplifies to

$$
\sum_{i=1}^n \nabla_\theta \log f_i(X_i, \widehat{\theta}) = 0,
\tag{2.2.27}
$$

which again enables us to analyze the behavior of $\widehat{\theta}$ using known properties of sums of independent random variables. Evidently, there may be solutions of (2.2.27) that are not maxima or only local maxima, and as we have seen in Example 2.2.5, situations with θ well defined but (2.2.27) doesn't make sense. Nevertheless, the dual point of view of (2.2.22) and (2.2.27) is very important and we shall explore it extensively in the natural and favorable setting of multiparameter exponential families in the next section.

Here are two simple examples with θ real.

Example 2.2.6. Consider a population with three kinds of individuals labeled 1, 2, and 3 and occurring in the Hardy–Weinberg proportions

$$p(1,\theta) = \theta^2, \; p(2,\theta) = 2\theta(1-\theta), \; p(3,\theta) = (1-\theta)^2$$

where $0 < \theta < 1$ (see Example 2.1.4). If we observe a sample of three individuals and obtain $x_1 = 1$, $x_2 = 2$, $x_3 = 1$, then

$$L_{\mathbf{x}}(\theta) = p(1,\theta)p(2,\theta)p(1,\theta) = 2\theta^5(1-\theta).$$

The likelihood equation is

$$\frac{\partial}{\partial\theta} l_{\mathbf{x}}(\theta) = \frac{5}{\theta} - \frac{1}{1-\theta} = 0,$$

which has the unique solution $\widehat{\theta} = \frac{5}{6}$. Because

$$\frac{\partial^2}{\partial\theta^2} l_{\mathbf{x}}(\theta) = -\frac{5}{\theta^2} - \frac{1}{(1-\theta)^2} < 0$$

for all $\theta \in (0,1)$, $\frac{5}{6}$ maximizes $L_{\mathbf{x}}(\theta)$. In general, let n_1, n_2, and n_3 denote the number of $\{x_1, \ldots, x_n\}$ equal to 1, 2 and 3, respectively. Then the same calculation shows that if $2n_1 + n_2$ and $n_2 + 2n_3$ are both positive, the maximum likelihood estimate exists and is given by

$$\widehat{\theta}(\mathbf{x}) = \frac{2n_1 + n_2}{2n}. \tag{2.2.28}$$

If $2n_1 + n_2$ is zero, the likelihood is $(1-\theta)^{2n}$, which is maximized by $\theta = 0$, so the MLE does not exist because $\Theta = (0,1)$. Similarly, the MLE does not exist if $n_2 + 2n_3 = 0$. $\quad\square$

Example 2.2.7. Let X denote the number of customers arriving at a service counter during n hours. If we make the usual simplifying assumption that the arrivals form a Poisson process, then X has a Poisson distribution with parameter $n\lambda$, where λ represents the expected number of arrivals in an hour or, equivalently, the rate of arrival. In practice, λ is an unknown positive constant and we wish to estimate λ using X. Here X takes on values $\{0, 1, 2, \ldots\}$ with probabilities,

$$p(x,\lambda) = \frac{e^{-\lambda n}(\lambda n)^x}{x!}, \; x = 0, 1, \ldots \tag{2.2.29}$$

The likelihood equation is

$$\frac{\partial}{\partial \lambda} l_x(\lambda) = -n + \frac{x}{\lambda} = 0,$$

which has the unique solution $\widehat{\lambda} = x/n$. If x is positive, this estimate is the MLE of λ (see (2.2.30)). If $x = 0$ the MLE does not exist. However, the maximum is approached as $\lambda \downarrow 0$.   □

To apply the likelihood equation successfully we need to know when a solution is an MLE. A sufficient condition, familiar from calculus, is that l be *concave* in θ. If l is twice differentiable, this is well known to be equivalent to

$$\frac{\partial^2}{\partial \theta^2} l_x(\theta) \leq 0, \tag{2.2.30}$$

for *all* θ. This is the condition we applied in Example 2.2.6. A similar condition applies for vector parameters.

Example 2.2.8. *Multinomial Trials.* As in Example 1.6.7, consider an experiment with n i.i.d. trials in which each trial can produce a result in one of k categories. Let $X_i = j$ if the ith trial produces a result in the jth category, let $\theta_j = P(X_i = j)$ be the probability of the jth category, and let $N_j = \sum_{i=1}^{n} 1[X_i = j]$ be the number of observations in the jth category. We assume that $n \geq k - 1$. Then, for an experiment in which we observe $n_j = \sum_{i=1}^{n} 1[X_i = j]$, $p(\mathbf{x}, \boldsymbol{\theta}) = \prod_{j=1}^{k} \theta_j^{n_j}$, and

$$l_{\mathbf{x}}(\boldsymbol{\theta}) = \sum_{j=1}^{k} n_j \log \theta_j, \ \boldsymbol{\theta} \in \Theta = \{\boldsymbol{\theta} : \theta_j \geq 0, \sum_{j=1}^{k} \theta_j = 1\}. \tag{2.2.31}$$

To obtain the MLE $\widehat{\boldsymbol{\theta}}$ we consider l as a function of $\theta_1, \ldots, \theta_{k-1}$ with

$$\theta_k = 1 - \sum_{j=1}^{k-1} \theta_j. \tag{2.2.32}$$

We first consider the case with all the n_j positive. Then $p(\mathbf{x}, \boldsymbol{\theta}) = 0$ if any of the θ_j are zero; thus, the MLE must have all $\widehat{\theta}_j > 0$, and must satisfy the likelihood equations

$$\frac{\partial}{\partial \theta_j} l_{\mathbf{x}}(\boldsymbol{\theta}) = \frac{\partial}{\partial \theta_j} \sum_{l=1}^{k} n_l \log \theta_l = \sum_{l=1}^{k} \frac{n_l}{\theta_l} \frac{\partial \theta_l}{\partial \theta_j} = 0, \ j = 1, \ldots, k - 1.$$

By (2.2.32), $\partial \theta_k / \partial \theta_j = -1$, and the equation becomes $(\widehat{\theta}_k / \widehat{\theta}_j) = (n_k / n_j)$. Next use (2.2.32) to find

$$\widehat{\theta}_j = \frac{n_j}{n}, \ j = 1, \ldots, k.$$

To show that this $\widehat{\boldsymbol{\theta}}$ maximizes $l_{\mathbf{x}}(\boldsymbol{\theta})$, we check the concavity of $l_{\mathbf{x}}(\boldsymbol{\theta})$: let $1 \leq r \leq k - 1$, $1 \leq j \leq k - 1$, then

$$
\begin{aligned}
\frac{\partial}{\partial \theta_r} \frac{\partial}{\partial \theta_j} l_{\mathbf{x}}(\boldsymbol{\theta}) &= \frac{\partial}{\partial \theta_r} \left(\frac{n_j}{\theta_j} - \frac{n_k}{\theta_k} \right) \\
&= - \left(\frac{n_r}{\theta_r^2} + \frac{n_k}{\theta_k^2} \right) < 0, \ r = j \\
&= - \frac{n_k}{\theta_k^2} < 0, \ r \neq j.
\end{aligned}
\tag{2.2.33}
$$

It follows that in this $n_j > 0$, $\theta_j > 0$ case, $l_{\mathbf{x}}(\boldsymbol{\theta})$ is strictly concave and $\widehat{\boldsymbol{\theta}}$ is the unique maximizer of $l_{\mathbf{x}}(\boldsymbol{\theta})$. Next suppose that $n_j = 0$ for some j. Then $\widehat{\boldsymbol{\theta}}$ with $\widehat{\theta}_j = n_j/n$, $j = 1, \ldots, k$, is still the unique MLE of $\boldsymbol{\theta}$. See Problem 2.2.30. The $0 < \theta_j < 1$, $1 \leq j \leq k$, version of this example will be considered in the exponential family case in Section 2.3.

Example 2.2.9. Suppose that $X_1, \ldots, X_n$ are i.i.d. $\mathcal{N}(\mu, \sigma^2)$ with μ and σ^2 both unknown. Using the concavity argument, we find that for $n \geq 2$ the unique MLEs of μ and σ^2 are $\widehat{\mu} = \bar{X}$ and $\widehat{\sigma}^2 = n^{-1} \sum_{i=1}^n (X_i - \bar{X})^2$ (Problem 2.2.11(a)).

Maximum likelihood and least squares

We conclude with the link between least squares and maximum likelihood. Suppose the model $\mathcal{P}_0$ (see (2.2.1)) of Example 2.1.1 holds and $\mathbf{X} = (Y_1, \ldots, Y_n)^T$. Then

$$
\begin{aligned}
l_{\mathbf{X}}(\boldsymbol{\beta}) &= \log \prod_{i=1}^n \frac{1}{\sigma_0} \varphi \left(\frac{Y_i - g(\boldsymbol{\beta}, \mathbf{z}_i)}{\sigma_0} \right) \\
&= -\frac{n}{2} \log(2\pi\sigma_0^2) - \frac{1}{2\sigma_0^2} \sum_{i=1}^n [Y_i - g(\boldsymbol{\beta}, \mathbf{z}_i)]^2.
\end{aligned}
\tag{2.2.34}
$$

Evidently maximizing $l_{\mathbf{X}}(\boldsymbol{\beta})$ is equivalent to minimizing $\sum_{i=1}^n [Y_i - g(\boldsymbol{\beta}, \mathbf{z}_i)]^2$. Thus, least squares estimates are maximum likelihood for the particular model $\mathcal{P}_0$. As we have seen and shall see more in Section 6.6, these estimates viewed as an algorithm applied to the set of data $\mathbf{X}$ make sense much more generally. It is easy to see that weighted least squares estimates are also maximum likelihood estimates of $\boldsymbol{\beta}$ for the model Y_i independent $\mathcal{N}(g(\boldsymbol{\beta}, \mathbf{z}_i), w_i \sigma_0^2)$, $1 \leq i \leq n$. More generally, we can consider $\widehat{\boldsymbol{\beta}}$ minimizing $\sum_{i,j} [Y_i - g(\boldsymbol{\beta}, \mathbf{z}_i)][Y_j - g(\boldsymbol{\beta}, \mathbf{z}_j)] w_{ij}$ where $W = \|w_{ij}\|_{n \times n}$ is a symmetric positive definite matrix, as maximum likelihood estimates for $\boldsymbol{\beta}$ when $\mathbf{Y}$ is distributed as $\mathcal{N}_n((g(\boldsymbol{\beta}, \mathbf{z}_1), \ldots, g(\boldsymbol{\beta}, \mathbf{z}_n))^T, \sigma_0^2 W^{-1})$, see Problem 2.2.28.

Summary. In Section 2.2.1 we consider *least squares estimators* (LSEs) obtained by minimizing a contrast of the form $\sum_{i=1}^n [Y_i - g_i(\boldsymbol{\beta})]^2$, where $E(Y_i) = g_i(\boldsymbol{\beta})$, g_i, $i = 1, \ldots, n$, are known functions and $\boldsymbol{\beta}$ is a parameter to be estimated from the independent observations $Y_1, \ldots, Y_n$, where $\mathrm{Var}(Y_i)$ does not depend on i. This approach is applied to experiments in which for the ith case in a study the mean of the response Y_i depends on a set of available covariate values $z_{i1}, \ldots, z_{id}$. In particular we consider the case with $g_i(\boldsymbol{\beta}) = \sum_{j=1}^{d} z_{ij}\beta_j$ and give the LSE of $\boldsymbol{\beta}$ in the case in which $\|z_{ij}\|_{n \times d}$ is of rank d. Extensions to weighted least squares, which are appropriate when $\text{Var}(Y_i)$ depends on i or the Y's are correlated, are given. In Section 2.2.2 we consider *maximum likelihood estimators* (MLEs) $\widehat{\boldsymbol{\theta}}$ that are defined as maximizers of the likelihood $L_x(\theta) = p(x, \boldsymbol{\theta})$. These estimates are shown to be equivalent to minimum contrast estimates based on a contrast function related to Shannon entropy and Kullback–Leibler information divergence. In the case of independent response variables Y_i that are modeled to have a $\mathcal{N}(g_i(\boldsymbol{\beta}), \sigma^2)$ distribution, it is shown that the MLEs coincide with the LSEs.

2.3   MAXIMUM LIKELIHOOD IN MULTIPARAMETER EXPONENTIAL FAMILIES

Questions of existence and uniqueness of maximum likelihood estimates in canonical exponential families can be answered completely and elegantly. This is largely a consequence of the strict concavity of the log likelihood in the natural parameter η, though the results of Theorems 1.6.3 and 1.6.4 and Corollaries 1.6.1 and 1.6.2 and other exponential family properties also play a role. Concavity also plays a crucial role in the analysis of algorithms in the next section. Properties that derive solely from concavity are given in Propositon 2.3.1.

We start with a useful general framework and lemma. Suppose $\Theta \subset R^p$ is an open set. Let $\partial\Theta = \bar{\Theta} - \Theta$ be the boundary of Θ, where $\bar{\Theta}$ denotes the closure of Θ in $[-\infty, \infty]^p$. That is, $\partial\Theta$ is the set of points outside of Θ that can be obtained as limits of points in Θ, including all points with $\pm\infty$ as a coordinate. For instance, if $X \sim \mathcal{N}(\theta_1, \theta_2), \Theta = R \times R^+$ and

$$\partial\Theta = \{(a,b) : a = \pm\infty, 0 \leq b \leq \infty\} \cup \{(a,b) : a \in R, b \in \{0,\infty\}\}.$$

In general, for a sequence $\{\boldsymbol{\theta}_m\}$ of points from Θ open, we define $\boldsymbol{\theta}_m \to \partial\Theta$ as $m \to \infty$ to mean that for any subsequence $\{\boldsymbol{\theta}_{m_k}\}$ either $\boldsymbol{\theta}_{m_k} \to \mathbf{t}$ with $\mathbf{t} \notin \Theta$, or $\boldsymbol{\theta}_{m_k}$ diverges with $|\boldsymbol{\theta}_{m_k}| \to \infty$, as $k \to \infty$, where $| \ |$ denotes the Euclidean norm. For instance, in the $\mathcal{N}(\theta_1, \theta_2)$ case, $(a, m^{-1}), (m, b), (-m, b), (a, m), (m, m^{-1})$ all tend to $\partial\Theta$ as $m \to \infty$.

Lemma 2.3.1. *Suppose we are given a function* $l : \Theta \to R$ *where* $\Theta \subset R^p$ *is open and* l *is continuous. Suppose also that*

$$\lim\{l(\boldsymbol{\theta}) : \boldsymbol{\theta} \to \partial\Theta\} = -\infty. \tag{2.3.1}$$

Then there exists $\widehat{\boldsymbol{\theta}} \in \Theta$ *such that*

$$l(\widehat{\boldsymbol{\theta}}) = \max\{l(\boldsymbol{\theta}) : \boldsymbol{\theta} \in \Theta\}.$$

Proof. See Problem 2.3.5.

Existence and unicity of the MLE in exponential families depend on the strict concavity of the log likelihood and the condition of Lemma 2.3.1 only. Formally,

Proposition 2.3.1. *Suppose* $X \sim \{P_{\boldsymbol{\theta}} : \boldsymbol{\theta} \in \Theta\}$, Θ *open* $\subset R^p$, *with corresponding densities* $p(x, \boldsymbol{\theta})$. *If further* $l_x(\boldsymbol{\theta}) \equiv \log p(x, \boldsymbol{\theta})$ *is strictly concave and* $l_x(\boldsymbol{\theta}) \to -\infty$ *as* $\boldsymbol{\theta} \to \partial\Theta$, *then the MLE* $\widehat{\boldsymbol{\theta}}(x)$ *exists and is unique.*

Proof. From (B.9) we know that $\boldsymbol{\theta} \to l_x(\boldsymbol{\theta})$ is continuous on Θ. By Lemma 2.3.1, $\widehat{\boldsymbol{\theta}}(x)$ exists. If $\widehat{\boldsymbol{\theta}}_1$ and $\widehat{\boldsymbol{\theta}}_2$ are distinct maximizers, then $l_x\left(\frac{1}{2}(\widehat{\boldsymbol{\theta}}_1 + \widehat{\boldsymbol{\theta}}_2)\right) > \frac{1}{2}(l_x(\widehat{\boldsymbol{\theta}}_1) + l_x(\widehat{\boldsymbol{\theta}}_2)) = l_x(\widehat{\boldsymbol{\theta}}_1)$, a contradiction.

Applications of this theorem are given in Problems 2.3.8 and 2.3.12. $\qquad\qquad\square$

We can now prove the following.

Theorem 2.3.1. *Suppose* $\mathcal{P}$ *is the canonical exponential family generated by* $(\mathbf{T}, h)$ *and that*

(i) *The natural parameter space,* $\mathcal{E}$, *is open.*
(ii) *The family is of rank* k.
Let x *be the observed data vector and set* $\mathbf{t}_0 = \mathbf{T}(x)$.
(a) *If* $\mathbf{t}_0 \in R^k$ *satisfies*[1]

$$P[\mathbf{c}^T \mathbf{T}(X) > \mathbf{c}^T \mathbf{t}_0] > 0 \qquad \forall \mathbf{c} \neq \mathbf{0} \tag{2.3.2}$$

then the MLE $\widehat{\boldsymbol{\eta}}$ *exists, is unique, and is a solution to the equation*

$$\dot{A}(\boldsymbol{\eta}) = E_{\boldsymbol{\eta}}(\mathbf{T}(X)) = \mathbf{t}_0. \tag{2.3.3}$$

(b) *Conversely, if* $\mathbf{t}_0$ *doesn't satisfy (2.3.2), then the MLE doesn't exist and (2.3.3) has no solution.*

We, thus, have a necessary and sufficient condition for existence and uniqueness of the MLE given the data.

Define the *convex support* of a probability P to be the smallest convex set C such that $P(C) = 1$.

Corollary 2.3.1. *Suppose the conditions of Theorem 2.3.1 hold. If* $C_{\mathbf{T}}$ *is the convex support of the distribution of* $\mathbf{T}(X)$, *then* $\widehat{\boldsymbol{\eta}}$ *exists and is unique iff* $\mathbf{t}_0 \in C_{\mathbf{T}}^0$ *where* $C_{\mathbf{T}}^0$ *is the interior of* $C_{\mathbf{T}}$.

Proof of Theorem 2.3.1. We give the proof for the continuous case.

Existence and Uniqueness of the MLE $\widehat{\boldsymbol{\eta}}$. Without loss of generality we can suppose $h(x) = p(x, \boldsymbol{\eta}_0)$ for some reference $\boldsymbol{\eta}_0 \in \mathcal{E}$ (see Problem 1.6.27). Furthermore, we may also assume that $\mathbf{t}_0 = \mathbf{T}(x) = 0$ because $\mathcal{P}$ is the same as the exponential family generated by $\mathbf{T}(x) - \mathbf{t}_0$. Then, if $l_x(\boldsymbol{\eta}) \equiv \log p(x, \boldsymbol{\eta})$ with $\mathbf{T}(x) = 0$,

$$l_x(\boldsymbol{\eta}) = -A(\boldsymbol{\eta}) + \log h(x).$$

We show that if $\{\boldsymbol{\eta}_m\}$ has no subsequence converging to a point in $\mathcal{E}$, then $l_x(\boldsymbol{\eta}_m) \to -\infty$, which implies existence of $\widehat{\boldsymbol{\eta}}$ by Lemma 2.3.1. Write $\boldsymbol{\eta}_m = \lambda_m \mathbf{u}_m$, $\mathbf{u}_m = \frac{\boldsymbol{\eta}_m}{\|\boldsymbol{\eta}_m\|}$, $\lambda_m =$

$\|\boldsymbol{\eta}_m\|$. So, $\|\mathbf{u}_m\| = 1$. Then, if $\{\boldsymbol{\eta}_m\}$ has no subsequence converging in $\mathcal{E}$ it must have a subsequence $\{\boldsymbol{\eta}_{m_k}\}$ that obeys either case 1 or 2 as follows.

Case 1: $\lambda_{m_k} \to \infty$, $\mathbf{u}_{m_k} \to \mathbf{u}$. Write E_0 for $E_{\boldsymbol{\eta}_0}$ and P_0 for $P_{\boldsymbol{\eta}_0}$. Then

$$
\begin{aligned}
\underline{\lim}_k \int e^{\boldsymbol{\eta}_{m_k}^T \mathbf{T}(x)} h(x) dx &= \underline{\lim}_k E_0 e^{\lambda_{m_k} \mathbf{u}_{m_k}^T \mathbf{T}(X)} \\
&\geq \underline{\lim} e^{\lambda_{m_k} \delta} P_0[\mathbf{u}_{m_k}^T \mathbf{T}(X) > \delta] \\
&\geq \underline{\lim} e^{\lambda_{m_k} \delta} P_0[\mathbf{u}^T \mathbf{T}(X) > \delta] = \infty
\end{aligned}
$$

because for some $\delta > 0$, $P_0[\mathbf{u}^T \mathbf{T}(X) > \delta] > 0$. So we have

$$
A(\boldsymbol{\eta}_{m_k}) = \log \int e^{\boldsymbol{\eta}_{m_k}^T \mathbf{T}(x)} h(x) dx \to \infty \text{ and } l_x(\boldsymbol{\eta}_{m_k}) \to -\infty.
$$

Case 2: $\lambda_{m_k} \to \lambda$, $\mathbf{u}_{m_k} \to \mathbf{u}$. Then $\lambda \mathbf{u} \notin \mathcal{E}$ by assumption. So

$$
lim_k E_0 e^{\lambda_{m_k} \mathbf{u}_{m_k}^T \mathbf{T}(X)} = E_0 e^{\lambda \mathbf{u}^T \mathbf{T}(X)} = \infty.
$$

In either case $\lim_{m_k} l_x(\boldsymbol{\eta}_{m_k}) = -\infty$. Because any subsequence of $\{\boldsymbol{\eta}_m\}$ has no subsequence converging in $\mathcal{E}$ we conclude $l_x(\boldsymbol{\eta}_m) \to -\infty$ and $\hat{\boldsymbol{\eta}}$ exists. It is unique and satisfies (2.3.3) by Theorem 1.6.4.

Nonexistence: If (2.3.2) fails, there exists $\mathbf{c} \neq \mathbf{0}$ such that $P_0[\mathbf{c}^T \mathbf{T} \leq 0] = 1 \Rightarrow E_{\boldsymbol{\eta}}(\mathbf{c}^T \mathbf{T}(X)) \leq 0$, for all $\boldsymbol{\eta}$. If $\hat{\boldsymbol{\eta}}$ exists then $E_{\boldsymbol{\eta}} \mathbf{T} = 0 \Rightarrow E_{\boldsymbol{\eta}}(\mathbf{c}^T \mathbf{T}) = 0 \Rightarrow P_{\boldsymbol{\eta}}[\mathbf{c}^T \mathbf{T} = 0] = 1$, contradicting the assumption that the family is of rank k. $\qquad\square$

Proof of Corollary 2.3.1. By (B.9.1) a point $\mathbf{t}_0$ belongs to the interior C of a convex set C iff there exist points in C^0 on either side of it, that is, iff, for every $\mathbf{d} \neq \mathbf{0}$, both $\{\mathbf{t} : \mathbf{d}^T \mathbf{t} > \mathbf{d}^T \mathbf{t}_0\} \cap C^0$ and $\{\mathbf{t} : \mathbf{d}^T \mathbf{t} < \mathbf{d}^T \mathbf{t}_0\} \cap C^0$ are nonempty open sets. The equivalence of (2.3.2) and Corollary 2.3.1 follow. $\qquad\square$

Example 2.3.1. *The Gaussian Model.* Suppose $X_1, \ldots, X_n$ are i.i.d. $\mathcal{N}(\mu, \sigma^2)$, $\mu \in R$, $\sigma^2 > 0$. As we observed in Example 1.6.5, this is the exponential family generated by $T(\mathbf{X}) \equiv (\sum_{i=1}^n X_i, \sum_{i=1}^n X_i^2)$ and 1. Evidently, $C_T = R \times R^+$. For $n = 2$, $T(\mathbf{X})$ has a density by Problem B.2.7. The $n > 2$ case follows because it involves convolution with the $n = 2$ case. Thus, $C_T = C_T^0$ and the MLE always exists. For $n = 1$, $C_T^0 = \emptyset$ because $T(X_1)$ is always a point on the parabola $T_2 = T_1^2$ and the MLE does not exist. This is equivalent to the fact that if $n = 1$ the formal solution to the likelihood equations gives $\hat{\sigma}^2 = 0$, which is impossible. $\qquad\square$

Existence of MLEs when $\mathbf{T}$ has a continuous case density is a general phenomenon.

Theorem 2.3.2. *Suppose the conditions of Theorem 2.3.1 hold and* $\mathbf{T}_{k \times 1}$ *has a continuous case density on* R^k. *Then the MLE* $\hat{\boldsymbol{\eta}}$ *exists with probability 1 and necessarily satisfies (2.3.3).*

Proof. The boundary of a convex set necessarily has volume 0 (Problem 2.3.9), thus, if $\mathbf{T}$ has a continuous case density $p_{\mathbf{T}}(\mathbf{t})$, then

$$P[\mathbf{T} \in \partial C_{\mathbf{T}}] = \int_{\partial C_{\mathbf{T}}} p_{\mathbf{T}}(\mathbf{t})d\mathbf{t} = 0$$

and the result follows from Corollary 2.3.1.   □

Remark 2.3.1. From Theorem 1.6.3 we know that $E_{\boldsymbol{\eta}}T(X) = \dot{A}(\boldsymbol{\eta})$. Thus, using (2.3.3), the MLE $\widehat{\boldsymbol{\eta}}$ in exponential families has an interpretation as a generalized method of moments estimate (see Problem 2.1.13 and the next example). When method of moments and frequency substitution estimates are not unique, the maximum likelihood principle in many cases selects the "best" estimate among them. For instance, in the Hardy–Weinberg examples 2.1.4 and 2.2.6, $\widehat{\theta}_1 = \sqrt{n_1/n}$, $\widehat{\theta}_2 = 1 - \sqrt{n_3/n}$ and $\widehat{\theta}_3 = (2n_1 + n_2)/2n$ are frequency substitution estimates (Problem 2.1.1), but only $\widehat{\theta}_3$ is a MLE. In Example 3.4.4 we will see that $\widehat{\theta}_3$ is, in a certain sense, the best estimate of θ.

A nontrivial application of Theorem 2.3.2 follows.

Example 2.3.2. *The Two-Parameter Gamma Family.* Suppose $X_1, \ldots, X_n$ are i.i.d. with density $g_{p,\lambda}(x) = \frac{\lambda^p}{\Gamma(p)}e^{-\lambda x}x^{p-1}$, $x > 0$, $p > 0$, $\lambda > 0$. This is a rank 2 canonical exponential family generated by $\mathbf{T} = (\sum \log X_i, \sum X_i)$, $h(x) = x^{-1}$, with

$$\eta_1 = p, \ \eta_2 = -\lambda, \ A(\eta_1, \eta_2) = n(\log \Gamma(\eta_1) - \eta_1 \log(-\eta_2))$$

by Problem 2.3.2(a). The likelihood equations are equivalent to (Problem 2.3.2(b))

$$\frac{\Gamma'}{\Gamma}(\widehat{p}) - \log \widehat{\lambda} = \overline{\log X} \tag{2.3.4}$$

$$\frac{\widehat{p}}{\widehat{\lambda}} = \bar{X} \tag{2.3.5}$$

where $\overline{\log X} \equiv \frac{1}{n}\sum_{i=1}^n \log X_i$. If $n = 2$, $\mathbf{T}$ has a density by Theorem B.2.2. The $n > 2$ case follows because it involves convolution with the $n = 2$ case. We conclude from Theorem 2.3.2 that (2.3.4) and (2.3.5) have a unique solution with probability 1. How to find such nonexplicit solutions is discussed in Section 2.4.   □

If $\mathbf{T}$ is discrete MLEs need not exist. Here is an example.

Example 2.3.3. *Multinomial Trials.* We follow the notation of Example 1.6.7. The statistic of rank $k - 1$ which generates the family is $\mathbf{T}_{(k-1)} = (T_1, \ldots, T_{k-1})^T$, where $T_j(\mathbf{X}) = \sum_{i=1}^n 1(X_i = j)$, $1 \leq j \leq k$. We assume $n \geq k - 1$ and verify using Theorem 2.3.1 that in this case MLEs of $\eta_j = \log(\lambda_j/\lambda_k)$, $1 \leq j \leq k - 1$, where $0 < \lambda_j \equiv P[X = j] < 1$, exist iff all $T_j > 0$. They are determined by $\widehat{\lambda}_j = T_j/n$, $1 \leq j \leq k$. To see this note that $T_j > 0$, $1 \leq j \leq k$ iff $0 < T_j < n$, $1 \leq j \leq k$. Thus, if we write $\mathbf{c}^T\mathbf{t}_0 = \sum\{c_j t_{j0} : c_j > 0\} + \sum\{c_j t_{j0} : c_j < 0\}$ we can increase $\mathbf{c}^T\mathbf{t}_0$ by replacing a t_{j0} by $t_{j0} + 1$ in the first sum or a t_{j0} by $t_{j0} - 1$ in the second. Because the resulting value of $\mathbf{t}$ is possible if $0 < t_{j0} < n$,

$1 \leq j \leq k$, and one of the two sums is nonempty because $\mathbf{c} \neq \mathbf{0}$, we see that (2.3.2) holds. On the other hand, if any $T_j = 0$ or n, $0 \leq j \leq k - 1$ we can obtain a contradiction to (2.3.2) by taking $c_i = -1(i = j)$, $1 \leq i \leq k - 1$. The remaining case $T_k = 0$ gives a contradiction if $\mathbf{c} = (1, 1, \ldots, 1)^T$. Alternatively we can appeal to Corollary 2.3.1 directly (Problem 2.3.10). □

Remark 2.3.1. In Example 2.2.8 we saw that in the multinomial case with the closed parameter set $\{\lambda_j : \lambda_j \geq 0, \sum_{j=1}^{k} \lambda_j = 1\}$, $n \geq k-1$, the MLEs of $\lambda_j, j = 1, \ldots, k$, exist and are unique. However, when we put the multinomial in canonical exponential family form, our parameter set is open. Similarly, note that in the Hardy–Weinberg Example 2.2.6, if $2n_1 + n_2 = 0$, the MLE does not exist if $\Theta = (0, 1)$, whereas if $\theta = [0, 1]$ it does exist and is unique. □

The argument of Example 2.3.3 can be applied to determine existence in cases for which (2.3.3) does not have a closed-form solution as in Example 1.6.8—see Problem 2.3.1 and Haberman (1974).

In some applications, for example, the bivariate normal case (Problem 2.3.13), the following corollary to Theorem 2.3.1 is useful.

Corollary 2.3.2. *Consider the exponential family*

$$p(x, \theta) = h(x) \exp \left\{ \sum_{j=1}^{k} c_j(\theta) T_j(x) - B(\theta) \right\}, \ x \in \mathcal{X}, \ \theta \in \Theta.$$

Let C^0 denote the interior of the range of $(c_1(\theta), \ldots, c_k(\theta))^T$ and let x be the observed data. If the equations

$$E_\theta T_j(X) = T_j(x), \ j = 1, \ldots, k$$

have a solution $\widehat{\theta}(x) \in C^0$, then it is the unique MLE of θ.

When $\mathcal{P}$ is not an exponential family both existence and unicity of MLEs become more problematic. The following result can be useful. Let $Q = \{P_{\boldsymbol{\theta}} : \boldsymbol{\theta} \in \Theta\}$, Θ open $\subset R^m$, $m \leq k - 1$, be a curved exponential family

$$p(x, \boldsymbol{\theta}) = \exp\{c^T(\boldsymbol{\theta}) \mathbf{T}(x) - A(c(\boldsymbol{\theta}))\} h(x). \tag{2.3.6}$$

Suppose $c : \Theta \rightarrow \mathcal{E} \subset R^k$ has a differential $\dot{c}(\boldsymbol{\theta}) \equiv \left\| \frac{\partial c_j}{\partial \theta_i}(\boldsymbol{\theta}) \right\|_{m \times k}$ on Θ. Here $\mathcal{E}$ is the natural parameter space of the exponential family $\mathcal{P}$ generated by $(\mathbf{T}, h)$. Then

Theorem 2.3.3. *If $\mathcal{P}$ above satisfies the condition of Theorem 2.3.1, $c(\Theta)$ is closed in $\mathcal{E}$ and $\mathbf{T}(\mathbf{x}) = \mathbf{t}_0$ satisfies (2.3.2) so that the MLE $\widehat{\boldsymbol{\eta}}$ in $\mathcal{P}$ exists, then so does the $MLE \ \widehat{\boldsymbol{\theta}}$ in Q and it satisfies the likelihood equation*

$$\dot{c}^T(\widehat{\boldsymbol{\theta}})(\mathbf{t}_0 - \dot{A}(c(\widehat{\boldsymbol{\theta}}))) = 0. \tag{2.3.7}$$

Note that $c(\widehat{\boldsymbol{\theta}}) \in c(\Theta)$ and is in general not $\widehat{\boldsymbol{\eta}}$. Unfortunately strict concavity of l_x is not inherited by curved exponential families, and unicity can be lost—take $\mathbf{c}$ not one-to-one for instance.

The proof is sketched in Problem 2.3.11.

Example 2.3.4. *Gaussian with Fixed Signal to Noise.* As in Example 1.6.9, suppose $X_1, \ldots, X_n$ are i.i.d. $\mathcal{N}(\mu, \sigma^2)$ with $\mu/\sigma = \lambda_0 > 0$ known. This is a curved exponential family with $c_1(\mu) = \frac{\lambda_0^2}{\mu}$, $c_2(\mu) = -\frac{\lambda_0^2}{2\mu^2}$, $\mu > 0$, corresponding to $\eta_1 = \frac{\mu}{\sigma^2}$, $\eta_2 = -\frac{1}{2\sigma^2}$. Evidently $c(\Theta) = \{(\eta_1, \eta_2) : \eta_2 = -\frac{1}{2}\eta_1^2\lambda_0^{-2}, \eta_1 > 0, \eta_2 < 0\}$, which is closed in $\mathcal{E} = \{(\eta_1, \eta_2) : \eta_1 \in R, \eta_2 < 0\}$. As a consequence of Theorems 2.3.2 and 2.3.3, we can conclude that an MLE $\widehat{\mu}$ always exists and satisfies (2.3.7) if $n \geq 2$. We find

$$\dot{c}(\theta) = \lambda_0^2(-\mu^{-2}, \mu^{-3})^T,$$

and from Example 1.6.5

$$\dot{A}(\boldsymbol{\eta}) = \frac{1}{2}n(-\eta_1/\eta_2, \eta_1^2/2\eta_2^2 - 1/\eta_2)^T.$$

Thus, with $t_1 = \sum x_i$ and $t_2 = \sum x_i^2$, Equation (2.3.7) becomes

$$\lambda_0^2(-\mu^{-2}, \mu^{-3})(t_1 - n\mu, t_2 - n(\mu^2 + \lambda_0^2\mu^2))^T = 0,$$

which with $\widehat{\mu}_2 = n^{-1}\sum x_i^2$ simplifies to

$$\mu^2 + \lambda_0^2\bar{x}\mu - \lambda_0^2\widehat{\mu}_2 = 0$$

$$\widehat{\mu}_\pm = \frac{1}{2}[\lambda_0^2\bar{x} \pm \lambda_0\sqrt{\lambda_0^2\bar{x}^2 + 4\widehat{\mu}_2}].$$

Note that $\widehat{\mu}_+\widehat{\mu}_- = -\lambda_0^2\widehat{\mu}_2 < 0$, which implies $\widehat{\mu}_+ > 0$, $\widehat{\mu}_- < 0$. Because $\mu > 0$, the solution we seek is $\widehat{\mu}_+$.   □

Example 2.3.5. *Location-Scale Regression.* Suppose that $Y_{j1}, \ldots, Y_{jm}$, $j = 1, \ldots, n$, are n independent random samples, where $Y_{jl} \sim \mathcal{N}(\mu_j, \sigma_j^2)$. Using Examples 1.6.5 and 1.6.10, we see that the distribution of $\{Y_{jl} : j = 1, \ldots, n, l = 1, \ldots, m\}$ is a $2n$-parameter canonical exponential family with $\eta_i = \mu_i/\sigma_i^2$, $\eta_{n+i} = -1/2\sigma_i^2$, $i = 1, \ldots, n$, generated by $h(\mathbf{Y}) = 1$ and

$$T(\mathbf{Y}) = \left(\sum_{l=1}^{m} Y_{1l}, \ldots, \sum_{l=1}^{m} Y_{nl}, \sum_{l=1}^{m} Y_{1l}^2, \ldots, \sum_{l=1}^{m} Y_{nl}^2\right)^T.$$

Next suppose, as in Example 1.6.10, that

$$\mu_i = \theta_1 + \theta_2 z_i, \quad \sigma_i^2 = \theta_3(\theta_1 + \theta_2 z_i)^2, \quad z_1 < \cdots < z_n$$

where $z_1, \ldots, z_n$ are given constants. Now $p(\mathbf{y}, \boldsymbol{\theta})$ is a curved exponential family of the form (2.3.6) with

$$c_i(\boldsymbol{\theta}) = \theta_3^{-1}(\theta_1 + \theta_2 z_i)^{-1}, \quad c_{n+i}(\boldsymbol{\theta}) = \frac{1}{2}\theta_3^{-1}(\theta_1 + \theta_2 z_i)^{-2}, \quad i = 1, \ldots, n.$$

If $m \geq 2$, then the full $2n$-parameter model satisfies the conditions of Theorem 2.3.1. Let $\mathcal{E}$ be the canonical parameter set for this full model and let

$$\Theta = \{\boldsymbol{\theta} : \theta_1 \in R, \theta_2 \in R, \theta_3 > 0\}.$$

Then $c(\Theta)$ is closed in $\mathcal{E}$ and we can conclude that for $m \geq 2$, an MLE $\hat{\boldsymbol{\theta}}$ of $\boldsymbol{\theta}$ exists and $\hat{\theta}$ satisfies (2.3.7).   □

Summary. In this section we derive necessary and sufficient conditions for existence of MLEs in canonical exponential families of full rank with $\mathcal{E}$ open (Theorem 2.3.1 and Corollary 2.3.1). These results lead to a necessary condition for existence of the MLE in curved exponential families but without a guarantee of unicity or sufficiency. Finally, the basic property making Theorem 2.3.1 work, strict concavity, is isolated and shown to apply to a broader class of models.

2.4   ALGORITHMIC ISSUES

As we have seen, even in the context of canonical multiparameter exponential families, such as the two-parameter gamma, MLEs may not be given explicitly by formulae but only implicitly as the solutions of systems of nonlinear equations. In fact, even in the classical regression model with design matrix $\mathbf{Z}_D$ of full rank d, the formula (2.1.10) for $\hat{\beta}$ is easy to write down symbolically but not easy to evaluate if d is at all large because inversion of $\mathbf{Z}_D^T \mathbf{Z}_D$ requires on the order of nd^2 operations to evaluate each of $d(d+1)/2$ terms with n operations to get $\mathbf{Z}_D^T \mathbf{Z}_D$ and then, if implemented as usual, order d^3 operations to invert. The packages that produce least squares estimates do not in fact use formula (2.1.10).

It is not our goal in this book to enter seriously into questions that are the subject of textbooks in numerical analysis. However, in this section, we will discuss three algorithms of a type used in different statistical contexts both for their own sakes and to illustrate what kinds of things can be established about the black boxes to which we all, at various times, entrust ourselves.

We begin with the bisection and coordinate ascent methods, which give a complete though slow solution to finding MLEs in the canonical exponential families covered by Theorem 2.3.1.

2.4.1   The Method of Bisection

The bisection method is the essential ingredient in the coordinate ascent algorithm that yields MLEs in k-parameter exponential families. Given f continuous on (a, b), $f \uparrow$ strictly, $f(a+) < 0 < f(b-)$, then, by the intermediate value theorem, there exists unique $x^* \epsilon(a, b)$ such that $f(x^*) = 0$. Here, in pseudocode, is the bisection algorithm to find x^*. Given tolerance $\epsilon > 0$ for $|x_{\text{final}} - x^*|$:

   Find $x_0 < x_1$, $f(x_0) < 0 < f(x_1)$ by taking $|x_0|, |x_1|$ large enough. Initialize $x_{\text{old}}^+ = x_1$, $x_{\text{old}}^- = x_0$.

(1) If $|x_{\text{old}}^{+} - x_{\text{old}}^{-}| < 2\epsilon$, $x_{\text{final}} = \frac{1}{2}(x_{\text{old}}^{+} + x_{\text{old}}^{-})$ and return x_{final}.

(2) Else, $x_{\text{new}} = \frac{1}{2}(x_{\text{old}}^{+} + x_{\text{old}}^{-})$.

(3) If $f(x_{\text{new}}) = 0$, $x_{\text{final}} = x_{\text{new}}$ and return x_{final}.

(4) If $f(x_{\text{new}}) < 0$, $x_{\text{old}}^{-} = x_{\text{new}}$.

(5) If $f(x_{\text{new}}) > 0$, $x_{\text{old}}^{+} = x_{\text{new}}$.

Go to (1).

End

Lemma 2.4.1. *The bisection algorithm stops at a solution* x_{final} *such that*

$$|x_{\text{final}} - x^{*}| \leq \epsilon.$$

Proof. If x_m is the mth iterate of x_{new}

$$(1) \qquad |x_m - x_{m-1}| \leq \frac{1}{2}|x_{m-1} - x_{m-2}| \leq \cdots \leq \cdots \frac{1}{2^{m-1}}|x_1 - x_0|.$$

Moreover, by the intermediate value theorem,

$$(2) \qquad\qquad\qquad x_m \leq x^{*} \leq x_{m+1} \text{ for all } m.$$

Therefore,

$$(3) \qquad\qquad\qquad |x_{m+1} - x^{*}| \leq 2^{-m}|x_1 - x_0|$$

and $x_m \to x^{*}$ as $m \to \infty$. Moreover, for $m = \log_2(|x_1 - x_0|/\epsilon)$, $|x_{m+1} - x^{*}| \leq \epsilon$. $\square$

If desired one could evidently also arrange it so that, in addition, $|f(x_{\text{final}})| \leq \epsilon$. From this lemma we can deduce the following.

Theorem 2.4.1. *Let* $p(x, \eta)$ *be a one-parameter canonical exponential family generated by* (T, h), *satisfying the conditions of Theorem 2.3.1 and* $T = t_0 \in C_T^0$, *the interior of the convex support of* p_T. *Then, the MLE* $\hat{\eta}$, *which exists and is unique by Theorem 2.3.1, may be found (to tolerance* ϵ) *by the method of bisection applied to*

$$f(\eta) \equiv E_\eta T(X) - t_0.$$

Proof. By Theorem 1.6.4, $f'(\eta) = \text{Var}_\eta T(X) > 0$ for all η so that f is strictly increasing and continuous and necessarily because $\hat{\eta}$ exists, $f(a+) < 0 < f(b-)$ if $\mathcal{E} = (a, b)$. $\square$

Example 2.4.1. *The Shape Parameter Gamma Family.* Let $X_1, \ldots, X_n$ be i.i.d. $\Gamma(\theta, 1)$,

$$p(x, \theta) = \Gamma^{-1}(\theta)x^{\theta-1}e^{-x}, \ x > 0, \ \theta > 0. \qquad\qquad (2.4.1)$$

Because $T(\mathbf{X}) = \sum_{i=1}^{n} \log X_i$ has a density for all n the MLE always exists. It solves the equation

$$\frac{\Gamma'(\theta)}{\Gamma(\theta)} = \frac{T(\mathbf{X})}{n},$$

which by Theorem 2.4.1 can be evaluated by bisection. This example points to another hidden difficulty. The function $\Gamma(\theta) = \int_0^\infty x^{\theta-1} e^{-x} dx$ needed for the bisection method can itself only be evaluated by numerical integration or some other numerical method. However, it is in fact available to high precision in standard packages such as NAG or MATLAB. In fact, bisection itself is a defined function in some packages.   □

2.4.2 Coordinate Ascent

The problem we consider is to solve numerically, for a canonical k-parameter exponential family,

$$E_{\boldsymbol{\eta}}(\mathbf{T}(\mathbf{X})) = \dot{A}(\boldsymbol{\eta}) = \mathbf{t}_0$$

when the MLE $\widehat{\boldsymbol{\eta}} \equiv \widehat{\boldsymbol{\eta}}(t_0)$ exists. Here is the algorithm, which is slow, but as we shall see, always converges to $\widehat{\boldsymbol{\eta}}$.

The case $k = 1$: See Theorem 2.4.1.

The general case: Initialize

$$\widehat{\boldsymbol{\eta}}^0 = (\widehat{\eta}_1^0, \ldots, \widehat{\eta}_k^0).$$

Solve

$$\text{for } \widehat{\eta}_1^1 : \frac{\partial}{\partial \eta_1} A(\eta_1, \widehat{\eta}_2^0, \ldots, \widehat{\eta}_k^0) = t_1$$

$$\text{for } \widehat{\eta}_2^1 : \frac{\partial}{\partial \eta_2} A(\widehat{\eta}_1^1, \eta_2, \widehat{\eta}_3^0, \ldots, \widehat{\eta}_k^0) = t_2$$

$$\vdots$$

$$\text{for } \widehat{\eta}_k^1 : \frac{\partial}{\partial \eta_k} A(\widehat{\eta}_1^1, \widehat{\eta}_2^1, \ldots, \eta_k) = t_k.$$

Set

$$\widehat{\boldsymbol{\eta}}^{01} \equiv (\widehat{\eta}_1^1, \widehat{\eta}_2^0, \ldots, \widehat{\eta}_k^0), \ \widehat{\boldsymbol{\eta}}^{02} \equiv (\widehat{\eta}_1^1, \widehat{\eta}_2^1, \widehat{\eta}_3^0, \ldots, \widehat{\eta}_k^0), \text{ and so on,}$$

and finally

$$\widehat{\boldsymbol{\eta}}^{0k} \equiv \widehat{\boldsymbol{\eta}}^{(1)} = (\widehat{\eta}_1^1, \ldots, \widehat{\eta}_k^1).$$

Repeat, getting $\widehat{\boldsymbol{\eta}}^{(r)}$, $r \geq 1$, eventually.

Notes:

(1) In practice, we would again set a tolerence to be, say ϵ, for each of the $\widehat{\boldsymbol{\eta}}^{jl}$, $1 \leq l \leq k$, in cycle j and stop possibly in midcycle as soon as

$$|\widehat{\boldsymbol{\eta}}^{jl} - \widehat{\boldsymbol{\eta}}^{j(l-1)}| \leq \epsilon.$$

(2) Notice that $\frac{\partial A}{\partial \eta_l}(\widehat{\eta}_1^j, \ldots, \widehat{\eta}_{l-2}^j, \eta_l, \widehat{\eta}_{l+1}^{j-1}, \ldots)$ is the expectation of $T_l(\mathbf{X})$ in the one-parameter exponential family model with all parameters save η_l assumed known. Thus, the algorithm may be viewed as successive fitting of one-parameter families. We pursue this discussion next.

Theorem 2.4.2. *If $\widehat{\eta}^{(r)}$ are as above, (i), (ii) of Theorem 2.3.1 hold and $t_0 \in C_T^0$,*

$$\widehat{\eta}^{(r)} \to \widehat{\eta} \text{ as } r \to \infty.$$

Proof. We give a series of steps. Let $l(\eta) = \mathbf{t}_0^T \eta - A(\eta) + \log h(\mathbf{x})$, the log likelihood.

(1) $l(\widehat{\eta}^{ij}) \uparrow$ in j for i fixed and in i. If $1 \leq j \leq k$, $\widehat{\eta}^{ij}$ and $\widehat{\eta}^{i(j+1)}$ differ in only one coordinate for which $\widehat{\eta}^{i(j+1)}$ maximizes l. Therefore, $\lim_{i,j} l(\widehat{\eta}^{ij}) = \lambda$ (say) exists and is $> -\infty$.

(2) The sequence $(\widehat{\eta}^{i1}, \ldots, \widehat{\eta}^{ik})$ has a convergent subsequence in $\bar{\mathcal{E}} \times \cdots \times \bar{\mathcal{E}}$

$$(\widehat{\eta}^{i_n 1}, \ldots, \widehat{\eta}^{i_n k}) \to (\eta^1, \ldots, \eta^k).$$

But $\eta^j \in \mathcal{E}$, $1 \leq j \leq k$. Else $\lim_i l(\widehat{\eta}^{ij}) = -\infty$ for some j.

(3) $l(\eta^j) = \lambda$ for all j because the sequence of likelihoods is monotone.

(4) $\frac{\partial l}{\partial \eta_j}(\eta^j) = 0$ because $\frac{\partial l}{\partial \eta_j}(\widehat{\eta}^{i_n j}) = 0$, $\forall n$.

(5) Because η^1, η^2 differ only in the second coordinate, (3) and (4) $\Rightarrow \eta^1 = \eta^2$. Continuing, $\eta^1 = \cdots = \eta^k$. Here we use the strict concavity of l.

(6) By (4) and (5), $\dot{A}(\eta^1) = \mathbf{t}_0$. Hence, η^1 is the unique MLE.

To complete the proof notice that if $\widehat{\eta}^{(r_k)}$ is any subsequence of $\widehat{\eta}^{(r)}$ that converges to $\widehat{\eta}^*$ (say) then, by (1), $l(\widehat{\eta}^*) = \lambda$. Because $l(\widehat{\eta}^1) = \lambda$ and the MLE is unique, $\widehat{\eta}^* = \widehat{\eta}^1 = \widehat{\eta}$. By a standard argument it follows that, $\widehat{\eta}^{(r)} \to \widehat{\eta}$. □

Example 2.4.2. *The Two-Parameter Gamma Family (continued).* We use the notation of Example 2.3.2. For $n \geq 2$ we know the MLE exists. We can initialize with the method of moments estimate from Example 2.1.2, $\widehat{\lambda}^{(0)} = \frac{\bar{X}}{\sigma^2}$, $\widehat{p}^{(0)} = \frac{\bar{X}^2}{\sigma^2}$. We now use bisection to get $\widehat{p}^{(1)}$ solving $\frac{\Gamma'}{\Gamma}(\widehat{p}^{(1)}) = \overline{\log X} + \log \widehat{\lambda}^{(0)}$ and then $\widehat{\lambda}^{(1)} = \frac{\widehat{p}^{(1)}}{\bar{X}}$, $\widehat{\eta}^1 = (\widehat{p}^{(1)}, -\widehat{\lambda}^{(1)})$. Continuing in this way we can get arbitrarily close to $\widehat{\eta}$. This two-dimensional problem is essentially no harder than the one-dimensional problem of Example 2.4.1 because the equation leading to $\widehat{\lambda}_{\text{new}}$ given $\widehat{p}_{\text{old}}$, (2.3.5), is computationally explicit and simple. Whenever we can obtain such steps in algorithms, they result in substantial savings of time. □

It is natural to ask what happens if, in fact, the MLE $\widehat{\eta}$ doesn't exist; that is, $\mathbf{t}_0 \notin C_T^0$. Fortunately in these cases the algorithm, as it should, refuses to converge (in η space!)—see Problem 2.4.2.

We note some important generalizations. Consider a point we noted in Example 2.4.2: For some coordinates l, $\widehat{\eta}_l^j$ can be explicit. Suppose that this is true for each l. Then each step of the iteration both within cycles and from cycle to cycle is quick. Suppose that we can write $\eta^T = (\eta_1^T, \ldots, \eta_r^T)$ where η_j has dimension d_j and $\sum_{j=1}^r d_j = k$ and the problem of obtaining $\widehat{\eta}_l(\mathbf{t}_0, \eta_j; \ j \neq l)$ can be solved in closed form. The case we have just discussed has $d_1 = \cdots = d_r = 1, r = k$. Then it is easy to see that Theorem 2.4.2 has a generalization with cycles of length r, each of whose members can be evaluated easily. A special case of this is the famous Deming–Stephan proportional fitting of contingency tables algorithm—see Bishop, Feinberg, and Holland (1975), for instance, and Problems 2.4.9–2.4.10.

Next consider the setting of Proposition 2.3.1 in which $l_{\mathbf{x}}(\boldsymbol{\theta})$, the log likelihood for $\boldsymbol{\theta} \in \Theta$ open $\subset R^p$, is strictly concave. If $\widehat{\boldsymbol{\theta}}(\mathbf{x})$ exists and $l_{\mathbf{x}}$ is differentiable, the method extends straightforwardly. Solve $\frac{\partial l_{\mathbf{x}}}{\partial \theta_j}(\theta_1^1, \ldots, \theta_{j-1}^1, \theta_j, \theta_{j+1}^0, \ldots, \theta_p^0) = 0$ by the method of bisection in θ_j to get θ_j^1 for $j = 1, \ldots, p$, iterate and proceed. Figure 2.4.1 illustrates the process. See also Problem 2.4.7.

The coordinate ascent algorithm can be slow if the contours in Figure 2.4.1 are not close to spherical. It can be speeded up at the cost of further computation by Newton's method, which we now sketch.

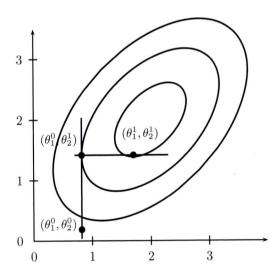

Figure 2.4.1. The coordinate ascent algorithm. The graph shows log likelihood contours, that is, values of $(\theta_1, \theta_2)^T$ where the log likelihood is constant. At each stage with one coordinate fixed, find that member of the family of contours to which the vertical (or horizontal) line is tangent. Change other coordinates accordingly.

2.4.3   The Newton–Raphson Algorithm

An algorithm that, in general, can be shown to be faster than coordinate ascent, when it converges, is the Newton–Raphson method. This method requires computation of the inverse of the Hessian, which may counterbalance its advantage in speed of convergence when it does converge. Here is the method: If $\widehat{\eta}_{old}$ is the current value of the algorithm, then

$$\widehat{\eta}_{new} = \widehat{\eta}_{old} - \ddot{A}^{-1}(\widehat{\eta}_{old})(\dot{A}(\widehat{\eta}_{old}) - t_0). \tag{2.4.2}$$

The rationale here is simple. If $\widehat{\eta}_{old}$ is close to the root $\widehat{\eta}$ of $\dot{A}(\widehat{\eta}) = t_0$, then by expanding $\dot{A}(\widehat{\eta})$ around $\widehat{\eta}_{old}$, we obtain

$$t_0 - \dot{A}(\widehat{\eta}_{old}) = \dot{A}(\widehat{\eta}) - \dot{A}(\widehat{\eta}_{old}) \simeq \ddot{A}(\widehat{\eta}_{old})(\widehat{\eta} - \widehat{\eta}_{old}).$$

$\widehat{\eta}_{new}$ is the solution for $\widehat{\eta}$ to the approximation equation given by the right- and left-hand sides. If $\widehat{\eta}_{old}$ is close enough to $\widehat{\eta}$, this method is known to converge to $\widehat{\eta}$ at a faster rate than coordinate ascent—see Dahlquist, Björk, and Anderson (1974). A hybrid of the two methods that always converges and shares the increased speed of the Newton–Raphson method is given in Problem 2.4.7.

Newton's method also extends to the framework of Proposition 2.3.1. In this case, if $l(\theta)$ denotes the log likelihood, the argument that led to (2.4.2) gives

$$\widehat{\theta}_{new} = \widehat{\theta}_{old} - \ddot{l}^{-1}(\widehat{\theta}_{old})\dot{l}(\widehat{\theta}_{old}). \tag{2.4.3}$$

Example 2.4.3. Let $X_1, \ldots, X_n$ be a sample from the logistic distribution with d.f.

$$F(x, \theta) = [1 + \exp\{-(x - \theta)\}]^{-1}.$$

The density is

$$f(x, \theta) = \frac{\exp\{-(x - \theta)\}}{[1 + \exp\{-(x - \theta)\}]^2}.$$

We find

$$\dot{l}(\theta) = n - 2 \sum_{i=1}^{n} \exp\{-(X_i - \theta)\} F(X_i, \theta)$$

$$\ddot{l}(\theta) = -2 \sum_{i=1}^{n} f(X_i, \theta) < 0.$$

The Newton–Raphson method can be implemented by taking $\widehat{\theta}_{old} = \widehat{\theta}_{MOM} = \bar{X}$.   □

The Newton–Raphson algorithm has the property that for large n, $\widehat{\eta}_{new}$ after only one step behaves approximately like the MLE. We return to this property in Problem 6.6.10.

When likelihoods are non-concave, methods such as bisection, coordinate ascent, and Newton–Raphson's are still employed, though there is a distinct possibility of nonconvergence or convergence to a local rather than global maximum. A one-dimensional problem in which such difficulties arise is given in Problem 2.4.13. Many examples and important issues and methods are discussed, for instance, in Chapter 6 of Dahlquist, Björk, and Anderson (1974).

2.4.4   The EM (Expectation/Maximization) Algorithm

There are many models that have the following structure. There are ideal observations, $X \sim P_\theta$ with density $p(x, \theta)$, $\theta \in \Theta \subset R^d$. Their log likelihood $l_{p,x}(\theta)$ is "easy" to maximize. Say there is a closed-form MLE or at least $l_{p,x}(\theta)$ is concave in θ. Unfortunately, we observe $S \equiv S(X) \sim Q_\theta$ with density $q(s, \theta)$ where $l_{q,s}(\theta) = \log q(s, \theta)$ is difficult to maximize; the function is not concave, difficult to compute, and so on. A fruitful way of thinking of such problems is in terms of S as representing part of X, the rest of X is "missing" and its "reconstruction" is part of the process of estimating θ by maximum likelihood. The algorithm was formalized with many examples in Dempster, Laird, and Rubin (1977), though an earlier general form goes back to Baum, Petrie, Soules, and Weiss (1970). We give a few examples of situations of the foregoing type in which it is used, and its main properties. For detailed discussion we refer to Little and Rubin (1987) and MacLachlan and Krishnan (1997). A prototypical example follows.

Example 2.4.4. *Lumped Hardy–Weinberg Data.* As in Example 2.2.6, let X_i, $i = 1, \ldots, n$, be a sample from a population in Hardy–Weinberg equilibrium for a two-allele locus, $X_i = (\epsilon_{i1}, \epsilon_{i2}, \epsilon_{i3})$, where $P_\theta[X = (1, 0, 0)] = \theta^2$, $P_\theta[X = (0, 1, 0)] = 2\theta(1 - \theta)$, $P_\theta[X = (0, 0, 1)] = (1 - \theta)^2$, $0 < \theta < 1$. What is observed, however, is not **X** but **S** where

$$\begin{aligned} S_i &= X_i, \ 1 \le i \le m \\ S_i &= (\epsilon_{i1} + \epsilon_{i2}, \epsilon_{i3}), \ m + 1 \le i \le n. \end{aligned} \tag{2.4.4}$$

Evidently, $\mathbf{S} = \mathbf{S}(\mathbf{X})$ where $\mathbf{S}(\mathbf{X})$ is given by (2.4.4). This could happen if, for some individuals, the homozygotes of one type ($\epsilon_{i1} = 1$) could not be distinguished from the heterozygotes ($\epsilon_{i2} = 1$). The log likelihood of **S** now is

$$\begin{aligned} l_{q,s}(\theta) &= \sum_{i=1}^{m} [2\epsilon_{i1} \log \theta + \epsilon_{i2} \log 2\theta(1 - \theta) + 2\epsilon_{i3} \log(1 - \theta)] \\ &+ \sum_{i=m+1}^{n} [(\epsilon_{i1} + \epsilon_{i2}) \log(1 - (1 - \theta)^2) + 2\epsilon_{i3} \log(1 - \theta)] \end{aligned} \tag{2.4.5}$$

a function that is of curved exponential family form. It does turn out that in this simplest case an explicit maximum likelihood solution is still possible, but the computation is clearly not as simple as in the original Hardy–Weinberg canonical exponential family example. If we suppose (say) that observations $S_1, \ldots, S_m$ are not X_i but $(\epsilon_{i1}, \epsilon_{i2} + \epsilon_{i3})$, then explicit solution is in general not possible. Yet the EM algorithm, with an appropriate starting point, leads us to an MLE if it exists in both cases.   □

Here is another important example.

Example 2.4.5. *Mixture of Gaussians.* Suppose $S_1, \ldots, S_n$ is a sample from a popu-
lation P whose density is modeled as a mixture of two Gaussian densities, $p(s, \theta) = (1 - \lambda)\varphi_{\sigma_1}(s - \mu_1) + \lambda\varphi_{\sigma_2}(s - \mu_2)$ where $\boldsymbol{\theta} = (\lambda, (\mu_i, \sigma_i), i = 1, 2)$ and $0 < \lambda < 1$,
$\sigma_1, \sigma_2 > 0$, $\mu_1, \mu_2 \in R$ and $\varphi_\sigma(s) = \frac{1}{\sigma}\varphi\left(\frac{s}{\sigma}\right)$. It is not obvious that this falls under our
scheme but let

$$X_i = (\Delta_i, S_i), \ 1 \le i \le n \qquad (2.4.6)$$

where Δ_i are independent identically distributed with $P_{\boldsymbol{\theta}}[\Delta_i = 1] = \lambda = 1 - P_{\boldsymbol{\theta}}[\Delta_i = 0]$.
Suppose that given $\boldsymbol{\Delta} = (\Delta_1, \ldots, \Delta_n)$, the S_i are independent with

$$\mathcal{L}_{\boldsymbol{\theta}}(S_i \mid \boldsymbol{\Delta}) = \mathcal{L}_{\boldsymbol{\theta}}(S_i \mid \Delta_i) = \mathcal{N}(\Delta_i\mu_1 + (1 - \Delta_i)\mu_2, \Delta_i\sigma_1^2 + (1 - \Delta_i)\sigma_2^2).$$

That is, Δ_i tells us whether to sample from $\mathcal{N}(\mu_1, \sigma_1^2)$ or $\mathcal{N}(\mu_2, \sigma_2^2)$. It is easy to see
(Problem 2.4.11), that under $\boldsymbol{\theta}$, $\mathbf{S}$ has the marginal distribution given previously. Thus, we
can think of $\mathbf{S}$ as $S(\mathbf{X})$ where $\mathbf{X}$ is given by (2.4.6).

This five-parameter model is very rich permitting up to two modes and scales. The log
likelihood similarly can have a number of local maxima and can tend to ∞ as $\boldsymbol{\theta}$ tends to
the boundary of the parameter space (Problem 2.4.12). Although MLEs do not exist in
these models, a local maximum close to the true $\boldsymbol{\theta}_0$ turns out to be a good "proxy" for the
nonexistent MLE. The EM algorithm can lead to such a local maximum.   □

The EM Algorithm. Here is the algorithm. Let

$$J(\theta \mid \theta_0) \equiv E_{\theta_0}\left(\log \frac{p(X, \theta)}{p(X, \theta_0)} \mid S(X) = s\right) \qquad (2.4.7)$$

where we suppress dependence on s.

Initialize with $\theta_{\text{old}} = \theta_0$.

The first (E) step of the algorithm is to compute $J(\theta \mid \theta_{\text{old}})$ for as many values of θ as
needed. If this is difficult, the EM algorithm is probably not suitable.

The second (M) step is to maximize $J(\theta \mid \theta_{\text{old}})$ as a function of θ. Again, if this step
is difficult, EM is not particularly appropriate.

Then we set $\theta_{\text{new}} = \arg\max J(\theta \mid \theta_{\text{old}})$, reset $\theta_{\text{old}} = \theta_{\text{new}}$ and repeat the process.

As we shall see in important situations, including the examples, we have given, the M
step is easy and the E step doable.

The rationale behind the algorithm lies in the following formulas, which we give for θ
real and which can be justified easily in the case that $\mathcal{X}$ is finite (Problem 2.4.12)

$$\frac{q(s, \theta)}{q(s, \theta_0)} = E_{\theta_0}\left(\frac{p(X, \theta)}{p(X, \theta_0)} \mid S(X) = s\right) \qquad (2.4.8)$$

and

$$\frac{\partial}{\partial \theta} \log q(s, \theta)\bigg|_{\theta=\theta_0} = E_{\theta_0}\left(\frac{\partial}{\partial \theta} \log p(X, \theta) \mid S(X) = s\right)\bigg|_{\theta=\theta_0} \qquad (2.4.9)$$

for all θ_0 (under suitable regularity conditions). Note that (2.4.9) follows from (2.4.8) by
taking logs in (2.4.8), differentiating and exchanging E_{θ_0} and differentiation with respect

to θ at θ_0. Because, formally,

$$\frac{\partial J(\theta \mid \theta_0)}{\partial \theta} = E_{\theta_0} \left(\frac{\partial}{\partial \theta} \log p(X, \theta) \mid S(X) = s \right) \tag{2.4.10}$$

and, hence,

$$\left. \frac{\partial J(\theta \mid \theta_0)}{\partial \theta} \right|_{\theta_0} = \frac{\partial}{\partial \theta} \log q(s, \theta_0) \tag{2.4.11}$$

it follows that a fixed point $\widetilde{\theta}$ of the algorithm satisfies the likelihood equation,

$$\frac{\partial}{\partial \theta} \log q(s, \widetilde{\theta}) = 0. \tag{2.4.12}$$

The main reason the algorithm behaves well follows.

Lemma 2.4.1. *If $\theta_{\mathrm{new}}, \theta_{\mathrm{old}}$ are as defined earlier and $S(X) = s$,*

$$q(s, \theta_{\mathrm{new}}) \geq q(s, \theta_{\mathrm{old}}). \tag{2.4.13}$$

Equality holds in (2.4.13) iff the conditional distribution of X given $S(X) = s$ is the same for θ_{new} as for θ_{old} and θ_{old} maximizes $J(\theta \mid \theta_{\mathrm{old}})$.

Proof. We give the proof in the discrete case. However, the result holds whenever the quantities in $J(\theta \mid \theta_0)$ can be defined in a reasonable fashion. In the discrete case we appeal to the product rule. For $x \in \mathcal{X}$, $S(x) = s$

$$p(x, \theta) = q(s, \theta) r(x \mid s, \theta) \tag{2.4.14}$$

where $r(\cdot \mid \cdot, \theta)$ is the conditional frequency function of X given $S(X) = s$. Then

$$J(\theta \mid \theta_0) = \log \frac{q(s, \theta)}{q(s, \theta_0)} + E_{\theta_0} \left\{ \log \frac{r(X \mid s, \theta)}{r(X \mid s, \theta_0)} \mid S(X) = s \right\}. \tag{2.4.15}$$

If $\theta_0 = \theta_{\mathrm{old}}$, $\theta = \theta_{\mathrm{new}}$,

$$\log \frac{q(s, \theta_{\mathrm{new}})}{q(s, \theta_{\mathrm{old}})} = J(\theta_{\mathrm{new}} \mid \theta_{\mathrm{old}}) - E_{\theta_{\mathrm{old}}} \left\{ \log \frac{r(X \mid s, \theta_{\mathrm{new}})}{r(X \mid s, \theta_{\mathrm{old}})} \mid S(X) = s \right\}. \tag{2.4.16}$$

Now, $J(\theta_{\mathrm{new}} \mid \theta_{\mathrm{old}}) \geq J(\theta_{\mathrm{old}} \mid \theta_{\mathrm{old}}) = 0$ by definition of θ_{new}. On the other hand,

$$-E_{\theta_{\mathrm{old}}} \left\{ \log \frac{r(X \mid s, \theta_{\mathrm{new}})}{r(X \mid s, \theta_{\mathrm{old}})} \mid S(X) = s \right\} \geq 0 \tag{2.4.17}$$

by Shannon's inequality, Lemma 2.2.1. $\qquad\square$

The most important and revealing special case of this lemma follows.

Theorem 2.4.3. *Suppose $\{P_\theta : \theta \in \Theta\}$ is a canonical exponential family generated by (T, h) satisfying the conditions of Theorem 2.3.1. Let $S(X)$ be any statistic, then*

(a) *The EM algorithm consists of the alternation*

$$\dot{A}(\theta_{\text{new}}) = E_{\theta_{\text{old}}}(T(X) \mid S(X) = s) \qquad (2.4.18)$$

$$\theta_{\text{old}} = \theta_{\text{new}}. \qquad (2.4.19)$$

If a solution of (2.4.18) exists it is necessarily unique.
   (b) *If the sequence of iterates $\{\theta_m\}$ so obtained is bounded and the equation*

$$\dot{A}(\theta) = E_{\theta}(T(X) \mid S(X) = s) \qquad (2.4.20)$$

has a unique solution, then it converges to a limit $\widehat{\theta}^*$, which is necessarily a local maximum of $q(s, \theta)$.

Proof. In this case,

$$
\begin{aligned}
J(\theta \mid \theta_0) &= E_{\theta_0}\{(\theta - \theta_0)^T T(X) - (A(\theta) - A(\theta_0)) \mid S(X) = s\} \\
&= (\theta - \theta_0)^T E_{\theta_0}(T(X) \mid S(X) = y) - (A(\theta) - A(\theta_0))
\end{aligned} \qquad (2.4.21)
$$

Part (a) follows.
   Part (b) is more difficult. A proof due to Wu (1983) is sketched in Problem 2.4.16.   □

Example 2.4.4 (continued). X is distributed according to the exponential family

$$p(\mathbf{x}, \theta) = \exp\{\eta(2N_{1n}(\mathbf{x}) + N_{2n}(\mathbf{x})) - A(\boldsymbol{\eta})\}h(\mathbf{x}) \qquad (2.4.22)$$

where

$$\eta = \log\left(\frac{\theta}{1 - \theta}\right), \ h(\mathbf{x}) = 2^{N_{2n}(\mathbf{x})}, \ A(\eta) = 2n\log(1 + e^{\eta})$$

and $N_{jn} = \sum_{i=1}^{n} \epsilon_{ij}(x_i), 1 \leq j \leq 3$. Now,

$$A'(\eta) = 2n\theta \qquad (2.4.23)$$

$$
\begin{aligned}
E_{\theta}(2N_{1n} + N_{2n} \mid \mathbf{S}) &= 2N_{1m} + N_{2m} \\
&\quad + E_{\theta}\left(\sum_{i=m+1}^{n}(2\epsilon_{i1} + \epsilon_{i2}) \mid \epsilon_{i1} + \epsilon_{i2}, \ m+1 \leq i \leq n\right).
\end{aligned} \qquad (2.4.24)
$$

Under the assumption that the process that causes lumping is independent of the values of the ϵ_{ij},

$$
\begin{aligned}
P_{\theta}[\epsilon_{ij} = 1 \mid \epsilon_{i1} + \epsilon_{i2} = 0] &= 0, \ 1 \leq j \leq 2 \\
P_{\theta}[\epsilon_{i1} = 1 \mid \epsilon_{i1} + \epsilon_{i2} = 1] &= \frac{\theta^2}{\theta^2 + 2\theta(1 - \theta)} = \frac{\theta^2}{1 - (1 - \theta)^2} \\
&= 1 - P_{\theta}[\epsilon_{i2} = 1 \mid \epsilon_{i1} + \epsilon_{i2} = 1].
\end{aligned}
$$

Thus, we see, after some simplification, that,

$$E_{\theta}(2N_{1n} + N_{2n} \mid \mathbf{S}) = 2N_{1m} + N_{2m} + \frac{2}{2 - \widehat{\theta}_{\text{old}}}M_n \qquad (2.4.25)$$

where

$$M_n = \sum_{i=m+1}^{n} (\epsilon_{i1} + \epsilon_{i2}).$$

Thus, the EM iteration is

$$\widehat{\theta}_{new} = \frac{2N_{1m} + N_{2m}}{2n} + \frac{M_n}{n}(2 - \widehat{\theta}_{old})^{-1}. \tag{2.4.26}$$

It may be shown directly (Problem 2.4.15) that if $2N_{1m} + N_{2m} > 0$ and $M_n > 0$, then $\widehat{\theta}_m$ converges to the unique root of

$$\theta^2 - \left(\frac{2 + 2N_{1m} + N_{2m}}{2n}\right)\theta + \frac{2N_{1m} + N_{2m} + M_n}{n} = 0$$

in $(0, 1)$, which is indeed the MLE when $\mathbf{S}$ is observed.   □

Example 2.4.6. Let $(Z_1, Y_1), \ldots, (Z_n, Y_n)$ be i.i.d. as (Z, Y), where $(Z, Y) \sim \mathcal{N}(\mu_1, \mu_2, \sigma_1^2, \sigma_2^2, \rho)$. Suppose that some of the Z_i and some of the Y_i are missing as follows: For $1 \leq i \leq n_1$ we observe both Z_i and Y_i, for $n_1 + 1 \leq i \leq n_2$, we oberve only Z_i, and for $n_2 + 1 \leq i \leq n$, we observe only Y_i. In this case a set of sufficient statistics is

$$T_1 = \bar{Z}, \; T_2 = \bar{Y}, \; T_3 = n^{-1}\sum_{i=1}^{n} Z_i^2, \; T_4 = n^{-1}\sum_{i=1}^{n} Y_i^2, \; T_5 = n^{-1}\sum_{i=1}^{n} Z_i Y_i.$$

The observed data are

$$S = \{(Z_i, Y_i) : 1 \leq i \leq n_1\} \cup \{Z_i : n_1 + 1 \leq i \leq n_2\} \cup \{Y_i : n_2 + 1 \leq i \leq n\}.$$

To compute $E_\theta(\mathbf{T} \mid S = s)$, where $\theta = (\mu_1, \mu_2, \sigma_1^2, \sigma_2^2, \theta)$, we note that for the cases with Z_i and/or Y_i observed, the conditional expected values equal their observed values. For other cases we use the properties of the bivariate normal distribution (Appendix B.4 and Section 1.4), to conclude

$$\begin{aligned}
E_\theta(Y_i \mid Z_i) &= \mu_2 + \rho\sigma_2(Z_i - \mu_1)/\sigma_1 \\
E_\theta(Y_i^2 \mid Z_i) &= [\mu_2 + \rho\sigma_2(Z_i - \mu_1)/\sigma_1]^2 + (1 - \rho^2)\sigma_2^2 \\
E_\theta(Z_i Y_i \mid Z_i) &= [\mu_2 + \rho\sigma_2(Z_i - \mu_1)/\sigma_1]Z_i
\end{aligned}$$

with the corresponding Z on Y regression equations when conditioning on Y_i (Problem 2.4.1). This completes the E-step. For the M-step, compute (Problem 2.4.1)

$$\dot{A}(\theta) = E_\theta \mathbf{T} = (\mu_1, \mu_2, \sigma_1^2 + \mu_1^2, \sigma_2^2 + \mu_2^2, \sigma_1\sigma_2\rho + \mu_1\mu_2).$$

We take $\widehat{\theta}_{old} = \widehat{\theta}_{MOM}$, where $\widehat{\theta}_{MOM}$ is the method of moment estimates $(\widehat{\mu}_1, \widehat{\mu}_2, \widehat{\sigma}_1^2, \widehat{\sigma}_2^2, r)$ (Problem 2.1.8) of θ based on the observed data. It may be shown directly (Problem 2.4.1) that the M-step produces

$$\begin{aligned}
\widehat{\mu}_{1,new} &= T_1(\widehat{\theta}_{old}), \; \widehat{\mu}_{2,new} = T_2(\widehat{\theta}_{old}), \; \widehat{\sigma}_{1,new}^2 = T_3(\widehat{\theta}_{old}) - \widehat{T}_1^2 \\
\widehat{\sigma}_{2,new}^2 &= T_4(\widehat{\theta}_{old}) - \widehat{T}_2^2, \\
\widehat{\rho}_{new} &= [T_5(\widehat{\theta}_{old}) - \widehat{T}_1\widehat{T}_2]/\{[T_3(\widehat{\theta}_{old}) - \widehat{T}_1][T_4(\widehat{\theta}_{old}) - \widehat{T}_2]\}^{\frac{1}{2}}
\end{aligned} \tag{2.4.27}$$

where $T_j(\theta)$ denotes T_j with missing values replaced by the values computed in the E-step and $\widehat{T}_j = T_j(\widehat{\theta}_{\text{old}})$, $j = 1, 2$. Now the process is repeated with $\widehat{\theta}_{\text{MOM}}$ replaced by $\widehat{\theta}_{\text{new}}$. $\square$

Because the E-step, in the context of Example 2.4.6, involves imputing missing values, the EM algorithm is a form of *multiple imputation*.

Remark 2.4.1. Note that if $S(X) = X$, then $J(\theta \mid \theta_0)$ is $\log[p(X, \theta)/p(X, \theta_0)]$, which as a function of θ is maximized where the contrast $-\log p(X, \theta)$ is minimized. Also note that, in general, $-E_{\theta_0}[J(\theta \mid \theta_0)]$ is the Kullback–Leibler divergence (2.2.23).

Summary. The basic bisection algorithm for finding roots of monotone functions is developed and shown to yield a rapid way of computing the MLE in all one-parameter canonical exponential families with $\mathcal{E}$ open (when it exists). We then, in Section 2.4.2, use this algorithm as a building block for the general coordinate ascent algorithm, which yields with certainty the MLEs in k-parameter canonical exponential families with $\mathcal{E}$ open when it exists. Important variants of and alternatives to this algorithm, including the Newton–Raphson method, are discussed and introduced in Section 2.4.3 and the problems. Finally in Section 2.4.4 we derive and discuss the important EM algorithm and its basic properties.

2.5 PROBLEMS AND COMPLEMENTS

Problems for Section 2.1

1. Consider a population made up of three different types of individuals occurring in the Hardy–Weinberg proportions θ^2, $2\theta(1 - \theta)$ and $(1 - \theta)^2$, respectively, where $0 < \theta < 1$.

(a) Show that $T_3 = N_1/n + N_2/2n$ is a frequency substitution estimate of θ.

(b) Using the estimate of (a), what is a frequency substitution estimate of the odds ratio $\theta/(1 - \theta)$?

(c) Suppose X takes the values $-1, 0, 1$ with respective probabilities p_1, p_2, p_3 given by the Hardy–Weinberg proportions. By considering the first moment of X, show that T_3 is a method of moment estimate of θ.

2. Consider n systems with failure times $X_1, \ldots, X_n$ assumed to be independent and identically distributed with exponential, $\mathcal{E}(\lambda)$, distributions.

(a) Find the method of moments estimate of λ based on the first moment.

(b) Find the method of moments estimate of λ based on the second moment.

(c) Combine your answers to (a) and (b) to get a method of moment estimate of λ based on the first two moments.

(d) Find the method of moments estimate of the probability $P(X_1 \geq 1)$ that one system will last at least a month.

3. Suppose that i.i.d. $X_1, \ldots, X_n$ have a beta, $\beta(\alpha_1, \alpha_2)$ distribution. Find the method of moments estimates of $\alpha = (\alpha_1, \alpha_2)$ based on the first two moments.

Hint: See Problem B.2.5.

4. Let $X_1, \ldots, X_n$ be the indicators of n Bernoulli trials with probability of success θ.

(a) Show that $\bar{X}$ is a method of moments estimate of θ.

(b) Exhibit method of moments estimates for $\mathrm{Var}_\theta \bar{X} = \theta(1 - \theta)/n$ first using only the first moment and then using only the second moment of the population. Show that these estimates coincide.

(c) Argue that in this case all frequency substitution estimates of $q(\theta)$ must agree with $q(\bar{X})$.

5. Let $X_1, \ldots, X_n$ and θ be as in Problem 2.1.4. Define $\psi : R^n \times (0, 1) \to R$ by

$$\psi(X_1, \ldots, X_n, \theta) = (S/\theta) - (n - S)/(1 - \theta)$$

where $S = \sum X_i$. Find V as defined by (2.1.3) and show that θ_0 is the unique solution of $V(\theta, \theta_0) = 0$. Find the estimating equation estimate of θ_0.

6. Let $X_{(1)} \leq \cdots \leq X_{(n)}$ be the order statistics of a sample $X_1, \ldots, X_n$. (See Problem B.2.8.) There is a one-to-one correspondence between the empirical distribution function $\widehat{F}$ and the order statistics in the sense that, given the order statistics we may construct $\widehat{F}$ and given $\widehat{F}$, we know the order statistics. Give the details of this correspondence.

7. The jth cumulant $\widehat{c}_j$ of the empirical distribution function is called the jth *sample cumulant* and is a method of moments estimate of the cumulant c_j. Give the first three sample cumulants. See A.12.

8. Let $(Z_1, Y_1), (Z_2, Y_2), \ldots, (Z_n, Y_n)$ be a set of independent and identically distributed random vectors with common distribution function F. The natural estimate of $F(s, t)$ is the *bivariate empirical* distribution function $\widehat{F}(s, t)$, which we define by

$$\widehat{F}(s, t) = \frac{\text{Number of vectors } (Z_i, Y_i) \text{ such that } Z_i \leq s \text{ and } Y_i \leq t}{n}.$$

(a) Show that $\widehat{F}(\cdot, \cdot)$ is the distribution function of a probability $\widehat{P}$ on R^2 assigning mass $1/n$ to each point (Z_i, Y_i).

(b) Define the sample product moment of order (i, j), the sample covariance, the sample correlation, and so on, as the corresponding characteristics of the distribution $\widehat{F}$. Show that the sample product moment of order (i, j) is given by

$$\frac{1}{n} \sum_{k=1}^{n} Z_k^i Y_k^j.$$

The sample covariance is given by

$$\frac{1}{n} \sum_{k=1}^{n} (Z_k - \bar{Z})(Y_k - \bar{Y}) = \frac{1}{n} \sum_{k=1}^{n} Z_k Y_k - \bar{Z}\bar{Y},$$

where $\bar{Z}, \bar{Y}$ are the sample means of the $Z_1, \ldots, Z_n$ and $Y_1, \ldots, Y_n$, respectively. The sample correlation coefficient is given by

$$r = \frac{\sum_{k=1}^n (Z_k - \bar{Z})(Y_k - \bar{Y})}{\sqrt{\sum_{k=1}^n (Z_k - \bar{Z})^2 \sum_{k=1}^n (Y_k - \bar{Y})^2}}.$$

All of these quantities are natural estimates of the corresponding population characteristics and are also called method of moments estimates. (See Problem 2.1.17.) Note that it follows from (A.11.19) that $-1 \leq r \leq 1$.

9. Suppose $\mathbf{X} = (X_1, \ldots, X_n)$ where the X_i are independent $\mathcal{N}(0, \sigma^2)$.

    (a) Find an estimate of σ^2 based on the second moment.

    (b) Construct an estimate of σ using the estimate of part (a) and the equation $\sigma = \sqrt{\sigma^2}$.

    (c) Use the empirical substitution principle to construct an estimate of σ using the relation $E(|X_1|) = \sigma\sqrt{2\pi}$.

10. In Example 2.1.1, suppose that $g(\beta, \mathbf{z})$ is continuous in β and that $|g(\beta, \mathbf{z})|$ tends to ∞ as $|\beta|$ tends to ∞. Show that the least squares estimate exists.
    Hint: Set $c = \rho(X, \mathbf{0})$. There exists a compact set K such that for β in the complement of K, $\rho(X, \beta) > c$. Since $\rho(X, \beta)$ is continuous on K, the result follows.

11. In Example 2.1.2 with $X \sim \Gamma(\alpha, \lambda)$, find the method of moments estimate based on $\hat{\mu}_1$ and $\hat{\mu}_3$.
    Hint: See Problem B.2.4.

12. Let $X_1, \ldots, X_n$ be i.i.d. as $X \sim P_\theta$, $\theta \in \Theta \subset R^d$, with θ identifiable. Suppose X has possible values $v_1, \ldots, v_k$ and that $q(\theta)$ can be written as

$$q(\theta) = h(\mu_1(\theta), \ldots, \mu_r(\theta))$$

for some R^k-valued function h. Show that the method of moments estimate $\hat{q} = h(\hat{\mu}_1, \ldots, \hat{\mu}_r)$ can be written as a frequency plug-in estimate.

13. *General method of moment estimates*[1]. Suppose $X_1, \ldots, X_n$ are i.i.d. as $X \sim P_\theta$, with $\theta \in \Theta \subset R^d$ and θ identifiable. Let $g_1, \ldots, g_r$ be given linearly independent functions and write

$$\mu_j(\theta) = E_\theta(g_j(X)), \quad \hat{\mu}_j = n^{-1} \sum_{i=1}^n g_j(X_i), \quad j = 1, \ldots, r.$$

Suppose that X has possible values $v_1, \ldots, v_k$ and that

$$q(\theta) = h(\mu_1(\theta), \ldots, \mu_r(\theta))$$

for some R^k-valued function h.

(a) Show that the method of moments estimate $\widehat{q} = h(\widehat{\mu}_1, \ldots, \widehat{\mu}_r)$ is a frequency plug-in estimate.

(b) Suppose $\{P_{\boldsymbol{\theta}} : \boldsymbol{\theta} \in \Theta\}$ is the k-parameter exponential family given by (1.6.10). Let $g_j(X) = T_j(X)$, $1 \leq j \leq k$. In the following cases, find the method of moments estimates

(i) Beta, $\beta(1, \theta)$

(ii) Beta, $\beta(\theta, 1)$

(iii) Rayleigh, $p(x, \theta) = (x/\theta^2) \exp(-x^2/2\theta^2)$, $x > 0$, $\theta > 0$

(iv) Gamma, $\Gamma(p, \theta)$, p fixed

(v) Inverse Gaussian, $IG(\mu, \lambda)$, $\boldsymbol{\theta} = (\mu, \lambda)$. See Problem 1.6.36.

Hint: Use Corollary 1.6.1.

14. When the data are not i.i.d., it may still be possible to express parameters as functions of moments and then use estimates based on replacing population moments with "sample" moments. Consider the Gaussian $AR(1)$ model of Example 1.1.5.

(a) Use $E(X_i)$ to give a method of moments estimate of μ.

(b) Suppose $\mu = \mu_0$ and $\beta = b$ are fixed. Use $E(U_i^2)$, where

$$U_i = (X_i - \mu_0) \bigg/ \left(\sum_{j=0}^{i-1} b^{2j} \right)^{1/2},$$

to give a method of moments estimate of σ^2.

(c) If μ and σ^2 are fixed, can you give a method of moments estimate of β?

15. *Hardy–Weinberg with six genotypes.* In a large natural population of plants (Mimulus guttatus) there are three possible alleles S, I, and F at one locus resulting in six genotypes labeled SS, II, FF, SI, SF, and IF. Let θ_1, θ_2, and θ_3 denote the probabilities of S, I, and F, respectively, where $\sum_{j=1}^{3} \theta_j = 1$. The Hardy–Weinberg model specifies that the six genotypes have probabilities

Genotype	1	2	3	4	5	6
Genotype	SS	II	FF	SI	SF	IF
Probability	θ_1^2	θ_2^2	θ_3^2	$2\theta_1\theta_2$	$2\theta_1\theta_3$	$2\theta_2\theta_3$

Let N_j be the number of plants of genotype j in a sample of n independent plants, $1 \leq j \leq 6$ and let $\widehat{p}_j = N_j/n$. Show that

$$\begin{aligned} \widehat{\theta}_1 &= \widehat{p}_1 + \tfrac{1}{2}\widehat{p}_4 + \tfrac{1}{2}\widehat{p}_5 \\ \widehat{\theta}_2 &= \widehat{p}_2 + \tfrac{1}{2}\widehat{p}_4 + \tfrac{1}{2}\widehat{p}_6 \\ \widehat{\theta}_3 &= \widehat{p}_3 + \tfrac{1}{2}\widehat{p}_5 + \tfrac{1}{2}\widehat{p}_6 \end{aligned}$$

are frequency plug-in estimates of θ_1, θ_2, and θ_3.

16. Establish $(2.1.6)$.
$\quad$ *Hint:* $[Y_i - g(\boldsymbol{\beta}, \boldsymbol{z}_i)] = [Y_i - g(\boldsymbol{\beta}_0, \boldsymbol{z}_i)] + [g(\boldsymbol{\beta}_0, \boldsymbol{z}_i) - g(\boldsymbol{\beta}, \boldsymbol{z}_i)]$.

17. *Multivariate method of moments.* For a vector $X = (X_1, \ldots, X_q)$, of observations, let
the moments be

$$m_{jkrs} = E(X_r^j X_s^k), \; j \geq 0, \; k \geq 0; \; r, s = 1, \ldots, q.$$

For independent identically distributed $X_i = (X_{i1}, \ldots, X_{iq})$, $i = 1, \ldots, n$, we define the
empirical or sample moment to be

$$\widehat{m}_{jkrs} = \frac{1}{n} \sum_{i=1}^{n} X_{ir}^j X_{is}^k, \; j \geq 0, \; k \geq 0; \; r, s = 1, \ldots, q.$$

If $\boldsymbol{\theta} = (\theta_1, \ldots, \theta_m)$ can be expressed as a function of the moments, the method of moments
estimate $\widehat{\boldsymbol{\theta}}$ of $\boldsymbol{\theta}$ is obtained by replacing m_{jkrs} by $\widehat{m}_{jkrs}$. Let $X = (Z, Y)$ and $\boldsymbol{\theta} = (a_1, b_1)$, where (Z, Y) and (a_1, b_1) are as in Theorem 1.4.3. Show that method of moments
estimators of the parameters b_1 and a_1 in the best linear predictor are

$$\widehat{b}_1 = \frac{n^{-1} \sum Z_i Y_i - \bar{Z}\bar{Y}}{n^{-1} \sum Z_i - (\bar{Z})^2}, \; \widehat{a}_1 = \bar{Y} - \widehat{b}_1 \bar{Z}.$$

Problems for Section 2.2

1. An object of unit mass is placed in a force field of unknown constant intensity θ. Read-
ings $Y_1, \ldots, Y_n$ are taken at times $t_1, \ldots, t_n$ on the position of the object. The reading Y_i
differs from the true position $(\theta/2)t_i^2$ by a random error ϵ_i. We suppose the ϵ_i to have mean
0 and be uncorrelated with constant variance. Find the LSE of θ.

2. Show that the formulae of Example 2.2.2 may be derived from Theorem 1.4.3, if we consider   the   distribution   assigning   mass $1/n$ to   each   of   the   points
$(z_1, y_1), \ldots, (z_n, y_n)$.

3. Suppose that observations $Y_1, \ldots, Y_n$ have been taken at times $z_1, \ldots, z_n$ and that the
linear regression model holds. A new observation Y_{n+1} is to be taken at time z_{n+1}. What
is the least squares estimate based on $Y_1, \ldots, Y_n$ of the best (MSPE) predictor of Y_{n+1}?

4. Show that the two sample regression lines coincide (when the axes are interchanged) if
and only if the points (z_i, y_i), $i = 1, \ldots, n$, in fact, all lie on a line.
$\quad$ *Hint:* Write the lines in the form

$$\frac{(z - \bar{z})}{\widehat{\sigma}} = \widehat{\rho}\frac{(y - \bar{y})}{\widehat{\tau}}.$$

5. The regression line minimizes the sum of the squared vertical distances from the points
$(z_1, y_1), \ldots, (z_n, y_n)$. Find the line that minimizes the sum of the squared *perpendicular*
distance to the same points.

Hint: The quantity to be minimized is

$$\frac{\sum_{i=1}^{n}(y_i - \theta_1 - \theta_2 z_i)^2}{1 + \theta_2^2}.$$

6. (a) Let $Y_1, \ldots, Y_n$ be independent random variables with equal variances such that $E(Y_i) = \alpha z_i$ where the z_i are known constants. Find the least squares estimate of α.

(b) Relate your answer to the formula for the best zero intercept linear predictor of Section 1.4.

7. Show that the least squares estimate is always defined and satisfies the equations (2.1.7) provided that g is differentiable with respect to β_i, $1 \leq i \leq d$, the range $\{g(\mathbf{z}_1, \boldsymbol{\beta}), \ldots, g(\mathbf{z}_n, \boldsymbol{\beta}), \boldsymbol{\beta} \in R^d\}$ is closed, and $\boldsymbol{\beta}$ ranges over R^d.

8. Find the least squares estimates for the model $Y_i = \theta_1 + \theta_2 z_i + \epsilon_i$ with ϵ_i as given by (2.2.4)–(2.2.6) under the restrictions $\theta_1 \geq 0, \theta_2 \leq 0$.

9. Suppose $Y_i = \theta_1 + \epsilon_i$, $i = 1, \ldots, n_1$ and $Y_i = \theta_2 + \epsilon_i$, $i = n_1 + 1, \ldots, n_1 + n_2$, where $\epsilon_1, \ldots, \epsilon_{n_1 + n_2}$ are independent $\mathcal{N}(0, \sigma^2)$ variables. Find the least squares estimates of θ_1 and θ_2.

10. Let $X_1, \ldots, X_n$ denote a sample from a population with one of the following densities or frequency functions. Find the MLE of θ.

(a) $f(x, \theta) = \theta e^{-\theta x}$, $x \geq 0; \theta > 0$. (exponential density)

(b) $f(x, \theta) = \theta c^{\theta} x^{-(\theta+1)}$, $x \geq c$; c constant $> 0; \theta > 0$. (Pareto density)

(c) $f(x, \theta) = c \theta^c x^{-(c+1)}$, $x \geq \theta$; c constant $> 0; \theta > 0$. (Pareto density)

(d) $f(x, \theta) = \sqrt{\theta} x^{\sqrt{\theta}-1}$, $0 \leq x \leq 1$, $\theta > 0$. (beta, $\beta(\sqrt{\theta}, 1)$, density)

(e) $f(x, \theta) = (x/\theta^2) \exp\{-x^2/2\theta^2\}$, $x > 0; \theta > 0$. (Rayleigh density)

(f) $f(x, \theta) = \theta c x^{c-1} \exp\{-\theta x^c\}$, $x \geq 0$; c constant $> 0; \theta > 0$. (Weibull density)

11. Suppose that $X_1, \ldots, X_n$, $n \geq 2$, is a sample from a $\mathcal{N}(\mu, \sigma^2)$ distribution.

(a) Show that if μ and σ^2 are unknown, $\mu \in R$, $\sigma^2 > 0$, then the unique MLEs are $\hat{\mu} = \bar{X}$ and $\hat{\sigma}^2 = n^{-1} \sum_{i=1}^{n}(X_i - \bar{X})^2$.

(b) Suppose μ and σ^2 are both known to be nonnegative but otherwise unspecified. Find maximum likelihood estimates of μ and σ^2.

12. Let $X_1, \ldots, X_n$, $n \geq 2$, be independently and identically distributed with density

$$f(x, \theta) = \frac{1}{\sigma} \exp\{-(x - \mu)/\sigma\}, \; x \geq \mu,$$

where $\theta = (\mu, \sigma^2)$, $-\infty < \mu < \infty$, $\sigma^2 > 0$.

(a) Find maximum likelihood estimates of μ and σ^2.

(b) Find the maximum likelihood estimate of $P_\theta[X_1 \geq t]$ for $t > \mu$.

Hint: You may use Problem 2.2.16(b).

13. Let $X_1, \ldots, X_n$ be a sample from a $\mathcal{U}[\theta - \frac{1}{2}, \theta + \frac{1}{2}]$ distribution. Show that any T such that $X_{(n)} - \frac{1}{2} \leq T \leq X_{(1)} + \frac{1}{2}$ is a maximum likelihood estimate of θ. (We write $\mathcal{U}[a, b]$ to make $p(a) = p(b) = (b-a)^{-1}$ rather than 0.)

14. If $n = 1$ in Example 2.1.5 show that no maximum likelihood estimate of $\theta = (\mu, \sigma^2)$ exists.

15. Suppose that $T(\mathbf{X})$ is sufficient for θ and that $\widehat{\theta}(\mathbf{X})$ is an MLE of θ. Show that $\widehat{\theta}$ depends on $\mathbf{X}$ through $T(\mathbf{X})$ only provided that $\widehat{\theta}$ is unique.

Hint: Use the factorization theorem (Theorem 1.5.1).

16. (a) Let $\mathbf{X} \sim P_\theta$, $\theta \in \Theta$ and let $\widehat{\theta}$ denote the MLE of θ. Suppose that h is a one-to-one function from Θ onto $h(\Theta)$. Define $\eta = h(\theta)$ and let $f(\mathbf{x}, \eta)$ denote the density or frequency function of $\mathbf{X}$ in terms of η (i.e., reparametrize the model using η). Show that the MLE of η is $h(\widehat{\theta})$ (i.e., MLEs are unaffected by reparametrization, they are *equivariant* under one-to-one transformations).

(b) Let $\mathcal{P} = \{P_{\boldsymbol{\theta}} : \boldsymbol{\theta} \in \Theta\}$, $\Theta \subset R^p$, $p \geq 1$, be a family of models for $\mathbf{X} \in \mathcal{X} \subset R^d$. Let $\mathbf{q}$ be a map from Θ onto Ω, $\Omega \subset R^k$, $1 \leq k \leq p$. Show that if $\widehat{\boldsymbol{\theta}}$ is a MLE of $\boldsymbol{\theta}$, then $\mathbf{q}(\widehat{\boldsymbol{\theta}})$ is an MLE of $\boldsymbol{\omega} = \mathbf{q}(\boldsymbol{\theta})$.

Hint: Let $\Theta(\boldsymbol{\omega}) = \{\boldsymbol{\theta} \in \Theta : \mathbf{q}(\boldsymbol{\theta}) = \boldsymbol{\omega}\}$, then $\{\Theta(\boldsymbol{\omega}) : \boldsymbol{\omega} \in \Omega\}$ is a partition of Θ, and $\widehat{\boldsymbol{\theta}}$ belongs to only one member of this partition, say $\Theta(\widehat{\boldsymbol{\omega}})$. Because $\mathbf{q}$ is onto Ω, for each $\boldsymbol{\omega} \in \Omega$ there is $\boldsymbol{\theta} \in \Theta$ such that $\boldsymbol{\omega} = \mathbf{q}(\boldsymbol{\theta})$. Thus, the MLE of $\boldsymbol{\omega}$ is by definition

$$\widehat{\boldsymbol{\omega}}_{MLE} = \arg \sup_{\boldsymbol{\omega} \in \Omega} \sup\{L_{\mathbf{X}}(\boldsymbol{\theta}) : \boldsymbol{\theta} \in \Theta(\boldsymbol{\omega})\}.$$

Now show that $\widehat{\boldsymbol{\omega}}_{MLE} = \widehat{\boldsymbol{\omega}} = \mathbf{q}(\widehat{\boldsymbol{\theta}})$.

17. *Censored Geometric Waiting Times.* If time is measured in discrete periods, a model that is often used for the time X to failure of an item is

$$P_\theta[X = k] = \theta^{k-1}(1 - \theta), \ k = 1, 2, \ldots$$

where $0 < \theta < 1$. Suppose that we only record the time of failure, if failure occurs on or before time r and otherwise just note that the item has lived at least $(r + 1)$ periods. Thus, we observe $Y_1, \ldots, Y_n$ which are independent, identically distributed, and have common frequency function,

$$f(k, \theta) = \theta^{k-1}(1 - \theta), \ k = 1, \ldots, r$$

$$f(r + 1, \theta) = 1 - P_\theta[X \leq r] = 1 - \sum_{k=1}^{r} \theta^{k-1}(1 - \theta) = \theta^r.$$

(We denote by "$r + 1$" survival for at least $(r + 1)$ periods.) Let M = number of indices i such that $Y_i = r + 1$. Show that the maximum likelihood estimate of θ based on $Y_1, \ldots, Y_n$

is

$$\hat{\theta}(\mathbf{Y}) = \frac{\sum_{i=1}^n Y_i - n}{\sum_{i=1}^n Y_i - M}.$$

18. Derive maximum likelihood estimates in the following models.

(a) The observations are indicators of Bernoulli trials with probability of success θ. We want to estimate θ and $\mathrm{Var}_\theta X_1 = \theta(1-\theta)$.

(b) The observations are $X_1 = $ the number of failures before the first success, $X_2 = $ the number of failures between the first and second successes, and so on, in a sequence of binomial trials with probability of success θ. We want to estimate θ.

19. Let $X_1, \ldots, X_n$ be independently distributed with X_i having a $\mathcal{N}(\theta_i, 1)$ distribution, $1 \leq i \leq n$.

(a) Find maximum likelihood estimates of the θ_i under the assumption that these quantities vary freely.

(b) Solve the problem of part (a) for $n = 2$ when it is known that $\theta_1 \leq \theta_2$. A general solution of this and related problems may be found in the book by Barlow, Bartholomew, Bremner, and Brunk (1972).

20. In the "life testing" problem 1.6.16(i), find the MLE of θ.

21. (Kiefer–Wolfowitz) Suppose $(X_1, \ldots, X_n)$ is a sample from a population with density

$$f(x, \theta) = \frac{9}{10\sigma}\varphi\left(\frac{x-\mu}{\sigma}\right) + \frac{1}{10}\varphi(x-\mu)$$

where φ is the standard normal density and $\theta = (\mu, \sigma^2) \in \Theta = \{(\mu, \sigma^2) : -\infty < \mu < \infty, 0 < \sigma^2 < \infty\}$. Show that maximum likelihood estimates do not exist, but that $\sup_\sigma p(\mathbf{x}, \hat{\mu}, \sigma^2) = \sup_{\mu,\sigma} p(\mathbf{x}, \mu, \sigma^2)$ if, and only if, $\hat{\mu}$ equals one of the numbers $x_1, \ldots, x_n$. Assume that $x_i \neq x_j$ for $i \neq j$ and that $n \geq 2$.

22. Suppose X has a hypergeometric, $\mathcal{H}(b, N, n)$, distribution. Show that the maximum likelihood estimate of b for N and n fixed is given by

$$\hat{b}(X) = \left[\frac{X}{n}(N+1)\right]$$

if $\frac{X}{n}(N+1)$ is *not* an integer, and

$$\hat{b}(X) = \frac{X}{n}(N+1) \text{ or } \frac{X}{n}(N+1) - 1$$

otherwise, where $[t]$ is the largest integer that is $\leq t$.
    Hint: Consider the ratio $L(b+1, x)/L(b, x)$ as a function of b.

23. Let $X_1, \ldots, X_m$ and $Y_1, \ldots, Y_n$ be two independent samples from $\mathcal{N}(\mu_1, \sigma^2)$ and $\mathcal{N}(\mu_2, \sigma^2)$ populations, respectively. Show that the MLE of $\theta = (\mu_1, \mu_2, \sigma^2)$ is $\hat{\theta} = (\bar{X}, \bar{Y}, \tilde{\sigma}^2)$ where

$$\tilde{\sigma}^2 = \left[\sum_{i=1}^{m}(X_i - \bar{X})^2 + \sum_{j=1}^{n}(Y_j - \bar{Y})^2 \right] / (m+n).$$

24. *Polynomial Regression.* Suppose $Y_i = \mu(\mathbf{z}_i) + \epsilon_i$, where ϵ_i satisfy (2.2.4)–(2.2.6). Set $\mathbf{z^j} = z_1^{j_1} \cdots z_p^{j_p}$ where $\mathbf{j} \in \mathcal{J}$ and $\mathcal{J}$ is a subset of $\{(j_1, \ldots, j_p) : 0 \leq j_k \leq J, \, 1 \leq k \leq p\}$, and assume that

$$\mu(\mathbf{z}) = \sum \{\alpha_{\mathbf{j}} \mathbf{z^j} : \mathbf{j} \in \mathcal{J}\}.$$

In an experiment to study tool life (in minutes) of steel-cutting tools as a function of cutting speed (in feet per minute) and feed rate (in thousands of an inch per revolution), the following data were obtained (from S. Weisberg, 1985).

TABLE 2.6.1. Tool life data

Feed	Speed	Life	Feed	Speed	Life
-1	-1	54.5	$-\sqrt{2}$	0	20.1
-1	-1	66.0	$\sqrt{2}$	0	2.9
1	-1	11.8	0	0	3.8
1	-1	14.0	0	0	2.2
-1	1	5.2	0	0	3.2
-1	1	3.0	0	0	4.0
1	1	0.8	0	0	2.8
1	1	0.5	0	0	3.2
0	$-\sqrt{2}$	86.5	0	0	4.0
0	$\sqrt{2}$	0.4	0	0	3.5

The researchers analyzed these data using

$$Y = \log \text{ tool life}, \ z_1 = (\text{feed rate} - 13)/6, \ z_2 = (\text{cutting speed} - 900)/300.$$

Two models are contemplated

(a) $Y = \beta_0 + \beta_1 z_1 + \beta_2 z_2 + \epsilon$

(b) $Y = \alpha_0 + \alpha_1 z_1 + \alpha_2 z_2 + \alpha_3 z_1^2 + \alpha_4 z_2^2 + \alpha_5 z_1 z_2 + \epsilon.$

Use a least squares computer package to compute estimates of the coefficients (β's and α's) in the two models. Use these estimated coefficients to compute the values of the contrast function (2.1.5) for (a) and (b). Both of these models are approximations to the true mechanism generating the data. Being larger, the second model provides a better approximation. However, this has to be balanced against greater variability in the estimated coefficients. This will be discussed in Volume II.

25. Consider the model (2.2.1), (2.2.4)–(2.2.6) with $g(\beta, \mathbf{z}) = \mathbf{z}^T\beta$. Show that the following are equivalent.

  (a) The parameterization $\beta \to \mathbf{Z}_D\beta$ is identifiable.

  (b) $\mathbf{Z}_D$ is of rank d.

  (c) $\mathbf{Z}_D^T\mathbf{Z}_D$ is of rank d.

26. Let (Z, Y) have joint probability P with joint density $f(z, y)$, let $v(z, y) \geq 0$ be a weight funciton such that $E(v(Z, Y)Z^2)$ and $E(v(Z, Y)Y^2)$ are finite. The best linear weighted mean squared prediction error predictor $\beta_1(P) + \beta_2(P)Z$ of Y is defined as the minimizer of

$$E\{v(Z, Y)[Y - (b_1 + b_2Z)]^2\}.$$

  (a) Let (Z^*, Y^*) have density $v(z, y)f(z, y)/c$ where $c = \int \int v(z, y)f(z, y)dzdy$. Show that $\beta_2(P) = \mathrm{Cov}(Z^*, Y^*)/\mathrm{Var}\ Z^*$ and $\beta_1(P) = E(Y^*) - \beta_2(P)E(Z^*)$.

  (b) Let $\widehat{P}$ be the empirical probability defined in Problem 2.1.8 and let $v(z, y) = 1/\mathrm{Var}(Y \mid Z = z)$. Show that $\beta_1(\widehat{P})$ and $\beta_2(\widehat{P})$ coincide with $\widehat{\beta}_1$ and $\widehat{\beta}_2$ of Example 2.2.3. That is, weighted least squares estimates are plug-in estimates.

27. Derive the weighted least squares normal equations (2.2.19).

28. Let $\mathbf{Z}_D = \|z_{ij}\|_{n \times d}$ be a design matrix and let $\mathbf{W}_{n \times n}$ be a known symmetric invertible matrix. Consider the model $\mathbf{Y} = \mathbf{Z}_D\beta + \epsilon$ where ϵ has covariance matrix $\sigma^2\mathbf{W}$, σ^2 unknown. Let $\mathbf{W}^{-\frac{1}{2}}$ be a square root matrix of $\mathbf{W}^{-1}$ (see (B.6.6)). Set $\widetilde{\mathbf{Y}} = \mathbf{W}^{-\frac{1}{2}}\mathbf{Y}$, $\widetilde{\mathbf{Z}}_D = \mathbf{W}^{-\frac{1}{2}}\mathbf{Z}_D$ and $\widetilde{\epsilon} = \mathbf{W}^{-\frac{1}{2}}\epsilon$.

  (a) Show that $\widetilde{\mathbf{Y}} = \widetilde{\mathbf{Z}}_D\beta + \widetilde{\epsilon}$ satisfy the linear regression model (2.2.1), (2.2.4)–(2.2.6) with $g(\beta, \mathbf{z}) = \widetilde{\mathbf{Z}}_D\beta$.

  (b) Show that if $\mathbf{Z}_D$ has rank d, then the $\widehat{\beta}$ that minimizes

$$(\widetilde{\mathbf{Y}} - \widetilde{\mathbf{Z}}_D\beta)^T(\widetilde{\mathbf{Y}} - \widetilde{\mathbf{Z}}_D\beta) = (\mathbf{Y} - \mathbf{Z}_D\beta)^T\mathbf{W}^{-1}(\mathbf{Y} - \mathbf{Z}_D\beta)$$

is given by (2.2.20).

29. Let $e_i = (\epsilon_i + \epsilon_{i+1})/2$, $i = 1, \ldots, n$, where $\epsilon_1, \ldots, \epsilon_{n+1}$ are i.i.d. with mean zero and variance σ^2. The e_i are called *moving average errors*.
  Consider the model $Y_i = \mu + e_i$, $i = 1, \ldots, n$.

  (a) Show that $E(Y_{i+1} \mid Y_1, \ldots, Y_i) = \frac{1}{2}(\mu + Y_i)$. That is, in this model the optimal MSPE predictor of the future Y_{i+1} given the past $Y_1, \ldots, Y_i$ is $\frac{1}{2}(\mu + Y_i)$.

  (b) Show that $\bar{Y}$ is a multivariate method of moments estimate of μ. (See Problem 2.1.17.)

(c) Find a matrix A such that $e_{n\times 1} = A_{n\times(n+1)}\epsilon_{(n+1)\times 1}$.

(d) Find the covariance matrix W of e.

(e) Find the weighted least squares estimate of μ.

(f) The following data give the elapsed times $Y_1, \ldots, Y_n$ spent above a fixed high level for a series of $n = 66$ consecutive wave records at a point on the seashore. Use a weighted least squares computer routine to compute the weighted least squares estimate $\hat{\mu}$ of μ. Is $\hat{\mu}$ different from $\bar{Y}$?

> TABLE 2.5.1. Elapsed times spent above a certain high level for a series
> of 66 wave records taken at San Francisco Bay. The data (courtesy
> S. J. Chou) should be read row by row.

2.968	2.097	1.611	3.038	7.921	5.476	9.858	1.397	0.155	1.301
9.054	1.958	4.058	3.918	2.019	3.689	3.081	4.229	4.669	2.274
1.971	10.379	3.391	2.093	6.053	4.196	2.788	4.511	7.300	5.856
0.860	2.093	0.703	1.182	4.114	2.075	2.834	3.968	6.480	2.360
5.249	5.100	4.131	0.020	1.071	4.455	3.676	2.666	5.457	1.046
1.908	3.064	5.392	8.393	0.916	9.665	5.564	3.599	2.723	2.870
1.582	5.453	4.091	3.716	6.156	2.039				

30. In the multinomial Example 2.2.8, suppose some of the n_j are zero. Show that the MLE of θ_j is $\hat{\theta}$ with $\hat{\theta}_j = n_j/n$, $j = 1, \ldots, k$.

Hint: Suppose without loss of generality that $n_1 = n_2 = \cdots = n_q = 0, n_{q+1} > 0, \ldots, n_k > 0$. Then

$$p(\mathbf{x}, \boldsymbol{\theta}) = \prod_{j=q+1}^{k} \theta_j^{n_j},$$

which vanishes if $\theta_j = 0$ for any $j = q + 1, \ldots, k$.

31. Suppose $Y_1, \ldots, Y_n$ are independent with Y_i uniformly distributed on $[\mu_i - \sigma, \mu_i + \sigma]$, $\sigma > 0$, where $\mu_i = \sum_{j=1}^{p} z_{ij}\beta_j$ for given covariate values $\{z_{ij}\}$. Show that the MLE of $(\beta_1, \ldots, \beta_p, \sigma)^T$ is obtained by finding $\hat{\beta}_1, \ldots, \hat{\beta}_p$ that minimizes the *maximum absolute value* contrast function $\max_i |y_i - \mu_i|$ and then setting $\hat{\sigma} = \max_i |y_i - \hat{\mu}_i|$, where $\hat{\mu}_i = \sum_{j=1}^{p} z_{ij}\hat{\beta}_j$.

32. Suppose $Y_1, \ldots, Y_n$ are independent with Y_i having the Laplace density

$$\frac{1}{2\sigma} \exp\{-|y_i - \mu_i|/\sigma\}, \ \sigma > 0$$

where $\mu_i = \sum_{j=1}^{p} z_{ij}\beta_j$ for given covariate values $\{z_{ij}\}$.

(a) Show that the MLE of $(\beta_1, \ldots, \beta_p, \sigma)$ is obtained by finding $\hat{\beta}_1, \ldots, \hat{\beta}_p$ that minimizes the *least absolute deviation* contrast function $\sum_{i=1}^{n} |y_i - \mu_i|$ and then setting $\hat{\sigma} =$

$n^{-1} \sum_{i=1}^{n} |y_i - \widehat{\mu}_i|$, where $\widehat{\mu}_i = \sum_{j=1}^{p} z_{ij} \widehat{\beta}_j$. These $\widehat{\beta}_1, \ldots, \widehat{\beta}_r$ and $\widehat{\mu}_1, \ldots, \widehat{\mu}_n$ are called *least absolute deviation estimates (LADEs)*.

(b) Suppose $\mu_i = \mu$ for each i. Show that the sample median $\widehat{y}$, as defined in Section 1.3, is the minimizer of $\sum_{i=1}^{n} |y_i - \mu|$.

Hint: Use Problem 1.4.7 with Y having the empirical distribution $\widehat{F}$.

33. The *Hodges–Lehmann (location)* estimate $\widetilde{x}_{HL}$ is defined to be the median of the $\frac{1}{2} n(n+1)$ pairwise averages $\frac{1}{2}(x_i + x_j)$, $i \leq j$. An asymptotically equivalent procedure $\widetilde{x}_{HL}$ is to take the median of the distribution placing mass $\frac{2}{n^2}$ at each point $\frac{x_i + x_j}{2}$, $i < j$ and mass $\frac{1}{n^2}$ at each x_i.

(a) Show that the Hodges–Lehmann estimate is the minimizer of the contrast function

$$\rho(x, \theta) = \sum_{i \leq j} |x_i + x_j - 2\theta|.$$

Hint: See Problem 2.2.32(b).

(b) Define θ_{HL} to be the minimizer of

$$\int |x - 2\theta| d(F * F)(x)$$

where $F * F$ denotes convolution. Show that $\widetilde{x}_{HL}$ is a plug-in estimate of θ_{HL}.

34. Let X_i be i.i.d. as $(Z, Y)^T$ where $Y = Z + \sqrt{\lambda} W$, $\lambda > 0$, Z and W are independent $\mathcal{N}(0, 1)$. Find the MLE of λ and give its mean and variance.

Hint: See Example 1.6.3.

35. Let $g(x) = 1/\pi(1 + x^2)$, $x \in R$, be the Cauchy density, let X_1 and X_2 be i.i.d. with density $g(x - \theta)$, $\theta \in R$. Let x_1 and x_2 be the observations and set $\Delta = \frac{1}{2}(x_1 - x_2)$. Let $\widehat{\theta} = \arg \max L_{\mathbf{x}}(\theta)$ be "the" MLE.

(a) Show that if $|\Delta| \leq 1$, then the MLE exists and is unique. Give the MLE when $|\Delta| \leq 1$.

(b) Show that if $|\Delta| > 1$, then the MLE is not unique. Find the values of θ that maximize the likelihood $L_{\mathbf{x}}(\theta)$ when $|\Delta| > 1$.

Hint: Factor out $(\bar{x} - \theta)$ in the likelihood equation.

36. Problem 35 can be generalized as follows (Dharmadhikari and Joag–Dev, 1985). Let g be a probability density on R satisfying the following three conditions:

   1. g is continuous, symmetric about 0, and positive everywhere.
   2. g is twice continuously differentiable everywhere except perhaps at 0.
   3. If we write $h = \log g$, then $h''(y) > 0$ for some nonzero y.

Let (X_1, X_2) be a random sample from the distribution with density $f(x, \theta) = g(x - \theta)$, where $x \in R$ and $\theta \in R$. Let x_1 and x_2 be the observed values of X_1 and X_2 and write $\bar{x} = (x_1 + x_2)/2$ and $\Delta = (x_1 - x_2)/2$. The likelihood function is given by

$$\begin{aligned} L_{\mathbf{x}}(\theta) &= g(x_1 - \theta)g(x_2 - \theta) \\ &= g(\bar{x} + \Delta - \theta)g(\bar{x} - \Delta - \theta). \end{aligned}$$

Let $\widehat{\theta} = \arg\max L_{\mathbf{x}}(\theta)$ be "the" MLE.
     Show that

   (a) The likelihood is symmetric about $\bar{x}$.

   (b) Either $\widehat{\theta} = \bar{x}$ or $\widehat{\theta}$ is not unique.

   (c) There is an interval (a, b), $a < b$, such that for every $y \in (a, b)$ there exists a $\delta > 0$ such that $h(y + \delta) - h(y) > h(y) - h(y - \delta)$.

   (d) Use (c) to show that if $\Delta \in (a, b)$, then $\widehat{\theta}$ is not unique.

37. Suppose $X_1, \ldots, X_n$ are i.i.d. $\mathcal{N}(0, \sigma^2)$ and let $p(\mathbf{x}, \theta)$ denote their joint density. Show that the entropy of $p(\mathbf{x}, \theta)$ is $\frac{1}{2}n$ and that the Kullback–Liebler divergence between $p(\mathbf{x}, \theta)$ and $p(\mathbf{x}, \theta_0)$ is $\frac{1}{2}n(\theta - \theta_0)^2/\sigma^2$.

38. Let $\mathbf{X} \sim P_\theta$, $\theta \in \Theta$. Suppose h is a 1-1 function from Θ onto $\Omega = h(\Theta)$. Define $\eta = h(\theta)$ and let $p^*(\mathbf{x}, \eta) = p(\mathbf{x}, h^{-1}(\eta))$ denote the density or frequency function of $\mathbf{X}$ for the η parametrization. Let $K(\theta_0, \theta_1)$ $(K^*(\eta_0, \eta_1))$ denote the Kullback–Leibler divergence between $p(\mathbf{x}, \theta_0)$ and $p(x, \theta_1)$ $(p^*(\mathbf{x}, \eta_0)$ and $p^*(\mathbf{x}, \eta_1))$. Show that

$$K^*(\eta_0, \eta_1) = K(h^{-1}(\eta_0), h^{-1}(\eta_1)).$$

39. Let X_i denote the number of hits at a certain Web site on day i, $i = 1, \ldots, n$. Assume that $S = \sum_{i=1}^{n} X_i$ has a Poisson, $\mathcal{P}(n\lambda)$, distribution. On day $n + 1$ the Web Master decides to keep track of two types of hits (money making and not money making). Let V_j and W_j denote the number of hits of type 1 and 2 on day j, $j = n+1, \ldots, n+m$. Assume that $S_1 = \sum_{j=n+1}^{n+m} V_j$ and $S_2 = \sum_{j=n+1}^{n+m} W_j$ have $\mathcal{P}(m\lambda_1)$ and $\mathcal{P}(m\lambda_2)$ distributions, where $\lambda_1 + \lambda_2 = \lambda$. Also assume that S, S_1, and S_2 are independent. Find the MLEs of λ_1 and λ_2 based on S, S_1, and S_2.

40. Let $X_1, \ldots, X_n$ be a sample from the generalized Laplace distribution with density

$$
\begin{aligned}
f(x, \theta_1, \theta_2) &= \frac{1}{\theta_1 + \theta_2} \exp\{-x/\theta_1\},\ x > 0, \\
&= \frac{1}{\theta_1 + \theta_2} \exp\{x/\theta_2\},\ x < 0
\end{aligned}
$$

where $\theta_j > 0$, $j = 1, 2$.

   (a) Show that $T_1 = \sum X_i 1[X_i > 0]$ and $T_2 = \sum -X_i 1[X_i < 0]$ are sufficient statistics.

   (b) Find the maximum likelihood estimates of θ_1 and θ_2 in terms of T_1 and T_2. Carefully check the "$T_1 = 0$ or $T_2 = 0$" case.

41. The mean relative growth of an organism of size y at time t is sometimes modeled by the equation (Richards, 1959; Seber and Wild, 1989)

$$\frac{1}{y}\frac{dy}{dt} = \beta\left[1 - \left(\frac{y}{\alpha}\right)^{\frac{1}{\delta}}\right],\ y > 0;\ \alpha > 0,\ \beta > 0,\ \delta > 0.$$

(a) Show that a solution to this equation is of the form $y = g(t; \boldsymbol{\theta})$, where $\boldsymbol{\theta} = (\alpha, \beta, \mu, \delta)$, $\mu \in R$, and

$$g(t, \boldsymbol{\theta}) = \frac{\alpha}{\{1 + \exp[-\beta(t - \mu)/\delta]\}^\delta}.$$

(b) Suppose we have observations $(t_1, y_1), \ldots, (t_n, y_n)$, $n \geq 4$, on a population of a large number of organisms. Variation in the population is modeled on the log scale by using the model

$$\log Y_i = \log \alpha - \delta \log\{1 + \exp[-\beta(t_i - \mu)/\delta]\} + \epsilon_i$$

where $\epsilon_1, \ldots, \epsilon_n$ are uncorrelated with mean 0 and variance σ^2. Give the least squares estimating equations (2.1.7) for estimating α, β, δ, and μ.

(c) Let Y_i denote the response of the ith organism in a sample and let z_{ij} denote the level of the jth covariate (stimulus) for the ith organism, $i = 1, \ldots, n$; $j = 1, \ldots, p$. An example of a *neural net model* is

$$Y_i = \sum_{j=1}^{p} h(z_{ij}; \boldsymbol{\lambda}_j) + \epsilon_i, \; i = 1, \ldots, n$$

where $\boldsymbol{\lambda} = (\alpha, \beta, \mu)$, $h(z; \boldsymbol{\lambda}) = g(z; \alpha, \beta, \mu, 1)$, and $\epsilon_1, \ldots, \epsilon_n$ are uncorrelated with mean zero and variance σ^2. For the case $p = 1$, give the least square estimating equations (2.1.7) for α, β, and μ.

42. Suppose $X_1, \ldots, X_n$ satisfy the autoregressive model of Example 1.1.5.

(a) If μ is known, show that the MLE of β is

$$\widehat{\beta} = \frac{-\sum_{i=2}^{n}(x_{i-1} - \mu)(x_i - \mu)}{\sum_{i=1}^{n-1}(x_i - \mu)^2}.$$

(b) If β is known, find the covariance matrix $\mathbf{W}$ of the vector $\boldsymbol{\epsilon} = (\epsilon_1, \ldots, \epsilon_n)^T$ of autoregression errors. (One way to do this is to find a matrix $\mathbf{A}$ such that $\boldsymbol{\epsilon}_{n \times 1} = \mathbf{A}_{n \times n} \boldsymbol{\epsilon}_{n \times 1}$.) Then find the weighted least square estimate of μ. Is this also the MLE of μ?

Problems for Section 2.3

1. Suppose $Y_1, \ldots, Y_n$ are independent

$$P[Y_i = 1] = p(x_i, \alpha, \beta) = 1 - P[Y_i = 0], \; 1 \leq i \leq n, \; n \geq 2,$$

$$\log \frac{p}{1 - p}(x, \alpha, \beta) = \alpha + \beta x, \; x_1 < \cdots < x_n.$$

Show that the MLE of α, β exists iff $(Y_1, \ldots, Y_n)$ is not a sequence of 1's followed by all 0's or the reverse.

Hint:

$$c_1 \sum_{i=1}^{n} y_i + c_2 \sum_{i=1}^{n} x_i y_i = \sum_{i=1}^{n} (c_1 + c_2 x_i) y_i \leq \sum_{i=1}^{n} (c_1 + c_2 x_i) 1(c_2 x_i + c_1 \geq 0).$$

If $c_2 > 0$, the bound is sharp and is attained only if $y_i = 0$ for $x_i \leq -\frac{c_1}{c_2}$, $y_i = 1$ for $x_i \geq -\frac{c_1}{c_2}$.

2. Let $X_1, \ldots, X_n$ be i.i.d. gamma, $\Gamma(\lambda, p)$.

(a) Show that the density of $\mathbf{X} = (X_1, \ldots, X_n)^T$ can be written as the rank 2 canonical exponential family generated by $\mathbf{T} = (\Sigma \log X_i, \Sigma X_i)$ and $h(x) = x^{-1}$ with $\eta_1 = p$, $\eta_2 = -\lambda$ and

$$A(\eta_1, \eta_2) = n[\log \Gamma(\eta_1) - \eta_1 \log(-\eta_2)],$$

where Γ denotes the gamma function.

(b) Show that the likelihood equations are equivalent to (2.3.4) and (2.3.5).

3. Consider the Hardy–Weinberg model with the six genotypes given in Problem 2.1.15. Let $\Theta = \{(\theta_1, \theta_2) : \theta_1 > 0, \theta_2 > 0, \theta_1 + \theta_2 < 1\}$ and let $\theta_3 = 1 - (\theta_1 + \theta_2)$. In a sample of n independent plants, write $x_i = j$ if the ith plant has genotype j, $1 \leq j \leq 6$. Under what conditions on $(x_1, \ldots, x_n)$ does the MLE exist? What is the MLE? Is it unique?

4. Give details of the proof of Corollary 2.3.1.

5. Prove Lemma 2.3.1.
   Hint: Let $c = l(\mathbf{0})$. There exists a compact set $K \subset \Theta$ such that $l(\boldsymbol{\theta}) < c$ for all $\boldsymbol{\theta}$ not in K. This set K will have a point where the max is attained.

6. In the heteroscedastic regression Example 1.6.10 with $n \geq 3, 0 < z_1 < \cdots < z_n$, show that the MLE exists and is unique.

7. Let $Y_1, \ldots, Y_n$ denote the duration times of n independent visits to a Web site. Suppose $\mathbf{Y}$ has an exponential, $\mathcal{E}(\lambda_i)$, distribution where

$$\mu_i = E(Y_i) = \lambda_i^{-1} = \exp\{\alpha + \beta z_i\}, \quad z_1 < \cdots < z_n$$

and z_i is the income of the person whose duration time is Y_i, $0 < z_1 < \cdots < z_n$, $n \geq 2$. Show that the MLE of $(\alpha, \beta)^T$ exists and is unique. See also Problem 1.6.40.

8. Let $X_1, \ldots, X_n \in R^p$ be i.i.d. with density,

$$f_{\boldsymbol{\theta}}(\mathbf{x}) = c(\alpha) \exp\{-|\mathbf{x} - \boldsymbol{\theta}|^{\alpha}\}, \quad \boldsymbol{\theta} \in R^p, \; \alpha \geq 1$$

where $c^{-1}(\alpha) = \int_{R_p} \exp\{-|\mathbf{x}|^{\alpha}\} d\mathbf{x}$ and $|\cdot|$ is the Euclidean norm.

(a) Show that if $\alpha > 1$, the MLE $\widehat{\boldsymbol{\theta}}$ exists and is unique.

(b) Show that if $\alpha = 1$ and $p = 1$, the MLE $\widehat{\boldsymbol{\theta}}$ exists but is not unique if n is even.

9. Show that the boundary ∂C of a convex C set in R^k has volume 0.

Hint: If ∂C has positive volume, then it must contain a sphere and the center of the sphere is an interior point by (B.9.1).

10. Use Corollary 2.3.1 to show that in the multinomial Example 2.3.3, MLEs of η_j exist iff all $T_j > 0, 1 \le j \le k - 1$.

Hint: The k points $(0, \dots, 0), (0, n, 0, \dots, 0), \dots, (0, 0, \dots, n)$ are the vertices of the convex set $\{(t_1, \dots, t_{k-1}) : t_j \ge 0, 1 \le j \le k - 1, \sum_{j=1}^{k-1} t_j \le n\}$.

11. Prove Theorem 2.3.3.

Hint: If it didn't there would exist $\eta_j = c(\theta_j)$ such that $\eta_j^T t_0 - A(\eta_j) \to \max\{\eta^T t_0 - A(\eta) : \eta \in c(\Theta)\} > -\infty$. Then $\{\eta_j\}$ has a subsequence that converges to a point $\eta^0 \in \mathcal{E}$. But $c(\Theta)$ is closed so that $\eta^0 = c(\theta^0)$ and θ^0 must satisfy the likelihood equations.

12. Let $X_1, \dots, X_n$ be i.i.d. $\frac{1}{\sigma} f_0 \left(\frac{x-\mu}{\sigma} \right), \sigma > 0, \mu \in R$, and assume for $w \equiv -\log f_0$ that $w'' > 0$ so that w is strictly convex, $w(\pm\infty) = \infty$.

 (a) Show that, if $n \ge 2$, the likelihood equations

$$\sum_{i=1}^{n} w' \left(\frac{X_i - \mu}{\sigma} \right) = 0$$

$$\sum_{i=1}^{n} \left\{ \frac{(X_i - \mu)}{\sigma} w' \left(\frac{X_i - \mu}{\sigma} \right) - 1 \right\} = 0$$

have a unique solution $(\widehat{\mu}, \widehat{\sigma})$.

 (b) Give an algorithm such that starting at $\widehat{\mu}^0 = 0, \widehat{\sigma}^0 = 1, \widehat{\mu}^{(i)} \to \widehat{\mu}, \widehat{\sigma}^{(i)} \to \widehat{\sigma}$.

 (c) Show that for the logistic distribution $F_0(x) = [1 + \exp\{-x\}]^{-1}$, w is strictly convex and give the likelihood equations for μ and σ. (See Example 2.4.3.)

Hint: **(a)** The function $D(a, b) = \sum_{i=1}^{n} w(aX_i - b) - n \log a$ is strictly convex in (a, b) and $\lim_{(a,b) \to (a_0, b_0)} D(a, b) = \infty$ if either $a_0 = 0$ or ∞ or $b_0 = \pm\infty$.

 (b) Reparametrize by $a = \frac{1}{\sigma}, b = \frac{\mu}{\sigma}$ and consider varying a, b successively.

Note: You may use without proof (see Appendix B.9).

 (i) If a strictly convex function has a minimum, it is unique.

 (ii) If $\frac{\partial^2 D}{\partial a^2} > 0, \frac{\partial^2 D}{\partial b^2} > 0$ and $\frac{\partial^2 D}{\partial a^2} \frac{\partial^2 D}{\partial b^2} > \left(\frac{\partial^2 D}{\partial a \partial b} \right)^2$, then D is strictly convex.

13. Let $(X_1, Y_1), \dots, (X_n, Y_n)$ be a sample from a $\mathcal{N}(\mu_1, \mu_2, \sigma_1^2, \sigma_2^2, \rho)$ population.

 (a) Show that the MLEs of σ_1^2, σ_2^2, and ρ when μ_1 and μ_2 are assumed to be known are $\widetilde{\sigma}_1^2 = (1/n) \sum_{i=1}^{n} (X_i - \mu_1)^2, \widetilde{\sigma}_2^2 = (1/n) \sum_{i=1}^{n} (Y_i - \mu_2)^2$, and

$$\widetilde{\rho} = \left[\sum_{i=1}^{n} (X_i - \mu_1)(Y_i - \mu_2)/n\widetilde{\sigma}_1\widetilde{\sigma}_2 \right]$$

respectively, provided that $n \geq 3$.

(b) If $n \geq 5$ and μ_1 and μ_2 are unknown, show that the estimates of $\mu_1, \mu_2, \sigma_1^2, \sigma_2^2, \rho$ coincide with the method of moments estimates of Problem 2.1.8.

Hint: (b) Because (X_1, Y_1) has a density you may assume that $\tilde{\sigma}_1^2 > 0$, $\tilde{\sigma}_2^2 > 0$, $|\tilde{\rho}| < 1$. Apply Corollary 2.3.2.

Problems for Section 2.4

1. EM for bivariate data.

(a) In the bivariate normal Example 2.4.6, complete the E-step by finding $E(Z_i \mid Y_i)$, $E(Z_i^2 \mid Y_i)$ and $E(Z_i Y_i \mid Y_i)$.

(b) In Example 2.4.6, verify the M-step by showing that

$$E_\theta \mathbf{T} = (\mu_1, \mu_2, \sigma_1^2 + \mu_1^2, \sigma_2^2 + \mu_2^2, \rho\sigma_1\sigma_2 + \mu_1\mu_2).$$

2. Show that if T is minimal and $\mathcal{E}$ is open and the MLE doesn't exist, then the coordinate ascent algorithm doesn't converge to a member of $\mathcal{E}$.

3. Describe in detail what the coordinate ascent algorithm does in estimation of the regression coefficients in the Gaussian linear model

$$\mathbf{Y} = \mathbf{Z}_D\boldsymbol{\beta} + \boldsymbol{\epsilon}, \ \text{rank}(\mathbf{Z}_D) = k, \ \epsilon_1, \ldots, \epsilon_n \ \text{i.i.d.} \ \mathcal{N}(0, \sigma^2).$$

(Check that you are describing the Gauss–Seidel iterative method for solving a system of linear equations. See, for example, Golub and Van Loan, 1985, Chapter 10.)

4. Let (I_i, Y_i), $1 \leq i \leq n$, be independent and identically distributed according to P_θ, $\theta = (\lambda, \mu) \in (0, 1) \times R$ where

$$P_\theta[I_1 = 1] = \lambda = 1 - P_\theta[I_1 = 0],$$

and given $I_1 = j$, $Y_1 \sim \mathcal{N}(\mu, \sigma_j^2)$, $j = 0, 1$ and $\sigma_0^2 \neq \sigma_1^2$ known.

(a) Show that $\mathbf{X} \equiv \{(I_i, Y_i) : 1 \leq i \leq n\}$ is distributed according to an exponential family with $\mathbf{T} = \left(\frac{1}{\sigma_1^2}\sum_i Y_i I_i + \frac{1}{\sigma_0^2}\sum_i Y_i(1 - I_i), \sum_i I_i\right)$, $\eta_1 = \mu$, $\eta_2 = \log\left(\frac{\lambda}{1-\lambda}\right) + \frac{\mu^2}{2}\left(\frac{1}{\sigma_0^2} - \frac{1}{\sigma_1^2}\right)$.

(b) Deduce that $\mathbf{T}$ is minimal sufficient.

(c) Give explicitly the maximum likelihood estimates of μ and λ, when they exist.

5. Suppose the I_i in Problem 4 are not observed.

(a) Justify the following crude estimates of μ and λ,

$$\begin{aligned}
\tilde{\mu} &= \bar{Y} \\
\tilde{\lambda} &= \left(\tfrac{1}{n}\sum_{i=1}^{n}(Y_i - \bar{Y})^2 - \sigma_0^2\right) / (\sigma_1^2 - \sigma_0^2).
\end{aligned}$$

Do you see any problems with $\widetilde{\lambda}$?

(b) Give as explicitly as possible the E- and M-steps of the EM algorithm for this problem.

Hint: Use Bayes rule.

6. Consider a genetic trait that is directly unobservable but will cause a disease among a certain proportion of the individuals that have it. For families in which one member has the disease, it is desired to estimate the proportion θ that has the genetic trait. Suppose that in a family of n members in which one has the disease (and, thus, also the trait), X is the number of members who have the trait. Because it is known that $X \geq 1$, the model often used for X is that it has the conditional distribution of a $\mathcal{B}(n, \theta)$ variable, $\theta \in [0, 1]$, given $X \geq 1$.

(a) Show that $P(X = x \mid X \geq 1) = \dfrac{\binom{n}{x} \theta^x (1 - \theta)^{n-x}}{1 - (1 - \theta)^n}$, $x = 1, \ldots, n$, and that the MLE exists and is unique.

(b) Use $(2.4.3)$ to show that the Newton–Raphson algorithm gives

$$\widehat{\theta}_1 = \widetilde{\theta} - \frac{\widetilde{\theta}(1 - \widetilde{\theta})[1 - (1 - \widetilde{\theta})^n]\{x - n\widetilde{\theta} - x(1 - \widetilde{\theta})^n\}}{n\widetilde{\theta}^2(1 - \widetilde{\theta})^n[n - 1 + (1 - \widetilde{\theta})^n] - [1 - (1 - \widetilde{\theta})^n]^2[(1 - 2\widetilde{\theta})x + n\widetilde{\theta}^2]},$$

where $\widetilde{\theta} = \widehat{\theta}_{\text{old}}$ and $\widehat{\theta}_1 = \widehat{\theta}_{\text{new}}$, as the first approximation to the maximum likelihood estimate of θ.

(c) If $n = 5$, $x = 2$, find $\widehat{\theta}_1$ of (b) above using $\widetilde{\theta} = x/n$ as a preliminary estimate.

7. Consider the following algorithm under the conditions of Theorem 2.4.2. Define $\widehat{\eta}^0$ as before. Let

$$\widehat{\eta}(\lambda) \equiv \widehat{\eta}_{\text{old}} + \lambda \ddot{A}^{-1}(\widehat{\eta}_{\text{old}})(\dot{A}(\widehat{\eta}_{\text{old}}) - \mathbf{t}_0)$$

and

$$\widehat{\eta}_{\text{new}} = \widehat{\eta}(\lambda^*)$$

where λ^* maximizes

$$\mathbf{t}_0^T \widehat{\eta}(\lambda) - A(\widehat{\eta}(\lambda)).$$

Show that the sequence defined by this algorithm converges to the MLE if it exists.

Hint: Apply the argument of the proof of Theorem 2.4.2 noting that the sequence of iterates $\{\widehat{\eta}_m\}$ is bounded and, hence, the sequence $(\widehat{\eta}_m, \widehat{\eta}_{m+1})$ has a convergent subsequence.

8. Let X_1, X_2, X_3 be independent observations from the Cauchy distribution about θ, $f(x, \theta) = \pi^{-1}(1 + (x - \theta)^2)^{-1}$. Suppose $X_1 = 0$, $X_2 = 1$, $X_3 = a$. Show that for a

sufficiently large the likelihood function has local maxima between 0 and 1 and between p and a.

(a) Deduce that depending on where bisection is started the sequence of iterates may converge to one or the other of the local maxima.

(b) Make a similar study of the Newton–Raphson method in this case.

9. Let $X_1, \ldots, X_n$ be i.i.d. where $X = (U, V, W)$, $P[U = a, V = b, W = c] \equiv p_{abc}$, $1 \le a \le A$, $1 \le b \le B$, $1 \le c \le C$ and $\sum_{a,b,c} p_{abc} = 1$.

(a) Suppose for all a, b, c,

$$(1) \; \log p_{abc} = \mu_{ac} + \nu_{bc} \text{ where } -\infty < \mu, \nu < \infty.$$

Show that this holds iff

$$P[U = a, V = b \mid W = c] = P[U = a \mid W = c]P[V = b \mid W = c],$$

i.e. iff U and V are independent given W.

(b) Show that the family of distributions obtained by letting μ, ν vary freely is an exponential family of rank $(C - 1) + C(A + B - 2) = C(A + B - 1) - 1$ generated by $N_{++c}, N_{a+c}, N_{+bc}$ where $N_{abc} = \#\{i : X_i = (a, b, c)\}$ and "+" indicates summation over the index.

(c) Show that the MLEs exist iff $0 < N_{a+c}, N_{+bc} < N_{++c}$ for all a, b, c and then are given by

$$\widehat{p}_{abc} = \frac{N_{++c}}{n} \frac{N_{a+c}}{N_{++c}} \frac{N_{+bc}}{N_{++c}}.$$

Hint:

(b) Consider $N_{a+c} - N_{++c}/A$, $N_{+bc} - N_{++c}/B$, N_{++c}.

(c) The model implies $\widehat{p}_{abc} = \widehat{p}_{+bc}\widehat{p}_{a+c}/\widehat{p}_{++c}$ and use the likelihood equations.

10. Suppose X is as in Problem 9, but now

$$(2) \; \log p_{abc} = \mu_{ac} + \nu_{bc} + \gamma_{ab} \text{ where } \mu, \nu, \gamma \text{ vary freely.}$$

(a) Show that this is an exponential family of rank

$$A + B + C - 3 + (A - 1)(C - 1) + (B - 1)(C - 1) + (A - 1)(B - 1)$$
$$= AB + AC + BC - (A + B + C).$$

(b) Consider the following "proportional fitting" algorithm for finding the maximum likelihood estimate in this model.

Initialize: $\widehat{p}_{abc}^{(0)} = \dfrac{N_{a++}}{n} \dfrac{N_{+b+}}{n} \dfrac{N_{++c}}{n}$

$$\widehat{p}_{abc}^{(1)} = \frac{N_{ab+}}{n} \frac{\widehat{p}_{abc}^{(0)}}{\widehat{p}_{ab+}^{(0)}}$$

$$\widehat{p}_{abc}^{(2)} = \frac{N_{a+c}}{n} \frac{\widehat{p}_{abc}^{(1)}}{\widehat{p}_{a+c}^{(1)}}$$

$$\widehat{p}_{abc}^{(3)} = \frac{N_{+bc}}{n} \frac{\widehat{p}_{abc}^{(2)}}{\widehat{p}_{+bc}^{(2)}}.$$

Reinitialize with $\widehat{p}_{abc}^{(3)}$. Show that the algorithm converges to the MLE if it exists and diverges otherwise.

Hint: Note that because $\{p_{abc}^{(0)}\}$ belongs to the model so do all subsequent iterates and that $\widehat{p}_{abc}^{(1)}$ is the MLE for the exponential family

$$p_{abc} = \frac{e^{\mu_{ab}} p_{abc}^{(0)}}{\displaystyle\sum_{a',b',c'} e^{\mu_{a'b'}} p_{a'b'c'}^{(0)}}$$

obtained by fixing the "b, c" and "a, c" parameters.

11. (a) Show that **S** in Example 2.4.5 has the specified mixture of Gaussian distribution.

(b) Give explicitly the E- and M-steps of the EM algorithm in this case.

12. Justify formula (2.4.8).
   Hint: $P_{\theta_0}[X = x \mid S(X) = s] = \frac{p(x,\theta_0)}{q(s,\theta_0)} 1(S(x) = s)$.

13. Let $f_\theta(x) = f_0(x - \theta)$ where

$$f_0(x) = \frac{1}{3}\varphi(x) + \frac{2}{3}\varphi(x - a)$$

and φ is the $\mathcal{N}(0, 1)$ density. Show for $n = 1$ that bisection may lead to a local maximum of the likelihood, if a is sufficiently large.

14. Establish the last claim in part (2) of the proof of Theorem 2.4.2.
   Hint: Use the canonical nature of the family and openness of $\mathcal{E}$.

15. Verify the formula given below (2.4.26) in Example 2.4.4 for the actual MLE in that example.
   Hint: Show that $\{(\theta_m, \theta_{m+1})\}$ has a subsequence converging to (θ^*, θ^*) and necessarily $\theta^* = \widehat{\theta}_0$.

16. Establish part (b) of Theorem 2.4.3.
   Hint: Show that $\{(\theta_m, \theta_{m+1})\}$ has a subsequence converging to (θ^*, θ^*) and, thus, necessarily θ^* is the global maximizer.

17. *Limitations of the missing value model of Example 2.4.6.* The assumption underlying Example 2.4.6 is that the conditional probability that a component X_j of the data vector X is missing given the rest of the data vector is not a function of X_j. That is, given $X - \{X_j\}$, the process determining whether X_j is missing is independent of X_j. This condition is called *missing at random.* For example, in Example 2.4.6, the probability that Y_i is missing may depend on Z_i, but not on Y_i. That is, given Z_i, the "missingness" of Y_i is independent of Y_i. If Y_i represents the seriousness of a disease, this assumption may not be satisfied. For instance, suppose all subjects with $Y_i \geq 2$ drop out of the study. Then using the E-step to impute values for the missing Y's would greatly underpredict the actual Y's because all the Y's in the imputation would have $Y \leq 2$. In Example 2.4.6, suppose Y_i is missing iff $Y_i \geq 2$. If $\mu_2 = 1.5$, $\sigma_1 = \sigma_2 = 1$ and $\rho = 0.5$, find the probability that $E(Y_i \mid Z_i)$ underpredicts Y_i.

18. *EM and Regression.* For $X = \{(Z_i, Y_i) : i = 1, \ldots, n\}$, consider the model

$$Y_i = \beta_1 + \beta_2 Z_i + \epsilon_i$$

where $\epsilon_1, \ldots, \epsilon_n$ are i.i.d. $\mathcal{N}(0, \sigma^2)$, $Z_1, \ldots, Z_n$ are i.i.d. $\mathcal{N}(\mu_1, \sigma_1^2)$ and independent of $\epsilon_1, \ldots, \epsilon_n$. Suppose that for $1 \leq i \leq m$ we observe both Z_i and Y_i and for $m + 1 \leq i \leq n$, we observe only Y_i. Complete the E- and M-steps of the EM algorithm for estimating $(\mu_1, \beta_1, \sigma_1^2, \sigma^2, \beta_2)$.

2.6 NOTES

Notes for Section 2.1

(1) "Natural" now was not so natural in the eighteenth century when the least squares principle was introduced by Legendre and Gauss. For a fascinating account of the beginnings of estimation in the context of astronomy see Stigler (1986).

(2) The frequency plug-in estimates are sometimes called *Fisher consistent.* R. A. Fisher (1922) argued that only estimates possessing the substitution property should be considered and the best of these selected. These considerations lead essentially to maximum likelihood estimates.

Notes for Section 2.2

(1) An excellent historical account of the development of least squares methods may be found in Eisenhart (1964).

(2) For further properties of Kullback–Leibler divergence, see Cover and Thomas (1991).

Note for Section 2.3

(1) Recall that in an exponential family, for any A, $P[\mathbf{T}(X) \in A] = 0$ for all or for no $P \in \mathcal{P}$.

Note for Section 2.5

(1) In the econometrics literature (e.g. Appendix A.2; Campbell, Lo, and MacKinlay, 1997), a multivariate version of minimum contrasts estimates are often called generalized method of moment estimates.

2.7 REFERENCES

BARLOW, R. E., D. J. BARTHOLOMEW, J. M. BREMNER, AND H. D. BRUNK, *Statistical Inference Under Order Restrictions* New York: Wiley, 1972.

BAUM, L. E., T. PETRIE, G. SOULES, AND N. WEISS, "A Maximization Technique Occurring in the Statistical Analysis of Probabilistic Functions of Markov Chains," *Ann. Math. Statist., 41*, 164–171 (1970).

BISHOP, Y. M. M., S. E. FEINBERG, AND P. W. HOLLAND, *Discrete Multivariate Analysis: Theory and Practice* Cambridge, MA: MIT Press, 1975.

CAMPBELL, J. Y., A. W. LO, AND A. C. MACKINLAY, *The Econometrics of Financial Markets* Princeton, NJ: Princeton University Press, 1997.

COVER, T. M., AND J. A. THOMAS, *Elements of Information Theory* New York: Wiley, 1991.

DAHLQUIST, G., A. BJÖRK, AND N. ANDERSON, *Numerical Analysis* New York: Prentice Hall, 1974.

DEMPSTER, A., M. M. LAIRD, AND D. B. RUBIN, "Maximum Likelihood Estimation from Incomplete Data via the EM Algorithm," *J. Roy. Statist. Soc. B*, 1–38 (1977).

DHARMADHIKARI, S., AND K. JOAG-DEV, "Examples of Nonunique Maximum Likelihood Estimators," *The American Statistician, 39*, 199–200 (1985).

EISENHART, C., "The Meaning of Least in Least Squares," *Journal Wash. Acad. Sciences, 54*, 24–33 (1964).

FAN, J., AND I. GIJBELS, *Local Polynomial Modelling and Its Applications* London: Chapman and Hall, 1996.

FISHER, R. A., "On the Mathematical Foundations of Theoretical Statistics," reprinted in *Contributions to Mathematical Statistics* (by R. A. Fisher 1950) New York: J. Wiley and Sons, 1922.

GOLUB, G. H., AND C. F. VAN LOAN, *Matrix Computations* Baltimore: John Hopkins University Press, 1985.

HABERMAN, S. J., *The Analysis of Frequency Data* Chicago: University of Chicago Press, 1974.

KOLMOGOROV, A. N., "On the Shannon Theory of Information Transmission in the Case of Continuous Signals," *IRE Transf. Inform. Theory, IT2*, 102–108 (1956).

LITTLE, R. J. A., AND D. B. RUBIN, *Statistical Analysis with Missing Data* New York: J. Wiley, 1987.

MACLACHLAN, G. J., AND T. KRISHNAN, *The EM Algorithm and Extensions* New York: Wiley, 1997.

MOSTELLER, F., "Association and Estimation in Contingency Tables," *J. Amer. Statist. Assoc., 63*, 1–28 (1968).

RICHARDS, F. J., "A Flexible Growth Function for Empirical Use," *J. Exp. Botany, 10*, 290–300 (1959).

RUPPERT, D., AND M. P. WAND, "Multivariate Locally Weighted Least Squares Regression," *Ann. Statist., 22*, 1346–1370 (1994).

SEBER, G. A. F., AND C.J. WILD, *Nonlinear Regression* New York: Wiley, 1989.

SHANNON, C. E., "A Mathematical Theory of Communication," *Bell System Tech. Journal, 27*, 379–243, 623–656 (1948).

SNEDECOR, G. W., AND W. COCHRAN, *Statistical Methods*, 6th ed. Ames, IA: Iowa State University Press, 1967.

STIGLER, S., *The History of Statistics* Cambridge, MA: Harvard University Press, 1986.

WEISBERG, S., *Applied Linear Regression*, 2nd ed. New York: Wiley, 1985.

WU, C. F. J., "On the Convergence Properties of the EM Algorithm," *Ann. Statist., 11*, 95–103 (1983).

Chapter 3

MEASURES OF PERFORMANCE, NOTIONS OF OPTIMALITY, AND OPTIMAL PROCEDURES

3.1 INTRODUCTION

Here we develop the theme of Section 1.3, which is how to appraise and select among decision procedures. In Sections 3.2 and 3.3 we show how the important Bayes and minimax criteria can in principle be implemented. However, actual implementation is limited. Our examples are primarily estimation of a real parameter. In Section 3.4, we study, in the context of estimation, the relation of the two major decision theoretic principles to the non-decision theoretic principle of maximum likelihood and the somewhat out of favor principle of unbiasedness. We also discuss other desiderata that strongly compete with decision theoretic optimality, in particular computational simplicity and robustness. We return to these themes in Chapter 6, after similarly discussing testing and confidence bounds, in Chapter 4 and developing in Chapters 5 and 6 the asymptotic tools needed to say something about the multiparameter case.

3.2 BAYES PROCEDURES

Recall from Section 1.3 that if we specify a parametric model $\mathcal{P} = \{P_\theta : \theta \in \Theta\}$, action space $\mathcal{A}$, loss function $l(\theta, a)$, then for data $X \sim P_\theta$ and any decision procedure δ randomized or not we can define its risk function, $R(\cdot, \delta) : \Theta \to R^+$ by

$$R(\theta, \delta) = E_\theta l(\theta, \delta(X)).$$

We think of $R(\cdot, \delta)$ as measuring a priori the performance of δ for this model. Strict comparison of δ_1 and δ_2 on the basis of the risks alone is not well defined unless $R(\theta, \delta_1) \leq R(\theta, \delta_2)$ for all θ or vice versa. However, by introducing a Bayes prior density (say) π for θ comparison becomes unambiguous by considering the scalar Bayes risk,

$$r(\pi, \delta) \equiv ER(\boldsymbol{\theta}, \delta) = El(\boldsymbol{\theta}, \delta(X)), \tag{3.2.1}$$

where $(\boldsymbol{\theta}, X)$ is given the joint distribution specified by (1.2.3). Recall also that we can define

$$R(\pi) = \inf\{r(\pi, \delta) : \delta \in \mathcal{D}\} \tag{3.2.2}$$

the *minimum Bayes risk* of the problem, and that in Section 1.3 we showed how in an example, we could identify the *Bayes rules* δ_π^* such that

$$r(\pi, \delta_\pi^*) = R(\pi). \tag{3.2.3}$$

In this section we shall show systematically how to construct Bayes rules. This exercise is interesting and important even if we do not view π as reflecting an implicitly believed in prior distribution on $\boldsymbol{\theta}$. After all, if π is a density and $\Theta \subset R$

$$r(\pi, \delta) = \int R(\theta, \delta)\pi(\theta)d\theta \tag{3.2.4}$$

and π may express that we care more about the values of the risk in some rather than other regions of Θ. For testing problems the hypothesis is often treated as more important than the alternative. We may have vague prior notions such as "$|\theta| \geq 5$ is physically implausible" if, for instance, θ denotes mean height of people in meters. If π is then thought of as a weight function roughly reflecting our knowledge, it is plausible that δ_π^* if computable will behave reasonably even if our knowledge is only roughly right. Clearly, $\pi(\theta) \equiv c$ plays a special role ("equal weight") though (Problem 3.2.4) the parametrization plays a crucial role here. It is in fact clear that prior and loss function cannot be separated out clearly either. Thus, considering $l_1(\theta, a)$ and $\pi_1(\theta)$ is equivalent to considering $l_2(\theta, a) = \pi_1(\theta)l_1(\theta, a)$ and $\pi_2(\theta) \equiv 1$. Issues such as these and many others are taken up in the fundamental treatises on Bayesian statistics such as Jeffreys (1948) and Savage (1954) and are reviewed in the modern works of Berger (1985) and Bernardo and Smith (1994). We don't pursue them further except in Problem 3.2.5, and instead turn to construction of Bayes procedures.

We first consider the problem of estimating $q(\theta)$ with quadratic loss, $l(\theta, a) = (q(\theta) - a)^2$, using a nonrandomized decision rule δ. Suppose $\boldsymbol{\theta}$ is a random variable (or vector) with (prior) frequency function or density $\pi(\theta)$. Our problem is to find the function δ of $\mathbf{X}$ that minimizes $r(\pi, \delta) = E(q(\boldsymbol{\theta}) - \delta(\mathbf{X}))^2$. This is just the problem of finding the best mean squared prediction error (MSPE) predictor of $q(\theta)$ given $\mathbf{X}$ (see Remark 1.4.5). Using our results on MSPE prediction, we find that either $r(\pi, \delta) = \infty$ for all δ or the Bayes rule δ^* is given by

$$\delta^*(\mathbf{X}) = E[q(\boldsymbol{\theta}) \mid \mathbf{X}]. \tag{3.2.5}$$

This procedure is called the *Bayes estimate for squared error loss*.

In view of formula (1.2.8) for the posterior density and frequency functions, we can give the Bayes estimate a more explicit form. In the continuous case with θ real valued and prior density π,

$$\delta^*(\mathbf{x}) = \frac{\int_{-\infty}^{\infty} q(\theta)p(x \mid \theta)\pi(\theta)d\theta}{\int_{-\infty}^{\infty} p(x \mid \theta)\pi(\theta)d\theta}. \tag{3.2.6}$$

In the discrete case, as usual, we just need to replace the integrals by sums. Here is an example.

Example 3.2.1. *Bayes Estimates for the Mean of a Normal Distribution with a Normal Prior.* Suppose that we want to estimate the mean θ of a normal distribution with known variance σ^2 on the basis of a sample $X_1, \ldots, X_n$. If we choose the conjugate prior $\mathcal{N}(\eta_0, \tau^2)$ as in Example 1.6.12, we obtain the posterior distribution

$$\mathcal{N}\left(\eta_0\left(\frac{\sigma^2}{n\tau^2 + \sigma^2}\right) + \bar{x}\left(\frac{n\tau^2}{n\tau^2 + \sigma^2}\right), \frac{\sigma^2}{n}\left(1 + \frac{\sigma^2}{n\tau^2}\right)^{-1}\right).$$

The Bayes estimate is just the mean of the posterior distribution

$$\delta^*(\mathbf{X}) = \eta_0\left[\frac{1/\tau^2}{n/\sigma^2 + 1/\tau^2}\right] + \bar{X}\left[\frac{n/\sigma^2}{n/\sigma^2 + 1/\tau^2}\right]. \qquad (3.2.7)$$

Its Bayes risk (the MSPE of the predictor) is just

$$\begin{aligned}
r(\pi, \delta^*) &= E(\theta - E(\theta \mid \mathbf{X}))^2 = E[E((\theta - E(\theta \mid \mathbf{X}))^2 \mid \mathbf{X})] \\
&= E\left[\frac{\sigma^2}{n} \Big/ \left(1 + \frac{\sigma^2}{n\tau^2}\right)\right] = \frac{1}{n/\sigma^2 + 1/\tau^2}.
\end{aligned}$$

No finite choice of η_0 and τ^2 will lead to $\bar{X}$ as a Bayes estimate. But $\bar{X}$ is the limit of such estimates as prior knowledge becomes "vague" ($\tau \to \infty$ with η_0 fixed). In fact, $\bar{X}$ is the estimate that (3.2.6) yields, if we substitute the prior "density" $\pi(\theta) \equiv 1$ (Problem 3.2.1). Such priors with $\int \pi(\theta) = \infty$ or $\sum \pi(\theta) = \infty$ are called *improper*. The resulting Bayes procedures are also called improper.

Formula (3.2.7) reveals the Bayes estimate in the proper case to be a weighted average

$$w\eta_0 + (1 - w)\bar{X}$$

of the estimate to be used when there are no observations, that is, η_0, and $\bar{X}$ with weights inversely proportional to the Bayes risks of these two estimates. Because the Bayes risk of $\bar{X}$, σ^2/n, tends to 0 as $n \to \infty$, the Bayes estimate corresponding to the prior density $\mathcal{N}(\eta_0, \tau^2)$ differs little from $\bar{X}$ for n large. In fact, $\bar{X}$ is approximately a Bayes estimate for any one of these prior distributions in the sense that $[r(\pi, \bar{X}) - r(\pi, \delta^*)]/r(\pi, \delta^*) \to 0$ as $n \to \infty$. For more on this, see Section 5.5.   □

We now turn to the problem of finding Bayes rules for general action spaces $\mathcal{A}$ and loss functions l. To begin with we consider only nonrandomized rules. If we look at the proof of Theorem 1.4.1, we see that the key idea is to consider what we should do given $\mathbf{X} = \mathbf{x}$. Thus, $E(Y \mid \mathbf{X})$ is the best predictor because $E(Y \mid \mathbf{X} = \mathbf{x})$ minimizes the conditional MSPE $E((Y - a)^2 \mid \mathbf{X} = \mathbf{x})$ as a function of the action a. Applying the same idea in the general Bayes decision problem, we form the *posterior risk*

$$r(a \mid \mathbf{x}) = E(l(\boldsymbol{\theta}, a) \mid \mathbf{X} = \mathbf{x}).$$

This quantity $r(a \mid \mathbf{x})$ is what we expect to lose, if $\mathbf{X} = \mathbf{x}$ and we use action a. Intuitively, we should, for each $\mathbf{x}$, take that action $a = \delta^*(\mathbf{x})$ that makes $r(a \mid \mathbf{x})$ as small as possible. This action need not exist nor be unique if it does exist. However,

Proposition 3.2.1. *Suppose that there exists a function* $\delta^*(\mathbf{x})$ *such that*

$$r(\delta^*(\mathbf{x}) \mid \mathbf{x}) = \inf\{r(a \mid \mathbf{x}) : a \in \mathcal{A}\}. \qquad (3.2.8)$$

Then δ^* *is a Bayes rule.*

Proof. As in the proof of Theorem 1.4.1, we obtain for any δ

$$r(\pi, \delta) = E[l(\boldsymbol{\theta}, \delta(\mathbf{X}))] = E[E(l(\boldsymbol{\theta}, \delta(\mathbf{X})) \mid \mathbf{X})]. \qquad (3.2.9)$$

But, by (3.2.8),

$$E[l(\theta, \delta(\mathbf{X})) \mid \mathbf{X} = \mathbf{x}] = r(\delta(\mathbf{x}) \mid \mathbf{x}) \geq r(\delta^*(\mathbf{x}) \mid \mathbf{x}) = E[l(\theta, \delta^*(\mathbf{X})) \mid \mathbf{X} = \mathbf{x}].$$

Therefore,

$$E[l(\theta, \delta(\mathbf{X})) \mid \mathbf{X}] \geq E[l(\theta, \delta^*(\mathbf{X})) \mid \mathbf{X}],$$

and the result follows from (3.2.9). $\qquad\qquad\qquad\qquad\qquad\qquad\qquad\qquad \square$

As a first illustration, consider the oil-drilling example (Example 1.3.5) with prior $\pi(\theta_1) = 0.2$, $\pi(\theta_2) = 0.8$. Suppose we observe $x = 0$. Then the posterior distribution of $\boldsymbol{\theta}$ is by (1.2.8)

$$\pi(\theta_1 \mid X = 0) = \frac{1}{9}, \ \pi(\theta_2 \mid X = 0) = \frac{8}{9}.$$

Thus, the posterior risks of the actions a_1, a_2, and a_3 are

$$
\begin{array}{rclcrcl}
r(a_1 \mid 0) & = & \dfrac{1}{9}l(\theta_1, a_1) & + & \dfrac{8}{9}l(\theta_2, a_1) & = & 10.67 \\[2mm]
r(a_2 \mid 0) & = & 2, & & r(a_3 \mid 0) & = & 5.89.
\end{array}
$$

Therefore, a_2 has the smallest posterior risk and, if δ^* is the Bayes rule,

$$\delta^*(0) = a_2.$$

Similarly,

$$r(a_1 \mid 1) = 8.35, \ r(a_2 \mid 1) = 3.74, \ r(a_3 \mid 1) = 5.70$$

and we conclude that

$$\delta^*(1) = a_2.$$

Therefore, $\delta^* = \delta_5$ as we found previously. The great advantage of our new approach is that it enables us to compute the Bayes procedure without undertaking the usually impossible calculation of the Bayes risks of all competing procedures.

More generally consider the following class of situations.

Example 3.2.2. *Bayes Procedures When* Θ *and* $\mathcal{A}$ *Are Finite.* Let $\Theta = \{\theta_0, \ldots, \theta_p\}$, $\mathcal{A} = \{a_0, \ldots, a_q\}$, let $w_{ij} \geq 0$ be given constants, and let the loss incurred when θ_i is true and action a_j is taken be given by

$$l(\theta_i, a_j) = w_{ij}.$$

Let $\pi(\theta)$ be a prior distribution assigning mass π_i to θ_i, so that $\pi_i \geq 0$, $i = 0, \ldots, p$, and $\sum_{i=0}^{p} \pi_i = 1$. Suppose, moreover, that $\mathbf{X}$ has density or frequency function $p(\mathbf{x} \mid \theta)$ for each θ. Then, by (1.2.8), the posterior probabilities are

$$P[\boldsymbol{\theta} = \theta_i \mid \mathbf{X} = \mathbf{x}] = \frac{\pi_i p(\mathbf{x} \mid \theta_i)}{\sum_j \pi_j p(\mathbf{x} \mid \theta_j)}$$

and, thus,

$$r(a_j \mid \mathbf{x}) = \frac{\sum_i \omega_{ij} \pi_i p(\mathbf{x} \mid \theta_i)}{\sum_i \pi_i p(\mathbf{x} \mid \theta_i)}. \tag{3.2.10}$$

The optimal action $\delta^*(\mathbf{x})$ has

$$r(\delta^*(\mathbf{x}) \mid \mathbf{x}) = \min_{0 \leq j \leq q} r(a_j \mid \mathbf{x}).$$

Here are two interesting specializations.

(a) *Classification:* Suppose that $p = q$, we identify a_j with θ_j, $j = 0, \ldots, p$, and let

$$\omega_{ij} = 1, \quad i \neq j$$
$$\omega_{ii} = 0.$$

This can be thought of as the *classification problem* in which we have $p + 1$ known disjoint populations and a new individual $\mathbf{X}$ comes along who is to be classified in one of these categories. In this case,

$$r(\theta_i \mid \mathbf{x}) = P[\boldsymbol{\theta} \neq \theta_i \mid \mathbf{X} = \mathbf{x}]$$

and minimizing $r(\theta_i \mid \mathbf{x})$ is equivalent to the reasonable procedure of maximizing the posterior probability,

$$P[\boldsymbol{\theta} = \theta_i \mid \mathbf{X} = \mathbf{x}] = \frac{\pi_i p(\mathbf{x} \mid \theta_i)}{\sum_j \pi_j p(\mathbf{x} \mid \theta_j)}.$$

(b) *Testing:* Suppose $p = q = 1$, $\pi_0 = \pi$, $\pi_1 = 1 - \pi$, $0 < \pi < 1$, a_0 corresponds to deciding $\theta = \theta_0$ and a_1 to deciding $\theta = \theta_1$. This is a special case of the testing formulation of Section 1.3 with $\Theta_0 = \{\theta_0\}$ and $\Theta_1 = \{\theta_1\}$. The Bayes rule is then to

$$\text{decide } \theta = \theta_1 \text{ if } (1 - \pi)p(\mathbf{x} \mid \theta_1) > \pi p(\mathbf{x} \mid \theta_0)$$
$$\text{decide } \theta = \theta_0 \text{ if } (1 - \pi)p(\mathbf{x} \mid \theta_1) < \pi p(\mathbf{x} \mid \theta_0)$$

and decide either a_0 or a_1 if equality occurs. See Sections 1.3 and 4.2 on the option of randomizing between a_0 and a_1 if equality occurs. As we let π vary between zero and one, we obtain what is called the class of *Neyman–Pearson tests*, which provides the solution to the problem of minimizing P (type II error) given P (type I error) $\leq \alpha$. This is treated further in Chapter 4. $\qquad \Box$

To complete our illustration of the utility of Proposition 3.2.1, we exhibit in "closed form" the Bayes procedure for an estimation problem when the loss is not quadratic.

Example 3.2.3. *Bayes Estimation of the Probability of Success in n Bernoulli Trials.* Suppose that we wish to estimate θ using $X_1, \ldots, X_n$, the indicators of n Bernoulli trials with probability of success θ. We shall consider the loss function l given by

$$l(\theta, a) = \frac{(\theta - a)^2}{\theta(1 - \theta)}, \quad 0 < \theta < 1, \ a \text{ real.} \tag{3.2.11}$$

This close relative of quadratic loss gives more weight to parameter values close to zero and one. Thus, for θ close to zero, this $l(\theta, a)$ is close to the relative squared error $(\theta - a)^2/\theta$. It makes $\bar{X}$ have constant risk, a property we shall find important in the next section. The analysis can also be applied to other loss functions. See Problem 3.2.5.

By sufficiency we need only consider the number of successes, S. Suppose now that we have a prior distribution. Then, if all terms on the right-hand side are finite,

$$
\begin{aligned}
r(a \mid k) \ &= \ E\left\{ \frac{(\theta - a)^2}{\theta(1 - \theta)} \,\Big|\, S = k \right\} = E\left\{ \frac{\theta}{(1 - \theta)} \,\Big|\, S = k \right\} \\
&- \ 2aE\left\{ \frac{1}{(1 - \theta)} \,\Big|\, S = k \right\} + a^2 E\left\{ \frac{1}{\theta(1 - \theta)} \,\Big|\, S = k \right\}.
\end{aligned}
\tag{3.2.12}
$$

Minimizing this parabola in a, we find our Bayes procedure is given by

$$\delta^*(k) = \frac{E(1/(1 - \theta) \mid S = k)}{E(1/\theta(1 - \theta) \mid S = k)} \tag{3.2.13}$$

provided the denominator is not zero. For convenience let us now take as prior density the density $b_{r,s}(\theta)$ of the beta distribution $\beta(r, s)$. In Example 1.2.1 we showed that this leads to a $\beta(k + r, n + s - k)$ posterior distribution for θ if $S = k$. If $1 \le k \le n - 1$ and $n \ge 2$, then all quantities in (3.2.12) and (3.2.13) are finite, and

$$
\begin{aligned}
\delta^*(k) \ &= \ \frac{\int_0^1 (1/(1 - \theta)) b_{k+r, n-k+s}(\theta) d\theta}{\int_0^1 (1/\theta(1 - \theta)) b_{k+r, n-k+s}(\theta) d\theta} \\
&= \ \frac{B(k + r, n - k + s - 1)}{B(k + r - 1, n - k + s - 1)} = \frac{k + r - 1}{n + s + r - 2},
\end{aligned}
\tag{3.2.14}
$$

where we are using the notation B.2.11 of Appendix B. If $k = 0$, it is easy to see that $a = 0$ is the only a that makes $r(a \mid k) < \infty$. Thus, $\delta^*(0) = 0$. Similarly, we get $\delta^*(n) = 1$. If we assume a uniform prior density, $(r = s = 1)$, we see that the Bayes procedure is the usual estimate, $\bar{X}$. This is *not* the case for quadratic loss (see Problem 3.2.2). □

"Real" computation of Bayes procedures

The closed forms of (3.2.6) and (3.2.10) make the computation of (3.2.8) appear straightforward. Unfortunately, this is far from true in general. Suppose, as is typically the case, that $\theta = (\theta_1, \ldots, \theta_p)$ has a hierarchically defined prior density,

$$\pi(\theta_1, \theta_2, \ldots, \theta_p) = \pi_1(\theta_1)\pi_2(\theta_2 \mid \theta_1) \ldots \pi_p(\theta_p \mid \theta_{p-1}). \tag{3.2.15}$$

Here is an example.

Example 3.2.4. The *random effects model* we shall study in Volume II has

$$X_{ij} = \mu + \Delta_i + \epsilon_{ij}, \ 1 \le i \le I, \ 1 \le j \le J \qquad (3.2.16)$$

where the ϵ_{ij} are i.i.d. $\mathcal{N}(0, \sigma_e^2)$ and μ and the vector $\boldsymbol{\Delta} = (\Delta_1, \dots, \Delta_I)$ are independent of $\{\epsilon_{ij} : 1 \le i \le I, \ 1 \le j \le J\}$ with $\Delta_1, \dots, \Delta_I$ i.i.d. $\mathcal{N}(0, \sigma_\Delta^2)$, $1 \le j \le J$, $\mu \sim \mathcal{N}(\mu_0, \sigma_\mu^2)$. Here the X_{ij} can be thought of as measurements on individual i and Δ_i is an "individual" effect. If we now put a prior distribution on $(\mu, \sigma_e^2, \sigma_\Delta^2)$ making them independent, we have a Bayesian model in the usual form. But it is more fruitful to think of this model as parametrized by $\theta = (\mu, \sigma_e^2, \sigma_\Delta^2, \Delta_1, \dots, \Delta_I)$ with the $X_{ij} \mid \theta$ independent $\mathcal{N}(\mu + \Delta_i, \sigma_e^2)$. Then $p(\mathbf{x} \mid \theta) = \prod_{i,j} \varphi_{\sigma_e}(x_{ij} - \mu - \Delta_i)$ and

$$\pi(\theta) = \pi_1(\mu)\pi_2(\sigma_e^2)\pi_3(\sigma_\Delta^2) \prod_{i=1}^{I} \varphi_{\sigma_\Delta}(\Delta_i) \qquad (3.2.17)$$

where φ_σ denotes the $\mathcal{N}(0, \sigma^2)$ density.

In such a context a loss function frequently will single out some single coordinate θ_s (e.g., Δ_1 in 3.2.17) and to compute $r(a \mid \mathbf{x})$ we will need the posterior distribution of $\Delta_1 \mid \mathbf{x}$. But this is obtainable from the posterior distribution of θ given $\mathbf{X} = \mathbf{x}$ only by integrating out θ_j, $j \ne s$, and if p is large this is intractable. In recent years so-called Markov Chain Monte Carlo (MCMC) techniques have made this problem more tractable and the use of Bayesian methods has spread. We return to the topic in Volume II. $\quad\square$

Linear Bayes estimates

When the problem of computing $r(\pi, \delta)$ and δ_π is daunting, an alternative is to consider a class $\widetilde{\mathcal{D}}$ of procedures for which $r(\pi, \delta)$ is easy to compute and then to look for $\widetilde{\delta}_\pi \in \widetilde{\mathcal{D}}$ that minimizes $r(\pi, \delta)$ for $\delta \in \widetilde{\mathcal{D}}$. An example is *linear Bayes* estimates where, in the case of squared error loss $[q(\theta) - a]^2$, the problem is equivalent to minimizing the mean squared prediction error among functions of the form $b_0 + \sum_{j=1}^{d} b_j X_j$. If in (1.4.14) we identify $q(\theta)$ with Y and $\mathbf{X}$ with $\mathbf{Z}$, the solution is

$$\widetilde{\delta}(\mathbf{X}) = Eq(\boldsymbol{\theta}) + [\mathbf{X} - E(\mathbf{X})]^T \boldsymbol{\beta}$$

where $\boldsymbol{\beta}$ is as defined in Section 1.4. For example, if in the model (3.2.16), (3.2.17) we set $q(\boldsymbol{\theta}) = \Delta_1$, we can find the linear Bayes estimate of Δ_1 by using 1.4.6 and Problem 1.4.21. We find from (1.4.14) that the best linear Bayes estimator of Δ_1 is

$$\delta_L(\mathbf{X}) = E(\Delta_1) + (\mathbf{X} - \boldsymbol{\mu})^T \boldsymbol{\beta} \qquad (3.2.18)$$

where $E(\Delta_1) = 0$, $\mathbf{X} = (X_{11}, \dots, X_{1J})^T$, $\boldsymbol{\mu} = E(\mathbf{X})$ and $\boldsymbol{\beta} = \Sigma_{\mathbf{XX}}^{-1}\Sigma_{\mathbf{X}\Delta_1}$. For the given model

$$E(X_{1j}) = EE(X_{1j} \mid \theta) = E(\mu + \Delta_1) = E(\mu)$$

$$\mathrm{Var}(X_{1j}) = E\,\mathrm{Var}(X_{1j} \mid \theta) + \mathrm{Var}\,E(X_{1j} \mid \theta) = E(\sigma_\epsilon^2) + \sigma_\mu^2 + \sigma_{\Delta_1}^2$$

$$
\begin{aligned}
\mathrm{Cov}(X_{1j}, X_{1k}) &= E\,\mathrm{Cov}(X_{1j}, X_{1k} \mid \theta) + \mathrm{Cov}(E(X_{1j} \mid \theta), E(X_{1k} \mid \theta)) \\
&= 0 + \mathrm{Cov}(\mu + \Delta_1, \mu + \Delta_1) = \sigma_\mu^2 + \sigma_{\Delta_1}^2
\end{aligned}
$$

$$\mathrm{Cov}(\Delta_1, X_{1j}) = E\,\mathrm{Cov}(X_{1j}, \Delta_1 \mid \theta) + \mathrm{Cov}(E(X_{1j} \mid \theta), E(\Delta_1 \mid \theta)) = 0 + \sigma_{\Delta_1}^2 = \sigma_{\Delta_1}^2.$$

From these calculations we find $\boldsymbol{\beta}$ and $\delta_L(\mathbf{X})$. We leave the details to Problem 3.2.10. Linear Bayes procedures are useful in actuarial science, for example, Bühlmann (1970) and Norberg (1986).

Bayes estimation, maximum likelihood, and equivariance

As we have noted earlier, the maximum likelihood estimate can be thought of as the mode of the Bayes posterior density when the prior density is (the usually improper) prior $\pi(\theta) \equiv c$. When modes and means coincide for the improper prior (as in the Gaussian case), the MLE is an improper Bayes estimate. In general, computing means is harder than modes and that again accounts in part for the popularity of maximum likelihood.

An important property of the MLE is equivariance: An estimating method M producing the estimate $\widehat{\theta}_M$ is said to be *equivariant* with respect to reparametrization if for every one-to-one function h from Θ to $\Omega = h(\Theta)$, the estimate of $\omega \equiv h(\theta)$ is $\widehat{\omega}_M = h(\widehat{\theta}_M)$; that is, $\widehat{(h(\theta))}_M = h(\widehat{\theta}_M)$. In Problem 2.2.16 we show that the MLE procedure is equivariant. If we consider squared error loss, then the Bayes procedure $\widehat{\theta}_B = E(\boldsymbol{\theta} \mid X)$ is not equivariant for nonlinear transformations because

$$E(h(\boldsymbol{\theta}) \mid X) \neq h(E(\boldsymbol{\theta} \mid X))$$

for nonlinear h (e.g., Problem 3.2.3).

The source of the lack of equivariance of the Bayes risk and procedure for squared error loss is evident from (3.2.9): In the discrete case the conditional Bayes risk is

$$r_\Theta(a \mid x) = \sum_{\theta \in \Theta} [\theta - a]^2 \pi(\theta \mid x). \tag{3.2.19}$$

If we set $\omega = h(\boldsymbol{\theta})$ for h one-to-one onto $\Omega = h(\Theta)$, then ω has prior $\lambda(\omega) \equiv \pi(h^{-1}(\omega))$ and in the ω parametrization, the posterior Bayes risk is

$$
\begin{aligned}
r_\Omega(a \mid x) &= \sum_{\omega \in \Omega} [\omega - a]^2 \lambda(\omega \mid x) \\
&= \sum_{\theta \in \Theta} [h(\theta) - a]^2 \pi(\theta \mid x).
\end{aligned}
\tag{3.2.20}
$$

Thus, the Bayes procedure for squared error loss is not equivariant because squared error loss is not equivariant and, thus, $r_\Omega(a \mid x) \neq r_\Theta(h^{-1}(a) \mid x)$.

Loss functions of the form $l(\theta, a) = Q(P_\theta, P_a)$ are necessarily equivariant. The Kullback–Leibler divergence $K(\theta, a)$, $\theta, a \in \Theta$, is an example of such a loss function. It satisfies $K_\Omega(\omega, a) = K_\Theta(\theta, h^{-1}(a))$, thus, with this loss function,

$$r_\Omega(a \mid \mathbf{x}) = r_\Theta(h^{-1}(a) \mid \mathbf{x}).$$

See Problem 2.2.38. In the discrete case using K means that the importance of a loss is measured in probability units, with a similar interpretation in the continuous case (see (A.7.10)). In the $\mathcal{N}(\theta, \sigma_0^2)$ case the KL (Kullback–Leibler) loss $K(\theta, a)$ is $\sigma^{-2}\frac{1}{2}n(a - \theta)^2$ (Problem 2.2.37), that is, equivalent to squared error loss. In canonical exponential families

$$K(\boldsymbol{\eta}, \mathbf{a}) = \sum_{j=1}^{k} [\eta_j - a_j] E_{\boldsymbol{\eta}} T_j + A(\boldsymbol{\eta}) - A(\mathbf{a}).$$

Moreover, if we can find the KL loss Bayes estimate $\widehat{\boldsymbol{\eta}}_{BKL}$ of the canonical parameter $\boldsymbol{\eta}$ and if $\boldsymbol{\eta} \equiv \mathbf{c}(\boldsymbol{\theta}) : \Theta \to \mathcal{E}$ is one-to-one, then the KL loss Bayes estimate of $\boldsymbol{\theta}$ in the general exponential family is $\widehat{\boldsymbol{\theta}}_{BKL} = \mathbf{c}^{-1}(\widehat{\boldsymbol{\eta}}_{BKL})$.

For instance, in Example 3.2.1 where μ is the mean of a normal distribution and the prior is normal, we found the squared error Bayes estimate $\widehat{\mu}_B = w\eta_0 + (1 - w)\bar{X}$, where η_0 is the prior mean and w is a weight. Because the KL loss is equivalent to squared error for the canonical parameter μ, then if $\omega = h(\mu)$, $\widehat{\omega}_{BKL} = h(\widehat{\mu}_{BKL})$, where $\widehat{\mu}_{BKL} = w\eta_0 + (1 - w)\bar{X}$.

Bayes procedures based on the Kullback–Leibler divergence loss function are important for their applications to model selection and their connection to "minimum description (message) length" procedures. See Rissanen (1987) and Wallace and Freeman (1987). More recent reviews are Shibata (1997), Dowe, Baxter, Oliver, and Wallace (1998), and Hansen and Yu (2001). We will return to this in Volume II.

Bayes methods and doing reasonable things

There is a school of Bayesian statisticians (Berger, 1985; DeGroot, 1969; Lindley, 1965; Savage, 1954) who argue on normative grounds that a decision theoretic framework and rational behavior force individuals to use only Bayes procedures appropriate to their personal prior π. This is not a view we espouse because we view a model as an imperfect approximation to imperfect knowledge. However, given that we view a model and loss structure as an adequate approximation, it is good to know that generating procedures on the basis of Bayes priors viewed as weighting functions is a reasonable thing to do. This is the conclusion of the discussion at the end of Section 1.3. It may be shown quite generally as we consider all possible priors that the class $\mathcal{D}_0$ of Bayes procedures and their limits is *complete* in the sense that for any $\delta \in \mathcal{D}$ there is a $\delta_0 \in \mathcal{D}_0$ such that $R(\theta, \delta_0) \leq R(\theta, \delta)$ for all θ.

Summary. We show how Bayes procedures can be obtained for certain problems by computing posterior risk. In particular, we present Bayes procedures for the important cases of classification and testing statistical hypotheses. We also show that for more complex problems, the computation of Bayes procedures require sophisticated statistical numerical techniques or approximations obtained by restricting the class of procedures.

3.3 MINIMAX PROCEDURES

In Section 1.3 on the decision theoretic framework we introduced minimax procedures as ones corresponding to a worst-case analysis; the true θ is one that is as "hard" as possible. That is, δ_1 is better than δ_2 from a minimax point of view if $\sup_\theta R(\theta, \delta_1) < \sup_\theta R(\theta, \delta_2)$ and δ^* is said to be *minimax* if

$$\sup_\theta R(\theta, \delta^*) = \inf_\delta \sup_\theta R(\theta, \delta).$$

Here θ and δ are taken to range over Θ and $\mathcal{D} = \{$all possible decision procedures (not randomized)$\}$ while $\mathcal{P} = \{P_\theta : \theta \in \Theta\}$. It is fruitful to consider proper subclasses of $\mathcal{D}$ and subsets of $\mathcal{P}$, but we postpone this discussion.

The nature of this criterion and its relation to Bayesian optimality is clarified by considering a so-called zero sum game played by two players N (Nature) and S (the statistician). The statistician has at his or her disposal the set $\mathcal{D}$ or the set $\overline{\mathcal{D}}$ of all randomized decision procedures whereas Nature has at her disposal all prior distributions π on Θ. For the basic game, S picks δ without N's knowledge, N picks π without S's knowledge and then all is revealed and S pays N

$$r(\pi, \delta) = \int R(\theta, \delta) d\pi(\theta)$$

where the notation $\int R(\theta, \delta) d\pi(\theta)$ stands for $\int R(\theta, \delta) \pi(\theta) d\theta$ in the continuous case and $\sum R(\theta_j, \delta) \pi(\theta_j)$ in the discrete case.

S tries to minimize his or her loss, N to maximize her gain. For simplicity, we assume in the general discussion that follows that all sup's and inf's are assumed. There are two related partial information games that are important.

I: N is told the choice δ of S before picking π and S knows the rules of the game. Then N naturally picks π_δ such that

$$r(\pi_\delta, \delta) = \sup_\pi r(\pi, \delta), \tag{3.3.1}$$

that is, π_δ is *least favorable against* δ. Knowing the rules of the game S naturally picks δ^* such that

$$r(\pi_{\delta^*}, \delta^*) = \sup_\pi r(\pi, \delta^*) = \inf_\delta \sup_\pi r(\pi, \delta). \tag{3.3.2}$$

We claim that δ^* is minimax. To see this we note first that,

$$r(\pi, \delta) = \int R(\theta, \delta) d\pi(\theta) \le \sup_\theta R(\theta, \delta)$$

for all π, δ. On the other hand, if $R(\theta_\delta, \delta) = \sup_\theta R(\theta, \delta)$, then if π_δ is point mass at θ_δ, $r(\pi_\delta, \delta) = R(\theta_\delta, \delta)$ and we conclude that

$$\sup_\pi r(\pi, \delta) = \sup_\theta R(\theta, \delta) \tag{3.3.3}$$

and our claim follows.

II: S is told the choice π of N before picking δ and N knows the rules of the game. Then S naturally picks δ_π such that

$$r(\pi, \delta_\pi) = \inf_\delta r(\pi, \delta).$$

That is, δ_π is a Bayes procedure for π. Then N should pick π^* such that

$$r(\pi^*, \delta_{\pi^*}) = \sup_\pi r(\pi, \delta_\pi) = \sup_\pi \inf_\delta r(\pi, \delta). \qquad (3.3.4)$$

For obvious reasons, π^* is called a *least favorable* (to S) *prior* distribution. As we shall see by example, although the right-hand sides of (3.3.2) and (3.3.4) are always defined, least favorable priors and/or minimax procedures may not exist and, if they exist, may not be unique.

The key link between the search for minimax procedures in the basic game and games I and II is the von Neumann minimax theorem of game theory, which we state in our language.

Theorem 3.3.1. *(von Neumann). If both Θ and $\mathcal{D}$ are finite, then:*

(*a*) $$\underline{v} \equiv \sup_\pi \inf_\delta r(\pi, \delta), \quad \bar{v} \equiv \inf_\delta \sup_\pi r(\pi, \delta)$$

are both assumed by (say) π^* (least favorable), δ^* minimax, respectively. Further,

$$\underline{v} = r(\pi^*, \delta^*) = \bar{v} \qquad (3.3.5)$$

and, hence, $\delta^* = \delta_{\pi^*}$, $\pi^* = \pi_{\delta^*}$.

$\underline{v}$ and $\bar{v}$ are called the *lower* and *upper values* of the basic game. When $\underline{v} = \bar{v} = v$ (say), v is called the *value* of the game.

Remark 3.3.1. Note (Problem 3.3.3) that von Neumann's theorem applies to classification and testing when $\Theta_0 = \{\theta_0\}$ and $\Theta_1 = \{\theta_1\}$ (Example 3.2.2) but is too restrictive in its assumption for the great majority of inference problems. A generalization due to Wald and Karlin—see Karlin (1959)—states that the conclusions of the theorem remain valid if Θ and $\mathcal{D}$ are compact subsets of Euclidean spaces. There are more far-reaching generalizations but, as we shall see later, without some form of compactness of Θ and/or $\mathcal{D}$, although equality of $\underline{v}$ and $\bar{v}$ holds quite generally, existence of least favorable priors and/or minimax procedures may fail.

The main practical import of minimax theorems is, in fact, contained in a converse and its extension that we now give. Remarkably these hold without essentially any restrictions on Θ and $\mathcal{D}$ and are easy to prove.

Proposition 3.3.1. *Suppose δ^{**}, π^{**} can be found such that*

$$\delta^{**} = \delta_{\pi^{**}}, \quad \pi^{**} = \pi_{\delta^{**}} \qquad (3.3.6)$$

that is, δ^{**} is Bayes against π^{**} and π^{**} is least favorable against δ^{**}. Then $\underline{v} = \bar{v} = r(\pi^{**}, \delta^{**})$. That is, π^{**} is least favorable and δ^{**} is minimax.

To utilize this result we need a characterization of π_δ. This is given by

Proposition 3.3.2. *π_δ is least favorable against δ iff*

$$\pi_\delta\{\theta : R(\theta, \delta) = \sup_{\theta'} R(\theta', \delta)\} = 1. \tag{3.3.7}$$

That is, π_δ assigns probability only to points θ at which the function $R(\cdot, \delta)$ is maximal.

Thus, combining Propositions 3.3.1 and 3.3.2 we have a simple criterion, "A Bayes rule with constant risk is minimax."

Note that π_δ may not be unique. In particular, if $R(\theta, \delta) \equiv$ constant, the rule has constant risk, then all π are least favorable.

We now prove Propositions 3.3.1 and 3.3.2.

Proof of Proposition 3.3.1. Note first that we always have

$$\underline{v} \leq \bar{v} \tag{3.3.8}$$

because, trivially,

$$\inf_\delta r(\pi, \delta) \leq r(\pi, \delta') \tag{3.3.9}$$

for all π, δ'. Hence,

$$\underline{v} = \sup_\pi \inf_\delta r(\pi, \delta) \leq \sup_\pi r(\pi, \delta') \tag{3.3.10}$$

for all δ' and $\underline{v} \leq \inf_{\delta'} \sup_\pi r(\pi, \delta') = \bar{v}$. On the other hand, by hypothesis,

$$\underline{v} \geq \inf_\delta r(\pi^{**}, \delta) = r(\pi^{**}, \delta^{**}) = \sup_\pi r(\pi, \delta^{**}) \geq \bar{v}. \tag{3.3.11}$$

Combining (3.3.8) and (3.3.11) we conclude that

$$\underline{v} = \inf_\delta r(\pi^{**}, \delta) = r(\pi^{**}, \delta^{**}) = \sup_\pi r(\pi, \delta^{**}) = \bar{v} \tag{3.3.12}$$

as advertised. $\qquad\square$

Proof of Proposition 3.3.2. π is least favorable for δ iff

$$E_\pi R(\boldsymbol{\theta}, \delta) = \int R(\theta, \delta) d\pi(\theta) = \sup_\pi r(\pi, \delta). \tag{3.3.13}$$

But by (3.3.3),

$$\sup_\pi r(\pi, \delta) = \sup_\theta R(\theta, \delta). \tag{3.3.14}$$

Because $E_\pi R(\boldsymbol{\theta}, \delta) = \sup_\theta R(\theta, \delta)$, (3.3.13) is possible iff (3.3.7) holds. $\qquad\square$

Putting the two propositions together we have the following.

Theorem 3.3.2. *Suppose δ^* has $\sup_\theta R(\theta, \delta^*) = r < \infty$. If there exists a prior π^* such that δ^* is Bayes for π^* and $\pi^*\{\theta : R(\theta, \delta^*) = r\} = 1$, then δ^* is minimax.*

Example 3.3.1. *Minimax Estimation in the Binomial Case.* Suppose S has a $\mathcal{B}(n, \theta)$ distribution and $\bar{X} = S/n$, as in Example 3.2.3. Let $l(\theta, a) = (\theta - a)^2/\theta(1-\theta), 0 < \theta < 1$. For this loss function,

$$R(\theta, \bar{X}) = \frac{E(\bar{X} - \theta)^2}{\theta(1 - \theta)} = \frac{\theta(1 - \theta)}{n\theta(1 - \theta)} = \frac{1}{n},$$

and $\bar{X}$ does have constant risk. Moreover, we have seen in Example 3.2.3 that $\bar{X}$ is Bayes, when θ is $\mathcal{U}(0, 1)$. By Theorem 3.3.2 we conclude that $\bar{X}$ is minimax and, by Proposition 3.3.2, the uniform distribution least favorable.

For the usual quadratic loss neither of these assertions holds. The minimax estimate is

$$\delta^*(S) = \frac{S + \frac{1}{2}\sqrt{n}}{n + \sqrt{n}} = \frac{\sqrt{n}}{\sqrt{n} + 1}\bar{X} + \frac{1}{\sqrt{n} + 1} \cdot \frac{1}{2}.$$

This estimate does have constant risk and is Bayes against a $\beta(\sqrt{n}/2, \sqrt{n}/2)$ prior (Problem 3.3.4). This is an example of a situation in which the minimax principle leads us to an unsatisfactory estimate. For quadratic loss, the limit as $n \to \infty$ of the ratio of the risks of δ^* and $\bar{X}$ is > 1 for every $\theta \neq \frac{1}{2}$. At $\theta = \frac{1}{2}$ the ratio tends to 1. Details are left to Problem 3.3.4. □

Example 3.3.2. *Minimax Testing. Satellite Communications.* A test to see whether a communications satellite is in working order is run as follows. A very strong signal is beamed from Earth. The satellite responds by sending a signal of intensity $v > 0$ for n seconds or, if it is not working, does not answer. Because of the general "noise" level in space the signals received on Earth vary randomly whether the satellite is sending or not. The mean voltage per second of the signal for each of the n seconds is recorded. Denote the mean voltage of the signal received through the ith second less expected mean voltage due to noise by X_i. We assume that the X_i are independently and identically distributed as $\mathcal{N}(\mu, \sigma^2)$ where $\mu = v$, if the satellite functions, and 0 otherwise. The variance σ^2 of the "noise" is assumed known. Our problem is to decide whether "$\mu = 0$" or "$\mu = v$." We view this as a decision problem with $0 - 1$ loss. If the number of transmissions is fixed, the minimax rule minimizes the maximum probability of error (see $(1.3.6)$). What is this risk?

A natural first step is to use the characterization of Bayes tests given in the preceding section. If we assign probability π to 0 and $1 - \pi$ to v, use $0 - 1$ loss, and set $L(\mathbf{x}, 0, v) = p(\mathbf{x} \mid v)/p(\mathbf{x} \mid 0)$, then the Bayes test decides $\mu = v$ if

$$L(\mathbf{x}, 0, v) = \exp\left\{\frac{v}{\sigma^2}\Sigma x_i - \frac{nv^2}{2\sigma^2}\right\} \geq \frac{\pi}{1 - \pi}$$

and decides $\mu = 0$ if

$$L(\mathbf{x}, 0, v) < \frac{\pi}{1 - \pi}.$$

This test is equivalent to deciding $\mu = v$ (Problem 3.3.1) if, and only if,

$$T = \frac{1}{\sigma\sqrt{n}}\Sigma x_i \geq t,$$

where,

$$t = \frac{\sigma}{v\sqrt{n}}\left[\log \frac{\pi}{1-\pi} + \frac{nv^2}{2\sigma^2}\right].$$

If we call this test δ_π,

$$R(0, \delta_\pi) = 1 - \Phi(t) = \Phi(-t)$$
$$R(v, \delta_\pi) = \Phi\left(t - \frac{v\sqrt{n}}{\sigma}\right).$$

To get a minimax test we must have $R(0, \delta_\pi) = R(v, \delta_\pi)$, which is equivalent to

$$-t = t - \frac{v\sqrt{n}}{\sigma}$$

or

$$t = \frac{v\sqrt{n}}{2\sigma}.$$

Because this value of t corresponds to $\pi = \frac{1}{2}$, the intuitive test, which decides $\mu = v$ if and only if $T \geq \frac{1}{2}[E_0(T) + E_v(T)]$, is indeed minimax. $\square$

If Θ is not bounded, minimax rules are often not Bayes rules but instead can be obtained as limits of Bayes rules. To deal with such situations we need an extension of Theorem 3.3.2.

Theorem 3.3.3. *Let δ^* be a rule such that $\sup_\theta R(\theta, \delta^*) = r < \infty$, and let $\{\pi_k\}$ denote a sequence of prior distributions. Let $r_k = \inf_\delta r(\pi_k, \delta)$, where $r(\pi_k, \delta)$ denotes the Bayes risk wrt π_k. If*

$$r_k \to r \text{ as } k \to \infty, \tag{3.3.15}$$

then δ^* is minimax.

Proof. By assumption

$$\sup_\theta R(\theta, \delta^*) = r_k + o(1)$$

where $o(1) \to 0$ as $k \to \infty$. For any competitor δ

$$\sup_\theta R(\theta, \delta) \geq E_{\pi_k}(R(\boldsymbol{\theta}, \delta)) \geq r_k = \sup_\theta R(\theta, \delta^*) - o(1). \tag{3.3.16}$$

If we let $k \to \infty$ the left-hand side of (3.3.16) is unchanged, whereas the right tends to $\sup_\theta R(\theta, \delta^*)$. $\square$

To apply this result, we need to find a sequence of priors π_k such that $\inf_\delta r(\pi_k, \delta) \to \sup_\theta R(\theta, \delta^*)$. Here are two examples.

Example 3.3.3. *Normal Mean.* We now show that $\bar{X}$ is minimax in Example 3.2.1. Identify π_k with the $\mathcal{N}(\eta_0, \tau^2)$ prior where $k = \tau^2$. We know that $R(\theta, \bar{X}) = \sigma^2/n$, whereas the Bayes risk of the Bayes rule of Example 3.2.1 is

$$\inf_\delta r(\pi_k, \delta) = \frac{\tau^2}{(\sigma^2/n) + \tau^2} \frac{\sigma^2}{n} = \frac{\sigma^2}{n} - \frac{1}{(\sigma^2/n) + \tau^2} \frac{\sigma^2}{n}.$$

Because $(\sigma^2/n)/((\sigma^2/n) + \tau^2) \to 0$ as $\tau^2 \to \infty$, we can conclude that $\bar{X}$ is minimax. $\square$

Example 3.3.4. *Minimax Estimation in a Nonparametric Setting (after Lehmann).* Suppose $X_1, \ldots, X_n$ are i.i.d. $F \in \mathcal{F}$

$$\mathcal{F} = \{F : \mathrm{Var}_F(X_1) \leq M\}.$$

Then $\bar{X}$ is minimax for estimating $\theta(F) \equiv E_F(X_1)$ with quadratic loss. This can be viewed as an extension of Example 3.3.3. Let π_k be a prior distribution on $\mathcal{F}$ constructed as follows:[1]

(i) $\pi_k\{F : \mathrm{Var}_F(X_1) \neq M\} = 0$.

(ii) $\pi_k\{F : F \neq \mathcal{N}(\mu, M) \text{ for some } \mu\} = 0$.

(iii) F is chosen by first choosing $\mu = \theta(F)$ from a $\mathcal{N}(0, k)$ distribution and then taking $F = \mathcal{N}(\theta(F), M)$.

The Bayes risk is now the same as in Example 3.3.3 with $\sigma^2 = M$. Because, evidently,

$$\max_{\mathcal{F}} R(F, \bar{X}) = \max_{\mathcal{F}} \frac{\mathrm{Var}_F(X_1)}{n} = \frac{M}{n},$$

Theorem 3.3.3 applies and the result follows. $\square$

Remark 3.3.2. If δ^* has constant risk and is Bayes with respect to some prior π, then $\inf_\delta r(\pi, \delta) = \sup_\theta R(\theta, \delta^*)$ is satisfied and δ^* is minimax. See Problem 3.3.4 for an example.

Minimax procedures and symmetry

As we have seen, minimax procedures have constant risk or at least constant risk on the "most difficult" θ. There is a deep connection between symmetries of the model and the structure of such procedures developed by Hunt and Stein, Lehmann, and others, which is discussed in detail in Chapter 9 of Lehmann (1986) and Chapter 5 of Lehmann and Casella (1998), for instance. We shall discuss this approach somewhat in Volume II but refer to Lehmann (1986) and Lehmann and Casella (1998) for further reading.

Summary. We introduce the minimax principle in the context of the theory of games. Using this framework we connect minimaxity and Bayes methods and develop sufficient conditions for a procedure to be minimax and apply them in several important examples. More specifically, we show how finding minimax procedures can be viewed as solving a *game* between a statistician S and nature N in which S selects a decision rule δ and N selects a prior π. The *lower (upper) value* $\underline{v}(\bar{v})$ of the game is the supremum (infimum) over priors (decision rules) of the infimum (supremum) over decision rules (priors) of the Bayes risk. A prior for which the Bayes risk of the Bayes procedure equals the lower value of the game is called *least favorable.* When $\underline{v} = \bar{v}$, the game is said to have a *value v.* Von Neumann's Theorem states that if Θ and $\mathcal{D}$ are both finite, then the game of S versus N has a value v, there is a least favorable prior π^* and a minimax rule δ^* such that δ^* is the Bayes rule for π^* and π^* maximizes the Bayes risk of δ^* over all priors. Moreover, v equals the Bayes risk of the Bayes rule δ^* for the prior π^*. We show that Bayes rules with constant risk, or more generally with constant risk over the support of some prior, are minimax. This result is extended to rules that are limits of Bayes rules with constant risk and we use it to show that $\bar{x}$ is a minimax rule for squared error loss in the $\mathcal{N}(\theta, \sigma_0^2)$ model.

3.4   UNBIASED ESTIMATION AND RISK INEQUALITIES

3.4.1   Unbiased Estimation, Survey Sampling

In the previous two sections we have considered two decision theoretic optimality principles, Bayes and minimaxity, for which it is possible to characterize and, in many cases, compute procedures (in particular estimates) that are best in the class of all procedures, $\mathcal{D}$, according to these criteria. An alternative approach is to specify a proper subclass of procedures, $\mathcal{D}_0 \subset \mathcal{D}$, on other grounds, computational ease, symmetry, and so on, and then see if within the $\mathcal{D}_0$ we can find $\delta^* \in \mathcal{D}_0$ that is best according to the "gold standard," $R(\theta, \delta) \geq R(\theta, \delta^*)$ for all θ, all $\delta \in \mathcal{D}_0$. Obviously, we can also take this point of view with humbler aims, for example, looking for the procedure $\delta_\pi^* \in \mathcal{D}_0$ that minimizes the Bayes risk with respect to a prior π among all $\delta \in \mathcal{D}_0$. This approach has early on been applied to parametric families $\mathcal{D}_0$. When $\mathcal{D}_0$ is the class of linear procedures and l is quadratic loss, the solution is given in Section 3.2.

In the non-Bayesian framework, if $\mathbf{Y}$ is postulated as following a linear regression model with $E(Y) = \mathbf{z}^T \beta$ as in Section 2.2.1, then in estimating a linear function of the β_j it is natural to consider the computationally simple class of linear estimates, $S(\mathbf{Y}) = \sum_{i=1}^n d_i Y_i$. This approach coupled with the principle of unbiasedness we now introduce leads to the famous Gauss–Markov theorem proved in Section 6.6.

We introduced, in Section 1.3, the notion of bias of an estimate $\delta(X)$ of a parameter $q(\theta)$ in a model $\mathcal{P} \equiv \{P_\theta : \theta \in \Theta\}$ as

$$\text{Bias}_\theta(\delta) \equiv E_\theta \delta(X) - q(\theta).$$

An estimate such that $\text{Bias}_\theta(\delta) \equiv 0$ is called *unbiased.* This notion has intuitive appeal, ruling out, for instance, estimates that ignore the data, such as $\delta(X) \equiv q(\theta_0)$, which can't be beat for $\theta = \theta_0$ but can obviously be arbitrarily terrible. The most famous unbiased estimates are the familiar estimates of μ and σ^2 when $X_1, \ldots, X_n$ are i.i.d. $\mathcal{N}(\mu, \sigma^2)$ given by (see Example 1.3.3 and Problem 1.3.8)

$$\hat{\mu} = \bar{X} \tag{3.4.1}$$

$$s^2 = \frac{1}{n-1} \sum_{i=1}^{n} (X_i - \bar{X})^2. \tag{3.4.2}$$

Because for unbiased estimates mean square error and variance coincide we call an unbiased estimate $\delta^*(X)$ of $q(\theta)$ that has minimum MSE among all unbiased estimates for all θ, UMVU (uniformly minimum variance unbiased). As we shall see shortly for $\bar{X}$ and in Volume 2 for s^2, these are both UMVU.

Unbiased estimates play a particularly important role in survey sampling.

Example 3.4.1. *Unbiased Estimates in Survey Sampling.* Suppose we wish to sample from a finite population, for instance, a census unit, to determine the average value of a variable (say) monthly family income during a time between two censuses and suppose that we have available a list of families in the unit with family incomes at the last census. Write $x_1, \ldots, x_N$ for the unknown current family incomes and correspondingly $u_1, \ldots, u_N$ for the known last census incomes. We ignore difficulties such as families moving. We let $X_1, \ldots, X_n$ denote the incomes of a sample of n families drawn at random without replacement. This leads to the model with $\mathbf{x} = (x_1, \ldots, x_N)$ as parameter

$$P_{\mathbf{x}}[X_1 = a_1, \ldots, X_n = a_n] = \frac{1}{\binom{N}{n}} \text{ if } \{a_1, \ldots, a_n\} \subset \{x_1, \ldots, x_N\}$$

$$\tag{3.4.3}$$

$$= 0 \text{ otherwise.}$$

We want to estimate the parameter $\bar{x} = \frac{1}{N} \sum_{j=1}^{N} x_j$. It is easy to see that the natural estimate $\bar{X} \equiv \frac{1}{n} \sum_{i=1}^{n} X_i$ is unbiased (Problem 3.4.14) and has

$$MSE(\bar{X}) = \text{Var}_{\mathbf{x}}(\bar{X}) = \frac{\sigma_{\mathbf{x}}^2}{n} \left(1 - \frac{n-1}{N-1}\right) \tag{3.4.4}$$

where

$$\sigma_{\mathbf{x}}^2 = \frac{1}{N} \sum_{i=1}^{N} (x_i - \bar{x})^2. \tag{3.4.5}$$

This method of sampling does not use the information contained in $u_1, \ldots, u_N$. One way to do this, reflecting the probable correlation between $(u_1, \ldots, u_N)$ and $(x_1, \ldots, x_N)$, is to estimate by a regression estimate

$$\hat{\bar{X}}_R \equiv \bar{X} - b(\bar{U} - \bar{u}) \tag{3.4.6}$$

where b is a prespecified positive constant, U_i is the last census income corresponding to X_i, and $\bar{u} = \frac{1}{N}\sum_{i=1}^{N} u_i$, $\bar{U} = \frac{1}{n}\sum_{i=1}^{n} U_i$. Clearly for each b, $\widehat{X}_R$ is also unbiased. If the correlation of U_i and X_i is positive and $b < 2\text{Cov}(\bar{U}, \bar{X})/\text{Var}(\bar{U})$, this will be a better estimate than $\bar{X}$ and the best choice of b is $b_{\text{opt}} \equiv \text{cov}(\bar{U}, \bar{X})/\text{Var}(\bar{U})$ (Problem 3.4.19). The value of b_{opt} is unknown but can be estimated by

$$\widehat{b}_{\text{opt}} = \frac{\frac{1}{n}\sum_{i=1}^{n}(X_i - \bar{X})(U_i - \bar{U})}{\frac{1}{N}\sum_{j=1}^{N}(U_j - \bar{U})^2}.$$

The resulting estimate is no longer unbiased but behaves well for large samples—see Problem 5.3.11.

An alternative approach to using the u_j is to not sample all units with the same probability. Specifically let $0 \le \pi_1, \ldots, \pi_N \le 1$ with $\sum_{j=1}^{N} \pi_j = n$. For each unit $1, \ldots, N$ toss a coin with probability π_j of landing heads and select x_j if the coin lands heads. The result is a sample $S = \{X_1, \ldots, X_M\}$ of random size M such that $E(M) = n$ (Problem 3.4.15). If the π_j are not all equal, $\bar{X}$ is not unbiased but the following estimate known as the *Horvitz–Thompson estimate* is:

$$\widehat{\bar{x}}_{HT} \equiv \frac{1}{N}\sum_{i=1}^{M} \frac{X_i}{\pi_{J_i}} \tag{3.4.7}$$

where J_i is defined by $X_i = x_{J_i}$. To see this write

$$\widehat{\bar{x}}_{HT} = \frac{1}{N}\sum_{j=1}^{N} \frac{x_j}{\pi_j} 1(x_j \in S).$$

Because $\pi_j = P[x_j \in S]$ by construction unbiasedness follows. A natural choice of π_j is $\frac{u_j}{\bar{u}}n$. This makes it more likely for big incomes to be included and is intuitively desirable. It is possible to avoid the undesirable random sample size of these schemes and yet have specified π_j. The Horvitz–Thompson estimate then stays unbiased. Further discussion of this and other sampling schemes and comparisons of estimates are left to the problems. □

Discussion. Unbiasedness is also used in stratified sampling theory (see Problem 1.3.4). However, outside of sampling, the unbiasedness principle has largely fallen out of favor for a number of reasons.

(i) Typically unbiased estimates do not exist—see Bickel and Lehmann (1969) and Problem 3.4.18, for instance.

(ii) Bayes estimates are necessarily biased—see Problem 3.4.20—and minimax estimates often are.

(iii) Unbiased estimates do not obey the attractive equivariance property. If $\tilde{\theta}$ is unbiased for θ, $q(\tilde{\theta})$ is biased for $q(\tilde{\theta})$ unless q is linear. They necessarily in general differ from maximum likelihood estimates except in an important special case we develop later.

Nevertheless, as we shall see in Chapters 5 and 6, good estimates in large samples are approximately unbiased. We expect that $|\text{Bias}_\theta(\widehat{\theta}_n)|/\text{Var}_\theta^{\frac{1}{2}}(\widehat{\theta}_n) \to 0$ or equivalently $\text{Var}_\theta(\widehat{\theta}_n)/MSE_\theta(\widehat{\theta}_n) \to 1$ as $n \to \infty$. In particular we shall show that maximum likelihood estimates are approximately unbiased and approximately best among all estimates. The arguments will be based on asymptotic versions of the important inequalities in the next subsection.

Finally, unbiased estimates are still in favor when it comes to estimating residual variances. For instance, in the linear regression model $\mathbf{Y} = \mathbf{Z}_D\boldsymbol{\beta} + \boldsymbol{\varepsilon}$ of Section 2.2, the variance $\sigma^2 = \text{Var}(\varepsilon_i)$ is estimated by the unbiased estimate $S^2 = \widehat{\boldsymbol{\varepsilon}}^T\widehat{\boldsymbol{\varepsilon}}/(n - p)$ where $\widehat{\boldsymbol{\varepsilon}} = (\mathbf{Y} - \mathbf{Z}_D\widehat{\boldsymbol{\beta}})$, $\widehat{\boldsymbol{\beta}}$ is the least squares estimate, and p is the number of coefficients in $\boldsymbol{\beta}$. This preference of S^2 over the MLE $\widehat{\sigma}^2 = \widehat{\boldsymbol{\varepsilon}}^T\boldsymbol{\varepsilon}/n$ is in accord with optimal behavior when both the number of observations and number of parameters are large. See Problem 3.4.9.

3.4.2 The Information Inequality

The one-parameter case

We will develop a lower bound for the variance of a statistic, which can be used to show that an estimate is UMVU. The lower bound is interesting in its own right, has some decision theoretic applications, and appears in the asymptotic optimality theory of Section 5.4.

We suppose throughout that we have a regular parametric model and further that Θ is an open subset of the line. From this point on we will suppose $p(x, \theta)$ is a density. The discussion and results for the discrete case are essentially identical and will be referred to in the future by the same numbers as the ones associated with the continuous-case theorems given later. We make two regularity assumptions on the family $\{P_\theta : \theta \in \Theta\}$.

(I) The set $A = \{x : p(x, \theta) > 0\}$ does not depend on θ. For all $x \in A, \theta \in \Theta$, $\partial/\partial\theta \log p(x, \theta)$ exists and is finite.

(II) If T is any statistic such that $E_\theta(|T|) < \infty$ for all $\theta \in \Theta$, then the operations of integration and differentiation by θ can be interchanged in $\int T(x)p(x, \theta)dx$. That is, for integration over R^q,

$$\frac{\partial}{\partial\theta}\left[\int T(x)p(x, \theta)dx\right] = \int T(x)\frac{\partial}{\partial\theta}p(x, \theta)dx \qquad (3.4.8)$$

whenever the right-hand side of (3.4.8) is finite.

Note that in particular (3.4.8) is assumed to hold if $T(x) = 1$ for all x, and we can interchange differentiation and integration in $\int p(x, \theta)dx$.

Assumption II is practically useless as written. What is needed are simple sufficient conditions on $p(x, \theta)$ for II to hold. Some classical conditions may be found in Apostol (1974), p. 167. Simpler assumptions can be formulated using Lebesgue integration theory. For instance, suppose I holds. Then II holds provided that for all T such that $E_\theta(|T|) < \infty$

for all θ, the integrals

$$\int T(x) \left[\frac{\partial}{\partial \theta} p(x, \theta) \right] dx \text{ and } \int \left| T(x) \left[\frac{\partial}{\partial \theta} p(x, \theta) \right] \right| dx$$

are continuous functions[3] of θ. It is not hard to check (using Laplace transform theory) that a one-parameter exponential family quite generally satisfies Assumptions I and II.

Proposition 3.4.1. *If $p(x, \theta) = h(x) \exp\{\eta(\theta)T(x) - B(\theta)\}$ is an exponential family and $\eta(\theta)$ has a nonvanishing continuous derivative on Θ, then I and II hold.*

For instance, suppose $X_1, \ldots, X_n$ is a sample from a $\mathcal{N}(\theta, \sigma^2)$ population, where σ^2 is known. Then (see Table 1.6.1) $\eta(\theta) = \theta/\sigma^2$ and I and II are satisfied. Similarly, I and II are satisfied for samples from gamma and beta distributions with one parameter fixed.

If I holds it is possible to define an important characteristic of the family $\{P_\theta\}$, the *Fisher information number*, which is denoted by $I(\theta)$ and given by

$$I(\theta) = E_\theta \left(\frac{\partial}{\partial \theta} \log p(X, \theta) \right)^2 = \int \left(\frac{\partial}{\partial \theta} \log p(x, \theta) \right)^2 p(x, \theta) dx. \tag{3.4.9}$$

Note that $0 \leq I(\theta) \leq \infty$.

Lemma 3.4.1. *Suppose that I and II hold and that*

$$E \left| \frac{\partial}{\partial \theta} \log p(X, \theta) \right| < \infty.$$

Then

$$E_\theta \left(\frac{\partial}{\partial \theta} \log p(X, \theta) \right) = 0 \tag{3.4.10}$$

and, thus,

$$I(\theta) = \text{Var} \left(\frac{\partial}{\partial \theta} \log p(X, \theta) \right). \tag{3.4.11}$$

Proof.

$$\begin{aligned}
E_\theta \left(\frac{\partial}{\partial \theta} \log p(X, \theta) \right) &= \int \left\{ \left[\frac{\partial}{\partial \theta} p(x, \theta) \right] \Big/ p(x, \theta) \right\} p(x, \theta) dx \\
&= \int \frac{\partial}{\partial \theta} p(x, \theta) dx = \frac{\partial}{\partial \theta} \int p(x, \theta) dx = 0.
\end{aligned}$$

$\square$

Example 3.4.2. Suppose $X_1, \ldots, X_n$ is a sample from a Poisson $\mathcal{P}(\theta)$ population. Then

$$\frac{\partial}{\partial \theta} \log p(\mathbf{x}, \theta) = \frac{\sum_{i=1}^n x_i}{\theta} - n \text{ and } I(\theta) = \text{Var} \left(\frac{\sum_{i=1}^n X_i}{\theta} \right) = \frac{1}{\theta^2} n\theta = \frac{n}{\theta}.$$

$\square$

Here is the main result of this section.

Theorem 3.4.1.   (Information Inequality).   *Let $T(X)$ be any statistic such that $\mathrm{Var}_\theta(T(X)) < \infty$ for all θ. Denote $E_\theta(T(X))$ by $\psi(\theta)$. Suppose that I and II hold and $0 < I(\theta) < \infty$. Then for all θ, $\psi(\theta)$ is differentiable and*

$$\mathrm{Var}_\theta(T(X)) \geq \frac{[\psi'(\theta)]^2}{I(\theta)}. \tag{3.4.12}$$

Proof. Using I and II we obtain,

$$\psi'(\theta) = \int T(x) \frac{\partial}{\partial \theta} p(x,\theta) dx = \int T(x) \left(\frac{\partial}{\partial \theta} \log p(x,\theta) \right) p(x,\theta) dx. \tag{3.4.13}$$

By (A.11.14) and Lemma 3.4.1,

$$\psi'(\theta) = \mathrm{Cov}\left(\frac{\partial}{\partial \theta} \log p(X,\theta), T(X) \right). \tag{3.4.14}$$

Now let us apply the correlation (Cauchy–Schwarz) inequality (A.11.16) to the random variables $\partial/\partial\theta \log p(X,\theta)$ and $T(X)$. We get

$$|\psi'(\theta)| \leq \sqrt{\mathrm{Var}(T(X))\mathrm{Var}\left(\frac{\partial}{\partial \theta} \log p(X,\theta) \right)}. \tag{3.4.15}$$

The theorem follows because, by Lemma 3.4.1, $\mathrm{Var}\left(\frac{\partial}{\partial \theta} \log p(X,\theta) \right) = I(\theta)$. $\qquad \square$

The lower bound given in the information inequality depends on $T(X)$ through $\psi(\theta)$. If we consider the class of unbiased estimates of $q(\theta) = \theta$, we obtain a universal lower bound given by the following.

Corollary 3.4.1. *Suppose the conditions of Theorem 3.4.1 hold and T is an unbiased estimate of θ. Then*

$$\mathrm{Var}_\theta(T(X)) \geq \frac{1}{I(\theta)}. \tag{3.4.16}$$

The number $1/I(\theta)$ is often referred to as the *information* or *Cramér–Rao lower bound* for the variance of an unbiased estimate of $\psi(\theta)$.

Here's another important special case.

Proposition 3.4.2. *Suppose that $\mathbf{X} = (X_1,\ldots,X_n)$ is a sample from a population with density $f(x,\theta)$, $\theta \in \Theta$, and that the conditions of Theorem 3.4.1 hold. Let $I_1(\theta) = E\left(\frac{\partial}{\partial \theta} \log f(X_1,\theta) \right)^2$, then*

$$I(\theta) = nI_1(\theta) \text{ and } \mathrm{Var}_\theta(T(\mathbf{X})) \geq \frac{[\psi'(\theta)]^2}{nI_1(\theta)}, \tag{3.4.17}$$

Proof. This is a consequence of Lemma 3.4.1 and

$$I(\theta) = \text{Var}\left[\frac{\partial}{\partial\theta}\log p(\mathbf{X},\theta)\right] = \text{Var}\left[\sum_{i=1}^{n}\frac{\partial}{\partial\theta}\log f(X_i,\theta)\right]$$

$$= \sum_{i=1}^{n}\text{Var}\left[\frac{\partial}{\partial\theta}\log f(X_i,\theta)\right] = nI_1(\theta).$$

$\square$

$I_1(\theta)$ is often referred to as the information contained in one observation. We have just shown that the information $I(\theta)$ in a sample of size n is $nI_1(\theta)$.

Next we note how we can apply the information inequality to the problem of unbiased estimation. If the family $\{P_\theta\}$ satisfies I and II and if there exists an unbiased estimate T^* of $\psi(\theta)$ such that $\text{Var}_\theta[T^*(X)] = [\psi'(\theta)]^2/I(\theta)$ for all $\theta \in \Theta$, then T^* is UMVU as an estimate of ψ.

Example 3.4.2. *(Continued).* For a sample from a $\mathcal{P}(\theta)$ distribution, the MLE is $\widehat\theta = \bar{X}$. Because $\bar{X}$ is unbiased and $\text{Var}(\bar{X}) = \theta/n$, $\bar{X}$ is UMVU.

Example 3.4.3. Suppose $X_1,\ldots,X_n$ is a sample from a normal distribution with unknown mean θ and known variance σ^2. As we previously remarked, the conditions of the information inequality are satisfied. By Corollary 3.4.1 we see that the conclusion that $\bar{X}$ is UMVU follows if

$$\text{Var}(\bar{X}) = \frac{1}{nI_1(\theta)}. \tag{3.4.18}$$

Now $\text{Var}(\bar{X}) = \sigma^2/n$, whereas if φ denotes the $\mathcal{N}(0,1)$ density, then

$$I_1(\theta) = E\left[\frac{\partial}{\partial\theta}\log\left\{\frac{1}{\sigma}\varphi\left(\frac{X_1-\theta}{\sigma}\right)\right\}\right]^2 = E\left(\frac{(X_1-\theta)}{\sigma^2}\right)^2 = \frac{1}{\sigma^2},$$

and (3.4.18) follows. Note that because $\bar{X}$ is UMVU whatever may be σ^2, we have in fact proved that $\bar{X}$ is UMVU even if σ^2 is unknown. $\square$

We can similarly show (Problem 3.4.1) that if $X_1,\ldots,X_n$ are the indicators of n Bernoulli trials with probability of success θ, then $\bar{X}$ is a UMVU estimate of θ. These are situations in which $\mathbf{X}$ follows a one-parameter exponential family. This is no accident.

Theorem 3.4.2. *Suppose that the family $\{P_\theta : \theta \in \Theta\}$ satisfies assumptions I and II and there exists an unbiased estimate T^* of $\psi(\theta)$, which achieves the lower bound of Theorem 3.4.1 for every θ. Then $\{P_\theta\}$ is a one-parameter exponential family with density or frequency function of the form*

$$p(x,\theta) = h(x)\exp[\eta(\theta)T^*(x) - B(\theta)]. \tag{3.4.19}$$

Conversely, if $\{P_\theta\}$ is a one-parameter exponential family of the form (1.6.1) with natural sufficient statistic $T(X)$ and $\eta(\theta)$ has a continuous nonvanishing derivative on Θ, then $T(X)$ achieves the information inequality bound and is a UMVU estimate of $E_\theta(T(X))$.

Proof. We start with the first assertion. Our argument is essentially that of Wijsman (1973). By (3.4.14) and the conditions for equality in the correlation inequality (A.11.16) we know that T^* achieves the lower bound for all θ if, and only if, there exist functions $a_1(\theta)$ and $a_2(\theta)$ such that

$$\frac{\partial}{\partial\theta}\log p(X,\theta) = a_1(\theta)T^*(X) + a_2(\theta) \qquad (3.4.20)$$

with P_θ probability 1 for each θ. From this equality of random variables we shall show that $P_\theta[X \in A^*] = 1$ for all θ where

$$A^* = \left\{ x : \frac{\partial}{\partial\theta}\log p(x,\theta) = a_1(\theta)T^*(x) + a_2(\theta) \text{ for all } \theta \in \Theta \right\}. \qquad (3.4.21)$$

Upon integrating both sides of (3.4.20) with respect to θ we get (3.4.19).

The passage from (3.4.20) to (3.4.19) is highly technical. However, it is necessary. Here is the argument. If A_θ denotes the set of x for which (3.4.20) hold, then (3.4.20) guarantees $P_\theta(A_\theta) = 1$ and assumption I guarantees $P_{\theta'}(A_\theta) = 1$ for all θ' (Problem 3.4.6). Let $\theta_1, \theta_2, \ldots$ be a denumerable dense subset of Θ. Note that if $A^{**} = \cap_m A_{\theta_m}$, $P_{\theta'}(A^{**}) = 1$ for all θ'. Suppose without loss of generality that $T(x_1) \neq T(x_2)$ for $x_1, x_2 \in A^{**}$. By solving for a_1, a_2 in

$$\frac{\partial}{\partial\theta}\log p(x_j,\theta) = a_1(\theta)T^*(x_j) + a_2(\theta) \qquad (3.4.22)$$

for $j = 1, 2$, we see that a_1, a_2 are linear combinations of $\partial \log p(x_j,\theta)/d\theta$, $j = 1, 2$ and, hence, continuous in θ. But now if x is such that

$$\frac{\partial}{\partial\theta}\log p(x,\theta) = a_1(\theta)T^*(x) + a_2(\theta) \qquad (3.4.23)$$

for all $\theta_1, \theta_2, \ldots$ and both sides are continuous in θ, then (3.4.23) must hold for all θ. Thus, $A^{**} = A^*$ and the result follows.

Conversely in the exponential family case (1.6.1) we assume without loss of generality (Problem 3.4.3) that we have the canonical case with $\eta(\theta) = \theta$ and $B(\theta) = A(\theta) = \log \int h(x)\exp\{\theta T(x)\}dx$. Then

$$\frac{\partial}{\partial\theta}\log p(X,\theta) = T(X) - A'(\theta) \qquad (3.4.24)$$

so that

$$I(\theta) = \text{Var}_\theta(T(X) - A'(\theta)) = \text{Var}_\theta T(X) = A''(\theta). \qquad (3.4.25)$$

But $\psi(\theta) = A'(\theta)$ and, thus, the information bound is $[A''(\theta)]^2/A''(\theta) = A''(\theta) = \text{Var}_\theta(T(X))$ so that $T(X)$ achieves the information bound as an estimate of $E_\theta T(X)$. $\square$

Example 3.4.4. In the Hardy-Weinberg model of Examples 2.1.4 and 2.2.6,

$$\begin{aligned} p(\mathbf{x},\theta) &= 2^{n_2}\exp\{(2n_1+n_2)\log\theta + (2n_3+n_3)\log(1-\theta)\} \\ &= 2^{n_2}\exp\{(2n_1+n_2)[\log\theta - \log(1-\theta)] + 2n\log(1-\theta)\} \end{aligned}$$

where we have used the identity $(2n_1 + n_2) + (2n_3 + n_2) = 2n$. Because this is an exponential family, Theorem 3.4.2 implies that $T = (2N_1 + N_2)/2n$ is UMVU for estimating $E(T) = (2n)^{-1}[2n\theta^2 + 2n\theta(1 - \theta)] = \theta$.

This T coincides with the MLE $\widehat{\theta}$ of Example 2.2.6. The variance of $\widehat{\theta}$ can be computed directly using the moments of the multinomial distribution of (N_1, N_2, N_3), or by transforming $p(\mathbf{x}, \theta)$ to canonical form by setting $\eta = \log[\theta/(1 - \theta)]$ and then using Theorem 1.6.2. A third method would be to use $\mathrm{Var}(\widehat{\theta}) = 1/I(\theta)$ and formula (3.4.25). We find (Problem 3.4.7) $\mathrm{Var}(\widehat{\theta}) = \theta(1 - \theta)/2n$. □

Note that by differentiating (3.4.24), we have

$$\frac{\partial^2}{\partial\theta^2} \log p(X, \theta) = -A''(\theta).$$

By (3.4.25) we obtain

$$I(\theta) = -E_\theta \frac{\partial^2}{\partial\theta^2} \log p(X, \theta). \tag{3.4.26}$$

It turns out that this identity also holds outside exponential families:

Proposition 3.4.3. *Suppose $p(\cdot, \theta)$ satisfies in addition to I and II: $p(\cdot, \theta)$ is twice differentiable and interchange between integration and differentiation is permitted. Then (3.4.26) holds.*

Proof. We need only check that

$$\frac{\partial^2}{\partial\theta^2} \log p(x, \theta) = \frac{1}{p(x, \theta)} \frac{\partial^2}{\partial\theta^2} p(x, \theta) - \left(\frac{\partial}{\partial\theta} \log p(x, \theta) \right)^2 \tag{3.4.27}$$

and integrate both sides with respect to $p(x, \theta)$. □

Example 3.4.2. *(Continued).* For a sample from a $\mathcal{P}(\theta)$ distribution

$$E_\theta \left(-\frac{\partial^2}{\partial\theta^2} \log p(\mathbf{X}, \theta) \right) = \theta^{-2} E \left(\sum_{i=1}^{n} X_i \right) = \frac{n}{\theta},$$

which equals $I(\theta)$. □

Discussion. It often happens, for instance, in the $\mathcal{U}(0, \theta)$ example, that I and II fail to hold, although UMVU estimates exist. See Volume II. Even worse, as Theorem 3.4.2 suggests, in many situations, assumptions I and II are satisfied and UMVU estimates of $\psi(\theta)$ exist, but the variance of the best estimate is not equal to the bound $[\psi'(\theta)]^2/I(\theta)$. Sharpenings of the information inequality are available but don't help in general.

Extensions to models in which θ is multidimensional are considered next.

The multiparameter case

We will extend the information lower bound to the case of several parameters, $\boldsymbol{\theta} = (\theta_1, \ldots, \theta_d)$. In particular, we will find a lower bound on the variance of an estimator

$\widehat{\theta}_1 = T$ of θ_1 when the parameters $\theta_2, \ldots, \theta_d$ are unknown. We assume that Θ is an open subset of R^d and that $\{p(x, \boldsymbol{\theta}) : \boldsymbol{\theta} \in \Theta\}$ is a *regular parametric model* with conditions I and II satisfied when differentiation is with respect θ_j, $j = 1, \ldots, d$. Let $p(x, \boldsymbol{\theta})$ denote the density or frequency function of X where $X \in \mathcal{X} \subset R^q$.

The *(Fisher) information matrix* is defined as

$$I_{p \times p}(\boldsymbol{\theta}) = (I_{jk}(\boldsymbol{\theta}))_{1 \le j \le d, 1 \le k \le d}, \tag{3.4.28}$$

where

$$I_{jk}(\boldsymbol{\theta}) = E\left(\frac{\partial}{\partial \theta_j} \log p(X, \boldsymbol{\theta}) \frac{\partial}{\partial \theta_k} \log p(X, \boldsymbol{\theta})\right). \tag{3.4.29}$$

Proposition 3.4.4. *Under the conditions in the opening paragraph,*
(a)

$$E_{\boldsymbol{\theta}}\left(\frac{\partial}{\partial \theta_j} \log p(X, \boldsymbol{\theta})\right) = 0 \tag{3.4.30}$$

$$I_{jk}(\boldsymbol{\theta}) = Cov_{\boldsymbol{\theta}}\left(\frac{\partial}{\partial \theta_j} \log p(X, \boldsymbol{\theta}), \frac{\partial}{\partial \theta_k} \log p(X, \boldsymbol{\theta})\right). \tag{3.4.31}$$

That is,

$$E_{\boldsymbol{\theta}}(\nabla_{\boldsymbol{\theta}} \log p(X, \boldsymbol{\theta})) = \mathbf{0},$$

and

$$I(\boldsymbol{\theta}) = Var(\nabla_{\boldsymbol{\theta}} \log p(X, \boldsymbol{\theta})).$$

(b) *If* $X_1, \ldots, X_n$ *are i.i.d. as* X, *then* $\mathbf{X} = (X_1, \ldots, X_n)^T$ *has information matrix* $nI_1(\boldsymbol{\theta})$ *where* I_1 *is the information matrix of* X.

(c) *If, in addition,* $p(\cdot, \boldsymbol{\theta})$ *is twice differentiable and double integration and differentiation under the integral sign can be interchanged,*

$$I(\boldsymbol{\theta}) = -\left\|E_{\boldsymbol{\theta}}\left(\frac{\partial^2}{\partial \theta_j \partial \theta_k} \log p(X, \boldsymbol{\theta})\right)\right\|, \quad 1 \le j \le d, \ 1 \le k \le d. \tag{3.4.32}$$

Proof. The arguments follow the $d = 1$ case and are left to the problems.

Example 3.4.5. Suppose $X \sim \mathcal{N}(\mu, \sigma^2)$, $\boldsymbol{\theta} = (\mu, \sigma^2)$. Then

$$\log p(x, \boldsymbol{\theta}) = -\frac{1}{2}\log(2\pi) - \frac{1}{2}\log \sigma^2 - \frac{1}{2\sigma^2}(x - \mu)^2$$

$$I_{11}(\boldsymbol{\theta}) = -E\left[\frac{\partial^2}{\partial \mu^2} \log p(x, \boldsymbol{\theta})\right] = E[\sigma^{-2}] = \sigma^{-2}$$

$$I_{12}(\boldsymbol{\theta}) = -E\left[\frac{\partial}{\partial \sigma^2}\frac{\partial}{\partial \mu} \log p(x, \boldsymbol{\theta})\right] = -\sigma^{-4}E(x - \mu) = 0 = I_{21}(\boldsymbol{\theta})$$

$$I_{22}(\boldsymbol{\theta}) = -E\left[\frac{\partial^2}{(\partial \sigma^2)^2} \log p(x, \boldsymbol{\theta})\right] = \sigma^{-4}/2.$$

Thus, in this case

$$I(\boldsymbol{\theta}) = \begin{pmatrix} \sigma^{-2} & 0 \\ 0 & \sigma^{-4}/2 \end{pmatrix}. \tag{3.4.33}$$

□

Example 3.4.6. *Canonical k-Parameter Exponential Family.* Suppose

$$p(x, \boldsymbol{\theta}) = \exp\{\sum_{j=1}^{k} T_j(x)\theta_j - A(\boldsymbol{\theta})\}h(x) \tag{3.4.34}$$

$\boldsymbol{\theta} \in \Theta$ open. The conditions I, II are easily checked and because

$$\nabla_{\boldsymbol{\theta}} \log p(x, \boldsymbol{\theta}) = \mathbf{T}(X) - \dot{A}(\boldsymbol{\theta}),$$

then

$$I(\boldsymbol{\theta}) = \text{Var}_{\boldsymbol{\theta}} \mathbf{T}(X).$$

By (3.4.30) and Corollary 1.6.1,

$$I(\boldsymbol{\theta}) = \text{Var}_{\boldsymbol{\theta}} \mathbf{T}(X) = \ddot{A}(\boldsymbol{\theta}). \tag{3.4.35}$$

□

Next suppose $\widehat{\theta}_1 = T$ is an estimate of θ_1 with $\theta_2, \ldots, \theta_d$ assumed unknown. Let $\psi(\boldsymbol{\theta}) = E_{\boldsymbol{\theta}} T(X)$ and let $\dot{\psi}(\boldsymbol{\theta}) = \nabla\psi(\boldsymbol{\theta})$ be the $d \times 1$ vector of partial derivatives. Then

Theorem 3.4.3. *Assume the conditions of the opening paragraph hold and suppose that the matrix $I(\boldsymbol{\theta})$ is nonsingular. Then for all $\boldsymbol{\theta}$, $\dot{\psi}(\boldsymbol{\theta})$ exists and*

$$\text{Var}_{\boldsymbol{\theta}}(T(X)) \geq \dot{\psi}(\boldsymbol{\theta})I^{-1}(\boldsymbol{\theta})[\dot{\psi}(\boldsymbol{\theta})]^T. \tag{3.4.36}$$

Proof. We will use the prediction inequality $\text{Var}(Y) \geq \text{Var}(\mu_L(\mathbf{Z}))$, where $\mu_L(\mathbf{Z})$ denotes the optimal MSPE linear predictor of Y; that is,

$$\mu_L(\mathbf{Z}) = \mu_Y + (\mathbf{Z} - \mu_{\mathbf{Z}})^T \Sigma_{\mathbf{ZZ}}^{-1} \Sigma_{\mathbf{Z}Y}. \tag{3.4.37}$$

Now set $Y = T(X)$, $\mathbf{Z} = \nabla_{\boldsymbol{\theta}} \log p(X, \boldsymbol{\theta})$. Then, by (B.5.3),

$$\text{Var}_{\boldsymbol{\theta}}(T(X)) \geq \Sigma_{\mathbf{Z}Y}^T I^{-1}(\boldsymbol{\theta}) \Sigma_{\mathbf{Z}Y}, \tag{3.4.38}$$

where $\Sigma_{\mathbf{Z}Y} = E_{\boldsymbol{\theta}}(T\nabla_{\boldsymbol{\theta}} \log p(X, \boldsymbol{\theta})) = \nabla_{\boldsymbol{\theta}} E_{\boldsymbol{\theta}}(T(X))$ and the last equality follows from the argument in (3.4.13).   □

Here are some consequences of this result.

Example 3.4.6. *(continued). UMVU Estimates in Canonical Exponential Families.* Suppose the conditions of Example 3.4.6 hold. We claim that each of $T_j(X)$ is a UMVU estimate of $E_{\boldsymbol{\theta}} T_j(X)$. This is a different claim than $T_j(X)$ is UMVU for $E_{\boldsymbol{\theta}} T_j(X)$ if θ_i, $i \neq j$, are known. To see our claim note that in our case

$$\psi(\boldsymbol{\theta}) = \frac{\partial A(\boldsymbol{\theta})}{\partial \theta_1}, \quad \dot{\psi}(\boldsymbol{\theta}) = \left(\frac{\partial^2 A}{\partial \theta_1^2}, \ldots, \frac{\partial^2 A}{\partial \theta_1 \partial \theta_k} \right) \tag{3.4.39}$$

where, without loss of generality, we let $j = 1$. We have already computed in Proposition 3.4.4

$$I^{-1}(\boldsymbol{\theta}) = \left(\frac{\partial^2 A}{\partial \theta_i \partial \theta_j} \right)_{k \times k}^{-1}. \tag{3.4.40}$$

We claim that in this case

$$\dot{\psi}(\boldsymbol{\theta}) I^{-1}(\boldsymbol{\theta}) \dot{\psi}^T(\boldsymbol{\theta}) = \frac{\partial^2 A}{\partial \theta_1^2} \tag{3.4.41}$$

because $\dot{\psi}(\boldsymbol{\theta})$ is the first row of $I(\boldsymbol{\theta})$ and, hence, $\dot{\psi}(\boldsymbol{\theta}) I^{-1}(\boldsymbol{\theta}) = (1, 0, \ldots, 0)$. But $\frac{\partial^2}{\partial \theta_1^2} A(\boldsymbol{\theta})$ is just $\mathrm{Var}_{\boldsymbol{\theta}} T_1(X)$.

Example 3.4.7. *Multinomial Trials.* In the multinomial Example 1.6.6 with $X_1, \ldots, X_n$ i.i.d. as X and $\lambda_j = P(X = j)$, $j = 1, \ldots, k$, we transformed the multinomial model $\mathcal{M}(n, \lambda_1, \ldots, \lambda_k)$ to the canonical form

$$p(\mathbf{x}, \boldsymbol{\theta}) = \exp\{\mathbf{T}^T(\mathbf{x})\boldsymbol{\theta} - A(\boldsymbol{\theta})\}$$

where $\mathbf{T}^T(\mathbf{x}) = (T_1(\mathbf{x}), \ldots, T_{k-1}(\mathbf{x}))$,

$$T_j(\mathbf{X}) = \sum_{j=1}^{n} 1[X_i = j], \quad \mathbf{X} = (X_1, \ldots, X_n)^T, \quad \boldsymbol{\theta} = (\theta_1, \ldots, \theta_{k-1})^T,$$

$\theta_j = \log(\lambda_j/\lambda_k)$, $j = 1, \ldots, k-1$, and

$$A(\boldsymbol{\theta}) = n \log \left(1 + \sum_{j=1}^{k-1} e^{\theta_j} \right).$$

Note that

$$\frac{\partial}{\partial \theta_j} A(\boldsymbol{\theta}) = \frac{n e^{\theta_j}}{1 + \sum_{l=1}^{k-1} e^{\theta_l}} = n\lambda_j = nE(T_j(\mathbf{X}))$$

$$\frac{\partial^2}{\partial \theta_j^2} A(\boldsymbol{\theta}) = \frac{n e^{\theta_j} \left(1 + \sum_{l=1}^{k-1} e^{\theta_l} - e^{\theta_j} \right)}{\left(1 + \sum_{l=1}^{k-1} e^{\theta_l} \right)^2} = n\lambda_j(1 - \lambda_j) = \mathrm{Var}(T_j(\mathbf{X})).$$

Thus, by Theorem 3.4.3, the lower bound on the variance of an unbiased estimator of $\psi_j(\theta) = E(n^{-1}T_j(\mathbf{X})) = \lambda_j$ is $\lambda_j(1 - \lambda_j)/n$. But because N_j/n is unbiased and has variance $\lambda_j(1 - \lambda_j)/n$, then N_j/n is UMVU for λ_j.   □

Example 3.4.8. *The Normal Case.* If $X_1, \ldots, X_n$ are i.i.d. $\mathcal{N}(\mu, \sigma^2)$ then $\bar{X}$ is UMVU for μ and $\frac{1}{n} \sum X_i^2$ is UMVU for $\mu^2 + \sigma^2$. But it does not follow that $\frac{1}{n-1} \sum (X_i - \bar{X})^2$ is UMVU for σ^2. These and other examples and the implications of Theorem 3.4.3 are explored in the problems. $\qquad\square$

Here is an important extension of Theorem 3.4.3 whose proof is left to Problem 3.4.21.

Theorem 3.4.4. *Suppose that the conditions of Theorem* 3.4.3 *hold and*

$$\mathbf{T}(X) = (T_1(X), \ldots, T_d(X))^T$$

is a d-dimensional statistic. Let

$$\boldsymbol{\psi}(\boldsymbol{\theta}) = E_{\boldsymbol{\theta}}(\mathbf{T}(X))_{d \times 1} = (\psi_1(\boldsymbol{\theta}), \ldots, \psi_d(\boldsymbol{\theta}))^T$$

and $\dot{\boldsymbol{\psi}}(\boldsymbol{\theta}) = \left(\frac{\partial \psi_i}{\partial \theta_j}(\boldsymbol{\theta}) \right)_{d \times d}$. *Then*

$$Var_{\boldsymbol{\theta}} \mathbf{T}(X) \geq \dot{\boldsymbol{\psi}}(\boldsymbol{\theta}) I^{-1}(\boldsymbol{\theta}) \dot{\boldsymbol{\psi}}^T(\boldsymbol{\theta}) \qquad (3.4.42)$$

where $A \geq B$ *means* $\mathbf{a}^T (A - B) \mathbf{a} \geq 0$ *for all* $\mathbf{a}_{d \times 1}$.

Note that both sides of (3.4.42) are $d \times d$ matrices. Also note that

$$\widehat{\boldsymbol{\theta}} \text{ unbiased} \ \Rightarrow \ Var_{\boldsymbol{\theta}} \widehat{\boldsymbol{\theta}} \geq I^{-1}(\boldsymbol{\theta}).$$

In Chapters 5 and 6 we show that in smoothly parametrized models, reasonable estimates are asymptotically unbiased. We establish analogues of the information inequality and use them to show that under suitable conditions the MLE is asymptotically optimal.

Summary. We study the important application of the unbiasedness principle in survey sampling. We derive the information inequality in one-parameter models and show how it can be used to establish that in a canonical exponential family, $T(\mathbf{X})$ is the UMVU estimate of its expectation. Using inequalities from prediction theory, we show how the information inequality can be extended to the multiparameter case. Asymptotic analogues of these inequalities are sharp and lead to the notion and construction of efficient estimates.

3.5 NONDECISION THEORETIC CRITERIA

In practice, even if the loss function and model are well specified, features other than the risk function are also of importance in selection of a procedure. The three principal issues we discuss are the speed and numerical stability of the method of computation used to obtain the procedure, interpretability of the procedure, and robustness to model departures.

3.5.1 Computation

Speed of computation and numerical stability issues have been discussed briefly in Section 2.4. They are dealt with extensively in books on numerical analysis such as Dahlquist,

Björk, and Anderson (1974). We discuss some of the issues and the subtleties that arise in the context of some of our examples in estimation theory.

Closed form versus iteratively computed estimates

At one level closed form is clearly preferable. For instance, a method of moments estimate of (λ, p) in Example 2.3.2 is given by

$$\widehat{\lambda} = \frac{\bar{X}}{\widehat{\sigma}^2}, \; \widehat{p} = \frac{\bar{X}^2}{\widehat{\sigma}^2}$$

where $\widehat{\sigma}^2$ is the empirical variance (Problem 2.2.11). It is clearly easier to compute than the MLE. Of course, with ever faster computers a difference at this level is irrelevant. But it reappears when the data sets are big and the number of parameters large.

On the other hand, consider the Gaussian linear model of Example 2.1.1. Then least squares estimates are given in closed form by equation (2.2.10). The closed form here is deceptive because inversion of a $d \times d$ matrix takes on the order of d^3 operations when done in the usual way and can be numerically unstable. It is in fact faster and better to solve equation (2.1.9) by, say, Gaussian elimination for the particular $\mathbf{Z}_D^T \mathbf{Y}$.

Faster versus slower algorithms

Consider estimation of the MLE $\widehat{\boldsymbol{\theta}}$ in a general canonical exponential family as in Section 2.3. It may be shown that, in the algorithm we discuss in Section 2.4, if we seek to take enough steps J so that $|\widehat{\boldsymbol{\theta}}^{(J)} - \theta| \leq \varepsilon < 1$ then J is of the order of $\log \frac{1}{\varepsilon}$ (Problem 3.5.1). On the other hand, at least if started close enough to $\widehat{\boldsymbol{\theta}}$, the Newton–Raphson method in which the jth iterate, $\widehat{\boldsymbol{\theta}}^{(j)} = \widehat{\boldsymbol{\theta}}^{(j-1)} - \ddot{A}^{-1}(\widehat{\boldsymbol{\theta}}^{(j-1)})(T(X) - \dot{A}(\widehat{\boldsymbol{\theta}}^{(j-1)}))$, takes on the order of $\log \log \frac{1}{\varepsilon}$ steps (Problem 3.5.2). The improvement in speed may however be spurious since $\ddot{A}^{-1}$ is costly to compute if d is large—though the same trick as in computing least squares estimates can be used.

The interplay between estimated variance and computation

As we have seen in special cases in Examples 3.4.3 and 3.4.4, estimates of parameters based on samples of size n have standard deviations of order $n^{-1/2}$. It follows that striving for numerical accuracy of order smaller than $n^{-1/2}$ is wasteful. Unfortunately it is hard to translate statements about orders into specific prescriptions without assuming at least bounds on the constants involved.

3.5.2   Interpretability

Suppose that in the normal $\mathcal{N}(\mu, \sigma^2)$ Example 2.1.5 we are interested in the parameter μ/σ. This parameter, the signal-to-noise ratio, for this population of measurements has a clear interpretation. Its maximum likelihood estimate $\bar{X}/\widehat{\sigma}$ continues to have the same intuitive interpretation as an estimate of μ/σ even if the data are a sample from a distribution with mean μ and variance σ^2 other than the normal. On the other hand, suppose we initially postulate a model in which the data are a sample from a gamma, $\mathcal{G}(p, \lambda)$, distribution. Then $E(X)/\sqrt{\text{Var}(X)} = (p/\lambda)(p/\lambda^2)^{-1/2} = p^{1/2}$. We can now use the MLE $\widehat{p}^{1/2}$, which as we shall see later (Section 5.4) is for n large a more precise estimate than $\bar{X}/\widehat{\sigma}$ if this model is correct. However, the form of this estimate is complex and if the model is incorrect it no longer is an appropriate estimate of $E(X)/[\text{Var}(X)]^{1/2}$. We return to this in Section 5.5.

3.5.3   Robustness

Finally, we turn to robustness.

This is an issue easy to point to in practice but remarkably difficult to formalize appropriately. The idea of robustness is that we want estimation (or testing) procedures to perform reasonably even when the model assumptions under which they were designed to perform excellently are not exactly satisfied. However, what reasonable means is connected to the choice of the parameter we are estimating (or testing hypotheses about). We consider three situations.

(a) The problem dictates the parameter. For instance, the Hardy–Weinberg parameter θ has a clear biological interpretation and is *the* parameter for the experiment described in Example 2.1.4. Similarly, economists often work with median housing prices, that is, the parameter ν that has half of the population prices on either side (formally, ν is any value such that $P(X \leq \nu) \geq \frac{1}{2}, P(X \geq \nu) \geq \frac{1}{2}$). Alternatively, they may be interested in total consumption of a commodity such as coffee, say $\theta = N\mu$, where N is the population size and μ is the expected consumption of a randomly drawn individual.

(b) We imagine that the random variable X^* produced by the random experiment we are interested in has a distribution that follows a "true" parametric model with an interpretable parameter θ, but we do not necessarily observe X^*. The actual observation X is X^* contaminated with "gross errors"—see the following discussion. But θ is still the target in which we are interested.

(c) We have a qualitative idea of what the parameter is, but there are several parameters that satisfy this qualitative notion. This idea has been developed by Bickel and Lehmann (1975a, 1975b, 1976) and Doksum (1975), among others. For instance, we may be interested in the center of a population, and both the mean μ and median ν qualify. See Problem 3.5.13.

We will consider situations (b) and (c).

Gross error models

Most measurement and recording processes are subject to *gross errors*, anomalous values that arise because of human error (often in recording) or instrument malfunction. To be a bit formal, suppose that if n measurements $\mathbf{X}^* \equiv (X_1^*, \ldots, X_n^*)$ could be taken without gross errors then $P^* \in \mathcal{P}^*$ would be an adequate approximation to the distribution of $\mathbf{X}^*$ (i.e., we could suppose $\mathbf{X}^* \sim P^* \in \mathcal{P}^*$). However, if gross errors occur, we observe not $\mathbf{X}^*$ but $\mathbf{X} = (X_1, \ldots, X_n)$ where most of the $X_i = X_i^*$, but there are a few wild values. Now suppose we want to estimate $\theta(P^*)$ and use $\widehat{\theta}(X_1, \ldots, X_n)$ knowing that $\widehat{\theta}(X_1^*, \ldots, X_n^*)$ is a good estimate. Informally $\widehat{\theta}(X_1, \ldots, X_n)$ will continue to be a good or at least reasonable estimate if its value is not greatly affected by the $X_i \neq X_i^*$, the gross errors. Again informally we shall call such procedures *robust*. Formal definitions require model specification, specification of the gross error mechanism, and definitions of insensitivity to gross errors. Most analyses require asymptotic theory and will have to be postponed to Chapters 5 and 6. However, two notions, the sensitivity curve and the break-down point, make sense for fixed n. The breakdown point will be discussed in Volume II. We next define and examine the sensitivity curve in the context of the Gaussian location model, Example 1.1.2, and then more generally.

Consider the one-sample symmetric *location* model $\mathcal{P}$ defined by

$$X_i = \mu + \varepsilon_i, \qquad i = 1, \ldots, n, \tag{3.5.1}$$

where the errors are independent, identically distributed, and symmetric about 0 with common density f and d.f. F. If the error distribution is normal, $\bar{X}$ is the best estimate in a variety of senses.

In our new formulation it is the X_i^* that obey (3.5.1). A reasonable formulation of a model in which the possibility of gross errors is acknowledged is to make the ε_i still i.i.d. but with common distribution function F and density f of the form

$$f(x) = (1 - \lambda)\frac{1}{\sigma}\varphi\left(\frac{x}{\sigma}\right) + \lambda h(x). \tag{3.5.2}$$

Here h is the density of the gross errors and λ is the probability of making a gross error. This corresponds to,

$$
\begin{aligned}
X_i &= X_i^* \text{ with probability } 1 - \lambda \\
&= Y_i \text{ with probability } \lambda
\end{aligned}
$$

where Y_i has density $h(y - \mu)$ and (X_i^*, Y_i) are i.i.d. Note that this implies the possibly unreasonable assumption that committing a gross error is independent of the value of X^*. Further assumptions that are commonly made are that h has a particular form, for example, $h = \frac{1}{K\sigma}\varphi\left(\frac{x}{K\sigma}\right)$ where $K \gg 1$ or more generally that h is an unknown density symmetric about 0. Then the gross error model is semiparametric, $\mathcal{P}_\delta \equiv \{f(\cdot - \mu) : f \text{ satisfies } (3.5.2)$ for some h such that $h(x) = h(-x)$ for all $x\}$. $P_{(\mu, f)} \in \mathcal{P}_\delta$ iff $X_1, \ldots, X_n$ are i.i.d. with common density $f(x - \mu)$, where f satisfies (3.5.2). The advantage of this formulation is that μ remains identifiable. That is, it is the center of symmetry of $P_{(\mu, f)}$ for all such P. Unfortunately, the assumption that h is itself symmetric about 0 seems patently untenable for gross errors. However, if we drop the symmetry assumption, we encounter one of the basic difficulties in formulating robustness in situation (b). Without h symmetric the quantity μ is not a parameter, so it is unclear what we are estimating. That is, it is possible to have $P_{(\mu_1, f_1)} = P_{(\mu_2, f_2)}$ for $\mu_1 \neq \mu_2$ (Problem 3.5.18). Is μ_1 or μ_2 our goal? On the other hand, in situation (c), we do not need the symmetry assumption. We return to these issues in Chapter 6.

The sensitivity curve

At this point we ask: Suppose that an estimate $T(X_1, \ldots, X_n) = \theta(\widehat{F})$, where $\widehat{F}$ is the empirical d.f., is appropriate for the symmetric location model, $\mathcal{P}$, in particular, has the plug-in property, $\theta(P_{(\mu,f)}) = \mu$ for all $P_{(\mu,f)} \in \mathcal{P}$. How sensitive is it to the presence of gross errors among $X_1, \ldots, X_n$? An interesting way of studying this due to Tukey (1972) and Hampel (1974) is the *sensitivity curve* defined as follows for plug-in estimates (which are well defined for all sample sizes n).

We start by defining the sensitivity curve for general plug-in estimates. Suppose that $X \sim P$ and that $\theta = \theta(P)$ is a parameter. The empirical plug-in estimate of θ is $\widehat{\theta} = \theta(\widehat{P})$ where $\widehat{P}$ is the empirical probability distribution. See Section 2.1.2. The *sensitivity curve* of $\widehat{\theta}$ is defined as

$$SC(x; \widehat{\theta}) = n[\widehat{\theta}(x_1, \ldots, x_{n-1}, x) - \widehat{\theta}(x_1, \ldots, x_{n-1})],$$

where $x_1, \ldots, x_{n-1}$ represents an observed sample of size $n-1$ from P and x represents an observation that (potentially) comes from a distribution different from P. We are interested in the shape of the sensitivity curve, not its location. In our examples we shall, therefore, shift the sensitivity curve in the horizontal or vertical direction whenever this produces more transparent formulas. Often this is done by fixing $x_1, \ldots, x_{n-1}$ as an "ideal" sample of size $n - 1$ for which the estimator $\widehat{\theta}$ gives us the right value of the parameter and then we see what the introduction of a potentially deviant nth observation x does to the value of $\widehat{\theta}$.

We return to the location problem with θ equal to the mean $\mu = E(X)$. Because the estimators we consider are location invariant, that is, $\widehat{\theta}(X_1, \ldots, X_n) - \mu = \widehat{\theta}(X_1 - \mu, \ldots, X_n - \mu)$, and because $E(X_j - \mu) = 0$, we take $\mu = 0$ without loss of generality. Now fix $x_1, \ldots, x_{n-1}$ so that their mean has the ideal value zero. This is equivalent to shifting the SC vertically to make its value at $x = 0$ equal to zero. See Problem 3.5.14. Then

$$SC(x; \bar{x}) = n \left(\frac{x_1 + \cdots + x_{n-1} + x}{n} \right) = x.$$

Thus, the sample mean is arbitrarily sensitive to gross error—a large gross error can throw the mean off entirely. Are there estimates that are less sensitive?

A classical estimate of location based on the order statistics is the *sample median* $\widehat{X}$ which we write as

$$\widehat{X} = X_{(k+1)} \qquad\qquad \text{if } n = 2k + 1$$
$$= \tfrac{1}{2}(X_{(k)} + X_{(k+1)}) \quad \text{if } n = 2k$$

where $X_{(1)}, \ldots, X_{(n)}$ are the order statistics, that is, $X_1, \ldots, X_n$ ordered from smallest to largest. See Section 1.3 and (2.1.17).

The sample median can be motivated as an estimate of location on various grounds.

(i) It is the empirical plug-in estimate of the population median ν (Problem 3.5.4), and it splits the sample into two equal halves.

(ii) In the symmetric location model (3.5.1), ν coincides with μ and $\widehat{x}$ is an empirical plug-in estimate of μ.

(iii) The sample median is the MLE when we assume the common density $f(x)$ of the errors $\{\varepsilon_i\}$ in (3.5.1) is the Laplace (double exponential) density

$$f(x) = \frac{1}{2\tau}\exp\{-|x|/\tau\},$$

a density having substantially heavier tails than the normal. See Problems 2.2.32 and 3.5.9.

The sensitivity curve of the median is as follows:
If, say, $n = 2k + 1$ is odd and the median of $x_1, \ldots, x_{n-1} = (x^{(k)} + x^{(k+1)})/2 = 0$, we obtain

$$
\begin{aligned}
SC(x; \widehat{x}) &= nx^{(k)} = -nx^{(k+1)} && \text{for } x < x^{(k)} \\
&= nx && \text{for } x^{(k)} \le x \le x^{(k+1)} \\
&= nx^{(k+1)} && \text{for } x > x^{(k+1)}
\end{aligned}
$$

where $x^{(1)} \le \cdots \le x^{(n-1)}$ are the ordered $x_1, \ldots, x_{n-1}$.

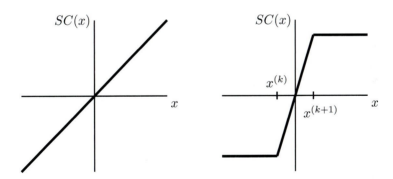

Figure 3.5.1. The sensitivity curves of the mean and median.

Although the median behaves well when gross errors are expected, its performance at the normal model is unsatisfactory in the sense that its variance is about 57% larger than the variance of $\bar{X}$. The sensitivity curve in Figure 3.5.1 suggests that we may improve matters by constructing estimates whose behavior is more like that of the mean when x is near μ. A class of estimates providing such intermediate behavior and including both the mean and the median has been known since the eighteenth century. Let $0 \leq \alpha < \frac{1}{2}$. We define the α trimmed mean, $\bar{X}_\alpha$, by

$$\bar{X}_\alpha = \frac{X_{([n\alpha]+1)} + \cdots + X_{(n-[n\alpha])}}{n - 2[n\alpha]} \qquad (3.5.3)$$

where $[n\alpha]$ is the largest integer $\leq n\alpha$ and $X_{(1)} < \cdots < X_{(n)}$ are the ordered observations. That is, we throw out the "outer" $[n\alpha]$ observations on either side and take the average of the rest. The estimates can be justified on plug-in grounds (see Problem 3.5.5). For more sophisticated arguments see Huber (1981). Note that if $\alpha = 0$, $\bar{X}_\alpha = \bar{X}$, whereas as $\alpha \uparrow \frac{1}{2}$, $\bar{X}_\alpha \to \widehat{X}$. For instance, suppose we take as our data the differences in Table 3.5.1.

If $[n\alpha] = [(n-1)\alpha]$ and the trimmed mean of $x_1, \ldots, x_{n-1}$ is zero, the sensitivity curve of an α trimmed mean is sketched in Figure 3.5.2. (The middle portion is the line $y = x(1 - 2[n\alpha]/n)^{-1}$.)

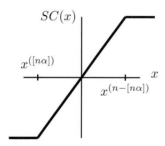

Figure 3.5.2. The sensitivity curve of the trimmed mean.

Intuitively we expect that if there are no gross errors, that is, $f = \varphi$, the mean is better than any trimmed mean with $\alpha > 0$ including the median, which corresponds approximately to $\alpha = \frac{1}{2}$. This can be verified in terms of asymptotic variances (MSEs)—see Problem 5.4.1. However, the sensitivity curve calculation points to an equally intuitive conclusion. If f is symmetric about 0 but has "heavier tails" (see Problem 3.5.8) than the Gaussian density, for example, the Laplace density, $f(x) = \frac{1}{2}e^{-|x|}$, or even more strikingly the Cauchy, $f(x) = 1/\pi(1+x^2)$, then the trimmed means for $\alpha > 0$ and even the median can be much better than the mean, infinitely better in the case of the Cauchy—see Problem 5.4.1 again.

Which α should we choose in the trimmed mean? There seems to be no simple answer. The range $0.10 \leq \alpha \leq 0.20$ seems to yield estimates that provide adequate protection against the proportions of gross errors expected and yet perform reasonably well when sampling is from the normal distribution. See Andrews, Bickel, Hampel, Haber, Rogers, and Tukey (1972). There has also been some research into procedures for which α is chosen using the observations. For a discussion of these and other forms of "adaptation," see Jaeckel (1971), Huber (1972), and Hogg (1974).

Gross errors or outlying data points affect estimates in a variety of situations. We next consider two estimates of the spread in the population as well as estimates of quantiles; other examples will be given in the problems. If we are interested in the spread of the values in a population, then the variance σ^2 or standard deviation σ is typically used. A fairly common quick and simple alternative is the IQR (interquartile range) defined as $\tau = x_{.75} - x_{.25}$, where x_α has 100α percent of the values in the population on its left (formally, x_α is any value such that $P(X \leq x_\alpha) \geq \alpha$, $P(X \geq x_\alpha) \geq 1 - \alpha$). x_α is called a αth *quantile* and $x_{.75}$ and $x_{.25}$ are called the *upper* and *lower quartiles*. The IQR is often calibrated so that it equals σ in the $\mathcal{N}(\mu, \sigma^2)$ model. Because $\tau = 2 \times (.674)\sigma$, the scale measure used is $0.742(x_{.75} - x_{.25})$.

Example 3.5.1. *Spread.* Let $\theta(P) = \text{Var}(X) = \sigma^2$ denote the variance in a population and let $X_1, \ldots, X_n$ denote a sample from that population. Then $\widehat{\sigma}_n^2 = n^{-1} \sum_{i=1}^n (X_i - \bar{X})^2$ is the empirical plug-in estimate of σ^2. To simplify our expression we shift the horizontal axis so that $\sum_{i=1}^{n-1} x_i = 0$. Write $\bar{x}_n = n^{-1} \sum_{i=1}^n x_i = n^{-1}x$, then

$$
\begin{aligned}
SC(x; \widehat{\sigma}^2) &= n(\widehat{\sigma}_n^2 - \widehat{\sigma}_{n-1}^2) \\
&= \sum_{i=1}^{n-1}(x_i - n^{-1}x)^2 + (x - n^{-1}x)^2 - n\widehat{\sigma}_{n-1}^2 \\
&= \sum_{i=1}^{n-1} x_i^2 + (n-1)(n^{-1}x)^2 + [(n-1)/n]^2 x^2 - n\widehat{\sigma}_{n-1}^2 \\
&= (n-1)\widehat{\sigma}_{n-1}^2 + \left\{ \left(\frac{n-1}{n}\right)^2 + \frac{n-1}{n^2} \right\} x^2 - n\widehat{\sigma}_{n-1}^2 \\
&= \left\{ \left(\frac{n-1}{n}\right)^2 + \frac{n-1}{n^2} \right\} x^2 - \widehat{\sigma}_{n-1}^2 \cong x^2 - \widehat{\sigma}_{n-1}^2.
\end{aligned}
$$

It is clear that $\widehat{\sigma}_n^2$ is very sensitive to large outlying $|x|$ values. Similarly,

$$
\begin{aligned}
SC(x; \widehat{\sigma}) &= n(\widehat{\sigma}_n - \widehat{\sigma}_{n-1}) \\
&= n[\{\widehat{\sigma}_{n-1}^2 + (\widehat{\sigma}_n^2 - \sigma_{n-1}^2)\}^{\frac{1}{2}} - \widehat{\sigma}_{n-1}] \\
&= n\widehat{\sigma}_{n-1}\left[\left\{1 + \frac{\widehat{\sigma}_n^2}{\widehat{\sigma}_{n-1}^2} - 1\right\}^{\frac{1}{2}} - 1\right] \qquad (3.5.4) \\
&\cong \widehat{\sigma}_{n-1}\frac{n}{2}\left(\frac{\widehat{\sigma}_n^2}{\widehat{\sigma}_{n-1}^2} - 1\right) \\
&= SC(x; \widehat{\sigma}^2)/2\widehat{\sigma}_{n-1}
\end{aligned}
$$

where the approximation is valid for x fixed, $n \to \infty$ (Problem 3.5.10).   □

Example 3.5.2. *Quantiles and the IQR.* Let $\theta(P) = x_\alpha$ denote a αth quantile of the distribution of X, $0 < \alpha < 1$, and let $\widehat{x}_\alpha$ denote the αth sample quantile (see 2.1.16).

If $n\alpha$ is an integer, say k, the αth sample quantile is $\widehat{x}_\alpha = \frac{1}{2}[x_{(k)} + x_{(k+1)}]$, and at sample size $n - 1$, $\widehat{x}_\alpha = x^{(k)}$, where $x^{(1)} \leq \cdots \leq x^{(n-1)}$ are the ordered $x_1, \ldots, x_{n-1}$, thus, for $2 \leq k \leq n - 2$,

$$
\begin{aligned}
SC(x; \widehat{x}_\alpha) &= \frac{1}{2}[x^{(k-1)} - x^{(k)}], \ x \leq x^{(k-1)} \\
&= \frac{1}{2}[x - x^{(k)}], \ x^{(k-1)} \leq x \leq x^{(k+1)} \qquad (3.5.5) \\
&= \frac{1}{2}[x^{(k+1)} - x^{(k)}], \ x \geq x^{(k+1)}.
\end{aligned}
$$

Clearly, $\widehat{x}_\alpha$ is not sensitive to outlying x's.

Next consider the *sample IQR*

$$
\widehat{\tau} = \widehat{x}_{.75} - \widehat{x}_{.25}.
$$

Then we can write

$$
SC(x; \widehat{\tau}) = SC(x; \widehat{x}_{.75}) - SC(x; \widehat{x}_{.25})
$$

and the sample IQR is robust with respect to outlying gross errors x.   □

Remark 3.5.1. The sensitivity of the parameter $\theta(F)$ to x can be measured by the *influence function*, which is defined by

$$
IF(x; \theta, F) = \lim_{\epsilon \downarrow 0} IF_\epsilon(x; \theta, F)
$$

where

$$
I_\epsilon(x; \theta, F) = \epsilon^{-1}[\theta((1 - \epsilon)F + \epsilon\Delta_x) - \theta(F)]
$$

and Δ_x is the distribution function of point mass at x ($\Delta_x(t) = 1[t \geq x]$). It is easy to see that (Problem 3.5.15)

$$
SC(x; \widehat{\theta}) = IF_{\frac{1}{n}}(x; \theta, \widehat{F}_{n-1})
$$

where $\widehat{F}_{n-1}$ denotes the empirical distribution based on $x_1, \ldots, x_{n-1}$. We will return to the influence function in Volume II. It plays an important role in functional expansions of estimates.

Discussion. Other aspects of robustness, in particular the breakdown point, have been studied extensively and a number of procedures proposed and implemented. Unfortunately these procedures tend to be extremely demanding computationally, although this difficulty appears to be being overcome lately. An exposition of this point of view and some of the earlier procedures proposed is in Hampel, Ronchetti, Rousseuw, and Stahel (1983).

Summary. We discuss briefly nondecision theoretic considerations for selecting procedures including interpretability, and computability. Most of the section focuses on robustness, discussing the difficult issues of identifiability. The rest of our very limited treatment focuses on the sensitivity curve as illustrated in the mean, trimmed mean, median, and other procedures.

3.6   PROBLEMS AND COMPLEMENTS

Problems for Section 3.2

1. Show that if $X_1, \ldots, X_n$ is a $\mathcal{N}(\theta, \sigma^2)$ sample and π is the improper prior $\pi(\theta) = 1$, $\theta \in \Theta = R$, then the improper Bayes rule for squared error loss is $\delta^*(\mathbf{x}) = \bar{x}$.

2. Let $X_1, \ldots, X_n$ be the indicators of n Bernoulli trials with success probability θ. Suppose $l(\theta, a)$ is the quadratic loss $(\theta - a)^2$ and that the prior $\pi(\theta)$ is the beta, $\beta(r, s)$, density. Find the Bayes estimate $\widehat{\theta}_B$ of θ and write it as a weighted average $w\theta_0 + (1 - w)\bar{X}$ of the mean θ_0 of the prior and the sample mean $\bar{X} = S/n$. Show that $\widehat{\theta}_B = (S+1)/(n+2)$ for the uniform prior.

3. In Problem 3.2.2 preceeding, give the MLE of the Bernoulli variance $q(\theta) = \theta(1 - \theta)$ and give the Bayes estimate of $q(\theta)$. Check whether $q(\widehat{\theta}_B) = E(q(\boldsymbol{\theta}) \mid \mathbf{x})$, where $\widehat{\theta}_B$ is the Bayes estimate of θ.

4. In the Bernoulli Problem 3.2.2 with uniform prior on the probabilility of success θ, we found that $(S + 1)/(n + 2)$ is the Bayes rule. In some studies (see Section 6.4.3), the parameter $\lambda = \theta/(1 - \theta)$, which is called the *odds ratio* (for success), is preferred to θ. If we put the (improper) uniform prior $\Pi(\lambda) = 1$ $(\lambda > 0)$ on λ and use quadratic loss $(\lambda - a)^2$, under what condition on S does the Bayes rule exist and what is the Bayes rule?

5. Suppose $\boldsymbol{\theta} \sim \pi(\theta)$, $(X \mid \boldsymbol{\theta} = \theta) \sim p(x \mid \theta)$.

(a) Show that the joint density of X and θ is

$$f(x, \theta) = p(x \mid \theta)\pi(\theta) = c(x)\pi(\theta \mid x)$$

where $c(x) = \int \pi(\theta)p(x \mid \theta)d\theta$.

(b) Let $l(\theta, a) = (\theta - a)^2/w(\theta)$ for some weight function $w(\theta) > 0$, $\theta \in \Theta$. Show that the Bayes rule is

$$\delta^* = E_{f_0}(\boldsymbol{\theta} \mid x)$$

where

$$f_0(x, \theta) = p(x \mid \theta)[\pi(\theta)/w(\theta)]/c$$

and

$$c = \int \int p(x \mid \theta)[\pi(\theta)/w(\theta)]d\theta dx$$

is assumed to be finite. That is, if π and l are changed to $a(\theta)\pi(\theta)$ and $l(\theta, a)/a(\theta)$, $a(\theta) > 0$, respectively, the Bayes rule does not change.

Hint: See Problem 1.4.24.

(c) In Example 3.2.3, change the loss function to $l(\theta, a) = (\theta - a)^2/\theta^\alpha(1 - \theta)^\beta$. Give the conditions needed for the posterior Bayes risk to be finite and find the Bayes rule.

6. Find the Bayes risk $r(\pi, \delta)$ of $\delta(\mathbf{x}) = \bar{X}$ in Example 3.2.1. Consider the relative risk $e(\delta, \pi) = R(\pi)/r(\pi, \delta)$, where $R(\pi)$ is the Bayes risk. Compute the limit of $e(\delta, \pi)$ as

(a) $\tau \to \infty$, (b) $n \to \infty$, (c) $\sigma^2 \to \infty$.

7. For the following problems, compute the posterior risks of the possible actions and give the optimal Bayes decisions when $x = 0$.

(a) Problem 1.3.1(d);

(b) Problem 1.3.2(d)(i) and (ii);

(c) Problem 1.3.19(c).

8. Suppose that $N_1, \ldots, N_r$ given $\boldsymbol{\theta} = \theta$ are multinomial $\mathcal{M}(n, \theta)$, $\theta = (\theta_1, \ldots, \theta_r)^T$, and that $\boldsymbol{\theta}$ has the Dirichlet distribution $\mathcal{D}(\alpha)$, $\alpha = (\alpha_1, \ldots, \alpha_r)^T$, defined in Problem 1.2.15. Let $q(\theta) = \sum_{j=1}^{r} c_j \theta_j$, where $c_1, \ldots, c_r$ are given constants.

(a) If $l(\theta, a) = [q(\theta) - a]^2$, find the Bayes decision rule δ^* and the minimum conditional Bayes risk $r(\delta^*(x) \mid x)$.
  Hint: If $\boldsymbol{\theta} \sim \mathcal{D}(\alpha)$, then $E(\boldsymbol{\theta}_j) = \alpha_j/\alpha_0$, $\mathrm{Var}(\boldsymbol{\theta}_j) = \alpha_j(\alpha_0 - \alpha_j)/\alpha_0^2(\alpha_0 + 1)$, and $\mathrm{Cov}(\boldsymbol{\theta}_i, \boldsymbol{\theta}_j) = -\alpha_i\alpha_j/\alpha_0^2(\alpha_0 + 1)$, where $\alpha_0 = \sum_{j=1}^{r} \alpha_j$. (Use these results, do not derive them.)

(b) When the loss function is $l(\theta, a) = (q(\theta) - a)^2/\prod_{j=1}^{r} \theta_j$, find necessary and sufficient conditions under which the Bayes risk is finite and under these conditions find the Bayes rule.

(c) We want to estimate the vector $(\theta_1, \ldots, \theta_r)$ with loss function $l(\theta, a) = \sum_{j=1}^{r} (\theta_j - a_j)^2$. Find the Bayes decision rule.

9. *Bioequivalence trials* are used to test whether a generic drug is, to a close approximation, equivalent to a name-brand drug. Let $\theta = \mu_G - \mu_B$ be the difference in mean effect of the generic and name-brand drugs. Suppose we have a sample $X_1, \ldots, X_n$ of differences in the effect of generic and name-brand effects for a certain drug, where $E(X) = \theta$. A regulatory agency specifies a number $\epsilon > 0$ such that if $\theta \in (-\epsilon, \epsilon)$, then the generic and brand-name drugs are, by definition, bioequivalent. On the basis of $\mathbf{X} = (X_1, \ldots, X_n)$ we want to decide whether or not $\theta \in (-\epsilon, \epsilon)$. Assume that given θ, $X_1, \ldots, X_n$ are i.i.d. $\mathcal{N}(\theta, \sigma_0^2)$, where σ_0^2 is known, and that $\boldsymbol{\theta}$ is random with a $\mathcal{N}(\eta_0, \tau_0^2)$ distribution.
  There are two possible actions:

$$a = 0 \Leftrightarrow \text{Bioequivalent}$$
$$a = 1 \Leftrightarrow \text{Not Bioequivalent}$$

with losses $l(\theta, 0)$ and $l(\theta, 1)$. Set

$$\lambda(\theta) = l(\theta, 0) - l(\theta, 1)$$

= difference in loss of acceptance and rejection of bioequivalence. Note that $\lambda(\theta)$ should be negative when $\theta \in (-\epsilon, \epsilon)$ and positive when $\theta \notin (-\epsilon, \epsilon)$. One such function (Lindley, 1998) is

$$\lambda(\theta) = r - \exp\left\{-\frac{1}{2c^2}\theta^2\right\}, \quad c^2 > 0$$

where $0 < r < 1$. Note that $\lambda(\pm\epsilon) = 0$ implies that r satisfies

$$\log r = -\frac{1}{2c^2}\epsilon^2.$$

This is an example with two possible actions 0 and 1 where $l(\theta, 0)$ and $l(\theta, 1)$ are not constant. Any two functions with difference $\lambda(\theta)$ are possible loss functions at $a = 0$ and 1.

(a) Show that the Bayes rule is equivalent to

$$\text{``Accept bioequivalence if } E(\lambda(\theta) \mid \mathbf{X} = \mathbf{x}) < 0\text{''} \qquad (3.6.1)$$

and show that $(3.6.1)$ is equivalent to

$$\text{``Accept bioequivalence if } [E(\theta \mid \mathbf{x})]^2 < (\tau_0^2(n) + c^2)\{\log(\frac{c^2}{\tau_0^2(n)+c^2}) + \frac{\epsilon^2}{c^2}\}\text{''}$$

where

$$E(\theta \mid \mathbf{x}) = w\eta_0 + (1-w)\bar{x}, \ w = \tau_0^2(n)/\tau_0^2\sigma_0^2, \ \tau_0^2(n) = \left(\frac{1}{\tau_0^2} + \frac{n}{\sigma_0^2}\right)^{-1}.$$

Hint: See Example 3.2.1.

(b) It is proposed that the preceding prior is "uninformative" if it has $\eta_0 = 0$ and τ_0^2 large ("$\tau_0^2 \to \infty$"). Discuss the preceding decision rule for this "prior."

(c) Discuss the behavior of the preceding decision rule for large n ("$n \to \infty$"). Consider the general case (a) and the specific case (b).

10. For the model defined by $(3.2.16)$ and $(3.2.17)$, find

(a) the linear Bayes estimate of Δ_1.

(b) the linear Bayes estimate of μ.

(c) Is the assumption that the Δ's are normal needed in (a) and (b)?

Problems for Section 3.3

1. In Example 3.3.2 show that $L(\mathbf{x}, 0, v) \geq \pi/(1 - \pi)$ is equivalent to $T \geq t$.

2. Suppose $g : S \times T \to R$. A point (x_0, y_0) is a *saddle point* of g if

$$g(x_0, y_0) = \sup_S g(x, y_0) = \inf_T g(x_0, y).$$

Suppose S and T are subsets of R^m, R^p, respectively, $(\mathbf{x}_0, \mathbf{y}_0)$ is in the interior of $S \times T$, and g is twice differentiable.

(a) Show that a necessary condition for $(\mathbf{x}_0, \mathbf{y}_0)$ to be a saddle point is that, representing $\mathbf{x} = (x_1, \ldots, x_m), \mathbf{y} = (y_1, \ldots, y_p)$,

$$\frac{\partial g}{\partial x_i}(\mathbf{x}_0, \mathbf{y}_0) = \frac{\partial g}{\partial y_j}(\mathbf{x}_0, \mathbf{y}_0) = 0,$$

and

$$\frac{\partial^2 g}{\partial x_a \partial x_b}(\mathbf{x}_0, \mathbf{y}_0) \le 0, \quad \frac{\partial^2 g(\mathbf{x}_0, \mathbf{y}_0)}{\partial y_c \partial y_d} \ge 0$$

for all $1 \le i, a, b \le m, 1 \le j, c, d \le p$.

(b) Suppose $S_m = \{\mathbf{x} : x_i \ge 0, \; 1 \le i \le m, \; \sum_{i=1}^m x_i = 1\}$, the simplex, and $g(\mathbf{x}, \mathbf{y}) = \sum_{i=1}^m \sum_{j=1}^p c_{ij} x_i y_j$ with $\mathbf{x} \in S_m, \mathbf{y} \in S_p$. Show that the von Neumann minimax theorem is equivalent to the existence of a saddle point for any g as above.

3. Suppose $\Theta = \{\theta_0, \theta_1\}, \mathcal{A} = \{0, 1\}$, and that the model is regular. Suppose

$$l(\theta_i, i) = 0, \; l(\theta_i, j) = w_{ij} > 0, \; i, j = 0, 1, \; i \ne j.$$

Let $L_X(\theta_0, \theta_1) = p(X, \theta_1)/p(X, \theta_0)$ and suppose that $L_X(\theta_0, \theta_1)$ has a continuous distribution under both P_{θ_0} and P_{θ_1}. Show that

(a) For every $0 < \pi < 1$, the test rule δ_π given by

$$\begin{aligned}
\delta_\pi(X) &= \; 1 \text{ if } L_X(\theta_0, \theta_1) \ge \frac{(1-\pi)w_{01}}{\pi w_{10}} \\
&= \; 0 \text{ otherwise}
\end{aligned}$$

is Bayes against a prior such that $P[\theta = \theta_1] = \pi = 1 - P[\theta = \theta_0]$, and

(b) There exists $0 < \pi^* < 1$ such that the prior π^* is least favorable against δ_{π^*}, that is, the conclusion of von Neumann's theorem holds.
Hint: Show that there exists (a unique) π^* so that

$$R(\theta_0, \delta_{\pi^*}) = R(\theta_1, \delta_{\pi^*}).$$

4. Let $S \sim \mathcal{B}(n, \theta), l(\theta, a) = (\theta - a)^2, \delta(S) = \bar{X} = S/n$, and

$$\delta^*(S) = (S + \tfrac{1}{2}\sqrt{n})/(n + \sqrt{n}).$$

(a) Show that δ^* has constant risk and is Bayes for the beta, $\beta(\sqrt{n}/2, \sqrt{n}/2)$, prior. Thus, δ^* is minimax.
Hint: See Problem 3.2.2.

(b) Show that $\lim_{n\to\infty}[R(\theta, \delta^*)/R(\theta, \delta)] > 1$ for $\theta \ne \tfrac{1}{2}$; and show that this limit equals 1 when $\theta = \tfrac{1}{2}$.

5. Let $X_1, \ldots, X_n$ be i.i.d. $\mathcal{N}(\mu, \sigma^2)$ and $l(\sigma^2, d) = \left(\frac{d}{\sigma^2} - 1\right)^2$.

(a) Show that if μ is known to be 0

$$\delta^*(X_1, \ldots, X_n) = \frac{1}{n+2}\sum X_i^2$$

is minimax.

(b) If $\mu = 0$, show that δ^* is uniformly best among all rules of the form $\delta_c(\mathbf{X}) = c \sum X_i^2$. Conclude that the MLE is inadmissible.

(c) Show that if μ is unknown, $\delta(\mathbf{X}) = \frac{1}{n+1} \sum (X_i - \bar{X})^2$ is best among all rules of the form $\delta_c(\mathbf{X}) = c \sum (X_i - \bar{X})^2$ and, hence, that both the MLE and the estimate $S^2 = (n-1)^{-1} \sum (X_i - \bar{X})^2$ are inadmissible.

Hint: (a) Consider a gamma prior on $\theta = 1/\sigma^2$. See Problem 1.2.12. (c) Use (B.3.29).

6. Let $X_1, \ldots, X_k$ be independent with means $\mu_1, \ldots, \mu_k$, respectively, where

$$(\mu_1, \ldots, \mu_k) = (\mu_{i_1}^0, \ldots, \mu_{i_k}^0), \ \mu_1^0 < \cdots < \mu_k^0$$

is a known set of values, and $i_1, \ldots, i_k$ is an arbitrary unknown permutation of $1, \ldots, k$. Let $\mathcal{A} = \{(j_1, \ldots, j_k) : \text{Permutations of } 1, \ldots, k\}$

$$l((i_1, \ldots, i_k), (j_1, \ldots, j_k)) = \sum_{l,m} 1(i_l < i_m, j_l > j_m).$$

Show that the minimax rule is to take

$$\delta(X_1, \ldots, X_k) = (R_1, \ldots, R_k)$$

where R_j is the rank of X_j, that is, $R_j = \sum_{l=1}^{k} 1(X_l \leq X_j)$.

Hint: Consider the uniform prior on permutations and compute the Bayes rule by showing that the posterior risk of a permutation $(i_1, \ldots, i_k)$ is smaller than that of $(i'_1, \ldots, i'_k)$, where $i'_j = i_j, j \neq a, b, a < b, i'_a = i_b, i'_b = i_a$, and $R_a < R_b$.

7. Show that X has a Poisson (λ) distribution and $l(\lambda, a) = (\lambda - a)^2/\lambda$. Then X is minimax.

Hint: Consider the gamma, $\Gamma(k^{-1}, 1)$, prior. Let $k \to \infty$.

8. Let X_i be independent $\mathcal{N}(\mu_i, 1)$, $1 \leq i \leq k$, $\boldsymbol{\mu} = (\mu_1, \ldots, \mu_k)^T$. Write $\mathbf{X} = (X_1, \ldots, X_k)^T$, $\mathbf{d} = (d_1, \ldots, d_k)^T$. Show that if

$$l(\boldsymbol{\mu}, \mathbf{d}) = \sum_{i=1}^{k} (d_i - \mu_i)^2$$

then $\boldsymbol{\delta}(\mathbf{X}) = \mathbf{X}$ is minimax.

Remark: Stein (1956) has shown that if $k \geq 3$, $\mathbf{X}$ is no longer unique minimax. For instance,

$$\delta^*(\mathbf{X}) = \left(1 - \frac{k-2}{|\mathbf{X}|^2}\right) \mathbf{X}$$

is also minimax and $R(\boldsymbol{\mu}, \delta^*) < R(\boldsymbol{\mu}, \delta)$ for all $\boldsymbol{\mu}$. See Volume II.

9. Show that if $(N_1, \ldots, N_k)$ has a multinomial, $\mathcal{M}(n, p_1, \ldots, p_k)$, distribution, $0 < p_j < 1$, $1 \leq j \leq k$, then $\frac{\mathbf{N}}{n}$ is minimax for the loss function

$$l(\mathbf{p}, \mathbf{d}) = \sum_{j=1}^{k} \frac{(d_j - p_j)^2}{p_j q_j}$$

where $q_j = 1 - p_j, 1 \leq j \leq k$.

Hint: Consider Dirichlet priors on $(p_1, \ldots, p_{k-1})$ with density defined in Problem 1.2.15. See also Problem 3.2.8.

10. Let $X_i (i = 1, \ldots, n)$ be i.i.d. with unknown distribution F. For a given x we want to estimate the proportion $F(x)$ of the population to the left of x. Show that

$$\delta = \frac{\text{No. of } X_i \leq x}{\sqrt{n}} \cdot \frac{1}{1 + \sqrt{n}} + \frac{1}{2(1 + \sqrt{n})}$$

is minimax for estimating $F(x) = P(X_i \leq x)$ with squared error loss.
Hint: Consider the risk function of δ. See Problem 3.3.4.

11. Let $X_1, \ldots, X_n$ be independent $\mathcal{N}(\mu, 1)$. Define

$$\begin{aligned}
\delta(\bar{X}) &= \bar{X} + \frac{d}{\sqrt{n}} \text{ if } \bar{X} < -\frac{d}{\sqrt{n}} \\
&= 0 \text{ if } |\bar{X}| \leq \frac{d}{\sqrt{n}} \\
&= \bar{X} - \frac{d}{\sqrt{n}} \text{ if } \bar{X} > \frac{d}{\sqrt{n}}.
\end{aligned}$$

(a) Show that the risk (for squared error loss) $E(\sqrt{n}(\delta(\bar{X}) - \mu))^2$ of these estimates is bounded for all n and μ.

(b) How does the risk of these estimates compare to that of $\bar{X}$?

12. Suppose that given $\boldsymbol{\theta} = \theta$, X has a binomial, $\mathcal{B}(n, \theta)$, distribution. Show that the Bayes estimate of θ for the Kullback–Leibler loss function $l_p(\theta, a)$ is the posterior mean $E(\theta \mid X)$.

13. Suppose that given $\boldsymbol{\theta} = \theta = (\theta_1, \ldots, \theta_k)^T$, $X = (X_1, \ldots, X_k)^T$ has a multinomial, $\mathcal{M}(n, \theta)$, distribution. Let the loss function be the Kullback–Leibler divergence $l_p(\theta, a)$ and let the prior be the uniform prior

$$\pi(\theta_1, \ldots, \theta_{k-1}) = (k-1)!, \ \theta_j \geq 0, \ \sum_{j=1}^{k-1} \theta_j = 1.$$

Show that the Bayes estimate is $(X_i + 1)/(n + k)$.

14. Let $K(p_\theta, q)$ denote the KLD (Kullback–Leibler divergence) between the densities p_θ and q and define the *Bayes KLD* between $\mathcal{P} = \{p_\theta : \theta \in \Theta\}$ and q as

$$k(q, \pi) = \int K(p_\theta, q) \pi(\theta) d\theta.$$

Show that the marginal density of X,

$$p(x) = \int p_\theta(x) \pi(\theta) d\theta,$$

minimizes $k(q, \pi)$ and that the minimum is

$$I_{\boldsymbol{\theta}, X} \equiv \int \left[E_\theta \left\{ \log \frac{p_\theta(X)}{p(X)} \right\} \right] \pi(\theta) d\theta.$$

$I_{\boldsymbol{\theta}, X}$ is called the *mutual information* between $\boldsymbol{\theta}$ and X.

    Hint: $k(q, \pi) - k(p, \pi) = \int \left[E_\theta \left\{ \log \frac{p(X)}{q(X)} \right\} \right] \pi(\theta) d\theta \geq 0$ by Jensen's inequality.

15. *Jeffrey's "Prior."* A density proportional to $\sqrt{I_p(\theta)}$ is called Jeffrey's prior. It is often improper. Show that in the $\mathcal{N}(\theta, \sigma_0^2)$, $\mathcal{N}(\mu_0, \theta)$ and $\mathcal{B}(n, \theta)$ cases, Jeffrey's priors are proportional to 1, θ^{-1}, and $\theta^{-\frac{1}{2}}(1 - \theta)^{-\frac{1}{2}}$, respectively. Give the Bayes rules for squared error in these three cases.

Problems for Section 3.4

1. Let $X_1, \dots, X_n$ be the indicators of n Bernoulli trials with success probability θ. Show that $\bar{X}$ is an UMVU estimate of θ.

2. Let $\mathcal{A} = R$. We shall say a loss function is *convex*, if $l(\theta, \alpha a_0 + (1 - \alpha) a_1) \leq \alpha l(\theta, a_0) + (1 - \alpha) l(\theta, a_1)$, for any $a_0, a_1, \theta, 0 < \alpha < 1$. Suppose that there is an unbiased estimate δ of $q(\theta)$ and that $T(\mathbf{X})$ is sufficient. Show that if $l(\theta, a)$ is convex and $\delta^*(\mathbf{X}) = E(\delta(\mathbf{X}) \mid T(\mathbf{X}))$, then $R(\theta, \delta^*) \leq R(\theta, \delta)$.

    Hint: Use *Jensen's inequality:* If g is a convex function and X is a random variable, then $E(g(X)) \geq g(E(X))$.

3. *Equivariance.* Let $X \sim p(x, \theta)$ with $\theta \in \Theta \subset R$, suppose that assumptions I and II hold and that h is a monotone increasing differentiable function from Θ onto $h(\Theta)$. Reparametrize the model by setting $\eta = h(\theta)$ and let $q(x, \eta) = p(x, h^{-1}(\eta))$ denote the model in the new parametrization.

    (a) Show that if $I_p(\theta)$ and $I_q(\eta)$ denote the Fisher information in the two parametrizations, then

$$I_q(\eta) = I_p(h^{-1}(\eta)) / [h'(h^{-1}(\eta))]^2.$$

That is, Fisher information is not equivariant under increasing transformations of the parameter.

    (b) *Equivariance of the Fisher Information Bound.* Let $B_p(\theta)$ and $B_q(\eta)$ denote the information inequality lower bound $(\psi')^2 / I$ as in (3.4.12) for the two parametrizations $p(x, \theta)$ and $q(x, \eta)$. Show that $B_q(\eta) = B_p(h^{-1}(\eta))$; that is, the Fisher information lower bound is equivariant.

4. Prove Proposition 3.4.4.

5. Suppose $X_1, \dots, X_n$ are i.i.d. $\mathcal{N}(\mu, \sigma^2)$ with $\mu = \mu_0$ known. Show that

    (a) $\hat{\sigma}_0^2 = n^{-1} \sum_{i=1}^{n} (X_i - \mu_0)^2$ is a UMVU estimate of σ^2.

    (b) $\hat{\sigma}_0^2$ is inadmissible under squared error loss.

Hint: See Problem 3.3.5(b).

(c) if μ_0 is not known and the true distribution of X_i is $\mathcal{N}(\mu, \sigma^2)$, $\mu \neq \mu_0$, find the bias of $\widehat{\sigma}_0^2$.

6. Show that assumption I implies that if $A \equiv \{x : p(\mathbf{x}, \theta) > 0\}$ doesn't depend on θ, then for any set B, $P_\theta(B) = 1$ for some θ if and only if $P_\theta(B) = 1$ for all θ.

7. In Example 3.4.4, compute $\mathrm{Var}(\widehat{\theta})$ using each of the three methods indicated.

8. Establish the claims of Example 3.4.8.

9. Show that $S^2 = (\mathbf{Y} - \mathbf{Z}_D\widehat{\boldsymbol{\beta}})^T (\mathbf{Y} - \mathbf{Z}_D\widehat{\boldsymbol{\beta}})/(n - p)$ is an unbiased estimate of σ^2 in the linear regression model of Section 2.2.

10. Suppose $\widehat{\theta}$ is UMVU for estimating θ. Let a and b be constants. Show that $\widehat{\lambda} = a + b\widehat{\theta}$ is UMVU for estimating $\lambda = a + b\theta$.

11. Suppose $Y_1, \ldots, Y_n$ are independent Poisson random variables with $E(Y_i) = \mu_i$ where $\mu_i = \exp\{\alpha + \beta z_i\}$ depends on the levels z_i of a covariate; $\alpha, \beta \in R$. For instance, z_i could be the level of a drug given to the ith patient with an infectious disease and Y_i could denote the number of infectious agents in a given unit of blood from the ith patient 24 hours after the drug was administered.

(a) Write the model for $Y_1, \ldots, Y_n$ in two-parameter canonical exponential form and give the sufficient statistic.

(b) Let $\boldsymbol{\theta} = (\alpha, \beta)^T$. Compute $I(\boldsymbol{\theta})$ for the model in (a) and then find the lower bound on the variances of unbiased estimators $\widehat{\alpha}$ and $\widehat{\beta}$ of α and β.

(c) Suppose that $z_i = \log[i/(n + 1)]$, $i = 1, \ldots, n$. Find $\lim n^{-1}I(\boldsymbol{\theta})$ as $n \rightarrow \infty$, and give the limit of n times the lower bound on the variances of $\widehat{\alpha}$ and $\widehat{\beta}$.
Hint: Use the integral approximation to sums.

12. Let $X_1, \ldots, X_n$ be a sample from the beta, $\mathcal{B}(\theta, 1)$, distribution.

(a) Find the MLE of $1/\theta$. Is it unbiased? Does it achieve the information inequality lower bound?

(b) Show that $\bar{X}$ is an unbiased estimate of $\theta/(\theta + 1)$. Does $\bar{X}$ achieve the information inequality lower bound?

13. Let $\mathcal{F}$ denote the class of densities with mean θ^{-1} and variance θ^{-2} ($\theta > 0$) that satisfy the conditions of the information inequality. Show that a density that minimizes the Fisher information over $\mathcal{F}$ is $f(x, \theta) = \theta e^{-\theta x}1(x > 0)$.
Hint: Consider $T(X) = X$ in Theorem 3.4.1.

14. Show that if $(X_1, \ldots, X_n)$ is a sample drawn without replacement from an unknown finite population $\{x_1, \ldots, x_N\}$, then

(a) $\bar{X}$ is an unbiased estimate of $\bar{x} = \frac{1}{N}\sum_{i=1}^{N} x_i$.

(b) The variance of $\bar{X}$ is given by (3.4.4).

15. Suppose $u_1, \ldots, u_N$ are as in Example 3.4.1 and u_j is retained independently of all other u_j with probability π_j where $\sum_{j=1}^{N} \pi_j = n$. Show that if M is the expected sample size, then

$$E(M) = \sum_{j=1}^{N} \pi_j = n.$$

16. Suppose the sampling scheme given in Problem 15 is employed with $\pi_j \equiv \frac{n}{N}$. Show that the resulting unbiased Horvitz–Thompson estimate for the population mean has variance strictly larger than the estimate obtained by taking the mean of a sample of size n taken without replacement from the population.

17. *Stratified Sampling.* (See also Problem 1.3.4.) Suppose the u_j can be relabeled into strata $\{x_{ki}\}$, $1 \leq i \leq I_k$, $k = 1, \ldots, K$, $\sum_{k=1}^{K} I_k = N$. Let $\pi_k = \frac{I_k}{N}$ and suppose $\pi_k = \frac{m_k}{n}$, $1 \leq k \leq K$.

(a) Take samples with replacement of size m_k from stratum $k = \{x_{k1}, \ldots, x_{kI_k}\}$ and form the corresponding sample averages $\bar{X}_1, \ldots, \bar{X}_K$. Define

$$\bar{\bar{X}} \equiv \frac{1}{K} \sum_{k=1}^{K} \pi_k \bar{X}_k.$$

Show that $\bar{\bar{X}}$ is unbiased and if $\bar{X}$ is the mean of a simple random sample without replacement from the population then

$$\text{Var } \bar{\bar{X}} \leq \text{Var } \bar{X}$$

with equality iff $x_{k\cdot} = I_k^{-1} \sum_{i=1}^{I_k} x_{ki}$ doesn't depend on k for all k such that $\pi_k > 0$.

(b) Show that the inequality between Var $\bar{\bar{X}}$ and Var $\bar{X}$ continues to hold if $\frac{m_k - 1}{I_k - 1} \geq \frac{n-1}{N-1}$ for all k, even for sampling without replacement in each stratum.

18. Let X have a binomial, $\mathcal{B}(n, p)$, distribution. Show that $\frac{p}{1-p}$ is not unbiasedly estimable. More generally only polynomials of degree n in p are unbiasedly estimable.

19. Show that $\hat{X}_k$ given by (3.4.6) is **(a)** unbiased and **(b)** has smaller variance than $\bar{X}$ if $b < 2 \text{Cov}(\bar{U}, \bar{X})/\text{Var}(\bar{U})$.

20. Suppose X is distributed according to $\{P_\theta : \theta \in \Theta \subset R\}$ and π is a prior distribution for θ such that $E(\theta^2) < \infty$.

(a) Show that $\delta(X)$ is both an unbiased estimate of θ and the Bayes estimate with respect to quadratic loss, if and only if, $P[\delta(X) = \theta] = 1$.

(b) Deduce that if $P_\theta = \mathcal{N}(0, \sigma_0^2)$, X is not a Bayes estimate for any prior π.

(c) Explain how it is possible if P_θ is binomial, $\mathcal{B}(n, \theta)$, that $\frac{X}{n}$ is a Bayes estimate for θ.

Hint: Given $E(\delta(X) \mid \boldsymbol{\theta}) = \boldsymbol{\theta}$, $E(\boldsymbol{\theta} \mid X) = \delta(X)$ compute $E(\delta(X) - \boldsymbol{\theta})^2$.

21. Prove Theorem 3.4.4.

   Hint: It is equivalent to show that, for all $\mathbf{a}_{d \times 1}$,

$$\begin{aligned} \text{Var}(\mathbf{a}^T \widehat{\boldsymbol{\theta}}) &\geq \mathbf{a}^T (\dot{\psi}(\boldsymbol{\theta}) I^{-1}(\boldsymbol{\theta}) \dot{\psi}^T(\boldsymbol{\theta})) \mathbf{a} \\ &= [\dot{\psi}^T(\boldsymbol{\theta}) \mathbf{a}]^T I^{-1}(\boldsymbol{\theta}) [\dot{\psi}^T(\boldsymbol{\theta}) \mathbf{a}]. \end{aligned}$$

Note that $\dot{\psi}^T(\boldsymbol{\theta}) \mathbf{a} = \nabla E_{\boldsymbol{\theta}}(\mathbf{a}^T \widehat{\boldsymbol{\theta}})$ and apply Theorem 3.4.3.

22. *Regularity Conditions are Needed for the Information Inequality.* Let $X \sim \mathcal{U}(0, \theta)$ be the uniform distribution on $(0, \theta)$. Note that $\log p(x, \theta)$ is differentiable for all $\theta > x$, that is, with probability 1 for each θ, and we can thus define moments of $T = \partial/\partial\theta \log p(X, \theta)$. Show that, however,

(i) $E\left(\dfrac{\partial}{\partial\theta} \log p(X, \theta)\right) = -\dfrac{1}{\theta} \neq 0$

(ii) $\text{Var}\left(\dfrac{\partial}{\partial\theta} \log p(X, \theta)\right) = 0$ and $I(\theta) = ET^2 = \text{Var}\, T + (ET)^2 = 1/\theta^2$.

(iii) $2X$ is unbiased for θ and has variance $(1/3)\theta^2 < (1/I(\theta)) = \theta^2$.

Problems for Section 3.5

1. If $n = 2k$ is even, give and plot the sensitivity curve of the median.

2. If $\alpha = 0.25$ and $n\alpha = k$ is an integer, use (3.5.5) to plot the sensitivity curve of the IQR.

3. If $\alpha = 0.25$ and $(n - 1)\alpha$ is an integer, give and plot the sensitivity curves of the lower quartile $\widehat{x}_{.25}$, the upper quartile $\widehat{x}_{.75}$, and the IQR.

4. Show that the sample median $\widehat{X}$ is an empirical plug-in estimate of the population median ν.

5. Show that the α trimmed mean $\bar{X}_\alpha$ is an empirical plug-in estimate of

$$\mu_\alpha = (1 - 2\alpha)^{-1} \int_{x_{1-\alpha}}^{x_\alpha} x dF(x).$$

Here $\int x dF(x)$ denotes $\int x p(x) dx$ in the continuous case and $\sum x p(x)$ in the discrete case.

6. An estimate $\delta(\mathbf{X})$ is said to be *shift* or *translation* equivariant if, for all $x_1, \ldots, x_n, c$,

$$\delta(x_1 + c, \ldots, x_n + c) = \delta(x_1, \ldots, x_n) + c.$$

It is *antisymmetric* if for all $x_1, \ldots, x_n$

$$\delta(x_1, \ldots, x_n) = -\delta(-x_1, \ldots, -x_n).$$

(a) Show that $\widehat{X}, \bar{X}, \bar{X}_\alpha$ are translation equivariant and antisymmetric.

(b) Suppose $X_1, \ldots, X_n$ is a sample from a population with d.f. $F(x - \mu)$ where μ is unknown and $X_i - \mu$ is symmetrically distributed about 0. Show that if δ is translation equivariant and antisymmetric and $E_0(\delta(\mathbf{X}))$ exists and is finite, then

$$E_\mu(\delta(\mathbf{X})) = \mu$$

(i.e., δ is an unbiased estimate of μ). Deduce that $\bar{X}, \bar{X}_\alpha, \widehat{X}$ are unbiased estimates of the center of symmetry of a symmetric distribution.

7. The *Hodges–Lehmann (location)* estimate $\widehat{x}_{HL}$ is defined to be the median of the $\frac{1}{2}n(n + 1)$ pairwise averages $\frac{1}{2}(x_i + x_j)$, $i \leq j$. Its properties are similar to those of the trimmed mean. It has the advantage that there is no trimming proportion α that needs to be subjectively specified.

(a) Suppose $n = 5$ and the "ideal" ordered sample of size $n - 1 = 4$ is $-1.03, -.30,$ $.30, 1.03$ (these are expected values of four $\mathcal{N}(0, 1)$-order statistics). For $x \geq .3$, plot the sensitivity curves of the mean, median, trimmed mean with $\alpha = 1/4$, and the Hodges–Lehmann estimate.

(b) Show that $\widehat{x}_{HL}$ is translation equivariant and antisymmetric. (See Problem 3.5.6.)

8. The *Huber estimate* $\widehat{X}_k$ is defined implicitly as the solution of the equation

$$\sum_{i=1}^{n} \psi_k \left(\frac{X_i - \widehat{X}_k}{\widehat{\sigma}} \right) = 0$$

where $0 \leq k \leq \infty$, $\widehat{\sigma}$ is an estimate of scale, and

$$
\begin{aligned}
\psi_k(x) &= x \text{ if } |x| \leq k \\
&= k \text{ if } x > k \\
&= -k \text{ if } x < -k.
\end{aligned}
$$

One reasonable choice for k is $k = 1.5$ and for $\widehat{\sigma}$ is,

$$\widehat{\sigma} = \underset{1 \leq i \leq n}{\text{med}} \ |X_i - \widehat{X}|/0.67.$$

Show that

(a) $k = \infty$ corresponds to $\bar{X}$, $k \to 0$ to the median.

(b) If $\widehat{\sigma}$ is replaced by a known σ_0, then $\widehat{X}_k$ is the MLE of θ when $X_1, \ldots, X_n$ are i.i.d. with density $f_0((x - \theta)/\sigma_0)$ where

$$
\begin{aligned}
f_0(x) &= \frac{1 - \varepsilon}{\sqrt{2\pi}} e^{-x^2/2}, && \text{for } |x| \leq k \\
&= \frac{1 - \varepsilon}{\sqrt{2\pi}} e^{k^2/2 - k|x|}, && \text{for } |x| > k,
\end{aligned}
$$

with k and ε connected through

$$
\frac{2\varphi(k)}{k} - 2\Phi(-k) = \frac{\varepsilon}{1 - \varepsilon}.
$$

(c) $\widehat{x}_k$ exists and is unique when $k > 0$. Use a fixed known σ_0 in place of $\widehat{\sigma}$.

(d) $\widehat{x}_k$ is translation equivariant and antisymmetric (see Problem 3.5.6).

(e) If $k < \infty$, then $\lim_{|x| \to \infty} SC(x; \widehat{x}_k)$ is a finite constant.

9. If $f(\cdot)$ and $g(\cdot)$ are two densities with medians ν zero and identical scale parameters τ, we say that $g(\cdot)$ has *heavier tails* than $f(\cdot)$ if $g(x)$ is above $f(x)$ for $|x|$ large. In the case of the Cauchy density, the standard deviation does not exist; thus, we will use the IQR scale parameter $\tau = x_{.75} - x_{.25}$. In what follows adjust f and g to have $\nu = 0$ and $\tau = 1$.

(a) Find the set of $|x|$ where $g(|x|) \geq \varphi(|x|)$ for g equal to the Laplace and Cauchy densities $g_L(x) = (2\eta)^{-1} \exp\{-|x|/\eta\}$ and $g_C(x) = b[b^2 + x^2]^{-1}/\pi$.

(b) Find the tail probabilities $P(|X| \geq 2)$, $P(|X| \geq 3)$ and $P(|X| \geq 4)$ for the normal, Laplace, and Cauchy distributions.

(c) Show that $g_C(x)/\varphi(x)$ is of order $\exp\{x^2\}$ as $|x| \to \infty$.

10. Suppose $\sum_{i=1}^{n-1} x_i = 0$. Show that $SC(x, \widehat{\sigma}_n) \overset{P}{\to} (2\sigma)^{-1}(x^2 - \sigma^2)$ as $n \to \infty$.

11. Let μ_0 be a hypothesized mean for a certain population. The (student) t-ratio is defined as $t = \sqrt{n}(\bar{x} - \mu_0)/s$, where $s^2 = (n - 1)^{-1} \sum_{i=1}^{n} (x_i - \bar{x})^2$. Let $\mu_0 = 0$ and choose the ideal sample $x_1, \ldots, x_{n-1}$ to have sample mean zero. Find the limit of the sensitivity curve of t as

(a) $|x| \to \infty$, n is fixed, and

(b) $n \to \infty$, x is fixed.

12. For the ideal sample of Problem 3.5.7(a), plot the sensitivity curve of

(a) $\widehat{\sigma}_n$, and

(b) the t-ratio of Problem 3.5.11.
This problem may be done on the computer.

13. *Location Parameters.* Let X be a random variable with continuous distribution function F. The functional $\theta = \theta_X = \theta(F)$ is said to be *scale* and *shift (translation) equivariant* if $\theta_{a+bX} = a + b\theta_X$. It is *antisymmetric* if $\theta_X = \theta_{-X}$. Let Y denote a random variable with continuous distribution function G. X is said to be *stochastically smaller* than Y if $F(t) = P(X \le t) \ge P(Y \le t) = G(t)$ for all $t \in R$. In this case we write $X \overset{st}{\le} Y$. θ is said to be order preserving if $X \overset{st}{\le} Y \Rightarrow \theta_X \le \theta_Y$. If θ is scale and shift equivariant, antisymmetric, and order preserving, it is called a *location parameter.*

(a) Show that if F is symmetric about c and θ is a location parameter, then $\theta(F) = c$.

(b) Show that the mean μ, median ν, and trimmed population mean μ_α (see Problem 3.5.5) are location parameters.

(c) Let $\mu^{(k)}$ be the solution to the equation $E\left(\psi_k\left(\frac{X-\mu}{\tau}\right)\right) = 0$, where τ is the median of the distribution of $|X - \nu|/0.67$ and ψ_k is defined in Problem 3.5.8. Show that $\mu^{(k)}$ is a location parameter.

(d) For $0 < \alpha < 1$, let $\nu_\alpha = \nu_\alpha(F) = \frac{1}{2}(x_\alpha + x_{1-\alpha})$, $\underline{\nu}(F) = \inf\{\nu_\alpha(F) : 0 < \alpha \le 1/2\}$ and $\bar{\nu}(F) = \sup\{\nu_\alpha(F) : 0 < \alpha \le 1/2\}$. Show that ν_α is a location parameter and show that any location parameter $\theta(F)$ satisfies $\underline{\nu}(F) \le \theta(F) \le \bar{\nu}(F)$.

Hint: For the second part, let $H(x)$ be the distribution function whose inverse is $H^{-1}(\alpha) = \frac{1}{2}[x_\alpha - x_{1-\alpha}], 0 < \alpha < 1$, and note that $H(x - \bar{\nu}(F)) \le F(x) \le H(x - \underline{\nu}(F))$. Also note that $H(x)$ is symmetric about zero.

(e) Show that if the support $S(F) = \{x : 0 < F(x) < 1\}$ of F is a finite interval, then $\underline{\nu}(F)$ and $\bar{\nu}(F)$ are location parameters. $([\underline{\nu}(F), \bar{\nu}(F)]$ is the *location parameter set* in the sense that for any continuous F the value $\theta(F)$ of any location parameter must be in $[\underline{\nu}(F), \bar{\nu}(F)]$ and, if F is also strictly increasing, any point in $[\underline{\nu}(F), \bar{\nu}(F)]$ is the value of some location parameter.)

14. An estimate $\hat{\theta}_n$ is said to be *shift and scale equivariant* if for all $x_1, \ldots, x_n, a, b > 0$,

$$\hat{\theta}_n(a + bx_1, \ldots, a + bx_n) = a + b\hat{\theta}_n(x_1, \ldots, x_n).$$

(a) Show that the sample mean, sample median, and sample trimmed mean are shift and scale equivariant.

(b) Write the SC as $SC(x, \hat{\theta}, \mathbf{x}_{n-1})$ to show its dependence on $\mathbf{x}_{n-1} = (x_1, \ldots, x_{n-1})$. Show that if $\hat{\theta}$ is shift and location equivariant, then for $a \in R, b > 0, c \in R, d > 0$,

$$SC(a + bx, c + d\hat{\theta}, a + b\mathbf{x}_{n-1}) = bd SC(x, \hat{\theta}, \mathbf{x}_{n-1}).$$

That is, the SC is shift invariant and scale equivariant.

15. In Remark 3.5.1:

(a) Show that $SC(x, \hat{\theta}) = IF_{\frac{1}{n}}(x; \theta, \hat{F}_{n-1})$.

(b) In the following cases, compare $SC(x; \theta, F) \equiv \lim_{n\to\infty} SC(x, \hat{\theta})$ and $IF(x; \theta, F)$.

(i) $\theta(F) = \mu_F = \int x dF(x)$.

(ii) $\theta(F) = \sigma_F^2 = \int (x - \mu_F)^2 dF(x)$.

(iii) $\theta(F) = x_\alpha$. Assume that F is strictly increasing.

(c) Does $n^{-\frac{1}{2}}[SC(x, \widehat{\theta}) - IF(x, \theta, F)] \xrightarrow{P} 0$ in the cases (i), (ii), and (iii) preceding?

16. Show that in the bisection method, in order to be certain that the Jth iterate $\widehat{\theta}_J$ is within ϵ of the desired $\widehat{\theta}$ such that $\psi(\widehat{\theta}) = 0$, we in general must take on the order of $\log \frac{1}{\epsilon}$ steps. This is, consequently, also true of the method of coordinate ascent.

17. Let $d = 1$ and suppose that ψ is twice continuously differentiable, $\psi' > 0$, and we seek the unique solution $\widehat{\theta}$ of $\psi(\theta) = 0$. The Newton–Raphson method in this case is

$$\widehat{\theta}^{(j)} = \widehat{\theta}^{(j-1)} - \frac{\psi(\widehat{\theta}^{(j-1)})}{\psi'(\widehat{\theta}^{(j-1)})}.$$

(a) Show by example that for suitable ψ and $|\widehat{\theta}^{(0)} - \widehat{\theta}|$ large enough, $\{\widehat{\theta}^{(j)}\}$ do not converge.

(b) Show that there exists, $C < \infty$, $\delta > 0$ (depending on ψ) such that if $|\widehat{\theta}^{(0)} - \widehat{\theta}| \leq \delta$, then $|\widehat{\theta}^{(j)} - \widehat{\theta}| \leq C|\widehat{\theta}^{(j-1)} - \widehat{\theta}|^2$.
Hint: (a) Try $\psi(x) = A \log x$ with $A > 1$.
(b)

$$|\widehat{\theta}^{(j)} - \widehat{\theta}| = \left| \widehat{\theta}^{(j-1)} - \widehat{\theta} - \frac{1}{\psi'(\widehat{\theta}^{(j-1)})} (\psi(\widehat{\theta}^{(j-1)}) - \psi(\widehat{\theta})) \right|.$$

18. In the gross error model $(3.5.2)$, show that

(a) If h is a density that is symmetric about zero, then μ is identifiable.

(b) If no assumptions are made about h, then μ is not identifiable.

3.7 NOTES

Note for Section 3.3

(1) A technical problem is to give the class $\mathcal{S}$ of subsets of $\mathcal{F}$ for which we can assign probability (the measurable sets). We define $\mathcal{S}$ as the σ-field generated by $\mathcal{S}_{A,B} = \{F \in \mathcal{F} : P_F(A) \in B\}$, $A, B \in \mathcal{B}$, where $\mathcal{B}$ is the class of Borel sets.

Notes for Section 3.4

(1) The result of Theorem 3.4.1 is commonly known as the Cramér–Rao inequality. Because priority of discovery is now given to the French mathematician M. Frêchet, we shall

follow the lead of Lehmann and call the inequality after the Fisher information number that appears in the statement.

(2) Note that this inequality is true but uninteresting if $I(\theta) = \infty$ (and $\psi'(\theta)$ is finite) or if $\text{Var}_\theta(T(X)) = \infty$.

(3) The continuity of the first integral ensures that

$$\frac{\partial}{\partial \theta}\left[\int_{\theta_0}^{\theta}\int_{-\infty}^{\infty}\cdots\int_{-\infty}^{\infty} T(x)\frac{\partial}{\partial \lambda}p(x, \lambda)dxd\lambda\right] = \int_{-\infty}^{\infty}\cdots\int_{-\infty}^{\infty} T(x)\left[\frac{\partial}{\partial \theta}p(x, \theta)\right]dx$$

for all θ whereas the continuity (or even boundedness on compact sets) of the second integral guarantees that we can interchange the order of integration in

$$\int_{\theta_0}^{\theta}\int_{-\infty}^{\infty}\cdots\int_{-\infty}^{\infty} T(x)\left[\frac{\partial}{\partial \lambda}p(x, \lambda)\right]dxd\lambda.$$

(4) The finiteness of $\text{Var}_\theta(T(X))$ and $I(\theta)$ imply that $\psi'(\theta)$ is finite by the covariance interpretation given in (3.4.8).

3.8   REFERENCES

ANDREWS, D. F., P. J. BICKEL, F. R. HAMPEL, P. J. HUBER, W. H. ROGERS, AND J. W. TUKEY, *Robust Estimates of Location: Survey and Advances* Princeton, NJ: Princeton University Press, 1972.

APOSTOL, T. M., *Mathematical Analysis*, 2nd ed. Reading, MA: Addison–Wesley, 1974.

BERGER, J. O., *Statistical Decision Theory and Bayesian Analysis* New York: Springer, 1985.

BERNARDO, J. M., AND A. F. M. SMITH, *Bayesian Theory* New York: Wiley, 1994.

BICKEL, P., AND E. LEHMANN, "Unbiased Estimation in Convex Families," *Ann. Math. Statist., 40,* 1523–1535 (1969).

BICKEL, P., AND E. LEHMANN, "Descriptive Statistics for Nonparametric Models. I. Introduction," *Ann. Statist., 3,* 1038–1044 (1975a).

BICKEL, P., AND E. LEHMANN, "Descriptive Statistics for Nonparametric Models. II. Location," *Ann. Statist., 3,* 1045–1069 (1975b).

BICKEL, P., AND E. LEHMANN, "Descriptive Statistics for Nonparametric Models. III. Dispersion," *Ann. Statist., 4,* 1139–1158 (1976).

BÜHLMANN, H., *Mathematical Methods in Risk Theory* Heidelberg: Springer Verlag, 1970.

DAHLQUIST, G., A. BJÖRK, AND N. ANDERSON, *Numerical Analysis* New York: Prentice Hall, 1974.

DE GROOT, M. H., *Optimal Statistical Decisions* New York: McGraw–Hill, 1969.

DOKSUM, K. A., "Measures of Location and Asymmetry," *Scand. J. of Statist., 2,* 11–22 (1975).

DOWE, D. L., R. A. BAXTER, J. J. OLIVER, AND C. S. WALLACE, *Point Estimation Using the Kullback–Leibler Loss Function and MML,* in *Proceedings of the Second Pacific Asian Conference on Knowledge Discovery and Data Mining* Melbourne: Springer–Verlag, 1998.

HAMPEL, F., "The Influence Curve and Its Role in Robust Estimation," *J. Amer. Statist. Assoc., 69*, 383–393 (1974).

HAMPEL, F., E. RONCHETTI, P. ROUSSEUW, AND W. STAHEL, *Robust Statistics: The Approach Based on Influence Functions* New York: J. Wiley & Sons, 1986.

HANSEN, M. H., AND B. YU, "Model Selection and the Principle of Minimum Description Length," *J. Amer. Statist. Assoc.*, **96**, 746–774 (2001).

HOGG, R., "Adaptive Robust Procedures," *J. Amer. Statist. Assoc., 69*, 909–927 (1974).

HUBER, P., *Robust Statistics* New York: Wiley, 1981.

HUBER, P., "Robust Statistics: A Review," *Ann. Math. Statist., 43*, 1041–1067 (1972).

JAECKEL, L. A., "Robust Estimates of Location," *Ann. Math. Statist., 42*, 1020–1034 (1971).

JEFFREYS, H., *Theory of Probability*, 2nd ed. London: Oxford University Press, 1948.

KARLIN, S., *Mathematical Methods and Theory in Games, Programming, and Economics* Reading, MA: Addison–Wesley, 1959.

LEHMANN, E. L., *Testing Statistical Hypotheses* New York: Springer, 1986.

LEHMANN, E. L., AND G. CASELLA, *Theory of Point Estimation*, 2nd ed. New York: Springer, 1998.

LINDLEY, D. V., *Introduction to Probability and Statistics from a Bayesian Point of View*, Part I: *Probability*; Part II: *Inference*, Cambridge University Press, London, 1965.

LINDLEY, D.V., "Decision Analysis and Bioequivalence Trials," *Statistical Science, 13*, 136–141 (1998).

NORBERG, R., "Hierarchical Credibility: Analysis of a Random Effect Linear Model with Nested Classification," *Scand. Actuarial J.*, 204–222 (1986).

RISSANEN, J., "Stochastic Complexity (With Discussions)," *J. Royal Statist. Soc. B, 49*, 223–239 (1987).

SAVAGE, L. J., *The Foundations of Statistics* New York: J. Wiley & Sons, 1954.

SHIBATA, R., "Boostrap Estimate of Kullback–Leibler Information for Model Selection," *Statistica Sinica, 7*, 375–394 (1997).

STEIN, C., "Inadmissibility of the Usual Estimator for the Mean of a Multivariate Distribution," *Proc. Third Berkeley Symposium on Math. Statist. and Probability, 1*, University of California Press, 197–206 (1956).

TUKEY, J. W., *Exploratory Data Analysis* Reading, MA: Addison–Wesley, 1972.

WALLACE, C. S., AND P. R. FREEMAN, "Estimation and Inference by Compact Coding (With Discussions)," *J. Royal Statist. Soc. B, 49*, 240–251 (1987).

WIJSMAN, R. A., "On the Attainment of the Cramér–Rao Lower Bound," *Ann. Math. Statist., 1*, 538–542 (1973).

Chapter 4

TESTING AND CONFIDENCE REGIONS: BASIC THEORY

4.1 INTRODUCTION

In Sections 1.3, 3.2, and 3.3 we defined the testing problem abstractly, treating it as a decision theory problem in which we are to decide whether $P \in \mathcal{P}_0$ or $\mathcal{P}_1$ or, parametrically, whether $\theta \in \Theta_0$ or Θ_1 if $\mathcal{P}_j = \{P_\theta : \theta \in \Theta_j\}$, where $\mathcal{P}_0, \mathcal{P}_1$ or Θ_0, Θ_1 are a partition of the model $\mathcal{P}$ or, respectively, the parameter space Θ.

This framework is natural if, as is often the case, we are trying to get a yes or no answer to important questions in science, medicine, public policy, and indeed most human activities, and we have data providing some evidence one way or the other.

As we have seen, in examples such as 1.1.3 the questions are sometimes simple and the type of data to be gathered under our control. Does a new drug improve recovery rates? Does a new car seat design improve safety? Does a new marketing policy increase market share? We can design a clinical trial, perform a survey, or more generally construct an experiment that yields data X in $\mathcal{X} \subset R^q$, modeled by us as having distribution P_θ, $\theta \in \Theta$, where Θ is partitioned into $\{\Theta_0, \Theta_1\}$ with Θ_0 and Θ_1 corresponding, respectively, to answering "no" or "yes" to the preceding questions.

Usually, the situation is less simple. The design of the experiment may not be under our control, what is an appropriate stochastic model for the data may be questionable, and what Θ_0 and Θ_1 correspond to in terms of the stochastic model may be unclear. Here are two examples that illustrate these issues.

Example 4.1.1. *Sex Bias in Graduate Admissions at Berkeley.* The Graduate Division of the University of California at Berkeley attempted to study the possibility that sex bias operated in graduate admissions in 1973 by examining admissions data. They initially tabulated N_{m1}, N_{f1}, the numbers of admitted male and female applicants, and the corresponding numbers N_{m0}, N_{f0} of denied applicants. If n is the total number of applicants, it might be tempting to model $(N_{m1}, N_{m0}, N_{f1}, N_{f0})$ by a multinomial, $\mathcal{M}(n, p_{m1}, p_{m0}, p_{f1}, p_{f0})$, distribution. But this model is suspect because in fact we are looking at the population of all applicants here, not a sample. Accepting this model provisionally, what does the

213

hypothesis of no sex bias correspond to? Again it is natural to translate this into

$$P[\text{Admit} \mid \text{Male}] = \frac{p_{m1}}{p_{m1} + p_{m0}} = P[\text{Admit} \mid \text{Female}] = \frac{p_{f1}}{p_{f1} + p_{f0}}.$$

But is this a correct translation of what absence of bias means? Only if admission is determined centrally by the toss of a coin with probability

$$\frac{p_{m1}}{p_{m1} + p_{m0}} = \frac{p_{f1}}{p_{f1} + p_{f0}}.$$

In fact, as is discussed in a paper by Bickel, Hammel, and O'Connell (1975), admissions are performed at the departmental level and rates of admission differ significantly from department to department. If departments "use different coins," then the data are naturally decomposed into $\mathbf{N} = (N_{m1d}, N_{m0d}, N_{f1d}, N_{f0d}, \ d = 1, \dots, D)$, where N_{m1d} is the number of male admits to department d, and so on. Our multinomial assumption now becomes $\mathbf{N}_d = (N_{m1d}, N_{m0d}, N_{f1d}, N_{f0d})$ are independent with corresponding distributions $\mathcal{M}(n_d, p_{m1d}, p_{m0d}, p_{f1d}, p_{f0d}), \ d = 1, \dots, D$. In these terms the hypothesis of "no bias" can now be translated into:

$$H : \frac{p_{m1d}}{p_{m1d} + p_{m0d}} = \frac{p_{f1d}}{p_{f1d} + p_{f0d}}$$

for $d = 1, \dots, D$. This is *not* the same as our previous hypothesis unless all departments have the same number of applicants or all have the same admission rate,

$$\frac{p_{m1} + p_{f1}}{p_{m1} + p_{f1} + p_{m0} + p_{f0}}.$$

In fact, the same data can lead to opposite conclusions regarding these hypotheses—a phenomenon called *Simpson's paradox*. The example illustrates both the difficulty of specifying a stochastic model and translating the question one wants to answer into a statistical hypothesis.   □

Example 4.1.2. *Mendel's Peas.* In one of his famous experiments laying the foundation of the quantitative theory of genetics, Mendel crossed peas heterozygous for a trait with two alleles, one of which was dominant. The progeny exhibited approximately the expected ratio of one homozygous dominant to two heterozygous dominants (to one recessive). In a modern formulation, if there were n dominant offspring (seeds), the natural model is to assume, if the inheritance ratio can be arbitrary, that N_{AA}, the number of homozygous dominants, has a binomial (n, p) distribution. The hypothesis of dominant inheritance corresponds to $H : p = \frac{1}{3}$ with the alternative $K : p \neq \frac{1}{3}$. It was noted by Fisher as reported in Jeffreys (1961) that in this experiment the observed fraction $\frac{m}{n}$ was much closer to $\frac{1}{3}$ than might be expected under the hypothesis that N_{AA} has a binomial, $\mathcal{B}\left(n, \frac{1}{3}\right)$, distribution,

$$P\left[\left|\frac{N_{AA}}{n} - \frac{1}{3}\right| \leq \left|\frac{m}{n} - \frac{1}{3}\right|\right] = 7 \times 10^{-5}.$$

Fisher conjectured that rather than believing that such a very extraordinary event occurred it is more likely that the numbers were made to "agree with theory" by an overzealous assistant. That is, either N_{AA} cannot really be thought of as stochastic or any stochastic model needs to permit distributions other than $\mathcal{B}(n, p)$, for instance, $(1 - \epsilon)\delta_{\frac{n}{3}} + \epsilon\mathcal{B}(n, p)$, where $1 - \epsilon$ is the probability that the assistant fudged the data and $\delta_{\frac{n}{3}}$ is point mass at $\frac{n}{3}$.   □

What the second of these examples suggests is often the case. The set of distributions corresponding to one answer, say Θ_0, is better defined than the alternative answer Θ_1. That a treatment has no effect is easier to specify than what its effect is; see, for instance, our discussion of constant treatment effect in Example 1.1.3. In science generally a theory typically closely specifies the type of distribution P of the data X as, say, $P = P_\theta, \theta \in \Theta_0$. If the theory is false, it's not clear what P should be as in the preceding Mendel example. These considerations lead to the asymmetric formulation that saying $P \in \mathcal{P}_0$ $(\theta \in \Theta_0)$ corresponds to *acceptance* of the *hypothesis* $H : P \in \mathcal{P}_0$ and $P \in \mathcal{P}_1$ corresponds to *rejection* sometimes written as $K : P \in \mathcal{P}_1$.[1]

As we have stated earlier, acceptance and rejection can be thought of as actions $a = 0$ or 1, and we are then led to the natural $0 - 1$ loss $l(\theta, a) = 0$ if $\theta \in \Theta_a$ and 1 otherwise. Moreover, recall that a decision procedure in the case of a *test* is described by a test function $\delta : x \to \{0, 1\}$ or *critical region* $C \equiv \{x : \delta(x) = 1\}$, the set of points for which we reject.

It is convenient to distinguish between two structural possibilities for Θ_0 and Θ_1: If Θ_0 consists of only one point, we call Θ_0 and H *simple*. When Θ_0 contains more than one point, Θ_0 and H are called *composite*. The same conventions apply to Θ_1 and K.

We illustrate these ideas in the following example.

Example 4.1.3. Suppose we have discovered a new drug that we believe will increase the rate of recovery from some disease over the recovery rate when an old established drug is applied. Our hypothesis is then the *null hypothesis* that the new drug does not improve on the old drug. Suppose that we know from past experience that a fixed proportion $\theta_0 = 0.3$ recover from the disease with the old drug. What our hypothesis means is that the chance that an individual randomly selected from the ill population will recover is the same with the new and old drug. To investigate this question we would have to perform a random experiment. Most simply we would sample n patients, administer the new drug, and then base our decision on the observed sample $\mathbf{X} = (X_1, \ldots, X_n)$, where X_i is 1 if the ith patient recovers and 0 otherwise. Thus, suppose we observe $S = \Sigma X_i$, the number of recoveries among the n randomly selected patients who have been administered the new drug.[2] If we let θ be the probability that a patient to whom the new drug is administered recovers and the population of (present and future) patients is thought of as infinite, then S has a $\mathcal{B}(n, \theta)$ distribution. If we suppose the new drug is at least as effective as the old, then $\Theta = [\theta_0, 1]$, where θ_0 is the probability of recovery using the old drug. Now $\Theta_0 = \{\theta_0\}$ and H is simple; Θ_1 is the interval $(\theta_0, 1]$ and K is composite. In situations such as this one we shall simplify notation and write $H : \theta = \theta_0$, $K : \theta > \theta_0$. If we allow for the possibility that the new drug is less effective than the old, then $\Theta_0 = [0, \theta_0]$ and Θ_0 is composite. It will turn out that in most cases the solution to testing problems with Θ_0 simple also solves the composite Θ_0 problem. See Remark 4.1.

In this example with $\Theta_0 = \{\theta_0\}$ it is reasonable to reject H if S is "much" larger than what would be expected by chance if H is true and the value of θ is θ_0. Thus, we reject H if S exceeds or equals some integer, say k, and accept H otherwise. That is, in the terminology of Section 1.3, our critical region C is $\{\mathbf{X} : S \geq k\}$ and the test function or rule is $\delta_k(\mathbf{X}) = 1\{S \geq k\}$ with

$$P_I = \text{probability of type I error} = P_{\theta_0}(S \geq k)$$
$$P_{II} = \text{probability of type II error} = P_\theta(S < k),\ \theta > \theta_0.$$

The constant k that determines the critical region is called the *critical value*.   □

In most problems it turns out that the tests that arise naturally have the kind of structure we have just described. There is a statistic T that "tends" to be small, if H is true, and large, if H is false. We call T a *test statistic*. (Other authors consider test statistics T that tend to be small, when H is false. $-T$ would then be a test statistic in our sense.) We select a number c and our test is to calculate $T(x)$ and then reject H if $T(x) \geq c$ and accept H otherwise. The value c that completes our specification is referred to as the *critical value* of the test. Note that a test statistic generates a family of possible tests as c varies. We will discuss the fundamental issue of how to choose T in Sections 4.2, 4.3, and later chapters.

We now turn to the prevalent point of view on how to choose c.

The Neyman Pearson Framework

As we discussed in Section 1.3, the Neyman Pearson approach rests on the idea that, of the two errors, one can be thought of as more important. By convention this is chosen to be the type I error and that in turn determines what we call H and what we call K. Given this position, how reasonable is this point of view?

In the medical setting of Example 4.1.3 this asymmetry appears reasonable. It has also been argued that, generally in science, announcing that a new phenomenon has been observed when in fact nothing has happened (the so-called null hypothesis) is more serious than missing something new that has in fact occurred. We do not find this persuasive, but if this view is accepted, it again reasonably leads to a Neyman Pearson formulation.

As we noted in Examples 4.1.1 and 4.1.2, asymmetry is often also imposed because one of Θ_0, Θ_1, is much better defined than its complement and/or the distribution of statistics T under Θ_0 is easy to compute. In that case rejecting the hypothesis at level α is interpreted as a measure of the weight of evidence we attach to the falsity of H. For instance, testing techniques are used in searching for regions of the genome that resemble other regions that are known to have significant biological activity. One way of doing this is to align the known and unknown regions and compute statistics based on the number of matches. To determine significant values of these statistics a (more complicated) version of the following is done. Thresholds (critical values) are set so that if the matches occur at random (i.e., matches at one position are independent of matches at other positions) and the probability of a match is $\frac{1}{2}$, then the probability of exceeding the threshold (type I error is smaller than α. No one really believes that H is true and possible types of alternatives are vaguely known at best, but computation under H is easy.

The Neyman Pearson framework is still valuable in these situations by at least making us think of possible alternatives and then, as we shall see in Sections 4.2 and 4.3, suggesting what test statistics it is best to use.

There is an important class of situations in which the Neyman Pearson framework is inappropriate, such as the quality control Example 1.1.1. Indeed, it is too limited in any situation in which, even though there are just two actions, we can attach, even nominally, numbers to the two losses that are not equal and/or depend on θ. See Problem 3.2.9. Finally, in the Bayesian framework with a prior distribution on the parameter, the approach of Example 3.2.2(b) is the one to take in all cases with Θ_0 and Θ_1 simple.

Here are the elements of the Neyman Pearson story. Begin by specifying a small number $\alpha > 0$ such that probabilities of type I error greater than α are undesirable. Then restrict attention to tests that in fact have the probability of rejection less than or equal to α for all $\theta \in \Theta_0$. As we have noted in Section 1.3, such tests are said to have *level (of significance)* α, and we speak of rejecting H at level α. The values $\alpha = 0.01$ and 0.05 are commonly used in practice. Because a test of level α is also of level $\alpha' > \alpha$, it is convenient to give a name to the smallest level of significance of a test. This quantity is called the *size* of the test and is the maximum probability of type I error. That is, if we have a test statistic T and use critical value c, our test has size $\alpha(c)$ given by

$$\alpha(c) = \sup\{P_\theta[T(X) \geq c] : \theta \in \Theta_0\}. \tag{4.1.1}$$

Now $\alpha(c)$ is nonincreasing in c and typically $\alpha(c) \uparrow 1$ as $c \downarrow -\infty$ and $\alpha(c) \downarrow 0$ as $c \uparrow \infty$. In that case, if $0 < \alpha < 1$, there exists a unique smallest c for which $\alpha(c) \leq \alpha$. This is the critical value we shall use, if our test statistic is T and we want level α. It is referred to as the *level α critical value*. In Example 4.1.3 with $\delta(\mathbf{X}) = 1\{S \geq k\}$, $\theta_0 = 0.3$ and $n = 10$, we find from binomial tables the level 0.05 critical value 6 and the test has size $\alpha(6) = P_{\theta_0}(S \geq 6) = 0.0473$.

Once the level or critical value is fixed, the probabilities of type II error as θ ranges over Θ_1 are determined. By convention $1 - P$ [type II error] is usually considered. Specifically,

Definition 4.1.1. The *power* of a test against the alternative θ is the probability of rejecting H when θ is true.

Thus, the power is 1 minus the probability of type II error. It can be thought of as the probability that the test will "detect" that the alternative θ holds. The power is a function of θ on Θ_1. If Θ_0 is composite as well, then the probability of type I error is also a function of θ. Both the power and the probability of type I error are contained in the *power function*, which is defined *for all $\theta \in \Theta$* by

$$\beta(\theta) = \beta(\theta, \delta) = P_\theta[\text{Rejection}] = P_\theta[\delta(X) = 1] = P_\theta[T(X) \geq c].$$

If $\theta \in \Theta_0$, $\beta(\theta, \delta)$ is just the probability of type I error, whereas if $\theta \in \Theta_1$, $\beta(\theta, \delta)$ is the power against θ.

Example 4.1.3 (continued). Here

$$\beta(\theta, \delta_k) = P(S \geq k) = \sum_{j=k}^{n} \binom{n}{j} \theta^j (1 - \theta)^{n-j}.$$

A plot of this function for $n = 10$, $\theta_0 = 0.3$, $k = 6$ is given in Figure 4.1.1.

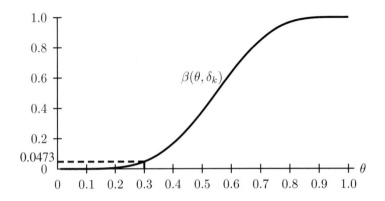

Figure 4.1.1. *Power function of the level 0.05 one-sided test δ_k of $H : \theta = 0.3$ versus $K : \theta > 0.3$ for the $\mathcal{B}(10, \theta)$ family of distributions. The power is plotted as a function of θ, $k = 6$ and the size is 0.0473.*

Note that in this example the power at $\theta = \theta_1 > 0.3$ is the probability that the level 0.05 test will detect an improvement of the recovery rate from 0.3 to $\theta_1 > 0.3$. When θ_1 is 0.5, a 67% improvement, this probability is only .3770. What is needed to improve on this situation is a larger sample size n. One of the most important uses of power is in the selection of sample sizes to achieve reasonable chances of detecting interesting alternatives. We return to this question in Section 4.3.   □

Remark 4.1. From Figure 4.1.1 it appears that the power function is increasing (a proof will be given in Section 4.3). It follows that the size of the test is unchanged if instead of $\Theta_0 = \{\theta_0\}$ we used $\Theta_0 = [0, \theta_0]$. That is,

$$\alpha(k) = \sup\{P_\theta[T(X) \geq k] : \theta \in \Theta_0\} = P_{\theta_0}[T(X) \geq k].$$

Example 4.1.4. *One-Sided Tests for the Mean of a Normal Distribution with Known Variance.* Suppose that $\mathbf{X} = (X_1, \ldots, X_n)$ is a sample from $\mathcal{N}(\mu, \sigma^2)$ population with σ^2 is known. (The σ^2 unknown case is treated in Section 4.5.) We want to test $H : \mu \leq 0$ versus $K : \mu > 0$. This problem arises when we want to compare two treatments or a treatment and control (nothing) and both treatments are administered to the same subject. For instance, suppose we want to see if a drug induces sleep. We might, for each of a group of n randomly selected patients, record sleeping time without the drug (or after the administration of a placebo) and then after some time administer the drug and record sleeping time again. Let X_i be the difference between the time slept after administration of the drug and time slept without administration of the drug by the ith patient. If we assume $X_1, \ldots, X_n$ are normally distributed with mean μ and variance σ^2, then the drug effect is measured by μ and H is the hypothesis that the drug has no effect or is detrimental, whereas K is the alternative that it has some positive effect.

Because $\bar{X}$ tends to be larger under K than under H, it is natural to reject H for large values of $\bar{X}$. It is convenient to replace $\bar{X}$ by the test statistic $T(\mathbf{X}) = \sqrt{n}\bar{X}/\sigma$, which generates the same family of critical regions. The power function of the test with critical value c is

$$\beta(\mu) = P_\mu[T(\mathbf{X}) \ge c] \;=\; P_\mu\left[\sqrt{n}\frac{(\bar{X} - \mu)}{\sigma} \ge c - \frac{\sqrt{n}\mu}{\sigma}\right]$$
$$= \;1 - \Phi\left(c - \frac{\sqrt{n}\mu}{\sigma}\right) = \Phi\left(-c + \frac{\sqrt{n}\mu}{\sigma}\right) \tag{4.1.2}$$

because $\Phi(z) = 1 - \Phi(-z)$. Because $\beta(\mu)$ is increasing,

$$\alpha(c) = \sup\{\beta(\mu) : \mu \le 0\} = \beta(0) = \Phi(-c).$$

The smallest c for which $\Phi(-c) \le \alpha$ is obtained by setting $\Phi(-c) = \alpha$ or

$$c = -z(\alpha)$$

where $-z(\alpha) = z(1 - \alpha)$ is the $(1 - \alpha)$ quantile of the $\mathcal{N}(0, 1)$ distribution. $\square$

The Heuristics of Test Construction

When hypotheses are expressed in terms of an estimable parameter $H : \theta \in \Theta_0 \subset \mathcal{R}^p$, and we have available a good estimate $\hat{\theta}$ of θ, it is clear that a reasonable test statistic is $d(\hat{\theta}, \Theta_0)$, where d is the Euclidean (or some other) distance and $d(x, S) \equiv \inf\{d(x, y) : y \in S\}$. This *minimum distance principle* is essentially what underlies Examples 4.1.2 and 4.1.3. In Example 4.1.2, $p = P[AA]$, $\frac{N_{AA}}{n}$ is the MLE of p and $d\left(\frac{N_{AA}}{n}, \Theta_0\right) = \left|\frac{N_{AA}}{n} - \frac{1}{3}\right|$. In Example 4.1.3, $\frac{X}{n}$ estimates θ and $d\left(\frac{X}{n}, [0, \theta_0]\right) = \left(\frac{X}{n} - \theta_0\right)_+$ where $y_+ = y1(y \ge 0)$. Rejecting for large values of this statistic is equivalent to rejecting for large values of X.

Given a test statistic $T(X)$ we need to determine critical values and eventually the power of the resulting tests. The task of finding a critical value is greatly simplified if $\mathcal{L}_\theta(T(\mathbf{X}))$ doesn't depend on θ for $\theta \in \Theta_0$. This occurs if Θ_0 is simple as in Example 4.1.3. But it occurs also in more interesting situations such as testing $\mu = \mu_0$ versus $\mu \ne \mu_0$ if we have $\mathcal{N}(\mu, \sigma^2)$ observations with both parameters unknown (the t tests of Example 4.5.1 and Example 4.1.5). In all of these cases, $\mathcal{L}_0$, the common distribution of $T(\mathbf{X})$ under $\theta \in \Theta_0$, has a closed form and is tabled. However, in any case, critical values yielding correct type I probabilities are easily obtained by *Monte Carlo methods*. That is, if we generate i.i.d. $T(\mathbf{X}^{(1)}), \ldots, T(\mathbf{X}^{(B)})$ from $\mathcal{L}_0$, then the test that rejects iff $T(\mathbf{X}) > T_{((B+1)(1-\alpha))}$, where $T_{(1)} \le \cdots \le T_{(B+1)}$ are the ordered $T(\mathbf{X}), T(\mathbf{X}^{(1)}), \ldots, T(\mathbf{X}^{(B)})$, has level α if $\mathcal{L}_0$ is continuous and $(B + 1)(1 - \alpha)$ is an integer (Problem 4.1.9).

The key feature of situations in which $\mathcal{L}_\theta(T_n) \equiv \mathcal{L}_0$ for $\theta \in \Theta_0$ is usually invariance under the action of a group of transformations. See Lehmann (1997) and Volume II for discussions of this property.

Here are two examples of testing hypotheses in a nonparametric context in which the minimum distance principle is applied and calculation of a critical value is straightforward.

Example 4.1.5. *Goodness of Fit Tests.* Let $X_1, \ldots, X_n$ be i.i.d. as $X \sim F$, where F is continuous. Consider the problem of testing $H : F = F_0$ versus $K : F \neq F_0$. Let $\widehat{F}$ denote the empirical distribution and consider the sup distance between the hypothesis F_0 and the plug-in estimate of F, the empirical distribution function $\widehat{F}$, as a test statistic

$$D_n = \sup_x |\widehat{F}(x) - F_0(x)|.$$

It can be shown (Problem 4.1.7) that D_n, which is called the *Kolmogorov statistic*, can be written as

$$D_n = \max_{i=1,\ldots,n} \max \left\{ \frac{i}{n} - F_0(x_{(i)}), \; F_0(x_{(i)}) - \frac{(i-1)}{n} \right\} \qquad (4.1.3)$$

where $x_{(1)} < \cdots < x_{(n)}$ is the ordered observed sample, that is, *the order statistics*. This statistic has the following *distribution-free* property:

Proposition 4.1.1. *The distribution of D_n under H is the same for all continuous F_0. In particular, $P_{F_0}(D_n \leq d) = P_U(D_n \leq d)$, where U denotes the $\mathcal{U}(0, 1)$ distribution.*

Proof. Set $U_i = F_0(X_i)$, then by Problem B.3.4, $U_i \sim \mathcal{U}(0, 1)$. Also

$$\begin{aligned} \widehat{F}(x) &= n^{-1}\Sigma 1\{X_i \leq x\} = n^{-1}\Sigma 1\{F_0(X_i) \leq F_0(x)\} \\ &= n^{-1}\Sigma 1\{U_i \leq F_0(x)\} = \widehat{U}(F_0(x)) \end{aligned}$$

where $\widehat{U}$ denotes the empirical distribution function of $U_1, \ldots, U_n$. As x ranges over R, $u = F_0(x)$ ranges over $(0, 1)$, thus,

$$D_n = \sup_{0<u<1} |\widehat{U}(u) - u|$$

and the result follows. $\qquad \square$

Note that the hypothesis here is simple so that for any one of these hypotheses $F = F_0$, the distribution can be simulated (or exhibited in closed form). What is remarkable is that it is independent of which F_0 we consider. This is again a consequence of invariance properties (Lehmann, 1997).

The distribution of D_n has been thoroughly studied for finite and large n. In particular, for $n > 80$, and

$$h_n(t) = t/(\sqrt{n} + 0.12 + 0.11/\sqrt{n})$$

close approximations to the size α critical values k_α are $h_n(1.628)$, $h_n(1.358)$, and $h_n(1.224)$ for $\alpha = .01, .05$, and $.10$ respectively. $\qquad \square$

Example 4.1.6. *Goodness of Fit to the Gaussian Family.* Suppose $X_1, \ldots, X_n$ are i.i.d. F and the hypothesis is $H : F = \Phi\left(\frac{\cdot - \mu}{\sigma}\right)$ for some μ, σ, which is evidently composite. We can proceed as in Example 4.1.5 rewriting $H : F(\mu + \sigma x) = \Phi(x)$ for all x where $\mu = E_F(X_1)$, $\sigma^2 = \mathrm{Var}_F(X_1)$. The natural estimate of the parameter $F(\mu + \sigma x)$ is

$\widehat{F}(\bar{X} + \widehat{\sigma}x)$ where $\bar{X}$ and $\widehat{\sigma}^2$ are the MLEs of μ and σ^2. Applying the sup distance again, we obtain the statistic

$$
\begin{aligned}
T_n &= \sup_x |\widehat{F}(\bar{X} + \widehat{\sigma}x) - \Phi(x)| \\
&= \sup_x |\widehat{G}(x) - \Phi(x)|
\end{aligned}
$$

where $\widehat{G}$ is the empirical distribution of $(\Delta_1, \ldots, \Delta_n)$ with $\Delta_i \equiv (X_i - \bar{X})/\widehat{\sigma}$. But, under H, the joint distribution of $(\Delta_1, \ldots, \Delta_n)$ doesn't depend on μ, σ^2 and is that of $(Z_i - \bar{Z}) / \left(\frac{1}{n} \sum_{i=1}^n (Z_i - \bar{Z})^2 \right)^{\frac{1}{2}}$, $1 \le i \le n$, where $Z_1, \ldots, Z_n$ are i.i.d. $\mathcal{N}(0,1)$. (See Section B.3.2.) Thus, T_n has the same distribution $\mathcal{L}_0$ under H, whatever be μ and σ^2, and the critical value may be obtained by simulating i.i.d. observations Z_i, $1 \le i \le n$, from $\mathcal{N}(0,1)$, then computing the T_n corresponding to those Z_i. We do this B times independently, thereby obtaining $T_{n1}, \ldots, T_{nB}$. Now the Monte Carlo critical value is the $[(B+1)(1-\alpha) + 1]$th order statistic among $T_n, T_{n1}, \ldots, T_{nB}$.[(3)]   □

The p-Value: The Test Statistic as Evidence

Different individuals faced with the same testing problem may have different criteria of size. Experimenter I may be satisfied to reject the hypothesis H using a test with size $\alpha = 0.05$, whereas experimenter II insists on using $\alpha = 0.01$. It is then possible that experimenter I rejects the hypothesis H, whereas experimenter II accepts H on the basis of the same outcome $\mathbf{x}$ of an experiment. If the two experimenters can agree on a common test statistic T, this difficulty may be overcome by reporting the outcome of the experiment in terms of the *observed size* or *p-value* or *significance probability* of the test. This quantity is a statistic that is defined as the smallest level of significance α at which an experimenter using T would reject on the *basis of the observed outcome* $\mathbf{x}$. That is, if the experimenter's critical value corresponds to a test of size less than the p-value, H is not rejected; otherwise, H is rejected.

Consider, for instance, Example 4.1.4. If we observe $\mathbf{X} = \mathbf{x} = (x_1, \ldots, x_n)$, we would reject H if, and only if, α satisfies

$$
T(\mathbf{x}) = \frac{\sqrt{n}\bar{x}}{\sigma} \ge -z(\alpha)
$$

or upon applying Φ to both sides if, and only if,

$$
\alpha \ge \Phi(-T(\mathbf{x})).
$$

Therefore, if $\mathbf{X} = \mathbf{x}$, the p-value is

$$
\Phi(-T(\mathbf{x})) = \Phi\left(\frac{-\sqrt{n}\bar{x}}{\sigma} \right). \tag{4.1.4}
$$

Considered as a statistic the p-value is $\Phi(-\sqrt{n}\bar{X}/\sigma)$.

In general, let X be a q dimensional random vector. We will show that we can express the p-value simply in terms of the function $\alpha(\cdot)$ defined in (4.1.1). Suppose that we observe $X = x$. Then if we use critical value c, we would reject H if, and only if,

$$T(x) \geq c.$$

Thus, the largest critical value c for which we would reject is $c = T(x)$. But the size of a test with critical value c is just $\alpha(c)$ and $\alpha(c)$ is decreasing in c. Thus, the smallest α for which we would reject corresponds to the largest c for which we would reject and is just $\alpha(T(x))$. We have proved the following.

Proposition 4.1.2. *The p-value is $\alpha(T(X))$.*

This is in agreement with (4.1.4). Similarly in Example 4.1.3,

$$\alpha(k) = \sum_{j=k}^{n} \binom{n}{j} \theta_0^j (1 - \theta_0)^{n-j}$$

and the p-value is $\alpha(s)$ where s is the observed value of X. The normal approximation is used for the p-value also. Thus, for $\min\{n\theta_0, n(1 - \theta_0)\} \geq 5$,

$$\alpha(s) = P_{\theta_0}(S \geq s) \simeq 1 - \Phi\left(\frac{s - \frac{1}{2} - n\theta_0}{[n\theta_0(1 - \theta_0)]^{\frac{1}{2}}}\right). \tag{4.1.5}$$

The p-value is used extensively in situations of the type we described earlier, when H is well defined, but K is not, so that type II error considerations are unclear. In this context, to quote Fisher (1958), "The actual value of p obtainable from the table by interpolation indicates the strength of the evidence against the null hypothesis" (p. 80).

The p-value can be thought of as a standardized version of our original statistic; that is, $\alpha(T)$ is on the unit interval and when H is simple and T has a continuous distribution, $\alpha(T)$ has a uniform, $\mathcal{U}(0, 1)$, distribution (Problem 4.1.5).

It is possible to use p-values to combine the evidence relating to a given hypothesis H provided by several different independent experiments producing different kinds of data. For example, if r experimenters use continuous test statistics $T_1, \ldots, T_r$ to produce p-values $\alpha(T_1), \ldots, \alpha(T_r)$, then if H is simple Fisher (1958) proposed using

$$\widehat{T} = -2 \sum_{j=1}^{r} \log \alpha(T_j) \tag{4.1.6}$$

to test H. The statistic $\widehat{T}$ has a chi-square distribution with $2r$ degrees of freedom (Problem 4.1.6). Thus, H is rejected if $\widehat{T} \geq x_{1-\alpha}$ where $x_{1-\alpha}$ is the $1 - \alpha$th quantile of the χ_{2r}^2 distribution. Various methods of combining the data from different experiments in this way are discussed by van Zwet and Osterhoff (1967). More generally, these kinds of issues are currently being discussed under the rubric of *data-fusion* and *meta-analysis* (e.g., see Hedges and Olkin, 1985).

The preceding paragraph gives an example in which the hypothesis specifies a distribution completely; that is, under H, $\alpha(T_i)$ has a $\mathcal{U}(0, 1)$ distribution. This is an instance of testing *goodness of fit*; that is, we test whether the distribution of X is different from a specified F_0.

Summary. We introduce the basic concepts and terminology of testing statistical hypotheses and give the Neyman–Pearson framework. In particular, we consider experiments in which important questions about phenomena can be turned into questions about whether a parameter θ belongs to Θ_0 or Θ_1, where Θ_0 and Θ_1 are disjoint subsets of the parameter space Θ. We introduce the basic concepts of simple and composite hypotheses, (null) hypothesis H and alternative (hypothesis) K, test functions, critical regions, test statistics, type I error, type II error, significance level, size, power, power function, and p-value. In the Neyman–Pearson framework, we specify a small number α and construct tests that have at most probability (significance level) α of rejecting H (deciding K) when H is true; then, subject to this restriction, we try to maximize the probability (power) of rejecting H when K is true.

4.2   CHOOSING A TEST STATISTIC: THE NEYMAN–PEARSON LEMMA

We have seen how a hypothesis-testing problem is defined and how performance of a given test δ, or equivalently, a given test statistic T, is measured in the Neyman–Pearson theory. Typically a test statistic is not given but must be chosen on the basis of its performance. In Sections 3.2 and 3.3 we derived test statistics that are best in terms of minimizing Bayes risk and maximum risk. In this section we will consider the problem of finding the level α test that has the highest possible power. Such a test and the corresponding test statistic are called *most powerful* (MP).

We start with the problem of testing a simple hypothesis $H : \theta = \theta_0$ versus a simple alternative $K : \theta = \theta_1$. In this case the Bayes principle led to procedures based on the *simple likelihood ratio statistic* defined by

$$L(x, \theta_0, \theta_1) = \frac{p(x, \theta_1)}{p(x, \theta_0)}$$

where $p(x, \theta)$ is the density or frequency function of the random vector X. The statistic L takes on the value ∞ when $p(x, \theta_1) > 0$, $p(x, \theta_0) = 0$; and, by convention, equals 0 when both numerator and denominator vanish.

The statistic L is reasonable for testing H versus K with large values of L favoring K over H. For instance, in the binomial example (4.1.3),

$$
\begin{aligned}
L(\mathbf{x}, \theta_0, \theta_1) &= (\theta_1/\theta_0)^S[(1 - \theta_1)/(1 - \theta_0)]^{n-S} \\
&= [\theta_1(1 - \theta_0)/\theta_0(1 - \theta_1)]^S[(1 - \theta_1)/(1 - \theta_0)]^n,
\end{aligned}
\tag{4.2.1}
$$

which is large when $S = \Sigma X_i$ is large, and S tends to be large when $K : \theta = \theta_1 > \theta_0$ is true.

We call φ_k a *likelihood ratio* or *Neyman–Pearson (NP) test (function)* if for some $0 \leq k \leq \infty$ we can write the test function φ_k as

$$\varphi_k(x) = \begin{array}{ll} 1 & \text{if } L(x, \theta_0, \theta_1) > k \\ 0 & \text{if } L(x, \theta_0, \theta_1) < k \end{array}$$

with $\varphi_k(x)$ any value in $(0, 1)$ if equality occurs. Note (Section 3.2) that φ_k is a Bayes rule with $k = \pi/(1 - \pi)$, where π denotes the prior probability of $\{\theta_0\}$. We show that in addition to being Bayes optimal, φ_k is MP for level $E_{\theta_0}\varphi_k(X)$.

Because we want results valid for all possible test sizes α in $[0, 1]$, we consider *randomized tests* φ, which are tests that may take values in $(0, 1)$. If $0 < \varphi(x) < 1$ for the observation vector x, the interpretation is that we toss a coin with probability of heads $\varphi(x)$ and reject H iff the coin shows heads. (See also Section 1.3.) For instance, if want size $\alpha = .05$ in Example 4.1.3 with $n = 10$ and $\theta_0 = 0.3$, we choose $\varphi(x) = 0$ if $S < 5$, $\varphi(x) = 1$ if $S > 5$, and

$$\varphi(x) = [0.05 - P(S > 5)]/P(S = 5) = .0262$$

if $S = 5$. Such randomized tests are not used in practice. They are only used to show that with randomization, likelihood ratio tests are unbeatable no matter what the size α is.

Theorem 4.2.1. (Neyman–Pearson Lemma).

(a) *If $\alpha > 0$ and φ_k is a size α likelihood ratio test, then φ_k is MP in the class of level α tests.*

(b) *For each $0 \leq \alpha \leq 1$ there exists an MP size α likelihood ratio test provided that randomization is permitted, $0 < \varphi(x) < 1$, for some x.*

(c) *If φ is an MP level α test, then it must be a level α likelihood ratio test; that is, there exists k such that*

$$P_\theta[\varphi(X) \neq \varphi_k(X), L(X, \theta_0, \theta_1) \neq k] = 0 \tag{4.2.2}$$

for $\theta = \theta_0$ and $\theta = \theta_1$.

Proof. (a) Let E_i denote E_{θ_i}, $i = 0, 1$, and suppose φ is a level α test, then

$$E_0\varphi_k(X) = \alpha, \quad E_0\varphi(X) \leq \alpha. \tag{4.2.3}$$

We want to show $E_1[\varphi_k(X) - \varphi(X)] \geq 0$. To this end consider

$$E_1[\varphi_k(X) - \varphi(X)] - kE_0[\varphi_k(X) - \varphi(X)]$$
$$= E_0[\varphi_k(X) - \varphi(X)] \left[\frac{p(X, \theta_1)}{p(X, \theta_0)} - k\right] + E_1[\varphi_k(X) - \varphi(X)]1\{p(X, \theta_0) = 0\}$$
$$= I + II \text{ (say), where}$$
$$I = E_0\{\varphi_k(X)[L(X, \theta_0, \theta_1) - k] - \varphi(X)[L(X, \theta_0, \theta_1) - k]\}.$$

Because $L(x, \theta_0, \theta_1) - k$ is < 0 or ≥ 0 according as $\varphi_k(x)$ is 0 or in $(0, 1]$, and because $0 \leq \varphi(x) \leq 1$, then $I \geq 0$. Note that $\alpha > 0$ implies $k < \infty$ and, thus, $\varphi_k(x) = 1$ if $p(x, \theta_0) = 0$. It follows that $II \geq 0$. Finally, using (4.2.3), we have shown that

$$E_1[\varphi_k(X) - \varphi(X)] \geq kE_0[\varphi_k(X) - \varphi(X)] \geq 0. \tag{4.2.4}$$

(b) If $\alpha = 0$, $k = \infty$ makes φ_k MP size α. If $\alpha = 1$, $k = 0$ makes $E_1 \varphi_k(X) = 1$ and φ_k is MP size α. Next consider $0 < \alpha < 1$. Let P_i denote P_{θ_i}, $i = 0, 1$. Because $P_0[L(X, \theta_0, \theta_1) = \infty] = 0$, then there exists $k < \infty$ such that

$$P_0[L(X, \theta_0, \theta_1) > k] \le \alpha \text{ and } P_0[L(X, \theta_0, \theta_1) \ge k] \ge \alpha.$$

If $P_0[L(X, \theta_0, \theta_1) = k] = 0$, then φ_k is MP size α. If not, define

$$\varphi_k(x) = \frac{\alpha - P_0[L(X, \theta_0, \theta_1) > k]}{P_0[L(X, \theta_0, \theta_1) = k]}$$

on the set $\{x : L(x, \theta_0, \theta_1) = k\}$. Now φ_k is MP size α.

(c) Let $x \in \{x : p(x, \theta_1) > 0\}$, then to have equality in (4.2.4) we need to have $\varphi(x) = \varphi_k(x) = 1$ when $L(x, \theta_0, \theta_1) > k$ and have $\varphi(x) = \varphi_k(x) = 0$ when $L(x, \theta_0, \theta_1) < k$. It follows that (4.2.2) holds for $\theta = \theta_1$. The same argument works for $x \in \{x : p(x, \theta_0) > 0\}$ and $\theta = \theta_0$.   □

It follows from the Neyman–Pearson lemma that an MP test has power at least as large as its level; that is,

Corollary 4.2.1. *If φ is an MP level α test, then $E_{\theta_1} \varphi(X) \ge \alpha$ with equality iff $p(\cdot, \theta_0) = p(\cdot, \theta_1)$.*

Proof. See Problem 4.2.7.

Remark 4.2.1. Let π denote the prior probability of θ_0 so that $(1 - \pi)$ is the prior probability of θ_1. Then the posterior probability of θ_1 is

$$\pi(\theta_1 \mid x) = \frac{(1 - \pi)p(x, \theta_1)}{(1 - \pi)p(x, \theta_1) + \pi p(x, \theta_0)} = \frac{(1 - \pi)L(x, \theta_0, \theta_1)}{(1 - \pi)L(x, \theta_0, \theta_1) + \pi}. \tag{4.2.5}$$

If δ_π denotes the Bayes procedure of Example 3.2.2(b), then, when $\pi = k/(k + 1)$, $\delta_\pi = \varphi_k$. Moreover, we conclude from (4.2.5) that this δ_π decides θ_1 or θ_0 according as $\pi(\theta_1 \mid x)$ is larger than or smaller than $1/2$.

Part (a) of the lemma can, for $0 < \alpha < 1$, also be easily argued from this Bayes property of φ_k (Problem 4.2.10).

Here is an example illustrating calculation of the most powerful level α test φ_k.

Example 4.2.1. Consider Example 3.3.2 where $\mathbf{X} = (X_1, \ldots, X_n)$ is a sample of n $N(\mu, \sigma^2)$ random variables with σ^2 known and we test $H : \mu = 0$ versus $K : \mu = v$, where v is a known signal. We found

$$L(\mathbf{X}, 0, v) = \exp \left\{ \frac{v}{\sigma^2} \sum_{i=1}^{n} X_i - \frac{nv^2}{2\sigma^2} \right\}.$$

Note that any strictly increasing function of an optimal statistic is optimal because the two statistics generate the same family of critical regions. Therefore,

$$T(\mathbf{X}) = \sqrt{n} \frac{\bar{X}}{\sigma} = \frac{\sigma}{v\sqrt{n}} \left[\log L(\mathbf{X}, 0, v) + \frac{nv^2}{2\sigma^2} \right]$$

is also optimal for this problem. But T is the test statistic we proposed in Example 4.1.4. From our discussion there we know that for any specified α, the test that rejects if, and only if,

$$T \geq z(1 - \alpha) \tag{4.2.6}$$

has probability of type I error α.

The power of this test is, by (4.1.2), $\Phi(z(\alpha) + (v\sqrt{n}/\sigma))$. By the Neyman–Pearson lemma this is the largest power available with a level α test. Thus, if we want the probability of detecting a signal v to be at least a preassigned value β (say, .90 or .95), then we solve $\Phi(z(\alpha)+(v\sqrt{n}/\sigma)) = \beta$ for n and find that we need to take $n = (\sigma/v)^2[z(1-\alpha)+z(\beta)]^2$. This is the smallest possible n for any size α test.   □

An interesting feature of the preceding example is that the test defined by (4.2.6) that is MP for a specified signal v does not depend on v: The same test maximizes the power for all possible signals $v > 0$. Such a test is called uniformly most powerful (UMP).

We will discuss the phenomenon further in the next section. The following important example illustrates, among other things, that the UMP test phenomenon is largely a feature of one-dimensional parameter problems.

Example 4.2.2. *Simple Hypothesis Against Simple Alternative for the Multivariate Normal: Fisher's Discriminant Function.* Suppose $\mathbf{X} \sim \mathcal{N}(\boldsymbol{\mu}_j, \boldsymbol{\Sigma}_j)$, $\boldsymbol{\theta}_j = (\boldsymbol{\mu}_j, \boldsymbol{\Sigma}_j)$, $j = 0, 1$. The likelihood ratio test for $H : \boldsymbol{\theta} = \boldsymbol{\theta}_0$ versus $K : \boldsymbol{\theta} = \boldsymbol{\theta}_1$ is based on

$$L(\boldsymbol{\theta}_0, \boldsymbol{\theta}_1) = \frac{\det^{\frac{1}{2}}(\Sigma_0) \exp\left\{-\frac{1}{2}(\mathbf{x} - \boldsymbol{\mu}_1)^T \Sigma_1^{-1}(\mathbf{x} - \boldsymbol{\mu}_1)\right\}}{\det^{\frac{1}{2}}(\Sigma_1) \exp\left\{-\frac{1}{2}(\mathbf{x} - \boldsymbol{\mu}_0)^T \Sigma_0^{-1}(\mathbf{x} - \boldsymbol{\mu}_0)\right\}}.$$

Rejecting H for L large is equivalent to rejecting for

$$Q \equiv (\mathbf{X} - \boldsymbol{\mu}_0)^T \Sigma_0^{-1}(\mathbf{X} - \boldsymbol{\mu}_0) - (\mathbf{X} - \boldsymbol{\mu}_1)^T \Sigma_1^{-1}(\mathbf{X} - \boldsymbol{\mu}_1)$$

large. Particularly important is the case $\Sigma_0 = \Sigma_1$ when "Q large" is equivalent to "$F \equiv (\boldsymbol{\mu}_1 - \boldsymbol{\mu}_0)\Sigma_0^{-1}\mathbf{X}$ large." The function F is known as the Fisher discriminant function. It is used in a classification context in which $\boldsymbol{\theta}_0, \boldsymbol{\theta}_1$ correspond to two known populations and we desire to classify a new observation $\mathbf{X}$ as belonging to one or the other. We return to this in Volume II. Note that in general the test statistic L depends intrinsically on $\boldsymbol{\mu}_0, \boldsymbol{\mu}_1$. However if, say, $\boldsymbol{\mu}_1 = \boldsymbol{\mu}_0 + \lambda\boldsymbol{\Delta}_0$, $\lambda > 0$ and $\Sigma_1 = \Sigma_0$, then, if $\boldsymbol{\mu}_0, \boldsymbol{\Delta}_0, \Sigma_0$ are known, a UMP (for all λ) test exists and is given by: Reject if

$$\boldsymbol{\Delta}_0^T \Sigma_0^{-1}(\mathbf{X} - \boldsymbol{\mu}_0) \geq c \tag{4.2.7}$$

where $c = z(1 - \alpha)[\boldsymbol{\Delta}_0^T \Sigma_0^{-1}\boldsymbol{\Delta}_0]^{\frac{1}{2}}$ (Problem 4.2.8). If $\boldsymbol{\Delta}_0 = (1, 0, \ldots, 0)^T$ and $\Sigma_0 = I$, then this test rule is to reject H if X_1 is large; however, if $\Sigma_0 \neq I$, this is no longer the case (Problem 4.2.9). In this example we have assumed that $\boldsymbol{\theta}_0$ and $\boldsymbol{\theta}_1$ for the two populations are known. If this is not the case, they are estimated with their empirical versions with sample means estimating population means and sample covariances estimating population covariances.

Summary. We introduce the simple likelihood ratio statistic and simple likelihood ratio (SLR) test for testing the simple hypothesis $H : \theta = \theta_0$ versus the simple alternative $K : \theta = \theta_1$. The Neyman–Pearson lemma, which states that the size α SLR test is uniquely most powerful (MP) in the class of level α tests, is established.

We note the connection of the MP test to the Bayes procedure of Section 3.2 for deciding between θ_0 and θ_1. Two examples in which the MP test does not depend on θ_1 are given. Such tests are said to be UMP (uniformly most powerful).

4.3   UNIFORMLY MOST POWERFUL TESTS AND MONOTONE LIKELIHOOD RATIO MODELS

We saw in the two Gaussian examples of Section 4.2 that UMP tests for one-dimensional parameter problems exist. This phenomenon is not restricted to the Gaussian case as the next example illustrates. Before we give the example, here is the general definition of UMP:

Definition 4.3.1. A level α test φ^* is *uniformly most powerful* (UMP) for $H : \theta \in \Theta_0$ versus $K : \theta \in \Theta_1$ if

$$\beta(\theta, \varphi^*) \geq \beta(\theta, \varphi) \text{ for all } \theta \in \Theta_1, \qquad (4.3.1)$$

for any other level α test φ.

Example 4.3.1. *Testing for a Multinomial Vector.* Suppose that $(N_1, \ldots, N_k)$ has a multinomial $\mathcal{M}(n, \theta_1, \ldots, \theta_k)$ distribution with frequency function,

$$p(n_1, \ldots, n_k, \theta) = \frac{n!}{n_1! \ldots n_k!} \theta_1^{n_1} \ldots \theta_k^{n_k}$$

where $n_1, \ldots, n_k$ are integers summing to n. With such data we often want to test a simple hypothesis $H : \theta_1 = \theta_{10}, \ldots, \theta_k = \theta_{k0}$. For instance, if a match in a genetic breeding experiment can result in k types, n offspring are observed, and N_i is the number of offspring of type i, then $(N_1, \ldots, N_k) \sim \mathcal{M}(n, \theta_1, \ldots, \theta_k)$. The simple hypothesis would correspond to the theory that the expected proportion of offspring of types $1, \ldots, k$ are given by $\theta_{10}, \ldots, \theta_{k0}$. Usually the alternative to H is composite. However, there is sometimes a simple alternative theory $K : \theta_1 = \theta_{11}, \ldots, \theta_k = \theta_{k1}$. In this case, the likelihood ratio L is

$$L = \prod_{i=1}^{k} \left(\frac{\theta_{i1}}{\theta_{i0}} \right)^{N_i}.$$

Here is an interesting special case: Suppose $\theta_{j0} > 0$ for all j, $0 < \epsilon < 1$ and for some fixed integer l with $1 \leq l \leq k$

$$\theta_{l1} = \epsilon \theta_{l0}; \quad \theta_{j1} = \rho \theta_{j0}, \; j \neq l, \qquad (4.3.2)$$

where

$$\rho = (1 - \theta_{l0})^{-1}(1 - \epsilon \theta_{l0}).$$

That is, under the alternative, type l is less frequent than under H and the conditional probabilities of the other types given that type l has not occurred are the same under K as they are under H. Then

$$L = \rho^{n-N_l}\epsilon^{N_l} = \rho^n(\epsilon/\rho)^{N_l}.$$

Because $\epsilon < 1$ implies that $\rho > \epsilon$, we conclude that the MP test rejects H, if and only if, $N_l \leq c$. Critical values for level α are easily determined because $N_l \sim \mathcal{B}(n,\theta_{l0})$ under H. Moreover, for $\alpha = P(N_l \leq c)$, this test is UMP for testing H versus $K : \theta \in \Theta_1 = \{\boldsymbol{\theta} : \boldsymbol{\theta}$ is of the form (4.3.2) with $0 < \epsilon < 1\}$. Note that because l can be any of the integers $1, \ldots, k$, we get radically different best tests depending on which θ_i we assume to be θ_{l0} under H.   □

Typically the MP test of $H : \theta = \theta_0$ versus $K : \theta = \theta_1$ depends on θ_1 and the test is not UMP. However, we have seen three models where, in the case of a real parameter, there is a statistic T such that the test with critical region $\{x : T(x) \geq c\}$ is UMP. This is part of a general phenomena we now describe.

Definition 4.3.2. The family of models $\{P_\theta : \theta \in \Theta\}$ with $\Theta \subset R$ is said to be a monotone likelihood ratio (MLR) family in T if for $\theta_1 < \theta_2$ the distributions P_{θ_1} and P_{θ_2} are distinct and there exists a statistic $T(x)$ such that the ratio $p(x,\theta_2)/p(x,\theta_1)$ is an increasing function of $T(x)$.   □

Example 4.3.2 (Example 4.1.3 continued). In this i.i.d. Bernoulli case, set $s = \sum_{i=1}^n x_i$, then

$$p(\mathbf{x},\theta) = \theta^s(1-\theta)^{n-s} = (1-\theta)^n[\theta/(1-\theta)]^s$$

and the model is by (4.2.1) MLR in s.   □

Example 4.3.3. Consider the one-parameter exponential family model

$$p(x,\theta) = h(x)\exp\{\eta(\theta)T(x) - B(\theta)\}.$$

If $\eta(\theta)$ is strictly increasing in $\theta \in \Theta$, then this family is MLR. Example 4.2.1 is of this form with $T(\mathbf{x}) = \sqrt{n}\bar{x}/\sigma$ and $\eta(\mu) = \sqrt{n}\mu/\sigma$, where σ is known.   □

Define the Neyman–Pearson (NP) test function

$$\delta_t(x) = \begin{array}{ll} 1 & \text{if } T(x) > t \\ 0 & \text{if } T(x) < t \end{array} \tag{4.3.3}$$

with $\delta_t(x)$ any value in $(0,1)$ if $T(x) = t$. Consider the problem of testing $H : \theta = \theta_0$ versus $K : \theta = \theta_1$ with $\theta_0 < \theta_1$. If $\{P_\theta : \theta \in \Theta\}$, $\Theta \subset R$, is an MLR family in $T(x)$, then $L(x,\theta_0,\theta_1) = h(T(x))$ for some increasing function h. Thus, δ_t equals the likelihood ratio test $\varphi_{h(t)}$ and is MP. Because δ_t does not depend on θ_1, it is UMP at level $\alpha = E_{\theta_0}\delta_t(x)$ for testing $H : \theta = \theta_0$ versus $K : \theta > \theta_0$, in fact.

Theorem 4.3.1. *Suppose* $\{P_\theta : \theta \in \Theta\}$, $\Theta \subset R$, *is an MLR family in* $T(x)$.
   (1) *For each* $t \in (0,\infty)$, *the power function* $\beta(\theta) = E_\theta\delta_t(X)$ *is increasing in* θ.

(2) If $E_{\theta_0}\delta_t(X) = \alpha > 0$, then δ_t is UMP level α for testing $H : \theta \leq \theta_0$ versus $K : \theta > \theta_0$.

Proof. (1) follows from $\delta_t = \varphi_{h(t)}$ and Corollary 4.2.1 by noting that for any $\theta_1 < \theta_2$, δ_t is MP at level $E_{\theta_1}\delta_t(X)$ for testing $H : \theta = \theta_1$ versus $K : \theta = \theta_2$. To show (2), recall that we have seen that δ_t maximizes the power for testing $H : \theta = \theta_0$ versus $K : \theta > \theta_0$ among the class of tests with level $\alpha = E_{\theta_0}\delta_t(X)$. If $\theta < \theta_0$, then by (1), $E_{\theta}\delta_t(X) \leq \alpha$ and δ_t is of level α for $H : \theta \leq \theta_0$. Because the class of tests with level α for $H : \theta \leq \theta_0$ is contained in the class of tests with level α for $H : \theta = \theta_0$, and because δ_t maximizes the power over this larger class, δ_t is UMP for $H : \theta \leq \theta_0$ versus $K : \theta > \theta_0$.   □

The following useful result follows immediately.

Corollary 4.3.1. *Suppose $\{P_\theta : \theta \in \Theta\}$, $\Theta \subset R$, is an MLR family in $T(x)$. If the distribution function F_0 of $T(X)$ under $X \sim P_{\theta_0}$ is continuous and if $t(1-\alpha)$ is a solution of $F_0(t) = 1 - \alpha$, then the test that rejects H if and only if $T(x) \geq t(1-\alpha)$ is UMP level α for testing $H : \theta \leq \theta_0$ versus $K : \theta > \theta_0$.*

Example 4.3.4. *Testing Precision.* Suppose $X_1, \ldots, X_n$ is a sample from a $\mathcal{N}(\mu, \sigma^2)$ population, where μ is a known standard, and we are interested in the precision σ^{-1} of the measurements $X_1, \ldots, X_n$. For instance, we could be interested in the precision of a new measuring instrument and test it by applying it to a known standard. Because the more serious error is to judge the precision adequate when it is not, we test $H : \sigma \geq \sigma_0$ versus $K : \sigma < \sigma_0$, where σ_0^{-1} represents the minimum tolerable precision. Let $S = \sum_{i=1}^{n}(X_i - \mu)^2$, then

$$p(\mathbf{x}, \theta) = \exp\left\{ -\frac{1}{2\sigma^2}S - \frac{n}{2}\log(2\pi\sigma^2) \right\}.$$

This is a one-parameter exponential family and is MLR in $T = -S$. The UMP level α test rejects H if and only if $S \leq s(\alpha)$ where $s(\alpha)$ is such that $P_{\sigma_0}(S \leq s(\alpha)) = \alpha$. If we write

$$\frac{S}{\sigma_0^2} = \sum_{i=1}^{n}\left(\frac{X_i - \mu}{\sigma_0}\right)^2$$

we see that S/σ_0^2 has a χ_n^2 distribution. Thus, the critical constant $s(\alpha)$ is $\sigma_0^2 x_n(\alpha)$, where $x_n(\alpha)$ is the αth quantile of the χ_n^2 distribution.   □

Example 4.3.5. *Quality Control.* Suppose that, as in Example 1.1.1, X is the observed number of defectives in a sample of n chosen at random without replacement from a lot of N items containing b defectives, where $b = N\theta$. If the inspector making the test considers lots with $b_0 = N\theta_0$ defectives or more unsatisfactory, she formulates the hypothesis H as $\theta \geq \theta_0$, the alternative K as $\theta < \theta_0$, and specifies an α such that the probability of rejecting H (keeping a bad lot) is at most α. If α is a value taken on by the distribution of X, we now show that the test δ^* which rejects H if, and only if, $X \leq h(\alpha)$, where $h(\alpha)$ is the αth quantile of the hypergeometric, $\mathcal{H}(N\theta_0, N, n)$, distribution, is UMP level α. For simplicity suppose that $b_0 \geq n$, $N - b_0 \geq n$. Then, if $N\theta_1 = b_1 < b_0$ and $0 \leq x \leq b_1$, (1.1.1) yields

$$L(x, \theta_0, \theta_1) = \frac{b_1(b_1 - 1) \ldots (b_1 - x + 1)(N - b_1) \ldots (N - b_1 - n + x + 1)}{b_0(b_0 - 1) \ldots (b_0 - x + 1)(N - b_0) \ldots (N - b_0 - n + x + 1)}.$$

Note that $L(x, \theta_0, \theta_1) = 0$ for $b_1 < x \leq n$. Thus, for $0 \leq x \leq b_1 - 1$,

$$\frac{L(x + 1, \theta_0, \theta_1)}{L(x, \theta_0, \theta_1)} = \left(\frac{b_1 - x}{b_0 - x}\right) \frac{(N - n + 1) - (b_0 - x)}{(N - n + 1) - (b_1 - x)} < 1.$$

Therefore, L is decreasing in x and the hypergeometric model is an MLR family in $T(x) = -x$. It follows that δ^* is UMP level α. The critical values for the hypergeometric distribution are available on statistical calculators and software.   □

Power and Sample Size

In the Neyman–Pearson framework we choose the test whose size is small. That is, we choose the critical constant so that the maximum probability of falsely rejecting the null hypothesis H is small. On the other hand, we would also like large power $\beta(\theta)$ when $\theta \in \Theta_1$; that is, we want the probability of correctly detecting an alternative K to be large. However, as seen in Figure 4.1.1 and formula (4.1.2), this is, in general, not possible for all parameters in the alternative Θ_1. In both these cases, H and K are of the form $H : \theta \leq \theta_0$ and $K : \theta > \theta_0$, and the powers are continuous increasing functions with $\lim_{\theta \downarrow \theta_0} \beta(\theta) = \alpha$. By Corollary 4.3.1, this is a general phenomenon in MLR family models with $p(\mathbf{x}, \theta)$ continuous in θ.

This continuity of the power shows that not too much significance can be attached to acceptance of H, if all points in the alternative are of equal significance: We can find $\theta > \theta_0$ sufficiently close to θ_0 so that $\beta(\theta)$ is arbitrarily close to $\beta(\theta_0) = \alpha$. For such θ the probability of falsely accepting H is almost $1 - \alpha$.

This is not serious in practice if we have an *indifference region*. This is a subset of the alternative on which we are willing to tolerate low power. In our normal example 4.1.4 we might be uninterested in values of μ in $(0, \Delta)$ for some small $\Delta > 0$ because such improvements are negligible. Thus, $(0, \Delta)$ would be our indifference region. Off the indifference region, we want guaranteed power as well as an upper bound on the probability of type I error. In our example this means that in addition to the indifference region and level α, we specify β close to 1 and would like to have $\beta(\mu) \geq \beta$ for all $\mu \geq \Delta$. This is possible for arbitrary $\beta < 1$ only by making the sample size n large enough. In Example 4.1.4 because $\beta(\mu)$ is increasing, the appropriate n is obtained by solving

$$\beta(\Delta) = \Phi(z(\alpha) + \sqrt{n}\Delta/\sigma) = \beta$$

for sample size n. This equation is equivalent to

$$z(\alpha) + \sqrt{n}\Delta/\sigma = z(\beta)$$

whose solution is

$$n = (\Delta/\sigma)^{-2}[z(1 - \alpha) + z(\beta)]^2.$$

Note that a small signal-to-noise ratio Δ/σ will require a large sample size n.

Dual to the problem of not having enough power is that of having too much. It is natural to associate statistical significance with practical significance so that a very low p-value is interpreted as evidence that the alternative that holds is physically significant, that is, far from the hypothesis. Formula (4.1.2) shows that, if n is very large and/or σ is small, we can have very great power for alternatives very close to 0. This problem arises particularly in goodness-of-fit tests (see Example 4.1.5), when we test the hypothesis that a very large sample comes from a particular distribution. Such hypotheses are often rejected even though for practical purposes "the fit is good enough." The reason is that n is so large that unimportant small discrepancies are picked up. There are various ways of dealing with this problem. They often reduce to adjusting the critical value so that the probability of rejection for parameter value at the boundary of some indifference region is α. In Example 4.1.4 this would mean rejecting H if, and only if,

$$\sqrt{n}\frac{\bar{X}}{\sigma} \geq z(1-\alpha) + \sqrt{n}\frac{\Delta}{\sigma}.$$

As a further example and precursor to Section 5.4.4, we next show how to find the sample size that will "approximately" achieve desired power β for the size α test in the binomial example.

Example 4.3.6 (Example 4.1.3 continued). Our discussion uses the classical normal approximation to the binomial distribution. First, to achieve approximate size α, we solve $\beta(\theta_0) = P_{\theta_0}(S \geq s)$ for s using (4.1.4) and find the approximate critical value

$$s_0 = n\theta_0 + \frac{1}{2} + z(1-\alpha)[n\theta_0(1-\theta_0)]^{1/2}.$$

Again using the normal approximation, we find

$$\beta(\theta) = P_\theta(S \geq s_0) = \Phi\left(\frac{n\theta + \frac{1}{2} - s_0}{[n\theta(1-\theta)]^{1/2}}\right).$$

Now consider the indifference region (θ_0, θ_1), where $\theta_1 = \theta_0 + \Delta$, $\Delta > 0$. We solve $\beta(\theta_1) = \beta$ for n and find the approximate solution

$$n = (\theta_1 - \theta_0)^{-2}\{z(1-\alpha)[\theta_0(1-\theta_0)]^{1/2} + z(\beta)[\theta_1(1-\theta_1)]^{1/2}\}^2.$$

For instance, if $\alpha = .05$, $\beta = .90$, $\theta_0 = 0.3$, and $\theta_1 = 0.35$, we need

$$n = (0.05)^{-2}\{1.645 \times 0.3(0.7) + 1.282 \times 0.35(0.65)\}^2 = 162.4.$$

Thus, the size .05 binomial test of $H : \theta = 0.3$ requires approximately 163 observations to have probability .90 of detecting the 17% increase in θ from 0.3 to 0.35. The power achievable (exactly, using the SPLUS package) for the level .05 test for $\theta = .35$ and $n = 163$ is 0.86. $\square$

Our discussion can be generalized. Suppose θ is a vector. Often there is a function $q(\theta)$ such that H and K can be formulated as $H : q(\theta) \leq q_0$ and $K : q(\theta) > q_0$. Now let

$q_1 > q_0$ be a value such that we want to have power $\beta(\theta)$ at least β when $q(\theta) \geq q_1$. The set $\{\theta : q_0 < q(\theta) < q_1\}$ is our indifference region. For each n suppose we have a level α test for H versus K based on a suitable test statistic T. Suppose that $\beta(\theta)$ depends on θ only through $q(\theta)$ and is a continuous increasing function of $q(\theta)$, and also increases to 1 for fixed $\theta \in \Theta_1$ as $n \to \infty$. To achieve level α and power at least β, first let c_0 be the smallest number c such that

$$P_{\theta_0}[T \geq c] \leq \alpha.$$

Then let n be the smallest integer such that

$$P_{\theta_1}[T \geq c_0] \geq \beta$$

where θ_0 is such that $q(\theta_0) = q_0$ and θ_1 is such that $q(\theta_1) = q_1$. This procedure can be applied, for instance, to the F test of the linear model in Section 6.1 by taking $q(\theta)$ equal to the noncentrality parameter governing the distribution of the statistic under the alternative.

Implicit in this calculation is the assumption that $P_{\theta_1}[T \geq c_0]$ is an increasing function of n.

We have seen in Example 4.1.5 that a particular test statistic can have a fixed distribution $\mathcal{L}_0$ under the hypothesis. It may also happen that the distribution of T_n as θ ranges over Θ_1 is determined by a one-dimensional parameter $\lambda(\theta)$ so that $\Theta_0 = \{\theta : \lambda(\theta) = 0\}$ and $\Theta_1 = \{\theta : \lambda(\theta) > 0\}$ and $\mathcal{L}_0(T_n) = \mathcal{L}_{\lambda(\theta)}(T_n)$ for all θ. The theory we have developed demonstrates that if $\mathcal{L}_\lambda(T_n)$ is an MLR family, then rejecting for large values of T_n is UMP among all tests based on T_n. Reducing the problem to choosing among such tests comes from invariance considerations that we do not enter into until Volume II. However, we illustrate what can happen with a simple example.

Example 4.3.7. *Testing Precision Continued.* Suppose that in the Gaussian model of Example 4.3.4, μ is unknown. Then the MLE of σ^2 is $\widehat{\sigma}^2 = \frac{1}{n}\sum_{i=1}^n (X_i - \bar{X})^2$ as in Example 2.2.9. Although $H : \sigma = \sigma_0$ is now composite, the distribution of $T_n \equiv n\widehat{\sigma}^2/\sigma_0^2$ is χ_{n-1}^2, independent of μ. Thus, the critical value for testing $H : \sigma = \sigma_0$ versus $K : \sigma < \sigma_0$ and rejecting H if T_n is small, is the α percentile of χ_{n-1}^2. It is evident from the argument of Example 4.3.3 that this test is UMP for $H : \sigma \geq \sigma_0$ versus $K : \sigma < \sigma_0$ *among* all tests depending on $\widehat{\sigma}^2$ only. $\qquad\square$

Complete Families of Tests

The Neyman–Pearson framework is based on using the 0-1 loss function. We may ask whether decision procedures other than likelihood ratio tests arise if we consider loss functions $l(\theta, a)$, $a \in \mathcal{A} = \{0, 1\}$, $\theta \in \Theta$, that are not 0-1. For instance, for $\Theta_1 = (\theta_0, \infty)$, we may consider $l(\theta, 0) = (\theta - \theta_0)$, $\theta \in \Theta_1$. In general, when testing $H : \theta \leq \theta_0$ versus $K : \theta > \theta_0$, a reasonable class of loss functions are those that satisfy

$$\begin{aligned} l(\theta, 1) - l(\theta, 0) &> 0 \quad \text{for } \theta < \theta_0 \\ l(\theta, 1) - l(\theta, 0) &< 0 \quad \text{for } \theta > \theta_0. \end{aligned} \tag{4.3.4}$$

The class $\mathcal{D}$ of decision procedures is said to be *complete*[(1),(2)] if for any decision rule φ there exists $\delta \in \mathcal{D}$ such that

$$R(\theta, \delta) \leq R(\theta, \varphi) \text{ for all } \theta \in \Theta. \qquad (4.3.5)$$

That is, if the model is correct and loss function is appropriate, then any procedure not in the complete class can be matched or improved *at all* θ by one in the complete class. Thus, it isn't worthwhile to look outside of complete classes. In the following the decision procedures are test functions.

Theorem 4.3.2. *Suppose $\{P_\theta : \theta \in \Theta\}$, $\Theta \subset R$, is an MLR family in $T(x)$ and suppose the loss function $l(\theta, a)$ satisfies (4.3.4), then the class of tests of the form (4.3.3) with $E\delta_t(X) = \alpha$, $0 \leq \alpha \leq 1$, is complete.*

Proof. The risk function of any test rule φ is

$$
\begin{aligned}
R(\theta, \varphi) &= E_\theta\{\varphi(X)l(\theta, 1) + [1 - \varphi(X)]l(\theta, 0)\} \\
&= E_\theta\{l(\theta, 0) + [l(\theta, 1) - l(\theta, 0)]\varphi(X)\}.
\end{aligned}
$$

Let $\delta_t(X)$ be such that, for some θ_0, $E_{\theta_0}\delta_t(X) = E_{\theta_0}\varphi(X) > 0$. If $E_\theta\varphi(X) \equiv 0$ for all θ then $\delta_\infty(X)$ clearly satisfies (4.3.5). Now δ_t is UMP for $H : \theta \leq \theta_0$ versus $K : \theta > \theta_0$ by Theorem 4.3.1 and, hence,

$$R(\theta, \delta_t) - R(\theta, \varphi) = (l(\theta, 1) - l(\theta, 0))(E_\theta(\delta_t(x)) - E_\theta(\varphi(X))) \leq 0 \text{ for } \theta > \theta_0. \qquad (4.3.6)$$

But $1 - \delta_t$ is similarly UMP for $H : \theta \geq \theta_0$ versus $K : \theta < \theta_0$ (Problem 4.3.12) and, hence, $E_\theta(1 - \delta_t(X)) = 1 - E_\theta\delta_t(X) \geq 1 - E_\theta\varphi(X)$ for $\theta < \theta_0$. Thus, (4.3.5) holds for all θ. $\qquad \square$

Summary. We consider models $\{P_\theta : \theta \in \Theta\}$ for which there exist tests that are most powerful for every θ in a composite alternative Θ_1 (UMP tests). For θ real, a model is said to be monotone likelihood ratio (MLR) if the simple likelihood ratio statistic for testing θ_0 versus θ_1 is an increasing function of a statistic $T(x)$ for every $\theta_0 < \theta_1$. For MLR models, the test that rejects $H : \theta \leq \theta_0$ for large values of $T(x)$ is UMP for $K : \theta > \theta_0$. In such situations we show how sample size can be chosen to guarantee minimum power for alternatives a given distance from H. Finally, we show that for MLR models, the class of most powerful Neyman Pearson (NP) tests is complete in the sense that for loss functions other than the 0-1 loss function, the risk of any procedure can be matched or improved by an NP test.

4.4   CONFIDENCE BOUNDS, INTERVALS, AND REGIONS

We have in Chapter 2 considered the problem of obtaining precise estimates of parameters and we have in this chapter treated the problem of deciding whether the parameter θ is a member of a specified set Θ_0. Now we consider the problem of giving confidence bounds, intervals, or sets that constrain the parameter with prescribed probability $1 - \alpha$. As an illustration consider Example 4.1.4 where $X_1, \ldots, X_n$ are i.i.d. $\mathcal{N}(\mu, \sigma^2)$ with σ^2 known. Suppose that μ represents the mean increase in sleep among patients administered a drug. Then we can use the experimental outcome $\mathbf{X} = (X_1, \ldots, X_n)$ to establish a lower bound $\underline{\mu}(\mathbf{X})$ for μ with a prescribed probability $(1 - \alpha)$ of being correct. In the non-Bayesian framework, μ is a constant, and we look for a statistic $\underline{\mu}(\mathbf{X})$ that satisfies $P(\underline{\mu}(\mathbf{X}) \leq \mu) = 1 - \alpha$ with $1 - \alpha$ equal to .95 or some other desired level of confidence. In our example this is achieved by writing

$$P\left(\frac{\sqrt{n}(\bar{X} - \mu)}{\sigma} \leq z(1 - \alpha)\right) = 1 - \alpha.$$

By solving the inequality inside the probability for μ, we find

$$P(\bar{X} - \sigma z(1 - \alpha)/\sqrt{n} \leq \mu) = 1 - \alpha$$

and

$$\underline{\mu}(\mathbf{X}) = \bar{X} - \sigma z(1 - \alpha)/\sqrt{n}$$

is a lower bound with $P(\underline{\mu}(\mathbf{X}) \leq \mu) = 1 - \alpha$. We say that $\underline{\mu}(\mathbf{X})$ is a *lower confidence bound with confidence level* $1 - \alpha$.

Similarly, as in (1.3.8), we may be interested in an upper bound on a parameter. In the $\mathcal{N}(\mu, \sigma^2)$ example this means finding a statistic $\bar{\mu}(\mathbf{X})$ such that $P(\bar{\mu}(\mathbf{X}) \geq \mu) = 1 - \alpha$; and a solution is

$$\bar{\mu}(\mathbf{X}) = \bar{X} + \sigma z(1 - \alpha)/\sqrt{n}.$$

Here $\bar{\mu}(\mathbf{X})$ is called an *upper level* $(1 - \alpha)$ *confidence bound* for μ.

Finally, in many situations where we want an indication of the accuracy of an estimator, we want both lower and upper bounds. That is, we want to find a such that the probability that the interval $[\bar{X} - a, \bar{X} + a]$ contains μ is $1 - \alpha$. We find such an interval by noting

$$P\left(\frac{\sqrt{n}|\bar{X} - \mu|}{\sigma} \leq z\left(1 - \tfrac{1}{2}\alpha\right)\right) = 1 - \alpha$$

and solving the inequality inside the probability for μ. This gives

$$P(\mu^-(\mathbf{X}) \leq \mu \leq \mu^+(\mathbf{X})) = 1 - \alpha$$

where

$$\mu^\pm(\mathbf{X}) = \bar{X} \pm \sigma z\left(1 - \tfrac{1}{2}\alpha\right)/\sqrt{n}.$$

We say that $[\mu^-(\mathbf{X}), \mu^+(\mathbf{X})]$ is a *level* $(1 - \alpha)$ *confidence interval* for μ.

In general, if $\nu = \nu(P)$, $P \in \mathcal{P}$, is a parameter, and $X \sim P$, $X \in R^q$, it may not be possible for a bound or interval to achieve exactly probability $(1 - \alpha)$ for a prescribed $(1 - \alpha)$ such as .95. In this case, we settle for a probability at least $(1 - \alpha)$. That is,

Definition 4.4.1. A statistic $\underline{\nu}(X)$ is called a *level* $(1 - \alpha)$ *lower confidence bound* for ν if for every $P \in \mathcal{P}$,

$$P[\underline{\nu}(X) \leq \nu] \geq 1 - \alpha.$$

Similarly, $\bar{\nu}(X)$ is called a *level* $(1 - \alpha)$ *upper confidence bound* for ν if for every $P \in \mathcal{P}$,

$$P[\bar{\nu}(X) \geq \nu] \geq 1 - \alpha.$$

Moreover, the random interval $[\underline{\nu}(X), \bar{\nu}(X)]$ formed by a pair of statistics $\underline{\nu}(X), \bar{\nu}(X)$ is a *level* $(1 - \alpha)$ or a $100(1 - \alpha)\%$ *confidence interval* for ν if, for all $P \in \mathcal{P}$,

$$P[\underline{\nu}(X) \leq \nu \leq \bar{\nu}(X)] \geq 1 - \alpha.$$

The quantities on the left are called the *probabilities of coverage* and $(1 - \alpha)$ is called a *confidence level*.

For a given bound or interval, the confidence level is clearly not unique because any number $(1 - \alpha') \leq (1 - \alpha)$ will be a confidence level if $(1 - \alpha)$ is. In order to avoid this ambiguity it is convenient to define the *confidence coefficient* to be the largest possible confidence level. Note that in the case of intervals this is just

$$\inf\{P[\underline{\nu}(X) \leq \nu \leq \bar{\nu}(X), P \in \mathcal{P}]\}$$

(i.e., the minimum probability of coverage). For the normal measurement problem we have just discussed the probability of coverage is independent of P and equals the confidence coefficient.

Example 4.4.1. *The (Student) t Interval and Bounds.* Let $X_1, \ldots, X_n$ be a sample from a $\mathcal{N}(\mu, \sigma^2)$ population, and assume initially that σ^2 is known. In the preceding discussion we used the fact that $Z(\mu) = \sqrt{n}(\bar{X} - \mu)/\sigma$ has a $\mathcal{N}(0, 1)$ distribution to obtain a confidence interval for μ by solving $-z\left(1 - \frac{1}{2}\alpha\right) \leq Z(\mu) \leq z\left(1 - \frac{1}{2}\alpha\right)$ for μ. In this process $Z(\mu)$ is called a *pivot*. In general, finding confidence intervals (or bounds) often involves finding appropriate pivots. Now we turn to the σ^2 unknown case and propose the pivot $T(\mu)$ obtained by replacing σ in $Z(\mu)$ by its estimate s, where

$$s^2 = \frac{1}{n - 1} \sum_{i=1}^{n} (X_i - \bar{X})^2.$$

That is, we will need the distribution of

$$T(\mu) = \frac{\sqrt{n}(\bar{X} - \mu)}{s}.$$

Now $Z(\mu) = \sqrt{n}(\bar{X} - \mu)/\sigma$ has a $\mathcal{N}(0, 1)$ distribution and is, by Theorem B.3.3, independent of $V = (n - 1)s^2/\sigma^2$, which has a χ_{n-1}^2 distribution. We conclude from the definition of the (Student) t distribution in Section B.3.1 that $Z(\mu)/\sqrt{V/(n - 1)} = T(\mu)$

has the t distribution $\mathcal{T}_{n-1}$ whatever be μ and σ^2. Let $t_k(p)$ denote the pth quantile of the $\mathcal{T}_k$ distribution. Then

$$P\left(-t_{n-1}\left(1 - \tfrac{1}{2}\alpha\right) \leq T(\mu) \leq t_{n-1}\left(1 - \tfrac{1}{2}\alpha\right)\right) = 1 - \alpha.$$

Solving the inequality inside the probability for μ, we find

$$P\left[\bar{X} - st_{n-1}\left(1 - \tfrac{1}{2}\alpha\right)/\sqrt{n} \leq \mu \leq \bar{X} + st_{n-1}\left(1 - \tfrac{1}{2}\alpha\right)/\sqrt{n}\right] = 1 - \alpha.$$

The shortest level $(1 - \alpha)$ confidence interval of the type $\bar{X} \pm sc/\sqrt{n}$ is, thus,

$$\left[\bar{X} - st_{n-1}\left(1 - \tfrac{1}{2}\alpha\right)/\sqrt{n}, \ \bar{X} + st_{n-1}\left(1 - \tfrac{1}{2}\alpha\right)/\sqrt{n}\right]. \tag{4.4.1}$$

Similarly, $\bar{X} - st_{n-1}(1 - \alpha)/\sqrt{n}$ and $\bar{X} + st_{n-1}(1 - \alpha)/\sqrt{n}$ are natural lower and upper confidence bounds with confidence coefficients $(1 - \alpha)$.

To calculate the coefficients $t_{n-1}\left(1 - \tfrac{1}{2}\alpha\right)$ and $t_{n-1}(1 - \alpha)$, we use a calculator, computer software, or Tables I and II. For instance, if $n = 9$ and $\alpha = 0.01$, we enter Table II to find that the probability that a $\mathcal{T}_{n-1}$ variable exceeds 3.355 is .005. Hence, $t_{n-1}\left(1 - \tfrac{1}{2}\alpha\right) = 3.355$ and

$$\left[\bar{X} - 3.355s/3, \ \bar{X} + 3.355s/3\right]$$

is the desired level 0.99 confidence interval.

From the results of Section B.7 (see Problem B.7.12), we see that as $n \to \infty$ the $\mathcal{T}_{n-1}$ distribution converges in law to the standard normal distribution. For the usual values of α, we can reasonably replace $t_{n-1}(p)$ by the standard normal quantile $z(p)$ for $n > 120$.

Up to this point, we have assumed that $X_1, \ldots, X_n$ are i.i.d. $\mathcal{N}(\mu, \sigma^2)$. It turns out that the distribution of the pivot $T(\mu)$ is fairly close to the $\mathcal{T}_{n-1}$ distribution if the X's have a distribution that is nearly symmetric and whose tails are not much heavier than the normal. In this case the interval (4.4.1) has confidence coefficient close to $1 - \alpha$. On the other hand, for very skew distributions such as the χ^2 with few degrees of freedom, or very heavy-tailed distributions such as the Cauchy, the confidence coefficient of (4.4.1) can be much smaller than $1 - \alpha$. The properties of confidence intervals such as (4.4.1) in non-Gaussian situations can be investigated using the asymptotic and Monte Carlo methods introduced in Chapter 5. See Figure 5.3.1. If we assume $\sigma^2 < \infty$, the interval will have probability $(1 - \alpha)$ in the limit as $n \to \infty$. $\quad\square$

Example 4.4.2. *Confidence Intervals and Bounds for the Variance of a Normal Distribution.* Suppose that $X_1, \ldots, X_n$ is a sample from a $\mathcal{N}(\mu, \sigma^2)$ population. By Theorem B.3.1, $V(\sigma^2) = (n-1)s^2/\sigma^2$ has a χ^2_{n-1} distribution and can be used as a pivot. Thus, if we let $x_{n-1}(p)$ denote the pth quantile of the χ^2_{n-1} distribution, and if $\alpha_1 + \alpha_2 = \alpha$, then

$$P(x(\alpha_1) \leq V(\sigma^2) \leq x(1 - \alpha_2)) = 1 - \alpha.$$

By solving the inequality inside the probability for σ^2 we find that

$$\left[(n-1)s^2/x(1 - \alpha_2), \ (n-1)s^2/x(\alpha_1)\right] \tag{4.4.2}$$

is a confidence interval with confidence coefficient $(1 - \alpha)$.

The length of this interval is random. There is a unique choice of α_1 and α_2, which uniformly minimizes expected length among all intervals of this type. It may be shown that for n large, taking $\alpha_1 = \alpha_2 = \frac{1}{2}\alpha$ is not far from optimal (Tate and Klett, 1959).

The pivot $V(\sigma^2)$ similarly yields the respective lower and upper confidence bounds $(n-1)s^2/x(1-\alpha)$ and $(n-1)s^2/x(\alpha)$.

In contrast to Example 4.4.1, if we drop the normality assumption, the confidence interval and bounds for σ^2 do not have confidence coefficient $1-\alpha$ even in the limit as $n \to \infty$. Asymptotic methods and Monte Carlo experiments as described in Chapter 5 have shown that the confidence coefficient may be arbitrarily small depending on the underlying true distribution, which typically is unknown. In Problem 4.4.16 we give an interval with correct limiting coverage probability.   □

The method of pivots works primarily in problems related to sampling from normal populations. If we consider "approximate" pivots, the scope of the method becomes much broader. We illustrate by an example.

Example 4.4.3. *Approximate Confidence Bounds and Intervals for the Probability of Success in n Bernoulli Trials.* If $X_1, \ldots, X_n$ are the indicators of n Bernoulli trials with probability of success θ, then $\bar{X}$ is the MLE of θ. There is no natural "exact" pivot based on $\bar{X}$ and θ. However, by the De Moivre–Laplace theorem, $\sqrt{n}(\bar{X} - \theta)/\sqrt{\theta(1-\theta)}$ has approximately a $\mathcal{N}(0,1)$ distribution. If we use this function as an "approximate" pivot and let $\approx$ denote "approximate equality," we can write

$$P\left[\left|\frac{\sqrt{n}(\bar{X} - \theta)}{\sqrt{\theta(1-\theta)}}\right| \leq z\left(1 - \tfrac{1}{2}\alpha\right)\right] \approx 1 - \alpha.$$

Let $k_\alpha = z\left(1 - \tfrac{1}{2}\alpha\right)$ and observe that this is equivalent to

$$P\left[(\bar{X} - \theta)^2 \leq \frac{k_\alpha^2}{n}\theta(1-\theta)\right] = P[g(\theta, \bar{X}) \leq 0] \approx 1 - \alpha$$

where

$$g(\theta, \bar{X}) = \left(1 + \frac{k_\alpha^2}{n}\right)\theta^2 - \left(2\bar{X} + \frac{k_\alpha^2}{n}\right)\theta + \bar{X}^2.$$

For fixed $0 \leq \bar{X} \leq 1$, $g(\theta, \bar{X})$ is a quadratic polynomial with two real roots. In terms of $S = n\bar{X}$, they are[1]

$$
\begin{aligned}
\underline{\theta}(\mathbf{X}) &= \left\{S + \frac{k_\alpha^2}{2} - k_\alpha\sqrt{[S(n-S)/n] + k_\alpha^2/4}\right\} \bigg/ (n + k_\alpha^2) \\
\bar{\theta}(\mathbf{X}) &= \left\{S + \frac{k_\alpha^2}{2} + k_\alpha\sqrt{[S(n-S)/n] + k_\alpha^2/4}\right\} \bigg/ (n + k_\alpha^2).
\end{aligned}
\tag{4.4.3}
$$

Because the coefficient of θ^2 in $g(\theta, \bar{X})$ is greater than zero,

$$[\theta : g(\theta, \bar{X}) \leq 0] = [\underline{\theta}(\mathbf{X}) \leq \theta \leq \bar{\theta}(\mathbf{X})], \tag{4.4.4}$$

so that $[\underline{\theta}(\mathbf{X}), \bar{\theta}(\mathbf{X})]$ is an approximate level $(1 - \alpha)$ confidence interval for θ. We can similarly show that the endpoints of the level $(1 - 2\alpha)$ interval are approximate upper and lower level $(1 - \alpha)$ confidence bounds. These intervals and bounds are satisfactory in practice for the usual levels, if the smaller of $n\theta, n(1 - \theta)$ is at least 6. For small n, it is better to use the exact level $(1 - \alpha)$ procedure developed in Section 4.5. A discussion is given in Brown, Cai, and Das Gupta (2001).

Note that in this example we can determine the sample size needed for desired accuracy. For instance, consider the market researcher whose interest is the proportion θ of a population that will buy a product. He draws a sample of n potential customers, calls willingness to buy success, and uses the preceding model. He can then determine how many customers should be sampled so that (4.4.4) has length 0.02 and is a confidence interval with confidence coefficient approximately 0.95. To see this, note that the length, say l, of the interval is

$$l = 2k_\alpha\{\sqrt{[S(n - S)/n] + k_\alpha^2/4}\}(n + k_\alpha^2)^{-1}.$$

Now use the fact that

$$S(n - S)/n = \tfrac{1}{4}n - n^{-1}\left(S - \tfrac{1}{2}n\right)^2 \leq \tfrac{1}{4}n \tag{4.4.5}$$

to conclude that

$$l \leq k_\alpha/\sqrt{n + k_\alpha^2}. \tag{4.4.6}$$

Thus, to bound l above by $l_0 = 0.02$, we choose n so that $k_\alpha(n + k_\alpha^2)^{-\frac{1}{2}} = l_0$. That is, we choose

$$n = \left(\frac{k_\alpha}{l_0}\right)^2 - k_\alpha^2.$$

In this case, $1 - \tfrac{1}{2}\alpha = 0.975$, $k_\alpha = z\left(1 - \tfrac{1}{2}\alpha\right) = 1.96$, and we can achieve the desired length 0.02 by choosing n so that

$$n = \left(\frac{1.96}{0.02}\right)^2 - (1.96)^2 = 9,600.16, \text{ or } n = 9,601.$$

This formula for the sample size is very crude because (4.4.5) is used and it is only good when θ is near $1/2$. Better results can be obtained if one has upper or lower bounds on θ such as $\theta \leq \theta_0 < \tfrac{1}{2}, \theta \geq \theta_1 > \tfrac{1}{2}$. See Problem 4.4.4.

Another approximate pivot for this example is $\sqrt{n}(\bar{X} - \theta)/\sqrt{\bar{X}(1 - \bar{X})}$. This leads to the simple interval

$$\bar{X} \pm k_\alpha\sqrt{\bar{X}(1 - \bar{X})}/\sqrt{n}. \tag{4.4.7}$$

See Brown, Cai, and Das Gupta (2001) for a discussion.   □

Confidence Regions for Functions of Parameters

We define a level $(1 - \alpha)$ confidence regions for a function $q(\theta)$ as random subsets of the range of q that cover the true value of $q(\theta)$ with probability at least $(1 - \alpha)$. Note that if $C(\mathbf{X})$ is a level $(1 - \alpha)$ confidence region for θ, then $q(C(\mathbf{X})) = \{q(\theta) : \theta \in C(\mathbf{X})\}$ is a level $(1 - \alpha)$ confidence region for $q(\theta)$.

If the distribution of $T_{\mathbf{X}}(\theta)$ does not invlove θ, then it is called a *pivot*. In this case, if $P_\theta(T_{\mathbf{X}}(\theta) \in A) \geq 1 - \alpha$, then $T_{\mathbf{X}}^{-1}(A)$ is a level $(1 - \alpha)$ confidence region for θ.

Example 4.4.4. Let $X_1, \ldots, X_n$ denote the number of hours a sample of Internet subscribers spend per week on the Internet. Suppose $X_1, \ldots, X_n$ is modeled as a sample from an exponential, $\mathcal{E}(\theta^{-1})$, distribution, and suppose we want a confidence interval for the population proportion $P(X \geq x)$ of subscribers that spend at least x hours per week on the Internet. Here $q(\theta) = 1 - F(x) = \exp\{-x/\theta\}$. By Problem B.3.4, $2n\bar{X}/\theta$ has a chi-square, χ^2_{2n}, distribution. By using $2n\bar{X}/\theta$ as a pivot we find the $(1 - \alpha)$ confidence interval

$$2n\bar{X}/x \left(1 - \tfrac{1}{2}\alpha\right) \leq \theta \leq 2n\bar{X}/x \left(\tfrac{1}{2}\alpha\right)$$

where $x(\beta)$ denotes the βth quantile of the χ^2_{2n} distribution. Let $\underline{\theta}$ and $\bar{\theta}$ denote the lower and upper boundaries of this interval, then

$$\exp\{-x/\underline{\theta}\} \leq q(\theta) \leq \exp\{-x/\bar{\theta}\}$$

is a confidence interval for $q(\theta)$ with confidence coefficient $(1 - \alpha)$.

If q is not $1 - 1$, this technique is typically wasteful. That is, we can find confidence regions for $q(\theta)$ entirely contained in $q(C(\mathbf{X}))$ with confidence level $(1-\alpha)$. For instance, we will later give confidence regions $C(\mathbf{X})$ for pairs $\theta = (\theta_1, \theta_2)^T$. In this case, if $q(\theta) = \theta_1$, $q(C(\mathbf{X}))$ is larger than the confidence set obtained by focusing on θ_1 alone.

Confidence Regions of Higher Dimension

We can extend the notion of a confidence interval for one-dimensional functions $q(\theta)$ to r-dimensional vectors $\mathbf{q}(\theta) = (q_1(\theta), \ldots, q_r(\theta))$. Suppose $\underline{q}_j(X)$ and $\bar{q}_j(X)$ are real-valued. Then the r-dimensional random rectangle

$$I(X) = \{\mathbf{q}(\theta) : \underline{q}_j(X) \leq q_j(\theta) \leq \bar{q}_j(X), \ j = 1, \ldots, r\}$$

is said to be a level $(1 - \alpha)$ confidence region, if the probability that it covers the unknown but fixed true $(q_1(\theta), \ldots, q_r(\theta))$ is at least $(1 - \alpha)$. We write this as

$$P[\mathbf{q}(\theta) \in I(X)] \geq 1 - \alpha.$$

Note that if $I_j(X) = [\underline{T}_j, \bar{T}_j]$ is a level $(1 - \alpha_j)$ confidence interval for $q_j(\theta)$ and if the pairs $(\underline{T}_1, \bar{T}_1), \ldots, (\underline{T}_r, \bar{T}_r)$ are independent, then the rectangle $I(X) = I_1(X) \times \cdots \times I_r(X)$ has level

$$\prod_{j=1}^{r} (1 - \alpha_j). \tag{4.4.8}$$

Thus, an r-dimensional confidence rectangle is in this case automatically obtained from the one-dimensional intervals. Moreover, if we choose $\alpha_j = 1 - (1 - \alpha)^{\frac{1}{r}}$, then $I(X)$ has confidence level $1 - \alpha$.

An approach that works even if the I_j are not independent is to use Bonferroni's inequality (A.2.7). According to this inequality,

$$P[\mathbf{q}(\theta) \in I(X)] \geq 1 - \sum_{j=1}^{r} P[q_j(\theta) \notin I_j(X)] \geq 1 - \sum_{j=1}^{r} \alpha_j.$$

Thus, if we choose $\alpha_j = \alpha/r$, $j = 1, \ldots, r$, then $I(X)$ has confidence level $(1 - \alpha)$.

Example 4.4.5. *Confidence Rectangle for the Parameters of a Normal Distribution.* Suppose $X_1, \ldots, X_n$ is a $\mathcal{N}(\mu, \sigma^2)$ sample and we want a confidence rectangle for (μ, σ^2). From Example 4.4.1

$$I_1(X) = \bar{X} \pm s t_{n-1} \left(1 - \tfrac{1}{4}\alpha\right)/\sqrt{n}$$

is a confidence interval for μ with confidence coefficient $\left(1 - \tfrac{1}{2}\alpha\right)$. From Example 4.4.2,

$$I_2(X) = \left[\frac{(n-1)s^2}{x_{n-1}\left(1 - \tfrac{1}{4}\alpha\right)}, \frac{(n-1)s^2}{x_{n-1}\left(\tfrac{1}{4}\alpha\right)} \right]$$

is a reasonable confidence interval for σ^2 with confidence coefficient $\left(1 - \tfrac{1}{2}\alpha\right)$. Thus, $I_1(X) \times I_2(X)$ is a level $(1 - \alpha)$ confidence rectangle for (μ, σ^2). The exact confidence coefficient is given in Problem 4.4.15. □

The method of pivots can also be applied to ∞-dimensional parameters such as F.

Example 4.4.6. Suppose $X_1, \ldots, X_n$ are i.i.d. as $X \sim P$, and we are interested in the distribution function $F(t) = P(X \leq t)$; that is, $\nu(P) = F(\cdot)$. We assume that F is continuous, in which case (Proposition 4.1.1) the distribution of

$$D_n(F) = \sup_{t \in R} |\widehat{F}(t) - F(t)|$$

does not depend on F and is known (Example 4.1.5). That is, $D_n(F)$ is a pivot. Let d_α be chosen such that $P_F(D_n(F) \leq d_\alpha) = 1 - \alpha$. Then by solving $D_n(F) \leq d_\alpha$ for F, we find that a simultaneous in t size $1 - \alpha$ confidence region $C(\mathbf{x})(\cdot)$ is the confidence band which, for each $t \in R$, consists of the interval

$$C(\mathbf{x})(t) = (\max\{0, \widehat{F}(t) - d_\alpha\}, \min\{1, \widehat{F}(t) + d_\alpha\}).$$

We have shown

$$P(C(\mathbf{X})(t) \supset F(t) \text{ for all } t \in R) = 1 - \alpha$$

for all $P \in \mathcal{P} = $ set of P with $P(-\infty, t]$ continuous in t. □

We can apply the notions studied in Examples 4.4.4 and 4.4.5 to give confidence regions for scalar or vector parameters in nonparametric models.

Example 4.4.7. *A Lower Confidence Bound for the Mean of a Nonnegative Random Variable.* Suppose $X_1, \ldots, X_n$ are i.i.d. as X and that X has a density $f(t) = F'(t)$, which is zero for $t < 0$ and nonzero for $t > 0$. By integration by parts, if $\mu = \mu(F) = \int_0^\infty t f(t) dt$ exists, then

$$\mu = \int_0^\infty [1 - F(t)] dt.$$

Let $\widehat{F}^-(t)$ and $\widehat{F}^+(t)$ be the lower and upper simultaneous confidence boundaries of Example 4.4.6. Then a $(1 - \alpha)$ lower confidence bound for μ is $\underline{\mu}$ given by

$$\underline{\mu} = \int_0^\infty [1 - \widehat{F}^+(t)] dt = \sum_{i \leq n(1-d_\alpha)} \left[1 - \left(\frac{i}{n} + d_\alpha \right) \right] [x_{(i+1)} - x_{(i)}] \qquad (4.4.9)$$

because for $C(\mathbf{X})$ as in Example 4.4.6, $\underline{\mu} = \inf\{\mu(F) : F \in C(\mathbf{X})\} = \mu(\widehat{F}^+)$ and $\sup\{\mu(F) : F \in C(\mathbf{X})\} = \mu(\widehat{F}^-) = \infty$—see Problem 4.4.19.

Intervals for the case of F supported on an interval (see Problem 4.4.18) arise in accounting practice (see Bickel, 1992, where such bounds are discussed and shown to be asymptotically strictly conservative). □

Summary. We define lower and upper confidence bounds (LCBs and UCBs), confidence intervals, and more generally confidence regions. In a parametric model $\{P_\theta : \theta \in \Theta\}$, a level $1 - \alpha$ confidence region for a parameter $q(\theta)$ is a set $C(x)$ depending only on the data x such that the probability under P_θ that $C(X)$ covers $q(\theta)$ is at least $1 - \alpha$ for all $\theta \in \Theta$. For a nonparametric class $\mathcal{P} = \{P\}$ and parameter $\nu = \nu(P)$, we similarly require $P(C(X) \supset \nu) \geq 1 - \alpha$ for all $P \in \mathcal{P}$. We derive the (Student) t interval for μ in the $\mathcal{N}(\mu, \sigma^2)$ model with σ^2 unknown, and we derive an exact confidence interval for the binomial parameter. In a nonparametric setting we derive a simultaneous confidence interval for the distribution function $F(t)$ and the mean of a positive variable X.

4.5   THE DUALITY BETWEEN CONFIDENCE REGIONS AND TESTS

Confidence regions are random subsets of the parameter space that contain the true parameter with probability at least $1 - \alpha$. Acceptance regions of statistical tests are, for a given hypothesis H, subsets of the sample space with probability of accepting H at least $1 - \alpha$ when H is true. We shall establish a duality between confidence regions and acceptance regions for families of hypotheses.

We begin by illustrating the duality in the following example.

Example 4.5.1. *Two-Sided Tests for the Mean of a Normal Distribution.* Suppose that an established theory postulates the value μ_0 for a certain physical constant. A scientist has reasons to believe that the theory is incorrect and measures the constant n times obtaining measurements $X_1, \ldots, X_n$. Knowledge of his instruments leads him to assume that the X_i are independent and identically distributed normal random variables with mean μ and variance σ^2. If any value of μ other than μ_0 is a possible alternative, then it is reasonable to formulate the problem as that of testing $H : \mu = \mu_0$ versus $K : \mu \neq \mu_0$.

We can base a size α test on the level $(1 - \alpha)$ confidence interval (4.4.1) we constructed for μ as follows. We accept H, if and only if, the postulated value μ_0 is a member of the level $(1 - \alpha)$ confidence interval

$$[\bar{X} - s t_{n-1} \left(1 - \tfrac{1}{2}\alpha\right) / \sqrt{n}, \ \bar{X} + s t_{n-1} \left(1 - \tfrac{1}{2}\alpha\right) / \sqrt{n}]. \tag{4.5.1}$$

If we let $T = \sqrt{n}(\bar{X} - \mu_0)/s$, then our test accepts H, if and only if, $-t_{n-1}\left(1 - \tfrac{1}{2}\alpha\right) \leq T \leq t_{n-1}\left(1 - \tfrac{1}{2}\alpha\right)$. Because $P_\mu[|T| = t_{n-1}\left(1 - \tfrac{1}{2}\alpha\right)] = 0$ the test is equivalently characterized by rejecting H when $|T| \geq t_{n-1}\left(1 - \tfrac{1}{2}\alpha\right)$. This test is called *two-sided* because it rejects for both large and small values of the statistic T. In contrast to the tests of Example 4.1.4, it has power against parameter values on either side of μ_0.

Because the same interval (4.5.1) is used for every μ_0 we see that we have, in fact, generated a family of level α tests $\{\delta(\mathbf{X}, \mu)\}$ where

$$\begin{aligned}\delta(\mathbf{X}, \mu) &= 1 \text{ if } \sqrt{n}\frac{|\bar{X}-\mu|}{s} \geq t_{n-1}\left(1 - \tfrac{1}{2}\alpha\right) \\ &= 0 \text{ otherwise.}\end{aligned} \tag{4.5.2}$$

These tests correspond to different hypotheses, $\delta(\mathbf{X}, \mu_0)$ being of size α only for the hypothesis $H : \mu = \mu_0$.

Conversely, by starting with the test (4.5.2) we obtain the confidence interval (4.5.1) by finding the set of μ where $\delta(\mathbf{X}, \mu) = 0$.

We achieve a similar effect, generating a family of level α tests, if we start out with (say) the level $(1 - \alpha)$ LCB $\bar{X} - t_{n-1}(1 - \alpha)s/\sqrt{n}$ and define $\delta^*(\mathbf{X}, \mu)$ to equal 1 if, and only if, $\bar{X} - t_{n-1}(1 - \alpha)s/\sqrt{n} \geq \mu$. Evidently,

$$P_{\mu_0}[\delta^*(\mathbf{X}, \mu_0) = 1] = P_{\mu_0}\left[\sqrt{n}\frac{(\bar{X} - \mu_0)}{s} \geq t_{n-1}(1 - \alpha)\right] = \alpha.$$

$\square$

These are examples of a general phenomenon. Consider the general framework where the random vector X takes values in the sample space $\mathcal{X} \subset R^q$ and X has distribution $P \in \mathcal{P}$. Let $\nu = \nu(P)$ be a parameter that takes values in the set $\mathcal{N}$. For instance, in Example 4.4.1, $\mu = \mu(P)$ takes values in $\mathcal{N} = (-\infty, \infty)$, in Example 4.4.2, $\sigma^2 = \sigma^2(P)$ takes values in $\mathcal{N} = (0, \infty)$, and in Example 4.4.5, (μ, σ^2) takes values in $\mathcal{N} = (-\infty, \infty) \times (0, \infty)$. For a function space example, consider $\nu(P) = F$, as in Example 4.4.6, where F is the distribution function of X_i. Here an example of $\mathcal{N}$ is the class of all continuous distribution functions. Let $S = S(X)$ be a map from $\mathcal{X}$ to subsets of $\mathcal{N}$, then S is a $(1 - \alpha)$ *confidence region* for ν if the probability that $S(X)$ contains ν is at least $(1 - \alpha)$, that is

$$P[\nu \in S(X)] \geq 1 - \alpha, \text{ all } P \in \mathcal{P}.$$

Next consider the testing framework where we test the hypothesis $H = H_{\nu_0} : \nu = \nu_0$ for some specified value ν_0. Suppose we have a test $\delta(X, \nu_0)$ with level α. Then the acceptance region

$$A(\nu_0) = \{x : \delta(x, \nu_0) = 0\}$$

is a subset of $\mathcal{X}$ with probability at least $1 - \alpha$. For some specified ν_0, H may be accepted, for other specified ν_0, H may be rejected. Consider the set of ν_0 for which H_{ν_0} is accepted; this is a random set contained in $\mathcal{N}$ with probability at least $1 - \alpha$ of containing the true value of $\nu(P)$ whatever be P. Conversely, if $S(X)$ is a level $1 - \alpha$ confidence region for ν, then the test that accepts H_{ν_0} if and only if ν_0 is in $S(\mathbf{X})$, is a level α test for H_{ν_0}.

Formally, let $\mathcal{P}_{\nu_0} = \{P : \nu(P) = \nu_0 : \nu_0 \in \mathcal{N}\}$. We have the following.

Duality Theorem. *Let* $S(X) = \{\nu_0 \in \mathcal{N} : X \in A(\nu_0)\}$, *then*

$$P[X \in A(\nu_0)] \geq 1 - \alpha \text{ for all } P \in \mathcal{P}_{\nu_0}$$

if and only if $S(X)$ is a $1 - \alpha$ confidence region for ν.

We next apply the duality theorem to MLR families:

Theorem 4.5.1. *Suppose $X \sim P_\theta$ where $\{P_\theta : \theta \in \Theta\}$ is MLR in $T = T(X)$ and suppose that the distribution function $F_\theta(t)$ of T under P_θ is continuous in each of the variables t and θ when the other is fixed. If the equation $F_\theta(t) = 1 - \alpha$ has a solution $\underline{\theta}_\alpha(t)$ in Θ, then $\underline{\theta}_\alpha(T)$ is a lower confidence bound for θ with confidence coefficient $1 - \alpha$. Similarly, any solution $\bar{\theta}_\alpha(T)$ of $F_\theta(T) = \alpha$ with $\bar{\theta}_\alpha \in \Theta$ is an upper confidence bound for θ with coefficient $(1 - \alpha)$. Moreover, if $\alpha_1 + \alpha_2 < 1$, then $[\underline{\theta}_{\alpha_1}, \bar{\theta}_{\alpha_2}]$ is confidence interval for θ with confidence coefficient $1 - (\alpha_1 + \alpha_2)$.*

Proof. By Corollary 4.3.1, the acceptance region of the UMP size α test of $H : \theta = \theta_0$ versus $K : \theta > \theta_0$ can be written

$$A(\theta_0) = \{x : T(x) \leq t_{\theta_0}(1 - \alpha)\}$$

where $t_{\theta_0}(1 - \alpha)$ is the $1 - \alpha$ quantile of F_{θ_0}. By the duality theorem, if

$$S(t) = \{\theta \in \Theta : t \leq t_\theta(1 - \alpha)\},$$

then $S(T)$ is a $1 - \alpha$ confidence region for θ. By applying F_θ to both sides of $t \leq t_\theta(1 - \alpha)$, we find

$$S(t) = \{\theta \in \Theta : F_\theta(t) \leq 1 - \alpha\}.$$

By Theorem 4.3.1, the power function $P_\theta(T \geq t) = 1 - F_\theta(t)$ for a test with critical constant t is increasing in θ. That is, $F_\theta(t)$ is decreasing in θ. It follows that $F_\theta(t) \leq 1 - \alpha$ iff $\theta \geq \underline{\theta}_\alpha(t)$ and $S(t) = [\underline{\theta}_\alpha, \infty)$. The proofs for the upper confidence bound and interval follow by the same type of argument. $\qquad\square$

We next give connections between confidence bounds, acceptance regions, and p-values for MLR families: Let t denote the observed value $t = T(x)$ of $T(X)$ for the datum x, let

$\alpha(t, \theta_0)$ denote the p-value for the UMP size α test of $H : \theta = \theta_0$ versus $K : \theta > \theta_0$, and let

$$A^*(\theta) = T(A(\theta)) = \{T(x) : x \in A(\theta)\}.$$

Corollary 4.5.1. *Under the conditions of Theorem* 4.4.1,

$$
\begin{aligned}
A^*(\theta) &= \{t : \alpha(t, \theta) \geq \alpha\} = (-\infty, t_\theta(1 - \alpha)] \\
S(t) &= \{\theta : \alpha(t, \theta) \geq \alpha\} = [\underline{\theta}_\alpha(t), \infty).
\end{aligned}
$$

Proof. The p-value is

$$\alpha(t, \theta) = P_\theta(T \geq t) = 1 - F_\theta(t).$$

We have seen in the proof of Theorem 4.3.1 that $1 - F_\theta(t)$ is increasing in θ. Because $F_\theta(t)$ is a distribution function, $1 - F_\theta(t)$ is decreasing in t. The result follows. □

In general, let $\alpha(t, \nu_0)$ denote the p-value of a level α test $\delta(T, \nu_0) = 1[T \geq c]$ of $H : \nu = \nu_0$ based on a statistic $T = T(X)$ with observed value $t = T(x)$. Then the set

$$C = \{(t, \nu) : \alpha(t, \nu) \geq \alpha\} = \{(t, \nu) : \delta(t, \nu) = 0\}$$

gives the pairs (t, ν) where, for the given t, ν will be accepted; and for the given ν, t is in the acceptance region. We call C the set of *compatible* (t, θ) points. In the (t, θ) plane, vertical sections of C are the confidence regions $S(t)$ whereas horizontal sections are the acceptance regions $A^*(\nu) = \{t : \delta(t, \nu) = 0\}$. We illustrate these ideas using the example of testing $H : \mu = \mu_0$ when $X_1, \ldots, X_n$ are i.i.d. $\mathcal{N}(\mu, \sigma^2)$ with σ^2 known. Let $T = \bar{X}$, then

$$C = \left\{(t, \mu) : |t - \mu| \leq \sigma z \left(1 - \tfrac{1}{2}\alpha\right) / \sqrt{n}\right\}.$$

Figure 4.5.1 shows the set C, a confidence region $S(t_0)$, and an acceptance set $A^*(\mu_0)$ for this example.

Example 4.5.2. *Exact Confidence Bounds and Intervals for the Probability of Success in n Bernoulli Trials.* Let $X_1, \ldots, X_n$ be the indicators of n Bernoulli trials with probability of success θ. For $\alpha \in (0, 1)$, we seek reasonable *exact* level $(1 - \alpha)$ upper and lower confidence bounds and confidence intervals for θ. To find a lower confidence bound for θ our preceding discussion leads us to consider level α tests for $H : \theta \leq \theta_0, \theta_0 \in (0, 1)$. We shall use some of the results derived in Example 4.1.3. Let $k(\theta_0, \alpha)$ denote the critical constant of a level α test of H. The corresponding level $(1 - \alpha)$ confidence region is given by

$$C(X_1, \ldots, X_n) = \{\theta : S \leq k(\theta, \alpha) - 1\},$$

where $S = \Sigma_{i=1}^n X_i$.

To analyze the structure of the region we need to examine $k(\theta, \alpha)$. We claim that

(i) $k(\theta, \alpha)$ is nondecreasing in θ.

(ii) $k(\theta, \alpha) \to k(\theta_0, \alpha)$ if $\theta \uparrow \theta_0$.

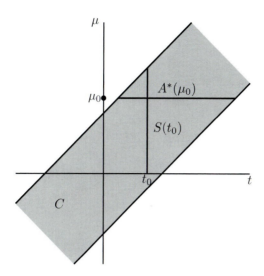

Figure 4.5.1. The shaded region is the compatibility set C for the two-sided test of $H_{\mu_0} : \mu = \mu_0$ in the normal model. $S(t_0)$ is a confidence interval for μ for a given value t_0 of T, whereas $A^*(\mu_0)$ is the acceptance region for H_{μ_0}.

(iii) $k(\theta, \alpha)$ increases by exactly 1 at its points of discontinuity.

(iv) $k(0, \alpha) = 1$ and $k(1, \alpha) = n + 1$.

   To prove (i) note that it was shown in Theorem 4.3.1(1) that $P_\theta[S \geq j]$ is nondecreasing in θ for fixed j. Clearly, it is also nonincreasing in j for fixed θ. Therefore, $\theta_1 < \theta_2$ and $k(\theta_1, \alpha) > k(\theta_2, \alpha)$ would imply that

$$\alpha \geq P_{\theta_2}[S \geq k(\theta_2, \alpha)] \geq P_{\theta_2}[S \geq k(\theta_2, \alpha) - 1] \geq P_{\theta_1}[S \geq k(\theta_1, \alpha) - 1] > \alpha,$$

a contradiction.

   The assertion (ii) is a consequence of the following remarks. If θ_0 is a discontinuity point of $k(\theta, \alpha)$, let j be the limit of $k(\theta, \alpha)$ as $\theta \uparrow \theta_0$. Then $P_\theta[S \geq j] \leq \alpha$ for all $\theta < \theta_0$ and, hence, $P_{\theta_0}[S \geq j] \leq \alpha$. On the other hand, if $\theta > \theta_0$, $P_\theta[S \geq j] > \alpha$. Therefore, $P_{\theta_0}[S \geq j] = \alpha$ and $j = k(\theta_0, \alpha)$. The claims (iii) and (iv) are left as exercises.

   From (i), (ii), (iii), and (iv) we see that, if we define

$$\underline{\theta}(S) = \inf\{\theta : k(\theta, \alpha) = S + 1\},$$

then

$$C(\mathbf{X}) = \begin{cases} (\underline{\theta}(S), 1] & \text{if } S > 0 \\ [0, 1] & \text{if } S = 0 \end{cases}$$

and $\underline{\theta}(S)$ is the desired level $(1 - \alpha)$ LCB for θ.[1] Figure 4.5.2 portrays the situation. From our discussion, when $S > 0$, then $k(\underline{\theta}(S), \alpha) = S$ and, therefore, we find $\underline{\theta}(S)$ as the unique solution of the equation,

$$\sum_{r=S}^{n} \binom{n}{r} \theta^r (1 - \theta)^{n-r} = \alpha.$$

When $S = 0$, $\underline{\theta}(S) = 0$.

Similarly, we define

$$\bar{\theta}(S) = \sup\{\theta : j(\theta, \alpha) = S - 1\}$$

where $j(\theta, \alpha)$ is given by,

$$\sum_{r=0}^{j(\theta, \alpha)} \binom{n}{r} \theta^r (1 - \theta)^{n-r} \leq \alpha < \sum_{r=0}^{j(\theta, \alpha)+1} \binom{n}{r} \theta^r (1 - \theta)^{n-r}.$$

Then $\bar{\theta}(S)$ is a level $(1 - \alpha)$ UCB for θ and when $S < n$, $\bar{\theta}(S)$ is the unique solution of

$$\sum_{r=0}^{S} \binom{n}{r} \theta^r (1 - \theta)^{n-r} = \alpha.$$

When $S = n$, $\bar{\theta}(S) = 1$. Putting the bounds $\underline{\theta}(S)$, $\bar{\theta}(S)$ together we get the confidence interval $[\underline{\theta}(S), \bar{\theta}(S)]$ of level $(1-2\alpha)$. These intervals can be obtained from computer packages that use algorithms based on the preceding considerations. As might be expected, if n is large, these bounds and intervals differ little from those obtained by the first approximate method in Example 4.4.3.

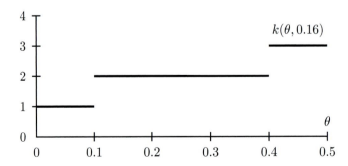

Figure 4.5.2. Plot of $k(\theta, 0.16)$ for $n = 2$.

Applications of Confidence Intervals to Comparisons and Selections

We have seen that confidence intervals lead naturally to two-sided tests. However, two-sided tests seem incomplete in the sense that if $H : \theta = \theta_0$ is rejected in favor of $H : \theta \neq \theta_0$, we usually want to know whether $H : \theta > \theta_0$ or $H : \theta < \theta_0$.

For instance, suppose θ is the expected difference in blood pressure when two treatments, A and B, are given to high blood pressure patients. Because we do not know whether A or B is to be preferred, we test $H : \theta = 0$ versus $K : \theta \neq 0$. If H is rejected, it is natural to carry the comparison of A and B further by asking whether $\theta < 0$ or $\theta > 0$. If we decide $\theta < 0$, then we select A as the better treatment, and vice versa.

The problem of deciding whether $\theta = \theta_0, \theta < \theta_0$, or $\theta > \theta_0$ is an example of a three-decision problem and is a special case of the decision problems in Section 1.4, and 3.1–3.3. Here we consider the simple solution suggested by the level $(1 - \alpha)$ confidence interval I:

1.   Make no judgment as to whether $\theta < \theta_0$ or $\theta > \theta_0$ if I contains θ_0;
2.   Decide $\theta < \theta_0$ if I is entirely to the left of θ_0; and   (4.5.3)
3.   Decide $\theta > \theta_0$ if I is entirely to the right of θ_0.

Example 4.5.2. Suppose $X_1, \ldots, X_n$ are i.i.d. $\mathcal{N}(\mu, \sigma^2)$ with σ^2 known. In Section 4.4 we considered the level $(1 - \alpha)$ confidence interval $\bar{X} \pm \sigma z(1 - \frac{1}{2}\alpha)/\sqrt{n}$ for μ. Using this interval and (4.5.3) we obtain the following three decision rule based on $T = \sqrt{n}(\bar{X} - \mu_0)/\sigma$:

$$\text{Do not reject } H : \mu = \mu_0 \text{ if } |T| \leq z(1 - \tfrac{1}{2}\alpha).$$
$$\text{Decide } \mu < \mu_0 \text{ if } T < -z(1 - \tfrac{1}{2}\alpha).$$
$$\text{Decide } \mu > \mu_0 \text{ if } T > z(1 - \tfrac{1}{2}\alpha).$$

Thus, the two-sided test can be regarded as the first step in the decision procedure where if H is not rejected, we make no claims of significance, but if H is rejected, we decide whether this is because μ is smaller or larger than μ_0. For this three-decision rule, the probability of falsely claiming significance of either $\mu < \mu_0$ or $\mu > \mu_0$ is bounded above by $\frac{1}{2}\alpha$. To see this consider first the case $\mu \geq \mu_0$. Then the wrong decision "$\mu < \mu_0$" is made when $T < -z(1 - \frac{1}{2}\alpha)$. This event has probability

$$P[T < -z(1 - \tfrac{1}{2}\alpha)] = \Phi\left(-z(1 - \tfrac{1}{2}\alpha) - \frac{\sqrt{n}(\mu - \mu_0)}{\sigma}\right) \leq \Phi(-z(1 - \tfrac{1}{2}\alpha)) = \tfrac{1}{2}\alpha.$$

Similarly, when $\mu \leq \mu_0$, the probability of the wrong decision is at most $\frac{1}{2}\alpha$. Therefore, by using this kind of procedure in a comparison or selection problem, we can control the probabilities of a wrong selection by setting the α of the parent test or confidence interval. We can use the two-sided tests and confidence intervals introduced in later chapters in similar fashions.

Summary. We explore the connection between tests of statistical hypotheses and confidence regions. If $\delta(x, \nu_0)$ is a level α test of $H : \nu = \nu_0$, then the set $S(x)$ of ν_0 where

$\delta(x, \nu_0) = 0$ is a level $(1 - \alpha)$ confidence region for ν_0. If $S(x)$ is a level $(1 - \alpha)$ confidence region for ν, then the test that accepts $H : \nu = \nu_0$ when $\nu_0 \in S(x)$ is a level α test. We give explicitly the construction of exact upper and lower confidence bounds and intervals for the parameter in the binomial distribution. We also give a connection between confidence intervals, two-sided tests, and the three-decision problem of deciding whether a parameter θ is θ_0, less than θ_0, or larger than θ_0, where θ_0 is a specified value.

4.6 UNIFORMLY MOST ACCURATE CONFIDENCE BOUNDS

In our discussion of confidence bounds and intervals so far we have not taken their accuracy into account. We next show that for a certain notion of accuracy of confidence bounds, which is connected to the power of the associated one-sided tests, optimality of the tests translates into accuracy of the bounds.

If $\underline{\theta}$ and $\underline{\theta}^*$ are two competing level $(1 - \alpha)$ lower confidence bounds for θ, they are both very likely to fall below the true θ. But we also want the bounds to be close to θ. Thus, we say that the bound with the smaller probability of being far below θ is more accurate. Formally, for $X \in \mathcal{X} \subset R^q$, the following is true.

Definition 4.6.1. A level $(1 - \alpha)$ LCB $\underline{\theta}^*$ of θ is said to be more *accurate* than a competing level $(1 - \alpha)$ LCB $\underline{\theta}$ if, and only if, for any fixed θ and all $\theta' < \theta$,

$$P_\theta[\underline{\theta}^*(X) \leq \theta'] \leq P_\theta[\underline{\theta}(X) \leq \theta']. \qquad (4.6.1)$$

Similarly, a level $(1 - \alpha)$ UCB $\overline{\theta}^*$ is more accurate than a competitor $\overline{\theta}$ if, and only if, for any fixed θ and all $\theta' > \theta$,

$$P_\theta[\overline{\theta}^*(X) \geq \theta'] \leq P_\theta[\overline{\theta}(X) \geq \theta']. \qquad (4.6.2)$$

Lower confidence bounds $\underline{\theta}^*$ satisfying $(4.6.1)$ for *all* competitors are called *uniformly most accurate* as are upper confidence bounds satisfying $(4.6.2)$ for all competitors. Note that $\underline{\theta}^*$ is a uniformly most accurate level $(1 - \alpha)$ LCB for θ, if and only if, $-\underline{\theta}^*$ is a uniformly most accurate level $(1 - \alpha)$ UCB for $-\theta$.

Example 4.6.1 (Examples 3.3.2 and 4.2.1 continued). Suppose $\mathbf{X} = (X_1, \ldots, X_n)$ is a sample of a $\mathcal{N}(\mu, \sigma^2)$ random variables with σ^2 known. A level α test of $H : \mu = \mu_0$ vs $K : \mu > \mu_0$ rejects H when $\sqrt{n}(\bar{X} - \mu_0)/\sigma \geq z(1 - \alpha)$. The dual lower confidence bound is $\underline{\mu}_1(\mathbf{X}) = \bar{X} - z(1 - \alpha)\sigma/\sqrt{n}$. Using Problem 4.5.6, we find that a competing lower confidence bound is $\underline{\mu}_2(\mathbf{X}) = X_{(k)}$, where $X_{(1)} \leq X_{(2)} \leq \cdots \leq X_{(n)}$ denotes the ordered $X_1, \ldots, X_n$ and k is defined to be the largest integer such that $P(S \geq k) \geq 1 - \alpha$ for a binomial, $B(n, \frac{1}{2})$, random variable S. Which lower bound is more accurate? It does turn out that $\underline{\mu}_1(\mathbf{X})$ is more accurate than $\underline{\mu}_2(\mathbf{X})$ and is, in fact, uniformly most accurate in the $\mathcal{N}(\mu, \sigma^2)$ model. This is a consequence of the following theorem, which reveals that $(4.6.1)$ is nothing more than a comparison of power functions.   □

Theorem 4.6.1. *Let $\underline{\theta}^*$ be a level $(1 - \alpha)$ LCB for θ, a real parameter, such that for each θ_0 the associated test whose critical function $\delta^*(x, \theta_0)$ is given by*

$$\delta^*(x, \theta_0) \;=\; 1 \text{ if } \underline{\theta}^*(x) > \theta_0$$
$$=\; 0 \text{ otherwise}$$

is UMP level α for $H : \theta = \theta_0$ versus $K : \theta > \theta_0$. Then $\underline{\theta}^*$ is uniformly most accurate at level $(1 - \alpha)$.

Proof. Let $\underline{\theta}$ be a competing level $(1 - \alpha)$ LCB θ_0. Defined $\delta(x, \theta_0)$ by

$$\delta(x, \theta_0) = 0 \text{ if, and only if, } \underline{\theta}(x) \le \theta_0.$$

Then $\delta(X, \theta_0)$ is a level α test for $H : \theta = \theta_0$ versus $K : \theta > \theta_0$. Because $\delta^*(X, \theta_0)$ is UMP level α for $H : \theta = \theta_0$ versus $K : \theta > \theta_0$, for $\theta_1 > \theta_0$ we must have

$$E_{\theta_1}(\delta(X, \theta_0)) \le E_{\theta_1}(\delta^*(X, \theta_0))$$

or

$$P_{\theta_1}[\underline{\theta}(X) > \theta_0] \le P_{\theta_1}[\underline{\theta}^*(X) > \theta_0].$$

Identify θ_0 with θ' and θ_1 with θ in the statement of Definition 4.4.2 and the result follows.   □

If we apply the result and Example 4.2.1 to Example 4.6.1, we find that $\bar{x} - z(1 - \alpha)\sigma/\sqrt{n}$ is uniformly most accurate. However, $X_{(k)}$ does have the advantage that we don't have to know σ or even the shape of the density f of X_i to apply it. Also, the robustness considerations of Section 3.5 favor $X_{(k)}$ (see Example 3.5.2).

Uniformly most accurate (UMA) bounds turn out to have related nice properties. For instance (see Problem 4.6.7 for the proof), they have the smallest expected "distance" to θ:

Corollary 4.6.1. *Suppose $\underline{\theta}^*(\mathbf{X})$ is a UMA level $(1 - \alpha)$ lower confidence bound for θ. Let $\underline{\theta}(\mathbf{X})$ be any other $(1 - \alpha)$ lower confidence bound, then*

$$E_\theta\{(\theta - \underline{\theta}^*(\mathbf{X}))^+\} \le E_\theta\{(\theta - \underline{\theta}(\mathbf{X}))^+\}$$

for all θ where $a^+ = a$, if $a \ge 0$, and 0 otherwise.

We can extend the notion of accuracy to confidence bounds for real-valued functions of an arbitrary parameter. We define $\underline{q}^*$ to be a uniformly most accurate level $(1 - \alpha)$ LCB for $q(\theta)$ if, and only if, for any other level $(1 - \alpha)$ LCB $\underline{q}$,

$$P_\theta[\underline{q}^* \le q(\theta')] \le P_\theta[\underline{q} \le q(\theta')]$$

whenever $q(\theta') < q(\theta)$. Most accurate upper confidence bounds are defined similarly.

Example 4.6.2. *Bounds for the Probability of Early Failure of Equipment.* Let $X_1, \ldots, X_n$ be the times to failure of n pieces of equipment where we assume that the X_i are independent $\mathcal{E}(\lambda)$ variables. We want a uniformly most accurate level $(1 - \alpha)$ upper confidence bound $\bar{q}^*$ for $q(\lambda) = 1 - e^{-\lambda t_0}$, the probability of early failure of a piece of equipment.

We begin by finding a uniformly most accurate level $(1 - \alpha)$ UCB $\bar{\lambda}^*$ for λ. To find $\bar{\lambda}^*$ we invert the family of UMP level α tests of $H : \lambda \geq \lambda_0$ versus $K : \lambda < \lambda_0$. By Problem 4.6.8, the UMP test accepts H if

$$\sum_{i=1}^{n} X_i < \chi_{2n}(1 - \alpha)/2\lambda_0 \tag{4.6.3}$$

or equivalently if

$$\lambda_0 < \frac{\chi_{2n}(1 - \alpha)}{2 \sum_{i=1}^{n} X_i}$$

where $\chi_{2n}(1 - \alpha)$ is the $(1 - \alpha)$ quantile of the χ_{2n}^2 distribution. Therefore, the confidence region corresponding to this test is $(0, \bar{\lambda}^*)$ where $\bar{\lambda}^*$ is by Theorem 4.6.1, a uniformly most accurate level $(1 - \alpha)$ UCB for λ and, because q is strictly increasing in λ, it follows that $q(\bar{\lambda}^*)$ is a uniformly most accurate level $(1 - \alpha)$ UCB for the probability of early failure.   □

Discussion

We have only considered confidence bounds. The situation with confidence intervals is more complicated. Considerations of accuracy lead us to ask that, subject to the requirement that the confidence level is $(1 - \alpha)$, the confidence interval be as short as possible. Of course, the length $\bar{T} - \underline{T}$ is random and it can be shown that in most situations there is no confidence interval of level $(1 - \alpha)$ that has uniformly minimum length among all such intervals. There are, however, some large sample results in this direction (see Wilks, 1962, pp. 374–376). If we turn to the expected length $E_\theta(\bar{T} - \underline{T})$ as a measure of precision, the situation is still unsatisfactory because, in general, there does not exist a member of the class of level $(1 - \alpha)$ intervals that has minimum expected length for all θ. However, as in the estimation problem, we can restrict attention to certain reasonable subclasses of level $(1 - \alpha)$ intervals for which members with uniformly smallest expected length exist. Thus, Neyman defines *unbiased* confidence intervals of level $(1 - \alpha)$ by the property that

$$P_\theta[\underline{T} \leq q(\theta) \leq \bar{T}] \geq P_\theta[\underline{T} \leq q(\theta') \leq \bar{T}]$$

for every θ, θ'. That is, the interval must be at least as likely to cover the true value of $q(\theta)$ as any other value. Pratt (1961) showed that in many of the classical problems of estimation there exist level $(1 - \alpha)$ confidence intervals that have uniformly minimum expected length among all level $(1 - \alpha)$ unbiased confidence intervals. In particular, the intervals developed in Example 4.5.1 have this property.

Confidence intervals obtained from two-sided tests that are uniformly most powerful within a restricted class of procedures can be shown to have optimality properties within restricted classes. These topics are discussed in Lehmann (1997).

Summary. By using the duality between one-sided tests and confidence bounds we show that confidence bounds based on UMP level α tests are uniformly most accurate (UMA) level $(1 - \alpha)$ in the sense that, in the case of lower bounds, they are less likely than other level $(1 - \alpha)$ lower confidence bounds to fall below any value θ' below the true θ.

4.7   FREQUENTIST AND BAYESIAN FORMULATIONS

We have so far focused on the frequentist formulation of confidence bounds and intervals where the data $X \in \mathcal{X} \subset R^q$ are random while the parameters are fixed but unknown. A consequence of this approach is that once a numerical interval has been computed from experimental data, no probability statement can be attached to this interval. Instead, the interpretation of a $100(1 - \alpha)\%$ confidence interval is that if we repeated an experiment indefinitely each time computing a $100(1 - \alpha)\%$ confidence interval, then $100(1 - \alpha)\%$ of the intervals would contain the true unknown parameter value.

In the Bayesian formulation of Sections 1.2 and 1.6.3, what are called level $(1 - \alpha)$ credible bounds and intervals are subsets of the parameter space which are given probability at least $(1 - \alpha)$ by the posterior distribution of the parameter given the data. Suppose that, given θ, X has distribution $P_\theta, \theta \in \Theta \subset R$, and that θ has the prior probability distribution Π.

Definition 4.7.1. Let $\Pi(\cdot|x)$ denote the posterior probability distribution of θ given $X = x$, then $\underline{\theta}$ and $\bar{\theta}$ are level $(1 - \alpha)$ *lower* and *upper credible* bounds for θ if they respectively satisfy

$$\Pi(\underline{\theta} \leq \theta|x) \geq 1 - \alpha, \quad \Pi(\theta \leq \bar{\theta}|x) \geq 1 - \alpha.$$

Turning to Bayesian credible intervals and regions, it is natural to consider the collection of θ that is "most likely" under the distribution $\Pi(\theta|x)$. Thus,

Definition 4.7.2. Let $\pi(\cdot|x)$ denote the density of θ given $X = x$, then

$$C_k = \{\theta : \pi(\theta|x) \geq k\}$$

is called a *level $(1 - \alpha)$ credible region* for θ if $\Pi(C_k|x) \geq 1 - \alpha$.

If $\pi(\theta|x)$ is unimodal, then C_k will be an interval of the form $[\underline{\theta}, \bar{\theta}]$. We next give such an example.

Example 4.7.1. Suppose that given μ, $X_1, \ldots, X_n$ are i.i.d. $\mathcal{N}(\mu, \sigma_0^2)$ with σ_0^2 known, and that $\mu \sim \mathcal{N}(\mu_0, \tau_0^2)$, with μ_0 and τ_0^2 known. Then, from Example 1.1.12, the posterior distribution of μ given $X_1, \ldots, X_n$ is $\mathcal{N}(\widehat{\mu}_B, \widehat{\sigma}_B^2)$, with

$$\widehat{\mu}_B = \frac{\frac{n}{\sigma_0^2}\bar{X} + \frac{1}{\tau_0^2}\mu_0}{\frac{n}{\sigma_0^2} + \frac{1}{\tau_0^2}}, \widehat{\sigma}_B^2 = \frac{1}{\frac{n}{\sigma_0^2} + \frac{1}{\tau_0^2}}$$

It follows that the level $1 - \alpha$ lower and upper credible bounds for μ are

$$\underline{\mu} = \widehat{\mu}_B - z_{1-\alpha}\frac{\sigma_0}{\sqrt{n}\left(1 + \frac{\sigma_0^2}{n\tau_0^2}\right)^{\frac{1}{2}}}$$

$$\bar{\mu} = \widehat{\mu}_B + z_{1-\alpha}\frac{\sigma_0}{\sqrt{n}\left(1 + \frac{\sigma_0^2}{n\tau_0^2}\right)^{\frac{1}{2}}}$$

while the level $(1 - \alpha)$ credible interval is $[\mu^-, \mu^+]$ with

$$\mu^{\pm} = \widehat{\mu}_B \pm z_{1-\frac{\alpha}{2}} \frac{\sigma_0}{\sqrt{n}\left(1 + \frac{\sigma_0^2}{n\tau_0^2}\right)^{\frac{1}{2}}}.$$

Compared to the frequentist interval $\bar{X} \pm z_{1-\frac{\alpha}{2}} \sigma_0 / \sqrt{n}$, the center $\widehat{\mu}_B$ of the Bayesian interval is pulled in the direction of μ_0, where μ_0 is a prior guess of the value of μ. See Example 1.3.4 for sources of such prior guesses. Note that as $\tau_0 \to \infty$, the Bayesian interval tends to the frequentist interval; however, the interpretations of the intervals are different.   □

Example 4.7.2. Suppose that given σ^2, $X_1, \ldots, X_n$ are i.i.d. $\mathcal{N}(\mu_0, \sigma^2)$ where μ_0 is known. Let $\lambda = \sigma^{-2}$ and suppose λ has the gamma $\Gamma(\frac{1}{2}a, \frac{1}{2}b)$ density

$$\pi(\lambda) \propto \lambda^{\frac{1}{2}(a-2)} \exp\left\{-\frac{1}{2}b\lambda\right\}, \ \lambda > 0$$

where $a > 0, b > 0$ are known parameters. Then, by Problem 1.2.12, given $x_1, \ldots, x_n$, $(t + b)\lambda$ has a χ_{a+n}^2 distribution, where $t = \sum(x_i - \mu_0)^2$. Let $x_{a+n}(\alpha)$ denote the αth quantile of the χ_{a+n}^2 distribution, then $\underline{\lambda} = x_{a+n}(\alpha)/(t + b)$ is a level $(1 - \alpha)$ lower credible bound for λ and

$$\bar{\sigma}_B^2 = (t + b)/x_{a+n}(\alpha)$$

is a level $(1-\alpha)$ upper credible bound for σ^2. Compared to the frequentist bound $t/x_n(\alpha)$, $\bar{\sigma}_B^2$ is shifted in the direction of the reciprocal b/a of the mean of $\pi(\lambda)$.

We shall analyze Bayesian credible regions further in Chapter 5.

Summary. In the Bayesian framework we define bounds and intervals, called level $(1 - \alpha)$ credible bounds and intervals, that determine subsets of the parameter space that are assigned probability at least $(1 - \alpha)$ by the posterior distribution of the parameter $\boldsymbol{\theta}$ given the data x. In the case of a normal prior $\pi(\theta)$ and normal model $p(x \mid \theta)$, the level $(1 - \alpha)$ credible interval is similar to the frequentist interval except it is pulled in the direction μ_0 of the prior mean and it is a little narrower. However, the interpretations are different: In the frequentist confidence interval, the probability of coverage is computed with the data X random and θ fixed, whereas in the Bayesian credible interval, the probability of coverage is computed with $X = x$ fixed and $\boldsymbol{\theta}$ random with probability distribution $\Pi(\theta \mid X = x)$.

4.8   PREDICTION INTERVALS

In Section 1.4 we discussed situations in which we want to predict the value of a random variable Y. In addition to point prediction of Y, it is desirable to give an interval $[\underline{Y}, \bar{Y}]$ that contains the unknown value Y with prescribed probability $(1 - \alpha)$. For instance, a doctor administering a treatment with delayed effect will give patients a time interval $[\underline{T}, \bar{T}]$ in which the treatment is likely to take effect. Similarly, we may want an interval for the future GPA of a student or a future value of a portfolio. We define a *level* $(1-\alpha)$ *prediction interval* as an interval $[\underline{Y}, \bar{Y}]$ based on data X such that $P(\underline{Y} \leq Y \leq \bar{Y}) \geq 1 - \alpha$. The problem of finding prediction intervals is similar to finding confidence intervals using a pivot:

Example 4.8.1. *The (Student) t Prediction Interval.* As in Example 4.4.1, let $X_1, \ldots, X_n$ be i.i.d. as $X \sim \mathcal{N}(\mu, \sigma^2)$. We want a prediction interval for $Y = X_{n+1}$, which is assumed to be also $\mathcal{N}(\mu, \sigma^2)$ and independent of $X_1, \ldots, X_n$. Let $\widehat{Y} = \widehat{Y}(\mathbf{X})$ denote a predictor based on $\mathbf{X} = (X_1, \ldots, X_n)$. Then $\widehat{Y}$ and Y are independent and the mean squared prediction error (MSPE) of $\widehat{Y}$ is

$$MSPE(\widehat{Y}) = E(\widehat{Y} - Y)^2 = E([\widehat{Y} - \mu] - [Y - \mu])^2 = E[\widehat{Y} - \mu]^2 + \sigma^2.$$

Note that $\widehat{Y}$ can be regarded as both a predictor of Y and as an estimate of μ, and when we do so, $MSPE(\widehat{Y}) = MSE(\widehat{Y}) + \sigma^2$, where MSE denotes the estimation theory mean squared error. It follows that, in this case, the optimal estimator when it exists is also the optimal predictor. In Example 3.4.8, we found that in the class of unbiased estimators, $\bar{X}$ is the optimal estimator. We define a predictor Y^* to be *prediction unbiased* for Y if $E(Y^* - Y) = 0$, and can conclude that in the class of prediction unbiased predictors, the optimal MSPE predictor is $\widehat{Y} = \bar{X}$.

We next use the prediction error $\widehat{Y} - Y$ to construct a pivot that can be used to give a prediction interval. Note that

$$\widehat{Y} - Y = \bar{X} - X_{n+1} \sim \mathcal{N}(0, [n^{-1} + 1]\sigma^2).$$

Moreover, $s^2 = (n-1)^{-1} \sum_1^n (X_i - \bar{X})$ is independent of $\bar{X}$ by Theorem B.3.3 and independent of X_{n+1} by assumption. It follows that

$$Z_p(Y) = \frac{\widehat{Y} - Y}{\sqrt{n^{-1} + 1}\sigma}$$

has a $\mathcal{N}(0, 1)$ distribution and is independent of $V = (n-1)s^2/\sigma^2$, which has a χ^2_{n-1} distribution. Thus, by the definition of the (Student) t distribution in Section B.3,

$$T_p(Y) \equiv \frac{Z_p(Y)}{\sqrt{\frac{V}{n-1}}} = \frac{\widehat{Y} - Y}{\sqrt{n^{-1} + 1}s}$$

has the t distribution, $\mathcal{T}_{n-1}$. By solving $-t_{n-1}\left(1 - \frac{1}{2}\alpha\right) \leq T_p(Y) \leq t_{n-1}\left(1 - \frac{1}{2}\alpha\right)$ for Y, we find the $(1 - \alpha)$ prediction interval

$$Y = \bar{X} \pm \sqrt{n^{-1} + 1}s t_{n-1}\left(1 - \frac{1}{2}\alpha\right). \tag{4.8.1}$$

Note that $T_p(Y)$ acts as a prediction interval pivot in the same way that $T(\mu)$ acts as a confidence interval pivot in Example 4.4.1. Also note that the prediction interval is much wider than the confidence interval (4.4.1). In fact, it can be shown using the methods of

Chapter 5 that the width of the confidence interval (4.4.1) tends to zero in probability at the rate $n^{-\frac{1}{2}}$, whereas the width of the prediction interval tends to $2\sigma z \left(1 - \frac{1}{2}\alpha\right)$. Moreover, the confidence level of (4.4.1) is approximately correct for large n even if the sample comes from a nonnormal distribution, whereas the level of the prediction interval (4.8.1) is not $(1 - \alpha)$ in the limit as $n \to \infty$ for samples from non-Gaussian distributions. $\qquad \square$

We next give a prediction interval that is valid for samples from any population with a continuous distribution.

Example 4.8.2. Suppose $X_1, \ldots, X_n$ are i.i.d. as $X \sim F$, where F is a continuous distribution function with positive density f on (a, b), $-\infty \leq a < b \leq \infty$. Let $X_{(1)} < \cdots < X_{(n)}$ denote the order statistics of $X_1, \ldots, X_n$. We want a prediction interval for $Y = X_{n+1} \sim F$, where X_{n+1} is independent of the data $X_1, \ldots, X_n$. Set $U_i = F(X_i)$, $i = 1, \ldots, n + 1$, then, by Problem B.2.12, $U_1, \ldots, U_{n+1}$ are i.i.d. uniform, $\mathcal{U}(0, 1)$. Let $U_{(1)} < \cdots < U_{(n)}$ be $U_1, \ldots, U_n$ ordered, then

$$
\begin{aligned}
P(X_{(j)} \leq X_{n+1} \leq X_{(k)}) &= P(U_{(j)} \leq U_{n+1} \leq U_{(k)}) \\
&= \int P(u \leq U_{n+1} \leq v \mid U_{(j)} = u, U_{(k)} = v) dH(u, v) \\
&= \int (v - u) dH(u, v) = E(U_{(k)}) - E(U_{(j)})
\end{aligned}
$$

where H is the joint distribution of $U_{(j)}$ and $U_{(k)}$. By Problem B.2.9, $E(U_{(i)}) = i/(n+1)$; thus,

$$
P(X_{(j)} \leq X_{n+1} \leq X_{(k)}) = \frac{k - j}{n + 1}. \tag{4.8.2}
$$

It follows that $[X_{(j)}, X_{(k)}]$ with $k = n + 1 - j$ is a level $\alpha = (n + 1 - 2j)/(n + 1)$ prediction interval for X_{n+1}. This interval is a *distribution-free* prediction interval. See Problem 4.8.5 for a simpler proof of (4.8.2).

Bayesian Predictive Distributions

Suppose that θ is random with $\theta \sim \pi$ and that given $\theta = \theta$, $X_1, \ldots, X_{n+1}$ are i.i.d. $p(X \mid \theta)$. Here $X_1, \ldots, X_n$ are observable and X_{n+1} is to be predicted. The *posterior predictive distribution* $Q(\cdot \mid x)$ of X_{n+1} is defined as the conditional distribution of X_{n+1} given $x = (x_1, \ldots, x_n)$; that is, $Q(\cdot \mid x)$ has in the continuous case density

$$
q(x_{n+1} \mid \mathbf{x}) = \int_\Theta \prod_{i=1}^{n+1} p(x_i \mid \theta) \pi(\theta) d\theta \Big/ \int_\Theta \prod_{i=1}^{n} p(x_i \mid \theta) \pi(\theta) d\theta
$$

with a sum replacing the integral in the discrete case. Now $[\underline{Y}_B, \bar{Y}_B]$ is said to be a level $(1 - \alpha)$ *Bayesian prediction interval* for $Y = X_{n+1}$ if

$$
Q(\underline{Y}_B \leq Y \leq \bar{Y}_B \mid \mathbf{x}) \geq 1 - \alpha.
$$

Example 4.8.3. Consider Example 3.2.1 where $(X_i \mid \theta) \sim \mathcal{N}(\theta, \sigma_0^2)$, σ_0^2 known, and $\pi(\theta)$ is $\mathcal{N}(\eta_0, \tau^2)$, τ^2 known. A sufficient statistic based on the observables $X_1, \ldots, X_n$ is $T = \bar{X} = n^{-1} \sum_{i=1}^{n} X_i$, and it is enough to derive the marginal distribution of $Y = X_{n+1}$ from the joint distribution of $\bar{X}$, X_{n+1} and θ, where $\bar{X}$ and X_{n+1} are independent. Note that

$$E[(X_{n+1} - \theta)\theta] = E\{E(X_{n+1} - \theta)\theta \mid \theta = \theta\} = 0.$$

Thus, $X_{n+1} - \theta$ and θ are uncorrelated and, by Theorem B.4.1, independent.

To obtain the predictive distribution, note that given $\bar{X} = t$, $X_{n+1} - \theta$ and θ are still uncorrelated and independent. Thus, if we let $\mathcal{L}$ denote "distribution of," then

$$\mathcal{L}\{X_{n+1} \mid \bar{X} = t\} = \mathcal{L}\{(X_{n+1} - \theta) + \theta \mid \bar{X} = t\} = \mathcal{N}(\widehat{\mu}_B, \sigma_0^2 + \widehat{\sigma}_B^2)$$

where, from Example 4.7.1,

$$\widehat{\sigma}_B^2 = \frac{1}{\frac{n}{\sigma_0^2} + \frac{1}{\tau^2}}, \quad \widehat{\mu}_B = (\widehat{\sigma}_B^2/\tau^2)\eta_0 + (n\widehat{\sigma}_B^2/\sigma_0^2)\bar{x}.$$

It follows that a level $(1 - \alpha)$ Bayesian prediction interval for Y is $[Y_B^-, Y_B^+]$ with

$$Y_B^\pm = \widehat{\mu}_B \pm z\left(1 - \tfrac{1}{2}\alpha\right)\sqrt{\sigma_0^2 + \widehat{\sigma}_B^2}. \tag{4.8.3}$$

To consider the frequentist properties of the Bayesian prediction interval (4.8.3) we compute its probability limit under the assumption that $X_1, \ldots, X_n$ are i.i.d. $\mathcal{N}(\theta, \sigma_0^2)$. Because $\widehat{\sigma}_B^2 \to 0$, $(n\widehat{\sigma}_B^2/\sigma_0^2) \to 1$, and $\bar{X} \xrightarrow{P} \theta$ as $n \to \infty$, we find that the interval (4.8.3) converges in probability to $\theta \pm z\left(1 - \tfrac{1}{2}\alpha\right)\sigma_0$ as $n \to \infty$. This is the same as the probability limit of the frequentist interval (4.8.1). $\square$

The posterior predictive distribution is also used to check whether the model and the prior give a reasonable description of the uncertainty in a study (see Box, 1983).

Summary. We consider intervals based on observable random variables that contain an unobservable random variable with probability at least $(1-\alpha)$. In the case of a normal sample of size $n+1$ with only n variables observable, we construct the Student t prediction interval for the unobservable variable. For a sample of size $n + 1$ from a continuous distribution we show how the order statistics can be used to give a distribution-free prediction interval. The Bayesian formulation is based on the posterior predictive distribution which is the conditional distribution of the unobservable variable given the observable variables. The Bayesian prediction interval is derived for the normal model with a normal prior.

4.9   LIKELIHOOD RATIO PROCEDURES

4.9.1   Introduction

Up to this point, the results and examples in this chapter deal mostly with one-parameter problems in which it sometimes is possible to find optimal procedures. However, even in the case in which θ is one-dimensional, optimal procedures may not exist. For instance, if $X_1, \ldots, X_n$ is a sample from a $\mathcal{N}(\mu, \sigma^2)$ population with σ^2 known, there is no UMP test for testing $H : \mu = \mu_0$ vs $K : \mu \neq \mu_0$. To see this, note that it follows from Example 4.2.1 that if $\mu_1 > \mu_0$, the MP level α test $\delta_\alpha(\mathbf{X})$ rejects H for $T > z(1 - \frac{1}{2}\alpha)$, where $T = \sqrt{n}(\bar{X} - \mu_0)/\sigma$. On the other hand, if $\mu_1 < \mu_0$, the MP level α test $\varphi_\alpha(\mathbf{X})$ rejects H if $T \leq z(\alpha)$. Because $\delta_\alpha(\mathbf{x}) \neq \varphi_\alpha(\mathbf{x})$, by the uniqueness of the NP test (Theorem 4.2.1(c)), there can be no UMP test of $H : \mu = \mu_0$ vs $H : \mu \neq \mu_0$.

In this section we introduce intuitive and efficient procedures that can be used when no optimal methods are available and that are natural for multidimensional parameters. The efficiency is in an approximate sense that will be made clear in Chapters 5 and 6. We start with a generalization of the Neyman-Pearson statistic $p(x, \theta_1)/p(x, \theta_0)$. Suppose that $\mathbf{X} = (X_1, \ldots, X_n)$ has density or frequency function $p(\mathbf{x}, \theta)$ and we wish to test $H : \theta \in \Theta_0$ vs $K : \theta \in \Theta_1$. The test statistic we want to consider is the *likelihood ratio* given by

$$L(\mathbf{x}) = \frac{\sup\{p(\mathbf{x}, \theta) : \theta \in \Theta_1\}}{\sup\{p(\mathbf{x}, \theta) : \theta \in \Theta_0\}}.$$

Tests that reject H for large values of $L(\mathbf{x})$ are called *likelihood ratio tests*.

To see that this is a plausible statistic, recall from Section 2.2.2 that we think of the likelihood function $L(\theta, \mathbf{x}) = p(\mathbf{x}, \theta)$ as a measure of how well θ "explains" the given sample $\mathbf{x} = (x_1, \ldots, x_n)$. So, if $\sup\{p(\mathbf{x}, \theta) : \theta \in \Theta_1\}$ is large compared to $\sup\{p(\mathbf{x}, \theta) : \theta \in \Theta_0\}$, then the observed sample is best explained by some $\theta \in \Theta_1$, and conversely. Also note that $L(\mathbf{x})$ coincides with the optimal test statistic $p(x, \theta_1)/p(x, \theta_0)$ when $\Theta_0 = \{\theta_0\}, \Theta_1 = \{\theta_1\}$. In particular cases, and for large samples, likelihood ratio tests have weak optimality properties to be discussed in Chapters 5 and 6.

In the cases we shall consider, $p(\mathbf{x}, \theta)$ is a continuous function of θ and Θ_0 is of smaller dimension than $\Theta = \Theta_0 \cup \Theta_1$ so that the likelihood ratio equals the test statistic

$$\lambda(\mathbf{x}) = \frac{\sup\{p(\mathbf{x}, \theta) : \theta \in \Theta\}}{\sup\{p(\mathbf{x}, \theta) : \theta \in \Theta_0\}} \tag{4.9.1}$$

whose computation is often simple. Note that in general

$$\lambda(\mathbf{x}) = \max(L(\mathbf{x}), 1).$$

We are going to derive likelihood ratio tests in several important testing problems. Although the calculations differ from case to case, the basic steps are always the same.

1. Calculate the MLE $\widehat{\theta}$ of θ.

2. Calculate the MLE $\widehat{\theta}_0$ of θ where θ may vary only over Θ_0.

3. Form $\lambda(\mathbf{x}) = p(\mathbf{x}, \widehat{\theta})/p(\mathbf{x}, \widehat{\theta}_0)$.

4. Find a function h that is strictly increasing on the range of λ such that $h(\lambda(\mathbf{X}))$ has a simple form and a tabled distribution under H. Because $h(\lambda(\mathbf{X}))$ is equivalent to $\lambda(\mathbf{X})$, we specify the size α likelihood ratio test through the test statistic $h(\lambda(\mathbf{X}))$ and its $(1 - \alpha)$th quantile obtained from the table.

We can also invert families of likelihood ratio tests to obtain what we shall call *likelihood confidence regions, bounds,* and so on. For instance, we can invert the family of size α likelihood ratio tests of the point hypothesis $H : \theta = \theta_0$ and obtain the level $(1 - \alpha)$ confidence region

$$C(\mathbf{x}) = \{\theta : p(\mathbf{x}, \theta) \geq [c(\theta)]^{-1} \sup_\theta p(\mathbf{x}, \theta)\} \tag{4.9.2}$$

where $\sup_\theta$ denotes sup over $\theta \in \Theta$ and the critical constant $c(\theta)$ satisfies

$$P_{\theta_0} \left[\frac{\sup_\theta p(\mathbf{X}, \theta)}{p(\mathbf{X}, \theta_0)} \geq c(\theta_0) \right] = \alpha.$$

It is often approximately true (see Chapter 6) that $c(\theta)$ is independent of θ. In that case, $C(\mathbf{x})$ is just the set of all θ whose likelihood is on or above some fixed value dependent on the data. An example is discussed in Section 4.9.2.

This section includes situations in which $\theta = (\theta_1, \theta_2)$ where θ_1 is the parameter of interest and θ_2 is a nuisance parameter. We shall obtain likelihood ratio tests for hypotheses of the form $H : \theta_1 = \theta_{10}$, which are composite because θ_2 can vary freely. The family of such level α likelihood ratio tests obtained by varying θ_{10} can also be inverted and yield confidence regions for θ_1. To see how the process works we refer to the specific examples in Sections 4.9.2–4.9.5.

4.9.2   Tests for the Mean of a Normal Distribution-Matched Pair Experiments

Suppose $X_1, \dots, X_n$ form a sample from a $\mathcal{N}(\mu, \sigma^2)$ population in which both μ and σ^2 are unknown. An important class of situations for which this model may be appropriate occurs in *matched pair experiments.* Here are some examples. Suppose we want to study the effect of a treatment on a population of patients whose responses are quite variable because the patients differ with respect to age, diet, and other factors. We are interested in expected differences in responses due to the treatment effect. In order to reduce differences due to the extraneous factors, we consider pairs of patients matched so that within each pair the patients are as alike as possible with respect to the extraneous factors. We can regard twins as being matched pairs. After the matching, the experiment proceeds as follows. In the ith pair one patient is picked at random (i.e., with probability $\frac{1}{2}$) and given the treatment, while the second patient serves as control and receives a placebo. Response measurements are taken on the treated and control members of each pair.

Studies in which subjects serve as their own control can also be thought of as matched pair experiments. That is, we measure the response of a subject when under treatment and when not under treatment. Examples of such measurements are hours of sleep when receiving a drug and when receiving a placebo, sales performance before and after a course in salesmanship, mileage of cars with and without a certain ingredient or adjustment, and so on.

Let X_i denote the difference between the treated and control responses for the ith pair. If the treatment and placebo have the same effect, the difference X_i has a distribution that is symmetric about zero. To this we have added the normality assumption. The test we derive will still have desirable properties in an approximate sense to be discussed in Chapter 5 if the normality assumption is not satisfied. Let $\mu = E(X_1)$ denote the mean difference between the response of the treated and control subjects. We think of μ as representing the treatment effect. Our null hypothesis of no treatment effect is then $H : \mu = 0$. However, for the purpose of referring to the duality between testing and confidence procedures, we test $H : \mu = \mu_0$, where we think of μ_0 as an established standard for an old treatment.

Two-Sided Tests

We begin by considering $K : \mu \neq \mu_0$. This corresponds to the alternative "The treatment has some effect, good or bad." However, as discussed in Section 4.5, the test can be modified into a three-decision rule that decides whether there is a significant positive or negative effect.

Form of the Two-Sided Tests

Let $\theta = (\mu, \sigma^2)$, $\Theta_0 = \{(\mu, \sigma^2) : \mu = \mu_0\}$.
Under our assumptions,

$$p(\mathbf{x}, \theta) = (2\pi\sigma^2)^{-n/2} \exp\left\{-\frac{1}{2\sigma^2}\sum_{i=1}^{n}(x_i - \mu)^2\right\}.$$

The problem of finding the supremum of $p(\mathbf{x}, \theta)$ was solved in Example 2.2.9. We found that

$$\sup\{p(\mathbf{x}, \theta) : \theta \in \Theta\} = p(\mathbf{x}, \widehat{\theta}),$$

where

$$\widehat{\theta} = (\bar{x}, \widehat{\sigma}^2) = \left(\frac{1}{n}\sum_{i=1}^{n}x_i, \frac{1}{n}\sum_{i=1}^{n}(x_i - \bar{x})^2,\right)$$

is the maximum likelihood estimate of θ.
     Finding $\sup\{p(\mathbf{x}, \theta) : \theta \in \Theta_0\}$ boils down to finding the maximum likelihood estimate $\widehat{\sigma}_0^2$ of σ^2 when $\mu = \mu_0$ is known and then evaluating $p(\mathbf{x}, \theta)$ at $(\mu_0, \widehat{\sigma}_0^2)$. The likelihood equation is

$$\frac{\partial}{\partial\sigma^2}\log p(\mathbf{x}, \theta) = \frac{1}{2}\left[\frac{1}{\sigma^4}\sum_{i=1}^{n}(x_i - \mu_0)^2 - \frac{n}{\sigma^2}\right] = 0,$$

which has the immediate solution

$$\widehat{\sigma}_0^2 = \frac{1}{n}\sum_{i=1}^{n}(x_i - \mu_0)^2.$$

By Theorem 2.3.1, $\widehat{\sigma}_0^2$ gives the maximum of $p(\mathbf{x}, \theta)$ for $\theta \in \Theta_0$. The test statistic $\lambda(\mathbf{x})$ is equivalent to $\log \lambda(\mathbf{x})$, which thus equals

$$
\begin{aligned}
\log \lambda(\mathbf{x}) &= \log p(\mathbf{x}, \widehat{\theta}) - \log p(\mathbf{x}, (\mu_0, \widehat{\sigma}_0^2)) \\
&= \left\{ -\frac{n}{2}[(\log 2\pi) + (\log \widehat{\sigma}^2)] - \frac{n}{2} \right\} - \left\{ -\frac{n}{2}[(\log 2\pi) + (\log \widehat{\sigma}_0^2)] - \frac{n}{2} \right\} \\
&= \frac{n}{2} \log(\widehat{\sigma}_0^2/\widehat{\sigma}^2).
\end{aligned}
$$

Our test rule, therefore, rejects H for large values of $(\widehat{\sigma}_0^2/\widehat{\sigma}^2)$. To simplify the rule further we use the following equation, which can be established by expanding both sides.

$$
\widehat{\sigma}_0^2 = \widehat{\sigma}^2 + (\bar{x} - \mu_0)^2
$$

Therefore,

$$
(\widehat{\sigma}_0^2/\widehat{\sigma}^2) = 1 + (\bar{x} - \mu_0)^2/\widehat{\sigma}^2.
$$

Because $s^2 = (n-1)^{-1} \sum (x_i - \bar{x})^2 = n\widehat{\sigma}^2/(n-1)$, $\widehat{\sigma}_0^2/\widehat{\sigma}^2$ is a monotone increasing function of $|T_n|$ where

$$
T_n = \frac{\sqrt{n}(\bar{x} - \mu_0)}{s}.
$$

Therefore, the likelihood ratio tests reject for large values of $|T_n|$. Because T_n has a T distribution under H (see Example 4.4.1), the size α critical value is $t_{n-1}(1 - \frac{1}{2}\alpha)$ and we can use calculators or software that gives quantiles of the t distribution, or Table III, to find the critical value. For instance, suppose $n = 25$ and we want $\alpha = 0.05$. Then we would reject H if, and only if, $|T_n| \geq 2.064$.

One-Sided Tests

The two-sided formulation is natural if two treatments, A and B, are considered to be equal before the experiment is performed. However, if we are comparing a treatment and control, the relevant question is whether the treatment creates an improvement. Thus, the testing problem $H : \mu \leq \mu_0$ versus $K : \mu > \mu_0$ (with $\mu_0 = 0$) is suggested. The statistic T_n is equivalent to the likelihood ratio statistic λ for this problem. A proof is sketched in Problem 4.9.2. In Problem 4.9.11 we argue that $P_\delta[T_n \geq t]$ is increasing in δ, where $\delta = (\mu - \mu_0)/\sigma$. Therefore, the test that rejects H for

$$
T_n \geq t_{n-1}(1 - \alpha),
$$

is of size α for $H : \mu \leq \mu_0$. Similarly, the size α likelihood ratio test for $H : \mu \geq \mu_0$ versus $K : \mu < \mu_0$ rejects H if, and only if,

$$
T_n \leq t_{n-1}(\alpha).
$$

Power Functions

To discuss the power of these tests, we need to introduce the *noncentral t distribution* with k degrees of freedom and noncentrality parameter δ. This distribution, denoted by $\mathcal{T}_{k,\delta}$, is by definition the distribution of $Z/\sqrt{V/k}$ where Z and V are independent and have $\mathcal{N}(\delta, 1)$ and χ^2_k distributions, respectively. The density of $Z/\sqrt{V/k}$ is given in Problem 4.9.12. To derive the distribution of T_n, note that from Section B.3, we know that $\sqrt{n}(\bar{X} - \mu)/\sigma$ and $(n-1)s^2/\sigma^2$ are independent and that $(n-1)s^2/\sigma^2$ has a χ^2_{n-1} distribution. Because $E[\sqrt{n}(\bar{X} - \mu_0)/\sigma] = \sqrt{n}(\mu - \mu_0)/\sigma$ and

$$\text{Var}(\sqrt{n}(\bar{X} - \mu_0)/\sigma) = 1,$$

$\sqrt{n}(\bar{X} - \mu_0)/\sigma$ has $\mathcal{N}(\delta, 1)$ distribution, with $\delta = \sqrt{n}(\mu - \mu_0)/\sigma$. Thus, the ratio

$$\frac{\sqrt{n}(\bar{X} - \mu_0)/\sigma}{\sqrt{\frac{(n-1)s^2/\sigma^2}{n-1}}} = T_n$$

has a $\mathcal{T}_{n-1,\delta}$ distribution, and the power can be obtained from computer software or tables of the noncentral t distribution. Note that the distribution of T_n depends on $\theta = (\mu, \sigma^2)$ only through δ.

The power functions of the one-sided tests are monotone in δ (Problem 4.9.11) just as the power functions of the corresponding tests of Example 4.2.1 are monotone in $\sqrt{n}\mu/\sigma$.

We can control both probabilities of error by selecting the sample size n large provided we consider alternatives of the form $|\delta| \geq \delta_1 > 0$ in the two-sided case and $\delta \geq \delta_1$ or $\delta \leq \delta_1$ in the one-sided cases. Computer software will compute n.

If we consider alternatives of the form $(\mu - \mu_0) \geq \Delta$, say, we can no longer control both probabilities of error by choosing the sample size. The reason is that, whatever be n, by making σ sufficiently large we can force the noncentrality parameter $\delta = \sqrt{n}(\mu - \mu_0)/\sigma$ as close to 0 as we please and, thus, bring the power arbitrarily close to α. We have met similar difficulties (Problem 4.4.7) when discussing confidence intervals. A Stein solution, in which we estimate σ for a first sample and use this estimate to decide how many more observations we need to obtain guaranteed power against all alternatives with $|\mu - \mu_0| \geq \Delta$, is possible (Lehmann, 1997, p. 260, Problem 17). With this solution, however, we may be required to take more observations than we can afford on the second stage.

Likelihood Confidence Regions

If we invert the two-sided tests, we obtain the confidence region

$$C(\mathbf{X}) = \{\mu : |\sqrt{n}(\bar{X} - \mu)/s| \leq t_{n-1}(1 - \tfrac{1}{2}\alpha)\}.$$

We recognize $C(\mathbf{X})$ as the confidence interval of Example 4.4.1. Similarly the one-sided tests lead to the lower and upper confidence bounds of Example 4.4.1.

Data Example

As an illustration of these procedures, consider the following data due to Cushny and Peebles (see Fisher, 1958, p. 121) giving the difference $B - A$ in sleep gained using drugs A and B on 10 patients. This is a matched pair experiment with each subject serving as its own control.

Patient i	1	2	3	4	5	6	7	8	9	10
A	0.7	-1.6	-0.2	-1.2	-0.1	3.4	3.7	0.8	0.0	2.0
B	1.9	0.8	1.1	0.1	-0.1	4.4	5.5	1.6	4.6	3.4
$B - A$	1.2	2.4	1.3	1.3	0.0	1.0	1.8	0.8	4.6	1.4

If we denote the difference as x's, then $\bar{x} = 1.58$, $s^2 = 1.513$, and $|T_n| = 4.06$. Because $t_9(0.995) = 3.25$, we conclude at the 1% level of significance that the two drugs are significantly different. The 0.99 confidence interval for the mean difference μ between treatments is [0.32, 2.84]. It suggests that not only are the drugs different but in fact B is better than A because no hypothesis $\mu = \mu' < 0$ is accepted at this level. (See also (4.5.3).)

4.9.3   Tests and Confidence Intervals for the Difference in Means of Two Normal Populations

We often want to compare two populations with distribution F and G on the basis of two independent samples $X_1, \ldots, X_{n_1}$ and $Y_1, \ldots, Y_{n_2}$, one from each population. For instance, suppose we wanted to test the effect of a certain drug on some biological variable (e.g., blood pressure). Then $X_1, \ldots, X_{n_1}$ could be blood pressure measurements on a sample of patients given a placebo, while $Y_1, \ldots, Y_{n_2}$ are the measurements on a sample given the drug. For quantitative measurements such as blood pressure, height, weight, length, volume, temperature, and so forth, it is usually assumed that $X_1, \ldots, X_{n_1}$ and $Y_1, \ldots, Y_{n_2}$ are independent samples from $\mathcal{N}(\mu_1, \sigma^2)$ and $\mathcal{N}(\mu_2, \sigma^2)$ populations, respectively.

The preceding assumptions were discussed in Example 1.1.3. A discussion of the consequences of the violation of these assumptions will be postponed to Chapters 5 and 6.

Tests

We first consider the problem of testing $H : \mu_1 = \mu_2$ versus $K : \mu_1 \neq \mu_2$. In the control versus treatment example, this is the problem of determining whether the treatment has any effect.

Let $\theta = (\mu_1, \mu_2, \sigma^2)$. Then $\Theta_0 = \{\theta : \mu_1 = \mu_2\}$ and $\Theta_1 = \{\theta : \mu_1 \neq \mu_2\}$. The log of the likelihood of $(\mathbf{X}, \mathbf{Y}) = (X_1, \ldots, X_{n_1}, Y_1, \ldots, Y_{n_2})$ is

$$\log p(\mathbf{x}, \mathbf{y}, \theta) = -(n/2)\log(2\pi\sigma^2) - \frac{1}{2\sigma^2}\left(\sum_{i=1}^{n_1}(x_i - \mu_1)^2 + \sum_{j=1}^{n_2}(y_j - \mu_2)^2\right)$$

where $n = n_1 + n_2$. As in Section 4.9.2 the likelihood function and its log are maximized over Θ by the maximum likelihood estimate $\hat{\theta}$. In Problem 4.9.6, it is shown that $\hat{\theta} =$

$(\bar{X}, \bar{Y}, \tilde{\sigma}^2)$, where

$$\tilde{\sigma}^2 = \frac{1}{n} \left[\sum_{i=1}^{n_1} (X_i - \bar{X})^2 + \sum_{j=1}^{n_2} (Y_j - \bar{Y})^2 \right]$$

When $\mu_1 = \mu_2 = \mu$, our model reduces to the one-sample model of Section 4.9.2. Thus, the maximum of p over Θ_0 is obtained for $\theta = (\hat{\mu}, \hat{\mu}, \tilde{\sigma}_0^2)$, where

$$\hat{\mu} = \frac{1}{n} \left[\sum_{i=1}^{n_1} X_i + \sum_{j=1}^{n_2} Y_j \right]$$

and

$$\tilde{\sigma}_0^2 = \frac{1}{n} \left[\sum_{i=1}^{n_1} (X_i - \hat{\mu})^2 + \sum_{j=1}^{n_2} (Y_j - \hat{\mu})^2 \right].$$

If we use the identities

$$\frac{1}{n} \sum_{i=1}^{n_1} (X_i - \hat{\mu})^2 = \frac{1}{n} \sum_{i=1}^{n_1} (X_i - \bar{X})^2 + \frac{n_1}{n} (\bar{X} - \hat{\mu})^2$$

$$\frac{1}{n} \sum_{i=1}^{n_2} (Y_i - \hat{\mu})^2 = \frac{1}{n} \sum_{i=1}^{n_1} (Y_i - \bar{Y})^2 + \frac{n_2}{n} (\bar{Y} - \hat{\mu})^2$$

obtained by writing $[X_i - \hat{\mu}]^2 = [(X_i - \bar{X}) + (\bar{X} - \hat{\mu})]^2$ and expanding, we find that the log likelihood ratio statistic

$$\log \lambda(\mathbf{x}, \mathbf{y}) = \frac{n}{2} \log(\tilde{\sigma}_0^2 / \tilde{\sigma}^2)$$

is equivalent to the test statistic $|T|$ where

$$T = \sqrt{\frac{n_1 n_2}{n}} \left(\frac{\bar{Y} - \bar{X}}{s} \right)$$

and

$$s^2 = n\tilde{\sigma}^2 / (n-2) = \frac{1}{n-2} \left[\sum_{i=1}^{n_1} (X_i - \bar{X})^2 + \sum_{j=1}^{n_2} (Y_j - \bar{Y})^2 \right].$$

To complete specification of the size α likelihood ratio test we show that T has $\mathcal{T}_{n-2}$ distribution when $\mu_1 = \mu_2$. By Theorem B.3.3

$$\frac{1}{\sigma} \bar{X}, \frac{1}{\sigma} \bar{Y}, \frac{1}{\sigma^2} \sum_{i=1}^{n_1} (X_i - \bar{X})^2 \text{ and } \frac{1}{\sigma^2} \sum_{j=1}^{n_2} (Y_j - \bar{Y})^2$$

are independent and distributed as $\mathcal{N}(\mu_1/\sigma, 1/n_1), \mathcal{N}(\mu_2/\sigma, 1/n_2), \chi^2_{n_1-1}, \chi^2_{n_2-1}$, respectively. We conclude from this remark and the additive property of the χ^2 distribution that $(n-2)s^2/\sigma^2$ has a χ^2_{n-2} distribution, and that $\sqrt{n_1 n_2/n}(\bar{Y} - \bar{X})/\sigma$ has a $\mathcal{N}(0, 1)$ distribution and is independent of $(n-2)s^2/\sigma^2$. Therefore, by definition, $T \sim \mathcal{T}_{n-2}$ under H and the resulting *two-sided, two-sample* t test rejects if, and only if,

$$|T| \geq t_{n-2}(1 - \frac{1}{2}\alpha)$$

As usual, corresponding to the two-sided test, there are two one-sided tests with critical regions,

$$T \geq t_{n-2}(1 - \alpha)$$

for $H : \mu_2 \leq \mu_1$ and

$$T \leq t_{n-2}(\alpha)$$

for $H : \mu_1 \leq \mu_2$.

We can show that these tests are likelihood ratio tests for these hypotheses. It is also true that these procedures are of size α for their respective hypotheses. As in the one-sample case, this follows from the fact that, if $\mu_1 \neq \mu_2$, T has a noncentral t distribution with noncentrality parameter,

$$\delta_2 = E\left(\sqrt{n_1 n_2/n}\frac{(\bar{Y} - \bar{X})}{\sigma}\right) = \sqrt{n_1 n_2/n}\frac{(\mu_2 - \mu_1)}{\sigma}.$$

Confidence Intervals

To obtain confidence intervals for $\mu_2 - \mu_1$ we naturally look at likelihood ratio tests for the family of testing problems $H : \mu_2 - \mu_1 = \Delta$ versus $K : \mu_2 - \mu_1 \neq \Delta$. As for the special case $\Delta = 0$, we find a simple equivalent statistic $|T(\Delta)|$ where

$$T(\Delta) = \sqrt{n_1 n_2/n}(\bar{Y} - \bar{X} - \Delta)/s.$$

If $\mu_2 - \mu_1 = \Delta$, $T(\Delta)$ has a $\mathcal{T}_{n-2}$ distribution and inversion of the tests leads to the interval

$$\bar{Y} - \bar{X} \pm t_{n-2}(1 - \frac{1}{2}\alpha)s\sqrt{n/n_1 n_2}, \qquad (4.9.3)$$

for $\mu_2 - \mu_1$. Similarly, one-sided tests lead to the upper and lower endpoints of the interval as $1 - \frac{1}{2}\alpha$ likelihood confidence bounds.

Data Example

As an illustration, consider the following experiment designed to study the permeability (tendency to leak water) of sheets of building material produced by two different machines. From past experience, it is known that the log of permeability is approximately normally distributed and that the variability from machine to machine is the same. The results in terms of logarithms were (from Hald, 1952, p. 472)

x (machine 1)	1.845	1.790	2.042
y (machine 2)	1.583	1.627	1.282

We test the hypothesis H of no difference in expected log permeability. H is rejected if $|T| \geq t_{n-2}(1 - \frac{1}{2}\alpha)$. Here $\bar{y} - \bar{x} = -0.395, s^2 = 0.0264$, and $T = -2.977$. Because $t_4(0.975) = 2.776$, we conclude at the 5% level of significance that there is a significant difference between the expected log permeability for the two machines. The level 0.95 confidence interval for the difference in mean log permeability is

$$-0.395 \pm 0.368.$$

On the basis of the results of this experiment, we would select machine 2 as producing the smaller permeability and, thus, the more waterproof material. Again we can show that the selection procedure based on the level $(1 - \alpha)$ confidence interval has probability at most $\frac{1}{2}\alpha$ of making the wrong selection.

4.9.4   The Two-Sample Problem with Unequal Variances

In two-sample problems of the kind mentioned in the introduction to Section 4.9.3, it may happen that the X's and Y's have different variances. For instance, a treatment that increases mean response may increase the variance of the responses. If normality holds, we are led to a model where $X_1, \ldots, X_{n_1}; Y_1, \ldots, Y_{n_2}$ are two independent $\mathcal{N}(\mu_1, \sigma_1^2)$ and $\mathcal{N}(\mu_2, \sigma_2^2)$ samples, respectively. As a first step we may still want to compare mean responses for the X and Y populations. This is the *Behrens-Fisher problem*.

Suppose first that σ_1^2 and σ_2^2 are known. The log likelihood, except for an additive constant, is

$$-\frac{1}{2\sigma_1^2} \sum_{i=1}^{n_1} (x_i - \mu_1)^2 - \frac{1}{2\sigma_2^2} \sum_{j=1}^{n_2} (y_j - \mu_2)^2.$$

The MLEs of μ_1 and μ_2 are, thus, $\widehat{\mu}_1 = \bar{x}$ and $\widehat{\mu}_2 = \bar{y}$ for $(\mu_1, \mu_2) \in R \times R$. When $\mu_1 = \mu_2 = \mu$, setting the derivative of the log likelihood equal to zero yields the MLE

$$\widehat{\mu} = \frac{\sum_{i=1}^{n_1} x_i + \gamma \sum_{j=1}^{n_2} y_j}{n_1 + \gamma n_2}$$

where $\gamma = \sigma_1^2/\sigma_2^2$. It follows that

$$
\begin{aligned}
\lambda(\mathbf{x}, \mathbf{y}) &= \frac{p(\mathbf{x}, \mathbf{y}, \bar{x}, \bar{y})}{p(\mathbf{x}, \mathbf{y}, \widehat{\mu}, \widehat{\mu})} \\
&= \exp\left\{\frac{1}{2\sigma_1^2}\left[\sum_{i=1}^{n_1}(x_i - \widehat{\mu})^2 - \sum_{i=1}^{n_1}(x_i - \bar{x})^2\right]\right. \\
&\quad + \left.\frac{1}{2\sigma_2^2}\left[\sum_{j=1}^{n_2}(y_j - \widehat{\mu})^2 - \sum_{j=1}^{n_2}(y_j - \bar{y})^2\right]\right\}.
\end{aligned}
$$

By writing

$$
\begin{aligned}
\sum_{i=1}^{n_1}(x_i - \widehat{\mu})^2 &= \sum_{i=1}^{n_1}(x_i - \bar{x})^2 + n_1(\widehat{\mu} - \bar{x})^2 \\
\sum_{j=1}^{n_2}(y_i - \widehat{\mu})^2 &= \sum_{j=1}^{n_2}(y_i - \bar{y})^2 + n_2(\widehat{\mu} - \bar{y})^2
\end{aligned}
$$

we obtain

$$
\lambda(\mathbf{x}, \mathbf{y}) = \exp\left\{\frac{n_1}{2\sigma_1^2}(\widehat{\mu} - \bar{x})^2 + \frac{n_2}{2\sigma_2^2}(\widehat{\mu} - \bar{y})^2\right\}.
$$

Next we compute

$$
\begin{aligned}
\widehat{\mu} - \bar{x} &= \frac{n_1\sum_{i=1}^{n_1}x_i + n_1\gamma\sum_{j=1}^{n_2}y_j - n_1\sum_{i=1}^{n_1}x_i - \gamma n_2\sum_{j=1}^{n_2}y_j}{n_1(n_1 + \gamma n_2)} \\
&= \gamma n_2(\bar{y} - \bar{x})/(n_1 + \gamma n_2).
\end{aligned}
$$

Similarly, $\widehat{\mu} - \bar{y} = n_2(\bar{x} - \bar{y})/(n_1 + \gamma n_2)$. It follows that the likelihood ratio test is equivalent to the statistic $|D|/\sigma_D$, where $D = \bar{Y} - \bar{X}$ and σ_D^2 is the variance of D, that is

$$
\sigma_D^2 = \text{Var}(\bar{X}) + \text{Var}(\bar{Y}) = \frac{\sigma_1^2}{n_1} + \frac{\sigma_2^2}{n_2}.
$$

Thus, D/σ_D has a $\mathcal{N}(\Delta, 1)$ distribution, where $\Delta = \mu_2 - \mu_1$. Because σ_D^2 is unknown, it must be estimated. An unbiased estimate is

$$
s_D^2 = \frac{s_1^2}{n_1} + \frac{s_2^2}{n_2}.
$$

It is natural to try and use D/s_D as a test statistic for the one-sided hypothesis $H : \mu_2 \le \mu_1$, $|D|/s_D$ as a test statistic for $H : \mu_1 = \mu_2$, and more generally $(D - \Delta)/s_D$ to generate confidence procedures. Unfortunately the distribution of $(D - \Delta)/s_D$ depends on σ_1^2/σ_2^2 for fixed n_1, n_2. For large n_1, n_2, by Slutsky's theorem and the central limit theorem, $(D - \Delta)/s_D$ has approximately a standard normal distribution (Problem 5.3.28). For small and moderate n_1, n_2 an approximation to the distribution of $(D - \Delta)/s_D$ due to Welch (1949) works well.

Let $c = s_1^2/n_1 s_D^2$. Then Welch's approximation is $\mathcal{T}_k$ where

$$k = \left[\frac{c^2}{n_1 - 1} + \frac{(1 - c)^2}{n_2 - 1} \right]^{-1}$$

When k is not an integer, the critical value is obtained by linear interpolation in the t tables or using computer software. The tests and confidence intervals resulting from this approximation are called Welch's solutions to the Behrens-Fisher problem. Wang (1971) has shown the approximation to be very good for $\alpha = 0.05$ and $\alpha = 0.01$, the maximum error in size being bounded by 0.003.

Note that Welch's solution works whether the variances are equal or not. The LR procedure derived in Section 4.9.3, which works well if the variances are equal or $n_1 = n_2$, can unfortunately be very misleading if $\sigma_1^2 \neq \sigma_2^2$ and $n_1 \neq n_2$. See Figure 5.3.3 and Problem 5.3.28.

4.9.5   Likelihood Ratio Procedures for Bivariate Normal Distributions

If n subjects (persons, machines, fields, mice, etc.) are sampled from a population and two numerical characteristics are measured on each case, then we end up with a bivariate random sample $(X_1, Y_1), \ldots, (X_n, Y_n)$. Empirical data sometimes suggest that a reasonable model is one in which the two characteristics (X, Y) have a joint bivariate normal distribution, $\mathcal{N}(\mu_1, \mu_2, \sigma_1^2, \sigma_2^2, \rho)$, with $\sigma_1^2 > 0, \sigma_2^2 > 0$.

Testing Independence, Confidence Intervals for ρ

The question "Are two random variables X and Y independent?" arises in many statistical studies. Some familiar examples are: $X = $ weight, $Y = $ blood pressure; $X = $ test score on mathematics exam, $Y = $ test score on English exam; $X = $ percentage of fat in diet, $Y = $ cholesterol level in blood; $X = $ average cigarette consumption per day in grams, $Y = $ age at death. If we have a sample as before and assume the bivariate normal model for (X, Y), our problem becomes that of testing $H : \rho = 0$.

Two-Sided Tests

If we are interested in all departures from H equally, it is natural to consider the two-sided alternative $K : \rho \neq 0$. We derive the likelihood ratio test for this problem. Let $\theta = (\mu_1, \mu_2, \sigma_1^2, \sigma_2^2, \rho)$. Then $\Theta_0 = \{\theta : \rho = 0\}$ and $\Theta_1 = \{\theta : \rho \neq 0\}$. From (B.4.9), the log of the likelihood function of $\mathbf{X} = ((X_1, Y_1), \ldots, (X_n, Y_n))$ is

$$\log p(\mathbf{x}, \theta) = -n[\log(2\pi\sigma_1\sigma_2\sqrt{1 - \rho^2})]$$
$$- \frac{1}{2(1 - \rho^2)} \left[\frac{\sum_{i=1}^n (x_i - \mu_1)^2}{\sigma_1^2} - \frac{2\rho \sum_{i=1}^n (x_i - \mu_1)(y_i - \mu_2)}{\sigma_1\sigma_2} + \frac{\sum_{i=1}^n (y_i - \mu_2)^2}{\sigma_2^2} \right].$$

The unrestricted maximum likelihood estimate $\widehat{\theta}$ was given in Problem 2.3.13 as $(\bar{x}, \bar{y}, \widehat{\sigma}_1^2, \widehat{\sigma}_2^2, \widehat{\rho})$, where

$$\widehat{\sigma}_1^2 = \frac{1}{n}\sum_{i=1}^{n}(x_i - \bar{x})^2, \widehat{\sigma}_2^2 = \frac{1}{n}\sum_{i=1}^{n}(y_i - \bar{y})^2,$$

$$\widehat{\rho} = \left[\sum_{i=1}^{n}(x_i - \bar{x})(y_i - \bar{y})\right] / n\widehat{\sigma}_1\widehat{\sigma}_2.$$

$\widehat{\rho}$ is called the *sample correlation coefficient* and satisfies $-1 \leq \widehat{\rho} \leq 1$ (Problem 2.1.8). When $\rho = 0$ we have two independent samples, and $\widehat{\theta}_0$ can be obtained by separately maximizing the likelihood of $X_1, \ldots, X_n$ and that of $Y_1, \ldots, Y_n$. We have $\widehat{\theta}_0 = (\bar{x}, \bar{y}, \widehat{\sigma}_1^2, \widehat{\sigma}_2^2, 0)$ and the log of the likelihood ratio statistic becomes

$$
\begin{aligned}
\log \lambda(\mathbf{x}) &= \{\log p(\mathbf{x}, \widehat{\theta})\} - \{\log p(\mathbf{x}, \widehat{\theta}_0)\} \\
&= \left\{-n[\log(2\pi\widehat{\sigma}_1\widehat{\sigma}_2)] - \frac{n}{2}[\log(1 - \widehat{\rho}^2)] - n\right\} \\
&\quad \{-n[\log(2\pi\widehat{\sigma}_1\widehat{\sigma}_2)] - n\} \\
&= -\frac{n}{2}\log(1 - \widehat{\rho}^2).
\end{aligned}
\tag{4.9.4}
$$

Thus, $\log \lambda(\mathbf{x})$ is an increasing function of $\widehat{\rho}^2$, and the likelihood ratio tests reject H for large values of $|\widehat{\rho}|$.

To obtain critical values we need the distribution of $\widehat{\rho}$ or an equivalent statistic under H. Now,

$$\widehat{\rho} = \sum_{i=1}^{n}(U_i - \bar{U})(V_i - \bar{V}) / \left[\sum_{i=1}^{n}(U_i - \bar{U})^2\right]^{1/2}\left[\sum_{i=1}^{n}(V_i - \bar{V})^2\right]^{1/2}$$

where $U_i = (X_i - \mu_1)/\sigma_1, V_i = (Y_i - \mu_2)/\sigma_2$. Because $(U_1, V_1), \ldots, (U_n, V_n)$ is a sample from the $\mathcal{N}(0, 0, 1, 1, \rho)$ distribution, the distribution of $\widehat{\rho}$ depends on ρ only. If $\rho = 0$, then by Problem B.4.8,

$$T_n = \frac{\sqrt{n - 2}\widehat{\rho}}{\sqrt{1 - \widehat{\rho}^2}} \tag{4.9.5}$$

has a $\mathcal{T}_{n-2}$ distribution. Because $|T_n|$ is an increasing function of $|\widehat{\rho}|$, the two-sided likelihood ratio tests can be based on $|T_n|$ and the critical values obtained from Table II.

When $\rho = 0$, the distribution of $\widehat{\rho}$ is available on computer packages. There is no simple form for the distribution of $\widehat{\rho}$ (or T_n) when $\rho \neq 0$. A normal approximation is available. See Example 5.3.6.

Qualitatively, for any α, the power function of the LR test is symmetric about $\rho = 0$ and increases continuously from α to 1 as ρ goes from 0 to 1. Therefore, if we specify indifference regions, we can control probabilities of type II error by increasing the sample size.

One-Sided Tests

In many cases, only one-sided alternatives are of interest. For instance, if we want to decide whether increasing fat in a diet significantly increases cholesterol level in the blood, we would test $H : \rho = 0$ versus $K : \rho > 0$ or $H : \rho \leq 0$ versus $K : \rho > 0$. It can be shown that $\widehat{\rho}$ is equivalent to the likelihood ratio statistic for testing $H : \rho \leq 0$ versus $K : \rho > 0$ and similarly that $-\widehat{\rho}$ corresponds to the likelihood ratio statistic for $H : \rho \geq 0$ versus $K : \rho < 0$. We can show that $P_\theta[\widehat{\rho} \geq c]$ is an increasing function of ρ for fixed c (Problem 4.9.15). Therefore, we obtain size α tests for each of these hypotheses by setting the critical value so that the probability of type I error is α when $\rho = 0$. The power functions of these tests are monotone.

Confidence Bounds and Intervals

Usually testing independence is not enough and we want bounds and intervals for ρ giving us an indication of what departure from independence is present. To obtain lower confidence bounds, we can start by constructing size α likelihood ratio tests of $H : \rho = \rho_0$ versus $K : \rho > \rho_0$. These tests can be shown to be of the form "Accept if, and only if, $\widehat{\rho} \leq c(\rho_0)$" where $P_{\rho_0}[\widehat{\rho} \leq c(\rho_0)] = 1 - \alpha$. We obtain $c(\rho)$ either from computer software or by the approximation of Chapter 5. Because c can be shown to be monotone increasing in ρ, inversion of this family of tests leads to level $1 - \alpha$ lower confidence bounds. We can similarly obtain $1 - \alpha$ upper confidence bounds and, by putting two level $1 - \frac{1}{2}\alpha$ bounds together, we obtain a commonly used confidence interval for ρ. These intervals do not correspond to the inversion of the size α LR tests of $H : \rho = \rho_0$ versus $K : \rho \neq \rho_0$ but rather of the "equal-tailed" test that rejects if, and only if $\widehat{\rho} \geq d(\rho_0)$ or $\widehat{\rho} \leq c(\rho_0)$ where $P_{\rho_0}[\widehat{\rho} \geq d(\rho_0)] = P_{\rho_0}[\widehat{\rho} \leq c(\rho_0)] = 1 - \frac{1}{2}\alpha$. However, for large n the equal tails and LR confidence intervals approximately coincide with each other.

Data Example

As an illustration, consider the following bivariate sample of weights x_i of young rats at a certain age and the weight increase y_i during the following week. We want to know whether there is a correlation between the initial weights and the weight increase and formulate the hypothesis $H : \rho = 0$.

x_i	383.2	356.4	362.5	397.4	356.0	387.6	385.1	346.6	370.7
y_i	27.3	41.0	38.4	24.4	25.9	21.9	13.1	28.5	18.1

Here $\widehat{\rho} = 0.18$ and $T_n = 0.48$. Thus, by using the two-sided test and referring to the T_7 tables, we find that there is no evidence of correlation: the p-value is bigger than 0.75.

Summary. The likelihood ratio test statistic λ is the ratio of the maximum value of the likelihood under the general model to the maximum value of the likelihood under the model specified by the hypothesis. We find the likelihood ratio tests and associated confidence procedures for four classical normal models:

(1) Matched pair experiments in which differences are modeled as $\mathcal{N}(\mu, \sigma^2)$ and we test the hypothesis that the mean difference μ is zero. The likelihood ratio test is equivalent to the one-sample (Student) t test.

(2) Two-sample experiments in which two independent samples are modeled as coming from $\mathcal{N}(\mu_1, \sigma^2)$ and $\mathcal{N}(\mu_2, \sigma^2)$ populations, respectively. We test the hypothesis that the means are equal and find that the likelihood ratio test is equivalent to the two-sample (Student) t test.

(3) Two-sample experiments in which two independent samples are modeled as coming from $\mathcal{N}(\mu_1, \sigma_1^2)$ and $\mathcal{N}(\mu_2, \sigma_2^2)$ populations, respectively. When σ_1^2 and σ_2^2 are known, the likelihood ratio test is equivalent to the test based on $|\bar{Y} - \bar{X}|$. When σ_1^2 and σ_2^2 are unknown, we use $(\bar{Y} - \bar{X})/s_D$, where s_D is an estimate of the standard deviation of $D = \bar{Y} - \bar{X}$. Approximate critical values are obtained using Welch's t distribution approximation.

(4) Bivariate sampling experiments in which we have two measurements X and Y on each case in a sample of n cases. We test the hypothesis that X and Y are independent and find that the likelihood ratio test is equivalent to the test based on $|\hat{\rho}|$, where $\hat{\rho}$ is the sample correlation coefficient. We also find that the likelihood ratio statistic is equivalent to a t statistic with $n - 2$ degrees of freedom.

4.10   PROBLEMS AND COMPLEMENTS

Problems for Section 4.1

1. Suppose that $X_1, \ldots, X_n$ are independently and identically distributed according to the uniform distribution $U(0, \theta)$. Let $M_n = \max(X_1, \ldots, X_n)$ and let

$$\delta_c(X) = 1 \text{ if } M_n \geq c$$
$$= 0 \text{ otherwise.}$$

(a) Compute the power function of δ_c and show that it is a monotone increasing function of θ.

(b) In testing $H : \theta \leq \frac{1}{2}$ versus $K : \theta > \frac{1}{2}$, what choice of c would make δ_c have size exactly 0.05?

(c) Draw a rough graph of the power function of δ_c specified in (b) when $n = 20$.

(d) How large should n be so that the δ_c specified in (b) has power 0.98 for $\theta = \frac{3}{4}$?

(e) If in a sample of size $n = 20$, $M_n = 0.48$, what is the p-value?

2. Let $X_1, \ldots, X_n$ denote the times in days to failure of n similar pieces of equipment. Assume the model where $\mathbf{X} = (X_1, \ldots, X_n)$ is an $\mathcal{E}(\lambda)$ sample. Consider the hypothesis H that the mean life $1/\lambda = \mu \leq \mu_0$.

(a) Use the result of Problem B.3.4 to show that the test with critical region

$$[\bar{X} \geq \mu_0 x(1 - \alpha)/2n],$$

where $x(1 - \alpha)$ is the $(1 - \alpha)$th quantile of the χ^2_{2n} distribution, is a size α test.

(b) Give an expression of the power in terms of the χ^2_{2n} distribution.

(c) Use the central limit theorem to show that $\Phi[(\mu_0 z(\alpha)/\mu) + \sqrt{n}(\mu - \mu_0)/\mu]$ is an approximation to the power of the test in part (a). Draw a graph of the approximate power function.

Hint: Approximate the critical region by $[\bar{X} \geq \mu_0(1 + z(1 - \alpha)/\sqrt{n})]$

(d) The following are days until failure of air monitors at a nuclear plant. If $\mu_0 = 25$, give a normal approximation to the significance probability. Days until failure:

$$3\ 150\ 40\ 34\ 32\ 37\ 34\ 2\ 31\ 6\ 5\ 14\ 150\ 27\ 4\ 6\ 27\ 10\ 30\ 37$$

Is H rejected at level $\alpha = 0.05$?

3. Let $X_1, \ldots, X_n$ be a $\mathcal{P}(\theta)$ sample.

(a) Use the MLE $\bar{X}$ of θ to construct a level α test for $H : \theta \leq \theta_0$ versus $K : \theta > \theta_0$.

(b) Show that the power function of your test is increasing in θ.

(c) Give an approximate expression for the critical value if n is large and θ not too close to 0 or ∞. (Use the central limit theorem.)

4. Let $X_1, \ldots, X_n$ be a sample from a population with the Rayleigh density

$$f(x, \theta) = (x/\theta^2) \exp\{-x^2/2\theta^2\}, \ x > 0, \theta > 0.$$

(a) Construct a test of $H : \theta = 1$ versus $K : \theta > 1$ with approximate size α using a sufficient statistic for this model.

Hint: Use the central limit theorem for the critical value.

(b) Check that your test statistic has greater expected value under K than under H.

5. Show that if H is simple and the test statistic T has a continuous distribution, then the p-value $\alpha(T)$ has a uniform, $\mathcal{U}(0, 1)$, distribution.

Hint: See Problem B.2.12.

6. Suppose that $T_1, \ldots, T_r$ are independent test statistics for the same simple H and that each T_j has a continuous distribution, $j = 1, \ldots, r$. Let $\alpha(T_j)$ denote the p-value for T_j, $j = 1, \ldots, r$.

Show that, under H, $\widehat{T} = -2 \sum_{j=1}^{r} \log \alpha(T_j)$ has a χ^2_{2r} distribution.

Hint: See Problem B.3.4.

7. Establish (4.1.3). Assume that F_0 and F are continuous.

8. (a) Show that the power $P_F[D_n \geq k_\alpha]$ of the Kolmogorov test is bounded below by

$$\sup_x P_F[|\widehat{F}(x) - F_0(x)| \geq k_\alpha].$$

Hint: $D_n \geq |\widehat{F}(x) - F_0(x)|$ for each x.

(b) Suppose F_0 is $\mathcal{N}(0,1)$ and $F(x) = (1+\exp(-x/\tau))^{-1}$ where $\tau = \sqrt{3}/\pi$ is chosen so that $\int_{-\infty}^{\infty} x^2 dF(x) = 1$. (This is the logistic distribution with mean zero and variance 1.) Evaluate the bound $P_F(|\widehat{F}(x) - F_0(x)| \geq k_\alpha)$ for $\alpha = 0.10$, $n = 80$ and $x = 0.5$, 1, and 1.5 using the normal approximation to the binomial distribution of $n\widehat{F}(x)$ and the approximate critical value in Example 4.1.5.

(c) Show that if F and F_0 are continuous and $F \neq F_0$, then the power of the Kolmogorov test tends to 1 as $n \to \infty$.

9. Let $X_1, \ldots, X_n$ be i.i.d. with distribution function F and consider $H : F = F_0$. Suppose that the distribution $\mathcal{L}_0$ of the statistic $T = T(\mathbf{X})$ is continuous under H and that H is rejected for large values of T. Let $T^{(1)}, \ldots, T^{(B)}$ be B independent Monte Carlo simulated values of T. (In practice these can be obtained by drawing B independent samples $\mathbf{X}^{(1)}, \ldots, \mathbf{X}^{(B)}$ from F_0 on the computer and computing $T^{(j)} = T(\mathbf{X}^{(j)})$, $j = 1, \ldots, B$. Here, to get X with distribution F_0, generate a $\mathcal{U}(0,1)$ variable on the computer and set $X = F_0^{-1}(U)$ as in Problem B.2.12(b).) Next let $T_{(1)}, \ldots, T_{(B+1)}$ denote $T, T^{(1)}, \ldots, T^{(B)}$ ordered. Show that the test that rejects H iff $T \geq T_{(B+2-m)}$ has level $\alpha = m/(B+1)$.

Hint: If H is true $T(\mathbf{X}), T(\mathbf{X}^{(1)}), \ldots, T(\mathbf{X}^{(B)})$ is a sample of size $B+1$ from $\mathcal{L}_0$. Use the fact that $T(\mathbf{X})$ is equally likely to be any particular order statistic.

10. (a) Show that the statistic T_n of Example 4.1.6 is invariant under location and scale. That is, if $X_i' = (X_i - a)/b$, $b > 0$, then $T_n(\mathbf{X}') = T_n(\mathbf{X})$.

(b) Use part (a) to conclude that $\mathcal{L}_{\mathcal{N}(\mu,\sigma^2)}(T_n) = \mathcal{L}_{\mathcal{N}(0,1)}(T_n)$.

11. In Example 4.1.5, let $\psi(u)$ be a function from $(0,1)$ to $(0,\infty)$, and let $\alpha > 0$. Define the statistics

$$S_{\psi,\alpha} = \sup_x \psi(F_0(x))|\widehat{F}(x) - F_0(x)|^\alpha$$

$$T_{\psi,\alpha} = \sup_x \psi(\widehat{F}(x))|\widehat{F}(x) - F_0(x)|^\alpha$$

$$U_{\psi,\alpha} = \int \psi(F_0(x))|\widehat{F}(x) - F_0(x)|^\alpha dF_0(x)$$

$$V_{\psi,\alpha} = \int \psi(\widehat{F}(x))|\widehat{F}(x) - F_0(x)|^\alpha d\widehat{F}(x).$$

(a) For each of these statistics show that the distribution under H does not depend on F_0.

(b) When $\psi(u) = 1$ and $\alpha = 2$, $V_{\psi,\alpha}$ is called the Cramér–von Mises statistic. Express the Cramer–von Mises statistic as a sum.

(c) Are any of the four statistics in (a) invariant under location and scale. (See Problem 4.1.10.)

12. *Expected p-values.* Consider a test with critical region of the form $\{T \geq c\}$ for testing $H : \theta = \theta_0$ versus $K : \theta > \theta_0$. Without loss of generality, take $\theta_0 = 0$. Suppose that T has a continuous distribution F_θ, then the p-value is

$$U = 1 - F_0(T).$$

(a) Show that if the test has level α, the power is

$$\beta(\theta) = P(U \leq \alpha) = 1 - F_\theta(F_0^{-1}(1 - \alpha))$$

where $F_0^{-1}(u) = \inf\{t : F_0(t) \geq u\}$.

(b) Define the expected p-value as $EPV(\theta) = E_\theta U$. Let T_0 denote a random variable with distribution F_0, which is independent of T. Show that $EPV(\theta) = P(T_0 \geq T)$.
Hint: $P(T_0 \geq T) = \int P(T_0 \geq t \mid T = t) f_\theta(t) dt$ where $f_\theta(t)$ is the density of $F_\theta(t)$.

(c) Suppose that for each $\alpha \in (0, 1)$, the UMP test is of the form $1\{T \geq c\}$. Show that the $EPV(\theta)$ for $1\{T \geq c\}$ is uniformly minimal in $\theta > 0$ when compared to the $EPV(\theta)$ for any other test.
Hint: $P(T \leq t_0 \mid T_0 = t_0)$ is 1 minus the power of a test with critical value t_0.

(d) Consider the problem of testing $H : \mu = \mu_0$ versus $K : \mu > \mu_0$ on the basis of the $\mathcal{N}(\mu, \sigma^2)$ sample $X_1, \ldots, X_n$, where σ is known. Let $T = \bar{X} - \mu_0$ and $\theta = \mu - \mu_0$. Show that $EPV(\theta) = \Phi(-\sqrt{n}\theta/\sqrt{2}\sigma)$, where Φ denotes the standard normal distribution. (For a recent review of expected p values see Sackrowitz and Samuel–Cahn, 1999.)

Problems for Section 4.2

1. Consider Examples 3.3.2 and 4.2.1. You want to buy one of two systems. One has signal-to-noise ratio $v/\sigma_0 = 2$, the other has $v/\sigma_0 = 1$. The first system costs $\$10^6$, the other $\$10^5$. One second of transmission on either system costs $\$10^3$ each. Whichever system you buy during the year, you intend to test the satellite 100 times. If each time you test, you want the number of seconds of response sufficient to ensure that both probabilities of error are ≤ 0.05, which system is cheaper on the basis of a year's operation?

2. Consider a population with three kinds of individuals labeled 1, 2, and 3 occuring in the Hardy–Weinberg proportions $f(1, \theta) = \theta^2$, $f(2, \theta) = 2\theta(1 - \theta)$, $f(3, \theta) = (1 - \theta)^2$. For a sample $X_1, \ldots, X_n$ from this population, let N_1, N_2, and N_3 denote the number of X_j equal to 1, 2, and 3, respectively. Let $0 < \theta_0 < \theta_1 < 1$.

(a) Show that $L(\mathbf{x}, \theta_0, \theta_1)$ is an increasing function of $2N_1 + N_2$.

(b) Show that if $c > 0$ and $\alpha \in (0, 1)$ satisfy $P_{\theta_0}[2N_1 + N_2 \geq c] = \alpha$, then the test that rejects H if, and only if, $2N_1 + N_2 \geq c$ is MP for testing $H : \theta = \theta_0$ versus $K : \theta = \theta_1$.

3. A gambler observing a game in which a single die is tossed repeatedly gets the impression that 6 comes up about 18% of the time, 5 about 14% of the time, whereas the other four numbers are equally likely to occur (i.e., with probability .17). Upon being asked to play, the gambler asks that he first be allowed to test his hypothesis by tossing the die n times.

(a) What test statistic should he use if the only alternative he considers is that the die is fair?

(b) Show that if $n = 2$ the most powerful level .0196 test rejects if, and only if, two 5's are obtained.

(c) Using the fact that if $(N_1, \ldots, N_k) \sim \mathcal{M}(n, \theta_1, \ldots, \theta_k)$, then $a_1 N_1 + \cdots + a_k N_k$ has approximately a $\mathcal{N}(n\mu, n\sigma^2)$ distribution, where $\mu = \sum_{i=1}^k a_i \theta_i$ and $\sigma^2 = \sum_{i=1}^k \theta_i (a_i - \mu)^2$, find an approximation to the critical value of the MP level α test for this problem.

4. A formulation of goodness of tests specifies that a test is best if the maximum probability of error (of either type) is as small as possible.

(a) Show that if in testing $H : \theta = \theta_0$ versus $K : \theta = \theta_1$ there exists a critical value c such that
$$P_{\theta_0}[L(\mathbf{X}, \theta_0, \theta_1) \geq c] = 1 - P_{\theta_1}[L(\mathbf{X}, \theta_0, \theta_1) \geq c]$$
then the likelihood ratio test with critical value c is best in this sense.

(b) Find the test that is best in this sense for Example 4.2.1.

5. A newly discovered skull has cranial measurements (X, Y) known to be distributed either (as in population 0) according to $\mathcal{N}(0, 0, 1, 1, 0.6)$ or (as in population 1) according to $\mathcal{N}(1, 1, 1, 1, 0.6)$ where all parameters are known. Find a statistic $T(X, Y)$ and a critical value c such that if we use the classification rule, (X, Y) belongs to population 1 if $T \geq c$, and to population 0 if $T < c$, then the maximum of the two *probabilities of misclassification* $P_0[T \geq c]$, $P_1[T < c]$ is as small as possible.
    Hint: Use Problem 4.2.4 and recall (Proposition B.4.2) that linear combinations of bivariate normal random variables are normally distributed.

6. Show that if randomization is permitted, MP-sized α likelihood ratio tests based on $X_1, \ldots, X_n$ with $0 < \alpha < 1$ have power nondecreasing in n.

7. Prove Corollary 4.2.1.
    Hint: The MP test has power at least that of the test with test function $\delta(x) = \alpha$.

8. In Examle 4.2.2, derive the UMP test defined by $(4.2.7)$.

9. In Example 4.2.2, if $\boldsymbol{\Delta}_0 = (1, 0, \ldots, 0)^T$ and $\Sigma_0 \neq I$, find the MP test for testing $H : \boldsymbol{\theta} = \boldsymbol{\theta}_0$ versus $K : \boldsymbol{\theta} = \boldsymbol{\theta}_1$.

10. For $0 < \alpha < 1$, prove Theorem 4.2.1(a) using the connection between likelihood ratio tests and Bayes tests given in Remark 4.2.1.

Problems for Section 4.3

1. Let X_i be the number of arrivals at a service counter on the ith of a sequence of n days. A possible model for these data is to assume that customers arrive according to a homogeneous Poisson process and, hence, that the X_i are a sample from a Poisson distribution with parameter θ, the expected number of arrivals per day. Suppose that if $\theta \leq \theta_0$ it is not worth keeping the counter open.

(a) Exhibit the optimal (UMP) test statistic for $H : \theta \leq \theta_0$ versus $K : \theta > \theta_0$.

(b) For what levels can you exhibit a UMP test?

(c) What distribution tables would you need to calculate the power function of the UMP test?

2. Consider the foregoing situation of Problem 4.3.1. You want to ensure that if the arrival rate is ≤ 10, the probability of your deciding to stay open is ≤ 0.01, but if the arrival rate is ≥ 15, the probability of your deciding to close is also ≤ 0.01. How many days must you observe to ensure that the UMP test of Problem 4.3.1 achieves this? (Use the normal approximation.)

3. In Example 4.3.4, show that the power of the UMP test can be written as

$$\beta(\sigma) = G_n(\sigma_0^2 x_n(\alpha)/\sigma^2)$$

where G_n denotes the χ_n^2 distribution function.

4. Let $X_1, \ldots, X_n$ be the times in months until failure of n similar pieces of equipment. If the equipment is subject to wear, a model often used (see Barlow and Proschan, 1965) is the one where $X_1, \ldots, X_n$ is a sample from a Weibull distribution with density $f(x, \lambda) = \lambda c x^{c-1} e^{-\lambda x^c}$, $x > 0$. Here c is a known positive constant and $\lambda > 0$ is the parameter of interest.

(a) Show that $\sum_{i=1}^{n} X_i^c$ is an optimal test statistic for testing $H : 1/\lambda \leq 1/\lambda_0$ versus $K : 1/\lambda > 1/\lambda_0$.

(b) Show that the critical value for the size α test with critical region $[\sum_{i=1}^{n} X_i^c \geq k]$ is $k = x_{2n}(1-\alpha)/2\lambda_0$ where $x_{2n}(1-\alpha)$ is the $(1-\alpha)$th quantile of the χ_{2n}^2 distribution and that the power function of the UMP level α test is given by

$$1 - G_{2n}(\lambda x_{2n}(1-\alpha)/\lambda_0)$$

where G_{2n} denotes the χ_{2n}^2 distribution function.
Hint: Show that $X_i^c \sim \mathcal{E}(\lambda)$.

(c) Suppose $1/\lambda_0 = 12$. Find the sample size needed for a level 0.01 test to have power at least 0.95 at the alternative value $1/\lambda_1 = 15$. Use the normal approximation to the critical value and the probability of rejection.

5. Show that if $X_1, \ldots, X_n$ is a sample from a *truncated binomial* distribution with

$$p(x, \theta) = \binom{k}{x} \theta^x (1-\theta)^{k-x}/[1 - (1-\theta)^k], \ x = 1, \ldots, k,$$

then $\sum_{i=1}^{n} X_i$ is an optimal test statistic for testing $H : \theta = \theta_0$ versus $K : \theta > \theta_0$.

6. Let $X_1, \ldots, X_n$ denote the incomes of n persons chosen at random from a certain population. Suppose that each X_i has the Pareto density

$$f(x, \theta) = c^\theta \theta x^{-(1+\theta)}, \ x > c$$

where $\theta > 1$ and $c > 0$.

(a) Express mean income μ in terms of θ.

(b) Find the optimal test statistic for testing $H : \mu = \mu_0$ versus $K : \mu > \mu_0$.

(c) Use the central limit theorem to find a normal approximation to the critical value of test in part (b).

Hint: Use the results of Theorem 1.6.2 to find the mean and variance of the optimal test statistic.

7. In the goodness-of-fit Example 4.1.5, suppose that $F_0(x)$ has a nonzero density on some interval (a, b), $-\infty \le a < b \le \infty$, and consider the alternative with distribution function $F(x, \theta) = F_0^\theta(x)$, $0 < \theta < 1$. Show that the UMP test for testing $H : \theta \ge 1$ versus $K : \theta < 1$ rejects H if $-2\Sigma \log F_0(X_i) \ge x_{1-\alpha}$, where $x_{1-\alpha}$ is the $(1 - \alpha)$th quantile of the χ^2_{2n} distribution. (See Problem 4.1.6.) It follows that Fisher's method for combining p-values (see 4.1.6) is UMP for testing that the p-values are uniformly distributed against $F(u) = u^\theta, 0 < \theta < 1$.

8. Let the distribution of survival times of patients receiving a standard treatment be the known distribution F_0, and let $Y_1, \ldots, Y_n$ be the i.i.d. survival times of a sample of patients receiving an experimental treatment.

(a) *Lehmann Alternative.* In Problem 1.1.12, we derived the model

$$G(y, \Delta) = 1 - [1 - F_0(y)]^\Delta, \ y > 0, \ \Delta > 0.$$

To test whether the new treatment is beneficial we test $H : \Delta \le 1$ versus $K : \Delta > 1$. Assume that F_0 has a density f_0. Find the UMP test. Show how to find critical values.

(b) *Nabeya–Miura Alternative.* For the purpose of modeling, imagine a sequence $X_1, X_2, \ldots$ of i.i.d. survival times with distribution F_0. Let N be a zero-truncated Poisson, $\mathcal{P}(\lambda)$, random variable, which is independent of $X_1, X_2, \ldots$.

(i) Show that if we model the distribution of Y as $\mathcal{L}(\max\{X_1, \ldots, X_N\})$, then

$$P(Y \le y) = \frac{e^{\lambda F_0(y)} - 1}{e^\lambda - 1}, \ y > 0, \ \lambda \ge 0.$$

(ii) Show that if we model the distribution of Y as $\mathcal{L}(\min\{X_1, \ldots, X_N\})$, then

$$P(Y \le y) = \frac{e^{-\lambda F_0(y)} - 1}{e^{-\lambda} - 1}, \ y > 0, \ \lambda \ge 0.$$

(iii) Consider the model

$$G(y, \theta) = \frac{e^{\theta F_0(y)} - 1}{e^{\theta} - 1}, \ \theta \neq 0$$

$$= F_0(y), \ \theta = 0.$$

To see whether the new treatment is beneficial, we test $H : \theta \leq 0$ versus $K : \theta > 0$. Assume that F_0 has a density $f_0(y)$. Show that the UMP test is based on the statistic $\sum_{i=1}^{n} F_0(Y_i)$.

9. Let $X_1, \ldots, X_n$ be i.i.d. with distribution function $F(x)$. We want to test whether F is exponential, $F(x) = 1 - \exp(-x)$, $x > 0$, or Weibull, $F(x) = 1 - \exp(-x^{\theta})$, $x > 0$, $\theta > 0$. Find the MP test for testing $H : \theta = 1$ versus $K : \theta = \theta_1 > 1$. Show that the test is not UMP.

10. Show that under the assumptions of Theorem 4.3.2 the class of all Bayes tests is complete.

Hint: Consider the class of all Bayes tests of $H : \theta = \theta_0$ versus $K : \theta = \theta_1$ where $\pi\{\theta_0\} = 1 - \pi\{\theta_1\}$ varies between 0 and 1.

11. Show that under the assumptions of Theorem 4.3.1 and 0-1 loss, every Bayes test for $H : \theta \leq \theta_0$ versus $K : \theta > \theta_1$ is of the form δ_t for some t.

Hint: A Bayes test rejects (accepts) H if

$$\int_{\theta_1}^{\infty} p(x, \theta) d\pi(\theta) / \int_{-\infty}^{\theta_0} p(x, \theta) d\pi(\theta) \overset{>}{(<)} 1.$$

The left-hand side equals

$$\frac{\int_{\theta_1}^{\infty} L(x, \theta, \theta_0) d\pi(\theta)}{\int_{-\infty}^{\theta_0} L(x, \theta, \theta_0) d\pi(\theta)}.$$

The numerator is an increasing function of $T(x)$, the denominator decreasing.

12. Show that under the assumptions of Theorem 4.3.1, $1 - \delta_t$ is UMP for testing $H : \theta \geq \theta_0$ versus $K : \theta < \theta_0$.

Problems for Section 4.4

1. Let $X_1, \ldots, X_n$ be a sample from a normal population with unknown mean μ and unknown variance σ^2. Using a pivot based on $\sum_{i=1}^{n}(X_i - \bar{X})^2$,

(a) Show how to construct level $(1 - \alpha)$ confidence intervals of fixed finite length for $\log \sigma^2$.

(b) Suppose that $\sum_{i=1}^{n}(X_i - \bar{X})^2 = 16.52$, $n = 2$, $\alpha = 0.01$. What would you announce as your level $(1 - \alpha)$ UCB for σ^2?

2. Let $X_i = (\theta/2)t_i^2 + \epsilon_i$, $i = 1, \ldots, n$, where the ϵ_i are independent normal random variables with mean 0 and known variance σ^2 (cf. Problem 2.2.1).

(a) Using a pivot based on the MLE $(2\Sigma_{i=1}^{n} t_i^2 X_i)/\Sigma_{i=1}^{n} t_i^4$ of θ, find a fixed length level $(1-\alpha)$ confidence interval for θ.

(b) If $0 \leq t_i \leq 1$, $i = 1, \ldots, n$, but we may otherwise choose the t_i freely, what values should we use for the t_i so as to make our interval as short as possible for given α?

3. Let $X_1, \ldots, X_n$ be as in Problem 4.4.1. Suppose that an experimenter thinking he knows the value of σ^2 uses a lower confidence bound for μ of the form $\mu(\mathbf{X}) = \bar{X} - c$, where c is chosen so that the confidence level under the assumed value of σ^2 is $1 - \alpha$. What is the actual confidence coefficient of μ, if σ^2 can take on all positive values?

4. Suppose that in Example 4.4.3 we know that $\theta \leq 0.1$.

(a) Justify the interval $[\underline{\theta}, \min(\bar{\theta}, 0.1)]$ if $\underline{\theta} < 0.1$, $[0.1, 0.1]$ if $\underline{\theta} \geq 0.1$, where $\underline{\theta}, \bar{\theta}$ are given by (4.4.3).

(b) Calculate the smallest n needed to bound the length of the 95% interval of part (a) by 0.02. Compare your result to the n needed for (4.4.3).

5. Show that if $\underline{q}(\mathbf{X})$ is a level $(1 - \alpha_1)$ LCB and $\bar{q}(\mathbf{X})$ is a level $(1 - \alpha_2)$ UCB for $q(\theta)$, then $[\underline{q}(\mathbf{X}), \bar{q}(\mathbf{X})]$ is a level $(1 - (\alpha_1 + \alpha_2))$ confidence interval for $q(\theta)$. (Define the interval arbitrarily if $\underline{q} > \bar{q}$.)

   Hint: Use (A.2.7).

6. Show that if $X_1, \ldots, X_n$ are i.i.d. $\mathcal{N}(\mu, \sigma^2)$ and $\alpha_1 + \alpha_2 \leq \alpha$, then the shortest level $(1-\alpha)$ interval of the form

$$\left[\bar{X} - z(1 - \alpha_1)\frac{\sigma}{\sqrt{n}}, \ \bar{X} + z(1 - \alpha_2)\frac{\sigma}{\sqrt{n}}\right]$$

is obtained by taking $\alpha_1 = \alpha_2 = \alpha/2$ (assume σ^2 known).

   Hint: Reduce to $\alpha_1 + \alpha_2 = \alpha$ by showing that if $\alpha_1 + \alpha_2 < \alpha$, there is a shorter interval with $\alpha_1 + \alpha_2 = \alpha$. Use calculus.

7. Suppose we want to select a sample size N such that the interval (4.4.1) based on $n = N$ observations has length at most l for some preassigned length $l = 2d$. Stein's (1945) two-stage procedure is the following. Begin by taking a fixed number $n_0 \geq 2$ of observations and calculate $\bar{X}_0 = (1/n_0)\Sigma_{i=1}^{n_0} X_i$ and

$$s_0^2 = (n_0 - 1)^{-1}\Sigma_{i=1}^{n_0}(X_i - \bar{X}_0)^2.$$

Then take $N - n_0$ further observations, with N being the smallest integer greater than n_0 and greater than or equal to

$$\left[s_0 t_{n_0-1}\left(1 - \tfrac{1}{2}\alpha\right)/d\right]^2.$$

Show that, although N is random, $\sqrt{N}(\bar{X} - \mu)/s_0$, with $\bar{X} = \Sigma_{i=1}^{N} X_i/N$, has a $\mathcal{T}_{n_0-1}$ distribution. It follows that

$$\left[\bar{X} - s_0 t_{n_0-1}\left(1 - \tfrac{1}{2}\alpha\right)/\sqrt{N}, \ \bar{X} + s_0 t_{n_0-1}\left(1 - \tfrac{1}{2}\alpha\right)/\sqrt{N}\right]$$

is a confidence interval with confidence coefficient $(1 - \alpha)$ for μ of length at most $2d$. (The sticky point of this approach is that we have no control over N, and, if σ is large, we may very likely be forced to take a prohibitively large number of observations. The reader interested in pursuing the study of sequential procedures such as this one is referred to the book of Wetherill and Glazebrook, 1986, and the fundamental monograph of Wald, 1947.)

Hint: Note that $\bar{X} = (n_0/N)\bar{X}_0 + (1/N)\Sigma_{i=n_0+1}^{N}X_i$. By Theorem B.3.3, s_0 is independent of $\bar{X}_0$. Because N depends only on s_0, given $N = k$, $\bar{X}$ has a $\mathcal{N}(\mu, \sigma^2/k)$ distribution. Hence, $\sqrt{N}(\bar{X} - \mu)$ has a $\mathcal{N}(0, \sigma^2)$ distribution and is independent of s_0.

8. (a) Show that in Problem 4.4.6, in order to have a level $(1 - \alpha)$ confidence interval of length at most $2d$ when σ^2 is known, it is necessary to take at least $z^2\left(1 - \frac{1}{2}\alpha\right)\sigma^2/d^2$ observations.

Hint: Set up an inequality for the length and solve for n.

(b) What would be the minimum sample size in part (a) if $\alpha = 0.001$, $\sigma^2 = 5$, $d = 0.05$?

(c) Suppose that σ^2 is not known exactly, but we are sure that $\sigma^2 \leq \sigma_1^2$. Show that $n \geq z^2\left(1 - \frac{1}{2}\alpha\right)\sigma_1^2/d^2$ observations are necessary to achieve the aim of part (a).

9. Let $S \sim \mathcal{B}(n, \theta)$ and $\bar{X} = S/n$.

(a) Use (A.14.18) to show that $\sin^{-1}(\sqrt{\bar{X}}) \pm z\left(1 - \frac{1}{2}\alpha\right)/2\sqrt{n}$ is an approximate level $(1 - \alpha)$ confidence interval for $\sin^{-1}(\sqrt{\theta})$.

(b) If $n = 100$ and $\bar{X} = 0.1$, use the result in part (a) to compute an approximate level 0.95 confidence interval for θ.

10. Let $X_1, \ldots, X_{n_1}$ and $Y_1, \ldots, Y_{n_2}$ be two independent samples from $\mathcal{N}(\mu, \sigma^2)$ and $\mathcal{N}(\eta, \tau^2)$ populations, respectively.

(a) If all parameters are unknown, find ML estimates of μ, ν, σ^2, τ^2. Show that this quadruple is sufficient.

(b) Exhibit a level $(1 - \alpha)$ confidence interval for τ^2/σ^2 using a pivot based on the statistics of part (a). Indicate what tables you would need to calculate the interval.

(c) If σ^2, τ^2 are known, exhibit a fixed length level $(1 - \alpha)$ confidence interval for $(\eta - \mu)$.

Such two sample problems arise in comparing the precision of two instruments and in determining the effect of a treatment.

11. Show that the endpoints of the approximate level $(1 - \alpha)$ interval defined by (4.4.3) are indeed approximate level $\left(1 - \frac{1}{2}\alpha\right)$ upper and lower bounds.

Hint: $[\underline{\theta}(\mathbf{X}) \leq \theta] = \left[\sqrt{n}(\bar{X} - \theta)/[\theta(1 - \theta)]^{\frac{1}{2}} \leq z\left(1 - \frac{1}{2}\alpha\right)\right]$.

12. Let $S \sim \mathcal{B}(n, \theta)$. Suppose that it is known that $\theta \leq \frac{1}{4}$.

(a) Show that $\bar{X} \pm \sqrt{3}z\left(1 - \frac{1}{2}\alpha\right)/4\sqrt{n}$ is an approximate level $(1 - \alpha)$ confidence interval for θ.

(b) What sample size is needed to guarantee that this interval has length at most 0.02?

13. Suppose that a new drug is tried out on a sample of 64 patients and that $S = 25$ cures are observed. If $S \sim B(64, \theta)$, give a 95% confidence interval for the true proportion of cures θ using (a) (4.4.3), and (b) (4.4.7).

14. Suppose that 25 measurements on the breaking strength of a certain alloy yield $\bar{x} = 11.1$ and $s = 3.4$. Assuming that the sample is from a $\mathcal{N}(\mu, \sigma^2)$ population, find

    (a) A level 0.9 confidence interval for μ.

    (b) A level 0.9 confidence interval for σ.

    (c) A level 0.9 confidence region for (μ, σ).

    (d) A level 0.9 confidence interval for $\mu + \sigma$.

15. Show that the confidence coefficient of the rectangle of Example 4.4.5 is

$$\int_a^b 2 \left[\Phi \left(\sqrt{\frac{1}{n-1}} \sqrt{\tau} \cdot c \right) - 1 \right] g(\tau) d\tau$$

where a is the $\frac{\alpha}{4}$th quantile of the χ_{n-1}^2 distribution; b is the $(1 - \frac{\alpha}{4})$th quantile of the χ_{n-1}^2 distribution; c is the $(1 - \frac{\alpha}{4})$th quantile of t-distribution; and $g(\cdot)$ is the density function of the χ_{n-1}^2 distribution.
    Hint: Use Theorem B.3.3 and (B.1.30). Condition on $[(n-1)s^2/\sigma^2] = \tau$.

16. In Example 4.4.2,

    (a) Show that $x(\alpha_1)$ and $x(1-\alpha_2)$ can be approximated by $x(\alpha_1) \cong (n-1) + \sqrt{2}(n-1)^{\frac{1}{2}} z(\alpha_1)$ and $x(1 - \alpha_2) \cong (n-1) + \sqrt{2}(n-1)^{\frac{1}{2}} z(1 - \alpha_2)$.
    Hint: By B.3.1, $V(\sigma^2)$ can be written as a sum of squares of $n-1$ independent $\mathcal{N}(0, 1)$ random variables, $\sum_{i=1}^{n-1} Z_i^2$. Now use the central limit theorem.

    (b) Suppose that X_i does not necessarily have a normal distribution, but assume that $\mu_4 = E(X_i - \mu)^4 < \infty$ and that $\kappa = \mathrm{Var}[(X_i - \mu)/\sigma]^2 = (\mu_4/\sigma^4) - 1$ is known. Find the limit of the distribution of $n^{-\frac{1}{2}}\{[(n-1)s^2/\sigma^2] - n\}$ and use this distribution to find an approximate $1 - \alpha$ confidence interval for σ^2. (In practice, κ is replaced by its MOM estimate. See Problem 5.3.30.)
    Hint: $(n-1)s^2 = \sum_{i=1}^n (X_i - \mu)^2 - n(\bar{X} - \mu)^2$. Now use the law of large numbers, Slutsky's theorem, and the central limit theorem as given in Appendix A.

    (c) Suppose X_i has a χ_k^2 distribution. Compute the (κ known) confidence intervals of part (b) when $k = 1, 10, 100$, and $10\,000$. Compare them to the approximate interval given in part (a). $\kappa - 2$ is known as the *kurtosis coefficient*. In the case where X_i is normal, it equals 0. See A.11.11.
    Hint: Use Problem B.2.4 and the fact that $\chi_k^2 = \Gamma\left(k, \frac{1}{2}\right)$.

17. Consider Example 4.4.6 with x fixed. That is, we want a level $(1 - \alpha)$ confidence interval for $F(x)$. In this case $n\widehat{F}(x) = \#[X_i \leq x]$ has a binomial distribution and

$$\frac{\sqrt{n}|\widehat{F}(x) - F(x)|}{\sqrt{F(x)[1 - F(x)]}}$$

is the approximate pivot given in Example 4.4.3 for deriving a confidence interval for $\theta = F(x)$.

(a) For $0 < a < b < 1$, define

$$A_n(F) = \sup\left\{\frac{\sqrt{n}|\widehat{F}(x) - F(x)|}{\sqrt{F(x)[1 - F(x)]}}, \; F^{-1}(a) \leq x \leq F^{-1}(b)\right\}.$$

Typical choices of a and b are .05 and .95. Show that for F continuous

$$P_F(A_n(F) \leq t) = P_U(A_n(U) \leq t)$$

where U denotes the uniform, $\mathcal{U}(0, 1)$, distribution function. It follows that the binomial confidence intervals for θ in Example 4.4.3 can be turned into simultaneous confidence intervals for $F(x)$ by replacing $z\left(1 - \frac{1}{2}\alpha\right)$ by the value u_α determined by $P_U(A_n(U) \leq u) = 1 - \alpha$.

(b) For $0 < a < b < 1$, define

$$B_n(F) = \sup\left\{\frac{\sqrt{n}|\widehat{F}(x) - F(x)|}{\sqrt{\widehat{F}(x)[1 - \widehat{F}(x)]}}, \; \widehat{F}^{-1}(a) \leq x \leq \widehat{F}^{-1}(b)\right\}.$$

Show that for F continuous,

$$P_F(B_n(F) \leq t) = P_U(B_n(U) \leq t).$$

(c) For testing $H_0 : F = F_0$ with F_0 continuous, indicate how critical values u_α and t_α for $A_n(F_0)$ and $B_n(F_0)$ can be obtained using the Monte Carlo method of Section 4.1.

18. Suppose $X_1, \ldots, X_n$ are i.i.d. as X and that X has density $f(t) = F'(t)$. Assume that $f(t) > 0$ iff $t \in (a, b)$ for some $-\infty < a \leq 0 < b < \infty$.

(a) Show that $\mu = - \int_a^0 F(x)dx + \int_0^b [1 - F(x)]dx$.

(b) Using Example 4.4.6, find a level $(1 - \alpha)$ confidence interval for μ.

19. In Example 4.4.7, verify the lower bounary $\underline{\mu}$ given by (4.4.9) and the upper boundary $\bar{\mu} = \infty$.

Problems for Section 4.5

1. Let $X_1, \ldots, X_{n_1}$ and $Y_1, \ldots, Y_{n_2}$ be independent exponential $\mathcal{E}(\theta)$ and $\mathcal{E}(\lambda)$ samples, respectively, and let $\Delta = \theta/\lambda$.

(a) If $f(\alpha)$ denotes the αth quantile of the $\mathcal{F}_{2n_1, 2n_2}$ distribution, show that $[\bar{Y}f(\frac{1}{2}\alpha)/\bar{X}, \bar{Y}f(1 - \frac{1}{2}\alpha)/\bar{X}]$ is a confidence interval for Δ with confidence coefficient $1 - \alpha$.
 Hint: Use the results of Problems B.3.4 and B.3.5.

(b) Show that the test with acceptance region $[f(\frac{1}{2}\alpha) \le \bar{X}/\bar{Y} \le f(1 - \frac{1}{2}\alpha)]$ has size α for testing $H : \Delta = 1$ versus $K : \Delta \neq 1$.

(c) The following are times until breakdown in days of air monitors operated under two different maintenance policies at a nuclear power plant. Experience has shown that the exponential assumption is warranted. Give a 90% confidence interval for the ratio Δ of mean life times.

x	3 150 40 34 32 37 34 2 31 6 5 14 150 27 4 6 27 10 30 37
y	8 26 10 8 29 20 10

Is $H : \Delta = 1$ rejected at level $\alpha = 0.10$?

2. Show that if $\bar{\theta}(\mathbf{X})$ is a level $(1 - \alpha)$ UCB for θ, then the test that accepts, if and only if $\bar{\theta}(\mathbf{X}) \ge \theta_0$, is of level α for testing $H : \theta \ge \theta_0$.
 Hint: If $\theta > \theta_0$, $[\bar{\theta}(\mathbf{X}) < \theta] \supset [\bar{\theta}(\mathbf{X}) < \theta_0]$.

3. (a) Deduce from Problem 4.5.2 that the tests of $H : \sigma^2 = \sigma_0^2$ based on the level $(1 - \alpha)$ UCBs of Example 4.4.2 are level α for $H : \sigma^2 \ge \sigma_0^2$.

(b) Give explicitly the power function of the test of part (a) in terms of the χ_{n-1}^2 distribution function.

(c) Suppose that $n = 16, \alpha = 0.05, \sigma_0^2 = 1$. How small must an alternative σ^2 be before the size α test given in part (a) has power 0.90?

4. (a) Find c such that δ_c of Problem 4.1.1 has size α for $H : \theta \le \theta_0$.

(b) Derive the level $(1 - \alpha)$ LCB corresponding to δ_c of part (a).

(c) Similarly derive the level $(1 - \alpha)$ UCB for this problem and exhibit the confidence intervals obtained by putting two such bounds of level $(1 - \alpha_1)$ and $(1 - \alpha_2)$ together.

(d) Show that $[M_n, M_n/\alpha^{1/n}]$ is the shortest such confidence interval.

5. Let X_1, X_2 be independent $\mathcal{N}(\theta_1, \sigma^2), \mathcal{N}(\theta_2, \sigma^2)$, respectively, and consider the problem of testing $H : \theta_1 = \theta_2 = 0$ versus $K : \theta_1^2 + \theta_2^2 > 0$ when σ^2 is known.

(a) Let $\delta_c(X_1, X_2) = 1$ if and only if $X_1^2 + X_2^2 \ge c$. What value of c gives size α?

(b) Using Problems B.3.12 and B.3.13 show that the power $\beta(\theta_1, \theta_2)$ is an increasing function of $\theta_1^2 + \theta_2^2$.

(c) Modify the test of part (a) to obtain a procedure that is level α for $H : \theta_1 = \theta_1^0, \theta_2 = \theta_2^0$ and exhibit the corresponding family of confidence circles for (θ_1, θ_2).

Hint: (c) $X_1 - \theta_1^0$, $X_2 - \theta_2^0$ are independent $\mathcal{N}(\theta_1 - \theta_1^0, \sigma^2), \mathcal{N}(\theta_2 - \theta_2^0, \sigma^2)$, respectively.

6. Let $X_1, \ldots, X_n$ be a sample from a population with density $f(t - \theta)$ where θ and f are unknown, but $f(t) = f(-t)$ for all t, and f is continuous and positive. Thus, we have a location parameter family.

(a) Show that testing $H : \theta \leq 0$ versus $K : \theta > 0$ is equivalent to testing

$$H' : P[X_1 \geq 0] \leq \frac{1}{2} \text{ versus } K' : P[X_1 \geq 0] > \frac{1}{2}.$$

(b) The *sign test* of H versus K is given by,

$$\delta_k(X) = 1 \text{ if } \left(\sum_{i=1}^{n} 1[X_i \geq 0] \right) \geq k$$
$$= 0 \text{ otherwise.}$$

Determine the smallest value $k = k(\alpha)$ such that $\delta_{k(\alpha)}$ is level α for H and show that for n large, $k \cong \frac{1}{2}n + \frac{1}{2}z(1 - \alpha)\sqrt{n}$.

(c) Show that $\delta_{k(\alpha)}(X_1 - \theta_0, \ldots, X_n - \theta_0)$ is a level α test of $H : \theta \leq \theta_0$ versus $K : \theta > \theta_0$.

(d) Deduce that $X_{(n-k(\alpha)+1)}$ (where $X_{(j)}$ is the jth order statistic of the sample) is a level $(1 - \alpha)$ LCB for θ *whatever be f satisfying our conditions*.

(e) Show directly that $P_\theta[X_{(j)} \leq \theta]$ and $P_\theta[X_{(j)} \leq \theta \leq X_{(k)}]$ do not depend on f or θ.

(f) Suppose that $\alpha = 2^{-(n-1)} \sum_{j=0}^{k-1} \binom{n}{j}$. Show that $P[X_{(k)} \leq \theta \leq X_{(n-k+1)}] = 1 - \alpha$.

(g) Suppose that we drop the assumption that $f(t) = f(-t)$ for all t and replace our assumptions by: $X_1, \ldots, X_n$ is a sample from a population with density f and median ν. Show that if we replace θ by ν, the conclusions of (a)–(f) still hold.

7. Suppose $\theta = (\eta, \tau)$ where η is a parameter of interest and τ is a nuisance parameter. We are given for each possible value η_0 of η a level α test $\delta(\mathbf{X}, \eta_0)$ of the composite hypothesis $H : \eta = \eta_0$. Let $C(\mathbf{X}) = \{\eta : \delta(\mathbf{X}, \eta) = 0\}$.

(a) Show that $C(\mathbf{X})$ is a level $(1 - \alpha)$ confidence region for the parameter η and conversely that any level $(1 - \alpha)$ confidence region for η is equivalent to a family of level α tests of these composite hypotheses.

(b) Find the family of tests corresponding to the level $(1 - \alpha)$ confidence interval for μ of Example 4.4.1 when σ^2 is unknown.

8. Suppose X, Y are independent and $X \sim \mathcal{N}(\nu, 1), Y \sim \mathcal{N}(\eta, 1)$. Let $\rho = \nu/\eta, \theta = (\rho, \eta)$. Define

$$\delta(X, Y, \rho) = 0 \text{ if } |X - \rho Y| \le (1 + \rho^2)^{\frac{1}{2}} z(1 - \frac{1}{2}\alpha)$$
$$= 1 \text{ otherwise.}$$

(a) Show that $\delta(X, Y, \rho_0)$ is a size α test of $H : \rho = \rho_0$.

(b) Describe the confidence region obtained by inverting the family $\{\delta(X, Y, \rho)\}$ as in Problem 4.5.7. Note that the region is not necessarily an interval or ray. This problem is a simplified version of that encountered in putting a confidence interval on the zero of a regression line.

9. Let $X \sim \mathcal{N}(\theta, 1)$ and $q(\theta) = \theta^2$.

(a) Show that the lower confidence bound for $q(\theta)$ obtained from the image under q of the ray $(X - z(1 - \alpha), \infty)$ is

$$\underline{q}(X) = (X - z(1 - \alpha))^2 \text{ if } X \ge z(1 - \alpha)$$
$$= 0 \text{ if } X < z(1 - \alpha).$$

(b) Show that

$$P_\theta[\underline{q}(X) \le \theta^2] = 1 - \alpha \text{ if } \theta \ge 0$$
$$= \Phi(z(1 - \alpha) - 2\theta) \text{ if } \theta < 0$$

and, hence, that $\sup_\theta P_\theta[\underline{q}(X) \le \theta^2] = 1$.

10. Let $\alpha(S, \theta_0)$ denote the p-value of the test of $H : \theta = \theta_0$ versus $K : \theta > \theta_0$ in Example 4.1.3 and let $[\underline{\theta}(S), \bar{\theta}(S)]$ be the exact level $(1 - 2\alpha)$ confidence interval for θ of Example 4.5.2. Show that as θ ranges from $\underline{\theta}(S)$ to $\bar{\theta}(S)$, $\alpha(S, \theta)$ ranges from α to a value no smaller than $1 - \alpha$. Thus, if $\theta_0 < \underline{\theta}(S)$ (S is inconsistent with $H : \theta = \theta_0$), the quantity $\Delta = \bar{\theta}(S) - \theta_0$ indicates how far we have to go from θ_0 before the value S is not at all surprising under H.

11. Establish (iii) and (iv) of Example 4.5.2.

12. Let η denote a parameter of interest, let τ denote a nuisance parameter, and let $\theta = (\eta, \tau)$. Then the level $(1 - \alpha)$ confidence interval $[\underline{\eta}(\mathbf{x}), \bar{\eta}(\mathbf{x})]$ for η is said to be *unbiased* if

$$P_\theta[\underline{\eta}(\mathbf{X}) \le \eta' \le \bar{\eta}(\mathbf{X})] \le 1 - \alpha \text{ for all } \eta' \ne \eta, \text{ all } \theta.$$

That is, the interval is unbiased if it has larger probability of covering the true value η than the wrong value η'. Show that the Student t interval (4.5.1) is unbiased.
   Hint: You may use the result of Problem 4.5.7.

13. *Confidence Regions for Quantiles.* Let $X_1, \ldots, X_n$ be a sample from a population with continuous and increasing distribution F. Let $x_p = F^{-1}(p), 0 < p < 1$, be the pth quantile of F. (See Section 3.5.) Suppose that p is specified. Thus, $x_{.95}$ could be the 95th percentile of the salaries in a certain profession, or $x_{.05}$ could be the fifth percentile of the duration time for a certain disease.

(a) Show that testing $H : x_p \leq 0$ versus $K : x_p > 0$ is equivalent to testing $H' :$ $P(X \geq 0) \leq (1 - p)$ versus $K' : P(X \geq 0) > (1 - p)$.

(b) The *quantile sign test* δ_k of H versus K has critical region $\{\mathbf{x} : \sum_{i=1}^{n} 1[X_i \geq 0] \geq k\}$. Determine the smallest value $k = k(\alpha)$ such that $\delta_{k(\alpha)}$ has level α for H and show that for n large, $k(\alpha) \cong h(\alpha)$, where

$$h(\alpha) \cong n(1 - p) + z_{1-\alpha}\sqrt{np(1 - p)}.$$

(c) Let x^* be a specified number with $0 < F(x^*) < 1$. Show that $\delta_k(X_1 - x^*, \dots, X_n - x^*)$ is a level α test for testing $H : x_p \leq x^*$ versus $K : x_p > x^*$.

(d) Deduce that $X_{(n-k(\alpha)+1)}$ ($X_{(j)}$ is the jth order statistic of the sample) is a level $(1 - \alpha)$ LCB for x_p whatever be f satisfying our conditions.

(e) Let S denote a $\mathcal{B}(n, p)$ variable and choose k and l such that $1 - \alpha = P(k \leq S \leq n - l + 1) = \sum_{j=k}^{n-l+1} p^j (1-p)^{n-j}$. Show that $P(X_{(k)} \leq x_p \leq X_{(n-l)}) = 1 - \alpha$. That is, $(X_{(k)}, X_{(n-l)})$ is a level $(1 - \alpha)$ confidence interval for x_p whatever be F satisfying our conditions. That is, it is distribution free.

(f) Show that k and l in part (e) can be approximated by $h\left(\frac{1}{2}\alpha\right)$ and $h\left(1 - \frac{1}{2}\alpha\right)$ where $h(\alpha)$ is given in part (b).

(g) Let $\widehat{F}(x)$ denote the empirical distribution. Show that the interval in parts (e) and (f) can be derived from the pivot

$$T(x_p) = \frac{\sqrt{n}[\widehat{F}(x_p) - F(x_p)]}{\sqrt{F(x_p)[1 - F(x_p)]}}.$$

Hint: Note that $F(x_p) = p$.

14. *Simultaneous Confidence Regions for Quantiles.* In Problem 13 preceding we gave a distribution-free confidence interval for the pth quantile x_p for p fixed. Suppose we want a distribution-free confidence region for x_p valid for all $0 < p < 1$. We can proceed as follows. Let F, $\widehat{F}^-(x)$, and $\widehat{F}^+(x)$ be as in Examples 4.4.6 and 4.4.7. Then

$$P(\widehat{F}^-(x) \leq F(x) \leq \widehat{F}^+(x)) \text{ for all } x \in (a, b) = 1 - \alpha.$$

(a) Show that this statement is equivalent to

$$P(\underline{x}_p \leq x_p \leq \bar{x}_p \text{ for all } p \in (0, 1)) = 1 - \alpha$$

where $\underline{x}_p = \sup\{x : a < x < b, \widehat{F}^+(x) \leq p\}$ and $\bar{x}_p = \inf\{x : a < x < b, \widehat{F}^-(x) \geq p\}$. That is, the desired confidence region is the band consisting of the collection of intervals $\{[\underline{x}_p, \bar{x}_p] : 0 < p < 1\}$. Note that $\underline{x}_p = -\infty$ for $p < d_\alpha$ and $\bar{x}_p = \infty$ for $p > 1 - d_\alpha$.

(b) Express $\underline{x}_p$ and $\bar{x}_p$ in terms of the critical value of the Kolmogorov statistic and the order statistics.

(c) Show how the statistic $A_n(F)$ of Problem 4.1.17(a) and (c) can be used to give another distribution-free simultaneous confidence band for x_p. Express the band in terms of critical values for $A_n(F)$ and the order statistics. Note the similarity to the interval in Problem 4.4.13(g) preceding.

15. Suppose X denotes the difference between responses after a subject has been given treatments A and B, where A is a placebo. Suppose that X has the continuous distribution F. We will write F_X for F when we need to distinguish it from the distribution F_{-X} of $-X$. The hypothesis that A and B are equally effective can be expressed as $H : F_{-X}(t) = F_X(t)$ for all $t \in R$. The alternative is that $F_{-X}(t) \neq F_X(t)$ for some $t \in R$. Let $\widehat{F}_X$ and $\widehat{F}_{-X}$ be the empirical distributions based on the i.i.d. $X_1, \ldots, X_n$ and $-X_1, \ldots, -X_n$.

(a) Consider the test statistic

$$D(\widehat{F}_X, \widehat{F}_{-X}) = \max\{|\widehat{F}_X(t) - \widehat{F}_{-X}(t)| : t \in R\}.$$

Show that if F_X is continuous and H holds, then $D(\widehat{F}_X, \widehat{F}_{-X})$ has the same distribution as $D(\widehat{F}_U, \widehat{F}_{1-U})$, where $\widehat{F}_U$ and $\widehat{F}_{1-U}$ are the empirical distributions of U and $1 - U$ with $U = F(X) \sim \mathcal{U}(0, 1)$.

Hint: $n\widehat{F}_X(x) = \sum_{i=1}^{n} 1[F_X(X_i) \leq F_X(x)] = n\widehat{F}_U(F(x))$ and

$$n\widehat{F}_{-X}(x) = \sum_{i=1}^{n} 1[-X_i \leq x] = \sum_{i=1}^{n} 1[F_{-X}(-X_i) \leq F_{-X}(x)]$$

$$= n\widehat{F}_{1-U}(F_{-X}(x)) = n\widehat{F}_{1-U}(F(x)) \text{ under } H.$$

See also Example 4.1.5.

(b) Suppose we measure the difference between the effects of A and B by $\frac{1}{2}$ the difference between the quantiles of X and $-X$, that is, $\nu_F(p) = \frac{1}{2}[x_p + x_{1-p}]$, where $p = F(x)$. Give a distribution-free level $(1 - \alpha)$ simultaneous confidence band for the curve $\{\nu_F(p) : 0 < p < 1\}$.

Hint: Let $\Delta(x) = F_{-X}^{-1}(F_X(x)) - x$, then

$$n\widehat{F}_{-X}(x + \Delta(x)) = \sum_{i=1}^{n} 1[-X_i \leq x + \Delta(x)]$$

$$= \sum_{i=1}^{n} 1[F_{-X}(-X_i) \leq F_X(x)] = n\widehat{F}_{1-U}(F_X(x)).$$

Moreover, $n\widehat{F}_X(x) = \sum_{i=1}^{n} 1[F_X(X_i) \leq F_X(x)] = n\widehat{F}_U(F_X(x))$. It follows that if we set $F_{-X,\Delta}^*(x) = \widehat{F}_{-X}(x + \Delta(x))$, then $D(\widehat{F}_X, F_{-X,\Delta}^*) \stackrel{\mathcal{L}}{=} D(\widehat{F}_U, \widehat{F}_{1-U})$, and by solving $D(\widehat{F}_X, \widehat{F}_{-X,\Delta}) \leq d_\alpha$ for Δ, where d_α is the αth quantile of the distribution of

$D(\widehat{F}_U, \widehat{F}_{1-U})$, we get a distribution-free level $(1 - \alpha)$ simultaneous confidence band for $\Delta(x) = F_{-X}^{-1}(F_X(x)) - x = -2\nu_F(F(x))$. Properties of this and other bands are given by Doksum, Fenstad and Aaberge (1977).

(c) *A Distribution and Parameter-Free Confidence Interval.* Let $\theta(\cdot) : \mathcal{F} \to R$, where $\mathcal{F}$ is the class of distribution functions with finite support, be a location parameter as defined in Problem 3.5.17. Let

$$\widehat{\nu}_F^- = \inf_{0<p<1} \widehat{\nu}_F^-(p), \quad \widehat{\nu}_F^+ = \sup_{0<p<1} \widehat{\nu}_F^+(p)$$

where $[\widehat{\nu}_F^-(p), \widehat{\nu}_F^+(p)]$ is the band in part (b). Show that for given $F \in \mathcal{F}$, the probability is $(1 - \alpha)$ that the interval $[\widehat{\nu}_F^-, \widehat{\nu}_F^+]$ contains the location set $L_F = \{\theta(F) : \theta(\cdot) \text{ is a location parameter}\}$ of *all* location parameter values at F.

Hint: Define H by $H^{-1}(p) = \frac{1}{2}[F_X^{-1}(p) - F_X^{-1}(1 - p)] = \frac{1}{2}[F_X^{-1}(p) + F_{-X}^{-1}(p)]$. Then H is symmetric about zero. Also note that

$$x = H^{-1}(F(x)) - \frac{1}{2}\Delta(x) = H^{-1}(F(x)) + \nu_F(F(x)).$$

It follows that X is stochastically between $X_S - \underline{\nu}_F$ and $X_S + \bar{\nu}_F$ where $X_S \equiv H^{-1}(F(X))$ has the symmetric distribution H. The result now follows from the properties of $\theta(\cdot)$.

16. As in Example 1.1.3, let $X_1, \ldots, X_n$ be i.i.d. treatment A (placebo) responses and let $Y_1, \ldots, Y_n$ be i.i.d. treatment B responses. We assume that the X's and Y's are independent and that they have respective continuous distributions F_X and F_Y. To test the hypothesis H that the two treatments are equally effective, we test $H : F_X(t) = F_Y(t)$ for all t versus $K : F_X(t) \neq F_Y(t)$ for some $t \in R$. Let $\widehat{F}_X$ and $\widehat{F}_Y$ denote the X and Y empirical distributions and consider the test statistic

$$D(\widehat{F}_X, \widehat{F}_Y) = \max_{t \in R} |\widehat{F}_Y(t) - \widehat{F}_X(t)|.$$

(a) Show that if H holds, then $D(\widehat{F}_X, \widehat{F}_Y)$ has the same distribution as $D(\widehat{F}_U, \widehat{F}_V)$, where $\widehat{F}_U$ and $\widehat{F}_V$ are independent $\mathcal{U}(0, 1)$ empirical distributions.

Hint: $n\widehat{F}_X(t) = \sum_{i=1}^n 1[F_X(X_i) \leq F_X(t)] = n\widehat{F}_U(F_X(t)); \quad n\widehat{F}_Y(t) = \sum_{i=1}^n 1[F_Y(Y_i) \leq F_Y(x_p)] = n\widehat{F}_V(F_X(t))$ under H.

(b) Consider the parameter $\delta_p(F_X, F_Y) = y_p - x_p$, where x_p and y_p are the pth quantiles of F_X and F_Y. Give a distribution-free level $(1 - \alpha)$ simultaneous confidence band $[\widehat{\delta}_p^-, \widehat{\delta}_p^+ : 0 < p < 1]$ for the curve $\{\delta_p(F_X, F_Y) : 0 < p < 1\}$.

Hint: Let $\Delta(x) = F_Y^{-1}(F_X(x)) - x$, then

$$n\widehat{F}_Y(x + \Delta(x)) = \sum_{i=1}^n 1[Y_i \leq F_Y^{-1}(F_X(x))] = \sum_{i=1}^n 1[F_Y(Y_i) \leq F_X(x)] = n\widehat{F}_V(F_X(x)).$$

Moreover, $n\widehat{F}_X(x) = n\widehat{F}_U(F_X(x))$. It follows that if we set $F^*_{Y,\Delta}(x) = \widehat{F}_Y(x + \Delta(x))$, then $D(\widehat{F}_X, F^*_{Y,\Delta}) \stackrel{\mathcal{L}}{=} D(\widehat{F}_U, \widehat{F}_V)$. Let d_α denote a size α critical value for $D(\widehat{F}_U, \widehat{F}_V)$, then by solving $D(\widehat{F}_X, F^*_{Y,\Delta}) \leq d_\alpha$ for Δ, we find a distribution-free level $(1 - \alpha)$ simultaneous confidence band for $\Delta(x_p) = F_X^{-1}(p) - F_Y^{-1}(p) = \delta_p(F_X, F_Y)$. Properties of this and other bands are given by Doksum and Sievers (1976).

(c) A parameter $\theta = \delta(\cdot, \cdot) : \mathcal{F} \times \mathcal{F} \to R$, where $\mathcal{F}$ is the class of distributions with finite support, is called a *shift parameter* if $\theta(F_X, F_{X+a}) = \theta(F_{X-a}, F_X) = a$ and

$$Y_1 \stackrel{st}{\geq} Y, \; X_1 \stackrel{st}{\geq} X \Rightarrow \theta(F_X, F_Y) \leq \theta(F_X, F_{Y_1}), \; \theta(F_X, F_Y) \geq \theta(F_{X_1}, F_Y).$$

Let $\underline{\delta} = \min_{0<p<1} \delta_p(F_X, F_Y)$ and $\bar{\delta} = \max_{0<p<1} \delta_p(F_X, F_Y)$. Show that if $\theta(\cdot, \cdot)$ is a shift parameter, then $\theta(F_X, F_Y)$ is in $[\underline{\delta}, \bar{\delta}]$.

Hint: Set $Y^* = X + \Delta(X)$, then $Y^* \stackrel{\mathcal{L}}{=} Y$, moreover $X + \underline{\delta} \leq Y^* \leq X + \bar{\delta}$. Now apply the axioms.

(d) Show that $E(Y) - E(X)$, $\delta_p(\cdot, \cdot)$, $0 < p < 1$, $\underline{\delta}$, and $\bar{\delta}$ are shift parameters.

(e) *A Distribution and Parameter-Free Confidence Interval.* Let $\widehat{\delta}^- = \min_{0<p<1} \widehat{\delta}^-(p)$, $\widehat{\delta}^+ = \max_{0<p<1} \widehat{\delta}^+(p)$. Show that for given $(F_X, F_Y) \in \mathcal{F} \times \mathcal{F}$, the probability is $(1 - \alpha)$ that the interval $[\widehat{\delta}^-, \widehat{\delta}^+]$ contains the *shift parameter set* $\{\theta(F_X, F_Y) : \theta(\cdot, \cdot) \text{ is a shift parameter}\}$ of the values of *all* shift parameters at (F_X, F_Y).

Problems for Section 4.6

1. Suppose $X_1, \ldots, X_n$ is a sample from a $\Gamma\left(p, \frac{1}{\theta}\right)$ distribution, where p is known and θ is unknown. Exhibit the UMA level $(1 - \alpha)$ UCB for θ.

2. (a) Consider the model of Problem 4.4.2. Show that

$$\underline{\theta}^* = \left(2 \sum_{i=1}^n t_i^2 X_i\right) / \sum_{i=1}^n t_i^4 - 2z(1 - \alpha)\sigma \left[\sum_{i=1}^n t_i^4\right]^{-\frac{1}{2}}$$

is a uniformly most accurate lower confidence bound for θ.

(b) Consider the unbiased estimate of θ, $T = \left(2 \sum_{i=1}^n X_i\right) / \sum_{i=1}^n t_i^2$. Show that

$$\underline{\theta} = \left(2 \sum_{i=1}^n X_i\right) / \sum_{i=1}^n t_i^2 - 2\sigma\sqrt{n}z(1 - \alpha) / \sum_{i=1}^n t_i^2$$

is also a level $(1 - \alpha)$ confidence bound for θ.

(c) Show that the statement that $\underline{\theta}^*$ is more accurate than $\underline{\theta}$ is equivalent to the assertion that $S = \left(2 \sum_{i=1}^n t_i^2 X_i\right) / \sum_{i=1}^n t_i^4$ has uniformly smaller variance than T.
Hint: Both $\underline{\theta}$ and $\underline{\theta}^*$ are normally distributed.

3. Show that for the model of Problem 4.3.4, if $\mu = 1/\lambda$, then $\bar{\mu} = 2 \sum_{i=1}^n X_i^c / x_{2n}(\alpha)$ is a uniformly most accurate level $1 - \alpha$ UCB for μ.

4. Construct uniformly most accurate level $1 - \alpha$ upper and lower confidence bounds for μ in the model of Problem 4.3.6 for c fixed, $n = 1$.

5. Establish the following result due to Pratt (1961). Suppose $[\underline{\theta}^*, \bar{\theta}^*]$, $[\underline{\theta}, \bar{\theta}]$ are two level $(1 - \alpha)$ confidence intervals such that

$$P_\theta[\underline{\theta}^* \le \theta' \le \bar{\theta}^*] \le P_\theta[\underline{\theta} \le \theta' \le \bar{\theta}] \text{ for all } \theta' \ne \theta.$$

Show that if $(\underline{\theta}, \bar{\theta})$, $(\underline{\theta}^*, \bar{\theta}^*)$ have joint densities, then $E_\theta(\bar{\theta}^* - \underline{\theta}^*) \le E_\theta(\bar{\theta} - \underline{\theta})$.

    Hint: $E_\theta(\bar{\theta} - \underline{\theta}) = \int_{-\infty}^{\infty} \int_{-\infty}^{\infty} \left(\int_s^t du \right) p(s,t) ds dt = \int_{-\infty}^{\infty} P_\theta[\underline{\theta} \le u \le \bar{\theta}] du$, where $p(s,t)$ is the joint density of $(\underline{\theta}, \bar{\theta})$.

6. Let U, V be random variables with d.f.'s F, G corresponding to densities f, g, respectively, satisfying the conditions of Problem B.2.12 so that F^{-1}, G^{-1} are well defined and strictly increasing. Show that if $F(x) \le G(x)$ for all x and $E(U), E(V)$ are finite, then $E(U) \ge E(V)$.

    Hint: By Problem B.2.12(b), $E(U) = \int_0^1 F^{-1}(t) dt$.

7. Suppose that $\underline{\theta}^*$ is a uniformly most accurate level $(1 - \alpha)$ LCB such that $P_\theta[\underline{\theta}^* \le \theta] = 1 - \alpha$. Prove Corollary 4.6.1.

    Hint: Apply Problem 4.6.6 to $V = (\theta - \underline{\theta}^*)^+$, $U = (\theta - \underline{\theta})^+$.

8. In Example 4.6.2, establish that the UMP test has acceptance region (4.6.3).

    Hint: Use Examples 4.3.3 and 4.4.4.

Problems for Section 4.7

1. (a) Show that if θ has a beta, $\beta(r, s)$, distribution with r and s positive integers, then $\lambda = s\theta/r(1 - \theta)$ has the F distribution $\mathcal{F}_{2r,2s}$.

    Hint: See Sections B.2 and B.3.

    (b) Suppose that given $\theta = \theta$, X has a binomial, $\mathcal{B}(n, \theta)$, distribution and that θ has beta, $\beta(r, s)$ distribution with r and s integers. Show how the quantiles of the F distribution can be used to find upper and lower credible bounds for λ and for θ.

2. Suppose that given $\lambda = \lambda$, $X_1, \ldots, X_n$ are i.i.d. Poisson, $\mathcal{P}(\lambda)$ and that λ is distributed as V/s_0, where s_0 is some constant and $V \sim \chi_k^2$. Let $T = \sum_{i=1}^n X_i$.

    (a) Show that $(\lambda \mid T = t)$ is distributed as W/s, where $s = s_0 + 2n$ and $W \sim \chi_m^2$ with $m = k + 2t$.

    (b) Show how quantiles of the χ^2 distribution can be used to determine level $(1 - \alpha)$ upper and lower credible bounds for λ.

3. Suppose that given $\theta = \theta$, $X_1, \ldots, X_n$ are i.i.d. uniform, $\mathcal{U}(0, \theta)$, and that θ has the Pareto, $Pa(c, s)$, density

$$\pi(t) = sc^s/t^{s-1}, \ t > c, \ s > 0, \ c > 0.$$

(a) Let $M = \max\{X_1, \ldots, X_n\}$. Show that $(\boldsymbol{\theta} \mid M = m) \sim Pa(c', s')$ with $c' = \max\{c, m\}$ and $s' = s + n$.

(b) Find level $(1 - \alpha)$ upper and lower credible bounds for θ.

(c) Give a level $(1 - \alpha)$ confidence interval for θ.

(d) Compare the level $(1 - \alpha)$ upper and lower credible bounds for θ to the level $(1 - \alpha)$ upper and lower confidence bounds for θ. In particular consider the credible bounds as $n \to \infty$.

4. Suppose that given $\boldsymbol{\theta} = (\boldsymbol{\mu}_1, \boldsymbol{\mu}_2, \boldsymbol{\tau}) = (\mu_1, \mu_2, \tau)$, $X_1, \ldots, X_m$ and $Y_1, \ldots, Y_n$ are two independent $\mathcal{N}(\mu_1, \tau)$ and $\mathcal{N}(\mu_2, \tau)$ samples, respectively. Suppose $\boldsymbol{\theta}$ has the improper prior $\pi(\theta) = 1/\tau, \tau > 0$.

(a) Let $s_0 = \Sigma(x_i - \bar{x})^2 + \Sigma(y_j - \bar{y})^2$. Show formally that the posterior $\pi(\theta \mid \mathbf{x}, \mathbf{y})$ is proportional to

$$\pi(\tau \mid s_0)\pi(\mu_1 \mid \tau, \bar{x})\pi(\mu_2 \mid \tau, \bar{y})$$

where $\pi(\tau \mid s_0)$ is the density of s_0/V with $V \sim \chi_{m+n-2}$, $\pi(\mu_1 \mid \tau, \bar{x})$ is a $\mathcal{N}(\bar{x}, \tau/m)$ density and $\pi(\mu_2 \mid \tau, \bar{y})$ is a $\mathcal{N}(\bar{y}, \tau/n)$ density.
  Hint: $p(\theta \mid \mathbf{x}, \mathbf{y})$ is proportional to

$$p(\boldsymbol{\theta})p(\mathbf{x} \mid \mu_1, \tau)p(\mathbf{y} \mid \mu_2, \tau).$$

(b) Show that given τ, μ_1 and μ_2 are independent in the posterior distribution $p(\theta \mid \mathbf{x}, \mathbf{y})$ and that the joint density of $\Delta = \mu_1 - \mu_2$ and τ is

$$\pi(\Delta, \tau \mid \mathbf{x}, \mathbf{y}) = \pi(\tau \mid s_0)\pi(\Delta \mid \bar{x} - \bar{y}, \tau)$$

where $\pi(\Delta \mid \bar{x} - \bar{y}, \varphi)$ is the $\mathcal{N}(\bar{x} - \bar{y}, \tau(m^{-1} + n^{-1}))$ distribution.

(c) Set $s^2 = s_0/(m + n - 2)$. Show that the posterior distribution $\pi(t \mid \mathbf{x}, \mathbf{y})$ of

$$t = \frac{\Delta - (\bar{x} - \bar{y})}{s\sqrt{\frac{1}{m} + \frac{1}{n}}}$$

is (Student) t with $m + n - 2$ degrees of freedom.
  Hint: $\pi(\Delta \mid \mathbf{x}, \mathbf{y})$ is obtained by integrating out τ in $\pi(\Delta, \tau \mid \mathbf{x}, \mathbf{y})$.

(d) Use part (c) to give level $(1 - \alpha)$ credible bounds and a level $(1 - \alpha)$ credible interval for Δ.

Problems for Section 4.8

1. Let $X_1, \ldots, X_{n+1}$ be i.i.d. as $X \sim \mathcal{N}(\mu, \sigma_0^2)$, where σ_0^2 is known. Here $X_1, \ldots, X_n$ is observable and X_{n+1} is to be predicted.

(a) Give a level $(1 - \alpha)$ prediction interval for X_{n+1}.

(b) Compare the interval in part (a) to the Bayesian prediction interval (4.8.3) by doing a frequentist computation of the probability of coverage. That is, suppose $X_1, \ldots, X_n$ are i.i.d. $\mathcal{N}(\mu, \sigma_0^2)$. Take $\sigma_0^2 = \tau^2 = 1$, $n = 100$, $\eta_0 = 10$, and $\alpha = .05$. Then the level of the frequentist interval is 95%. Find the probability that the Bayesian interval covers X_{n+1} for $\mu = 5, 8, 9, 9.5, 10, 10.5, 11, 12, 15$. Present the results in a table and a graph.

2. Let $X_1, \ldots, X_{n+1}$ be i.i.d. as $X \sim F$, where $X_1, \ldots, X_n$ are observable and X_{n+1} is to be predicted. A level $(1 - \alpha)$ *lower (upper) prediction* bound on $Y = X_{n+1}$ is defined to be a function $\underline{Y}(\bar{Y})$ of $X_1, \ldots, X_n$ such that $P(\underline{Y} \leq Y) \geq 1 - \alpha$ $(P(Y \leq \bar{Y}) \geq 1 - \alpha)$.

(a) If F is $\mathcal{N}(\mu, \sigma_0^2)$ with σ_0^2 known, give level $(1 - \alpha)$ lower and upper prediction bounds for X_{n+1}.

(b) If F is $\mathcal{N}(\mu, \sigma^2)$ with σ^2 unknown, give level $(1 - \alpha)$ lower and upper prediction bounds for X_{n+1}.

(c) If F is continuous with a positive density f on (a, b), $-\infty \leq a < b \leq \infty$, give level $(1 - \alpha)$ distribution free lower and upper prediction bounds for X_{n+1}.

3. Suppose $X_1, \ldots, X_{n+1}$ are i.i.d. as X where X has the exponential distribution

$$F(x \mid \theta) = 1 - e^{-x/\theta}, \ x > 0, \ \theta > 0.$$

Suppose $X_1, \ldots, X_n$ are observable and we want to predict X_{n+1}. Give a level $(1 - \alpha)$ prediction interval for X_{n+1}.
    Hint: $2X_i/\theta$ has a χ_2^2 distribution and $nX_{n+1}/\sum_{i=1}^{n} X_i$ has an $\mathcal{F}_{2,2n}$ distribution.

4. Suppose that given $\boldsymbol{\theta} = \theta$, X is a binomial, $\mathcal{B}(n, \theta)$, random variable, and that $\boldsymbol{\theta}$ has a beta, $\beta(r, s)$, distribution. Suppose that Y, which is not observable, has a $\mathcal{B}(m, \theta)$ distribution given $\boldsymbol{\theta} = \theta$. Show that the conditional (predictive) distribution of Y given $X = x$ is

$$q(y \mid x) = \binom{m}{y} B(r + x + y, s + n - x + m - y)/B(r + x, s + n - x)$$

where $B(\cdot, \cdot)$ denotes the beta function. (This $q(y \mid x)$ is sometimes called the Pólya distribution.)
    Hint: First show that

$$q(y \mid x) = \int p(y \mid \theta) \pi(\theta \mid x) d\theta.$$

5. In Example 4.8.2, let $U^{(1)} < \cdots < U^{(n+1)}$ denote $U_1, \ldots, U_{n+1}$ ordered. Establish (4.8.2) by using the observation that U_{n+1} is equally likely to be any of the values $U^{(1)}, \ldots, U^{(n+1)}$.

Problems for Section 4.9

1. Let X have a binomial, $\mathcal{B}(n, \theta)$, distribution. Show that the likelihood ratio statistic for testing $H : \theta = \frac{1}{2}$ versus $K : \theta \neq \frac{1}{2}$ is equivalent to $|2X - n|$.

Hint: Show that for $x \leq \frac{1}{2}n$, $\lambda(x)$ is an increasing function of $-(2x - n)$ and $\lambda(x) = \lambda(n - x)$.

In Problems 2–4, let $X_1, \ldots, X_n$ be a $\mathcal{N}(\mu, \sigma^2)$ sample with both μ and σ^2 unknown.

2. In testing $H : \mu \leq \mu_0$ versus $K : \mu > \mu_0$ show that the one-sided, one-sample t test is the likelihood ratio test (for $\alpha < \frac{1}{2}$).

Hint: Note that $\hat{\mu}_0 = \bar{X}$ if $\bar{X} \leq \mu_0$ and $= \mu_0$ otherwise. Thus, $\log \lambda(\mathbf{x}) = 0$, if $T_n \leq 0$ and $= (n/2) \log(1 + T_n^2/(n-1))$ for $T_n > 0$, where T_n is the t statistic.

3. *One-Sided Tests for Scale.* We want to test $H : \sigma^2 \leq \sigma_0^2$ versus $K : \sigma^2 > \sigma_0^2$. Show that

    (a) Likelihood ratio tests are of the form: Reject if, and only if,

$$\frac{n\hat{\sigma}^2}{\sigma_0^2} = \frac{1}{\sigma_0^2} \sum_{i=1}^{n}(X_i - \bar{X})^2 \geq c.$$

Hint: $\log \lambda(\mathbf{x}) = 0$, if $\hat{\sigma}^2/\sigma_0^2 \leq 1$ and $= (n/2)[\hat{\sigma}^2/\sigma_0^2 - 1 - \log(\hat{\sigma}^2/\sigma_0^2)]$ otherwise.

    (b) To obtain size α for H we should take $c = x_{n-1}(1 - \alpha)$.
    Hint: Recall Theorem B.3.3.

    (c) These tests coincide with the tests obtained by inverting the family of level $(1 - \alpha)$ lower confidence bounds for σ^2.

4. *Two-Sided Tests for Scale.* We want to test $H : \sigma = \sigma_0$ versus $K : \sigma \neq \sigma_0$.

    (a) Show that the size α likelihood ratio test accepts if, and only if,

$$c_1 \leq \frac{1}{\sigma_0^2} \sum_{i=1}^{n}(X_i - \bar{X})^2 \leq c_2 \text{ where } c_1 \text{ and } c_2 \text{ satisfy,}$$

    (i) $F(c_2) - F(c_1) = 1 - \alpha$, where F is the d.f. of the χ_{n-1}^2 distribution.

    (ii) $c_1 - c_2 = n \log c_1/c_2$.

    (b) Use the normal approximation to check that

$$\begin{aligned} c_{1n} &= n - \sqrt{2nz}(1 - \tfrac{1}{2}\alpha) \\ c_{2n} &= n + \sqrt{2nz}(1 - \tfrac{1}{2}\alpha) \end{aligned}$$

approximately satisfy (i) and also (ii) in the sense that the ratio

$$\frac{c_{1n} - c_{2n}}{n \log c_{1n}/c_{2n}} \to 1 \text{ as } n \to \infty.$$

    (c) Deduce that the critical values of the commonly used equal-tailed test, $x_{n-1}(\tfrac{1}{2}\alpha)$, $x_{n-1}(1 - \tfrac{1}{2}\alpha)$ also approximately satisfy (i) and (ii) of part (a).

5. The following blood pressures were obtained in a sample of size $n = 5$ from a certain population: 124, 110, 114, 100, 190. Assume the one-sample normal model.

  (a) Using the size $\alpha = 0.05$ one-sample t test, can we conclude that the mean blood pressure in the population is significantly larger than 100?

  (b) Compute a level 0.95 confidence interval for σ^2 corresponding to inversion of the equal-tailed tests of Problem 4.9.4.

  (c) Compute a level 0.90 confidence interval for the mean blood pressure μ.

6. Let $X_1, \ldots, X_{n_1}$ and $Y_1, \ldots, Y_{n_2}$ be two independent $\mathcal{N}(\mu_1, \sigma^2)$ and $\mathcal{N}(\mu_2, \sigma^2)$ samples, respectively.

  (a) Show that the MLE of $\theta = (\mu_1, \mu_2, \sigma^2)$ is $(\bar{X}, \bar{Y}, \tilde{\sigma}^2)$, where $\tilde{\sigma}^2$ is as defined in Section 4.9.3.

  (b) Consider the problem of testing $H : \mu_1 \leq \mu_2$ versus $K : \mu_1 > \mu_2$. Assume $\alpha \leq \frac{1}{2}$. Show that the likelihood ratio statistic is equivalent to the two-sample t statistic T.

  (c) Using the normal approximation $\Phi(z(\alpha) + \sqrt{n_1 n_2/n}(\mu_1 - \mu_2)/\sigma)$ to the power, find the sample size n needed for the level 0.01 test to have power 0.95 when $n_1 = n_2 = \frac{1}{2}n$ and $(\mu_1 - \mu_2)/\sigma = \frac{1}{2}$.

7. The following data are from an experiment to study the relationship between forage production in the spring and mulch left on the ground the previous fall. The control measurements (x's) correspond to 0 pounds of mulch per acre, whereas the treatment measurements (y's) correspond to 500 pounds of mulch per acre. Forage production is also measured in pounds per acre.

x	794	1800	576	411	897
y	2012	2477	3498	2092	1808

Assume the two-sample normal model with equal variances.

  (a) Find a level 0.95 confidence interval for $\mu_2 - \mu_1$.

  (b) Can we conclude that leaving the indicated amount of mulch on the ground significantly improves forage production? Use $\alpha = 0.05$.

  (c) Find a level 0.90 confidence interval for σ by using the pivot s^2/σ^2.

8. Suppose $\mathbf{X}$ has density $p(\mathbf{x}, \theta)$, $\theta \in \Theta$, and that T is sufficient for θ. Show that $\lambda(\mathbf{X}, \Theta_0, \Theta_1)$ depends on $\mathbf{X}$ only through T.

9. The normally distributed random variables $X_1, \ldots, X_n$ are said to be *serially correlated* or to follow an autoregressive model if we can write

$$X_i = \theta X_{i-1} + \epsilon_i, \quad i = 1, \ldots, n,$$

where $X_0 = 0$ and $\epsilon_1, \ldots, \epsilon_n$ are independent $\mathcal{N}(0, \sigma^2)$ random variables.

(a) Show that the density of $\mathbf{X} = (X_1, \ldots, X_n)$ is

$$p(\mathbf{x}, \theta) = (2\pi\sigma^2)^{-\frac{1}{2}n} \exp\{-(1/2\sigma^2) \sum_{i=1}^{n} (x_i - \theta x_{i-1})^2\}$$

for $-\infty < x_i < \infty$, $i = 1, \ldots, n$, $x_0 = 0$.

(b) Show that the likelihood ratio statistic of $H : \theta = 0$ (independence) versus $K : \theta \neq 0$ (serial correlation) is equivalent to $-(\sum_{i=2}^{n} X_i X_{i-1})^2 / \sum_{i=1}^{n-1} X_i^2$.

10. (An example due to C. Stein). Consider the following model. Fix $0 < \alpha < \frac{1}{2}$ and $\alpha/[2(1 - \alpha)] < c < \alpha$. Let Θ consist of the point -1 and the interval $[0, 1]$. Define the frequency functions $p(x, \theta)$ by the following table.

θ \ x	-2	-1	0	1	2
-1	$\frac{1}{2}\alpha$	$\frac{1}{2} - \alpha$	α	$\frac{1}{2} - \alpha$	$\frac{1}{2}\alpha$
$\neq -1$	θc	$\left(\frac{1-c}{1-\alpha}\right)\left(\frac{1}{2} - \alpha\right)$	$\left(\frac{1-c}{1-\alpha}\right)\alpha$	$\left(\frac{1-c}{1-\alpha}\right)\left(\frac{1}{2} - \alpha\right)$	$(1 - \theta)c$

(a) What is the size α likelihood ratio test for testing $H : \theta = -1$ versus $K : \theta \neq -1$?

(b) Show that the test that rejects if, and only if, $X = 0$, has level α and is strictly more powerful whatever be θ.

11. *The power functions of one- and two-sided t tests.* Suppose that T has a noncentral t, $\mathcal{T}_{k,\delta}$, distribution. Show that,

(a) $P_\delta[T \geq t]$ is an increasing function of δ.

(b) $P_\delta[|T| \geq t]$ is an increasing function of $|\delta|$.
Hint: Let Z and V be independent and have $\mathcal{N}(\delta, 1)$, χ_k^2 distributions respectively. Then, for each $v > 0$, $P_\delta[Z \geq t\sqrt{v/k}]$ is increasing in δ, $P_\delta[|Z| \geq t\sqrt{v/k}]$ is increasing in $|\delta|$. Condition on V and apply the double expectation theorem.

12. Show that the noncentral t distribution, $\mathcal{T}_{k,\delta}$, has density

$$f_{k,\delta}(t) = \frac{1}{\sqrt{\pi k}(\frac{1}{2}k)2^{\frac{1}{2}(k+1)}} \int_0^\infty x^{\frac{1}{2}(k-1)} e^{-\frac{1}{2}\{x + (t\sqrt{x/k} - \delta)^2\}} dx.$$

Hint: Let Z and V be as in the preceding hint. From the joint distribution of Z and V, get the joint distribution of $Y_1 = Z/\sqrt{V/k}$ and $Y_2 = V$. Then use $p_{Y_1}(y_1) = \int p_{Y_1, Y_2}(y_1, y_2) dy_2$.

13. *The F Test for Equality of Scale.* Let $X_1, \ldots, X_{n_1}$, $Y_1, \ldots, Y_{n_2}$ be two independent samples from $\mathcal{N}(\mu_1, \sigma_1^2)$, $\mathcal{N}(\mu_2, \sigma_2^2)$, respectively, with all parameters assumed unknown.

(a) Show that the LR test of $H : \sigma_1^2 = \sigma_2^2$ versus $K : \sigma_2^2 > \sigma_1^2$ is of the form: Reject if, and only if, $F = [(n_1 - 1)/(n_2 - 1)]\Sigma(Y_i - \bar{Y})^2/\Sigma(X_i - \bar{X})^2 \geq C$.

(b) Show that $(\sigma_1^2/\sigma_2^2)F$ has an $\mathcal{F}_{n_2-1,n_1-1}$ distribution and that critical values can be obtained from the $\mathcal{F}$ table.

(c) Justify the two-sided F test: Reject H if, and only if, $F \geq f(1 - \alpha/2)$ or $F \leq f(\alpha/2)$, where $f(t)$ is the tth quantile of the $\mathcal{F}_{n_2-1,n_1-1}$ distribution, as an approximation to the LR test of $H : \sigma_1 = \sigma_2$ versus $K : \sigma_1 \neq \sigma_2$. Argue as in Problem 4.9.4.

(d) Relate the two-sided test of part (c) to the confidence intervals for σ_2^2/σ_1^2 obtained in Problem 4.4.10.

14. The following data are the blood cholesterol levels (x's) and weight/height ratios (y's) of 10 men involved in a heart study.

x	254	240	279	284	315	250	298	384	310	337
y	2.71	2.96	2.62	2.19	2.68	2.64	2.37	2.61	2.12	1.94

Using the likelihood ratio test for the bivariate normal model, can you conclude at the 10% level of significance that blood cholesterol level is correlated with weight/height ratio?

15. Let $(X_1, Y_1), \ldots, (X_n, Y_n)$ be a sample from a bivariate $\mathcal{N}(0, 0, \sigma_1^2, \sigma_2^2, \rho)$ distribution. Consider the problem of testing $H : \rho = 0$ versus $K : \rho \neq 0$.

(a) Show that the likelihood ratio statistic is equivalent to $|r|$ where

$$r = \sum_{i=1}^{n} X_i Y_i \Bigg/ \sqrt{\sum_{i=1}^{n} X_i^2 \sum_{j=1}^{n} Y_j^2}.$$

(b) Show that if we have a sample from a bivariate $\mathcal{N}(\mu_1, \mu_2, \sigma_1^2, \sigma_2^2, \rho)$ distribution, then $P[\hat{\rho} \geq c]$ is an increasing function of ρ for fixed c.

Hint: Use the transformations and Problem B.4.7 to conclude that $\hat{\rho}$ has the same distribution as $S_{12}/S_1 S_2$, where

$$S_1^2 = \sum_{i=2}^{n} U_i^2, \ S_2^2 = \sum_{i=2}^{n} V_i^2, \ S_{12} = \sum_{i=2}^{n} U_i V_i$$

and $(U_2, V_2), \ldots, (U_n, V_n)$ is a sample from a $\mathcal{N}(0, 0, 1, 1, \rho)$ distribution. Let $R = S_{12}/S_1 S_2$, $T = \sqrt{n - 2} R/\sqrt{1 - R^2}$, and using the arguments of Problems B.4.7 and B.4.8, show that given $U_2 = u_2, \ldots, U_n = u_n$, T has a noncentral $\mathcal{T}_{n-2}$ distribution with noncentrality parameter ρ. Because this conditional distribution does not depend on $(u_2, \ldots, u_n)$, the continuous version of (B.1.24) implies that this is also the unconditional distribution. Finally, note that $\hat{\rho}$ has the same distribution as R, that T is an increasing function of R, and use Problem 4.8.11(a).

16. Let $\lambda(\mathbf{X})$ denote the likelihood ratio statistic for testing $H : \rho = 0$ versus $K : \rho \neq 0$ in the bivariate normal model. Show, using (4.9.4) and (4.9.5) that $2 \log \lambda(\mathbf{X}) \xrightarrow{\mathcal{L}} V$, where V has a χ_1^2 distribution.

17. Consider the bioequivalence example in Problem 3.2.9.

(a) Find the level α LR test for testing $H : \theta \in [-\epsilon, \epsilon]$ versus $K : \theta \notin [-\epsilon, \epsilon]$.

(b) In this example it is reasonable to use a level α test of $H: \theta \notin [-\epsilon, \epsilon]$ versus $K:$ $\theta \in [-\epsilon, \epsilon]$. Propose such a test and compare your solution to the Bayesian solution based on a continuous loss function given in Problem 3.2.9. Consider the cases $\eta_0 = 0$, $\tau_0^2 \to \infty$, and $\eta_0 = 0$, $n \to \infty$.

4.11   NOTES

Notes for Section 4.1

(1) The point of view usually taken in science is that of Karl Popper [1968]. Acceptance of a hypothesis is only provisional as an adequate current approximation to what we are interested in understanding. Rejection is more definitive.

(2) We ignore at this time some real-life inadequacies of this experiment such as the placebo effect (see Example 1.1.3).

(3) A good approximation (Durbin, 1973; Stephens, 1974) to the critical value is $c_n(t) = t/(\sqrt{n} - 0.01 + 0.85/\sqrt{n})$ where $t = 1.035, 0.895$ and $t = 0.819$ for $\alpha = 0.01, .05$ and 0.10, respectively.

Notes for Section 4.3

(1) Such a class is sometimes called essentially complete. The term *complete* is then reserved for the class where strict inequality in $(4.3.3)$ holds for some θ if $\varphi \notin \mathcal{D}$.

(2) The theory of complete and essentially complete families is developed in Wald (1950), see also Ferguson (1967). Essentially, if the parameter space is compact and loss functions are bounded, the class of Bayes procedures is complete. More generally the closure of the class of Bayes procedures (in a suitable metric) is complete.

Notes for Section 4.4

(1) If the continuity correction discussed in Section A.15 is used here, S in $\bar{\theta}(\mathbf{X})$ would be replaced by $S + \frac{1}{2}$, and S in $\underline{\theta}(\mathbf{X})$ is replaced by $S - \frac{1}{2}$.

Notes for Section 4.5

(1) In using $\underline{\theta}(S)$ as a confidence bound we are using the region $[\underline{\theta}(S), 1]$. Because the region contains $C(\mathbf{X})$, it also has confidence level $(1 - \alpha)$.

4.12   REFERENCES

BARLOW, R. AND F. PROSCHAN, *Mathematical Theory of Reliability* New York: J. Wiley & Sons, 1965.

BICKEL, P., "Inference and auditing: the Stringer bound." *Internat. Statist. Rev., 60,* 197–209 (1992).

BICKEL, P., E. HAMMEL, AND J. W. O'CONNELL, "Is there a sex bias in graduate admissions?" *Science, 187*, 398–404 (1975).

BOX, G. E. P., *Apology for Ecumenism in Statistics and Scientific Inference*, Data Analysis and Robustness, G. E. P. Box, T. Leonard, and C. F. Wu, Editors New York: Academic Press, 1983.

BROWN, L. D., T. CAI, AND A. DAS GUPTA, "Interval estimation for a binomial proportion," *Statistical Science*, 101–128 (2001).

DOKSUM, K. A. AND G. SIEVERS, "Plotting with confidence: Graphical comparisons of two populations," *Biometrika, 63*, 421–434 (1976).

DOKSUM, K. A., G. FENSTAD, AND R. AABERGE, "Plots and tests for symmetry," *Biometrika, 64*, 473–487 (1977).

DURBIN, J., "Distribution theory for tests based on the sample distribution function," *Regional Conference Series in Applied Math., 9*, SIAM, Philadelphia, Pennsylvania (1973).

FERGUSON, T., *Mathematical Statistics. A Decision Theoretic Approach* New York: Academic Press, 1967.

FISHER, R. A., *Statistical Methods for Research Workers*, 13th ed. New York: Hafner Publishing Company, 1958.

HALD, A., *Statistical Theory with Engineering Applications* New York: J. Wiley & Sons, 1952.

HEDGES, L. V. AND I. OLKIN, *Statistical Methods for Meta-Analysis* Orlando, FL: Academic Press, 1985.

JEFFREYS, H., *The Theory of Probability* Oxford: Oxford University Press, 1961.

LEHMANN, E. L., *Testing Statistical Hypotheses*, 2nd ed. New York: Springer, 1997.

POPPER, K. R., *Conjectures and Refutations; the Growth of Scientific Knowledge*, 3rd ed. New York: Harper and Row, 1968.

PRATT, J., "Length of confidence intervals," *J. Amer. Statist. Assoc., 56*, 549–567 (1961).

SACKROWITZ, H. AND E. SAMUEL–CAHN, "P values as random variables—Expected P values," *The American Statistician, 53*, 326–331 (1999).

STEPHENS, M., "EDF statistics for goodness of fit," *J. Amer. Statist., 69*, 730–737 (1974).

STEIN, C., "A two-sample test for a linear hypothesis whose power is independent of the variance," *Ann. Math. Statist., 16*, 243–258 (1945).

TATE, R. F. AND G. W. KLETT, "Optimal confidence intervals for the variance of a normal distribution," *J. Amer. Statist. Assoc., 54*, 674–682 (1959).

VAN ZWET, W. R. AND J. OSTERHOFF, "On the combination of independent test statistics," *Ann. Math. Statist., 38*, 659–680 (1967).

WALD, A., *Sequential Analysis* New York: Wiley, 1947.

WALD, A., *Statistical Decision Functions* New York: Wiley, 1950.

WANG, Y., "Probabilities of the type I errors of the Welch tests," *J. Amer. Statist. Assoc., 66*, 605–608 (1971).

WELCH, B., "Further notes on Mrs. Aspin's tables," *Biometrika, 36*, 243–246 (1949).

WETHERILL, G. B. AND K. D. GLAZEBROOK, *Sequential Methods in Statistics* New York: Chapman and Hall, 1986.

Chapter 5

ASYMPTOTIC APPROXIMATIONS

5.1 INTRODUCTION: THE MEANING AND USES OF ASYMPTOTICS

Despite the many simple examples we have dealt with, closed form computation of risks in terms of known functions or simple integrals is the exception rather than the rule. Even if the risk is computable for a specific P by numerical integration in one dimension, the qualitative behavior of the risk as a function of parameter and sample size is hard to ascertain. Worse, computation even at a single point may involve high-dimensional integrals. In particular, consider a sample $X_1, \ldots, X_n$ from a distribution F, our setting for this section and most of this chapter. If we want to estimate $\mu(F) \equiv E_F X_1$ and use $\bar{X}$ we can write,

$$MSE_F(\bar{X}) = \frac{\sigma^2(F)}{n}. \tag{5.1.1}$$

This is a highly informative formula, telling us exactly how the MSE behaves as a function of n, and calculable for any F and all n by a single one-dimensional integration. However, consider $\mathrm{med}(X_1, \ldots, X_n)$ as an estimate of the population median $\nu(F)$. If n is odd, $\nu(F) = F^{-1}\left(\frac{1}{2}\right)$, and F has density f we can write

$$MSE_F(\mathrm{med}(X_1, \ldots, X_n)) = \int_{-\infty}^{\infty} \left(x - F^{-1}\left(\tfrac{1}{2}\right)\right)^2 g_n(x) dx \tag{5.1.2}$$

where, from Problem (B.2.13), if $n = 2k + 1$,

$$g_n(x) = n \binom{2k}{k} F^k(x)(1 - F(x))^k f(x). \tag{5.1.3}$$

Evaluation here requires only evaluation of F and a one-dimensional integration, but a different one for each n (Problem 5.1.1). Worse, the qualitative behavior of the risk as a function of n and simple parameters of F is not discernible easily from (5.1.2) and (5.1.3). To go one step further, consider evaluation of the power function of the one-sided t test of Chapter 4. If $X_1, \ldots, X_n$ are i.i.d. $\mathcal{N}(\mu, \sigma^2)$ we have seen in Section 4.9.2 that $\sqrt{n}\bar{X}/s$ has a noncentral t distribution with parameter μ/σ and $n - 1$ degrees of freedom. This distribution may be evaluated by a two-dimensional integral using classical functions

(Problem 5.1.2) and its qualitative properties are reasonably transparent. But suppose F is not Gaussian. It seems impossible to determine explicitly what happens to the power function because the distribution of $\sqrt{n}\bar{X}/s$ requires the joint distribution of $(\bar{X}, s)$ and in general this is only representable as an n-dimensional integral;

$$P\left[\sqrt{n}\frac{\bar{X}}{s} \leq t\right] = \int_A f(x_1)\dots f(x_n)d\mathbf{x}$$

where

$$A = \left\{(x_1, \dots, x_n) : \sum_{i=1}^{n} x_i \leq \frac{\sqrt{n}t}{n-1}\left(\sum_{i=1}^{n} x_i^2 - \frac{(\Sigma x_i)^2}{n}\right)\right\}.$$

There are two complementary approaches to these difficulties. The first, which occupies us for most of this chapter, is to approximate the risk function under study

$$R_n(F) \equiv E_F l(F, \delta(X_1, \dots, X_n)),$$

by a qualitatively simpler to understand and easier to compute function, $\widetilde{R}_n(F)$. The other, which we explore further in later chapters, is to use the Monte Carlo method. In its simplest form, Monte Carlo is described as follows. Draw B independent "samples" of size n, $\{X_{1j}, \dots, X_{nj}\}$, $1 \leq j \leq B$ from F using a random number generator and an explicit form for F. Approximately evaluate $R_n(F)$ by

$$\widehat{R}_B = \frac{1}{B}\sum_{j=1}^{B} l(F, \delta(X_{1j}, \dots, X_{nj})). \tag{5.1.4}$$

By the law of large numbers as $B \to \infty$, $\widehat{R}_B \xrightarrow{P} R_n(F)$. Thus, save for the possibility of a very unlikely event, just as in numerical integration, we can approximate $R_n(F)$ arbitrarily closely. We now turn to a detailed discussion of asymptotic approximations but will return to describe Monte Carlo and show how it complements asymptotics briefly in Example 5.3.3.

Asymptotics in statistics is usually thought of as the study of the limiting behavior of statistics or, more specifically, of distributions of statistics, based on observing n i.i.d. observations $X_1, \dots, X_n$ as $n \to \infty$. We shall see later that the scope of asymptotics is much greater, but for the time being let's stick to this case as we have until now.

Asymptotics, in this context, always refers to a sequence of statistics

$$\{T_n(X_1, \dots, X_n)\}_{n\geq 1},$$

for instance the sequence of means $\{\bar{X}_n\}_{n\geq 1}$, where $\bar{X}_n \equiv \frac{1}{n}\sum_{i=1}^{n} X_i$, or the sequence of medians, or it refers to the sequence of their distributions

$$\{\mathcal{L}_F(T_n(X_1, \dots, X_n))\}_{n\geq 1}.$$

Asymptotic statements are always statements about *the sequence*. The classical examples are, $\bar{X}_n \xrightarrow{P} E_F(X_1)$ or

$$\mathcal{L}_F(\sqrt{n}(\bar{X}_n - E_F(X_1))) \to \mathcal{N}(0, \text{Var}_F(X_1)).$$

In theory these limits say nothing about any particular $T_n(X_1, \ldots, X_n)$ but in practice we act as if they do because the $T_n(X_1, \ldots, X_n)$ we consider are closely related as functions of n so that we expect the limit to *approximate* $T_n(X_1, \ldots, X_n)$ or $\mathcal{L}_F(T_n(X_1, \ldots, X_n))$ (in an appropriate sense). For instance, the weak law of large numbers tells us that, if $E_F|X_1| < \infty$, then

$$\bar{X}_n \xrightarrow{P} \mu \equiv E_F(X_1). \tag{5.1.5}$$

That is, (see A.14.1)

$$P_F[|\bar{X}_n - \mu| \geq \epsilon] \to 0 \tag{5.1.6}$$

for all $\epsilon > 0$. We interpret this as saying that, for n sufficiently large, $\bar{X}_n$ is approximately equal to its expectation. The trouble is that for any specified degree of approximation, say, $\epsilon = .01$, (5.1.6) does not tell us how large n has to be for the chance of the approximation not holding to this degree (the left-hand side of (5.1.6)) to fall, say, below .01. Is $n \geq 100$ enough or does it have to be $n \geq 100,000$? Similarly, the central limit theorem tells us that if $E_F|X_1^2| < \infty$, μ is as above and $\sigma^2 \equiv \text{Var}_F(X_1)$, then

$$P_F\left[\sqrt{n}\frac{(\bar{X}_n - \mu)}{\sigma} \leq z\right] \to \Phi(z) \tag{5.1.7}$$

where Φ is the standard normal d.f.

As an approximation, this reads

$$P_F[\bar{X}_n \leq x] \approx \Phi\left(\sqrt{n}\frac{(x - \mu)}{\sigma}\right). \tag{5.1.8}$$

Again we are faced with the questions of how good the approximation is for given n, x, and P_F. What we in principle prefer are bounds, which are available in the classical situations of (5.1.6) and (5.1.7). Thus, by Chebychev's inequality, if $E_F X_1^2 < \infty$,

$$P_F[|\bar{X}_n - \mu| \geq \epsilon] \leq \frac{\sigma^2}{n\epsilon^2}. \tag{5.1.9}$$

As a bound this is typically far too conservative. For instance, if $|X_1| \leq 1$, the much more delicate Hoeffding bound (B.9.6) gives

$$P_F[|\bar{X}_n - \mu| \geq \epsilon] \leq 2\exp\left\{-\tfrac{1}{2}n\epsilon^2\right\}. \tag{5.1.10}$$

Because $|X_1| \leq 1$ implies that $\sigma^2 \leq 1$ with $\sigma^2 = 1$ possible (Problem 5.1.3), the right-hand side of (5.1.9) when σ^2 is unknown becomes $1/n\epsilon^2$. For $\epsilon = .1$, $n = 400$, (5.1.9) is .25 whereas (5.1.10) is .14.

Further qualitative features of these bounds and relations to approximation (5.1.8) are given in Problem 5.1.4. Similarly, the celebrated Berry–Esséen bound (A.15.11) states that if $E_F|X_1|^3 < \infty$,

$$\sup_x \left|P_F\left[\sqrt{n}\frac{(\bar{X}_n - \mu)}{\sigma} \leq x\right] - \Phi(x)\right| \leq C\frac{E_F|X_1|^3}{\sigma^3 n^{1/2}} \tag{5.1.11}$$

where C is a universal constant known to be $\leq 33/4$. Although giving us some idea of how much (5.1.8) differs from the truth, (5.1.11) is again much too conservative generally.[1] The approximation (5.1.8) is typically much better than (5.1.11) suggests.

Bounds for the goodness of approximations have been available for $\bar{X}_n$ and its distribution to a much greater extent than for nonlinear statistics such as the median. Yet, as we have seen, even here they are not a very reliable guide. Practically one proceeds as follows:

(a) Asymptotic approximations are derived.

(b) Their validity for the given n and T_n for some plausible values of F is tested by numerical integration if possible or Monte Carlo computation.

If the agreement is satisfactory we use the approximation even though the agreement for the true but unknown F generating the data may not be as good.

Asymptotics has another important function beyond suggesting numerical approximations for specific n and F. If they are simple, asymptotic formulae suggest qualitative properties that may hold even if the approximation itself is not adequate. For instance, (5.1.7) says that the behavior of the distribution of $\bar{X}_n$ is for large n governed (approximately) only by μ and σ^2 in a precise way, although the actual distribution depends on P_F in a complicated way. It suggests that qualitatively the risk of $\bar{X}_n$ as an estimate of μ, for any loss function of the form $l(F, d) = \lambda(|\mu - d|)$ where $\lambda(0) = 0$, $\lambda'(0) > 0$, behaves like $\lambda'(0)(\sigma/\sqrt{n})(\sqrt{2\pi})$ (Problem 5.1.5) and quite generally that risk increases with σ and decreases with n, which is reasonable.

As we shall see, quite generally, good estimates $\widehat{\theta}_n$ of parameters $\theta(F)$ will behave like $\bar{X}_n$ does in relation to μ. The estimates $\widehat{\theta}_n$ will be *consistent*, $\widehat{\theta}_n \overset{P}{\to} \theta(F)$, for all F in the model, and *asymptotically normal*,

$$\mathcal{L}_F \left(\frac{\sqrt{n}[\widehat{\theta}_n - \theta(F)]}{\sigma(\theta, F)} \right) \to \mathcal{N}(0, 1) \tag{5.1.12}$$

where $\sigma(\theta, F)$ typically is the standard deviation (SD) of $\sqrt{n}\widehat{\theta}_n$ or an approximation to this SD. Consistency will be pursued in Section 5.2 and asymptotic normality via the delta method in Section 5.3. The qualitative implications of results such as these are very important when we consider comparisons between competing procedures. Note that this feature of simple asymptotic approximations using the normal distribution is not replaceable by Monte Carlo.

We now turn to specifics. As we mentioned, Section 5.2 deals with consistency of various estimates including maximum likelihood. The arguments apply to vector-valued estimates of Euclidean parameters. In particular, consistency is proved for the estimates of canonical parameters in exponential families. Section 5.3 begins with asymptotic computation of moments and asymptotic normality of functions of a scalar mean and include as an application asymptotic normality of the maximum likelihood estimate for one-parameter exponential families. The methods are then extended to vector functions of vector means and applied to establish asymptotic normality of the MLE $\widehat{\eta}$ of the canonical parameter η

in exponential families among other results. Section 5.4 deals with optimality results for likelihood-based procedures in one-dimensional parameter models. Finally in Section 5.5 we examine the asymptotic behavior of Bayes procedures. The notation we shall use in the rest of this chapter conforms closely to that introduced in Sections A.14, A.15, and B.7. We will recall relevant definitions from that appendix as we need them, but we shall use results we need from A.14, A.15, and B.7 without further discussion.

Summary. Asymptotic statements refer to the behavior of sequences of procedures as the sequence index tends to ∞. In practice, asymptotics are methods of approximating risks, distributions, and other statistical quantities that are not realistically computable in closed form, by quantities that can be so computed. Most asymptotic theory we consider leads to approximations that in the i.i.d. case become increasingly valid as the sample size increases. We also introduce Monte Carlo methods and discuss the interaction of asymptotics, Monte Carlo, and probability bounds.

5.2 CONSISTENCY

5.2.1 Plug-In Estimates and MLEs in Exponential Family Models

Suppose that we have a sample $X_1, \ldots, X_n$ from $P_{\boldsymbol{\theta}}$ where $\boldsymbol{\theta} \in \Theta$ and want to estimate a real or vector $q(\boldsymbol{\theta})$. The least we can ask of our estimate $\widehat{q}_n(X_1, \ldots, X_n)$ is that as $n \to \infty$, $\widehat{q}_n \xrightarrow{P_{\boldsymbol{\theta}}} q(\boldsymbol{\theta})$ for all $\boldsymbol{\theta}$. That is, in accordance with (A.14.1) and (B.7.1), for all $\boldsymbol{\theta} \in \Theta, \epsilon > 0$,

$$P_{\boldsymbol{\theta}}[|\widehat{q}_n(X_1, \ldots, X_n) - q(\boldsymbol{\theta})| \geq \epsilon] \to 0. \tag{5.2.1}$$

where $|\cdot|$ denotes Euclidean distance. A stronger requirement is

$$\sup_{\boldsymbol{\theta}} \left\{ P_{\boldsymbol{\theta}} \left[|\widehat{q}_n(X_1, \ldots, X_n) - q(\boldsymbol{\theta})| \geq \epsilon \right] : \boldsymbol{\theta} \in \Theta \right\} \to 0. \tag{5.2.2}$$

Bounds $b(n, \epsilon)$ for $\sup_{\boldsymbol{\theta}} P_{\boldsymbol{\theta}} [|\widehat{q}_n - q(\boldsymbol{\theta})| \geq \epsilon]$ that yield (5.2.2) are preferable and we shall indicate some of qualitative interest when we can. But, with all the caveats of Section 5.1, (5.2.1), which is called *consistency* of $\widehat{q}_n$ and can be thought of as 0'th order asymptotics, remains central to all asymptotic theory. The stronger statement (5.2.2) is called *uniform consistency*. If Θ is replaced by a smaller set K, we talk of uniform consistency over K.

Example 5.2.1. *Means.* The simplest example of consistency is that of the mean. If $X_1, \ldots, X_n$ are i.i.d. P where P is unknown but $E_P|X_1| < \infty$ then, by the WLLN,

$$\bar{X} \xrightarrow{P} \mu(P) \equiv E(X_1)$$

and $\mu(\widehat{P}) = \bar{X}$, where $\widehat{P}$ is the empirical distribution, is a consistent estimate of $\mu(P)$. For $\mathcal{P}$ this large it is not uniformly consistent. (See Problem 5.2.2.) However, if, for instance, $\mathcal{P} \equiv \{P : E_P X_1^2 \leq M < \infty\}$, then $\bar{X}$ is uniformly consistent over $\mathcal{P}$ because by Chebyshev's inequality, for all $P \in \mathcal{P}$,

$$P[|\bar{X} - \mu(P)| \geq \epsilon]| \leq \frac{\mathrm{Var}(\bar{X})}{\epsilon^2} \leq \frac{M}{n\epsilon^2}$$

$\square$

Example 5.2.2. *Binomial Variance.* Let $X_1, \dots, X_n$ be the indicators of binomial trials with $P[X_1 = 1] = p$. Then $N = \sum X_i$ has a $\mathcal{B}(n, p)$ distribution, $0 \leq p \leq 1$, and $\hat{p} = \bar{X} = N/n$ is a uniformly consistent estimate of p. But further, consider the plug-in estimate $\hat{p}(1-\hat{p})/n$ of the variance of $\hat{p}$, which is $\frac{1}{n}q(\hat{p})$, where $q(p) = p(1-p)$. Evidently, by A.14.6, $q(\hat{p})$ is consistent. Other moments of X_1 can be consistently estimated in the same way. $\square$

To some extent the plug-in method was justified by consistency considerations and it is not surprising that consistency holds quite generally for frequency plug-in estimates.

Theorem 5.2.1. *Suppose that* $\mathcal{P} = S = \{(p_1, \dots, p_k) : 0 \leq p_j \leq 1, 1 \leq j \leq k, \sum_{j=1}^{k} p_j = 1\}$, *the k-dimensional simplex, where* $p_j = P[X_1 = x_j]$, $1 \leq j \leq k$, *and* $\{x_1, \dots, x_k\}$ *is the range of* X_1. *Let* $N_j \equiv \sum_{i=1}^{n} 1(X_i = x_j)$ *and* $\hat{p}_j \equiv N_j/n$, $\mathbf{p}_n \equiv (\hat{p}_1, \dots, \hat{p}_k) \in S$ *be the empirical distribution. Suppose that* $q : S \to R^p$ *is continuous. Then* $\hat{q}_n \equiv q(\hat{\mathbf{p}}_n)$ *is a uniformly consistent estimate of* $q(\mathbf{p})$.

Proof. By the weak law of large numbers for all $\mathbf{p}, \delta > 0$

$$P_{\mathbf{p}}[|\hat{\mathbf{p}}_n - \mathbf{p}| \geq \delta] \to 0.$$

Because q is continuous and S is compact, it is uniformly continuous on S. Thus, for every $\epsilon > 0$, there exists $\delta(\epsilon) > 0$ such that $\mathbf{p}, \mathbf{p}' \in S$, $|\mathbf{p}' - \mathbf{p}| \leq \delta(\epsilon)$, implies $|q(\mathbf{p}') - q(\mathbf{p})| \leq \epsilon$. Then

$$P_{\mathbf{p}}[|\hat{q}_n - q| \geq \epsilon] \leq P_{\mathbf{p}}[|\hat{\mathbf{p}}_n - \mathbf{p}| \geq \delta(\epsilon)]$$

But, $\sup\{P_{\mathbf{p}}[|\hat{\mathbf{p}}_n - \mathbf{p}| \geq \delta] : \mathbf{p} \in S\} \leq k/4n\delta^2$ (Problem 5.2.1) and the result follows. $\square$

In fact, in this case, we can go further. Suppose the *modulus of continuity* of q, $\omega(q, \delta)$ is defined by

$$\omega(q, \delta) = \sup\{|q(\mathbf{p}) - q(\mathbf{p}')| : |\mathbf{p} - \mathbf{p}'| \leq \delta\}. \tag{5.2.3}$$

Evidently, $\omega(q, \cdot)$ is increasing in δ and has the range $[a, b)$ say. If q is continuous $\omega(q, \delta) \downarrow 0$ as $\delta \downarrow 0$. Let $\omega^{-1} : [a, b] \leq R^+$ be defined as the inverse of ω,

$$\omega^{-1}(\epsilon) = \inf\{\delta : \omega(q, \delta) \geq \epsilon\} \tag{5.2.4}$$

It easily follows (Problem 5.2.3) that

$$\sup\{P[|\hat{q}_n - q(\mathbf{p})| \geq \epsilon] : P \in \mathcal{P}\} \leq n^{-1}[\omega^{-1}(\epsilon)]^{-2}\frac{k}{4}. \tag{5.2.5}$$

A simple and important result for the case in which $X_1, \dots, X_n$ are i.i.d. with $X_i \in \mathcal{X}$ is the following:

Proposition 5.2.1. *Let* $\mathbf{g} \equiv (g_1, \ldots, g_d)$ *map* $\mathcal{X}$ *onto* $\mathcal{Y} \subset R^d$. *Suppose* $E_\theta |g_j(X_1)| < \infty$, $1 \le j \le d$, *for all* θ; *let* $m_j(\theta) \equiv E_\theta g_j(X_1)$, $1 \le j \le d$, *and let* $q(\theta) = h(\mathbf{m}(\theta))$, *where* $h : \mathcal{Y} \to R^p$. *Then, if* h *is continuous,*

$$\widehat{q} \equiv h(\bar{\mathbf{g}}) \equiv h\left(\frac{1}{n} \sum_{i=1}^{n} \mathbf{g}(X_i)\right)$$

is a consistent estimate of $q(\theta)$. *More generally if* $\nu(P) = h(E_P \mathbf{g}(X_1))$ *and* $\mathcal{P} = \{P : E_P |\mathbf{g}(X_1)| < \infty\}$, *then* $\nu(\widehat{P}) \equiv h(\bar{\mathbf{g}})$, *where* $\widehat{P}$ *is the empirical distribution, is consistent for* $\nu(P)$.

Proof. We need only apply the general weak law of large numbers (for vectors) to conclude that

$$\frac{1}{n} \sum_{i=1}^{n} \mathbf{g}(X_i) \xrightarrow{P} E_P \mathbf{g}(X_1) \tag{5.2.6}$$

if $E_p |\mathbf{g}(X_1)| < \infty$. For consistency of $h(\bar{\mathbf{g}})$ apply Proposition B.7.1: $\mathbf{U}_n \xrightarrow{P} \mathbf{U}$ implies that $h(\mathbf{U}_n) \xrightarrow{P} h(\mathbf{U})$ for all continuous h. $\square$

Example 5.2.3. *Variances and Correlations.* Let $X_i = (U_i, V_i)$, $1 \le i \le n$ be i.i.d. $\mathcal{N}_2(\mu_1, \mu_2, \sigma_1^2, \sigma_2^2, \rho)$, $\sigma_i^2 > 0$, $|\rho| < 1$. Let $\mathbf{g}(u, v) = (u, v, u^2, v^2, uv)$ so that $\sum_{i=1}^{n} \mathbf{g}(U_i, V_i)$ is the statistic generating this 5-parameter exponential family. If we let $\boldsymbol{\theta} \equiv (\mu_1, \mu_2, \sigma_1^2, \sigma_2^2, \rho)$, then

$$\mathbf{m}(\boldsymbol{\theta}) = (\mu_1, \mu_2, \sigma_1^2 + \mu_1^2, \sigma_2^2 + \mu_2^2, \rho\sigma_1\sigma_2 + \mu_1\mu_2).$$

If $h = \mathbf{m}^{-1}$, then

$$h(m_1, \ldots, m_5)$$
$$= (m_1, m_2, m_3 - m_1^2, m_4 - m_2^2, (m_5 - m_1 m_2)(m_3 - m_1^2)^{-1/2}(m_4 - m_2^2)^{-1/2}),$$

which is well defined and continuous at all points of the range of $\mathbf{m}$. We may, thus, conclude by Proposition 5.2.1 that the empirical means, variances, and correlation coefficient are all consistent. Questions of uniform consistency and consistency when $\mathcal{P} = \{$ Distributions such that $EU_1^2 < \infty$, $EV_1^2 < \infty$, $\mathrm{Var}(U_1) > 0$, $\mathrm{Var}(V_1) > 0$, $|\mathrm{Corr}(U_1, V_1)| < 1\}$ are discussed in Problem 5.2.4. $\square$

Here is a general consequence of Proposition 5.2.1 and Theorem 2.3.1.

Theorem 5.2.2. *Suppose* $\mathcal{P}$ *is a canonical exponential family of rank d generated by* $\mathbf{T}$. *Let* $\boldsymbol{\eta}, \mathcal{E}$ *and* $A(\cdot)$ *correspond to* $\mathcal{P}$ *as in Section 1.6. Suppose* $\mathcal{E}$ *is open. Then, if* $X_1, \ldots, X_n$ *are a sample from* $P_{\boldsymbol{\eta}} \in \mathcal{P}$,

(i) $P_{\boldsymbol{\eta}}[\text{The MLE } \widehat{\boldsymbol{\eta}} \text{ exists}] \to 1$.

(ii) $\widehat{\boldsymbol{\eta}}$ *is consistent.*

Proof. Recall from Corollary 2.3.1 to Theorem 2.3.1 that $\hat{\eta}(X_1, \ldots, X_n)$ exists iff $\frac{1}{n}\sum_{i=1}^{n} \mathbf{T}(X_i) = \overline{\mathbf{T}}_n$ belongs to the interior $C_{\mathbf{T}}^{\circ}$ of the convex support of the distribution of $\overline{\mathbf{T}}_n$. Note that, if $\boldsymbol{\eta}_0$ is true, $E_{\boldsymbol{\eta}_0}(\mathbf{T}(X_1))$ must by Theorem 2.3.1 belong to the interior of the convex support because the equation $\dot{A}(\boldsymbol{\eta}) = \mathbf{t}_0$, where $\mathbf{t}_0 = \dot{A}(\boldsymbol{\eta}_0) = E_{\boldsymbol{\eta}_0}\mathbf{T}(X_1)$, is solved by $\boldsymbol{\eta}_0$. By definition of the interior of the convex support there exists a ball $S_\delta \equiv \{\mathbf{t} : |\mathbf{t} - E_{\boldsymbol{\eta}_0}\mathbf{T}(X_1)| < \delta\} \subset C_{\mathbf{T}}^{\circ}$. By the law of large numbers,

$$\frac{1}{n}\sum_{i=1}^{n} \mathbf{T}(X_i) \xrightarrow{P\boldsymbol{\eta}_0} E_{\boldsymbol{\eta}_0}\mathbf{T}(X_1).$$

Hence,

$$P_{\boldsymbol{\eta}_0}[\frac{1}{n}\sum_{i=1}^{n} \mathbf{T}(X_i) \in C_{\mathbf{T}}^{\circ}] \rightarrow 1. \tag{5.2.7}$$

But $\hat{\boldsymbol{\eta}}$, which solves

$$\dot{A}(\boldsymbol{\eta}) = \frac{1}{n}\sum_{i=1}^{n} \mathbf{T}(X_i),$$

exists iff the event in (5.2.7) occurs and (i) follows. We showed in Theorem 2.3.1 that on $C_{\mathbf{T}}^{\circ}$ the map $\boldsymbol{\eta} \rightarrow \dot{A}(\boldsymbol{\eta})$ is 1-1 and continuous on $\mathcal{E}$. By a classical result, see, for example, Rudin (1987), the inverse $\dot{A}^{-1} : \dot{A}(\mathcal{E}) \rightarrow \mathcal{E}$ is continuous on S_δ and the result follows from Proposition 5.2.1. $\square$

5.2.2 Consistency of Minimum Contrast Estimates

The argument of the the previous subsection in which a minimum contrast estimate, the MLE, is a continuous function of a mean of i.i.d. vectors evidently used exponential family properties. A more general argument is given in the following simple theorem whose conditions are hard to check. Let $X_1, \ldots, X_n$ be i.i.d. P_θ, $\theta \in \Theta \subset R^d$. Let $\hat{\boldsymbol{\theta}}$ be a minimum contrast estimate that minimizes

$$\rho_n(\mathbf{X}, \boldsymbol{\theta}) = \frac{1}{n}\sum_{i=1}^{n} \rho(X_i, \boldsymbol{\theta})$$

where, as usual, $D(\boldsymbol{\theta}_0, \boldsymbol{\theta}) \equiv E_{\boldsymbol{\theta}_0}\rho(X_1, \boldsymbol{\theta})$ is uniquely minimized at $\boldsymbol{\theta}_0$ for all $\boldsymbol{\theta}_0 \in \Theta$.

Theorem 5.2.3. *Suppose*

$$\sup\{|\frac{1}{n}\sum_{i=1}^{n}[\rho(X_i, \boldsymbol{\theta}) - D(\boldsymbol{\theta}_0, \boldsymbol{\theta})]| : \boldsymbol{\theta} \in \Theta\} \xrightarrow{P_{\boldsymbol{\theta}_0}} 0 \tag{5.2.8}$$

and

$$\inf\{D(\boldsymbol{\theta}_0, \boldsymbol{\theta}) : |\boldsymbol{\theta} - \boldsymbol{\theta}_0| \geq \epsilon\} > D(\boldsymbol{\theta}_0, \boldsymbol{\theta}_0) \quad \text{for every } \epsilon > 0. \tag{5.2.9}$$

Then $\hat{\boldsymbol{\theta}}$ is consistent.

Proof. Note that,

$$P_{\theta_0}[|\widehat{\theta} - \theta_0| \geq \epsilon] \leq P_{\theta_0}[\inf\{\frac{1}{n}\sum_{i=1}^{n}[\rho(X_i, \theta) - \rho(X_i, \theta_0)] : |\theta - \theta_0| \geq \epsilon\} \leq 0]$$

(5.2.10)

By hypothesis, for all $\delta > 0$,

$$P_{\theta_0}[\inf\{\frac{1}{n}\sum_{i=1}^{n}(\rho(X_i, \theta) - \rho(X_i, \theta_0)) : |\theta - \theta_0| \geq \epsilon\}$$
$$- \inf\{D(\theta_0, \theta) - D(\theta_0, \theta_0)) : |\theta - \theta_0| \geq \epsilon\} < -\delta] \to 0 \qquad (5.2.11)$$

because the event in (5.2.11) implies that

$$\sup\{|\frac{1}{n}\sum_{i=1}^{n}[\rho(X_i, \theta) - D(\theta_0, \theta)]| : \theta \in \Theta\} > \frac{\delta}{2}, \qquad (5.2.12)$$

which has probability tending to 0 by (5.2.8). But for $\epsilon > 0$ let

$$\delta = \frac{1}{4}\inf\{D(\theta, \theta_0) - D(\theta_0, \theta_0) : |\theta - \theta_0| \geq \epsilon\}.$$

Then (5.2.11) implies that the right-hand side of (5.2.10) tends to 0.   □

A simple and important special case is given by the following.

Corollary 5.2.1. *If Θ is finite, $\Theta = \{\theta_1, \ldots, \theta_d\}$, $E_{\theta_0}|\log p(X_1, \theta)| < \infty$ and the parameterization is identifiable, then, if $\widehat{\theta}$ is the MLE, $P_{\theta_j}[\widehat{\theta} \neq \theta_j] \to 0$ for all j.*

Proof. Note that for some $\epsilon > 0$,

$$P_{\theta_0}[\widehat{\theta} \neq \theta_j] = P_{\theta_0}[|\widehat{\theta} - \theta_0| \geq \epsilon]. \qquad (5.2.13)$$

By Shannon's Lemma 2.2.1 we need only check that (5.2.8) and (5.2.9) hold for $\rho(x, \theta) = \log p(x, \theta)$. But because Θ is finite, (5.2.8) follows from the WLLN and

$$P_{\theta_0}[\max\{|\frac{1}{n}\sum_{i=1}^{n}(\rho(X_i, \theta_j) - D(\theta_0, \theta_j)) : 1 \leq j \leq d\} \geq \epsilon]$$
$$\leq d\max\{P_{\theta_0}[|\frac{1}{n}\sum_{i=1}^{n}(\rho(X_i, \theta_j) - D(\theta_0, \theta_j))| \geq \epsilon] : 1 \leq j \leq d\} \to 0.$$

and (5.2.9) follows from Shannon's lemma.   □

Condition (5.2.8) can often fail—see Problem 5.2.5. An alternative condition that is readily seen to work more widely is the replacement of (5.2.8) by

(i)   For all compact $K \subset \Theta$,

$$\sup\left\{\left|\frac{1}{n}\sum_{i=1}^{n}(\rho(X_i, \theta) - D(\theta_0, \theta))\right| : \theta \in K\right\} \xrightarrow{P_{\theta_0}} 0.$$

(ii)   For some compact $K \subset \Theta$,

(5.2.14)

$$P_{\theta_0}\left[\inf\left\{\frac{1}{n}\sum_{i=1}^{n}(\rho(X_i, \theta) - \rho(X_i, \theta_0)) : \theta \in K^c\right\} > 0\right] \to 1.$$

We shall see examples in which this modification works in the problems. Unfortunately checking conditions such as (5.2.8) and (5.2.14) is in general difficult. A general approach due to Wald and a similar approach for consistency of generalized estimating equation solutions are left to the problems. When the observations are independent but not identically distributed, consistency of the MLE may fail if the number of parameters tends to infinity, see Problem 5.3.33.

Summary. We introduce the minimal property we require of any estimate (strictly speaking, sequence of estimates), consistency. If $\widehat{\theta}_n$ is an estimate of $\theta(P)$, we require that $\widehat{\theta}_n \xrightarrow{P} \theta(P)$ as $n \to \infty$. Uniform consistency for $\mathcal{P}$ requires more, that $\sup\{P[|\widehat{\theta}_n - \theta(P)| \geq \epsilon] : P \in \mathcal{P}\} \to 0$ for all $\epsilon > 0$. We show how consistency holds for continuous functions of vector means as a consequence of the law of large numbers and derive consistency of the MLE in canonical multiparameter exponential families. We conclude by studying consistency of the MLE and more generally MC estimates in the case Θ finite and Θ Euclidean. Sufficient conditions are explored in the problems.

5.3   FIRST- AND HIGHER-ORDER ASYMPTOTICS: THE DELTA METHOD WITH APPLICATIONS

We have argued in Section 5.1 that the principal use of asymptotics is to provide quantitatively or qualitatively useful approximations to risk.

5.3.1   The Delta Method for Moments

We begin this section by deriving approximations to moments of smooth functions of scalar means and even provide crude bounds on the remainders. We then sketch the extension to functions of vector means.

As usual let $X_1, \ldots, X_n$ be i.i.d. $\mathcal{X}$ valued and for the moment take $\mathcal{X} = R$. Let $h : R \to R$, let $\|g\|_\infty = \sup\{|g(t)| : t \in R\}$ denote the sup norm, and assume

   (i)   (a) h is m times differentiable on R, $m \geq 2$. We denote the jth derivative of h by $h^{(j)}$ and assume

          (b) $\|h^{(m)}\|_\infty \equiv \sup_\mathcal{X} |h^{(m)}(x)| \leq M < \infty$

  (ii) $E|X_1|^m < \infty$

Let $E(X_1) = \mu$, $\text{Var}(X_1) = \sigma^2$. We have the following.

Theorem 5.3.1. *If (i) and (ii) hold, then*

$$Eh(\bar{X}) = h(\mu) + \sum_{j=1}^{m-1} \frac{h^{(j)}(\mu)}{j!} E(\bar{X} - \mu)^j + R_m \qquad (5.3.1)$$

where

$$|R_m| \leq M \frac{E|X_1|^m}{m!} n^{-m/2}.$$

The proof is an immediate consequence of Taylor's expansion,

$$h(\bar{X}) = h(\mu) + \sum_{k=1}^{m-1} \frac{h^{(k)}(\mu)}{k!}(\bar{X} - \mu)^k + \frac{h^{(m)}(\bar{X}^*)}{m!}(\bar{X} - \mu)^m \qquad (5.3.2)$$

where $|\bar{X}^* - \mu| \leq |\bar{X} - \mu|$, and the following lemma.

Lemma 5.3.1. *If $E|X_1|^j < \infty$, $j \geq 2$, then there are constants $C_j > 0$ and $D_j > 0$ such that*

$$E|\bar{X} - \mu|^j \leq C_j E|X_1|^j n^{-j/2} \qquad (5.3.3)$$

$$|E(\bar{X} - \mu)^j| \leq D_j E|X_1|^j n^{-(j+1)/2}, \; j \; odd. \qquad (5.3.4)$$

Note that for j even, $E|\bar{X} - \mu|^j = E(\bar{X} - \mu)^j$.

Proof. We give the proof of (5.3.4) for all j and (5.3.3) for j even. The more difficult argument needed for (5.3.3) and j odd is given in Problem 5.3.2.

Let $\mu = E(X_1) = 0$, then

(a)
$$\begin{aligned} E(\bar{X}^j) &= n^{-j} E(\textstyle\sum_{i=1}^n X_i)^j \\ &= n^{-j} \sum_{1 \leq i_1, \dots, i_j \leq n} E(X_{i_1} \dots X_{i_j}) \end{aligned}$$

But $E(X_{i_1} \dots X_{i_j}) = 0$ unless each integer that appears among $\{i_1, \dots, i_j\}$ appears at least twice. Moreover,

(b)
$$\sup_{i_1, \dots, i_j} |E(X_{i_1} \dots X_{i_j})| = E|X_1|^j$$

by Problem 5.3.5, so the number d of nonzero terms in (a) is

(c)
$$\sum_{r=1}^{[j/2]} \frac{n}{r} \sum_{\substack{i_1 + \dots + i_r = j \\ i_k \geq 2 \text{ all } k}} \frac{j}{i_1, \dots, i_r}$$

where $\frac{n}{i_1, \dots, i_r} = \frac{n!}{i_1! \dots i_r!}$ and $[t]$ denotes the greatest integer $\leq t$. The expression in (c) is, for $j \leq n/2$, bounded by

(d)
$$\frac{C_j}{[\frac{j}{2}]!} n(n-1) \dots (n - [j/2] + 1)$$

where

$$C_j = \max_{1 \leq r \leq [j/2]} \left\{ \sum \left\{ \frac{j}{i_1, \dots, i_r} : i_1 + \dots + i_r = j, i_k \geq 2, 1 \leq k \leq r \right\} \right\}.$$

But

(e) $$n^{-j}n(n-1)\dots(n-[j/2]+1) \leq n^{[j/2]-j}$$

and (c), (d), and (e) applied to (a) imply (5.3.4) for j odd and (5.3.3) for j even, if $\mu = 0$. In general by considering $X_i - \mu$ as our basic variables we obtain the lemma but with $E|X_1|^j$ replaced by $E|X_1 - \mu|^j$. By Problem 5.3.6, $E|X_1 - \mu|^j \leq 2^j E|X_1|^j$ and the lemma follows.   □

The two most important corollaries of Theorem 5.3.1, respectively, give approximations to the bias of $h(\bar{X})$ as an estimate of $h(\mu)$ and its variance and MSE.

Corollary 5.3.1.

(a) If $E|X_1|^3 < \infty$ and $||h^{(3)}||_\infty < \infty$, then

$$E h(\bar{X}) = h(\mu) + \frac{h^{(2)}(\mu)\sigma^2}{2n} + O(n^{-3/2}). \tag{5.3.5}$$

(b) If $E(X_1^4) < \infty$ and $||h^{(4)}||_\infty < \infty$ then $O(n^{-3/2})$ in (5.3.5) can be replaced by $O(n^{-2})$.

Proof. For (5.3.5) apply Theorem 5.3.1 with $m = 3$. Because $E(\bar{X} - \mu)^2 = \sigma^2/n$, (5.3.5) follows. If the conditions of (b) hold, apply Theorem 5.3.1 with $m = 4$. Then $R_m = O(n^{-2})$ and also $E(\bar{X} - \mu)^3 = O(n^{-2})$ by (5.3.4).   □

Corollary 5.3.2. If

(a) $||h^{(j)}||_\infty < \infty$, $1 \leq j \leq 3$ and $E|X_1|^3 < \infty$, then

$$\operatorname{Var} h(\bar{X}) = \frac{\sigma^2 [h^{(1)}(\mu)]^2}{n} + O(n^{-3/2}) \tag{5.3.6}$$

(b) If $||h^{(j)}||_\infty < \infty$, $1 \leq j \leq 3$, and $E X_1^4 < \infty$, then $O(n^{-3/2})$ in (5.3.6) can be replaced by $O(n^{-2})$.

Proof. (a) Write

$$
\begin{aligned}
E h^2(\bar{X}) &= h^2(\mu) + 2h(\mu)h^{(1)}(\mu)E(\bar{X} - \mu) + \{h^{(2)}(\mu)h(\mu) + [h^{(1)}]^2(\mu)\}E(\bar{X} - \mu)^2 \\
&\quad + \frac{1}{6}E[h^2]^{(3)}(\bar{X}^*)(\bar{X} - \mu)^3 \\
&= h^2(\mu) + \{h^{(2)}(\mu)h(\mu) + [h^{(1)}]^2(\mu)\}\frac{\sigma^2}{n} + O(n^{-3/2}).
\end{aligned}
$$

(b) Next, using Corollary 5.3.1,

$$
\begin{aligned}
[E h(\bar{X})]^2 &= (h(\mu) + \frac{h^{(2)}(\mu)}{2}\frac{\sigma^2}{n} + O(n^{-3/2}))^2 \\
&= h^2(\mu) + h(\mu)h^{(2)}(\mu)\frac{\sigma^2}{n} + O(n^{-3/2}).
\end{aligned}
$$

Subtracting (a) from (b) we get (5.3.6). To get part (b) we need to expand $Eh^2(\bar{X})$ to four terms and similarly apply the appropriate form of (5.3.5). $\qquad\square$

Clearly the statements of the corollaries as well can be turned to expansions as in Theorem 5.3.1 with bounds on the remainders.

Note an important qualitative feature revealed by these approximations. If $h(\bar{X})$ is viewed, as we normally would, as the plug-in estimate of the parameter $h(\mu)$ then, for large n, the bias of $h(\bar{X})$ defined by $Eh(\bar{X}) - h(\mu)$ is $O(n^{-1})$, which is neglible compared to the standard deviation of $h(\bar{X})$, which is $O(n^{-1/2})$ unless $h^{(1)}(\mu) = 0$. A qualitatively simple explanation of this important phenonemon will be given in Theorem 5.3.3.

Example 5.3.1. If $X_1, \dots, X_n$ are i.i.d. $\mathcal{E}(\lambda)$ the MLE of λ is $\bar{X}^{-1}$. If the X_i represent the lifetimes of independent pieces of equipment in hundreds of hours and the warranty replacement period is (say) 200 hours, then we may be interested in the warranty failure probability

$$P_\lambda[X_1 \le 2] = 1 - e^{-2\lambda}. \qquad (5.3.7)$$

If $h(t) = 1 - \exp(-2/t)$, then $h(\bar{X})$ is the MLE of $1 - \exp(-2\lambda) = h(\mu)$, where $\mu = E_\lambda X_1 = 1/\lambda$.

We can use the two corollaries to compute asymptotic approximations to the means and variance of $h(\bar{X})$. Thus, by Corollary 5.3.1,

$$\begin{aligned}
\text{Bias}_\lambda(h(\bar{X})) &\equiv E_\lambda(h(\bar{X}) - h(\mu)) \\
&= \frac{h^{(2)}(\mu)}{2}\frac{\sigma^2}{n} + O(n^{-2}) \qquad (5.3.8) \\
&= 2e^{-2\lambda}\lambda^3(1 - \lambda)/n + O(n^{-2})
\end{aligned}$$

because $h^{(2)}(t) = 4(t^{-3} - t^{-4})\exp(-2/t)$, $\sigma^2 = 1/\lambda^2$, and, by Corollary 5.3.2 (Problem 5.3.1)

$$\text{Var}_\lambda h(\bar{X}) = 4\lambda^2 e^{-4\lambda}/n + O(n^{-2}). \qquad (5.3.9)$$

$\qquad\square$

Further expansion can be done to increase precision of the approximation to $\text{Var}\, h(\bar{X})$ for large n. Thus, by expanding $Eh^2(\bar{X})$ and $Eh(\bar{X})$ to six terms we obtain the approximation

$$\begin{aligned}
\text{Var}(h(\bar{X})) &= \tfrac{1}{n}[h^{(1)}(\mu)]^2\sigma^2 + \tfrac{1}{n^2}\left\{h^{(1)}(\mu)h^{(2)}(\mu)\mu_3 \right. \\
&\quad \left. + \tfrac{1}{2}[h^{(2)}(\mu)]^2\sigma^4\right\} + R'_n
\end{aligned} \qquad (5.3.10)$$

with R'_n tending to zero at the rate $1/n^3$. Here μ_k denotes the kth central moment of X_i and we have used the facts that (see Problem 5.3.3)

$$E(\bar{X} - \mu)^3 = \frac{\mu_3}{n^2}, \quad E(\bar{X} - \mu)^4 = \frac{\mu_4}{n^3} + \frac{3(n-1)\sigma^4}{n^3}. \qquad (5.3.11)$$

Example 5.3.2. *Bias and Variance of the MLE of the Binomial Variance.* We will compare $E(h(\bar{X}))$ and $\text{Var}\, h(\bar{X})$ with their approximations, when $h(t) = t(1-t)$ and $X_i \sim$

$\mathcal{B}(1,p)$, and will illustrate how accurate (5.3.10) is in a situation in which the approximation can be checked.

First calculate

$$Eh(\bar{X}) = E(\bar{X}) - E(\bar{X}^2) = p - [\text{Var}(\bar{X}) + (E(\bar{X}))^2]$$

$$= p(1-p) - \frac{1}{n}p(1-p) = \frac{n-1}{n}p(1-p).$$

Because $h^{(1)}(t) = 1 - 2t$, $h^{(2)} = -2$, (5.3.5) yields

$$E(h(\bar{X})) = p(1-p) - \frac{1}{n}p(1-p)$$

and in this case (5.3.5) is exact as it should be. Next compute

$$\text{Var}\, h(\bar{X}) = \frac{p(1-p)}{n}\left\{(1-2p)^2 + \frac{2p(1-p)}{n-1}\right\}\left(\frac{n-1}{n}\right)^2.$$

Because $\mu_3 = p(1-p)(1-2p)$, (5.3.10) yields

$$\text{Var}\, h(\bar{X}) = \frac{1}{n}(1-2p)^2 p(1-p) + \frac{1}{n^2}\{-2(1-2p)p(1-p)(1-2p)$$

$$+ 2p^2(1-p)^2\} + R'_n$$

$$= \frac{p(1-p)}{n}\{(1-2p)^2 + \frac{1}{n}[2p(1-p) - 2(1-2p)^2]\} + R'_n.$$

Thus, the error of approximation is

$$R'_n = \frac{p(1-p)}{n^3}[(1-2p)^2 - 2p(1-p)]$$

$$= \frac{p(1-p)}{n^3}[1 - 6p(1-p)] = O(n^{-3}).$$

$\square$

The generalization of this approach to approximation of moments for functions of vector means is formally the same but computationally not much used for d larger than 2.

Theorem 5.3.2. *Suppose* $\mathbf{g} : \mathcal{X} \to R^d$ *and let* $\mathbf{Y}_i = \mathbf{g}(X_i) = (g_1(X_i), \dots, g_d(X_i))^T$,
$i = 1, \dots, n$, *where* $X_1, \dots, X_n$ *are i.i.d. Let* $h : R^d \to R$, *assume that h has continuous
partial derivatives of order up to m, and that*

(i) $\|D^m(h)\|_\infty < \infty$ *where* $D^m h(\mathbf{x})$ *is the array (tensor)*

$$\left\{\frac{\partial^m h}{\partial x_1^{i_1} \dots \partial x_d^{i_d}}(\mathbf{x}) : i_1 + \dots + i_d = m, 0 \le i_j \le m, 1 \le j \le d\right\}$$

and $\|D^m h\|_\infty$ *is the sup over all x and $i_1, \dots, i_d$ of* $\left|\frac{\partial^m h}{\partial x_1^{i_1} \dots \partial x_d^{i_d}}(\mathbf{x})\right|$.

(ii) $E|Y_{ij}|^m < \infty$, $1 \le i \le n$, $1 \le j \le d$, where $Y_{ij} \equiv g_j(X_i)$.

Then, if $\bar{Y}_k = \frac{1}{n}\sum_{i=1}^n Y_{ik}$, $\bar{\mathbf{Y}} = \frac{1}{n}\sum_{i=1}^n \mathbf{Y}_i$, and $\boldsymbol{\mu} = E\mathbf{Y}_1$, then

$$Eh(\bar{\mathbf{Y}}) = h(\boldsymbol{\mu}) + \sum_{j=1}^{m-1}\sum\{\frac{\partial^m h}{\partial x_1^{i_1}\dots\partial x_d^{i_d}}(\boldsymbol{\mu})(i_1!\dots i_d!)^{-1}$$
$$E\prod_{k=1}^d(\bar{Y}_k - \mu_k)^{i_k} : i_1 + \dots + i_d = j, 0 \le i_k \le j\} + O(n^{-m/2}).$$
$$(5.3.12)$$

This is a consequence of Taylor's expansion in d variables, B.8.11, and the appropriate generalization of Lemma 5.3.1. The proof is outlined in Problem 5.3.4. The most interesting application, as for the case $d = 1$, is to $m = 3$. We get, for $d = 2$, $E|Y_1|^3 < \infty$

$$Eh(\bar{\mathbf{Y}}) = h(\boldsymbol{\mu}) + \frac{1}{n}\left\{\frac{1}{2}\frac{\partial^2 h}{\partial x_1^2}(\boldsymbol{\mu})\operatorname{Var}(Y_{11}) + \frac{\partial^2 h}{\partial x_1\partial x_2}(\boldsymbol{\mu})\operatorname{Cov}(Y_{11}, Y_{12})\right.$$
$$+ \left.\frac{1}{2}\frac{\partial^2 h}{\partial x_2^2}(\boldsymbol{\mu})\operatorname{Var}(Y_{12})\right\} + O(n^{-3/2}).$$
$$(5.3.13)$$

Moreover, by (5.3.3), if $E|\mathbf{Y}_1|^4 < \infty$, then $O(n^{-3/2})$ in (5.3.13) can be replaced by $O(n^{-2})$. Similarly, under appropriate conditions (Problem 5.3.12), for $d = 2$,

$$\operatorname{Var} h(\bar{\mathbf{Y}}) = \frac{1}{n}\left[\left(\frac{\partial h}{\partial x_1}(\boldsymbol{\mu})\right)^2 \operatorname{Var}(Y_{11}) + 2\frac{\partial h}{\partial x_1}(\boldsymbol{\mu})\frac{\partial h}{\partial x_2}(\boldsymbol{\mu})\operatorname{Cov}(Y_{11}, Y_{12})\right.$$
$$+ \left.\left(\frac{\partial h}{\partial x_2}(\boldsymbol{\mu})\right)^2 \operatorname{Var}(Y_{12})\right] + O(n^{-2})$$
$$(5.3.14)$$

Approximations (5.3.5), (5.3.6), (5.3.13), and (5.3.14) do not help us to approximate risks for loss functions other than quadratic (or some power of $(d - h(\mu))$). The results in the next subsection go much further and "explain" the form of the approximations we already have.

5.3.2   The Delta Method for In Law Approximations

As usual we begin with $d = 1$.

Theorem 5.3.3. *Suppose that $\mathcal{X} = R$, $h : R \to R$, $EX_1^2 < \infty$ and h is differentiable at $\mu = E(X_1)$. Then*

$$\mathcal{L}(\sqrt{n}(h(\bar{X}) - h(\mu))) \to \mathcal{N}(0, \sigma^2(h))$$
$$(5.3.15)$$

where

$$\sigma^2(h) = [h^{(1)}(\mu)]^2\sigma^2$$

and $\sigma^2 = \operatorname{Var}(X_1)$.

The result follows from the more generally useful lemma.

Lemma 5.3.2. *Suppose $\{U_n\}$ are real random variables and that for a sequence $\{a_n\}$ of constants with $a_n \to \infty$ as $n \to \infty$,*

(i) $a_n(U_n - u) \xrightarrow{\mathcal{L}} V$ for some constant u.

(ii) $g : R \to R$ is differentiable at u with derivative $g^{(1)}(u)$.

Then

$$a_n(g(U_n) - g(u)) \xrightarrow{\mathcal{L}} g^{(1)}(u)V. \tag{5.3.16}$$

Proof. By definition of the derivative, for every $\epsilon > 0$ there exists a $\delta > 0$ such that

(a) $\qquad\qquad |v - u| \le \delta \Rightarrow |g(v) - g(u) - g^{(1)}(u)(v - u)| \le \epsilon |v - u|$

Note that (i) $\Rightarrow$

(b) $\qquad\qquad\qquad\qquad a_n(U_n - u) = O_p(1)$

$\Rightarrow$

(c) $\qquad\qquad\qquad\qquad U_n - u = O_p(a_n^{-1}) = o_p(1).$

Using (c), for every $\delta > 0$

(d) $\qquad\qquad\qquad\qquad P[|U_n - u| \le \delta] \to 1$

and, hence, from (a), for every $\epsilon > 0$,

(e) $\qquad\qquad P[|g(U_n) - g(u) - g^{(1)}(u)(U_n - u)| \le \epsilon |U_n - u|] \to 1.$

But (e) implies

(f) $\qquad a_n[g(U_n) - g(u) - g^{(1)}(u)(U_n - u)] = o_p(a_n(U_n - u)) = o_p(1)$

from (b). Therefore,

(g) $\qquad\qquad a_n[g(U_n) - g(u)] = g^{(1)}(u)a_n(U_n - u) + o_p(1).$

But, by hypothesis, $a_n(U_n - u) \xrightarrow{\mathcal{L}} V$ and the result follows. $\qquad\qquad\square$

The theorem follows from the central limit theorem letting $U_n = \bar{X}$, $a_n = n^{1/2}$, $u = \mu$, $V \sim \mathcal{N}(0, \sigma^2)$. $\qquad\qquad\square$

Note that (5.3.15) "explains" Lemma 5.3.1. Formally we expect that if $V_n \xrightarrow{\mathcal{L}} V$, then $EV_n^j \to EV^j$ (although this need not be true, see Problems 5.3.32 and B.7.8). Consider $V_n = \sqrt{n}(\bar{X} - \mu) \xrightarrow{\mathcal{L}} V \sim \mathcal{N}(0, \sigma^2)$. Thus, we expect

$$E[\sqrt{n}(\bar{X} - \mu)]^j \to \sigma^j EZ^j \tag{5.3.17}$$

where $Z \sim \mathcal{N}(0, 1)$. But if j is even, $EZ^j > 0$, else $EZ^j = 0$. Then (5.3.17) yields

$$E(\bar{X} - \mu)^j = O(\sigma^j EZ^j n^{-j/2}) = O(n^{-j/2}), \ j \text{ even}$$
$$= o(n^{-j/2}), j \text{ odd.}$$

Example 5.3.3. *"t" Statistics.*

(a) *The One-Sample Case.* Let $X_1, \ldots, X_n$ be i.i.d. $F \in \mathcal{F}$ where $E_F(X_1) = \mu$, $\text{Var}_F(X_1) = \sigma^2 < \infty$. A statistic for testing the hypothesis $H : \mu = 0$ versus $K : \mu > 0$ is

$$T_n = \sqrt{n}\frac{\bar{X}}{s}$$

where

$$s^2 = \frac{1}{n-1} \sum_{i=1}^{n}(X_i - \bar{X})^2.$$

If $\mathcal{F} = \{\text{Gaussian distributions}\}$, we can obtain the critical value $t_{n-1}(1 - \alpha)$ for T_n from the $\mathcal{T}_{n-1}$ distribution. In general we claim that if $F \in \mathcal{F}$ and H is true, then

$$T_n \overset{\mathcal{L}}{\to} \mathcal{N}(0, 1). \tag{5.3.18}$$

In particular this implies not only that $t_{n-1}(1-\alpha) \to z_{1-\alpha}$ but that the $t_{n-1}(1-\alpha)$ critical value (or $z_{1-\alpha}$) is approximately correct if H is true and F is not Gaussian. For the proof note that

$$U_n \equiv \sqrt{n}\frac{(\bar{X} - \mu)}{\sigma} \overset{\mathcal{L}}{\to} \mathcal{N}(0, 1)$$

by the central limit theorem, and

$$s^2 = \frac{n}{n-1} \left(\frac{1}{n}\sum_{i=1}^{n} X_i^2 - (\bar{X})^2 \right) \overset{P}{\to} \sigma^2$$

by Proposition 5.2.1 and Slutsky's theorem. Now Slutsky's theorem yields (5.3.18) because $T_n = U_n/(s_n/\sigma) = g(U_n, s_n/\sigma)$, where $g(u, v) = u/v$.

(b) *The Two-Sample Case.* Let $X_1, \ldots, X_{n_1}$ and $Y_1, \ldots, Y_{n_2}$ be two independent samples with $\mu_1 = E(X_1)$, $\sigma_1^2 = \text{Var}(X_1)$, $\mu_2 = E(Y_1)$ and $\sigma_2^2 = \text{Var}(Y_1)$. Consider testing $H : \mu_1 = \mu_2$ versus $K : \mu_2 > \mu_1$. In Example 4.9.3 we saw that the two sample t statistic

$$S_n = \sqrt{\frac{n_1 n_2}{n}} \left(\frac{\bar{Y} - \bar{X}}{s} \right), \ n = n_1 + n_2$$

has a $\mathcal{T}_{n-2}$ distribution under H when the X's and Y's are normal with $\sigma_1^2 = \sigma_2^2$. Using the central limit theorem, Slutsky's theorem, and the foregoing arguments, we find (Problem 5.3.28) that if $n_1/n \to \lambda, 0 < \lambda < 1$, then

$$S_n \overset{\mathcal{L}}{\to} \mathcal{N} \left(0, \frac{(1 - \lambda)\sigma_1^2 + \lambda\sigma_2^2}{\lambda\sigma_1^2 + (1 - \lambda)\sigma_2^2} \right).$$

It follows that if $n_1 = n_2$ or $\sigma_1^2 = \sigma_2^2$, then the critical value $t_{n-2}(1 - \alpha)$ for S_n is approximately correct if H is true and the X's and Y's are not normal.

Monte Carlo Simulation

As mentioned in Section 5.1, approximations based on asymptotic results should be checked by Monte Carlo simulations. We illustrate such simulations for the preceding t tests by generating data from the χ_d^2 distribution M times independently, each time computing the value of the t statistics and then giving the proportion of times out of M that the t statistics exceed the critical values from the t table. Here we use the χ_d^2 distribution because for small to moderate d it is quite different from the normal distribution. Other distributions should also be tried. Figure 5.3.1 shows that for the one-sample t test, when $\alpha = 0.05$, the asymptotic result gives a good approximation when $n \geq 10^{1.5} \cong 32$, and the true distribution F is χ_d^2 with $d \geq 10$. The χ_2^2 distribution is extremely skew, and in this case the $t_{n-1}(0.95)$ approximation is only good for $n \geq 10^{2.5} \cong 316$.

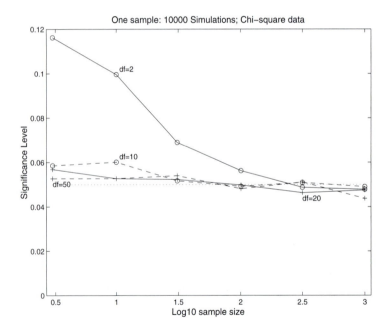

Figure 5.3.1. Each plotted point represents the results of 10,000 one-sample t tests using χ_d^2 data, where d is either 2, 10, 20, or 50, as indicated in the plot. The simulations are repeated for different sample sizes and the observed significance levels are plotted.

For the two-sample t tests, Figure 5.3.2 shows that when $\sigma_1^2 = \sigma_2^2$ and $n_1 = n_2$, the $t_{n-2}(1-\alpha)$ critical value is a very good approximation even for small n and for $X, Y \sim \chi_2^2$.

This is because, in this case, $\bar{Y} - \bar{X} = \frac{1}{n_1}\sum_{i=1}^{n_1}(Y_i - X_i)$, and $Y_i - X_i$ have a symmetric distribution. Other Monte Carlo runs (not shown) with $\sigma_1^2 \neq \sigma_2^2$ show that as long as $n_1 = n_2$, the $t_{n-2}(0.95)$ approximation is good for $n_1 \geq 100$, even when the X's and Y's have different χ_ν^2 distributions, scaled to have the same means, and $\sigma_2^2 = 12\sigma_1^2$. Moreover, the $t_{n-2}(1 - \alpha)$ approximation is good when $n_1 \neq n_2$ and $\sigma_1^2 = \sigma_2^2$. However, as we see from the limiting law of S_n and Figure 5.3.3, when both $n_1 \neq n_2$ and $\sigma_1^2 \neq \sigma_2^2$, then the two-sample t tests with critical region $1\{S_n \geq t_{n-2}(1 - \alpha)\}$ do not have approximate level α. In this case Monte Carlo studies have shown that the test in Section 4.9.4 based on Welch's approximation works well.

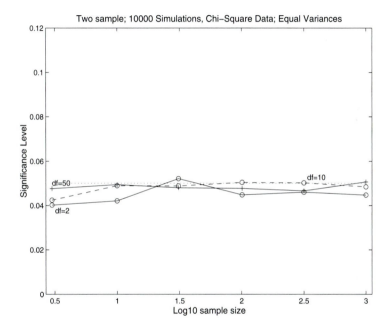

Figure 5.3.2. Each plotted point represents the results of 10,000 two-sample t tests. For each simulation the two samples are the same size (the size indicated on the x-axis), $\sigma_1^2 = \sigma_2^2$, and the data are χ_d^2 where d is one of 2, 10, or 50.

□

Next, in the one-sample situation, let $h(\bar{X})$ be an estimate of $h(\mu)$ where h is continuously differentiable at μ, $h^{(1)}(\mu) \neq 0$. By Theorem 5.3.3, $\sqrt{n}[h(\bar{X}) - h(\mu)] \xrightarrow{\mathcal{L}} \mathcal{N}(0, \sigma^2[h^{(1)}(\mu)]^2)$. To test the hypothesis $H : h(\mu) = h_0$ versus $K : h(\mu) > h_0$ the natural test statistic is

$$T_n = \frac{\sqrt{n}[h(\bar{X}) - h_0]}{s|h^{(1)}(\bar{X})|}.$$

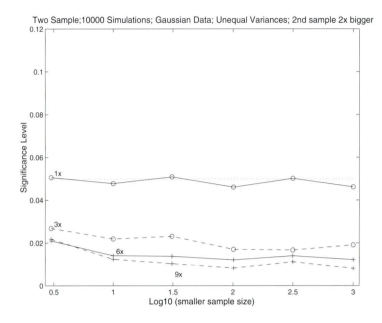

Figure 5.3.3. Each plotted point represents the results of 10,000 two-sample t tests. For each simulation the two samples differ in size: The second sample is two times the size of the first. The x-axis denotes the size of the smaller of the two samples. The data in the first sample are $\mathcal{N}(0, 1)$ and in the second they are $\mathcal{N}(0, \sigma^2)$ where σ^2 takes on the values 1, 3, 6, and 9, as indicated in the plot.

Combining Theorem 5.3.3 and Slutsky's theorem, we see that here, too, if H is true

$$T_n \xrightarrow{\mathcal{L}} \mathcal{N}(0, 1)$$

so that $z_{1-\alpha}$ is the asymptotic critical value.

Variance Stabilizing Transformations

Example 5.3.4. In Appendices A and B we encounter several important families of distributions, such as the binomial, Poisson, gamma, and beta, which are indexed by one or more parameters. If we take a sample from a member of one of these families, then the sample mean $\bar{X}$ will be approximately normally distributed with variance σ^2/n depending on the parameters indexing the family considered. We have seen that smooth transformations $h(\bar{X})$ are also approximately normally distributed. It turns out to be useful to know transformations h, called *variance stabilizing*, such that $\operatorname{Var} h(\bar{X})$ is approximately independent of the parameters indexing the family we are considering. From (5.3.6) and

(5.3.13) we see that a first approximation to the variance of $h(\bar{X})$ is $\sigma^2[h^{(1)}(\mu)]^2/n$. Thus, finding a variance stabilizing transformation is equivalent to finding a function h such that

$$\sigma^2[h^{(1)}(\mu)]^2 \equiv c$$

for all μ and σ appropriate to our family. Such a function can usually be found if σ depends only on μ, which varies freely. In this case (5.3.19) is an ordinary differential equation. As an example, suppose that $X_1, \ldots, X_n$ is a sample from a $\mathcal{P}(\lambda)$ family. In this case $\sigma^2 = \lambda$ and $\text{Var}(\bar{X}) = \lambda/n$. To have $\text{Var}\, h(\bar{X})$ approximately constant in λ, h must satisfy the differential equation $[h^{(1)}(\lambda)]^2\lambda = c > 0$ for some arbitrary $c > 0$. If we require that h is increasing, this leads to $h^{(1)}(\lambda) = \sqrt{c}/\sqrt{\lambda}$, $\lambda > 0$, which has as its solution $h(\lambda) = 2\sqrt{c\lambda} + d$, where d is arbitrary. Thus, $h(t) = \sqrt{t}$ is a variance stabilizing transformation of $\bar{X}$ for the Poisson family of distributions. Substituting in (5.3.6) we find $\text{Var}(\bar{X})^{\frac{1}{2}} \cong 1/4n$ and $\sqrt{n}((\bar{X})^{\frac{1}{2}} - (\lambda)^{\frac{1}{2}})$ has approximately a $\mathcal{N}(0, 1/4)$ distribution.   □

One application of variance stabilizing transformations, by their definition, is to exhibit monotone functions of parameters of interest for which we can give fixed length (independent of the data) confidence intervals. Thus, in the preceding $\mathcal{P}(\lambda)$ case,

$$\sqrt{\bar{X}} \pm \frac{2z(1 - \frac{1}{2}\alpha)}{\sqrt{n}}$$

is an approximate $1 - \alpha$ confidence interval for $\sqrt{\lambda}$. A second application occurs for models where the families of distribution for which variance stabilizing transformations exist are used as building blocks of larger models. Major examples are the generalized linear models of Section 6.5. The comparative roles of variance stabilizing and canonical transformations as link functions are discussed in Volume II. Some further examples of variance stabilizing transformations are given in the problems.

The notion of such transformations can be extended to the following situation. Suppose, $\widehat{\gamma}_n(X_1, \ldots, X_n)$ is an estimate of a real parameter γ indexing a family of distributions from which $X_1, \ldots, X_n$ are an i.i.d. sample. Suppose further that

$$\mathcal{L}_\gamma(\sqrt{n}(\widehat{\gamma}_n - \gamma)) \to \mathcal{N}(0, \sigma^2(\gamma)).$$

Then again, a variance stabilizing transformation h is such that

$$\mathcal{L}_\gamma\left[\sqrt{n}(h(\widehat{\gamma}_n) - h(\gamma))\right] \to \mathcal{N}(0, c) \tag{5.3.19}$$

for all γ. See Example 5.3.6. Also closely related but different are so-called normalizing transformations. See Problems 5.3.15 and 5.3.16.

Edgeworth Approximations

The normal approximation to the distribution of $\bar{X}$ utilizes only the first two moments of $\bar{X}$. Under general conditions (Bhattacharya and Rao, 1976, p. 538) one can improve on the normal approximation by utilizing the third and fourth moments. Let F_n denote the distribution of $T_n = \sqrt{n}(\bar{X} - \mu)/\sigma$ and let γ_{1n} and γ_{2n} denote the coefficient of skewness and kurtosis of T_n. Then under some conditions,[1]

$$F_n(x) = \Phi(x) - \varphi(x)[\frac{1}{6}\gamma_{1n}H_2(x) + \frac{1}{24}\gamma_{2n}H_3(x) + \frac{1}{72}\gamma_{1n}^2 H_5(x)] + r_n \qquad (5.3.20)$$

where r_n tends to zero at a rate faster than $1/n$ and H_2, H_3, and H_5 are *Hermite polynomials* defined by

$$H_2(x) = x^2 - 1, H_3(x) = x^3 - 3x, H_5(x) = x^5 - 10x^3 + 15x. \qquad (5.3.21)$$

The expansion (5.3.20) is called the *Edgeworth expansion* for F_n.

Example 5.3.5. *Edgeworth Approximations to the χ^2 Distribution.* Suppose $V \sim \chi_n^2$. According to Theorem B.3.1, V has the same distribution as $\sum_{i=1}^{n} X_i^2$, where the X_i are independent and $X_i \sim \mathcal{N}(0, 1), i = 1, \dots, n$. It follows from the central limit theorem that $T_n = (\sum_{i=1}^{n} X_i^2 - n)/\sqrt{2n} = (V - n)/\sqrt{2n}$ has approximately a $\mathcal{N}(0, 1)$ distribution. To improve on this approximation, we need only compute γ_{1n} and γ_{2n}. We can use Problem B.2.4 to compute

$$\gamma_{1n} = \frac{E(V - n)^3}{(2n)^{\frac{3}{2}}} = \frac{2\sqrt{2}}{\sqrt{n}}, \gamma_{2n} = \frac{E(V - n)^4}{(2n)^2} - 3 = \frac{12}{n}.$$

Therefore,

$$F_n(x) = \Phi(x) - \varphi(x)\left[\frac{\sqrt{2}}{3\sqrt{n}}(x^2 - 1) + \frac{1}{2n}(x^3 - 3x) + \frac{1}{9n}(x^5 - 10x^3 + 15x)\right] + r_n.$$

Table 5.3.1 gives this approximation together with the exact distribution and the normal approximation when $n = 10$. □

x	-2.04	-1.95	-1.91	-1.75	-1.66	-1.51	-1.35	
Exact	0.0001	0.0005	0.0010	0.0050	0.0100	0.0250	0.0500	
EA	0	0	0	0.0032	0.0105	0.0287	0.0553	
NA	0.0208	0.0254	0.0284	0.0397	0.0481	0.0655	0.0877	

x	-1.15	-0.85	-0.61	-0.38	-0.15	0.11	0.40	0.77
Exact	0.1000	0.2000	0.3000	0.4000	0.5000	0.6000	0.7000	0.8000
EA	0.1051	0.2024	0.3006	0.4000	0.4999	0.5999	0.6999	0.8008
NA	0.1254	0.1964	0.2706	0.3513	0.4415	0.5421	0.6548	0.7792

x	1.34	1.86	2.34	2.95	3.40	4.38	4.79	5.72
Exact	0.9000	0.9500	0.9750	0.9900	0.9950	0.9990	0.9995	0.9999
EA	0.9029	0.9506	0.9724	0.9876	0.9943	0.9996	0.9999	1.0000
NA	0.9097	0.9684	0.9905	0.9984	0.9997	1.0000	1.0000	1.0000

TABLE 5.3.1. Edgeworth[2] and normal approximations EA and NA to the χ_{10}^2 distribution, $P(T_n \le x)$, where T_n is a standardized χ_{10}^2 random variable.

The Multivariate Case

Lemma 5.3.2 extends to the d-variate case.

Lemma 5.3.3. *Suppose* $\{\mathbf{U}_n\}$ *are d-dimensional random vectors and that for some sequence of constants* $\{a_n\}$ *with* $a_n \to \infty$ *as* $n \to \infty$,

(i) $a_n(\mathbf{U}_n - \mathbf{u}) \overset{\mathcal{L}}{\to} \mathbf{V}_{d \times 1}$ *for some* $d \times 1$ *vector of constants* $\mathbf{u}$.

(ii) $\mathbf{g} : R^d \to R^p$ *has a differential* $\mathbf{g}^{(1)}_{p \times d}(\mathbf{u})$ *at* $\mathbf{u}$. *Then*

$$a_n[\mathbf{g}(\mathbf{U}_n) - \mathbf{g}(\mathbf{u})] \overset{\mathcal{L}}{\to} \mathbf{g}^{(1)}(\mathbf{u})\mathbf{V}.$$

Proof. The proof follows from the arguments of the proof of Lemma 5.3.2.

Example 5.3.6. Let $(X_1, Y_1), \ldots, (X_n, Y_n)$ be i.i.d. as (X, Y) where $0 < EX^4 < \infty$, $0 < EY^4 < \infty$. Let $\rho^2 = \text{Cov}^2(X, Y)/\sigma_1^2\sigma_2^2$ where $\sigma_1^2 = \text{Var } X$, $\sigma_2^2 = \text{Var } Y$; and let $r^2 = \widehat{C}^2/\widehat{\sigma}_1^2\widehat{\sigma}_2^2$ where

$$\widehat{C} = n^{-1}\Sigma(X_i - \bar{X})(Y_i - \bar{Y}), \ \widehat{\sigma}_1^2 = n^{-1}\Sigma(X_i - \bar{X})^2, \ \widehat{\sigma}_2^2 = n^{-1}\Sigma(Y_i - \bar{Y})^2.$$

Recall from Section 4.9.5 that in the bivariate normal case the sample correlation coefficient r is the MLE of the population correlation coefficient ρ and that the likelihood ratio test of $H : \rho = 0$ is based on $|r|$. We can write $r^2 = g(\widehat{C}, \widehat{\sigma}_1^2, \widehat{\sigma}_2^2) : R^3 \to R$, where $g(u_1, u_2, u_3) = u_1^2/u_2u_3$. Because of the location and scale invariance of ρ and r, we can use the transformations $\widetilde{X}_i = (X_i - \mu_1)/\sigma_1$, and $\widetilde{Y}_j = (Y_j - \mu_2)/\sigma_2$ to conclude that without loss of generality we may assume $\mu_1 = \mu_2 = 0, \sigma_1^2 = \sigma_2^2 = 1, \rho = E(XY)$. Using the central limit and Slutsky's theorems, we can show (Problem 5.3.9) that $\sqrt{n}(\widehat{C} - \rho)$, $\sqrt{n}(\widehat{\sigma}_1^2 - 1)$ and $\sqrt{n}(\widehat{\sigma}_2^2 - 1)$ jointly have the same asymptotic distribution as $\sqrt{n}(\mathbf{U}_n - \mathbf{u})$ where

$$\mathbf{U}_n = (n^{-1}\Sigma X_iY_i, n^{-1}\Sigma X_i^2, n^{-1}\Sigma Y_i^2)$$

and $\mathbf{u} = (\rho, 1, 1)$. Let $\tau_{k,j}^2 = \text{Var}(\widetilde{X}^k\widetilde{Y}^j)$ and $\lambda_{k,j,m,l} = \text{Cov}(\widetilde{X}^k\widetilde{Y}^j, \widetilde{X}^m\widetilde{Y}^l)$, then by the central limit theorem

$$\sqrt{n}(\mathbf{U} - \mathbf{u}) \to \mathcal{N}(0, \Sigma), \ \Sigma = \begin{pmatrix} \tau_{1,1}^2 & \lambda_{1,1,2,0} & \lambda_{1,1,0,2} \\ \lambda_{1,1,2,0} & \tau_{2,0}^2 & \lambda_{2,0,2,0} \\ \lambda_{1,1,0,2} & \lambda_{2,0,2,0} & \tau_{0,2}^2 \end{pmatrix}. \tag{5.3.22}$$

Next we compute

$$g^{(1)}(\mathbf{u}) = (2u_1/u_2u_3, -u_1^2/u_2^2u_3, -u_1^2/u_2u_3^2) = (2\rho, -\rho^2, -\rho^2).$$

It follows from Lemma 5.3.3 and (B.5.6) that $\sqrt{n}(r^2 - \rho^2)$ is asymptotically normal, $\mathcal{N}(0, \sigma_0^2)$, with

$$\begin{aligned} \sigma_0^2 = g^{(1)}(u)\Sigma[g^{(1)}(u)]^T &= 4\rho^2\tau_{11}^2 + \rho^4\tau_{20}^2 + \rho^4\tau_{02}^2 \\ &+ 2\{-2\rho^3\lambda_{1,1,2,0} - 2\rho^3\lambda_{1,1,0,2} + \rho^4\lambda_{2,0,2,0}\}. \end{aligned}$$

When $(X, Y) \sim \mathcal{N}(\mu_1, \mu_2, \sigma_1^2, \sigma_2^2, \rho)$, then $\sigma_0^2 = 4\rho^2(1 - \rho^2)^2$, and (Problem 5.3.9) $\sqrt{n}(r - \rho) \overset{\mathcal{L}}{\to} \mathcal{N}(0, (1 - \rho^2)^2)$.

Referring to (5.3.19), we see (Problem 5.3.10) that in the bivariate normal case a variance stabilizing transformation $h(r)$ with $\sqrt{n}[h(r) - h(\rho)] \overset{\mathcal{L}}{\to} \mathcal{N}(0, 1)$ is achieved by choosing

$$h(\rho) = \frac{1}{2} \log \left(\frac{1 + \rho}{1 - \rho} \right).$$

The approximation based on this transformation, which is called *Fisher's z*, has been studied extensively and it has been shown (e.g., David, 1938) that

$$\mathcal{L}(\sqrt{n - 3}(h(r) - h(\rho))$$

is closely approximated by the $\mathcal{N}(0, 1)$ distribution, that is,

$$P(r \leq c) \approx \Phi(\sqrt{n - 3}[h(c) - h(\rho)]), \quad c \in (-1, 1).$$

This expression provides approximations to the critical value of tests of $H : \rho = 0$, it gives approximations to the power of these tests, and it provides the approximate $100(1 - \alpha)\%$ confidence interval of fixed length,

$$\rho = \tanh \left\{ h(r) \pm z \left(1 - \tfrac{1}{2}\alpha \right) / \sqrt{n - 3} \right\}$$

where tanh is the hyperbolic tangent. □

Here is an extension of Theorem 5.3.3.

Theorem 5.3.4. *Suppose* $\mathbf{Y}_1, \dots, \mathbf{Y}_n$ *are independent identically distributed d vectors with* $E|\mathbf{Y}_1|^2 < \infty$, $E\mathbf{Y}_1 = \mathbf{m}$, $\mathrm{Var}\, \mathbf{Y}_1 = \Sigma$ *and* $\mathbf{h} : \mathcal{O} \to R^p$ *where* $\mathcal{O}$ *is an open subset of* R^d, $\mathbf{h} = (h_1, \dots, h_p)$ *and* $\mathbf{h}$ *has a total differential* $\mathbf{h}^{(1)}(\mathbf{m}) = \left\| \frac{\partial h_i}{\partial x_j}(\mathbf{m}) \right\|_{p \times d}$. *Then*

$$\mathbf{h}(\bar{\mathbf{Y}}) = \mathbf{h}(\mathbf{m}) + \mathbf{h}^{(1)}(\mathbf{m})(\bar{\mathbf{Y}} - \mathbf{m}) + o_p(n^{-1/2}) \tag{5.3.23}$$

$$\sqrt{n}[\mathbf{h}(\bar{\mathbf{Y}}) - \mathbf{h}(\mathbf{m})] \overset{\mathcal{L}}{\to} \mathcal{N}(0, \mathbf{h}^{(1)}(\mathbf{m})\Sigma[\mathbf{h}^{(1)}(\mathbf{m})]^T) \tag{5.3.24}$$

Proof. Argue as before using B.8.5

(a) $$\mathbf{h}(\mathbf{y}) = \mathbf{h}(\mathbf{m}) + \mathbf{h}^{(1)}(\mathbf{m})(\mathbf{y} - \mathbf{m}) + o(|\mathbf{y} - \mathbf{m}|)$$

and

(b) $$\sqrt{n}(\bar{\mathbf{Y}} - \mathbf{m}) \overset{\mathcal{L}}{\to} \mathcal{N}(0, \Sigma)$$

so that

(c)
$$\sqrt{n}(\mathbf{h}(\bar{\mathbf{Y}}) - \mathbf{h}(\mathbf{m})) = \sqrt{n}\mathbf{h}^{(1)}(\mathbf{m})(\bar{\mathbf{Y}} - \mathbf{m}) + o_p(1).$$

$\square$

Example 5.3.7. χ_1^2 *and Normal Approximation to the Distribution of $\mathcal{F}$ Statistics.* Suppose that $X_1, \ldots, X_n$ is a sample from a $\mathcal{N}(0, 1)$ distribution. Then according to Corollary B.3.1, the $\mathcal{F}$ statistic

$$T_{k,m} = \frac{(1/k) \sum_{i=1}^{k} X_i^2}{(1/m) \sum_{i=k+1}^{k+m} X_i^2} \tag{5.3.25}$$

has an $\mathcal{F}_{k,m}$ distribution, where $k + m = n$. Suppose that $n \geq 60$ so that Table IV cannot be used for the distribution of $T_{k,m}$. When k is fixed and m (or equivalently $n = k+m$) is large, we can use Slutsky's theorem (A.14.9) to find an approximation to the distribution of $T_{k,m}$. To show this, we first note that $(1/m) \sum_{i=k+1}^{k+m} X_i^2$ is the average of m independent χ_1^2 random variables. By Theorem B.3.1, the mean of a χ_1^2 variable is $E(Z^2)$, where $Z \sim \mathcal{N}(0, 1)$. But $E(Z^2) = \text{Var}(Z) = 1$. Now the weak law of large numbers (A.15.7) implies that as $m \to \infty$,

$$\frac{1}{m} \sum_{i=k+1}^{k+m} X_i^2 \xrightarrow{P} 1.$$

Using the (b) part of Slutsky's theorem, we conclude that for fixed k,

$$T_{k,m} \xrightarrow{\mathcal{L}} \frac{1}{k} \sum_{i=1}^{k} X_i^2$$

as $m \to \infty$. By Theorem B.3.1, $\sum_{i=1}^{k} X_i^2$ has a χ_k^2 distribution. Thus, when the number of degrees of freedom in the denominator is large, the $\mathcal{F}_{k,m}$ distribution can be approximated by the distribution of V/k, where $V \sim \chi_k^2$.

To get an idea of the accuracy of this approximation, check the entries of Table IV against the last row. This row, which is labeled $m = \infty$, gives the quantiles of the distribution of V/k. For instance, if $k = 5$ and $m = 60$, then $P[T_{5,60} \leq 2.37] = P[(V/k) \leq 2.21] = 0.05$ and the respective 0.05 quantiles are 2.37 for the $\mathcal{F}_{5,60}$ distribution and 2.21 for the distribution of V/k. See also Figure B.3.1 in which the density of V/k, when $k = 10$, is given as the $\mathcal{F}_{10,\infty}$ density.

Next we turn to the normal approximation to the distribution of $T_{k,m}$. Suppose for simplicity that $k = m$ and $k \to \infty$. We write T_k for $T_{k,k}$. The case $m = \lambda k$ for some $\lambda > 0$ is left to the problems. We do not require the X_i to be normal, only that they be i.i.d. with $EX_1 = 0$, $EX_1^2 > 0$ and $EX_1^4 < \infty$. Then, if $\sigma^2 = \text{Var}(X_1)$, we can write,

$$T_k = \frac{1}{k} \sum_{i=1}^{k} Y_{i1} \bigg/ \frac{1}{k} \sum_{i=1}^{k} Y_{i2} \tag{5.3.26}$$

where $Y_{i1} = X_i^2/\sigma^2$ and $Y_{i2} = X_{k+i}^2/\sigma^2$, $i = 1, \ldots, k$. Equivalently $T_k = h(\bar{\mathbf{Y}})$ where $\mathbf{Y}_i = (Y_{i1}, Y_{i2})^T$, $E(\mathbf{Y}_i) = (1, 1)^T$ and $h(u, v) = \frac{u}{v}$. By Theorem 5.3.4,

$$\sqrt{n}(T_k - 1) \xrightarrow{\mathcal{L}} \mathcal{N}(0, h^{(1)}(\mathbf{1})\mathbf{\Sigma}[h^{(1)}(\mathbf{1})]^T) \tag{5.3.27}$$

where $\mathbf{1} = (1, 1)^T$, $h^{(1)}(u, v) = (\frac{1}{v}, -\frac{u}{v^2})^T$ and $\mathbf{\Sigma} = \mathrm{Var}(Y_{11})\mathbf{J}$, where $\mathbf{J}$ is the 2×2 identity. We conclude that

$$\sqrt{n}(T_k - 1) \xrightarrow{\mathcal{L}} \mathcal{N}(0, 2\,\mathrm{Var}(Y_{11})).$$

In particular if $X_1 \sim \mathcal{N}(0, \sigma^2)$, as $k \to \infty$,

$$\sqrt{n}(T_k - 1) \xrightarrow{\mathcal{L}} \mathcal{N}(0, 4).$$

In general, when $\min\{k, m\} \to \infty$, the distribution of $\sqrt{\frac{mk}{m+k}}(T_{k,m} - 1)$ can be approximated by a $\mathcal{N}(0, 2)$ distribution. Thus (Problem 5.3.7), when $X_i \sim \mathcal{N}(0, \sigma^2)$,

$$\begin{aligned} P[T_{k,m} \le t] &= P[\sqrt{\tfrac{mk}{m+k}}(T_{k,m} - 1) \le \sqrt{\tfrac{mk}{m+k}}(t - 1)] \\ &\approx \Phi(\sqrt{\tfrac{mk}{m+k}}(t - 1)/\sqrt{2}). \end{aligned} \tag{5.3.28}$$

An interesting and important point (noted by Box, 1953) is that unlike the t test, the F test for equality of variances (Problem 5.3.8(a)) does not have robustness of level. Specifically, if $\mathrm{Var}(X_1^2) \ne 2\sigma^4$, the upper $\mathcal{F}_{k,m}$ critical value $f_{k,m}(1 - \alpha)$, which by (5.3.28) satisfies

$$z_{1-\alpha} \approx \sqrt{\frac{mk}{m+k}}(f_{k,m}(1 - \alpha) - 1)/\sqrt{2}$$

or

$$f_{k,m}(1 - \alpha) \approx 1 + \sqrt{\frac{2(m+k)}{mk}}z_{1-\alpha}$$

is asymptotically incorrect. In general (Problem 5.3.8(c)) one has to use the critical value

$$c_{k,m} = \left(1 + \sqrt{\frac{\kappa(m+k)}{mk}}z_{1-\alpha}\right) \tag{5.3.29}$$

where $\kappa = \mathrm{Var}[(X_1 - \mu_1)/\sigma_1]^2$, $\mu_1 = E(X_1)$, and $\sigma_1^2 = \mathrm{Var}(X_1)$. When κ is unknown, it can be estimated by the method of moments (Problem 5.3.8(d)). $\qquad \Box$

5.3.3 Asymptotic Normality of the Maximum Likelihood Estimate in Exponential Families

Our final application of the δ-method follows.

Theorem 5.3.5. *Suppose $\mathcal{P}$ is a canonical exponential family of rank d generated by $\mathbf{T}$ with $\mathcal{E}$ open. Then if $X_1, \ldots, X_n$ are a sample from $P_{\boldsymbol{\eta}} \in \mathcal{P}$ and $\hat{\boldsymbol{\eta}}$ is defined as the MLE if it exists and equal to $\mathbf{c}$ (some fixed value) otherwise,*

(i) $\widehat{\boldsymbol{\eta}} = \boldsymbol{\eta} + \frac{1}{n}\sum_{i=1}^{n}\ddot{A}^{-1}(\boldsymbol{\eta})(\mathbf{T}(X_i) - \dot{A}(\boldsymbol{\eta})) + o_{P_{\boldsymbol{\eta}}}(n^{-\frac{1}{2}})$

(ii) $\mathcal{L}_{\boldsymbol{\eta}}(\sqrt{n}(\widehat{\boldsymbol{\eta}} - \boldsymbol{\eta})) \to \mathcal{N}_d(\mathbf{0}, \ddot{A}^{-1}(\boldsymbol{\eta}))$.

Proof. The result is a consequence of Theorems 5.2.2 and 5.3.4. We showed in the proof of Theorem 5.2.2 that, if $\bar{\mathbf{T}} \equiv \frac{1}{n}\sum_{i=1}^{n}\mathbf{T}(X_i)$, $P_{\boldsymbol{\eta}}[\bar{\mathbf{T}} \in \dot{A}(\mathcal{E})] \to 1$ and, hence, $P_{\boldsymbol{\eta}}[\widehat{\boldsymbol{\eta}} = \dot{A}^{-1}(\bar{\mathbf{T}})] \to 1$. Identify $\mathbf{h}$ in Theorem 5.3.4 with $\dot{A}^{-1}$ and $\mathbf{m}$ with $\dot{A}(\boldsymbol{\eta})$. Note that by B.8.14, if $\mathbf{t} = \dot{A}(\boldsymbol{\eta})$,

$$D\dot{A}^{-1}(\mathbf{t}) = [D\dot{A}(\boldsymbol{\eta})]^{-1}. \tag{5.3.30}$$

But $D\dot{A} = \ddot{A}$ by definition and, thus, in our case,

$$\mathbf{h}^{(1)}(\mathbf{m}) = \ddot{A}^{-1}(\boldsymbol{\eta}). \tag{5.3.31}$$

Thus, (i) follows from (5.3.23). For (ii) simply note that, in our case, by Corollary 1.6.1,

$$\boldsymbol{\Sigma} = \mathrm{Var}(\mathbf{T}(X_1)) = \ddot{A}(\boldsymbol{\eta})$$

and, therefore,

$$\mathbf{h}^{(1)}(\mathbf{m})\boldsymbol{\Sigma}[\mathbf{h}^{(1)}(\mathbf{m})]^T = \ddot{A}^{-1}\ddot{A}\ddot{A}^{-1}(\boldsymbol{\eta}) = \ddot{A}^{-1}(\boldsymbol{\eta}). \tag{5.3.32}$$

Hence, (ii) follows from (5.3.24). $\qquad\square$

Remark 5.3.1. Recall that
$$\ddot{A}(\boldsymbol{\eta}) = \mathrm{Var}_{\boldsymbol{\eta}}(\mathbf{T}) = I(\boldsymbol{\eta})$$
is the Fisher information. Thus, the asymptotic variance matrix $I^{-1}(\boldsymbol{\eta})$ of $\sqrt{n}(\widehat{\boldsymbol{\eta}} - \boldsymbol{\eta})$ equals the lower bound (3.4.38) on the variance matrix of $\sqrt{n}(\widetilde{\boldsymbol{\eta}} - \boldsymbol{\eta})$ for any unbiased estimator $\widetilde{\boldsymbol{\eta}}$. This is an "asymptotic efficiency" property of the MLE we return to in Section 6.2.1.

Example 5.3.8. Let $X_1, \dots, X_n$ be i.i.d. as X with $X \sim N(\mu, \sigma^2)$. Then $T_1 = \bar{X}$ and $T_2 = n^{-1}\Sigma X_i^2$ are sufficient statistics in the canonical model. Now

$$\sqrt{n}[T_1 - \mu, T_2 - (\mu^2 + \sigma^2)] \xrightarrow{\mathcal{L}} N(0, 0, I(\boldsymbol{\eta})) \tag{5.3.33}$$

where, by Example 2.3.4,

$$I(\boldsymbol{\eta}) = \ddot{A}(\boldsymbol{\eta}) = \frac{1}{2\eta_2^2}\begin{pmatrix} -\eta_2 & \eta_1 \\ \eta_1 & 1 - \eta_1^2(4\eta_2)^{-1} \end{pmatrix}.$$

Here $\eta_1 = \mu/\sigma^2$, $\eta_2 = -1/2\sigma^2$, $\widehat{\eta}_1 = \bar{X}/\widehat{\sigma}^2$, and $\widehat{\eta}_2 = -1/2\widehat{\sigma}^2$ where $\widehat{\sigma}^2 = T_2 - (T_1)^2$. By Theorem 5.3.5,

$$\sqrt{n}(\widehat{\eta}_1 - \eta_1, \widehat{\eta}_2 - \eta_2) \xrightarrow{\mathcal{L}} N(0, 0, I^{-1}(\boldsymbol{\eta})).$$

Because $\bar{X} = T_1$ and $\widehat{\sigma}^2 = T_2 - (T_1)^2$, we can use (5.3.33) and Theorem 5.3.4 to find (Problem 5.3.26)

$$\sqrt{n}(\bar{X} - \mu, \widehat{\sigma}^2 - \sigma^2) \xrightarrow{\mathcal{L}} \mathcal{N}(0, 0, \Sigma_0)$$

where $\Sigma_0 = \operatorname{diag}(\sigma^2, 2\sigma^4)$.

Summary. Consistency is 0th-order asymptotics. First-order asymptotics provides approximations to the difference between a quantity tending to a limit and the limit, for instance, the difference between a consistent estimate and the parameter it estimates. Second-order asymptotics provides approximations to the difference between the error and its first-order approximation, and so on. We begin in Section 5.3.1 by studying approximations to moments and central moments of estimates. Fundamental asymptotic formulae are derived for the bias and variance of an estimate first for smooth function of a scalar mean and then a vector mean. These "δ method" approximations based on Taylor's formula and elementary results about moments of means of i.i.d. variables are explained in terms of similar stochastic approximations to $h(\bar{\mathbf{Y}}) - h(\boldsymbol{\mu})$ where $\mathbf{Y}_1, \ldots, \mathbf{Y}_n$ are i.i.d. as $\mathbf{Y}$, $E\mathbf{Y} = \boldsymbol{\mu}$, and h is smooth. These stochastic approximations lead to Gaussian approximations to the laws of important statistics. The moment and in law approximations lead to the definition of variance stabilizing transformations for classical one-dimensional exponential families. Higher-order approximations to distributions (Edgeworth series) are discussed briefly. Finally, stochastic approximations in the case of vector statistics and parameters are developed, which lead to a result on the asymptotic normality of the MLE in multiparameter exponential families.

5.4   ASYMPTOTIC THEORY IN ONE DIMENSION

In this section we define and study asymptotic optimality for estimation, testing, and confidence bounds, under i.i.d. sampling, when we are dealing with one-dimensional smooth parametric models. Specifically we shall show that important likelihood based procedures such as MLE's are asymptotically optimal. In Chapter 6 we sketch how these ideas can be extended to multi-dimensional parametric families.

5.4.1   Estimation: The Multinomial Case

Following Fisher (1958),[1] we develop the theory first for the case that $X_1, \ldots, X_n$ are i.i.d. taking values $\{x_0, \ldots, x_k\}$ only so that P is defined by $\mathbf{p} \equiv (p_0, \ldots, p_k)$ where

$$p_j \equiv P[X_1 = x_j], \ 0 \leq j \leq k \tag{5.4.1}$$

and $\mathbf{p} \in \mathcal{S}$, the $(k+1)$-dimensional simplex (see Example 1.6.7). Thus, $\mathbf{N} = (N_0, \ldots, N_k)$ where $N_j \equiv \sum_{i=1}^{n} 1(X_i = x_j)$ is sufficient. We consider one-dimensional parametric submodels of $\mathcal{S}$ defined by $\mathcal{P} = \{(p(x_0, \theta), \ldots, p(x_k, \theta)) : \theta \in \Theta\}$, Θ open $\subset R$ (e.g., see Example 2.1.4 and Problem 2.1.15). We focus first on estimation of θ. Assume

$$A : \theta \to p(x_j, \theta), \ 0 < p_j < 1, \ \text{is twice differentiable for } 0 \leq j \leq k.$$

Note that A implies that

$$l(X_1, \theta) \equiv \log p(X_1, \theta) = \sum_{j=0}^{k} \log p(x_j, \theta) 1(X_1 = x_j) \tag{5.4.2}$$

is twice differentiable and $\frac{\partial l}{\partial \theta}(X_1, \theta)$ is a well-defined, bounded random variable

$$\frac{\partial l}{\partial \theta}(X_1, \theta) = \sum_{j=0}^{k} \left(\frac{\partial p}{\partial \theta}(x_j, \theta) \right) \frac{1}{p(x_j, \theta)} \cdot 1(X_1 = x_j). \tag{5.4.3}$$

Furthermore (Section 3.4.2),

$$E_\theta \frac{\partial l}{\partial \theta}(X_1, \theta) = 0 \tag{5.4.4}$$

and $\frac{\partial^2 l}{\partial \theta^2}(X_1, \theta)$ is similarly bounded and well defined with

$$I(\theta) \equiv \mathrm{Var}_\theta \left(\frac{\partial l}{\partial \theta}(X_1, \theta) \right) = -E_\theta \frac{\partial^2}{\partial \theta^2} l(X_1, \theta). \tag{5.4.5}$$

As usual we call $I(\theta)$ the *Fisher information.*
   Next suppose we are given a plug-in estimator $h\left(\frac{N}{n} \right)$ (see (2.1.11)) of θ where

$$h : \mathcal{S} \to R$$

satisfies

$$h(\mathbf{p}(\theta)) = \theta \text{ for all } \theta \in \Theta \tag{5.4.6}$$

where $\mathbf{p}(\theta) = (p(x_0, \theta), \ldots, p(x_k, \theta))^T$. Many such h exist if $k > 1$. Consider Example 2.1.4, for instance. Assume

$$H : h \text{ is differentiable.}$$

Then we have the following theorem.

Theorem 5.4.1. *Under H, for all θ,*

$$\mathcal{L}_\theta \left(\sqrt{n} \left(h\left(\frac{N}{n} \right) - \theta \right) \right) \to \mathcal{N}(0, \sigma^2(\theta, h)) \tag{5.4.7}$$

where $\sigma^2(\theta, h)$ is given by (5.4.11). Moreover, if A also holds,

$$\sigma^2(\theta, h) \geq I^{-1}(\theta) \tag{5.4.8}$$

with equality if and only if,

$$\frac{\partial h}{\partial p_j}(\mathbf{p}(\theta)) \bigg|_{\mathbf{p}(\theta)} = I^{-1}(\theta) \frac{\partial l}{\partial \theta}(x_j, \theta), \ 0 \leq j \leq k. \tag{5.4.9}$$

Proof. Apply Theorem 5.3.2 noting that

$$\sqrt{n}\left(h\left(\frac{\mathbf{N}}{n}\right) - h(\mathbf{p}(\theta))\right) = \sqrt{n}\sum_{j=0}^{k}\frac{\partial h}{\partial p_j}(\mathbf{p}(\theta))\left(\frac{N_j}{n} - p(x_j,\theta)\right) + o_p(1).$$

Note that, using the definition of N_j,

$$\sum_{j=0}^{k}\frac{\partial h}{\partial p_j}(\mathbf{p}(\theta))\left(\frac{N_j}{n} - p(x_j,\theta)\right) = n^{-1}\sum_{i=1}^{n}\sum_{j=0}^{k}\frac{\partial h}{\partial p_j}(\mathbf{p}(\theta))(1(X_i = x_j) - p(x_j,\theta)).$$

$$(5.4.10)$$

Thus, by (5.4.10), not only is $\sqrt{n}\left\{h\left(\frac{\mathbf{N}}{n}\right) - h(\mathbf{p}(\theta))\right\}$ asymptotically normal with mean 0, but also its asymptotic variance is

$$\sigma^2(\theta,h) = \operatorname{Var}_\theta\left(\sum_{j=0}^{k}\frac{\partial h}{\partial p_j}(\mathbf{p}(\theta))1(X_1 = x_j)\right)$$

$$(5.4.11)$$

$$= \sum_{j=0}^{k}\left(\frac{\partial h}{\partial p_j}(\mathbf{p}(\theta))\right)^2 p(x_j,\theta) - \left(\sum_{j=0}^{k}\frac{\partial h}{\partial p_j}(\mathbf{p}(\theta))p(x_j,\theta)\right)^2.$$

Note that by differentiating (5.4.6), we obtain

$$\sum_{j=0}^{k}\frac{\partial h}{\partial p_j}(\mathbf{p}(\theta))\frac{\partial p}{\partial \theta}(x_j,\theta) = 1$$

$$(5.4.12)$$

or equivalently, by noting $\frac{\partial p}{\partial \theta}(x_j,\theta) = \left[\frac{\partial}{\partial \theta}l(x_j,\theta)\right]p(x_j,\theta)$,

$$\operatorname{Cov}_\theta\left(\sum_{j=0}^{k}\frac{\partial h}{\partial p_j}(\mathbf{p}(\theta))1(X_1 = x_j), \frac{\partial l}{\partial \theta}(X_1,\theta)\right) = 1.$$

$$(5.4.13)$$

By (5.4.13), using the correlation inequality (A.11.16) as in the proof of the information inequality (3.4.12), we obtain

$$1 \leq \sigma^2(\theta,h)\operatorname{Var}_\theta\frac{\partial l}{\partial \theta}(X_1,\theta) = \sigma^2(\theta,h)I(\theta)$$

$$(5.4.14)$$

with equality iff,

$$\sum_{j=0}^{k}\frac{\partial h}{\partial p_j}(\mathbf{p}(\theta))(1(X_1 = x_j) - p(x_j,\theta)) = a(\theta)\frac{\partial l}{\partial \theta}(X_1,\theta) + b(\theta)$$

$$(5.4.15)$$

for some $a(\theta) \neq 0$ and some $b(\theta)$ with probability 1. Taking expectations we get $b(\theta) = 0$. Noting that the covariance of the right- and left-hand sides is $a(\theta)$, while their common variance is $a^2(\theta)I(\theta) = \sigma^2(\theta,h)$, we see that equality in (5.4.8) gives

$$a^2(\theta)I^2(\theta) = 1,$$

$$(5.4.16)$$

which implies (5.4.9).   □

We shall see in Section 5.4.3 that the information bound (5.4.8) is, if it exists and under regularity conditions, achieved by $\widehat{\theta} = \widehat{h}\left(\frac{\mathbf{N}}{n}\right)$, the MLE of θ where $\widehat{h}$ is defined implicitly by: $\widehat{h}(\mathbf{p})$ is the value of θ, which

(i) maximizes $\sum_{j=0}^{k} N_j \log p(x_j, \theta)$

and

(ii) solves $\sum_{j=0}^{k} N_j \frac{\partial l}{\partial \theta}(x_j, \theta) = 0$.

Example 5.4.1. *One-Parameter Discrete Exponential Families.* Suppose $p(x, \theta) = \exp\{\theta T(x) - A(\theta)\} h(x)$ where $h(x) = 1(x \in \{x_0, \ldots, x_k\})$, $\theta \in \Theta$, is a canonical one-parameter exponential family (supported on $\{x_0, \ldots, x_k\}$) and Θ is open. Then Theorem 5.3.5 applies to the MLE $\widehat{\theta}$ and

$$\mathcal{L}_\theta(\sqrt{n}(\widehat{\theta} - \theta)) \rightarrow N\left(0, \frac{1}{I(\theta)}\right) \tag{5.4.17}$$

with the asymptotic variance achieving the information bound $I^{-1}(\theta)$. Note that because $\bar{T} = n^{-1} \sum_{i=1}^{n} T(X_i) = \sum_{j=0}^{k} T(x_j) \frac{N_j}{n}$, then, by (2.3.3)

$$\widehat{\theta} = [\dot{A}]^{-1}(\bar{T}), \tag{5.4.18}$$

and

$$\widehat{h}(\mathbf{p}) = [\dot{A}]^{-1}\left(\sum_{j=0}^{k} T(x_j) p_j\right). \tag{5.4.19}$$

The binomial (n, p) and Hardy–Weinberg models can both be put into this framework with canonical parameters such as $\theta = \log\left(\frac{p}{1-p}\right)$ in the first case.   □

Both the asymptotic variance bound and its achievement by the MLE are much more general phenomena. In the next two subsections we consider some more general situations.

5.4.2   Asymptotic Normality of Minimum Contrast and M-Estimates

We begin with an asymptotic normality theorem for minimum contrast estimates. As in Theorem 5.2.3 we give this result under conditions that are themselves implied by more technical sufficient conditions that are easier to check.

Suppose i.i.d. $X_1, \ldots, X_n$ are tentatively modeled to be distributed according to P_θ, $\theta \in \Theta$ open $\subset R$ and corresponding density/frequency functions $p(\cdot, \theta)$. Write $\mathcal{P} = \{P_\theta : \theta \in \Theta\}$. Let $\rho : X \times \Theta \rightarrow R$ where

$$D(\theta, \theta_0) = E_{\theta_0}(\rho(X_1, \theta) - \rho(X_1, \theta_0))$$

is uniquely minimized at θ_0. Let $\bar{\theta}_n$ be the minimum contrast estimate

$$\bar{\theta}_n = \text{argmin} \; \frac{1}{n} \sum_{i=1}^{n} \rho(X_i, \theta).$$

Suppose

A0: $\psi = \frac{\partial \rho}{\partial \theta}$ is well defined.

Then

$$\frac{1}{n} \sum_{i=1}^{n} \psi(X_i, \bar{\theta}_n) = 0. \tag{5.4.20}$$

In what follows we let P, rather than P_θ, denote the distribution of X_i. This is because, as pointed out later in Remark 5.4.3, under regularity conditions the properties developed in this section are valid for $P \notin \{P_\theta : \theta \in \Theta\}$. We need only that $\theta(P)$ is a well defined parameter as defined in Section 1.1.2. As we saw in Section 2.1, parameters and their estimates can often be extended to larger classes of distributions than they originally were defined for. Suppose

A1: The parameter $\theta(P)$ given by the solution of

$$\int \psi(x, \theta) dP(x) = 0 \tag{5.4.21}$$

is well defined on $\mathcal{P}$. That is,

$$\int |\psi(x, \theta)| dP(x) < \infty, \; \theta \in \Theta, \; P \in \mathcal{P}$$

and $\theta(P)$ is the unique solution of (5.4.21) and, hence, $\theta(P_\theta) = \theta$.

A2: $E_P \psi^2(X_1, \theta(P)) < \infty$ for all $P \in \mathcal{P}$.

A3: $\psi(\cdot, \theta)$ is differentiable, $\frac{\partial \psi}{\partial \theta}(X_1, \theta)$ has a finite expectation and

$$E_P \frac{\partial \psi}{\partial \theta}(X_1, \theta(P)) \neq 0.$$

A4: $\sup_t \left\{ \left| \frac{1}{n} \sum_{i=1}^{n} \left(\frac{\partial \psi}{\partial \theta}(X_i, t) - \frac{\partial \psi}{\partial \theta}(X_i, \theta(P)) \right) \right| : |t - \theta(P)| \leq \epsilon_n \right\} \xrightarrow{P} 0$ if $\epsilon_n \to 0$.

A5: $\bar{\theta}_n \xrightarrow{P} \theta(P)$. That is, $\bar{\theta}_n$ is consistent on $\mathcal{P} = \{P_\theta : \theta \in \Theta\}$.

Theorem 5.4.2. *Under* A0–A5,

$$\bar{\theta}_n = \theta(P) + \frac{1}{n} \sum_{i=1}^{n} \tilde{\psi}(X_i, \theta(P)) + o_p(n^{-1/2}) \tag{5.4.22}$$

where

$$\tilde{\psi}(x, P) = \psi(x, \theta(P)) \Big/ \left(-E_P \frac{\partial \psi}{\partial \theta}(X_1, \theta(P)) \right). \tag{5.4.23}$$

Hence,

$$\mathcal{L}_P(\sqrt{n}(\bar{\theta}_n - \theta(P))) \rightarrow \mathcal{N}(0, \sigma^2(\psi, P))$$

where

$$\sigma^2(\psi, P) = \frac{E_P \psi^2(X_1, \theta(P))}{\left(E_P \frac{\partial \psi}{\partial \theta}(X_1, \theta(P))\right)^2}. \tag{5.4.24}$$

Proof. Claim (5.4.24) follows from the central limit theorem and Slutsky's theorem, applied to (5.4.22) because

$$\sqrt{n}(\bar{\theta}_n - \theta(P)) = n^{-1/2} \sum_{i=1}^{n} \tilde{\psi}(X_i, P) + o_p(1)$$

and

$$\begin{aligned}
E_P \tilde{\psi}(X_1, P) &= E_P \psi(X_1, \theta(P)) \Big/ \left(-E_P \frac{\partial \psi}{\partial \theta}(X_1, \theta(P))\right) \\
&= 0
\end{aligned}$$

while

$$E_P \tilde{\psi}^2(X_1, P) = \sigma^2(\psi, P) < \infty$$

by A1, A2, and A3. Next we show that (5.4.22) follows by a Taylor expansion of the equations (5.4.20) and (5.4.21). Let $\bar{\theta}_n = \theta(\widehat{P})$ where $\widehat{P}$ denotes the empirical probability. By expanding $n^{-1} \sum_{i=1}^{n} \psi(X_i, \bar{\theta}_n)$ around $\theta(P)$, we obtain, using (5.4.20),

$$-\frac{1}{n} \sum_{i=1}^{n} \psi(X_i, \theta(P)) = \frac{1}{n} \sum_{i=1}^{n} \frac{\partial \psi}{\partial \theta}(X_i, \theta_n^*)(\bar{\theta}_n - \theta(P)) \tag{5.4.25}$$

where $|\theta_n^* - \theta(P)| \leq |\bar{\theta}_n - \theta(P)|$. Apply A5 and A4 to conclude that

$$\frac{1}{n} \sum_{i=1}^{n} \frac{\partial \psi}{\partial \theta}(X_i, \theta_n^*) = \frac{1}{n} \sum_{i=1}^{n} \frac{\partial}{\partial \theta} \psi(X_i, \theta(P)) + o_p(1) \tag{5.4.26}$$

and A3 and the WLLN to conclude that

$$\frac{1}{n} \sum_{i=1}^{n} \frac{\partial \psi}{\partial \theta}(X_i, \theta(P)) = E_P \frac{\partial \psi}{\partial \theta}(X_1, \theta(P)) + o_p(1). \tag{5.4.27}$$

Combining (5.4.25)–(5.4.27) we get,

$$(\bar{\theta}_n - \theta(P)) \left(-E_P \frac{\partial \psi}{\partial \theta}(X_1, \theta(P)) + o_p(1)\right) = \frac{1}{n} \sum_{i=1}^{n} \psi(X_i, \theta(P)). \tag{5.4.28}$$

But by the central limit theorem and A1,

$$\frac{1}{n} \sum_{i=1}^{n} \psi(X_i, \theta(P)) = O_p(n^{-1/2}). \tag{5.4.29}$$

Dividing by the second factor in (5.4.28) we finally obtain

$$\bar{\theta}_n - \theta(P) = \frac{1}{n}\sum_{i=1}^{n}\tilde{\psi}(X_i, \theta(P)) + o_p\left(\frac{1}{n}\sum_{i=1}^{n}\psi(X_i, \theta(P))\right)$$

and (5.4.22) follows from the foregoing and (5.4.29). $\square$

Remark 5.4.1. An additional assumption A6 gives a slightly different formula for $E_P \frac{\partial \psi}{\partial \theta}(X_1, \theta(P))$ if $P = P_\theta$.

A6: Suppose $P = P_\theta$ so that $\theta(P) = \theta$, and that the model $\mathcal{P}$ is regular and let $l(x, \theta) = \log p(x, \theta)$ where $p(\cdot, \theta)$ is as usual a density or frequency function. Suppose l is differentiable and assume that

$$\begin{aligned}E_\theta \frac{\partial \psi}{\partial \theta}(X_1, \theta(P)) &= -E_\theta \frac{\partial l}{\partial \theta}(X_1, \theta)\psi(X_1, \theta) \\ &= -\text{Cov}_\theta\left(\frac{\partial l}{\partial \theta}(X_1, \theta), \psi(X_1, \theta)\right).\end{aligned} \quad (5.4.30)$$

Note that (5.4.30) is formally obtained by differentiating the equation (5.4.21), written as

$$\int \psi(x, \theta)p(x, \theta)d\mu(x) = 0 \quad (5.4.31)$$

for all θ. If an unbiased estimate $\delta(X_1)$ of θ exists and we let $\psi(x, \theta) = \delta(x) - \theta$, it is easy to see that A6 is the same as (3.4.8). If further X_1 takes on a finite set of values, $\{x_0, \ldots, x_k\}$, and we define $h(\mathbf{p}) = \sum_{j=0}^{k}\delta(x_j)p_j$, we see that A6 corresponds to (5.4.12).

Identity (5.4.30) suggests that if $\mathcal{P}$ is regular the conclusion of Theorem 5.4.2 may hold even if ψ is not differentiable provided that $-E_\theta \frac{\partial \psi}{\partial \theta}(X_1, \theta)$ is replaced by $\text{Cov}_\theta(\psi(X_1, \theta), \frac{\partial l}{\partial \theta}(X_1, \theta))$ and a suitable replacement for A3, A4 is found. This is in fact true—see Problem 5.4.1.

Remark 5.4.2. Solutions to (5.4.20) are called M-estimates as well as estimating equation estimates—see Section 2.2.1. Our arguments apply to M-estimates. Nothing in the arguments require that $\bar{\theta}_n$ be a minimum contrast as well as an M-estimate (i.e., that $\psi = \frac{\partial \rho}{\partial \theta}$ for some ρ).

Remark 5.4.3. Our arguments apply even if $X_1, \ldots, X_n$ are i.i.d. P but $P \notin \mathcal{P} = \{P_\theta : \theta \in \Theta\}$. $\theta(P)$ in A1–A5 is then replaced by

(1) $\theta(P) = \text{argmin } E_P \rho(X_1, \theta)$

or more generally, for M-estimates,

(2) $\theta(P)$ solves $E_P \psi(X_1, \theta) = 0$.

Theorem 5.4.2 is valid with $\theta(P)$ as in (1) or (2). This extension will be pursued in Volume 2.

We conclude by stating some sufficient conditions, essentially due to Cramer (1946), for A4 and A6. Conditions A0, A1, A2, and A3 are readily checkable whereas we have given conditions for A5 in Section 5.2.

A4':

(a) $\theta \to \frac{\partial \psi}{\partial \theta}(x_1, \theta)$ is a continuous function of θ for all x.

(b) There exists $\delta(\theta) > 0$ such that

$$\sup \left\{ \left| \frac{\partial \psi}{\partial \theta}(X_1, \theta') - \frac{\partial \psi}{\partial \theta}(X_1, \theta) \right| : |\theta - \theta'| \leq \delta(\theta) \right\} \leq M(X_1, \theta),$$

where $E_\theta M(X_1, \theta) < \infty$.

A6': $\frac{\partial \psi}{\partial \theta}(x, \theta')$ is defined for all x, $|\theta' - \theta| \leq \delta(\theta)$ and $\int_{\theta-\delta}^{\theta+\delta} \int \left| \frac{\partial \psi}{\partial s}(x, s) \right| d\mu(x) ds < \infty$ for some $\delta = \delta(\theta) > 0$, where $\mu(x)$ is the dominating measure for $P(x)$ defined in (A.10.13). That is, "$dP(x) = p(x)d\mu(x)$."

Details of how A4' (with A0–A3) implies A4 and A6' implies A6 are given in the problems. We also indicate by example in the problems that some conditions are needed (Problem 5.4.4) but A4' and A6' are not necessary (Problem 5.4.1).

5.4.3 Asymptotic Normality and Efficiency of the MLE

The most important special case of (5.4.30) occurs when $\rho(x, \theta) = -l(x, \theta) \equiv -\log p(x, \theta)$ and $\psi(x, \theta) \equiv \frac{\partial l}{\partial \theta}(x, \theta)$ obeys A0–A6. In this case $\bar{\theta}_n$ is the MLE $\hat{\theta}_n$ and we obtain an identity of Fisher's,

$$
\begin{aligned}
-E_\theta \frac{\partial^2 l}{\partial \theta^2}(X_1, \theta) &= E_\theta \left(\frac{\partial l}{\partial \theta}(X_1, \theta) \right)^2 \\
&= \mathrm{Var}_\theta \left(\frac{\partial l}{\partial \theta}(X_1, \theta) \right) \equiv I(\theta),
\end{aligned}
\tag{5.4.32}
$$

where $I(\theta)$ is the Fisher information introduced in Section 3.4. We can now state the basic result on asymptotic normality and efficiency of the MLE.

Theorem 5.4.3. *If A0–A6 apply to* $\rho(x, \theta) = -l(x, \theta)$ *and* $P = P_\theta$, *then the MLE* $\hat{\theta}_n$ *satisfies*

$$\hat{\theta}_n = \theta + \frac{1}{n} \sum_{i=1}^{n} \frac{1}{I(\theta)} \frac{\partial l}{\partial \theta}(X_1, \theta) + o_p(n^{-1/2}) \tag{5.4.33}$$

so that

$$\mathcal{L}_\theta(\sqrt{n}(\hat{\theta}_n - \theta)) \to N\left(0, \frac{1}{I(\theta)}\right). \tag{5.4.34}$$

Furthermore, if $\bar{\theta}_n$ *is a minimum contrast estimate whose corresponding* ρ *and* ψ *satisfy A0–A6, then*

$$\sigma^2(\psi, P_\theta) \geq \frac{1}{I(\theta)} \tag{5.4.35}$$

with equality iff $\psi = a(\theta)\frac{\partial l}{\partial \theta}$ *for some* $a \neq 0$.

Proof. Claims (5.4.33) and (5.4.34) follow directly by Theorem 5.4.2. By (5.4.30) and (5.4.35), claim (5.4.35) is equivalent to

$$\frac{E_\theta \psi^2(X_1, \theta)}{\left[\text{Cov}_\theta \left(\psi(X_1, \theta), \frac{\partial l}{\partial \theta}(X_1, \theta)\right)\right]^2} \geq \frac{1}{\text{Var}_\theta \left(\frac{\partial l}{\partial \theta}(X_1, \theta)\right)}. \tag{5.4.36}$$

Because $E_\theta \psi(X_1, \theta) = 0$, cross multiplication shows that (5.4.36) is just the correlation inequality and the theorem follows because equality holds iff ψ is a nonzero multiple $a(\theta)$ of $\frac{\partial l}{\partial \theta}(X_1, \theta)$. $\qquad\square$

Note that Theorem 5.4.3 generalizes Example 5.4.1 once we identify $\psi(x, \theta)$ with $T(x) - A'(\theta)$.

The optimality part of Theorem 5.4.3 is not valid without some conditions on the estimates being considered.

Example 5.4.2. *Hodges's Example.* Let $X_1, \ldots, X_n$ be i.i.d. $\mathcal{N}(\theta, 1)$. Then $\bar{X}$ is the MLE of θ and it is trivial to calculate $I(\theta) \equiv 1$.

Consider the following competitor to $\bar{X}$:

$$\begin{aligned}
\widetilde{\theta}_n &= \quad 0 \text{ if } |\bar{X}| \leq n^{-1/4} \\
&= \quad \bar{X} \text{ if } |\bar{X}| > n^{-1/4}.
\end{aligned} \tag{5.4.37}$$

We can interpret this estimate as first testing $H : \theta = 0$ using the test "Reject iff $|\bar{X}| \geq n^{-1/4}$" and using $\bar{X}$ as our estimate if the test rejects and 0 as our estimate otherwise. We next compute the limiting distribution of $\sqrt{n}(\widetilde{\theta}_n - \theta)$. Let $Z \sim \mathcal{N}(0, 1)$. Then

$$\begin{aligned}
P_\theta[|\bar{X}| \leq n^{-1/4}] &= \quad P[|Z + \sqrt{n}\theta| \leq n^{1/4}] \\
&= \quad \Phi(n^{1/4} - \sqrt{n}\theta) - \Phi(-n^{1/4} - \sqrt{n}\theta).
\end{aligned} \tag{5.4.38}$$

Therefore, if $\theta \neq 0$, $P_\theta[|\bar{X}| \leq n^{-1/4}] \to 0$ because $n^{1/4} - \sqrt{n}\theta \to -\infty$, and, thus, $P_\theta[\widetilde{\theta}_n = \bar{X}] \to 1$. If $\theta = 0$, $P_\theta[|\bar{X}| \leq n^{1/4}] \to 1$, and $P_\theta[\widetilde{\theta}_n = 0] \to 1$. Therefore,

$$\mathcal{L}_\theta(\sqrt{n}(\widetilde{\theta}_n - \theta)) \to \mathcal{N}(0, \sigma^2(\theta)) \tag{5.4.39}$$

where $\sigma^2(\theta) = 1 = \frac{1}{I(\theta)}, \theta \neq 0, \sigma^2(0) = 0 < \frac{1}{I(\theta)}$.

The phenomenon (5.4.39) with $\sigma^2(\theta) \leq I^{-1}(\theta)$ for all $\theta \in \Theta$ and $\sigma^2(\theta_0) < I^{-1}(\theta_0)$, for some $\theta_0 \in \Theta$ is known as *superefficiency*. For this estimate superefficiency implies poor behavior of $\widehat{\theta}_n$ at values close to 0, see Lehmann and Casella, 1998, p. 442. However, for higher-dimensional θ, the phenomenon becomes more disturbing and has important practical consequences. We discuss this further in Volume II. $\qquad\square$

5.4.4   Testing

The major testing problem if θ is one-dimensional is $H : \theta \leq \theta_0$ versus $K : \theta > \theta_0$. If $p(\cdot, \theta)$ is an MLR family in $T(X)$, we know that all likelihood ratio tests for simple θ_1

versus simple θ_2, $\theta_1 < \theta_2$, as well as the likelihood ratio test for H versus K, are of the form "Reject H for $T(X)$ large" with the critical value specified by making the probability of type I error α at θ_0. If $p(\cdot, \theta)$ is a one-parameter exponential family in θ generated by $T(X)$, this test can also be interpreted as a test of $H : \lambda \leq \lambda_0$ versus $K : \lambda > \lambda_0$, where $\lambda = \dot{A}(\theta)$ because $\dot{A}$ is strictly increasing. The test is then precisely, "Reject H for large values of the MLE $T(X)$ of λ." It seems natural in general to study the behavior of the test, "Reject H if $\widehat{\theta}_n \geq c(\alpha, \theta_0)$" where $P_{\theta_0}[\widehat{\theta}_n \geq c(\alpha, \theta_0)] = \alpha$ and $\widehat{\theta}_n$ is the MLE of θ. We will use asymptotic theory to study the behavior of this test when we observe i.i.d. $X_1, \ldots, X_n$ distributed according to P_θ, $\theta \in (a, b)$, $a < \theta_0 < b$, derive an optimality property, and then directly and through problems exhibit other tests with the same behavior.

Let $c_n(\alpha, \theta_0)$ denote the critical value of the test using the MLE $\widehat{\theta}_n$ based on n observations.

Theorem 5.4.4. *Suppose the model* $\mathcal{P} = \{P_\theta : \theta \in \Theta\}$ *is such that the conditions of Theorem 5.4.2 apply to* $\psi = \frac{\partial l}{\partial \theta}$ *and* $\widehat{\theta}_n$, *the MLE. That is,*

$$\mathcal{L}_\theta(\sqrt{n}(\widehat{\theta}_n - \theta)) \to \mathcal{N}(0, I^{-1}(\theta)) \tag{5.4.40}$$

where $I(\theta) > 0$ *for all* θ. *Then*

$$c_n(\alpha, \theta_0) = \theta_0 + z_{1-\alpha}/\sqrt{nI(\theta_0)} + o(n^{-1/2}) \tag{5.4.41}$$

where $z_{1-\alpha}$ *is the* $1 - \alpha$ *quantile of the* $\mathcal{N}(0, 1)$ *distribution.*

Suppose (A4′) holds as well as (A6) and $I(\theta) < \infty$ *for all* θ. *Then If* $\theta > \theta_0$,

$$P_\theta[\widehat{\theta}_n > c_n(\alpha, \theta_0)] \to 1. \tag{5.4.42}$$

If $\theta < \theta_0$,

$$P_\theta[\widehat{\theta}_n > c_n(\alpha, \theta_0)] \to 0. \tag{5.4.43}$$

Property (5.4.42) is sometimes called *consistency* of the test against a fixed alternative.

Proof. The proof is straightforward:

$$P_{\theta_0}[\sqrt{nI(\theta_0)}(\widehat{\theta}_n - \theta_0) \geq z] \to 1 - \Phi(z)$$

by (5.4.40). Thus,

$$P_{\theta_0}[\widehat{\theta}_n \geq \theta_0 + z_{1-\alpha}/\sqrt{nI(\theta_0)}] = P_{\theta_0}[\sqrt{nI(\theta_0)}(\widehat{\theta}_n - \theta_0) \geq z_{1-\alpha}] \to \alpha. \tag{5.4.44}$$

But Polya's theorem (A.14.22) guarantees that

$$\sup_z |P_{\theta_0}[\sqrt{n}(\widehat{\theta}_n - \theta_0) \geq z] - (1 - \Phi(z))| \to 0, \tag{5.4.45}$$

which implies that $\sqrt{nI(\theta_0)}(c_n(\alpha, \theta_0) - \theta_0) - z_{1-\alpha} \to 0$, and (5.4.41) follows. On the other hand,

$$P_\theta[\widehat{\theta}_n \geq c_n(\alpha, \theta_0)] = P_\theta[\sqrt{nI(\theta)}(\widehat{\theta}_n - \theta) \geq \sqrt{nI(\theta)}(c_n(\alpha, \theta_0) - \theta)]. \tag{5.4.46}$$

By (5.4.41),

$$\sqrt{nI(\theta)}(c_n(\alpha, \theta_0) - \theta) = \sqrt{nI(\theta)}(\theta_0 - \theta + z_{1-\alpha}/\sqrt{nI(\theta_0)} + o(n^{-1/2}))$$
$$= \sqrt{nI(\theta)}(\theta_0 - \theta) + O(1) \to -\infty \text{ if } \theta > \theta_0$$

and $\to \infty$ if $\theta < \theta_0$. Claims (5.4.42) and (5.4.43) follow.   □

Theorem 5.4.4 tells us that the test under discussion is consistent and that for n large the power function of the test rises steeply to α from the left at θ_0 and continues rising steeply to 1 to the right of θ_0. Optimality claims rest on a more refined analysis involving a reparametrization from θ to $\gamma \equiv \sqrt{n}(\theta - \theta_0)$.

Theorem 5.4.5. *Suppose the conditions of Theorem 5.4.2 and (5.4.40) hold uniformly for θ in a neighborhood of θ_0. That is, assume*

$$\sup\{|P_\theta[\sqrt{nI(\theta)}(\widehat{\theta}_n - \theta) \le z] - (1 - \Phi(z))| : |\theta - \theta_0| \le \epsilon(\theta_0)\} \to 0, \qquad (5.4.47)$$

for some $\epsilon(\theta_0) > 0$. Let $Q_\gamma \equiv P_\theta$, $\gamma = \sqrt{n}(\theta - \theta_0)$, then

$$Q_\gamma[\widehat{\theta}_n \ge c_n(\alpha, \theta_0)] \to 1 - \Phi(z_{1-\alpha} - \gamma\sqrt{I(\theta_0)}) \qquad (5.4.48)$$

uniformly in γ. Furthermore, if $\varphi_n(X_1, \dots, X_n)$ is any sequence of (possibly randomized) critical (test) functions such that

$$E_{\theta_0}\varphi_n(X_1, \dots, X_n) \to \alpha, \qquad (5.4.49)$$

then

$$\overline{\lim}_n E_{\theta_0 + \frac{\gamma}{\sqrt{n}}} \varphi_n(X_1, \dots, X_n) \quad \begin{aligned} &\le \quad 1 - \Phi(z_{1-\alpha} - \gamma\sqrt{I(\theta_0)}) \text{ if } \gamma > 0 \\ &\ge \quad 1 - \Phi(z_{1-\alpha} - \gamma\sqrt{I(\theta_0)}) \text{ if } \gamma < 0. \end{aligned} \qquad (5.4.50)$$

Note that (5.4.48) and (5.4.50) can be interpreted as saying that among all tests that are asymptotically level α (obey (5.4.49)) the test based on rejecting for large values of $\widehat{\theta}_n$ is asymptotically uniformly most powerful (obey (5.4.50)) and has asymptotically smallest probability of type I error for $\theta \le \theta_0$. In fact, these statements can only be interpreted as valid in a small neighborhood of θ_0 because γ fixed means $\theta \to \theta_0$. On the other hand, if $\sqrt{n}(\theta - \theta_0)$ tends to zero, then by (5.4.50), the power of tests with asymptotic level α tend to α. If $\sqrt{n}(\theta - \theta_0)$ tends to infinity, the power of the test based on $\widehat{\theta}_n$ tends to 1 by (5.4.48). In either case, the test based on $\widehat{\theta}_n$ is still asymptotically MP.

Proof. Write

$$P_\theta[\widehat{\theta}_n \ge c_n(\alpha, \theta_0)] = P_\theta[\sqrt{nI(\theta)}(\widehat{\theta}_n - \theta) \ge \sqrt{nI(\theta)}(c_n(\alpha, \theta_0) - \theta)]$$
$$= P_\theta[\sqrt{nI(\theta)}(\widehat{\theta}_n - \theta) \ge \sqrt{nI(\theta)}(\theta_0 - \theta + z_{1-\alpha}/\sqrt{nI(\theta_0)}$$
$$+ o(n^{-1/2}))].$$

$$(5.4.51)$$

If $\gamma = \sqrt{n}(\theta - \theta_0)$ is fixed, $I(\theta) = I\left(\theta_0 + \frac{\gamma}{\sqrt{n}}\right) \to I(\theta_0)$ because our uniformity assumption implies that $\theta \to I(\theta)$ is continuous (Problem 5.4.7). Thus,

$$
\begin{aligned}
Q_\gamma[\widehat{\theta}_n \geq c_n(\alpha, \theta_0)] &= 1 - \Phi(z_{1-\alpha}(1 + o(1)) + \sqrt{n(I(\theta_0) + o(1))}(\theta_0 - \theta) + o(1)) \\
&= 1 - \Phi(z_{1-\alpha} - \gamma\sqrt{I(\theta_0)}) + o(1))
\end{aligned}
$$

$$(5.4.52)$$

and (5.4.48) follows.

To prove (5.4.50) note that by the Neyman–Pearson lemma, if $\gamma > 0$,

$$
\begin{aligned}
E_{\theta_0 + \frac{\gamma}{\sqrt{n}}}\varphi_n(X_1, \ldots, X_n) &\leq P_{\theta_0 + \frac{\gamma}{\sqrt{n}}}\left[\sum_{i=1}^n \log \frac{p\left(X_i, \theta_0 + \frac{\gamma}{\sqrt{n}}\right)}{p(X_i, \theta_0)} \geq d_n(\alpha, \theta_0)\right] \\
&\quad + \epsilon_n P_{\theta_0 + \frac{\gamma}{\sqrt{n}}}\left[\sum_{i=1}^n \log \frac{p\left(X_i, \theta_0 + \frac{\gamma}{\sqrt{n}}\right)}{p(X_i, \theta_0)} = d_n(\alpha, \theta_0)\right],
\end{aligned}
$$

$$(5.4.53)$$

where $p(x, \theta)$ denotes the density of X_i and d_n, ϵ_n are uniquely chosen so that the right-hand side of (5.4.53) is α if γ is 0.

Further Taylor expansion and probabilistic arguments of the type we have used show that the right-hand side of (5.4.53) tends to the right-hand side of (5.4.50) for all γ. The details are in Problem 5.4.5. □

The asymptotic results we have just established do *not* establish that the test that rejects for large values of $\widehat{\theta}_n$ is necessarily good for all alternatives for any n.

The test $1[\widehat{\theta}_n \geq c_n(\alpha, \theta_0)]$ of Theorems 5.4.4 and 5.4.5 in the future will be referred to as a *Wald* test. There are two other types of test that have the same asymptotic behavior. These are the likelihood ratio test and the *score* or *Rao* test.

It is easy to see that the likelihood ratio test for testing $H : \theta \leq \theta_0$ versus $K : \theta > \theta_0$ is of the form

$$
\text{``Reject if } \sum_{i=1}^n \log[p(X_i, \widehat{\theta}_n)/p(X_i, \theta_0)]1(\widehat{\theta}_n > \theta_0) \geq k_n(\theta_0, \alpha).\text{''}
$$

It may be shown (Problem 5.4.8) that, for $\alpha \leq \frac{1}{2}$, $k_n(\theta_0, \alpha) = z_{1-\alpha}^2 + o(1)$ and that if $\delta_{Wn}^*(X_1, \ldots, X_n)$ is the critical function of the Wald test and $\delta_{Ln}^*(X_1, \ldots, X_n)$ is the critical function of the LR test then, for all γ,

$$
P_{\theta_0 + \frac{\gamma}{\sqrt{n}}}[\delta_{Ln}^*(X_1, \ldots, X_n) = \delta_{Wn}^*(X_1, \ldots, X_n)] \to 1. \tag{5.4.54}
$$

Assertion (5.4.54) establishes that the test δ_{Ln}^* yields equality in (5.4.50) and, hence, is asymptotically most powerful as well.

Finally, note that the Neyman Pearson LR test for $H : \theta = \theta_0$ versus $K : \theta_0 + \epsilon, \epsilon > 0$ rejects for large values of

$$
\frac{1}{\epsilon}[\log p_n(X_1, \ldots, X_n, \theta_0 + \epsilon) - \log p_n(X_1, \ldots, X_n, \theta_0)]
$$

where $p_n(X_1, \ldots, X_n, \theta)$ is the joint density of $X_1, \ldots, X_n$. For ϵ small, n fixed, this is approximately the same as rejecting for large values of $\frac{\partial}{\partial \theta_0} \log p_n(X_1, \ldots, X_n, \theta_0)$.

The preceding argument doesn't depend on the fact that $X_1, \ldots, X_n$ are i.i.d. with common density or frequency function $p(x, \theta)$ and the test that rejects H for large values of $\frac{\partial}{\partial \theta_0} \log p_n(X_1, \ldots, X_n, \theta_0)$ is, in general, called the *score* or *Rao* test. For the case we are considering it simplifies, becoming

$$\text{``Reject } H \text{ iff } \sum_{i=1}^{n} \frac{\partial}{\partial \theta_0} \log p(X_i, \theta_0) \geq r_n(\alpha, \theta_0).\text{''}$$

It is easy to see (Problem 5.4.15) that

$$r_n(\alpha, \theta_0) = z_{1-\alpha} \sqrt{nI(\theta_0)} + o(n^{1/2}) \tag{5.4.55}$$

and that again if $\delta_{Rn}^*(X_1, \ldots, X_n)$ is the critical function of the Rao test then

$$P_{\theta_0 + \frac{\gamma}{\sqrt{n}}}[\delta_{Rn}^*(X_1, \ldots, X_n) = \delta_{Wn}^*(X_1, \ldots, X_n)] \to 1, \tag{5.4.56}$$

(Problem 5.4.8) and the Rao test is asymptotically optimal.

Note that for all these tests and the confidence bounds of Section 5.4.5, $I(\theta_0)$, which may require numerical integration, can be replaced by $-n^{-1} \frac{d^2}{d\theta^2} l_n(\widehat{\theta}_n)$ (Problem 5.4.10).

5.4.5   Confidence Bounds

We define an *asymptotic level* $1 - \alpha$ *lower confidence bound (LCB)* $\underline{\theta}_n$ by the requirement that

$$P_\theta[\underline{\theta}_n \leq \theta] \to 1 - \alpha \tag{5.4.57}$$

for all θ and similarly define asymptotic level $1 - \alpha$ UCBs and confidence intervals.

We can approach obtaining asymptotically optimal confidence bounds in two ways:

  (i) By using a natural pivot.

  (ii) By inverting the testing regions derived in Section 5.4.4.

Method (i) is easier: If the assumptions of Theorem 5.4.4 hold, that is, (A0)–(A6), (A4'), and $I(\theta)$ finite for all θ, it follows (Problem 5.4.9) that

$$\mathcal{L}_\theta(\sqrt{nI(\widehat{\theta}_n)}(\widehat{\theta}_n - \theta)) \to \mathcal{N}(0, 1) \tag{5.4.58}$$

for all θ and, hence, an asymptotic level $1 - \alpha$ lower confidence bound is given by

$$\underline{\theta}_n^* = \widehat{\theta}_n - z_{1-\alpha}/\sqrt{nI(\widehat{\theta}_n)}. \tag{5.4.59}$$

Turning to method (ii), inversion of δ_{Wn}^* gives formally

$$\underline{\theta}_{n1}^* = \inf\{\theta : c_n(\alpha, \theta) \geq \widehat{\theta}_n\} \tag{5.4.60}$$

or if we use the approximation $\tilde{c}_n(\alpha, \theta) = \theta + z_{1-\alpha}/\sqrt{nI(\theta)}$, (5.4.41),

$$\underline{\theta}_{n2}^* = \inf\{\theta : \tilde{c}_n(\alpha, \theta) \geq \hat{\theta}_n\}. \tag{5.4.61}$$

In fact neither $\underline{\theta}_{n1}^*$, or $\underline{\theta}_{n2}^*$ properly inverts the tests unless $c_n(\alpha, \theta)$ and $\tilde{c}_n(\alpha, \theta)$ are increasing in θ. The three bounds are different as illustrated by Examples 4.4.3 and 4.5.2.

If it applies and can be computed, $\underline{\theta}_{n1}^*$ is preferable because this bound is not only approximately but genuinely level $1 - \alpha$. But computationally it is often hard to implement because $c_n(\alpha, \theta)$ needs, in general, to be computed by simulation for a grid of θ values. Typically, (5.4.59) or some equivalent alternatives (Problem 5.4.10) are preferred but can be quite inadequate (Problem 5.4.11).

These bounds $\underline{\theta}_n^*, \underline{\theta}_{n1}^*, \underline{\theta}_{n2}^*$, are in fact asymptotically equivalent and optimal in a suitable sense (Problems 5.4.12 and 5.4.13).

Summary. We have defined asymptotic optimality for estimates in one-parameter models. In particular, we developed an asymptotic analogue of the information inequality of Chapter 3 for estimates of θ in a one-dimensional subfamily of the multinomial distributions, showed that the MLE formally achieves this bound, and made the latter result sharp in the context of one-parameter discrete exponential families. In Section 5.4.2 we developed the theory of minimum contrast and M-estimates, generalizations of the MLE, along the lines of Huber (1967). The asymptotic formulae we derived are applied to the MLE both under the model that led to it and under an arbitrary P. We also delineated the limitations of the optimality theory for estimation through Hodges's example. We studied the optimality results parallel to estimation in testing and confidence bounds. Results on asymptotic properties of statistical procedures can also be found in Ferguson (1996), Le Cam and Yang (1990), Lehmann (1999), Rao (1973), and Serfling (1980).

5.5   ASYMPTOTIC BEHAVIOR AND OPTIMALITY OF THE POSTERIOR DISTRIBUTION

Bayesian and frequentist inferences merge as $n \to \infty$ in a sense we now describe. The framework we consider is the one considered in Sections 5.2 and 5.4, i.i.d. observations from a regular model in which Θ is open $\subset R$ or $\Theta = \{\theta_1, \ldots, \theta_k\}$ finite, and θ is identifiable.

Most of the questions we address and answer are under the assumption that $\boldsymbol{\theta} = \theta$, an arbitrary specified value, or in frequentist terms, that θ is true.

Consistency

The first natural question is whether the Bayes posterior distribution as $n \to \infty$ concentrates all mass more and more tightly around θ. Intuitively this means that the data that are coming from P_θ eventually wipe out any prior belief that parameter values not close to θ are likely.

Formalizing this statement about the posterior distribution, $\Pi(\cdot \mid X_1, \ldots, X_n)$, which is a function-valued statistic, is somewhat subtle in general. But for $\Theta = \{\theta_1, \ldots, \theta_k\}$ it is straightforward. Let

$$\pi(\theta \mid X_1, \ldots, X_n) \equiv P[\boldsymbol{\theta} = \theta \mid X_1, \ldots, X_n]. \tag{5.5.1}$$

Then we say that $\Pi(\cdot \mid X_1, \ldots, X_n)$ is *consistent* iff for all $\theta \in \Theta$,

$$P_\theta\big[|\pi(\theta \mid X_1, \ldots, X_n) - 1| \geq \epsilon\big] \to 0 \tag{5.5.2}$$

for all $\epsilon > 0$. There is a slightly stronger definition: $\Pi(\cdot \mid X_1, \ldots, X_n)$ is *a.s. consistent* iff for all $\theta \in \Theta$,

$$\pi(\theta \mid X_1, \ldots, X_n) \to 1 \text{ a.s. } P_\theta. \tag{5.5.3}$$

General a.s. consistency is not hard to formulate:

$$\pi(\cdot \mid X_1, \ldots, X_n) \Rightarrow \delta_{\{\theta\}} \text{ a.s. } P_\theta \tag{5.5.4}$$

where $\Rightarrow$ denotes convergence in law and $\delta_{\{\theta\}}$ is point mass at θ. There is a completely satisfactory result for Θ finite.

Theorem 5.5.1. *Let* $\pi_j \equiv P[\boldsymbol{\theta} = \theta_j]$, $j = 1, \ldots, k$ *denote the prior distribution of* $\boldsymbol{\theta}$. *Then* $\Pi(\cdot \mid X_1, \ldots, X_n)$ *is consistent (a.s. consistent) iff* $\pi_j > 0$ *for* $j = 1, \ldots, k$.

Proof. Let $p(\cdot, \theta)$ denote the frequency or density function of X. The necessity of the condition is immediate because $\pi_j = 0$ for some j implies that $\pi(\theta_j \mid X_1, \ldots, X_n) = 0$ for all $X_1, \ldots, X_n$ because, by (1.2.8),

$$
\begin{aligned}
\pi(\theta_j \mid X_1, \ldots, X_n) &= P[\boldsymbol{\theta} = \theta_j \mid X_1, \ldots, X_n] \\
&= \frac{\pi_j \prod_{i=1}^n p(X_i, \theta_j)}{\sum_{a=1}^k \pi_a \prod_{i=1}^n p(X_i, \theta_a)}.
\end{aligned}
\tag{5.5.5}
$$

Intuitively, no amount of data can convince a Bayesian who has decided a priori that θ_j is impossible.

On the other hand, suppose all π_j are positive. If the true θ is θ_j or equivalently $\boldsymbol{\theta} = \theta_j$, then

$$\log \frac{\pi(\theta_a \mid X_1, \ldots, X_n)}{\pi(\theta_j \mid X_1, \ldots, X_n)} = n\left(\frac{1}{n}\log\frac{\pi_a}{\pi_j} + \frac{1}{n}\sum_{i=1}^n \log\frac{p(X_i, \theta_a)}{p(X_i, \theta_j)}\right).$$

By the weak (respectively strong) LLN, under P_{θ_j},

$$\frac{1}{n}\sum_{i=1}^n \log\frac{p(X_i, \theta_a)}{p(X_i, \theta_j)} \to E_{\theta_j}\left(\log\frac{p(X_1, \theta_a)}{p(X_1, \theta_j)}\right)$$

in probability (respectively a.s.). But $E_{\theta_j}\left(\log\frac{p(X_1,\theta_a)}{p(X_1,\theta_j)}\right) < 0$, by Shannon's inequality, if $\theta_a \neq \theta_j$. Therefore,

$$\log\frac{\pi(\theta_a \mid X_1, \ldots, X_n)}{\pi(\theta_j \mid X_1, \ldots, X_n)} \to -\infty,$$

in the appropriate sense, and the theorem follows. $\square$

Remark 5.5.1. We have proved more than is stated. Namely, that for each $\theta \in \Theta$, $P_\theta[\boldsymbol{\theta} \neq \theta \mid X_1, \ldots, X_n] \to 0$ exponentially.   □

As this proof suggests, consistency of the posterior distribution is very much akin to consistency of the MLE. The appropriate analogues of Theorem 5.2.3 are valid. Next we give a much stronger connection that has inferential implications:

Asymptotic normality of the posterior distribution

Under conditions A0–A6 for $\rho(x, \theta) = l(x, \theta) \equiv \log p(x, \theta)$, we showed in Section 5.4 that if $\widehat{\theta}$ is the MLE,

$$\mathcal{L}_\theta(\sqrt{n}(\widehat{\theta} - \theta)) \to \mathcal{N}(0, I^{-1}(\theta)). \tag{5.5.6}$$

Consider $\mathcal{L}(\sqrt{n}(\boldsymbol{\theta} - \widehat{\theta}) \mid X_1, \ldots, X_n)$, the posterior probability distribution of $\sqrt{n}(\boldsymbol{\theta} - \widehat{\theta}(X_1, \ldots, X_n))$, where we emphasize that $\widehat{\theta}$ depends only on the data and is a constant given $X_1, \ldots, X_n$. For conceptual ease we consider A4(a.s.) and A5(a.s.), assumptions that strengthen A4 and A5 by replacing convergence in P_θ probability by convergence a.s. P_θ. We also add,

A7: For all θ, and all $\delta > 0$ there exists $\epsilon(\delta, \theta) > 0$ such that

$$P_\theta\left[\sup\left\{\frac{1}{n}\sum_{i=1}^n [l(X_i, \theta') - l(X_i, \theta)] : |\theta' - \theta| \geq \delta\right\} \leq -\epsilon(\delta, \theta)\right] \to 1.$$

A8: The prior distribution has a density $\pi(\cdot)$ on Θ such that $\pi(\cdot)$ is continuous and positive at all θ.

Remarkably,

Theorem 5.5.2 ("Bernstein/von Mises"). *If conditions* A0–A3, A4(a.s.), A5(a.s.), A6, A7, *and* A8 *hold, then*

$$\mathcal{L}(\sqrt{n}(\boldsymbol{\theta} - \widehat{\theta}) \mid X_1, \ldots, X_n) \to \mathcal{N}(0, I^{-1}(\theta)) \tag{5.5.7}$$

a.s. under P_θ *for all* θ.

We can rewrite (5.5.7) more usefully as

$$\sup_x |P[\sqrt{n}(\boldsymbol{\theta} - \widehat{\theta}) \leq x \mid X_1, \ldots, X_n] - \Phi(x\sqrt{I(\theta)})| \to 0 \tag{5.5.8}$$

for all θ a.s. P_θ and, of course, the statement holds for our usual and weaker convergence in P_θ probability also. From this restatement we obtain the important corollary.

Corollary 5.5.1. *Under the conditions of Theorem* 5.5.2,

$$\sup_x |P[\sqrt{n}(\boldsymbol{\theta} - \widehat{\theta}) \leq x \mid X_1, \ldots, X_n] - \Phi(x\sqrt{I(\widehat{\theta})})| \to 0 \tag{5.5.9}$$

a.s. P_θ *for all* θ.

Remarks

(1) Statements (5.5.4) and (5.5.7)–(5.5.9) are, in fact, frequentist statements about the asymptotic behavior of certain function-valued statistics.

(2) Claims (5.5.8) and (5.5.9) hold with a.s. replaced by in P_θ probability if A4 and A5 are used rather than their strong forms—see Problem 5.5.7.

(3) Condition A7 is essentially equivalent to (5.2.8), which coupled with (5.2.9) and identifiability guarantees consistency of $\widehat{\theta}$ in a regular model.

Proof. We compute the posterior density of $\sqrt{n}(\theta - \widehat{\theta})$ as

$$q_n(t) = c_n^{-1}\pi\left(\widehat{\theta} + \frac{t}{\sqrt{n}}\right)\prod_{i=1}^{n}p\left(X_i, \widehat{\theta} + \frac{t}{\sqrt{n}}\right) \qquad (5.5.10)$$

where $c_n = c_n(X_1, \ldots, X_n)$ is given by

$$c_n(X_1, \ldots, X_n) = \int_{-\infty}^{\infty}\pi\left(\widehat{\theta} + \frac{s}{\sqrt{n}}\right)\prod_{i=1}^{n}p\left(X_i, \widehat{\theta} + \frac{s}{\sqrt{n}}\right)ds.$$

Divide top and bottom of (5.5.10) by $\prod_{i=1}^{n}p(X_i, \widehat{\theta})$ to obtain

$$q_n(t) = d_n^{-1}\pi\left(\widehat{\theta} + \frac{t}{\sqrt{n}}\right)\exp\left\{\sum_{i=1}^{n}\left(l\left(X_i, \widehat{\theta} + \frac{t}{\sqrt{n}}\right) - l(X_i, \widehat{\theta})\right)\right\} \qquad (5.5.11)$$

where $l(x, \theta) = \log p(x, \theta)$ and

$$d_n = \int_{-\infty}^{\infty}\pi\left(\widehat{\theta} + \frac{s}{\sqrt{n}}\right)\exp\left\{\sum_{i=1}^{n}\left(l\left(X_i, \widehat{\theta} + \frac{s}{\sqrt{n}}\right) - l(X_i, \widehat{\theta})\right)\right\}ds.$$

We claim that

$$P_\theta\left[d_n q_n(t) \to \pi(\theta)\exp\left\{-\frac{t^2 I(\theta)}{2}\right\} \text{ for all } t\right] = 1 \qquad (5.5.12)$$

for all θ. To establish this note that

(a) $\sup\left\{\left|\pi\left(\widehat{\theta} + \frac{t}{\sqrt{n}}\right) - \pi(\theta)\right| : |t| \le M\right\} \to 0$ a.s. for all M because $\widehat{\theta}$ is a.s. consistent and π is continuous.

(b) Expanding,

$$\sum_{i=1}^{n}\left(l\left(X_i, \widehat{\theta} + \frac{t}{\sqrt{n}}\right) - l(X_i, \widehat{\theta})\right) = \frac{t^2}{2}\frac{1}{n}\sum_{i=1}^{n}\frac{\partial^2 l}{\partial\theta^2}(X_i, \theta^*(t)) \qquad (5.5.13)$$

where $|\hat{\theta} - \theta^*_{(t)}| \leq \frac{t}{\sqrt{n}}$. We use $\sum_{i=1}^{n} \frac{\partial l}{\partial \theta}(X_i, \hat{\theta}) = 0$ here. By A4(a.s.), A5(a.s.),

$$\sup \left\{ \left| \frac{1}{n} \sum_{i=1}^{n} \frac{\partial^2 l}{\partial \theta^2}(X_i, \theta^*(t)) - \frac{1}{n} \sum_{i=1}^{n} \frac{\partial^2 l}{\partial \theta^2}(X_i, \theta) \right| : |t| \leq M \right\} \to 0,$$

for all M, a.s. P_θ. Using (5.5.13), the strong law of large numbers (SLLN) and A8, we obtain (Problem 5.5.3),

$$P_\theta \left[d_n q_n(t) \to \pi(\theta) \exp \left\{ E_\theta \frac{\partial^2 l}{\partial \theta^2}(X_1, \theta) \frac{t^2}{2} \right\} \text{ for all } t \right] = 1. \qquad (5.5.14)$$

Using A6 we obtain (5.5.12).

Now consider

$$\begin{aligned}
d_n &= \int_{-\infty}^{\infty} \pi \left(\hat{\theta} + \frac{s}{\sqrt{n}} \right) \exp \left\{ \sum_{i=1}^{n} l \left(X_i, \hat{\theta} + \frac{s}{\sqrt{n}} \right) - l(X_i, \hat{\theta}) \right\} ds \\
&= \int_{|s| \leq \delta \sqrt{n}} d_n q_n(s) ds \qquad\qquad (5.5.15) \\
&\quad + \sqrt{n} \int \pi(t) \exp \left\{ \sum_{i=1}^{n} (l(X_i, t) - l(X_i, \hat{\theta})) \right\} 1(|t - \hat{\theta}| > \delta) dt
\end{aligned}$$

By A5 and A7,

$$P_\theta \left[\sup \left\{ \exp \left\{ \sum_{i=1}^{n} (l(X_i, t) - l(X_i, \hat{\theta})) \right\} : |t - \hat{\theta}| > \delta \right\} \leq e^{-n\epsilon(\delta, \theta)} \right] \to 1 \quad (5.5.16)$$

for all δ so that the second term in (5.5.14) is bounded by $\sqrt{n} e^{-n\epsilon(\delta, \theta)} \to 0$ a.s. P_θ for all $\delta > 0$. Finally note that (Problem 5.5.4) by arguing as for (5.5.14), there exists $\delta(\theta) > 0$ such that

$$P_\theta \left[d_n q_n(t) \leq 2\pi(\theta) \exp \left\{ \frac{1}{4} E_\theta \left(\frac{\partial^2 l}{\partial \theta^2}(X_1, \theta) \right) \frac{t^2}{2} \right\} \text{ for all } |t| \leq \delta(\theta) \sqrt{n} \right] \to 1.$$

$$(5.5.17)$$

By (5.5.15) and (5.5.16), for all $\delta > 0$,

$$P_\theta \left[d_n - \int_{|s| \leq \delta \sqrt{n}} d_n q_n(s) ds \to 0 \right] = 1. \qquad (5.5.18)$$

Finally, apply the dominated convergence theorem, Theorem B.7.5, to $d_n q_n(s1(|s| \leq \delta(\theta)\sqrt{n}))$, using (5.5.14) and (5.5.17) to conclude that, a.s. P_θ,

$$d_n \to \pi(\theta) \int_{-\infty}^{\infty} \exp \left\{ -\frac{s^2 I(\theta)}{2} \right\} ds = \frac{\pi(\theta) \sqrt{2\pi}}{\sqrt{I(\theta)}}. \qquad (5.5.19)$$

Hence, a.s. P_θ,

$$q_n(t) \to \sqrt{I(\theta)}\varphi(t\sqrt{I(\theta)})$$

where φ is the standard Gaussian density and the theorem follows from Scheffé's Theorem B.7.6 and Proposition B.7.2. □

Example 5.5.1. *Posterior Behavior in the Normal Translation Model with Normal Prior.* (Example 3.2.1 continued). Suppose as in Example 3.2.1 we have observations from a $\mathcal{N}(\theta, \sigma^2)$ distribution with σ^2 known and we put a $\mathcal{N}(\eta, \tau^2)$ prior on θ. Then the posterior distribution of θ is $\mathcal{N}\left(w_{1n}\eta + w_{2n}\bar{X}, \left(\frac{n}{\sigma^2} + \frac{1}{\tau^2}\right)^{-1}\right)$ where

$$w_{1n} = \frac{\sigma^2}{n\tau^2 + \sigma^2}, \quad w_{2n} = 1 - w_{1n}. \tag{5.5.20}$$

Evidently, as $n \to \infty$, $w_{1n} \to 0$, $\bar{X} \to \theta$, a.s., if $\theta = \theta$, and $\left(\frac{n}{\sigma^2} + \frac{1}{\tau^2}\right)^{-1} \to 0$. That is, the posterior distribution has mean approximately θ and variance approximately 0, for n large, or equivalently the posterior is close to point mass at θ as we expect from Theorem 5.5.1. Because $\hat{\theta} = \bar{X}$, $\sqrt{n}(\theta - \hat{\theta})$ has posterior distribution $\mathcal{N}\left(\sqrt{n}w_{1n}(\eta - \bar{X}), n\left(\frac{n}{\sigma^2} + \frac{1}{\tau^2}\right)^{-1}\right)$. Now, $\sqrt{n}w_{1n} = O(n^{-1/2}) = o(1)$ and $n\left(\frac{n}{\sigma^2} + \frac{1}{\tau^2}\right)^{-1} \to \sigma^2 = I^{-1}(\theta)$ and we have directly established the conclusion of Theorem 5.5.2. □

Example 5.5.2. *Posterior Behavior in the Binomial-Beta Model.* (Example 3.2.3 continued). If we observe S_n with a binomial, $B(n, \theta)$, distribution, or equivalently we observe $X_1, \ldots, X_n$ i.i.d. Bernoulli $(1, \theta)$ and put a beta, $\beta(r, s)$ prior on θ, then, as in Example 3.2.3, θ has posterior $\beta(S_n + r, n + s - S_n)$. We have shown in Problem 5.3.20 that if $U_{a,b}$ has a $\beta(a, b)$ distribution, then as $a \to \infty$, $b \to \infty$,

$$\left[\frac{(a+b)^3}{ab}\right]^{\frac{1}{2}}\left(U_{a,b} - \frac{a}{a+b}\right) \xrightarrow{\mathcal{L}} \mathcal{N}(0, 1). \tag{5.5.21}$$

If $0 < \theta < 1$ is true, $S_n/n \xrightarrow{\text{a.s.}} \theta$ so that $S_n + r \to \infty$, $n + s - S_n \to \infty$ a.s. P_θ. By identifying a with $S_n + r$ and b with $n + s - S_n$ we conclude after some algebra that because $\hat{\theta} = \bar{X}$,

$$\sqrt{n}(\theta - \bar{X}) \xrightarrow{\mathcal{L}} \mathcal{N}(0, \theta(1 - \theta))$$

a.s. P_θ, as claimed by Theorem 5.5.2. □

Bayesian optimality of optimal frequentist procedures and frequentist optimality of Bayesian procedures

Theorem 5.5.2 has two surprising consequences.

(a) Bayes estimates for a wide variety of loss functions and priors are asymptotically efficient in the sense of the previous section.

(b) The maximum likelihood estimate is asymptotically equivalent in a Bayesian sense to the Bayes estimate for a variety of priors and loss functions.

As an example of this phenomenon consider the following.

Theorem 5.5.3. *Suppose the conditions of Theorem 5.5.2 are satisfied. Let $\widehat{\theta}$ be the MLE of θ and let $\widehat{\theta}^*$ be the median of the posterior distribution of $\boldsymbol{\theta}$. Then*
(i)

$$\sqrt{n}(\widehat{\theta}^* - \widehat{\theta}) \to 0 \tag{5.5.22}$$

a.s. P_θ for all θ. Consequently,

$$\widehat{\theta}^* = \theta + \frac{1}{n}\sum_{i=1}^{n} I^{-1}(\theta)\frac{\partial l}{\partial \theta}(X_i, \theta) + o_{P_\theta}(n^{-1/2}) \tag{5.5.23}$$

and $\mathcal{L}_\theta(\sqrt{n}(\widehat{\theta}^* - \theta)) \to \mathcal{N}(0, I^{-1}(\theta))$.
(ii)

$$E(\sqrt{n}(|\boldsymbol{\theta} - \widehat{\theta}| - |\boldsymbol{\theta} - \widehat{\theta}^*|) \mid X_1, \ldots, X_n) = o_P(1) \tag{5.5.24}$$

$$E(\sqrt{n}(|\boldsymbol{\theta} - \widehat{\theta}| - |\boldsymbol{\theta}| \mid X_1, \ldots, X_n) = \min_d E(\sqrt{n}(|\boldsymbol{\theta} - d| - |\boldsymbol{\theta}|) \mid X_1, \ldots, X_n) + o_P(1). \tag{5.5.25}$$

Thus, (i) corresponds to claim (a) whereas (ii) corresponds to claim (b) for the loss functions $l_n(\theta, d) = \sqrt{n}(|\theta - d| - |\theta|)$. But the Bayes estimates for l_n and for $l(\theta, d) = |\theta - d|$ must agree whenever $E(|\boldsymbol{\theta}| \mid X_1, \ldots, X_n) < \infty$. (Note that if $E(|\boldsymbol{\theta}| \mid X_1, \ldots, X_n) = \infty$, then the posterior Bayes risk under l is infinite and all estimates are equally poor.) Hence, (5.5.25) follows. The proof of a corresponding claim for quadratic loss is sketched in Problem 5.5.5.

Proof. By Theorem 5.5.2 and Polya's theorem (A.14.22)

$$\sup_x |P[\sqrt{n}(\boldsymbol{\theta} - \widehat{\theta}) \leq x \mid X_1, \ldots, X_n] - \Phi(x\sqrt{I(\theta)})| \to 0 \text{ a.s. } P_\theta. \tag{5.5.26}$$

But uniform convergence of distribution functions implies convergence of quantiles that are unique for the limit distribution (Problem B.7.11). Thus, any median of the posterior distribution of $\sqrt{n}(\boldsymbol{\theta} - \widehat{\theta})$ tends to 0, the median of $\mathcal{N}(0, I^{-1}(\theta))$, a.s. P_θ. But the median of the posterior of $\sqrt{n}(\boldsymbol{\theta} - \widehat{\theta})$ is $\sqrt{n}(\widehat{\theta}^* - \widehat{\theta})$, and (5.5.22) follows. To prove (5.5.24) note that

$$|\sqrt{n}(|\boldsymbol{\theta} - \widehat{\theta}| - |\boldsymbol{\theta} - \widehat{\theta}^*|)| \leq \sqrt{n}|\widehat{\theta} - \widehat{\theta}^*|$$

and, hence, that

$$E(\sqrt{n}(|\boldsymbol{\theta} - \widehat{\theta}| - |\boldsymbol{\theta} - \widehat{\theta}^*|) \mid X_1, \ldots, X_n) \leq \sqrt{n}|\widehat{\theta} - \widehat{\theta}^*| \to 0 \tag{5.5.27}$$

a.s. P_θ, for all θ. Because a.s. convergence P_θ for all θ implies a.s. convergence P (B.?), claim (5.5.24) follows and, hence,

$$E(\sqrt{n}(|\boldsymbol{\theta} - \widehat{\theta}| - |\boldsymbol{\theta}|) \mid X_1, \ldots, X_n) = E(\sqrt{n}(|\boldsymbol{\theta} - \widehat{\theta}^*| - |\boldsymbol{\theta}|) \mid X_1, \ldots, X_n) + o_P(1). \tag{5.5.28}$$

Because by Problem 1.4.7 and Proposition 3.2.1, $\widehat{\theta}^*$ is the Bayes estimate for $l_n(\theta, d)$, (5.5.25) and the theorem follows.   □

Remark. In fact, Bayes procedures can be efficient in the sense of Sections 5.4.3 and 6.2.3 even if MLEs do not exist. See Le Cam and Yang (1990).

Bayes credible regions

There is another result illustrating that the frequentist inferential procedures based on $\widehat{\theta}$ agree with Bayesian procedures to first order.

Theorem 5.5.4. *Suppose the conditions of Theorem 5.5.2 are satisfied. Let*

$$C_n(X_1, \ldots, X_n) = \{\theta : \pi(\theta \mid X_1, \ldots, X_n) \geq c_n\},$$

where c_n is chosen so that $\pi(C_n \mid X_1, \ldots, X_n) = 1 - \alpha$, be the Bayes credible region defined in Section 4.7. Let $I_n(\gamma)$ be the asymptotically level $1 - \gamma$ optimal interval based on $\widehat{\theta}$, given by

$$I_n(\gamma) = [\widehat{\theta} - d_n(\gamma), \widehat{\theta} + d_n(\gamma)]$$

where $d_n(\gamma) = z\left(1 - \frac{\gamma}{2}\right)\sqrt{\frac{I(\theta)}{n}}$. Then, for every $\epsilon > 0$, θ,

$$P_\theta[I_n(\alpha + \epsilon) \subset C_n(X_1, \ldots, X_n) \subset I_n(\alpha - \epsilon)] \to 1. \tag{5.5.29}$$

The proof, which uses a strengthened version of Theorem 5.5.2 by which the posterior density of $\sqrt{n}(\theta - \widehat{\theta})$ converges to the $\mathcal{N}(0, I^{-1}(\theta))$ density uniformly over compact neighborhoods of θ for each fixed θ, is sketched in Problem 5.5.6. The message of the theorem should be clear. Bayesian and frequentist coverage statements are equivalent to first order. A finer analysis both in this case and in estimation reveals that any approximations to Bayes procedures on a scale finer than $n^{-1/2}$ do involve the prior. A particular choice, the Jeffrey's prior, makes agreement between frequentist and Bayesian confidence procedures valid even to the higher n^{-1} order (see Schervisch, 1995).

Testing

Bayes and frequentist inferences diverge when we consider testing a point hypothesis. For instance, in Problem 5.5.1, the posterior probability of θ_0 given $X_1, \ldots, X_n$ if H is false is of a different magnitude than the p-value for the same data. For more on this so-called Lindley paradox see Berger (1985), Schervisch (1995) and Problem 5.5.1. However, if instead of considering hypotheses specifying one points θ_0 we consider indifference regions where H specifies $[\theta_0 + \Delta)$ or $(\theta_0 - \Delta, \theta_0 + \Delta)$, then Bayes and frequentist testing procedures agree in the limit. See Problem 5.5.2.

Summary. Here we established the frequentist consistency of Bayes estimates in the finite parameter case, if all parameter values are a prior possible. Second, we established the so-called Bernstein–von Mises theorem actually dating back to Laplace (see Le Cam and Yang, 1990), which establishes frequentist optimality of Bayes estimates and Bayes optimality of the MLE for large samples and priors that do not rule out any region of the parameter space. Finally, the connection between the behavior of the posterior given by the so-called Bernstein–von Mises theorem and frequentist confidence regions is developed.

5.6 PROBLEMS AND COMPLEMENTS

Problems for Section 5.1

1. Suppose $X_1, \ldots, X_n$ are i.i.d. as $X \sim F$, where F has median $F^{-1}\left(\frac{1}{2}\right)$ and a continuous case density.

(a) Show that, if $n = 2k + 1$,

$$E_F \text{med}(X_1, \ldots, X_n) = n \binom{2k}{k} \int_0^1 F^{-1}(t) t^k (1-t)^k \, dt$$

$$E_F \text{med}^2(X_1, \ldots, X_n) = n \binom{2k}{k} \int_0^1 [F^{-1}(t)]^2 t^k (1-t)^k \, dt$$

(b) Suppose F is uniform, $\mathcal{U}(0,1)$. Find the MSE of the sample median for $n = 1, 3$, and 5.

2. Suppose $Z \sim \mathcal{N}(\mu, 1)$ and V is independent of Z with distribution χ^2_m. Then $T \equiv Z / \left(\frac{V}{m}\right)^{\frac{1}{2}}$ is said to have a noncentral t distribution with noncentrality μ and m degrees of freedom. See Section 4.9.2.

(a) Show that

$$P[T \le t] = 2m \int_0^\infty \Phi(tw - \mu) f_m(mw^2) w \, dw$$

where $f_m(w)$ is the χ^2_m density, and Φ is the normal distribution function.

(b) If $X_1, \ldots, X_n$ are i.i.d. $\mathcal{N}(\mu, \sigma^2)$ show that $\sqrt{n}\bar{X} \left/ \left(\frac{1}{n-1} \sum (X_i - \bar{X})^2\right)^{\frac{1}{2}} \right.$ has a noncentral t distribution with noncentrality parameter $\sqrt{n}\mu/\sigma$ and $n-1$ degrees of freedom.

(c) Show that T^2 in (a) has a noncentral $\mathcal{F}_{1,m}$ distribution with noncentrality parameter μ^2. Deduce that the density of T is

$$p(t) = 2 \sum_{i=0}^\infty P[R = i] \cdot f_{2i+1}(t^2) [\varphi(t - \mu) 1(t > 0) + \varphi(t + \mu) 1(t < 0)]$$

where R is given in Problem B.3.12.

Hint: Condition on $|T|$.

3. Show that if $P[|X| \leq 1] = 1$, then $\mathrm{Var}(X) \leq 1$ with equality iff $X = \pm 1$ with probability $\frac{1}{2}$.
    Hint: $\mathrm{Var}(X) \leq EX^2$.

4. *Comparison of Bounds:* Both the Hoeffding and Chebychev bounds are functions of n and ϵ through $\sqrt{n}\epsilon$.

    (a) Show that the ratio of the Hoeffding function $h(\sqrt{n}\epsilon)$ to the Chebychev function $c(\sqrt{n}\epsilon)$ tends to 0 as $\sqrt{n}\epsilon \to \infty$ so that $h(\cdot)$ is arbitrarily better than $c(\cdot)$ in the tails.

    (b) Show that the normal approximation $2\Phi\left(\frac{\sqrt{n}\epsilon}{\sigma}\right) - 1$ gives lower results than h in the tails if $P[|X| \leq 1] = 1$ because, if $\sigma^2 \leq 1, 1 - \Phi(t) \sim \varphi(t)/t$ as $t \to \infty$.
    Note: Hoeffding (1963) exhibits better bounds for known σ^2.

5. Suppose $\lambda : R \to R$ has $\lambda(0) = 0$, is bounded, and has a bounded second derivative λ''. Show that if $X_1, \ldots, X_n$ are i.i.d., $EX_1 = \mu$ and $\mathrm{Var}\, X_1 = \sigma^2 < \infty$, then

$$E\lambda(\bar{X} - \mu) = \lambda'(0)\frac{\sigma}{\sqrt{n}}\sqrt{\frac{2}{\pi}} + O\left(\frac{1}{n}\right) \quad \text{as } n \to \infty.$$

    Hint: $\sqrt{n}E(\lambda(|\bar{X} - \mu|) - \lambda(0)) = E\lambda'(0)\sqrt{n}|\bar{X} - \mu| + E\left(\frac{\lambda''}{2}(\tilde{X} - \mu)(\bar{X} - \mu)^2\right)$
where $|\tilde{X} - \mu| \leq |\bar{X} - \mu|$. The last term is $\leq \sup_x |\lambda''(x)|\sigma^2/2n$ and the first tends to $\lambda'(0)\sigma \int_{-\infty}^{\infty} |z|\varphi(z)dz$ by Remark B.7.1.

Problems for Section 5.2

1. Using the notation of Theorem 5.2.1, show that

$$\sup\{P_{\mathbf{p}}(|\hat{p}_n - \mathbf{p}| \geq \delta) : \mathbf{p} \in \mathcal{S}\} \leq k/4n\delta^2.$$

2. Let $X_1, \ldots, X_n$ be i.i.d. $\mathcal{N}(\mu, \sigma^2)$. Show that for all $n \geq 1$, all $\epsilon > 0$

$$\sup_{\sigma} P_{(\mu,\sigma)}[|\bar{X} - \mu| \geq \epsilon] = 1.$$

    Hint: Let $\sigma \to \infty$.

3. Establish (5.2.5).
    Hint: $|\hat{q}_n - q(\mathbf{p})| \geq \epsilon \Rightarrow |\hat{p}_n - \mathbf{p}| \geq \omega^{-1}(\epsilon)$.

4. Let $(U_i, V_i), 1 \leq i \leq n$, be i.i.d. $\sim P \in \mathcal{P}$.

    (a) Let $\gamma(P) = P[U_1 > 0, V_1 > 0]$. Show that if $P = \mathcal{N}(0, 0, 1, 1, \rho)$, then

$$\rho = \sin 2\pi \left(\gamma(P) - \frac{1}{4}\right).$$

(b) Deduce that if P is the bivariate normal distribution, then

$$\tilde{\rho} \equiv \sin\left\{2\pi\left(\frac{1}{n}\sum_{i=1}^{n}1(X_i > \bar{X})1(Y_i > \bar{Y})\right)\right\}$$

is a consistent estimate of ρ.

(c) Suppose $\rho(P)$ is defined generally as $\mathrm{Cov}_P(U,V)/\sqrt{\mathrm{Var}_P U \,\mathrm{Var}_P V}$ for $P \in \mathcal{P} = \{P : E_P U^2 + E_P V^2 < \infty, \mathrm{Var}_P U \,\mathrm{Var}_P V > 0\}$. Show that the sample correlation coefficient continues to be a consistent estimate of $\rho(P)$ but $\tilde{\rho}$ is no longer consistent.

5. Suppose $X_1,\dots,X_n$ are i.i.d. $\mathcal{N}(\mu,\sigma_0^2)$ where σ_0 is known and $\rho(\mathbf{x},\mu) = -\log p(\mathbf{x},\mu)$.

(a) Show that condition (5.2.8) fails even in this simplest case in which $\bar{X} \overset{P}{\to} \mu$ is clear.
Hint: $\sup_\mu \left|\frac{1}{n}\sum_{i=1}^{n}\left(\frac{(X_i-\mu)^2}{\sigma_0^2} - \left(1 + \frac{(\mu-\mu_0)^2}{\sigma_0^2}\right)\right)\right| = \infty.$

(b) Show that condition (5.2.14)(i),(ii) holds.
Hint: K can be taken as $[-A, A]$, where A is an arbitrary positive and finite constant.

6. Prove that (5.2.14)(i) and (ii) suffice for consistency.

7. (Wald) Suppose $\theta \to \rho(X,\theta)$ is continuous, $\theta \in R$ and

(i) For some $\epsilon(\theta_0) > 0$

$$E_{\theta_0}\sup\{|\rho(X,\theta') - \rho(X,\theta)| : |\theta - \theta'| \le \epsilon(\theta_0)\} < \infty.$$

(ii) $E_{\theta_0}\inf\{\rho(X,\theta) - \rho(X,\theta_0) : |\theta - \theta_0| \ge A\} > 0$ for some $A < \infty$.

Show that the minimum contrast estimate $\widehat{\theta}$ is consistent.
Hint: From continuity of ρ, (i), and the dominated convergence theorem,

$$\lim_{\delta \to 0} E_{\theta_0}\sup\{|\rho(X,\theta') - \rho(X,\theta)| : \theta' \in S(\theta,\delta)\} = 0$$

where $S(\theta,\delta)$ is the δ ball about θ. Therefore, by the basic property of maximum contrast estimates, for each $\theta \ne \theta_0$, and $\epsilon > 0$ there is $\delta(\theta) > 0$ such that

$$E_{\theta_0}\inf\{\rho(X,\theta') - \rho(X,\theta_0) : \theta' \in S(\theta,\delta(\theta))\} > \epsilon.$$

By compactness there is a finite number $\theta_1,\dots,\theta_r$ of sphere centers such that

$$K \cap \{\theta : |\theta - \theta_0| \ge \lambda\} \subset \bigcup_{j=1}^{r} S(\theta_j,\delta(\theta_j)).$$

Now

$$\inf\left\{\frac{1}{n}\sum_{i=1}^{n}\{\rho(X_i,\theta) - \rho(X_i,\theta_0)\} : \theta \in K \cap \{\theta : |\theta - \theta_0| \ge \lambda\}\right\}$$

$$\geq \min_{1\leq j\leq r}\left\{\frac{1}{n}\sum_{i=1}^{n}\inf\{\rho(X_i,\theta')-\rho(X_i,\theta_0)\}:\theta'\in S(\theta_j,\delta(\theta_j))\right\}.$$

For r fixed apply the law of large numbers.

8. The condition of Problem 7(ii) can also fail. Let X_i be i.i.d. $\mathcal{N}(\mu,\sigma^2)$. Compact sets K can be taken of the form $\{|\mu|\leq A, \epsilon\leq\sigma\leq 1/\epsilon, \epsilon>0\}$. Show that the log likelihood tends to ∞ as $\sigma\to 0$ and the condition fails.

9. Indicate how the conditions of Problem 7 have to be changed to ensure uniform consistency on K.

10. Extend the result of Problem 7 to the case $\theta\in R^p, p>1$.

Problems for Section 5.3

1. Establish (5.3.9) in the exponential model of Example 5.3.1.

2. Establish (5.3.3) for j odd as follows:

(i) Suppose $X_1',\ldots,X_n'$ are i.i.d. with the same distribution as $X_1,\ldots,X_n$ but independent of them, and let $\bar{X}'=n^{-1}\Sigma X_i'$. Then $E|\bar{X}-\mu|^j\leq E|\bar{X}-\bar{X}'|^j$.

(ii) If ϵ_i are i.i.d. and take the values ± 1 with probability $\frac{1}{2}$, and if $c_1,\ldots,c_n$ are constants, then by Jensen's inequality, for some constants M_j,

$$E\left|\sum_{i=1}^{n}c_i\epsilon_i\right|^j\leq E^{\frac{j}{j+1}}\left(\sum_{i=1}^{n}c_i\epsilon_i\right)^{j+1}\leq M_j\left(\sum_{i=1}^{n}c_i^2\right)^{\frac{j}{2}}.$$

(iii) Condition on $|X_i-X_i'|, i=1,\ldots,n$, in (i) and apply (ii) to get

(iv) $E|\sum_{i=1}^{n}(X_i-X_i')|^j\leq M_jE\left[\sum_{i=1}^{n}(X_i-X_i')^2\right]^{\frac{j}{2}}\leq$
$M_jn^{\frac{j}{2}}E\left[\frac{1}{n}\sum_{i=1}^{n}(X_i-X_i')^2\right]^{\frac{j}{2}}\leq M_jn^{\frac{j}{2}}E\left(\frac{1}{n}\sum|X_i-X_i'|^j\right)\leq M_jn^{\frac{j}{2}}E|X_1-\mu|^j.$

3. Establish (5.3.11).
   Hint: See part (a) of the proof of Lemma 5.3.1.

4. Establish Theorem 5.3.2. *Hint:* Taylor expand and note that if $i_1+\cdots+i_d=m$

$$E\left|\prod_{k=1}^{d}(\bar{Y}_k-\mu_k)^{i_k}\right|\leq m^{m+1}\sum_{k=1}^{d}E|\bar{Y}_k-\mu_k|^m$$
$$\leq C_mn^{-m/2}.$$

Suppose $a_j \geq 0$, $1 \leq j \leq d$, $\sum_{j=1}^{d} i_j = m$, then

$$a_1^{i_1}, \ldots, a_d^{i_d} \leq [\max(a_1, \ldots, a_d)]^m \leq \left(\sum_{j=1}^{d} a_j \right)^m$$

$$\leq m^{m-1} \sum_{j=1}^{d} a_j^m.$$

5. Let $X_1, \ldots, X_n$ be i.i.d. R valued with $EX_1 = 0$ and $E|X_1|^j < \infty$. Show that

$$\sup\{|E(X_{i_1}, \ldots, X_{i_j})| : 1 \leq i_k \leq n; k = 1, \ldots, n\} = E|X_1|^j.$$

6. Show that if $E|X_1|^j < \infty$, $j \geq 2$, then $E|X_1 - \mu|^j \leq 2^j E|X_1|^j$.

Hint: By the iterated expectation theorem

$$E|X_1 - \mu|^j = E\{|X_1 - \mu|^j \mid |X_1| \geq |\mu|\}P(|X_1| \geq |\mu|)$$
$$+E\{|X_1 - \mu|^j \mid |X_1| < |\mu|\}P(|X_1| < |\mu|).$$

7. Establish 5.3.28.

8. Let $X_1, \ldots, X_{n_1}$ be i.i.d. F and $Y_1, \ldots, Y_{n_2}$ be i.i.d. G, and suppose the X's and Y's are independent.

(a) Show that if F and G are $\mathcal{N}(\mu_1, \sigma_1^2)$ and $\mathcal{N}(\mu_2, \sigma_2^2)$, respectively, then the LR test of $H : \sigma_1^2 = \sigma_2^2$ versus $K : \sigma_1^2 \neq \sigma_2^2$ is based on the statistic s_1^2/s_2^2, where $s_1^2 = (n_1 - 1)^{-1} \sum_{i=1}^{n_1}(X_i - \bar{X})^2$, $s_2^2 = (n_2 - 1)^{-1} \sum_{j=1}^{n_2}(Y_j - \bar{Y})^2$.

(b) Show that when F and G are normal as in part (a), then $(s_1^2/\sigma_1^2)/(s_2^2/\sigma_2^2)$ has an $\mathcal{F}_{k,m}$ distribution with $k = n_1 - 1$ and $m = n_2 - 1$.

(c) Now suppose that F and G are not necessarily normal but that

$$G \in \mathcal{G} = \left\{ F\left(\frac{\cdot - a}{b} \right) : a \in R, \ b > 0 \right\}$$

and that $0 < \text{Var}(X_1^2) < \infty$. Show that if $m = \lambda k$ for some $\lambda > 0$ and

$$c_{k,m} = 1 + \sqrt{\frac{\kappa(k + m)}{km}} z_{1-\alpha}, \quad \kappa = \text{Var}[(X_1 - \mu_1)/\sigma_1]^2, \quad \mu_1 = E(X_1), \quad \sigma_1^2 = \text{Var}(X_1).$$

Then, under $H : \text{Var}(X_1) = \text{Var}(Y_1)$, $P(s_1^2/s_2^2 \leq c_{k,m}) \to 1 - \alpha$ as $k \to \infty$.

(d) Let $\hat{c}_{k,m}$ be $c_{k,m}$ with κ replaced by its method of moments estimate. Show that under the assumptions of part (c), if $0 < EX_1^8 < \infty$, $P_H(s_1^2/s_2^2 \leq \hat{c}_{k,m}) \to 1 - \alpha$ as $k \to \infty$.

(e) Next drop the assumption that $G \in \mathcal{G}$. Instead assume that $0 < \text{Var}(Y_1^2) < \infty$. Under the assumptions of part (c), use a normal approximation to find an approximate critical value $q_{k,m}$ (depending on $\kappa_1 = \text{Var}[(X_1 - \mu_1)/\sigma_1]^2$ and $\kappa_2 = \text{Var}[(X_2 - \mu_2)/\sigma_2]^2$ such that $P_H(s_1^2/s_2^2 \le q_{k,m}) \to 1 - \alpha$ as $k \to \infty$.

(f) Let $\widehat{q}_{k,m}$ be $q_{k,m}$ with κ_1 and κ_2 replaced by their method of moment estimates. Show that under the assumptions of part (e), if $0 < EX_1^8 < \infty$ and $0 < EY_1^8 < \infty$, then $P(s_1^2/s_2^2 \le \widehat{q}_{k,m}) \to 1 - \alpha$ as $k \to \infty$.

9. In Example 5.3.6, show that

(a) If $\mu_1 = \mu_2 = 0$, $\sigma_1 = \sigma_2 = 1$, $\sqrt{n}((\widehat{C} - \rho), (\widehat{\sigma}_1^2 - 1), (\widehat{\sigma}_2^2 - 1))^T$ has the same asymptotic distribution as $n^{\frac{1}{2}}[n^{-1}\Sigma X_i Y_i - \rho, n^{-1}\Sigma X_i^2 - 1, n^{-1}\Sigma Y_i^2 - 1]^T$.

(b) If $(X, Y) \sim \mathcal{N}(\mu_1, \mu_2, \sigma_1^2, \sigma_2^2, \rho)$, then $\sqrt{n}(r^2 - \rho^2) \overset{\mathcal{L}}{\to} \mathcal{N}(0, 4\rho^2(1 - \rho^2)^2)$ and, if $\rho \ne 0$, then $\sqrt{n}(r - \rho) \to \mathcal{N}(0, (1 - \rho^2)^2)$.

(c) Show that if $\rho = 0$, $\sqrt{n}(r - \rho) \to \mathcal{N}(0, 1)$.
Hint: Use the central limit theorem and Slutsky's theorem. Without loss of generality, $\mu_1 = \mu_2 = 0$, $\sigma_1^2 = \sigma_2^2 = 1$.

10. Show that $\frac{1}{2} \log \left(\frac{1+\rho}{1-\rho} \right)$ is the variance stabilizing transformation for the correlation coefficient in Example 5.3.6.
Hint: Write $\frac{1}{(1-\rho)^2} = \frac{1}{2} \left(\frac{1}{1-\rho} + \frac{1}{1+\rho} \right)$.

11. In survey sampling the *model-based* approach postulates that the population $\{x_1, \ldots, x_N\}$ or $\{(u_1, x_1), \ldots, (u_N, x_N)\}$ we are interested in is itself a sample from a superpopulation that is known up to parameters; that is, there exists $T_1, \ldots, T_N$ i.i.d. P_θ, $\theta \in \Theta$ such that $T_i = t_i$ where $t_i \equiv (u_i, x_i)$, $i = 1, \ldots, N$. In particular, suppose in the context of Example 3.4.1 that we use $T_{i_1}, \ldots, T_{i_n}$, which we have sampled at random from $\{t_1, \ldots, t_N\}$, to estimate $\bar{x} \equiv \frac{1}{N}\sum_{i=1}^{N} x_i$. Without loss of generality, suppose $i_j = j$, $1 \le j \le n$. Consider as estimates

(i) $\bar{X} = \frac{X_1 + \cdots + X_n}{n}$ when $T_i \equiv (U_i, X_i)$.

(ii) $\widehat{\bar{X}}_R = \bar{X} - b_{\text{opt}}(\bar{U} - \bar{u})$ as in Example 3.4.1.

Show that, if $\frac{n}{N} \to \lambda$ as $N \to \infty$, $0 < \lambda < 1$, and if $EX_1^2 < \infty$ (in the supermodel), then

(a) $\sqrt{n}(\bar{X} - \bar{x}) \overset{\mathcal{L}}{\to} \mathcal{N}(0, \tau^2(1 - \lambda))$ where $\tau^2 = \text{Var}(X_1)$.

(b) Suppose P_θ is such that $X_i = bU_i + \epsilon_i$, $i = 1, \ldots, N$ where the ϵ_i are i.i.d. and independent of the U_i, $E\epsilon_i = 0$, $\text{Var}(\epsilon_i) = \sigma^2 < \infty$ and $\text{Var}(U_i) > 0$. Show that $\sqrt{n}(\widehat{\bar{X}}_R - \bar{x}) \overset{\mathcal{L}}{\to} \mathcal{N}(0, (1 - \lambda)\sigma^2)$, $\sigma^2 < \tau^2$.
Hint: (a) $\bar{X} - \bar{x} = \left(1 - \frac{n}{N}\right)(\bar{X} - \bar{X}^c)$ where $\bar{X}^c = \frac{1}{N-n}\sum_{i=n+1}^{N} X_i$.
(b) Use the multivariate delta method and note $(\widehat{b}_{\text{opt}} - b)(\bar{U} - \bar{u}) = o_p(n^{-\frac{1}{2}})$.

12. (a) Let $\bar{Y}_k$ be as defined in Theorem 5.3.2. Suppose that $E|\mathbf{Y}_1|^3 < \infty$. Show that
$|E(\bar{Y}_a - \mu_a)(\bar{Y}_b - \mu_b)(\bar{Y}_c - \mu_c)| \le Mn^{-2}; a, b, c \in \{1, \dots, d\}$.

(b) Assume $EY_1^4 < \infty$. Deduce formula (5.3.14).
Hint: **(a)** If U is independent of (V, W), $EU = 0$, $E(WV) < \infty$, then $E(UVW) = 0$.

13. Let S_n have a χ_n^2 distribution.

(a) Show that if n is large, $\sqrt{S_n} - \sqrt{n}$ has approximately a $\mathcal{N}(0, \frac{1}{2})$ distribution. This is known as *Fisher's approximation.*

(b) From (a) deduce the approximation $P[S_n \le x] \approx \Phi(\sqrt{2x} - \sqrt{2n})$.

(c) Compare the approximation of (b) with the central limit approximation $P[S_n \le x] = \Phi((x - n)/\sqrt{2n})$ and the exact values of $P[S_n \le x]$ from the χ^2 table for $x = x_{0.90}$, $x = x_{0.99}$, $n = 5, 10, 25$. Here x_q denotes the qth quantile of the χ_n^2 distribution.

14. Suppose $X_1, \dots, X_n$ is a sample from a population with mean μ, variance σ^2, and third central moment μ_3. Justify formally

$$E[h(\bar{X}) - E(h(\bar{X}))]^3 = \frac{1}{n^2}[h'(\mu)]^3 \mu_3 + \frac{3}{n^2}h''(\mu)[h'(\mu)]^2\sigma^4 + O(n^{-3}).$$

15. It can be shown (under suitable conditions) that the normal approximation to the distribution of $h(\bar{X})$ improves as the coefficient of skewness γ_{1n} of $h(\bar{X})$ diminishes.

(a) Use this fact and Problem 5.3.14 to explain the numerical results of Problem 5.3.13(c).

(b) Let $S_n \sim \chi_n^2$. The following approximation to the distribution of S_n (due to Wilson and Hilferty, 1931) is found to be excellent

$$P[S_n \le x] \approx \Phi\left\{\left[\left(\frac{x}{n}\right)^{1/3} - 1 + \frac{2}{9n}\right]\sqrt{\frac{9n}{2}}\right\}.$$

Use (5.3.6) to explain why.

16. *Normalizing Transformation for the Poisson Distribution.* Suppose $X_1, \dots, X_n$ is a sample from a $\mathcal{P}(\lambda)$ distribution.

(a) Show that the only transformations h that make $E[h(\bar{X}) - E(h(\bar{X}))]^3 = 0$ to terms up to order $1/n^2$ for all $\lambda > 0$ are of the form $h(t) = ct^{2/3} + d$.

(b) Use (a) to justify the approximation

$$P\left[\bar{X} \le \frac{k}{n}\right] \approx \Phi\left\{\sqrt{n}\left[\left(\frac{k + \frac{1}{2}}{n}\right)^{2/3} - \lambda^{2/3}\right]\bigg/ \frac{2}{3}\lambda^{1/6}\right\}.$$

17. Suppose $X_1, \dots, X_n$ are independent, each with Hardy–Weinberg frequency function f given by

x	0	1	2
$f(x)$	θ^2	$2\theta(1-\theta)$	$(1-\theta)^2$

where $0 < \theta < 1$.

(a) Find an approximation to $P[\bar{X} \leq t]$ in terms of θ and t.

(b) Find an approximation to $P[\sqrt{\bar{X}} \leq t]$ in terms of θ and t.

(c) What is the approximate distribution of $\sqrt{n}(\bar{X} - \mu) + \bar{X}^2$, where $\mu = E(X_1)$?

18. *Variance Stabilizing Transformation for the Binomial Distribution.* Let $X_1, \ldots, X_n$ be the indicators of n binomial trials with probability of success θ. Show that the only variance stabilizing transformation h such that $h(0) = 0$, $h(1) = 1$, and $h'(t) \geq 0$ for all t, is given by $h(t) = (2/\pi) \sin^{-1}(\sqrt{t})$.

19. Justify formally the following expressions for the moments of $h(\bar{X}, \bar{Y})$ where $(X_1, Y_1), \ldots, (X_n, Y_n)$ is a sample from a bivariate population with $E(X) = \mu_1$, $E(Y) = \mu_2$, $\text{Var}(X) = \sigma_1^2$, $\text{Var}(Y) = \sigma_2^2$, $\text{Cov}(X, Y) = \rho\sigma_1\sigma_2$.

(a)
$$E(h(\bar{X}, \bar{Y})) = h(\mu_1, \mu_2) + O(n^{-1}).$$

(b)
$$\text{Var}(h(\bar{X}, \bar{Y})) \cong \frac{1}{n}\{[h_1(\mu_1, \mu_2)]^2\sigma_1^2$$
$$+2h_1(\mu_1, \mu_2)h_2(\mu_1, \mu_2)\rho\sigma_1\sigma_2 + [h_2(\mu_1, \mu_2)]^2\sigma_2^2\} + 0(n^{-2})$$

where
$$h_1(x, y) = \frac{\partial}{\partial x}h(x, y), \quad h_2(x, y) = \frac{\partial}{\partial y}h(x, y).$$

Hint: $h(\bar{X}, \bar{Y}) - h(\mu_1, \mu_2) = h_1(\mu_1, \mu_2)(\bar{X} - \mu_1) + h_2(\mu_1, \mu_2)(\bar{Y} - \mu_2) + 0(n^{-1}).$

20. Let $B_{m,n}$ have a beta distribution with parameters m and n, which are integers. Show that if m and n are both tending to ∞ in such a way that $m/(m + n) \to \alpha$, $0 < \alpha < 1$, then

$$P\left[\sqrt{m + n}\frac{(B_{m,n} - m/(m + n))}{\sqrt{\alpha(1 - \alpha)}} \leq x\right] \to \Phi(x).$$

Hint: Use $B_{m,n} = (m\bar{X}/n\bar{Y})[1 + (m\bar{X}/n\bar{Y})]^{-1}$ where $X_1, \ldots, X_m, Y_1, \ldots, Y_n$ are independent standard exponentials.

21. Show directly using Problem B.2.5 that under the conditions of the previous problem, if $m/(m + n) - \alpha$ tends to zero at the rate $1/(m + n)^2$, then

$$E(B_{m,n}) = \frac{m}{m + n}, \quad \text{Var } B_{m,n} = \frac{\alpha(1 - \alpha)}{m + n} + O((m + n)^{-2}).$$

22. Let $S_n \sim \chi_n^2$. Use Stirling's approximation and Problem B.2.4 to give a direct justification of

$$E(\sqrt{S_n}) = \sqrt{n} + R_n$$

where $R_n/\sqrt{n} \to 0$ as in $n \to \infty$. Recall *Stirling's approximation:*

$$\Gamma(p+1)/(\sqrt{2\pi}e^{-p}p^{p+\frac{1}{2}}) \to 1 \text{ as } p \to \infty.$$

(It may be shown but is not required that $|\sqrt{n}R_n|$ is bounded.)

23. Suppose that $X_1, \ldots, X_n$ is a sample from a population and that h is a real-valued function of $\bar{X}$ whose derivatives of order k are denoted by $h^{(k)}$, $k > 1$. Suppose $|h^{(4)}(x)| \leq M$ for all x and some constant M and suppose that μ_4 is finite. Show that $Eh(\bar{X}) = h(\mu) + \frac{1}{2}h^{(2)}(\mu)\frac{\sigma^2}{n} + R_n$ where $|R_n| \leq h^{(3)}(\mu)|\mu_3|/6n^2 + M(\mu_4 + 3\sigma^2)/24n^2$.
Hint:

$$\left| h(x) - h(\mu) - h^{(1)}(\mu)(x - \mu) - \frac{h^{(2)}(\mu)}{2}(x - \mu)^2 - \frac{h^{(3)}(\mu)}{6}(x - \mu)^3 \right| \leq \frac{M}{24}(x - \mu)^4.$$

Therefore,

$$\left| Eh(\bar{X}) - h(\mu) - h^{(1)}(\mu)E(\bar{X} - \mu) - \frac{h^{(2)}(\mu)}{2}E(\bar{X} - \mu)^2 \right|$$

$$\leq \frac{|h^{(3)}(\mu)|}{6}|E(\bar{X} - \mu)^3| + \frac{M}{24}E(\bar{X} - \mu)^4$$

$$\leq \frac{|h^{(3)}(\mu)|\,|\mu_3|}{6\,n^2} + \frac{M}{24}\frac{(\mu_4 + 3\sigma^4)}{n^2}.$$

24. Let $X_1, \ldots, X_n$ be a sample from a population with mean μ and variance $\sigma^2 < \infty$. Suppose h has a second derivative $h^{(2)}$ continuous at μ and that $h^{(1)}(\mu) = 0$.

(a) Show that $\sqrt{n}[h(\bar{X}) - h(\mu)] \to 0$ while $n[h(\bar{X} - h(\mu)]$ is asymptotically distributed as $\frac{1}{2}h^{(2)}(\mu)\sigma^2 V$ where $V \sim \chi_1^2$.

(b) Use part (a) to show that when $\mu = \frac{1}{2}$, $n[\bar{X}(1 - \bar{X}) - \mu(1 - \mu)] \xrightarrow{\mathcal{L}} -\sigma^2 V$ with $V \sim \chi_1^2$. Give an approximation to the distribution of $\bar{X}(1 - \bar{X})$ in terms of the χ_1^2 distribution function when $\mu = \frac{1}{2}$.

25. Let $X_1, \ldots, X_n$ be a sample from a population with $\sigma^2 = \text{Var}(X) < \infty$, $\mu = E(X)$ and let $T = \bar{X}^2$ be an estimate of μ^2.

(a) When $\mu \neq 0$, find the asymptotic distribution of $\sqrt{n}(T - \mu^2)$ using the delta method.

(b) When $\mu = 0$, find the asymptotic distriution of nT using $P(nT \leq t) = P(-\sqrt{t} \leq \sqrt{n}\bar{X} \leq \sqrt{t})$. Compare your answer to the answer in part (a).

(c) Find the limiting laws of $\sqrt{n}(\bar{X} - \mu)^2$ and $n(\bar{X} - \mu)^2$.

26. Show that if $X_1, \ldots, X_n$ are i.i.d. $\mathcal{N}(\mu, \sigma^2)$, then

$$\sqrt{n}(\bar{X} - \mu, \hat{\sigma}^2 - \sigma^2) \xrightarrow{\mathcal{L}} \mathcal{N}(0, 0, \Sigma_0)$$

where $\Sigma_0 = \mathrm{diag}(\sigma^2, 2\sigma^4)$.

Hint: Use (5.3.33) and Theorem 5.3.4.

27. Suppose $(X_1, Y_1), \ldots, (X_n, Y_n)$ are n sets of control and treatment responses in a matched pair experiment. Assume that the observations have a common $\mathcal{N}(\mu_1, \mu_2, \sigma_1^2, \sigma_2^2, \rho)$ distribution. We want to obtain confidence intervals on $\mu_2 - \mu_1 = \Delta$. Suppose that instead of using the one-sample t intervals based on the differences $Y_i - X_i$ we treat $X_1, \ldots, X_n$, $Y_1, \ldots, Y_n$ as separate samples and use the two-sample t intervals (4.9.3). What happens? Analysis for fixed n is difficult because $T(\Delta)$ no longer has a $\mathcal{T}_{2n-2}$ distribution. Let $n \to \infty$ and

(a) Show that $P[T(\Delta) \leq t] \to \Phi\left(t\left[1 - \frac{2\sigma_1\sigma_2\rho}{(\sigma_1^2 + \sigma_2^2)}\right]^{-\frac{1}{2}}\right)$.

(b) Deduce that if $\rho > 0$ and I_n is given by (4.9.3), then $\lim_n P[\Delta \in I_n] > 1 - \alpha$.

(c) Show that if $|I_n|$ is the length of the interval I_n,

$$\sqrt{n}|I_n| \to 2\sqrt{\sigma_1^2 + \sigma_2^2} z(1 - \frac{1}{2}\alpha) > 2(\sqrt{\sigma_1^2 + \sigma_2^2 - 2\rho\sigma_1\sigma_2}) z(1 - \frac{1}{2}\alpha)$$

where the right-hand side is the limit of $\sqrt{n}$ times the length of the one-sample t interval based on the differences.

Hint: (a), (c) Apply Slutsky's theorem.

28. Suppose $X_1, \ldots, X_{n_1}$ and $Y_1, \ldots, Y_{n_2}$ are as in Section 4.9.3 independent samples with $\mu_1 = E(X_1)$, $\sigma_1^2 = \mathrm{Var}(X_1)$, $\mu_2 = E(Y_1)$, and $\sigma_2^2 = \mathrm{Var}(Y_1)$. We want to study the behavior of the two-sample pivot $T(\Delta)$ of Example 4.9.3, if $n_1, n_2 \to \infty$, so that $n_1/n \to \lambda, 0 < \lambda < 1$.

(a) Show that $P[T(\Delta) \leq t] \to \Phi(t[(\lambda\sigma_1^2 + (1 - \lambda)\sigma_2^2)/((1 - \lambda)\sigma_1^2 + \lambda\sigma_2^2)]^{\frac{1}{2}})$.

(b) Deduce that if $\lambda = \frac{1}{2}$ or $\sigma_1 = \sigma_2$, the intervals (4.9.3) have correct asymptotic probability of coverage.

(c) Show that if $\sigma_2^2 > \sigma_1^2$ and $\lambda > 1 - \lambda$, the interval (4.9.3) has asymptotic probability of coverage $< 1 - \alpha$, whereas the situation is reversed if the sample size inequalities and variance inequalities agree.

(d) Make a comparison of the asymptotic length of (4.9.3) and the intervals based on the pivot $|D - \Delta|/s_D$ where D and s_D are as in Section 4.9.4.

29. Let $T = (D - \Delta)/s_D$ where D, Δ and s_D are as defined in Section 4.9.4. Suppose that $E(X_i^4) < \infty$ and $E(Y_j^4) < \infty$.

(a) Show that T has asymptotically a standard normal distribution as $n_1 \to \infty$ and $n_2 \to \infty$.

(b) Let k be the Welch degrees of freedom defined in Section 4.9.4. Show that $k \xrightarrow{P} \infty$ as $n_1 \to \infty$ and $n_2 \to \infty$.

(c) Show using parts (a) and (b) that the tests that reject $H : \mu_1 = \mu_2$ in favor of $K : \mu_2 > \mu_1$ when $T \geq t_k(1 - \alpha)$, where $t_k(1 - \alpha)$ is the critical value using the Welch approximation, has asymptotic level α.

(d) Find or write a computer program that carries out the Welch test. Carry out a Monte Carlo study such as the one that led to Figure 5.3.3 using the Welch test based on T rather than the two-sample t test based on S_n. Plot your results.

30. Generalize Lemma 5.3.1 by showing that if $\mathbf{Y}_1, \ldots, \mathbf{Y}_n \in R^d$ are i.i.d. vectors with zero means and $E|\mathbf{Y}_1|^k < \infty$, where $|\cdot|$ is the Euclidean norm, then for all integers $k \geq 2$:

$$E|\bar{\mathbf{Y}}|^k \leq Cn^{-k/2}$$

where C depends on d, $E|\mathbf{Y}_1|^k$ and k only.

Hint: If $|\mathbf{x}|_1 = \sum_{j=1}^d |x_j|$, $\mathbf{x} = (x_1, \ldots, x_d)^T$ and $|\mathbf{x}|$ is Euclidean distance, then there exist universal constants $0 < c_d < C_d < \infty$ such that $c_d |\mathbf{x}|_1 \leq |\mathbf{x}| \leq C_d |\mathbf{x}|_1$.

31. Let $X_1, \ldots, X_n$ be i.i.d. as $X \sim F$ and let $\mu = E(X)$, $\sigma^2 = \text{Var}(X)$, $\kappa = \text{Var}[(X - \mu)/\sigma]^2$, $s^2 = (n-1)^{-1} \sum_{i=1}^n (X_i - \bar{X})^2$. Then by Theorem B.3.1, $V_n = (n-1)s^2/\sigma^2$ has a χ^2_{n-1} distribution when F is the $\mathcal{N}(\mu, \sigma^2)$ distribution.

(a) Suppose $E(X^4) < \infty$.

(b) Let $x_{n-1}(\alpha)$ be the αth quantile of χ^2_{n-1}. Find approximations to $P(V_n \leq x_{n-1}(\alpha))$ and $P(V_n \leq x_{n-1}(1 - \alpha))$ and evaluate the approximations when F is T_5.
Hint: See Problems B.3.9 and 4.4.16.

(c) Let $\hat{\kappa}$ be the method of moment estimate of κ and let

$$\hat{v}_\alpha = (n - 1) + \sqrt{\hat{\kappa}(n - 1)} z(\alpha).$$

Show that if $0 < EX^8 < \infty$, then $P(V_n \leq \hat{v}_\alpha) \to \alpha$ as $n \to \infty$.

32. It may be shown that if T_n is any sequence of random variables such that $T_n \xrightarrow{\mathcal{L}} T$ and if the variances of T and T_n exist, then $\liminf_n \text{Var}(T_n) \geq \text{Var}(T)$. Let

$$T_n = X1[|X| \leq 1 - n^{-1}] + n1[|X| > 1 - n^{-1}]$$

where X is uniform, $\mathcal{U}(-1, 1)$. Show that as $n \to \infty$, $T_n \xrightarrow{\mathcal{L}} X$, but $\text{Var}(T_n) \to \infty$.

33. Let $X_{ij}(i = 1, \ldots, p; \ j = 1, \ldots, k)$ be independent with $X_{ij} \sim \mathcal{N}(\mu_i, \sigma^2)$.

(a) Show that the MLEs of μ_i and σ^2 are

$$\hat{\mu}_i = k^{-1} \sum_{j=1}^k X_{ij} \text{ and } \hat{\sigma}^2 = (kp)^{-1} \sum_{i=1}^p \sum_{j=1}^k (X_{ij} - \hat{\mu}_i)^2.$$

(b) Show that if k is fixed and $p \to \infty$, then $\widehat{\sigma}^2 \xrightarrow{P} (k-1)\sigma^2/k$. That is the MLE $\widehat{\mu}^2$ is not consistent (Neyman and Scott, 1948).

(c) Give a consistent estimate of σ^2.

Problems for Section 5.4

1. Let $X_1, \ldots, X_n$ be i.i.d. random variables distributed according to $P \in \mathcal{P}$. Suppose $\psi : R \to R$

(i) is monotone nondecreasing

(ii) $\psi(-\infty) < 0 < \psi(\infty)$

(iii) $|\psi|(x) \le M < \infty$ for all x.

(a) Show that (i), (ii), and (iii) imply that $\theta(P)$ defined (not uniquely) by

$$E_P \psi(X_1 - \theta(P)) \ge 0 \ge E_P \psi(X_1 - \theta'), \text{ all } \theta' > \theta(P)$$

is finite.

(b) Suppose that for all $P \in \mathcal{P}$, $\theta(P)$ is the unique solution of $E_P \psi(X_1 - \theta) = 0$. Let $\widehat{\theta}_n = \theta(\widehat{P})$, where $\widehat{P}$ is the empirical distribution of $X_1, \ldots, X_n$. Show that $\widehat{\theta}_n$ is consistent for $\theta(P)$ over $\mathcal{P}$. Deduce that the sample median is a consistent estimate of the population median if the latter is unique. (Use $\psi(x) = \text{sgn}(x)$.)
Hint: Show that $E_P \psi(X_1 - \theta)$ is nonincreasing in θ. Use the bounded convergence theorem applied to $\psi(X_1 - \theta) \xrightarrow{P} \psi(-\infty)$ as $\theta \to \infty$.

(c) Assume the conditions in (a) and (b). Set $\lambda(\theta) = E_P \psi(X_1 - \theta)$ and $\tau^2(\theta) = \text{Var}_P \psi(X_1 - \theta)$. Assume that $\lambda'(\theta) < 0$ exists and that

$$\frac{1}{\sqrt{n}\tau(\theta)} \sum_{i=1}^{n} [\psi(X_i - \theta_n) - \lambda(\theta_n)] \xrightarrow{\mathcal{L}} \mathcal{N}(0, 1)$$

for every sequence $\{\theta_n\}$ with $\theta_n = \theta + t/\sqrt{n}$ for $t \in R$. Show that

$$\sqrt{n}(\widehat{\theta}_n - \theta) \xrightarrow{\mathcal{L}} \mathcal{N}\left(0, \frac{\tau^2(\theta)}{[\lambda'(\theta)]^2}\right).$$

Hint: $P(\sqrt{n}(\widehat{\theta}_n - \theta)) < t) = P(\widehat{\theta} < \theta_n) = P(-\sum_{i=1}^{n} \psi(X_i - \theta_n) < 0)$.

(d) Assume part (c) and A6. Show that if $f'(x) = F''(x)$ exists, then $\lambda'(\theta) = \text{Cov}(\psi(X_1 - \theta), \frac{\partial}{\partial \theta} \log f(X_1 - \theta))$.

(e) Suppose that the d.f. $F(x)$ of X_1 is continuous and that $f(\theta) = F'(\theta) > 0$ exists. Let $\widehat{X}$ denote the sample median. Show that, under the conditions of (c), $\sqrt{n}(\widehat{X} - \theta) \xrightarrow{\mathcal{L}} \mathcal{N}(0, 1/4f^2(\theta))$.

(f) For two estimates $\widehat{\theta}_1$ and $\widehat{\theta}_2$ with $\sqrt{n}(\widehat{\theta}_j - \theta) \xrightarrow{\mathcal{L}} \mathcal{N}(0, \sigma_j^2)$, $j = 1, 2$, the *asymptotic relative efficiency* of $\widehat{\theta}_1$ with respect to $\widehat{\theta}_2$ is defined as $e_P(\widehat{\theta}_1, \widehat{\theta}_2) = \sigma_2^2/\sigma_1^2$. Show that if P is $\mathcal{N}(\mu, \sigma^2)$, then $e_P(\bar{X}, \widehat{X}) = \pi/2$.

(g) Suppose X_1 has the gross error density $f_\epsilon(x - \theta)$ (see Section 3.5) where

$$f_\epsilon(x) = (1 - \epsilon)\varphi_\sigma(x) + \epsilon\varphi_\tau(x),\ 0 \le \epsilon \le 0.5$$

and φ_σ denotes the $\mathcal{N}(0, \sigma^2)$ density. Find the efficiency $e_P(\bar{X}, \widehat{X})$ as defined in (f). If $\sigma = 1$, $\tau = 4$, evaluate the efficiency for $\epsilon = .05, 0.10, 0.15$ and 0.20 and note that $\widehat{X}$ is more efficient than $\bar{X}$ for these gross error cases.

(h) Suppose that X_1 has the Cauchy density $f(x) = 1/\pi(1 + x^2)$, $x \in R$. Show that $e_P(\bar{X}, \widehat{X}) = 0$.

2. Show that assumption A4′ of this section coupled with A0–A3 implies assumption A4.
   Hint:

$$E \sup \left\{ \left| \frac{1}{n} \sum_{i=1}^{n} \left(\frac{\partial \psi}{\partial \psi}(X_i, t) - \frac{\partial \psi}{\partial \theta}(X_i, \theta(p)) \right) \right| : |t - \theta(p)| \le \epsilon_n \right\}$$

$$\le E \sup \left\{ \left| \frac{\partial \psi}{\partial \theta}(X_1, t) - \frac{\partial \psi}{\partial \theta}(X_1, \theta(p)) \right| : |t - \theta(p)| \le \epsilon_n \right\}.$$

Apply A4 and the dominated convergence theorem B.7.5.

3. Show that A6′ implies A6.
   Hint: $\frac{\partial}{\partial \theta} \int \psi(x, \theta)p(x, \theta)d\mu(x) = \int \frac{\partial}{\partial \theta}(\psi(x, \theta)p(x, \theta))d\mu(x)$ if for all $-\infty < a < b < \infty$,

$$\int_a^b \int \frac{\partial}{\partial \theta}(\psi(x, \theta)p(x, \theta)d\mu(x)) = \int \psi(x, b)p(x, b)d\mu(x) - \int \psi(x, a)p(x, a)d\mu(x).$$

Condition A6′ permits interchange of the order of integration by Fubini's theorem (Billingsley, 1979) which you may assume.

4. Let $X_1, \ldots, X_n$ be i.i.d. $\mathcal{U}(0, \theta)$, $\theta > 0$.

   (a) Show that $\frac{\partial l}{\partial \theta}(x, \theta) = -\frac{1}{\theta}$ for $\theta > x$ and is undefined for $\theta \le x$. Conclude that $\frac{\partial l}{\partial \theta}(X, \theta)$ is defined with P_θ probability 1 but

$$E_\theta \frac{\partial l}{\partial \theta}(X, \theta) = -\frac{1}{\theta} \neq 0$$

   (b) Show that if $\widehat{\theta} = \max(X_1, \ldots, X_n)$ is the MLE, then $\mathcal{L}_\theta(n(\theta - \widehat{\theta})) \to \mathcal{E}(1/\theta)$. Thus, not only does asymptotic normality not hold but $\widehat{\theta}$ converges to θ faster than at rate $n^{-1/2}$. See also Problem 3.4.22.
   Hint: $P_\theta[n(\theta - \widehat{\theta}) \le x] = 1 - \left(1 - \frac{x}{n\theta}\right)^n \to 1 - \exp(-x/\theta)$.

5. (a) Show that in Theorem 5.4.5

$$\sum_{i=1}^{n} \log \frac{p\left(X_i, \theta_0 + \frac{\gamma}{\sqrt{n}}\right)}{p(X_i, \theta_0)} = \frac{\gamma}{\sqrt{n}} \sum_{i=1}^{n} \frac{\partial}{\partial \theta} \log p(X_i, \theta_0)$$

$$+ \frac{\gamma^2}{2} \frac{1}{n} \sum_{i=1}^{n} \frac{\partial^2}{\partial \theta^2} \log p(X_i, \theta_0) + o_p(1)$$

under P_{θ_0}, and conclude that

$$\mathcal{L}_{\theta_0} \left(\sum_{i=1}^{n} \log \frac{p\left(X_i, \theta_0 + \frac{\gamma}{\sqrt{n}}\right)}{p(X_i, \theta_0)} \right) \to \mathcal{N} \left(-\frac{\gamma^2}{2} I(\theta_0), \gamma^2 I(\theta_0) \right).$$

(b) Show that

$$\mathcal{L}_{\theta_0 + \frac{\gamma}{\sqrt{n}}} \left(\sum_{i=1}^{n} \log \frac{p\left(X_i, \theta_0 + \frac{\gamma}{\sqrt{n}}\right)}{p(X_i, \theta_0)} \right) \to \mathcal{N} \left(\frac{\gamma^2}{2} I(\theta_0), \gamma^2 I(\theta) \right).$$

(c) Prove (5.4.50).
Hint: (b) Expand as in (a) but around $\theta_0 + \frac{\gamma}{\sqrt{n}}$.

(d) Show that $P_{\theta_0 + \frac{\gamma}{\sqrt{n}}} \left[\sum_{i=1}^{n} \log \frac{p\left(X_i, \theta_0 + \frac{\gamma}{\sqrt{n}}\right)}{p(X_i, \theta_0)} = c_n \right] \to 0$ for any sequence $\{c_n\}$ by using (b) and Polyà's theorem (A.14.22).

6. Show that for the log likelihood ratio statistic

$$\log \Lambda_n = \sum_{i=1}^{n} \log \frac{p(X_i, \widehat{\theta}_n)}{p(X_i, \theta_0)} 1(\widehat{\theta}_n > \theta_0),$$

(a) $\mathcal{L}_{\theta_0} (\log \Lambda_n) \to \mathcal{L}(U)$ where $U \sim Z^2 1 (Z > 0)$ with $Z \sim \mathcal{N}(0, 1)$.

(b) $\mathcal{L}_{\theta_0 + \frac{\gamma}{\sqrt{n}}} (\log \Lambda_n) \to \mathcal{L} \left(\frac{1}{2} \left(Z + \gamma I^{\frac{1}{2}}(\theta_0) \right)^2 1[Z > -\gamma I^{\frac{1}{2}}(\theta_0)] \right)$

(c) Show that the asymptotic power function of the level α likelihood ratio test achieves equality in (5.4.50).

7. Suppose A4′, A2, and A6 hold for $\psi = \partial l / \partial \theta$ so that $E_\theta \frac{\partial^2 l}{\partial \theta^2}(X, \theta) = -I(\theta)$ and $I(\theta) < \infty$. Show that $\theta \to I(\theta)$ is continuous.
Hint: $\theta \to \frac{\partial^2 l}{\partial \theta^2}(X, \theta)$ is continuous and

$$\sup \left\{ \frac{\partial^2 l}{\partial \theta^2}(X, \theta') : |\theta - \theta'| \leq \epsilon_n \right\} \xrightarrow{P} 0$$

if $\epsilon_n \to 0$. Apply the dominated convergence theorem (B.7.5) to $\frac{\partial^2 l}{\partial \theta^2}(x, \theta) p(x, \theta)$.

8. Establish $(5.4.54)$ and $(5.4.56)$.

   Hint: Use Problems 5.4.5 and 5.4.6.

9. (a) Establish $(5.4.58)$.

   Hint: Use Problem 5.4.7, A.14.6 and Slutsky's theorem.

   (b) Suppose the conditions of Theorem 5.4.5 hold. Let $\underline{\theta}^*_n$ be as in $(5.4.59)$. Then $(5.4.57)$ can be strengthened to: For each $\theta_0 \in \Theta$, there is a neighborhood $V(\theta_0)$ of θ_0 such that $\varlimsup_n \sup\{P_\theta[\underline{\theta}^*_n \le \theta] : \theta \in V(\theta_0)\} \to 1 - \alpha$.

10. Let

$$\widehat{I} = -\frac{1}{n}\sum_{i=1}^{n}\frac{\partial^2 l}{\partial \theta^2}(X_i, \widehat{\theta}).$$

   (a) Show that under assumptions (A0)–(A6) for $\psi = \frac{\partial l}{\partial \theta}$ at all θ and (A4'), $\widehat{I}$ is a consistent estimate of $I(\theta)$.

   (b) Deduce that

$$\underline{\theta}^{**} = \widehat{\theta} - z_{1-\alpha}/\sqrt{n\widehat{I}}$$

is an asymptotic lower confidence bound for θ.

   (c) Show that if P_θ is a one parameter exponential family the bound of (b) and $(5.4.59)$ coincide.

11. Consider Example 4.4.3, setting a lower confidence bound for binomial θ.

   (a) Show that, if $\bar{X} = 0$ or 1, the bound $(5.4.59)$, which is just $(4.4.7)$, gives $\underline{\theta}^* = X$. Compare with the exact bound of Example 4.5.2.

   (b) Compare the bounds in (a) with the bound $(5.4.61)$, which agrees with $(4.4.3)$, and give the behavior of $(5.4.61)$ for $\bar{X} = 0$ and 1.

12. (a) Show that under assumptions (A0)–(A6) for all θ and (A4'),

$$\underline{\theta}^*_{nj} = \underline{\theta}^*_n + o_p(n^{-1/2})$$

for $j = 1, 2$.

13. Let $\widetilde{\underline{\theta}}_{n1}, \widetilde{\underline{\theta}}_{n2}$ be two asymptotic level $1 - \alpha$ lower confidence bounds. We say that $\widetilde{\underline{\theta}}_{n1}$ is asymptotically at least as good as $\widetilde{\underline{\theta}}_{n2}$ if, for all $\gamma > 0$

$$\varlimsup_n P_\theta\left[\widetilde{\underline{\theta}}_{n1} \le \theta - \frac{\gamma}{\sqrt{n}}\right] \le \varliminf_n P_\theta\left[\widetilde{\underline{\theta}}_{n2} \le \theta - \frac{\gamma}{\sqrt{n}}\right].$$

Show that $\underline{\theta}^*_{n1}$ and, hence, all the $\underline{\theta}^*_{nj}$ are at least as good as any competitors.

   Hint: Compare Theorem 4.4.2.

14. Suppose that $X_1, \ldots, X_n$ are i.i.d. inverse Gaussian with parameters μ and λ, where μ is known. That is, each X_i has density

$$\left(\frac{\lambda}{2\pi x^3}\right)^{1/2} \exp\left\{-\frac{\lambda x}{2\mu^2} + \frac{\lambda}{\mu} - \frac{\lambda}{2x}\right\}; \ x > 0; \ \mu > 0; \ \lambda > 0.$$

(a) Find the Neyman–Pearson (NP) test for testing $H : \lambda = \lambda_0$ versus $K : \lambda < \lambda_0$.

(b) Show that the NP test is UMP for testing $H : \lambda \geq \lambda_0$ versus $K : \lambda < \lambda_0$.

(c) Find the approximate critical value of the Neyman–Pearson test using a normal approximation.

(d) Find the Wald test for testing $H : \lambda = \lambda_0$ versus $K : \lambda < \lambda_0$.

(e) Find the Rao score test for testing $H : \lambda = \lambda_0$ versus $K : \lambda < \lambda_0$.

15. Establish $(5.4.55)$.
  Hint: By $(3.4.10)$ and $(3.4.11)$, the test statistic is a sum of i.i.d. variables with mean zero and variance $I(\theta)$. Now use the central limit theorem.

Problems for Section 5.5

1. Consider testing $H : \mu = 0$ versus $K : \mu \neq 0$ given $X_1, \ldots, X_n$ i.i.d. $\mathcal{N}(\mu, 1)$. Consider the Bayes test when μ is distributed according to π such that

$$1 > \pi(\{0\}) = \lambda > 0, \ \pi(\mu \neq 0) = 1 - \lambda$$

and given $\mu \neq 0$, μ has a $\mathcal{N}(0, \tau^2)$ distribution with density $\varphi_\tau(\mu)$.

(a) Show that the posterior probability of $\{0\}$ is

$$\tilde{\beta} \equiv \lambda\varphi(\sqrt{n}\bar{X})(\lambda\varphi(\sqrt{n}\bar{X}) + (1 - \lambda)m_n(\sqrt{n}\bar{X}))^{-1}$$

where $m_n(\sqrt{n}\bar{X}) = (1 + n\tau^2)^{-1/2}\varphi\left(\frac{\sqrt{n}\bar{X}}{(1+n\tau^2)^{1/2}}\right)$.
  Hint: Set $T = \bar{X}$. We want $\pi(\{0\}|t) = \lambda p_T(t|0)/p_T(t)$ where $p_T(t) = \lambda p_T(t|0) + (1 - \lambda)\int_{-\infty}^{\infty} \varphi_\tau(\mu)p_T(t|\mu)d\mu$.

(b) Suppose that $\mu = 0$. By Problem 4.1.5, the p-value $\hat{\beta} \equiv 2[1 - \Phi(\sqrt{n}|\bar{X}|)]$ has a $\mathcal{U}(0, 1)$ distribution. Show that $\tilde{\beta} \xrightarrow{P} 1$ as $n \to \infty$.

(c) Suppose that $\mu = \delta > 0$. Show that $\tilde{\beta}/\hat{\beta} \xrightarrow{P} \infty$. That is, if H is false, the evidence against H as measured by the smallness of the p-value is much greater than the evidence measured by the smallness of the posterior probability of the hypothesis (Lindley's "paradox").

2. Let $X_1, \ldots, X_n$ be i.i.d. $\mathcal{N}(\mu, 1)$. Consider the problem of testing $H : \mu \in [0, \Delta]$ versus $K : \mu > \Delta$, where Δ is a given number.

(a) Show that the test that rejects H for large values of $\sqrt{n}(\bar{X} - \Delta)$ has p-value $\hat{p} = \Phi(-\sqrt{n}(\bar{X} - \Delta))$ and that when $\mu = \Delta$, $\hat{p}$ has a $\mathcal{U}(0, 1)$ distribution.

(b) Suppose that μ has a $\mathcal{N}(0, 1)$ prior. Show that the posterior probability of H is

$$\tilde{p} = \Phi\left(\frac{-\sqrt{n}(a_n\bar{X} - \Delta)}{\sqrt{a_n}}\right) - \Phi\left(\frac{-\sqrt{n}a_n\bar{X}}{\sqrt{a_n}}\right)$$

where $a_n = n/(n+1)$.

(c) Show that when $\mu = \Delta$, $-\sqrt{n}(a_n\bar{X} - \Delta)/\sqrt{a_n} \overset{\mathcal{L}}{\to} \mathcal{N}(0, 1)$ and $\tilde{p} \overset{\mathcal{L}}{\to} \mathcal{U}(0, 1)$. (Lindley's "paradox" of Problem 5.1.1 is not in effect.)

(d) Compute $p \lim_{n \to \infty} \tilde{p}/\hat{p}$ for $\mu \neq \Delta$.

(e) Verify the following table giving posterior probabilities of $[0, \Delta]$ when $\sqrt{n}\bar{X} = 1.645$ and $\hat{p} = 0.05$.

n	10	20	50	100
$\Delta = 0.1$	.029	.034	.042	.046
$\Delta = 1.0$	.058	.054	.052	.050

3. Establish (5.5.14).
   Hint: By (5.5.13) and the SLLN,

$$\log d_n q_n(t) =$$
$$-\frac{t^2}{2}\left\{I(\theta) - \frac{1}{n}\sum_{i=1}^{n}\left(\frac{\partial^2 l}{\partial \theta^2}(X_i, \theta^*(t)) - \frac{\partial^2 l}{\partial \theta^2}(X_i, \theta_0)\right) + \log \pi\left(\hat{\theta} + \frac{t}{\sqrt{n}}\right)\right\}.$$

Apply the argument used for Theorem 5.4.2 and the continuity of $\pi(\theta)$.

4. Establish (5.5.17).
   Hint:

$$\left|\frac{1}{n}\sum_{i=1}^{n}\frac{\partial^2 l}{\partial \theta^2}(X_i, \theta^*(t))\right| \leq \frac{1}{n}\sum_{i=1}^{n}\sup\left\{\left|\frac{\partial^2}{\partial \theta^2}l(X_i, \theta')\right| : |\theta' - \theta| \leq \delta\right\}$$

if $|t| \leq \delta\sqrt{n}$. Apply the SLLN and $\delta \to E_\theta \sup\left\{\left|\frac{\partial^2}{\partial \theta^2}l(X_i, \theta')\right| : |\theta - \theta'| \leq \delta\right\}$ continuous at $\delta = 0$.

5. Suppose that in addition to the conditions of Theorem 5.5.2, $\int \theta^2 \pi(\theta)d\theta < \infty$. Then $\sqrt{n}(E(\theta \mid \mathbf{X}) - \hat{\theta}) \to 0$ a.s. P_θ.
   Hint: In view of Theorem 5.5.2 it is equivalent to show that

$$d_n \int t q_n(t)dt \to 0$$

a.s. P_θ. By Theorem 5.5.2, $\int_{-M}^{M} tq_n(t)dt \to 0$ a.s. for all $M < \infty$. By (5.5.17), $\int_{M}^{\delta(\theta)\sqrt{n}} |t|q_n(t)dt \leq \int_{M}^{\delta(\theta)\sqrt{n}} |t| \exp\left\{-\frac{1}{4}I(\theta)\frac{t^2}{2}\right\} dt \leq \epsilon$ for $M(\epsilon)$ sufficiently large, all $\epsilon > 0$. Finally,

$$d_n \int_{\delta(\theta)\sqrt{n}}^{\infty} tq_n(t)dt = \int_{\theta+\delta(\theta)}^{\infty} \sqrt{n}(t - \widehat{\theta}) \exp\left\{\sum_{i=1}^{n}(l(X_i, t) - l(X_i, \widehat{\theta}))\right\} \pi(t)dt.$$

Apply (5.5.16) noting that $\sqrt{n}e^{-n\epsilon(\delta,\theta)} \to 0$ and $\int |t|\pi(t)dt < \infty$.

6. (a) Show that $\sup\{|q_n(t) - I^{\frac{1}{2}}(\theta)\varphi(tI^{\frac{1}{2}}(\theta))| : |t| \leq M\} \to 0$ a.s. for all θ.

(b) Deduce (5.5.29).
    Hint: $\{t : \sqrt{I(\theta)}\varphi(t\sqrt{I(\theta)}) \geq c(d)\} = [-d, d]$ for some $c(d)$, all d and $c(d) \nearrow$ in d. The sets $C_n(c) \equiv \{t : q_n(t) \geq c\}$ are monotone increasing in c. Finally, to obtain

$$P[\theta \in C_n(X_1, \ldots, X_n) \mid X_1, \ldots, X_n] = 1 - \alpha$$

we must have $c_n = c(z_{1-\frac{\alpha}{2}}[I(\widehat{\theta})n]^{-1/2})(1 + o_p(1))$ by Theorem 5.5.1.

7. Suppose that in Theorem 5.5.2 we replace the assumptions A4(a.s.) and A5(a.s.) by A4 and A5. Show that (5.5.8) and (5.5.9) hold with a.s. convergence replaced by convergence in P_θ probability.

5.7   NOTES

Notes for Section 5.1

(1) The bound is actually known to be essentially attained for $X_i = 0$ with probability p_n and 1 with probability $1 - p_n$ where $p_n \to 0$ or 1. For n large these do not correspond to distributions one typically faces. See Bhattacharya and Ranga Rao (1976) for further discussion.

Notes for Section 5.3

(1) If the right-hand side is negative for some x, $F_n(x)$ is taken to be 0.

(2) Computed by Winston Chow.

Notes for Section 5.4

(1) This result was first stated by R. A. Fisher (1925). A proof was given by Cramer (1946).

Notes for Section 5.5

(1) This famous result appears in Laplace's work and was rediscovered by S. Bernstein and R. von Mises—see Stigler (1986) and Le Cam and Yang (1990).

5.8   REFERENCES

BERGER, J., *Statistical Decision Theory and Bayesian Analysis* New York: Springer–Verlag, 1985.

BHATTACHARYA, R. H. AND R. RANGA RAO, *Normal Approximation and Asymptotic Expansions* New York: Wiley, 1976.

BILLINGSLEY, P., *Probability and Measure* New York: Wiley, 1979.

BOX, G. E. P., "Non-normality and tests on variances," *Biometrika, 40*, 318–324 (1953).

CRAMER, H., *Mathematical Methods of Statistics* Princeton, NJ: Princeton University Press, 1946.

DAVID, F. N., *Tables of the Correlation Coefficient*, Cambridge University Press, reprinted in *Biometrika Tables for Statisticians* (1966), Vol. I, 3rd ed., H. O. Hartley and E. S. Pearson, Editors Cambridge: Cambridge University Press, 1938.

FERGUSON, T. S., *A Course in Large Sample Theory* New York: Chapman and Hall, 1996.

FISHER, R. A., "Theory of statistical estimation," *Proc. Camb. Phil. Soc., 22*, 700–725 (1925).

FISHER, R. A., *Statistical Inference and Scientific Method*, Vth Berkeley Symposium, 1958.

HAMMERSLEY, J. M. AND D. C. HANSCOMB, *Monte Carlo Methods* London: Methuen & Co., 1964.

HOEFFDING, W., "Probability inequalities for sums of bounded random variables," *J. Amer. Statist. Assoc., 58*, 13–80 (1963).

HUBER, P. J., *The Behavior of the Maximum Likelihood Estimator Under Non-Standard Conditions*, Proc. Vth Berk. Symp. Math. Statist. Prob., Vol. 1 Berkeley, CA: University of California Press, 1967.

LE CAM, L. AND G. L. YANG, *Asymptotics in Statistics, Some Basic Concepts* New York: Springer, 1990.

LEHMANN, E. L., *Elements of Large-Sample Theory* New York: Springer–Verlag, 1999.

LEHMANN, E. L. AND G. CASELLA, *Theory of Point Estimation* New York: Springer–Verlag, 1998.

NEYMAN, J. AND E. L. SCOTT, "Consistent estimates based on partially consistent observations," *Econometrica, 16*, 1–32 (1948).

RAO, C. R., *Linear Statistical Inference and Its Applications*, 2nd ed. New York: J. Wiley & Sons, 1973.

RUDIN, W., *Mathematical Analysis*, 3rd ed. New York: McGraw Hill, 1987.

SCHERVISCH, M., *Theory of Statistics* New York: Springer, 1995.

SERFLING, R. J., *Approximation Theorems of Mathematical Statistics* New York: J. Wiley & Sons, 1980.

STIGLER, S., *The History of Statistics: The Measurement of Uncertainty Before 1900* Cambridge, MA: Harvard University Press, 1986.

WILSON, E. B. AND M. M. HILFERTY, "The distribution of chi square," *Proc. Nat. Acad. Sci., U.S.A., 17*, p. 684 (1931).

Chapter 6

INFERENCE IN THE
MULTIPARAMETER CASE

6.1   INFERENCE FOR GAUSSIAN LINEAR MODELS

Most modern statistical questions involve large data sets, the modeling of whose stochastic structure involves complex models governed by several, often many, real parameters and frequently even more semi- or nonparametric models. In this final chapter of Volume I we develop the analogues of the asymptotic analyses of the behaviors of estimates, tests, and confidence regions in regular one-dimensional parametric models for d-dimensional models $\{P_{\boldsymbol{\theta}} : \boldsymbol{\theta} \in \Theta\}$, $\Theta \subset R^d$. We have presented several such models already, for instance, the multinomial (Examples 1.6.7, 2.3.3), multiple regression models (Examples 1.1.4, 1.4.3, 2.1.1) and more generally have studied the theory of multiparameter exponential families (Sections 1.6.2, 2.2, 2.3). However, with the exception of Theorems 5.2.2 and 5.3.5, in which we looked at asymptotic theory for the MLE in multiparameter exponential families, we have not considered asymptotic inference, testing, confidence regions, and prediction in such situations. We begin our study with a thorough analysis of the Gaussian linear model with known variance in which exact calculations are possible. We shall show how the exact behavior of likelihood procedures in this model correspond to limiting behavior of such procedures in the unknown variance case and more generally in large samples from regular d-dimensional parametric models and shall illustrate our results with a number of important examples.

This chapter is a lead-in to the more advanced topics of Volume II in which we consider the construction and properties of procedures in non- and semiparametric models. The approaches and techniques developed here will be successfully extended in our discussions of the delta method for function-valued statistics, the properties of nonparametric MLEs, curve estimates, the bootstrap, and efficiency in semiparametric models. There is, however, an important aspect of practical situations that is not touched by the approximation, the fact that d, the number of parameters, and n, the number of observations, are often both large and commensurate or nearly so. The inequalities of Vapnik–Chervonenkis, Talagrand type and the modern empirical process theory needed to deal with such questions will also appear in the later chapters of Volume II.

Notational Convention: In this chapter we will, when there is no ambiguity, let expressions such as $\boldsymbol{\theta}$ refer to both column and row vectors.

6.1.1   The Classical Gaussian Linear Model

Many of the examples considered in the earlier chapters fit the framework in which the ith measurement Y_i among n independent observations has a distribution that depends on known constants $z_{i1}, \ldots, z_{ip}$. In the classical Gaussian (normal) linear model this dependence takes the form

$$Y_i = \sum_{j=1}^{p} z_{ij}\beta_j + \epsilon_i, \ i = 1, \ldots, n \tag{6.1.1}$$

where $\epsilon_1, \ldots, \epsilon_n$ are i.i.d. $\mathcal{N}(0, \sigma^2)$. In vector and matrix notation, we write

$$Y_i = \mathbf{z}_i^T \boldsymbol{\beta} + \epsilon_i, \ i = 1, \ldots, n \tag{6.1.2}$$

and

$$\mathbf{Y} = \mathbf{Z}\boldsymbol{\beta} + \boldsymbol{\epsilon}, \ \boldsymbol{\epsilon} \sim \mathcal{N}(\mathbf{0}, \sigma^2 \mathbf{J}) \tag{6.1.3}$$

where $\mathbf{z}_i = (z_{i1}, \ldots, z_{ip})^T$, $\mathbf{Z} = (z_{ij})_{n \times p}$, and $\mathbf{J}$ is the $n \times n$ identity matrix.

Here Y_i is called the *response* variable, the z_{ij} are called the *design values*, and $\mathbf{Z}$ is called the *design matrix*.

In this section we will derive exact statistical procedures under the assumptions of the model (6.1.3). These are among the most commonly used statistical techniques. In Section 6.6 we will investigate the sensitivity of these procedures to the assumptions of the model. It turns out that these techniques are sensible and useful outside the narrow framework of model (6.1.3).

Here is Example 1.1.2(4) in this framework.

Example 6.1.1. *The One-Sample Location Problem.* We have n independent measurements $Y_1, \ldots, Y_n$ from a population with mean $\beta_1 = E(Y)$. The model is

$$Y_i = \beta_1 + \epsilon_i, \ i = 1, \ldots, n \tag{6.1.4}$$

where $\epsilon_1, \ldots, \epsilon_n$ are i.i.d. $\mathcal{N}(0, \sigma^2)$. Here $p = 1$ and $\mathbf{Z}_{n \times 1} = (1, \ldots, 1)^T$. $\quad\square$

The regression framework of Examples 1.1.4 and 2.1.1 is also of the form (6.1.3):

Example 6.1.2. *Regression.* We consider experiments in which n cases are sampled from a population, and for each case, say the ith case, we have a response Y_i and a set of $p - 1$ covariate measurements denoted by $z_{i2}, \ldots, z_{ip}$. We are interested in relating the mean of the response to the covariate values. The normal linear regression model is

$$Y_i = \beta_1 + \sum_{j=2}^{p} z_{ij}\beta_j + \epsilon_i, \ i = 1, \ldots, n \tag{6.1.5}$$

where β_1 is called the regression intercept and $\beta_2, \ldots, \beta_p$ are called the regression coefficients. If we set $z_{i1} = 1$, $i = 1, \ldots, n$, then the notation (6.1.2) and (6.1.3) applies.

We treat the covariate values z_{ij} as fixed (nonrandom). In this case, (6.1.5) is called the
fixed design normal linear regression model. The random design Gaussian linear regression
model is given in Example 1.4.3. We can think of the fixed design model as a conditional
version of the random design model with the inference developed for the conditional distribution of Y given a set of observed covariate values.   □

Example 6.1.3. *The p-Sample Problem or One-Way Layout.* In Example 1.1.3 and Section 4.9.3 we considered experiments involving the comparisons of two population means
when we had available two independent samples, one from each population. Two-sample
models apply when the design values represent a qualitative factor taking on only two values. Frequently, we are interested in qualitative factors taking on several values. If we are
comparing pollution levels, we want to do so for a variety of locations; we often have more
than two competing drugs to compare, and so on.

To fix ideas suppose we are interested in comparing the performance of $p \geq 2$ treatments on a population and that we administer only one treatment to each subject and a
sample of n_k subjects get treatment k, $1 \leq k \leq p$, $n_1 + \cdots + n_p = n$. If the control and
treatment responses are independent and normally distributed with the same variance σ^2,
we arrive at the *one-way layout* or *p-sample* model,

$$Y_{kl} = \beta_k + \epsilon_{kl}, \ 1 \leq l \leq n_k, \ 1 \leq k \leq p \tag{6.1.6}$$

where Y_{kl} is the response of the lth subject in the group obtaining the kth treatment, β_k
is the mean response to the kth treatment, and the ϵ_{kl} are independent $\mathcal{N}(0, \sigma^2)$ random
variables.

To see that this is a linear model we relabel the observations as $Y_1, \ldots, Y_n$, where
$Y_1, \ldots, Y_{n_1}$ correspond to the group receiving the first treatment, $Y_{n_1+1}, \ldots, Y_{n_1+n_2}$ to
that getting the second, and so on. Then for $1 \leq j \leq p$, if $n_0 = 0$, the design matrix has
elements:

$$z_{ij} = 1 \text{ if } \sum_{k=1}^{j-1} n_k + 1 \leq i \leq \sum_{k=1}^{j} n_k$$

$$= 0 \text{ otherwise}$$

and

$$\mathbf{Z} = \begin{pmatrix} \mathbf{I}_1 & 0 & \cdots & 0 \\ 0 & \mathbf{I}_2 & \cdots & 0 \\ \vdots & \vdots & & \vdots \\ 0 & 0 & \cdots & \mathbf{I}_p \end{pmatrix}$$

where $\mathbf{I}_j$ is a column vector of n_j ones and the 0 in the "row" whose jth member is $\mathbf{I}_j$ is a
column vector of n_j zeros. The model (6.1.6) is an example of what is often called *analysis
of variance models*. Generally, this terminology is commonly used when the design values
are qualitative.

The model (6.1.6) is often reparametrized by introducing $\alpha = p^{-1} \sum_{k=1}^{p} \beta_k$ and $\delta_k = \beta_k - \alpha$ because then δ_k represents the difference between the kth and average treatment effects, $k = 1, \ldots, p$. In terms of the new parameter $\beta^* = (\alpha, \delta_1, \ldots, \delta_p)^T$, the linear model is

$$\mathbf{Y} = \mathbf{Z}^* \boldsymbol{\beta}^* + \boldsymbol{\epsilon}, \ \boldsymbol{\epsilon} \sim \mathcal{N}(\mathbf{0}, \sigma^2 \mathbf{J})$$

where $\mathbf{Z}^*_{n \times (p+1)} = (\mathbf{1}, \mathbf{Z})$ and $\mathbf{1}_{n \times 1}$ is the vector with n ones. Note that $\mathbf{Z}^*$ is of rank p and that $\boldsymbol{\beta}^*$ is not identifiable for $\boldsymbol{\beta}^* \in R^{p+1}$. However, $\boldsymbol{\beta}^*$ is identifiable in the p-dimensional linear subspace $\{\boldsymbol{\beta}^* \in R^{p+1} : \sum_{k=1}^{p} \delta_k = 0\}$ of R^{p+1} obtained by adding the linear restriction $\sum_{k=1}^{p} \delta_k = 0$ forced by the definition of the δ_k's. This type of linear model with the number of columns d of the design matrix larger than its rank r, and with the parameters identifiable only once $d - r$ additional linear restrictions have been specified, is common in analysis of variance models.   □

Even if $\boldsymbol{\beta}$ is not a parameter (is unidentifiable), the vector of means $\boldsymbol{\mu} = (\mu_1, \ldots, \mu_n)^T$ of $\mathbf{Y}$ always is. It is given by

$$\boldsymbol{\mu} = \mathbf{Z}\boldsymbol{\beta} = \sum_{j=1}^{p} \beta_j \mathbf{c}_j$$

where the $\mathbf{c}_j$ are the columns of the design matrix,

$$\mathbf{c}_j = (z_{1j}, \ldots, z_{nj})^T, \ j = 1, \ldots, p.$$

The parameter set for $\boldsymbol{\beta}$ is R^p and the parameter set for $\boldsymbol{\mu}$ is

$$\omega = \{\boldsymbol{\mu} = \mathbf{Z}\boldsymbol{\beta}; \ \boldsymbol{\beta} \in R^p\}.$$

Note that ω is the linear space spanned by the columns $\mathbf{c}_j$, $j = 1, \ldots, n$, of the design matrix. Let r denote the number of linearly independent $\mathbf{c}_j$, $j = 1, \ldots, p$, then r is the rank of $\mathbf{Z}$ and ω has dimension r. It follows that the parametrization $(\boldsymbol{\beta}, \sigma^2)$ is identifiable if and only if $r = p$ (Problem 6.1.17). We assume that $n \geq r$.

The Canonical Form of the Gaussian Linear Model

The linear model can be analyzed easily using some geometry. Because $\dim \omega = r$, there exists (e.g., by the Gram–Schmidt process) (see Section B.3.2), an orthonormal basis $\mathbf{v}_1, \ldots, \mathbf{v}_n$ for R^n such that $\mathbf{v}_1, \ldots, \mathbf{v}_r$ span ω. Recall that orthonormal means $\mathbf{v}_i^T \mathbf{v}_j = 0$ for $i \neq j$ and $\mathbf{v}_i^T \mathbf{v}_i = 1$. When $\mathbf{v}_i^T \mathbf{v}_j = 0$, we call $\mathbf{v}_i$ and $\mathbf{v}_j$ *orthogonal*. Note that any $\mathbf{t} \in R^n$ can be written

$$\mathbf{t} = \sum_{i=1}^{n} (\mathbf{v}_i^T \mathbf{t}) \mathbf{v}_i \tag{6.1.7}$$

and that

$$\mathbf{t} \in \omega \Leftrightarrow \mathbf{t} = \sum_{i=1}^{r} (\mathbf{v}_i^T \mathbf{t}) \mathbf{v}_i \Leftrightarrow \mathbf{v}_i^T \mathbf{t} = 0, \ i = r+1, \ldots, n.$$

We now introduce the *canonical variables* and *means*

$$U_i = \mathbf{v}_i^T \mathbf{Y}, \ \eta_i = E(U_i) = \mathbf{v}_i^T \boldsymbol{\mu}, \ i = 1, \ldots, n.$$

Theorem 6.1.1. *The U_i are independent and $U_i \sim \mathcal{N}(\eta_i, \sigma^2)$, $i = 1, \ldots, n$, where*

$$\eta_i = 0, \ i = r+1, \ldots, n,$$

while $(\eta_1, \ldots, \eta_r)^T$ varies freely over R^r.

Proof. Let $\mathbf{A}_{n \times n}$ be the orthogonal matrix with rows $\mathbf{v}_1^T, \ldots, \mathbf{v}_n^T$. Then we can write $\mathbf{U} = \mathbf{A}\mathbf{Y}$, $\boldsymbol{\eta} = \mathbf{A}\boldsymbol{\mu}$, and by Theorem B.3.2, $U_1, \ldots, U_n$ are independent normal with variance σ^2 and $E(U_i) = \mathbf{v}_i^T \boldsymbol{\mu} = 0$ for $i = r+1, \ldots, n$ because $\boldsymbol{\mu} \in \omega$. $\qquad\square$

Note that

$$\mathbf{Y} = \mathbf{A}^{-1}\mathbf{U}. \tag{6.1.8}$$

So, observing $\mathbf{U}$ and $\mathbf{Y}$ is the same thing. $\boldsymbol{\mu}$ and $\boldsymbol{\eta}$ are equivalently related,

$$\boldsymbol{\mu} = \mathbf{A}^{-1}\boldsymbol{\eta}. \tag{6.1.9}$$

whereas

$$\text{Var}(\mathbf{Y}) = \text{Var}(\mathbf{U}) = \sigma^2 \mathbf{J}_{n \times n}. \tag{6.1.10}$$

It will be convenient to obtain our statistical procedures for the canonical variables $\mathbf{U}$, which are sufficient for $(\boldsymbol{\mu}, \sigma^2)$ using the parametrization $(\boldsymbol{\eta}, \sigma^2)^T$, and then translate them to procedures for $\boldsymbol{\mu}$, $\boldsymbol{\beta}$, and σ^2 based on $\mathbf{Y}$ using (6.1.8)–(6.1.10). We start by considering the log likelihood $l_{\mathbf{u}}(\eta)$ based on $\mathbf{U}$

$$
\begin{aligned}
l_{\mathbf{u}}(\eta) &= -\frac{1}{2\sigma^2} \sum_{i=1}^{n} (u_i - \eta_i)^2 - \frac{n}{2} \log(2\pi\sigma^2) \\
&= -\frac{1}{2\sigma^2} \sum_{i=1}^{n} u_i^2 + \frac{1}{\sigma^2} \sum_{i=1}^{r} \eta_i u_i - \sum_{i=1}^{r} \frac{\eta_i^2}{2\sigma^2} - \frac{n}{2} \log(2\pi\sigma^2).
\end{aligned}
\tag{6.1.11}
$$

6.1.2   Estimation

We first consider the σ^2 known case, which is the guide to asymptotic inference in general.

Theorem 6.1.2. *In the canonical form of the Gaussian linear model with σ^2 known*

(i) $\mathbf{T} = (U_1, \ldots, U_r)^T$ *is sufficient for $\boldsymbol{\eta}$.*

(ii) $U_1, \ldots, U_r$ *is the MLE of $\eta_1, \ldots, \eta_r$.*

(iii) U_i *is the UMVU estimate of η_i, $i = 1, \ldots, r$.*

(iv) *If $c_1, \ldots, c_r$ are constants, then the MLE of $\alpha = \sum_{i=1}^{r} c_i \eta_i$ is $\widehat{\alpha} = \sum_{i=1}^{r} c_i U_i$. $\widehat{\alpha}$ is also UMVU for α.*

(v) *The MLE of $\boldsymbol{\mu}$ is $\widehat{\boldsymbol{\mu}} = \sum_{i=1}^{r} \mathbf{v}_i U_i$ and $\widehat{\mu}_i$ is UMVU for μ_i, $i = 1, \ldots, n$. Moreover, $U_i = \mathbf{v}_i^T \widehat{\boldsymbol{\mu}}$, making $\widehat{\boldsymbol{\mu}}$ and $\mathbf{U}$ equivalent.*

Proof. (i) By observation, (6.1.11) is an exponential family with sufficient statistic $\mathbf{T}$.

(ii) $U_1, \ldots, U_r$ are the MLEs of $\eta_1, \ldots, \eta_r$ because, by observation, (6.1.11) is a function of $\eta_1, \ldots, \eta_r$ only through $\sum_{i=1}^{r} (u_i - \eta_i)^2$ and is minimized by setting $\eta_i = u_i$. (We could also apply Theorem 2.3.1.)

(iii) By Theorem 3.4.4 and Example 3.4.6, U_i is UMVU for $E(U_i) = \eta_i$, $i = 1, \ldots, r$.

(iv) By the invariance of the MLE (Section 2.2.2), the MLE of $q(\theta) = \sum_{i=1}^{r} c_i \eta_i$ is $q(\widehat{\theta}) = \sum_{i=1}^{r} c_i U_i$. If all the c's are zero, $\widehat{\alpha}$ is UMVU. Assume that at least one c is different from zero. By Problem 3.4.10, we can assume without loss of generality that $\sum_{i=1}^{r} c_i^2 = 1$. By Gram–Schmidt orthogonalization, there exists an orthonormal basis $\mathbf{v}_1, \ldots, \mathbf{v}_n$ of R^n with $\mathbf{v}_1 = \mathbf{c} = (c_1, \ldots, c_r, 0, \ldots, 0)^T \in R^n$. Let $W_i = \mathbf{v}_i^T \mathbf{U}$, $\xi_i = \mathbf{v}_i \boldsymbol{\eta}$, $i = 1, \ldots, n$, then $\mathbf{W} \sim \mathcal{N}(\boldsymbol{\xi}, \sigma^2 \mathbf{J})$ by Theorem B.3.2, where $\mathbf{J}$ is the $n \times n$ identity matrix. The distribution of $\mathbf{W}$ is an exponential family, $W_1 = \widehat{\alpha}$ is sufficient for $\xi_1 = \alpha$, and is UMVU for its expectation $E(W_1) = \alpha$.

(v) Follows from (iv).   □

Next we consider the case in which σ^2 is unknown and assume $n \geq r + 1$.

Theorem 6.1.3. *In the canonical Gaussian linear model with σ^2 unknown,*

(i) $\widetilde{\mathbf{T}} = \left(U_1, \ldots, U_r, \sum_{i=r+1}^{n} U_i^2 \right)^T$ *is sufficient for* $(\eta_1, \ldots, \eta_r, \sigma^2)^T$.

(ii) *The MLE of σ^2 is* $n^{-1} \sum_{i=r+1}^{n} U_i^2$.

(iii) $s^2 \equiv (n - r)^{-1} \sum_{i=r+1}^{n} U_i^2$ *is an unbiased estimator of σ^2.*

(iv) *The conclusions of Theorem 6.1.2 (ii),\ldots,(v) are still valid.*

Proof. By (6.1.11), $\left(U_1, \ldots, U_r, \sum_{i=1}^{n} U_i^2 \right)^T$ is sufficient. But because $\sum_{i=1}^{n} U_i^2 = \sum_{i=1}^{r} U_i^2 + \sum_{i=r+1}^{n} U_i^2$, this statistic is equivalent to $\widetilde{\mathbf{T}}$ and (i) follows. To show (ii), recall that the maximum of (6.1.11) has $\eta_i = U_i$, $i = 1, \ldots, r$. That is, we need to maximize

$$-\frac{1}{2\sigma^2} \sum_{i=r+1}^{n} U_i^2 - \frac{n}{2} (\log 2\pi\sigma^2)$$

as a function of σ^2. The maximizer is easily seen to be $n^{-1} \sum_{i=r+1}^{n} U_i^2$ (Problem 6.1.1). (iii) is clear because $EU_i^2 = \sigma^2$, $i \geq r+1$. To show (iv), apply Theorem 3.4.3 and Example 3.4.6 to the canonical exponential family obtained from (6.1.11) by setting $T_j = U_j$, $\theta_j = \eta_j / \sigma^2$, $j = 1, \ldots, r$, $T_{r+1} = \sum_{i=1}^{n} U_i^2$ and $\theta_{r+1} = -1/2\sigma^2$.

Projections

We next express $\widehat{\boldsymbol{\mu}}$ in terms of $\mathbf{Y}$, obtain the MLE $\widehat{\boldsymbol{\beta}}$ of $\boldsymbol{\beta}$, and give a geometric interpretation of $\widehat{\boldsymbol{\mu}}$, $\widehat{\boldsymbol{\beta}}$, and s^2. To this end, define the norm $|\mathbf{t}|$ of a vector $\mathbf{t} \in R_n$ by $|\mathbf{t}|^2 = \sum_{i=1}^{n} t_i^2$.

Definition 6.1.1. The *projection* $\mathbf{y}_0 = \pi(\mathbf{y} \mid \omega)$ of a point $\mathbf{y} \in R^n$ on ω is the point

$$\mathbf{y}_0 = \arg \min\{|\mathbf{y} - \mathbf{t}|^2 : \mathbf{t} \in \omega\}.$$

The maximum likelihood estimate $\widehat{\boldsymbol{\beta}}$ of $\boldsymbol{\beta}$ maximizes

$$\log p(\mathbf{y}, \boldsymbol{\beta}, \sigma) = -\frac{1}{2\sigma^2}|\mathbf{y} - \mathbf{Z}\boldsymbol{\beta}|^2 - \frac{n}{2}\log(2\pi\sigma^2)$$

or, equivalently,

$$\widehat{\boldsymbol{\beta}} = \arg \min\{|\mathbf{y} - \mathbf{Z}\boldsymbol{\beta}|^2 : \boldsymbol{\beta} \in R^p\}.$$

That is, the MLE of $\boldsymbol{\beta}$ equals the least squares estimate (LSE) of $\boldsymbol{\beta}$ defined in Example 2.1.1 and Section 2.2.1. We have

Theorem 6.1.4. *In the Gaussian linear model*

(i) $\widehat{\boldsymbol{\mu}}$ *is the unique projection of* $\mathbf{Y}$ *on* ω *and is given by*

$$\widehat{\boldsymbol{\mu}} = \mathbf{Z}\widehat{\boldsymbol{\beta}}. \tag{6.1.12}$$

(ii) $\widehat{\boldsymbol{\mu}}$ *is orthogonal to* $\mathbf{Y} - \widehat{\boldsymbol{\mu}}$.

(iii)

$$s^2 = |\mathbf{Y} - \widehat{\boldsymbol{\mu}}|^2/(n - r) \tag{6.1.13}$$

(iv) *If* $p = r$, *then* $\boldsymbol{\beta}$ *is identifiable,* $\boldsymbol{\beta} = (\mathbf{Z}^T\mathbf{Z})^{-1}\mathbf{Z}^T\boldsymbol{\mu}$, *the MLE = LSE of* $\boldsymbol{\beta}$ *is unique and given by*

$$\widehat{\boldsymbol{\beta}} = (\mathbf{Z}^T\mathbf{Z})^{-1}\mathbf{Z}^T\widehat{\boldsymbol{\mu}} = (\mathbf{Z}^T\mathbf{Z})^{-1}\mathbf{Z}^T\mathbf{Y}. \tag{6.1.14}$$

(v) *If* $p = r$, *then* $\widehat{\beta}_j$ *is the UMVU estimate of* β_j, $j = 1, \ldots, p$, *and* $\widehat{\mu}_i$ *is the UMVU estimate of* μ_i, $i = 1, \ldots, n$.

Proof. (i) is clear because $\mathbf{Z}\boldsymbol{\beta}$, $\boldsymbol{\beta} \in R^p$, spans ω. (ii) and (iii) are also clear from Theorem 6.1.3 because $\widehat{\boldsymbol{\mu}} = \sum_{i=1}^r v_i U_i$ and $\mathbf{Y} - \widehat{\boldsymbol{\mu}} = \sum_{j=r+1}^n v_j U_j$. To show (iv), note that $\boldsymbol{\mu} = \mathbf{Z}\boldsymbol{\beta}$ and (6.1.12) implies $\mathbf{Z}^T\boldsymbol{\mu} = \mathbf{Z}^T\mathbf{Z}\boldsymbol{\beta}$ and $\mathbf{Z}^T\widehat{\boldsymbol{\mu}} = \mathbf{Z}^T\mathbf{Z}\widehat{\boldsymbol{\beta}}$ and, because $\mathbf{Z}$ has full rank, $\mathbf{Z}^T\mathbf{Z}$ is nonsingular, and $\boldsymbol{\beta} = (\mathbf{Z}^T\mathbf{Z})^{-1}\mathbf{Z}^T\boldsymbol{\mu}$, $\widehat{\boldsymbol{\beta}} = (\mathbf{Z}^T\mathbf{Z})^{-1}\mathbf{Z}^T\widehat{\boldsymbol{\mu}}$. To show $\widehat{\boldsymbol{\beta}} = (\mathbf{Z}^T\mathbf{Z})^{-1}\mathbf{Z}^T\mathbf{Y}$, note that the space $\omega^\perp$ of vectors $\mathbf{s}$ orthogonal to ω can be written as

$$\omega^\perp = \{\mathbf{s} \in R^n : \mathbf{s}^T(\mathbf{Z}\boldsymbol{\beta}) = 0 \text{ for all } \boldsymbol{\beta} \in R^p\}.$$

It follows that $\boldsymbol{\beta}^T(\mathbf{Z}^T\mathbf{s}) = 0$ for all $\boldsymbol{\beta} \in R^p$, which implies $\mathbf{Z}^T\mathbf{s} = 0$ for all $\mathbf{s} \in \omega^\perp$. Thus, $\mathbf{Z}^T(\mathbf{Y} - \widehat{\boldsymbol{\mu}}) = 0$ and the second equality in (6.1.14) follows.

$\widehat{\beta}_j$ and $\widehat{\mu}_i$ are UMVU because, by (6.1.9), any linear combination of Y's is also a linear combination of U's, and by Theorems 6.1.2(iv) and 6.1.3(iv), any linear combination of U's is a UMVU estimate of its expectation. $\quad\square$

Note that in Example 2.1.1 we give an alternative derivation of $(6.1.14)$ and the normal equations $(\mathbf{Z}^T \mathbf{Z})\boldsymbol{\beta} = \mathbf{Z}^T \mathbf{Y}$.

The estimate $\widehat{\boldsymbol{\mu}} = \mathbf{Z}\widehat{\boldsymbol{\beta}}$ of $\boldsymbol{\mu}$ is called the *fitted value* and $\widehat{\boldsymbol{\epsilon}} = \mathbf{Y} - \widehat{\boldsymbol{\mu}}$ is called the *residual* from this fit. The goodness of the fit is measured by the *residual sum of squares* (RSS) $|\mathbf{Y} - \widehat{\boldsymbol{\mu}}|^2 = \sum_{i=1}^{n} \widehat{\epsilon}_i^2$. Example 2.2.2 illustrates this terminology in the context of Example 6.1.2 with $p = 2$. There the points $\widehat{\mu}_i = \widehat{\beta}_1 + \widehat{\beta}_2 z_i$, $i = 1, \ldots, n$ lie on the regression line fitted to the data $\{(z_i, y_i), \ i = 1, \ldots, n\}$; moreover, the residuals $\widehat{\epsilon}_i = [y_i - (\widehat{\beta}_1 + \widehat{\beta}_2 z_i)]$ are the vertical distances from the points to the fitted line.

Suppose we are given a value of the covariate $\mathbf{z}$ at which a value Y following the linear model $(6.1.3)$ is to be taken. By Theorem 1.4.1, the best MSPE predictor of Y if $\boldsymbol{\beta}$ is known as well as $\mathbf{z}$ is $E(Y) = \mathbf{z}^T \boldsymbol{\beta}$ and its best (UMVU) estimate not knowing $\boldsymbol{\beta}$ is $\widehat{Y} \equiv \mathbf{z}^T \widehat{\boldsymbol{\beta}}$. Taking $\mathbf{z} = \mathbf{z}_i$, $1 \leq i \leq n$, we obtain $\widehat{\mu}_i = \mathbf{z}_i \widehat{\boldsymbol{\beta}}$, $1 \leq i \leq n$. In this method of "prediction" of Y_i, it is common to write $\widehat{Y}_i$ for $\widehat{\mu}_i$, the ith component of the fitted value $\widehat{\boldsymbol{\mu}}$. That is, $\widehat{\mathbf{Y}} = \widehat{\boldsymbol{\mu}}$. Note that by $(6.1.12)$ and $(6.1.14)$, when $p = r$,

$$\widehat{\mathbf{Y}} = \mathbf{H}\mathbf{Y}$$

where

$$\mathbf{H} = \mathbf{Z}(\mathbf{Z}^T \mathbf{Z})^{-1} \mathbf{Z}^T.$$

The matrix $\mathbf{H}$ is the *projection matrix* mapping R^n into ω, see also Section B.10. In statistics it is also called the *hat matrix* because it "puts the hat on $\mathbf{Y}$." As a projection matrix $\mathbf{H}$ is necessarily symmetric and idempotent,

$$\mathbf{H}^T = \mathbf{H}, \ \mathbf{H}^2 = \mathbf{H}.$$

It follows from this and (B.5.3) that if $\mathbf{J} = \mathbf{J}_{n \times n}$ is the identity matrix, then

$$\mathrm{Var}(\widehat{\mathbf{Y}}) = \mathbf{H}(\sigma^2 \mathbf{J})\mathbf{H}^T = \sigma^2 \mathbf{H}. \tag{6.1.15}$$

Next note that the residuals can be written as

$$\widehat{\boldsymbol{\epsilon}} = \mathbf{Y} - \widehat{\mathbf{Y}} = (\mathbf{J} - \mathbf{H})\mathbf{Y}.$$

The residuals are the projection of $\mathbf{Y}$ on the orthocomplement of ω and

$$\mathrm{Var}(\widehat{\boldsymbol{\epsilon}}) = \sigma^2(\mathbf{J} - \mathbf{H}). \tag{6.1.16}$$

We can now conclude the following.

Corollary 6.1.1. *In the Gaussian linear model*

(i) *the fitted values $\widehat{\mathbf{Y}} = \widehat{\boldsymbol{\mu}}$ and the residual $\widehat{\boldsymbol{\epsilon}}$ are independent,*

(ii) $\widehat{\mathbf{Y}} \sim \mathcal{N}(\boldsymbol{\mu}, \sigma^2 \mathbf{H})$,

(iii) $\widehat{\boldsymbol{\epsilon}} \sim \mathcal{N}(\mathbf{0}, \sigma^2(\mathbf{J} - \mathbf{H}))$, *and*

(iv) *if $p = r$, $\widehat{\boldsymbol{\beta}} \sim \mathcal{N}(\boldsymbol{\beta}, \sigma^2(\mathbf{Z}^T \mathbf{Z})^{-1})$.*

Proof. $(\widehat{\mathbf{Y}}, \widehat{\boldsymbol{\epsilon}})$ is a linear transformation of $\mathbf{U}$ and, hence, joint Gaussian. The independence follows from the identification of $\widehat{\boldsymbol{\mu}}$ and $\widehat{\boldsymbol{\epsilon}}$ in terms of the U_i in the theorem. $\operatorname{Var}(\widehat{\boldsymbol{\beta}}) = \sigma^2 (\mathbf{Z}^T \mathbf{Z})^{-1}$ follows from (B.5.3). $\qquad\square$

We now return to our examples.

Example 6.1.1. *One Sample (continued).* Here $\mu = \beta_1$ and $\widehat{\mu} = \widehat{\beta}_1 = \bar{Y}$. Moreover, the unbiased estimator s^2 of σ^2 is $\sum_{i=1}^{n} (Y_i - \bar{Y})^2 / (n-1)$, which we have seen before in Problem 1.3.8 and (3.4.2). $\qquad\square$

Example 6.1.2. *Regression (continued).* If the design matrix $\mathbf{Z}$ has rank p, then the MLE = LSE estimate is $\widehat{\boldsymbol{\beta}} = (\mathbf{Z}^T \mathbf{Z})^{-1} \mathbf{Z}^T \mathbf{Y}$ as seen before in Example 2.1.1 and Section 2.2.1. We now see that the MLE of $\boldsymbol{\mu}$ is $\widehat{\boldsymbol{\mu}} = \mathbf{Z}\widehat{\boldsymbol{\beta}}$ and that $\widehat{\beta}_j$ and $\widehat{\mu}_i$ are UMVU for β_j and μ_i respectively, $j = 1, \ldots, p, i = 1, \ldots, n$. The variances of $\widehat{\boldsymbol{\beta}}, \widehat{\boldsymbol{\mu}} = \widehat{\mathbf{Y}}$, and $\widehat{\boldsymbol{\epsilon}} = \mathbf{Y} - \widehat{\mathbf{Y}}$ are given in Corollary 6.1.1. In the Gaussian case $\widehat{\boldsymbol{\beta}}, \widehat{\mathbf{Y}}$, and $\widehat{\boldsymbol{\epsilon}}$ are normally distributed with $\widehat{\mathbf{Y}}$ and $\widehat{\boldsymbol{\epsilon}}$ independent. The error variance $\sigma^2 = \operatorname{Var}(\epsilon_1)$ can be unbiasedly estimated by $s^2 = (n-p)^{-1} |\mathbf{Y} - \widehat{\boldsymbol{\mu}}|^2$. $\qquad\square$

Example 6.1.3. *The One-Way Layout (continued).* In this example the normal equations $(\mathbf{Z}^T \mathbf{Z})\boldsymbol{\beta} = \mathbf{Z}\mathbf{Y}$ become

$$n_k \beta_k = \sum_{l=1}^{n_k} Y_{kl}, \ k = 1, \ldots, p.$$

At this point we introduce an important notational convention in statistics. If $\{c_{ijk} \ldots\}$ is a multiple-indexed sequence of numbers or variables, then replacement of a subscript by a dot indicates that we are considering the average over that subscript. Thus,

$$Y_{k\cdot} = \frac{1}{n_k} \sum_{l=1}^{n_k} Y_{kl}, \ Y_{\cdot\cdot} = \frac{1}{n} \sum_{k=1}^{p} \sum_{l=1}^{n_k} Y_{kl}$$

where $n = n_1 + \cdots + n_p$ and we can write the least squares estimates as

$$\widehat{\beta}_k = Y_{k\cdot}, \ k = 1, \ldots, p.$$

By Theorem 6.1.3, in the Gaussian model, the UMVU estimate of the average effect of all the treatments, $\alpha = \beta_{\cdot}$, is

$$\widehat{\alpha} = \frac{1}{p} \sum_{k=1}^{p} Y_{k\cdot} \text{ (not } Y_{\cdot\cdot} \text{ in general)}$$

and the UMVU estimate of the incremental effect $\delta_k = \beta_k - \alpha$ of the kth treatment is

$$\widehat{\delta}_k = Y_{k\cdot} - \widehat{\alpha}, \ k = 1, \ldots, p.$$

$\qquad\square$

Remark 6.1.1. An alternative approach to the MLEs for the normal model and the associated LSEs of this section is an approach based on MLEs for the model in which the errors $\epsilon_1, \ldots, \epsilon_n$ in (6.1.1) have the Laplace distribution with density

$$\frac{1}{2\sigma} \exp\left\{-\frac{1}{\sigma}|t|\right\}$$

and the estimates of $\boldsymbol{\beta}$ and $\boldsymbol{\mu}$ are least absolute deviation estimates (LADEs) obtained by minimizing the absolute deviation distance $\sum_{i=1}^{n} |y_i - \mathbf{z}_i^T \boldsymbol{\beta}|$. The LADEs were introduced by Laplace before Gauss and Legendre introduced the LSEs—see Stigler (1986). The LSEs are preferred because of ease of computation and their geometric properties. However, the LADEs are obtained fairly quickly by modern computing methods; see Koenker and D'Orey (1987) and Portnoy and Koenker (1997). For more on LADEs, see Problems 1.4.7 and 2.2.31.

6.1.3   Tests and Confidence Intervals

The most important hypothesis-testing questions in the context of a linear model correspond to restriction of the vector of means $\boldsymbol{\mu}$ to a linear subspace of the space ω, which together with σ^2 specifies the model. For instance, in a study to investigate whether a drug affects the mean of a response such as blood pressure we may consider, in the context of Example 6.1.2, a regression equation of the form

$$\text{mean response} = \beta_1 + \beta_2 z_{i2} + \beta_3 z_{i3}, \tag{6.1.17}$$

where z_{i2} is the dose level of the drug given the ith patient, z_{i3} is the age of the ith patient, and the matrix $\|z_{ij}\|_{n \times 3}$ with $z_{i1} = 1$ has rank 3. Now we would test $H : \beta_2 = 0$ versus $K : \beta_2 \neq 0$. Thus, under H, $\{\boldsymbol{\mu} : \mu_i = \beta_1 + \beta_3 z_{i3}, \ i = 1, \ldots, n\}$ is a two-dimensional linear subspace of the full model's three-dimensional linear subspace of R^n given by (6.1.17).

Next consider the p-sample model of Example 1.6.3 with β_k representing the mean response for the kth population. The first inferential question is typically "Are the means equal or not?" Thus we test $H : \beta_1 = \cdots = \beta_p = \beta$ for some $\beta \in R$ versus K: "the β's are not all equal." Now, under H, the mean vector is an element of the space $\{\boldsymbol{\mu} : \mu_i = \beta \in R, \ i = 1, \ldots, p\}$, which is a one-dimensional subspace of R^n, whereas for the full model $\boldsymbol{\mu}$ is in a p-dimensional subspace of R^n.

In general, we let ω correspond to the full model with dimension r and let ω_0 be a q-dimensional linear subspace over which $\boldsymbol{\mu}$ can range under the null hypothesis H; $0 \leq q < r$.

We first consider the σ^2 known case and consider the likelihood ratio statistic

$$\lambda(\mathbf{y}) = \frac{\sup\{p(\mathbf{y}, \boldsymbol{\mu}) : \boldsymbol{\mu} \in \omega\}}{\sup\{p(\mathbf{y}, \boldsymbol{\mu}) : \boldsymbol{\mu} \in \omega_0\}}$$

for testing $H : \boldsymbol{\mu} \in \omega_0$ versus $K : \boldsymbol{\mu} \in \omega - \omega_0$. Because

$$p(\mathbf{Y}, \boldsymbol{\mu}) = (2\pi\sigma^2)^{-\frac{n}{2}} \exp\left\{ -\frac{1}{2\sigma^2} |\mathbf{Y} - \boldsymbol{\mu}|^2 \right\} \tag{6.1.18}$$

then, by Theorem 6.1.4,

$$\lambda(\mathbf{Y}) = \exp\left\{ -\frac{1}{2\sigma^2} \left(|\mathbf{Y} - \widehat{\boldsymbol{\mu}}|^2 - |\mathbf{Y} - \widehat{\boldsymbol{\mu}}_0|^2 \right) \right\}$$

where $\widehat{\boldsymbol{\mu}}$ and $\widehat{\boldsymbol{\mu}}_0$ are the projections of $\mathbf{Y}$ on ω and ω_0, respectively.

But if we let $\mathbf{A}_{n \times n}$ be an orthogonal matrix with rows $\mathbf{v}_1^T, \ldots, \mathbf{v}_n^T$ such that $\mathbf{v}_1, \ldots, \mathbf{v}_q$ span ω_0 and $\mathbf{v}_1, \ldots, \mathbf{v}_r$ span ω and set

$$\mathbf{U} = \mathbf{A}\mathbf{Y}, \ \boldsymbol{\eta} = \mathbf{A}\boldsymbol{\mu} \tag{6.1.19}$$

then, by Theorem 6.1.2(v),

$$\lambda(\mathbf{Y}) = \exp\left\{ \frac{1}{2\sigma^2} \sum_{i=q+1}^{r} U_i^2 \right\} = \exp\left\{ \frac{1}{2\sigma^2} |\widehat{\boldsymbol{\mu}} - \widehat{\boldsymbol{\mu}}_0|^2 \right\}. \tag{6.1.20}$$

It follows that

$$2 \log \lambda(\mathbf{Y}) = \sum_{i=q+1}^{r} (U_i/\sigma)^2.$$

Note that (U_i/σ) has a $\mathcal{N}(\theta_i, 1)$ distribution with $\theta_i = \eta_i/\sigma$. In this case the distribution of $\sum_{i=q+1}^{r}(U_i/\sigma)^2$ is called a *chi-square distribution with $r - q$ degrees of freedom* and *noncentrality parameter* $\theta^2 = |\boldsymbol{\theta}|^2 = \sum_{i=q+1}^{r} \theta_i^2$, where $\boldsymbol{\theta} = (\theta_{q+1}, \ldots, \theta_r)^T$ (see Problem B.3.12). We write $\chi_{r-q}^2(\theta^2)$ for this distribution. We have shown the following.

Proposition 6.1.1. *In the Gaussian linear model with σ^2 known, $2\log \lambda(\mathbf{Y})$ has a $\chi_{r-q}^2(\theta^2)$ distribution with*

$$\theta^2 = \sigma^{-2} \sum_{i=q+1}^{r} \eta_i^2 = \sigma^{-2}|\boldsymbol{\mu} - \boldsymbol{\mu}_0|^2 \tag{6.1.21}$$

where $\boldsymbol{\mu}_0$ is the projection of $\boldsymbol{\mu}$ on ω_0. In particular, when H holds, $2\log \lambda(\mathbf{Y}) \sim \chi_{r-q}^2$.

Proof. We only need to establish the second equality in (6.1.21). Write $\boldsymbol{\eta} = \mathbf{A}\boldsymbol{\mu}$ where $\mathbf{A}$ is as defined in (6.1.19), then

$$\sum_{i=q+1}^{r} \eta_i^2 = |\boldsymbol{\mu} - \boldsymbol{\mu}_0|^2.$$

$\square$

Next consider the case in which σ^2 is unknown. We know from Problem 6.1.1 that the MLEs of σ^2 for $\mu \in \omega$ and $\mu \in \omega_0$ are

$$\widehat{\sigma}^2 = \frac{1}{n}|\mathbf{Y} - \widehat{\boldsymbol{\mu}}|^2 \text{ and } \widehat{\sigma}_0^2 = \frac{1}{n}|\mathbf{Y} - \widehat{\boldsymbol{\mu}}_0|^2,$$

respectively. Substituting $\widehat{\boldsymbol{\mu}}, \widehat{\boldsymbol{\mu}}_0, \widehat{\sigma}^2$, and $\widehat{\sigma}_0^2$ into the likelihood ratio statistic, we obtain

$$\lambda(\mathbf{y}) = \frac{p(\mathbf{y}, \widehat{\boldsymbol{\mu}}, \widehat{\sigma}^2)}{p(\mathbf{y}, \widehat{\boldsymbol{\mu}}_0, \widehat{\sigma}_0^2)} = \left\{ \frac{|\mathbf{y} - \widehat{\boldsymbol{\mu}}_0|^2}{|\mathbf{y} - \widehat{\boldsymbol{\mu}}|^2} \right\}^{\frac{n}{2}}$$

where $p(\mathbf{y}, \boldsymbol{\mu}, \sigma^2)$ denotes the right-hand side of (6.1.18).

The resulting test is intuitive. It consists of rejecting H when the fit, as measured by the residual sum of squares under the model specified by H, is poor compared to the fit under the general model. For the purpose of finding critical values it is more convenient to work with a statistic equivalent to $\lambda(\mathbf{Y})$,

$$T = \frac{n - r}{r - q} \frac{|\mathbf{Y} - \widehat{\boldsymbol{\mu}}_0|^2 - |\mathbf{Y} - \widehat{\boldsymbol{\mu}}|^2}{|\mathbf{Y} - \widehat{\boldsymbol{\mu}}|^2} = \frac{(r - q)^{-1}|\widehat{\boldsymbol{\mu}} - \widehat{\boldsymbol{\mu}}_0|^2}{(n - r)^{-1}|\mathbf{Y} - \widehat{\boldsymbol{\mu}}|^2}. \tag{6.1.22}$$

Because $T = (n - r)(r - q)^{-1}\{[\lambda(\mathbf{Y})]^{2/n} - 1\}$, T is an increasing function of $\lambda(\mathbf{Y})$ and the two test statistics are equivalent. T is called *the F statistic* for the general linear hypothesis.

We have seen in Proposition 6.1.1 that $\sigma^{-2}|\widehat{\boldsymbol{\mu}} - \widehat{\boldsymbol{\mu}}_0|^2$ have a $\chi_{r-q}^2(\theta^2)$ distribution with $\theta^2 = \sigma^{-2}|\boldsymbol{\mu} - \boldsymbol{\mu}_0|^2$. By the canonical representation (6.1.19), we can write $\sigma^{-2}|\mathbf{Y} - \widehat{\boldsymbol{\mu}}|^2 = \sum_{i=r+1}^{n}(U_i/\sigma)^2$, which has a χ_{n-r}^2 distribution and is independent of $\sigma^{-2}|\widehat{\boldsymbol{\mu}} - \widehat{\boldsymbol{\mu}}_0|^2 = \sum_{i=q+1}^{r}(U_i/\sigma)^2$. Thus, T has the representation

$$T = \frac{(\text{noncentral } \chi_{r-q}^2 \text{ variable})/df}{(\text{central } \chi_{n-r}^2 \text{ variable})/df}$$

with the numerator and denominator independent. The distribution of such a variable is called the *noncentral F distribution with noncentrality parameter θ^2 and $r - q$ and $n - r$ degrees of freedom* (see Problem B.3.14). We write $\mathcal{F}_{k,m}(\theta^2)$ for this distribution where $k = r - q$ and $m = n - r$. We have shown the following.

Proposition 6.1.2. *In the Gaussian linear model the F statistic defined by (6.1.22), which is equivalent to the likelihood ratio statistic for $H : \boldsymbol{\mu} \in \omega_0$ for $K : \boldsymbol{\mu} \in \omega - \omega_0$, has the noncentral F distribution $\mathcal{F}_{r-q,n-r}(\theta^2)$ where $\theta^2 = \sigma^{-2}|\boldsymbol{\mu} - \boldsymbol{\mu}_0|^2$. In particular, when H holds, T has the (central) $\mathcal{F}_{r-q,n-r}$ distribution.*

Remark 6.1.2. In Proposition 6.1.1 suppose the assumption "σ^2 is known" is replaced by "σ^2 is the same under H and K and estimated by the MLE $\widehat{\sigma}^2$ for $\mu \in \omega$." In this case, it can be shown (Problem 6.1.5) that if we introduce the *variance equal likelihood ratio statistic*

$$\widetilde{\lambda}(\mathbf{y}) = \frac{\max\{p(\mathbf{y}, \boldsymbol{\mu}, \widehat{\sigma}^2) : \boldsymbol{\mu} \in \omega\}}{\max\{p(\mathbf{y}, \boldsymbol{\mu}, \widehat{\sigma}^2) : \boldsymbol{\mu} \in \omega_0\}} \tag{6.1.23}$$

then $\widetilde{\lambda}(\mathbf{Y})$ equals the likelihood ratio statistic for the σ^2 known case with σ^2 replaced by $\widehat{\sigma}^2$. It follows that

$$2 \log \widetilde{\lambda}(\mathbf{Y}) = \frac{r-q}{(n-r)/n} T = \frac{\text{noncentral } \chi^2_{r-q}}{\text{central } \chi^2_{n-r}/n} \tag{6.1.24}$$

where T is the F statistic (6.1.22).

Remark 6.1.3. The canonical representation (6.1.19) made it possible to recognize the identity

$$|\mathbf{Y} - \widehat{\boldsymbol{\mu}}_0|^2 = |\mathbf{Y} - \widehat{\boldsymbol{\mu}}|^2 + |\widehat{\boldsymbol{\mu}} - \widehat{\boldsymbol{\mu}}_0|^2, \tag{6.1.25}$$

which we exploited in the preceding derivations. This is the *Pythagorean identity*. See Figure 6.1.1 and Section B.10.

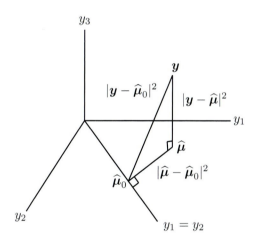

Figure 6.1.1. The projections $\widehat{\boldsymbol{\mu}}$ and $\widehat{\boldsymbol{\mu}}_0$ of $\mathbf{Y}$ on ω and ω_0; and the Pythagorean identity.

We next return to our examples.

Example 6.1.1. *One Sample (continued).* We test $H : \beta_1 = \mu_0$ versus $K : \beta \neq \mu_0$. In this case $\omega_0 = \{\mu_0\}$, $q = 0$, $r = 1$ and

$$T = \frac{(\bar{Y} - \mu_0)^2}{(n-1)^{-1}\Sigma(Y_i - \bar{Y})^2},$$

which we recognize as t^2/n, where t is the one-sample Student t statistic of Section 4.9.2.   □

Example 6.1.2. *Regression (continued).* We consider the possibility that a subset of $p - q$ covariates does not affect the mean response. Without loss of generality we ask whether the last $p - q$ covariates in multiple regression have an effect after fitting the first q. To formulate this question, we partition the design matrix $\mathbf{Z}$ by writing it as $\mathbf{Z} = (\mathbf{Z}_1, \mathbf{Z}_2)$ where $\mathbf{Z}_1$ is $n \times q$ and $\mathbf{Z}_2$ is $n \times (p - q)$, and we partition $\boldsymbol{\beta}$ as $\boldsymbol{\beta}^T = (\boldsymbol{\beta}_1^T, \boldsymbol{\beta}_2^T)$ where $\boldsymbol{\beta}_2$ is a $(p - q) \times 1$ vector of main (e.g., treatment) effect coefficients and $\boldsymbol{\beta}_1$ is a $q \times 1$ vector of "nuisance" (e.g., age, economic status) coefficients. Now the linear model can be written as

$$\mathbf{Y} = \mathbf{Z}_1\boldsymbol{\beta}_1 + \mathbf{Z}_2\boldsymbol{\beta}_2 + \boldsymbol{\epsilon}. \tag{6.1.26}$$

We test $H : \boldsymbol{\beta}_2 = \mathbf{0}$ versus $K : \boldsymbol{\beta}_2 \neq \mathbf{0}$. In this case $\widehat{\boldsymbol{\beta}} = (\mathbf{Z}^T\mathbf{Z})^{-1}\mathbf{Z}^T\mathbf{Y}$ and $\widehat{\boldsymbol{\beta}}_1 = (\mathbf{Z}_1^T\mathbf{Z}_1)^{-1}\mathbf{Z}_1^T\mathbf{Y}$ are the MLEs under the full model (6.1.26) and H, respectively. Using (6.1.22) we can write the F statistic version of the likelihood ratio test in the intuitive form

$$F = \frac{(RSS_H - RSS_F)/(df_H - df_F)}{RSS_F/df_F}$$

where $RSS_F = |\mathbf{Y} - \widehat{\boldsymbol{\mu}}|^2$ and $RSS_H = |\mathbf{Y} - \widehat{\boldsymbol{\mu}}_0|^2$ are the residual sums of squares under the full model and H, respectively; and $df_F = n - p$ and $df_H = n - q$ are the corresponding degrees of freedom. The F test rejects H if F is large when compared to the αth quantile of the $\mathcal{F}_{p-q,n-p}$ distribution.

Under the alternative F has a noncentral $\mathcal{F}_{p-q,n-p}(\theta^2)$ distribution with noncentrality parameter (Problem 6.1.7)

$$\theta^2 = \sigma^{-2}\boldsymbol{\beta}_2^T\{\mathbf{Z}_2^T\mathbf{Z}_2 - \mathbf{Z}_2^T\mathbf{Z}_1(\mathbf{Z}_1^T\mathbf{Z}_1)^{-1}\mathbf{Z}_1^T\mathbf{Z}_2\}\boldsymbol{\beta}_2. \tag{6.1.27}$$

In the special case that $\mathbf{Z}_1^T\mathbf{Z}_2 = \mathbf{0}$ so the variables in $\mathbf{Z}_1$ are orthogonal to the variables in $\mathbf{Z}_2$, θ^2 simplifies to $\sigma^{-2}\boldsymbol{\beta}_2^T(\mathbf{Z}_2^T\mathbf{Z}_2)\boldsymbol{\beta}_2$, which only depends on the second set of variables and coefficients. However, in general θ^2 depends on the sample correlations between the variables in $\mathbf{Z}_1$ and those in $\mathbf{Z}_2$. This issue is discussed further in Example 6.2.1. □

Example 6.1.3. *The One-Way Layout (continued).* Recall that the least squares estimates of $\beta_1, \ldots, \beta_p$ are $Y_{1\cdot}, \ldots, Y_{p\cdot}$. As we indicated earlier, we want to test $H : \beta_1 = \cdots = \beta_p$. Under H all the observations have the same mean so that,

$$\widehat{\boldsymbol{\mu}}_0 = (Y_{\cdot\cdot}, \ldots, Y_{\cdot\cdot})^T.$$

Thus,

$$|\widehat{\boldsymbol{\mu}} - \boldsymbol{\mu}_0|^2 = \sum_{k=1}^{p}\sum_{l=1}^{n_k}(Y_{k\cdot} - Y_{\cdot\cdot})^2 = \sum_{k=1}^{p}n_k(Y_{k\cdot} - Y_{\cdot\cdot})^2.$$

Substituting in (6.1.22) we obtain the F statistic for the hypothesis H in the one-way layout

$$T = \frac{n-p}{p-1}\frac{\sum_{k=1}^{p}n_k(Y_{k\cdot} - Y_{\cdot\cdot})^2}{\sum_{k=1}^{p}\sum_{l=1}^{n_k}(Y_{kl} - Y_{k\cdot})^2}.$$

When H holds, T has a $\mathcal{F}_{p-1,n-p}$ distribution. If the β_i are not all equal, T has a noncentral $\mathcal{F}_{p-1,n-p}$ distribution with noncentrality parameter

$$\delta^2 = \frac{1}{\sigma^2} \sum_{k=1}^{p} n_k (\beta_k - \bar{\beta})^2, \tag{6.1.28}$$

where $\bar{\beta} = n^{-1} \sum_{i=1}^{p} n_i \beta_i$. To derive δ^2, compute $\sigma^{-2}|\boldsymbol{\mu} - \boldsymbol{\mu}_0|^2$ for the vector $\boldsymbol{\mu} = (\beta_1, \ldots, \beta_1, \beta_2, \ldots, \beta_2, \ldots, \beta_p, \ldots, \beta_p)^T$ and its projection $\boldsymbol{\mu}_0 = (\bar{\beta}, \ldots, \bar{\beta})^T$.

There is an interesting way of looking at the pieces of information summarized by the F statistic. The sum of squares in the numerator,

$$SS_B = \sum_{k=1}^{p} n_k (Y_{k\cdot} - Y_{\cdot\cdot})^2$$

is a measure of variation *between* the p samples $Y_{11}, \ldots, Y_{1n_1}, \ldots, Y_{p1}, \ldots, Y_{pn_p}$. The sum of squares in the denominator,

$$SS_W = \sum_{k=1}^{p} \sum_{l=1}^{n_k} (Y_{kl} - Y_{k\cdot})^2,$$

measures variation *within* the samples. If we define the *total* sum of squares as

$$SS_T = \sum_{k=1}^{p} \sum_{l=1}^{n_k} (Y_{kl} - Y_{\cdot\cdot})^2,$$

which measures the variability of the pooled samples, then by the Pythagorean identity (6.1.25)

$$SS_T = SS_B + SS_W. \tag{6.1.29}$$

Thus, we have a decomposition of the variability of the whole set of data, SS_T, the *total* sum of squares, into two constituent components, SS_B, the *between groups* (or treatment) sum of squares and SS_W, the *within groups* (or *residual*) sum of squares. SS_T/σ^2 is a (noncentral) χ^2 variable with $(n-1)$ degrees of freedom and noncentrality parameter δ^2. Because SS_B/σ^2 and SS_W/σ^2 are independent χ^2 variables with $(p-1)$ and $(n-p)$ degrees of freedom, respectively, we see that the decomposition (6.1.29) can also be viewed stochastically, identifying δ^2 and $(p-1)$ degrees of freedom as "coming" from SS_B/σ^2 and the remaining $(n-p)$ of the $(n-1)$ degrees of freedom of SS_T/σ^2 as "coming" from SS_W/σ^2.

This information as well as $SS_B/(p-1)$ and $SS_W/(n-p)$, the unbiased estimates of δ^2 and σ^2, and the F statistic, which is their ratio, are often summarized in what is known as an *analysis of variance* (ANOVA) table. See Tables 6.1.1 and 6.1.3.

As an illustration, consider the following data[1] giving blood cholesterol levels of men in three different socioeconomic groups labeled I, II, and III with I being the "high" end. We assume the one-way layout is valid. Note that this implies the possibly unrealistic

TABLE 6.1.1. ANOVA table for the one-way layout

	Sum of squares	d.f.	Mean squares	F-value
Between samples	$SS_B = \sum_{k=1}^{p} n_k(Y_{k\cdot} - Y_{\cdot\cdot})^2$	$p-1$	$MS_B = \frac{SS_B}{p-1}$	$\frac{MS_B}{MS_W}$
Within samples	$SS_W = \sum_{k=1}^{p} \sum_{l=1}^{n-k}(Y_{kl} - Y_{k\cdot})^2$	$n-p$	$MS_W = \frac{SS_W}{n-p}$	
Total	$SS_T = \sum_{k=1}^{p} \sum_{l=1}^{n_k}(Y_{kl} - Y_{\cdot\cdot})^2$	$n-1$		

TABLE 6.1.2. Blood cholesterol levels

I	403	311	269	336	259				
II	312	222	302	420	420	386	353	210	286	290
III	403	244	353	235	319	260			

assumption that the variance of the measurement is the same in the three groups (not to speak of normality). But see Section 6.6 for "robustness" to these assumptions.

We want to test whether there is a significant difference among the mean blood cholesterol of the three groups. Here $p = 3$, $n_1 = 5$, $n_2 = 10$, $n_3 = 6$, $n = 21$, and we compute

TABLE 6.1.3. ANOVA table for the cholesterol data

	SS	d.f.	MS	F-value
Between groups	1202.5	2	601.2	0.126
Within groups	85,750.5	18	4763.9	
Total	86,953.0	20		

From $\mathcal{F}$ tables, we find that the p-value corresponding to the F-value 0.126 is 0.88. Thus, there is no evidence to indicate that mean blood cholesterol is different for the three socioeconomic groups.   □

Remark 6.1.4. Decompositions such as (6.1.29) of the response total sum of squares SS_T into a variety of sums of squares measuring variability in the observations corresponding to variation of covariates are referred to as *analysis of variance*. They can be formulated in any linear model including regression models. See Scheffé (1959, pp. 42–45) and Weisberg (1985, p. 48). Originally such decompositions were used to motivate F statistics and to establish the distribution theory of the components via a device known as Cochran's theorem (Graybill, 1961, p. 86). Their principal use now is in the motivation of the convenient summaries of information we call ANOVA tables.

Confidence Intervals and Regions

We next use our distributional results and the method of pivots to find confidence intervals for μ_i, $1 \leq i \leq n$, β_j, $1 \leq j \leq p$, and in general, any linear combination

$$\psi = \psi(\boldsymbol{\mu}) = \sum_{i=1}^{n} a_i \mu_i = \mathbf{a}^T \boldsymbol{\mu}$$

of the μ's. If we set $\widehat{\psi} = \sum_{i=1}^{n} a_i \widehat{\mu}_i = \mathbf{a}^T \widehat{\boldsymbol{\mu}}$ and

$$\sigma^2(\widehat{\psi}) = \text{Var}(\widehat{\psi}) = \mathbf{a}^T \text{Var}(\widehat{\boldsymbol{\mu}})\mathbf{a} = \sigma^2 \mathbf{a}^T \mathbf{H}\mathbf{a},$$

where H is the hat matrix, then $(\widehat{\psi} - \psi)/\sigma(\widehat{\psi})$ has a $\mathcal{N}(0,1)$ distribution. Moreover,

$$(n-r)s^2/\sigma^2 = |\mathbf{Y} - \widehat{\boldsymbol{\mu}}|^2/\sigma^2 = \sum_{i=r+1}^{n} (U_i/\sigma)^2$$

has a χ^2_{n-r} distribution and is independent of $\widehat{\psi}$. Let

$$\widehat{\sigma}(\widehat{\psi}) = (s^2 \mathbf{a}^T \mathbf{H}\mathbf{a})^{\frac{1}{2}}$$

be an estimate of the standard deviation $\sigma(\widehat{\psi})$ of $\widehat{\psi}$. This estimated standard deviation is called *the standard error* of $\widehat{\psi}$. By referring to the definition of the t distribution, we find that the pivot

$$T(\psi) = \frac{(\widehat{\psi} - \psi)/\sigma(\widehat{\psi})}{(s/\sigma)} = (\widehat{\psi} - \psi)/\widehat{\sigma}(\widehat{\psi})$$

has a $\mathcal{T}_{n-r}$ distribution. Let $t_{n-r}\left(1 - \frac{1}{2}\alpha\right)$ denote the $1 - \frac{1}{2}\alpha$ quantile of the $\mathcal{T}_{n-r}$ distribution, then by solving $|T(\psi)| \leq t_{n-r}\left(1 - \frac{1}{2}\alpha\right)$ for ψ, we find that

$$\psi = \widehat{\psi} \pm t_{n-r}\left(1 - \tfrac{1}{2}\alpha\right)\widehat{\sigma}(\widehat{\psi})$$

is, in the Gaussian linear model, a $100(1 - \alpha)\%$ confidence interval for ψ.

Example 6.1.1. *One Sample (continued).* Consider $\psi = \mu$. We obtain the interval

$$\mu = \bar{Y} \pm t_{n-1}\left(1 - \tfrac{1}{2}\alpha\right)s/\sqrt{n},$$

which is the same as the interval of Example 4.4.1 and Section 4.9.2.

Example 6.1.2. *Regression (continued).* Assume that $p = r$. First consider $\psi = \beta_j$ for some specified regression coefficient β_j. The $100(1 - \alpha)\%$ confidence interval for β_j is

$$\beta_j = \widehat{\beta}_j \pm t_{n-p}\left(1 - \tfrac{1}{2}\alpha\right)s\{[(\mathbf{Z}^T\mathbf{Z})^{-1}]_{jj}\}^{\frac{1}{2}}$$

where $[(\mathbf{Z}^T\mathbf{Z})^{-1}]_{jj}$ is the jth diagonal element of $(\mathbf{Z}^T\mathbf{Z})^{-1}$. Computer software computes $(\mathbf{Z}^T\mathbf{Z})^{-1}$ and labels $s\{[(\mathbf{Z}^T\mathbf{Z})^{-1}]_{jj}\}^{\frac{1}{2}}$ as the *standard error* of the (estimated) jth regression coefficient. Next consider $\psi = \mu_i = $ mean response for the ith case, $1 \leq i \leq n$. The level $(1-\alpha)$ confidence interval is

$$\mu_i = \widehat{\mu}_i \pm t_{n-p}\left(1 - \tfrac{1}{2}\alpha\right) s\sqrt{h_{ii}}$$

where h_{ii} is the ith diagonal element of the hat matrix $\mathbf{H}$. Here $s\sqrt{h_{ii}}$ is called the standard error of the (estimated) mean of the ith case.

Next consider the special case in which $p = 2$ and

$$Y_i = \beta_1 + \beta_2 z_{i2} + \epsilon_i, \ i = 1, \ldots, n.$$

If we use the identity

$$\sum_{i=1}^{n}(z_{i2} - z_{\cdot 2})(Y_i - \bar{Y}) = \sum(z_{i2} - z_{\cdot 2})Y_i - \bar{Y}\sum(z_{i2} - z_{\cdot 2})$$

$$= \sum(z_{i2} - z_{\cdot 2})Y_i,$$

we obtain from Example 2.2.2 that

$$\widehat{\beta}_2 = \frac{\sum_{i=1}^{n}(z_{i2} - z_{\cdot 2})Y_i}{\sum_{i=1}^{n}(z_{i2} - z_{\cdot 2})^2}. \tag{6.1.30}$$

Because $\text{Var}(Y_i) = \sigma^2$, we obtain

$$\text{Var}(\widehat{\beta}_2) = \sigma^2 / \sum_{i=1}^{n}(z_{i2} - z_{\cdot 2})^2,$$

and the $100(1-\alpha)\%$ confidence interval for β_2 has the form

$$\beta_2 = \widehat{\beta}_2 \pm t_{n-p}\left(1 - \tfrac{1}{2}\alpha\right) s / \sqrt{\sum(z_{i2} - z_{\cdot 2})^2}.$$

The confidence interval for β_1 is given in Problem 6.1.10.

Similarly, in the $p = 2$ case, it is straightforward (Problem 6.1.10) to compute

$$h_{ii} = \frac{1}{n} + \frac{(z_{i2} - z_{\cdot 2})^2}{\sum_{i=1}^{n}(z_{i2} - z_{\cdot 2})^2}$$

and the confidence interval for the mean response μ_i of the ith case has a simple explicit form.   □

Example 6.1.3. *One-Way Layout (continued).* We consider $\psi = \beta_k, 1 \leq k \leq p$. Because $\widehat{\beta}_k = Y_{k\cdot} \sim \mathcal{N}(\beta_k, \sigma^2/n_k)$, we find the $100(1-\alpha)\%$ confidence interval

$$\beta_k = \widehat{\beta}_k \pm t_{n-p}\left(1 - \tfrac{1}{2}\alpha\right) s / \sqrt{n_k}$$

where $s^2 = SS_W/(n-p)$. The intervals for $\mu = \beta_\cdot$ and the incremental effect $\delta_k = \beta_k - \mu$ are given in Problem 6.1.11.   □

Joint Confidence Regions

We have seen how to find confidence intervals for each individual β_j, $1 \leq j \leq p$. We next consider the problem of finding a confidence region C in R^p that covers the vector β with prescribed probability $(1 - \alpha)$. This can be done by inverting the likelihood ratio test or equivalently the F test. That is, we let C be the collection of β_0 that is accepted when the level $(1 - \alpha)$ F test is used to test $H : \beta = \beta_0$. Under H, $\mu = \mu_0 = \mathbf{Z}\beta_0$; and the numerator of the F statistic (6.1.22) is based on

$$|\widehat{\mu} - \mu_0|^2 = |\mathbf{Z}\widehat{\beta} - \mathbf{Z}\beta_0|^2 = (\widehat{\beta} - \beta_0)^T (\mathbf{Z}^T \mathbf{Z})(\widehat{\beta} - \beta_0).$$

Thus, using (6.1.22), the simultaneous confidence region for β is the ellipse

$$C = \left\{ \beta_0 : \frac{(\widehat{\beta} - \beta_0)^T (\mathbf{Z}^T \mathbf{Z})(\widehat{\beta} - \beta_0)}{rs^2} \leq f_{r,n-r}\left(1 - \tfrac{1}{2}\alpha\right) \right\} \qquad (6.1.31)$$

where $f_{r,n-r}\left(1 - \tfrac{1}{2}\alpha\right)$ is the $1 - \tfrac{1}{2}\alpha$ quantile of the $\mathcal{F}_{r,n-r}$ distribution.

Example 6.1.2. *Regression (continued).* We consider the case $p = r$ and as in (6.1.26) write $\mathbf{Z} = (\mathbf{Z}_1, \mathbf{Z}_2)$ and $\beta^T = (\beta_1^T, \beta_2^T)$, where β_2 is a vector of main effect coefficients and β_1 is a vector of "nuisance" coefficients. Similarly, we partition $\widehat{\beta}$ as $\widehat{\beta}^T = (\widehat{\beta}_1^T, \widehat{\beta}_2^T)$ where $\widehat{\beta}_1$ is $q \times 1$ and $\widehat{\beta}_2$ is $(p - q) \times 1$. By Corollary 6.1.1, $\sigma^2 (\mathbf{Z}^T \mathbf{Z})^{-1}$ is the variance-covariance matrix of $\widehat{\beta}$. It follows that if we let $\mathbf{S}$ denote the lower right $(p - q) \times (p - q)$ corner of $(\mathbf{Z}^T \mathbf{Z})^{-1}$, then $\sigma^2 \mathbf{S}$ is the variance-covariance matrix of $\widehat{\beta}_2$. Thus, a joint $100(1 - \alpha)\%$ confidence region for β_2 is the $p - q$ dimensional ellipse

$$C = \left\{ \beta_{02} : \frac{(\widehat{\beta}_2 - \beta_{02})^T \mathbf{S}^{-1}(\widehat{\beta}_2 - \beta_{02})}{(p - q)s^2} \leq f_{p-q,n-p}\left(1 - \tfrac{1}{2}\alpha\right) \right\}.$$

$\square$

Summary. We consider the classical Gaussian linear model in which the resonse Y_i for the ith case in an experiment is expressed as a linear combination $\mu_i = \sum_{j=1}^{p} \beta_j z_{ij}$ of covariates plus an error ϵ_i, where $\epsilon_i, \ldots, \epsilon_n$ are i.i.d. $\mathcal{N}(0, \sigma^2)$. By introducing a suitable orthogonal transformation, we obtain a canonical model in which likelihood analysis is straightforward. The inverse of the orthogonal transformation gives procedures and results in terms of the original variables. In particular we obtain maximum likelihood estimates, likelihood ratio tests, and confidence procedures for the regression coefficients $\{\beta_j\}$, the response means $\{\mu_i\}$, and linear combinations of these.

6.2   ASYMPTOTIC ESTIMATION THEORY IN p DIMENSIONS

In this section we largely parallel Section 5.4 in which we developed the asymptotic properties of the MLE and related tests and confidence bounds for one-dimensional parameters. We leave the analogue of Theorem 5.4.1 to the problems and begin immediately generalizing Section 5.4.2.

6.2.1   Estimating Equations

Our assumptions are as before save that everything is made a vector: $X_1, \ldots, X_n$ are i.i.d. P where $P \in \mathcal{Q}$, a model containing $\mathcal{P} \equiv \{P_{\boldsymbol{\theta}} : \boldsymbol{\theta} \in \Theta\}$ such that

(i) Θ open $\subset R^p$.

(ii) Densities of P_θ are $p(\cdot, \boldsymbol{\theta})$, $\theta \in \Theta$.

The following result gives the general asymptotic behavior of the solution of estimating equations. Let $\boldsymbol{\Psi} \equiv (\psi_1, \ldots, \psi_p)^T$ where, in the case of minimum contrast estimates based on $\rho(X, \theta)$, $\psi_j = \frac{\partial \rho}{\partial \theta_j}$ is well defined.

A0. We assume that $\bar{\boldsymbol{\theta}}_n$ satisfies

$$\frac{1}{n} \sum_{i=1}^{n} \boldsymbol{\Psi}(X_i, \bar{\boldsymbol{\theta}}_n) = \mathbf{0}. \tag{6.2.1}$$

A solution to $(6.2.1)$ is called an *estimating equation estimate* or an *M-estimate*.

A1. The parameter $\boldsymbol{\theta}(P)$ given by the solution of (the nonlinear system of p equations in p unknowns):

$$\int \boldsymbol{\Psi}(x, \boldsymbol{\theta}) dP(x) = \mathbf{0} \tag{6.2.2}$$

is well defined on $\mathcal{Q}$ so that $\boldsymbol{\theta}(P)$ is the unique solution of $(6.2.2)$. If $\psi_j(x, \boldsymbol{\theta}) = \frac{\partial}{\partial \theta_j} \log p(x, \boldsymbol{\theta})$, then $\boldsymbol{\theta}(P_{\boldsymbol{\theta}}) = \boldsymbol{\theta}$ because $\mathcal{Q} \supset \mathcal{P}$. See Section 1.1.2.

A2. $E_P|\boldsymbol{\Psi}(X_1, \boldsymbol{\theta}(P))|^2 < \infty$ where $|\cdot|$ is the Euclidean norm.

A3. $\psi_i(\cdot, \boldsymbol{\theta})$, $1 \leq i \leq p$, have first-order partials with respect to all coordinates and using the notation of Section B.8,

$$E_P|D\boldsymbol{\Psi}(X_1, \boldsymbol{\theta})| < \infty$$

where

$$E_P D\boldsymbol{\Psi}(X_1, \boldsymbol{\theta}) = \left\| E_P \frac{\partial \psi_i}{\partial \theta_j}(X_1, \boldsymbol{\theta}) \right\|_{p \times p}$$

is nonsingular.

A4. $\sup \left\{ \left| \frac{1}{n} \sum_{i=1}^{n} (D\boldsymbol{\Psi}(X_i, \mathbf{t}) - D\boldsymbol{\Psi}(X_i, \boldsymbol{\theta}(P))) \right| : |\mathbf{t} - \boldsymbol{\theta}(P)| \leq \epsilon_n \right\} \xrightarrow{P} 0$ if $\epsilon_n \to 0$.

A5. $\bar{\boldsymbol{\theta}}_n \xrightarrow{P} \boldsymbol{\theta}(P)$ for all $P \in \mathcal{Q}$.

Theorem 6.2.1. *Under* A0–A5 *of this section*

$$\bar{\boldsymbol{\theta}}_n = \boldsymbol{\theta}(P) + \frac{1}{n} \sum_{i=1}^{n} \widetilde{\boldsymbol{\Psi}}(X_i, \boldsymbol{\theta}(P)) + o_p(n^{-1/2}) \tag{6.2.3}$$

where

$$\widetilde{\boldsymbol{\Psi}}(x, \boldsymbol{\theta}(P)) = -[E_P D\boldsymbol{\Psi}(X_1, \boldsymbol{\theta}(P))]^{-1} \boldsymbol{\Psi}(x, \boldsymbol{\theta}(P)). \tag{6.2.4}$$

Hence,

$$\mathcal{L}_p(\sqrt{n}(\bar{\boldsymbol{\theta}}_n - \boldsymbol{\theta}(P))) \to \mathcal{N}_p(\mathbf{0}, \Sigma(\boldsymbol{\Psi}, P)) \tag{6.2.5}$$

where

$$\Sigma(\boldsymbol{\Psi}, P) = J(\boldsymbol{\theta}, P)E\boldsymbol{\Psi}\boldsymbol{\Psi}^T(X_1, \boldsymbol{\theta}(P))J^T(\boldsymbol{\theta}, P) \tag{6.2.6}$$

and

$$J^{-1}(\boldsymbol{\theta}, P) = -E_P D\boldsymbol{\Psi}(X_1, \boldsymbol{\theta}(P)) = \left\| -E_P \frac{\partial \psi_i}{\partial \theta_j}(X_1, \boldsymbol{\theta}(P)) \right\|.$$

The proof of this result follows precisely that of Theorem 5.4.2 save that we need multivariate calculus as in Section B.8. Thus,

$$-\frac{1}{n}\sum_{i=1}^{n}\boldsymbol{\Psi}(X_i, \boldsymbol{\theta}(P)) = \frac{1}{n}\sum_{i=1}^{n}D\boldsymbol{\Psi}(X_i, \boldsymbol{\theta}_n^*)(\bar{\boldsymbol{\theta}}_n - \boldsymbol{\theta}(P)). \tag{6.2.7}$$

Note that the left-hand side of (6.2.7) is a $p \times 1$ vector, the right is the product of a $p \times p$ matrix and a $p \times 1$ vector.

The rest of the proof follows essentially exactly as in Section 5.4.2 save that we need the observation that the set of nonsingular $p \times p$ matrices, when viewed as vectors, is an open subset of R^{p^2}, representable, for instance, as the set of vectors for which the determinant, a continuous function of the entries, is different from zero. We use this remark to conclude that A3 and A4 guarantee that with probability tending to 1, $\frac{1}{n}\sum_{i=1}^{n}D\boldsymbol{\Psi}(X_i, \boldsymbol{\theta}_n^*)$ is nonsingular.

Note. This result goes beyond Theorem 5.4.2 in making it clear that although the definition of $\bar{\boldsymbol{\theta}}_n$ is motivated by $\mathcal{P}$, the behavior in (6.2.3) is guaranteed for $P \in \mathcal{Q}$, which can include $P \notin \mathcal{P}$. In fact, typically $\mathcal{Q}$ is essentially the set of P's for which $\boldsymbol{\theta}(P)$ can be defined uniquely by (6.2.2).

We can again extend the assumptions of Section 5.4.2 to:

A6. If $l(\cdot, \boldsymbol{\theta})$ is differentiable

$$\begin{aligned} E_{\boldsymbol{\theta}}D\boldsymbol{\Psi}(X_1, \boldsymbol{\theta}) &= -E_{\boldsymbol{\theta}}\boldsymbol{\Psi}(X_1, \boldsymbol{\theta})Dl(X_1, \boldsymbol{\theta}) \\ &= -\text{Cov}_{\boldsymbol{\theta}}(\boldsymbol{\Psi}(X_1, \boldsymbol{\theta}), Dl(X_1, \boldsymbol{\theta})) \end{aligned} \tag{6.2.8}$$

defined as in B.5.2. The heuristics and conditions behind this identity are the same as in the one-dimensional case. Remarks 5.4.2, 5.4.3, and Assumptions **A4′** and **A6′** extend to the multivariate case readily.

Note that consistency of $\bar{\boldsymbol{\theta}}_n$ is assumed. Proving consistency usually requires different arguments such as those of Section 5.2. It may, however, be shown that with probability tending to 1, a root-finding algorithm starting at a consistent estimate $\boldsymbol{\theta}_n^*$ will find a solution $\bar{\boldsymbol{\theta}}_n$ of (6.2.1) that satisfies (6.2.3) (Problem 6.2.10).

6.2.2   Asymptotic Normality and Efficiency of the MLE

If we take $\rho(x, \boldsymbol{\theta}) = -l(x, \boldsymbol{\theta}) \equiv -\log p(x, \boldsymbol{\theta})$, and $\boldsymbol{\Psi}(x, \boldsymbol{\theta})$ obeys A0–A6, then (6.2.8) becomes

$$
\begin{aligned}
-\|E_{\boldsymbol{\theta}} D^2 l(X_1, \boldsymbol{\theta})\| &= E_{\boldsymbol{\theta}} Dl(X_1, \boldsymbol{\theta}) D^T l(X_1, \boldsymbol{\theta})) \\
&= \mathrm{Var}_{\boldsymbol{\theta}} Dl(X_1, \boldsymbol{\theta})
\end{aligned}
\tag{6.2.9}
$$

where

$$
\mathrm{Var}_{\boldsymbol{\theta}} Dl(X_1, \boldsymbol{\theta}) = \left\| E_{\boldsymbol{\theta}} \left(\frac{\partial l}{\partial \theta_i}(X_1, \boldsymbol{\theta}) \frac{\partial l}{\partial \theta_j}(X_1, \boldsymbol{\theta}) \right) \right\|
$$

is the *Fisher information matrix* $I(\boldsymbol{\theta})$ introduced in Section 3.4. If $\rho : \boldsymbol{\theta} \to R, \boldsymbol{\theta} \subset R^d$, is a scalar function, the matrix $\left\| \frac{\partial^2 \rho}{\partial \theta_i \partial \theta_j}(\boldsymbol{\theta}) \right\|$ is known as the *Hessian* or curvature matrix of the surface ρ. Thus, (6.2.9) states that the expected value of the Hessian of l is the negative of the Fisher information.

We also can immediately state the generalization of Theorem 5.4.3.

Theorem 6.2.2. *If A0–A6 hold for $\rho(x, \boldsymbol{\theta}) \equiv -\log p(x, \boldsymbol{\theta})$, then the MLE $\widehat{\boldsymbol{\theta}}_n$ satisfies*

$$
\widehat{\boldsymbol{\theta}}_n = \boldsymbol{\theta} + \frac{1}{n} \sum_{i=1}^{n} I^{-1}(\boldsymbol{\theta}) Dl(X_i, \boldsymbol{\theta}) + o_p(n^{-1/2})
\tag{6.2.10}
$$

so that

$$
\mathcal{L}(\sqrt{n}(\widehat{\boldsymbol{\theta}}_n - \boldsymbol{\theta})) \to \mathcal{N}(\mathbf{0}, I^{-1}(\boldsymbol{\theta})).
\tag{6.2.11}
$$

If $\bar{\boldsymbol{\theta}}_n$ is a minimum contrast estimate with ρ and ψ satisfying A0–A6 and corresponding asymptotic variance matrix $\Sigma(\boldsymbol{\Psi}, P_{\boldsymbol{\theta}})$, then

$$
\Sigma(\boldsymbol{\Psi}, P_{\boldsymbol{\theta}}) \geq I^{-1}(\boldsymbol{\theta})
\tag{6.2.12}
$$

in the sense of Theorem 3.4.4 with equality in (6.2.12) for $\boldsymbol{\theta} = \boldsymbol{\theta}_0$ iff, under $\boldsymbol{\theta}_0$,

$$
\bar{\boldsymbol{\theta}}_n = \widehat{\boldsymbol{\theta}}_n + o_p(n^{-1/2}).
\tag{6.2.13}
$$

Proof. The proofs of (6.2.10) and (6.2.11) parallel those of (5.4.33) and (5.4.34) exactly. The proof of (6.2.12) parallels that of Theorem 3.4.4. For completeness we give it. Note that by (6.2.6) and (6.2.8)

$$
\Sigma(\boldsymbol{\Psi}, P_{\boldsymbol{\theta}}) = \mathrm{Cov}_{\boldsymbol{\theta}}^{-1}(\mathbf{U}, \mathbf{V}) \mathrm{Var}_{\boldsymbol{\theta}}(\mathbf{U}) \mathrm{Cov}_{\boldsymbol{\theta}}^{-1}(\mathbf{V}, \mathbf{U})
\tag{6.2.14}
$$

where $\mathbf{U} \equiv \boldsymbol{\Psi}(X_1, \boldsymbol{\theta})$, $\mathbf{V} = Dl(X_1, \boldsymbol{\theta})$. But by (B.10.8), for any $\mathbf{U}, \mathbf{V}$ with $\mathrm{Var}(\mathbf{U}^T, \mathbf{V}^T)^T$ nonsingular

$$
\mathrm{Var}(\mathbf{V}) \geq \mathrm{Cov}(\mathbf{U}, \mathbf{V}) \mathrm{Var}^{-1}(\mathbf{U}) \mathrm{Cov}(\mathbf{V}, \mathbf{U}).
\tag{6.2.15}
$$

Taking inverses of both sides yields

$$
I^{-1}(\boldsymbol{\theta}) = \mathrm{Var}_{\boldsymbol{\theta}}^{-1}(\mathbf{V}) \leq \Sigma(\boldsymbol{\Psi}, \boldsymbol{\theta}).
\tag{6.2.16}
$$

Equality holds in (6.2.15) by (B.10.2.3) iff for some $\mathbf{b} = b(\boldsymbol{\theta})$

$$\mathbf{U} = \mathbf{b} + \mathrm{Cov}(\mathbf{U}, \mathbf{V})\mathrm{Var}^{-1}(\mathbf{V})\mathbf{V} \qquad (6.2.17)$$

with probability 1. This means in view of $E_{\boldsymbol{\theta}}\boldsymbol{\Psi} = E_{\boldsymbol{\theta}}Dl = \mathbf{0}$ that

$$\boldsymbol{\Psi}(X_1, \boldsymbol{\theta}) = \mathbf{b}(\boldsymbol{\theta})Dl(X_1, \boldsymbol{\theta}).$$

In the case of identity in (6.2.16) we must have

$$-[E_{\boldsymbol{\theta}}D\boldsymbol{\Psi}(X_1, \boldsymbol{\theta})]^{-1}\boldsymbol{\Psi}(X_1, \boldsymbol{\theta}) = I^{-1}(\boldsymbol{\theta})Dl(X_1, \boldsymbol{\theta}). \qquad (6.2.18)$$

Hence, from (6.2.3) and (6.2.10) we conclude that (6.2.13) holds. $\qquad \square$

We see that, by the theorem, the MLE is *efficient* in the sense that for any $\mathbf{a}_{p\times 1}$, $\mathbf{a}^T\widehat{\boldsymbol{\theta}}_n$ has asymptotic bias $o(n^{-1/2})$ and asymptotic variance $n^{-1}\mathbf{a}^T I^{-1}(\boldsymbol{\theta})\mathbf{a}$, which is no larger than that of any competing minimum contrast or M estimate. Further any competitor $\bar{\boldsymbol{\theta}}_n$ such that $\mathbf{a}^T\bar{\boldsymbol{\theta}}_n$ has the same asymptotic behavior as $\mathbf{a}^T\widehat{\boldsymbol{\theta}}_n$ for *all* $\mathbf{a}$ in fact agrees with $\widehat{\boldsymbol{\theta}}_n$ to order $n^{-1/2}$.

A special case of Theorem 6.2.2 that we have already established is Theorem 5.3.6 on the asymptotic normality of the MLE in canonical exponential families. A number of important new statistical issues arise in the multiparameter case. We illustrate with an example.

Example 6.2.1. *The Linear Model with Stochastic Covariates.* Let $X_i = (\mathbf{Z}_i^T, Y_i)^T$, $1 \le i \le n$, be i.i.d. as $X = (\mathbf{Z}^T, Y)^T$ where $\mathbf{Z}$ is a $p \times 1$ vector of explanatory variables and Y is the response of interest. This model is discussed in Section 2.2.1 and Example 1.4.3. We specialize in two ways:

(i)

$$Y = \alpha + \mathbf{Z}^T\boldsymbol{\beta} + \epsilon \qquad (6.2.19)$$

where ϵ is distributed as $\mathcal{N}(0, \sigma^2)$ independent of $\mathbf{Z}$ and $E(\mathbf{Z}) = \mathbf{0}$. That is, given $\mathbf{Z}$, Y has a $\mathcal{N}(\alpha + \mathbf{Z}^T\boldsymbol{\beta}, \sigma^2)$ distribution.

(ii) The distribution H_0 of $\mathbf{Z}$ is known with density h_0 and $E(\mathbf{Z}\mathbf{Z}^T)$ is nonsingular.

The second assumption is unreasonable but easily dispensed with. It readily follows (Problem 6.2.6) that the MLE of $\boldsymbol{\beta}$ is given by (with probability 1)

$$\widehat{\boldsymbol{\beta}} = [\widetilde{\mathbf{Z}}_{(n)}^T\widetilde{\mathbf{Z}}_{(n)}]^{-1}\widetilde{\mathbf{Z}}_{(n)}^T\mathbf{Y}. \qquad (6.2.20)$$

Here $\widetilde{\mathbf{Z}}_{(n)}$ is the $n \times p$ matrix $\|Z_{ij} - Z_{\cdot j}\|$ where $Z_{\cdot j} = \frac{1}{n}\sum_{i=1}^n Z_{ij}$. We used subscripts (n) to distinguish the use of $\mathbf{Z}$ as a vector in this section and as a matrix in Section 6.1. In the present context, $\mathbf{Z}_{(n)} = (\mathbf{Z}_1, \dots, \mathbf{Z}_n)^T$ is referred to as the *random design matrix*. This example is called the *random design case* as opposed to the *fixed design case* of Section 6.1. Also the MLEs of α and σ^2 are

$$\widehat{\alpha} = \bar{Y} - \sum_{j=1}^p Z_{\cdot j}\widehat{\beta}_j, \ \widehat{\sigma}^2 = \frac{1}{n}|\mathbf{Y} - (\widehat{\alpha} + \mathbf{Z}_{(n)}\widehat{\boldsymbol{\beta}})|^2. \qquad (6.2.21)$$

Note that although given $\mathbf{Z}_1, \ldots, \mathbf{Z}_n$, $\widehat{\boldsymbol{\beta}}$ is Gaussian, this is not true of the marginal distribution of $\widehat{\boldsymbol{\beta}}$.

It is not hard to show that A0–A6 hold in this case because if H_0 has density h_0 and if $\boldsymbol{\theta}$ denotes $(\alpha, \boldsymbol{\beta}^T, \sigma^2)^T$, then

$$
\begin{aligned}
l(X, \boldsymbol{\theta}) &= -\frac{1}{2\sigma^2}[Y - (\alpha + \mathbf{Z}^T \boldsymbol{\beta})]^2 - \frac{1}{2}(\log \sigma^2 + \log 2\pi) + \log h_0(\mathbf{z}) \\
Dl(X, \boldsymbol{\theta}) &= \left(\frac{\epsilon}{\sigma^2}, \ \mathbf{Z}\frac{\epsilon}{\sigma^2}, \ \frac{1}{2\sigma^4}(\epsilon^2 - 1) \right)
\end{aligned}
\tag{6.2.22}
$$

and

$$
I(\boldsymbol{\theta}) = \begin{pmatrix} \sigma^{-2} & \mathbf{0} & 0 \\ \mathbf{0} & \sigma^{-2} E(\mathbf{Z}\mathbf{Z}^T) & \mathbf{0} \\ 0 & \mathbf{0} & \frac{1}{2\sigma^4} \end{pmatrix}
\tag{6.2.23}
$$

so that by Theorem 6.2.2

$$
\mathcal{L}(\sqrt{n}(\widehat{\alpha} - \alpha, \widehat{\boldsymbol{\beta}} - \boldsymbol{\beta}, \widehat{\sigma}^2 - \sigma^2)) \to \mathcal{N}(\mathbf{0}, \mathrm{diag}(\sigma^2, \sigma^2[E(\mathbf{Z}\mathbf{Z}^T)]^{-1}, 2\sigma^4)). \tag{6.2.24}
$$

This can be argued directly as well (Problem 6.2.8). It is clear that the restriction of H_0 known plays no role in the limiting result for $\widehat{\alpha}, \widehat{\boldsymbol{\beta}}, \widehat{\sigma}^2$. Of course, these will only be the MLEs if H_0 depends only on parameters other than $(\alpha, \boldsymbol{\beta}, \sigma^2)$. In this case we can estimate $E(\mathbf{Z}\mathbf{Z}^T)$ by $\frac{1}{n}\sum_{i=1}^n \mathbf{Z}_i \mathbf{Z}_i^T$ and give approximate confidence intervals for $\beta_j, j = 1, \ldots, p$.

An interesting feature of (6.2.23) is that because $I(\boldsymbol{\theta})$ is a block diagonal matrix so is $I^{-1}(\boldsymbol{\theta})$ and, consequently, $\widehat{\boldsymbol{\beta}}$ and $\widehat{\sigma}^2$ are asymptotically independent. In the classical linear model of Section 6.1 where we perform inference conditionally given $\mathbf{Z}_i = \mathbf{z}_i, 1 \le i \le n$, we have noted this is exactly true.

This is an example of the phenomenon of *adaptation*. If we knew σ^2, the MLE would still be $\widehat{\boldsymbol{\beta}}$ and its asymptotic variance optimal for this model. If we knew α and $\boldsymbol{\beta}$, $\widehat{\sigma}^2$ would no longer be the MLE. But its asymptotic variance would be the same as that of the MLE and, by Theorem 6.2.2, $\widehat{\sigma}^2$ would be asymptotically equivalent to the MLE. To summarize, estimating either parameter with the other being a nuisance parameter is no harder than when the nuisance parameter is known. Formally, in a model $\mathcal{P} = \{P_{(\theta,\eta)} : \theta \in \Theta, \ \eta \in \mathcal{E}\}$ we say we can estimate θ *adaptively* at η_0 if the asymptotic variance of the MLE $\widehat{\theta}$ (or more generally, an efficient estimate of θ) in the pair $(\widehat{\theta}, \widehat{\eta})$ is the same as that of $\widehat{\theta}(\eta_0)$, the efficient estimate for $\mathcal{P}_{\eta_0} = \{P_{(\theta,\eta_0)} : \theta \in \Theta\}$. The possibility of adaptation is in fact rare, though it appears prominently in this way in the Gaussian linear model. In particular consider estimating β_1 in the presence of $\alpha, (\beta_2 \ldots, \beta_p)$ with

(i) $\alpha, \beta_2, \ldots, \beta_p$ known.

(ii) $\boldsymbol{\beta}$ arbitrary.

In case (i), we take, without loss of generality, $\alpha = \beta_2 = \cdots = \beta_p = 0$. Let $\mathbf{Z}_i = (Z_{i1}, \ldots, Z_{ip})^T$, then the efficient estimate in case (i) is

$$
\widehat{\beta}_1^0 = \frac{\sum_{i=1}^n Z_{i1} Y_i}{\sum_{i=1}^n Z_{i1}^2} \tag{6.2.25}
$$

with asymptotic variance $\sigma^2[EZ_1^2]^{-1}$. On the other hand, $\widehat{\beta}_1$ is the first coordinate of $\widehat{\beta}$ given by (6.2.20). Its asymptotic variance is the $(1,1)$ element of $\sigma^2[E\mathbf{Z}\mathbf{Z}^T]^{-1}$, which is strictly bigger than $\sigma^2[EZ_1^2]^{-1}$ unless $[E\mathbf{Z}\mathbf{Z}^T]^{-1}$ is a diagonal matrix (Problem 6.2.3). So in general we cannot estimate β_1 adaptively if $\beta_2, \ldots, \beta_p$ are regarded as nuisance parameters. What is happening can be seen by a representation of $[\mathbf{Z}_{(n)}^T\mathbf{Z}_{(n)}]^{-1}\mathbf{Z}_{(n)}^T\mathbf{Y}$ and $I^{11}(\boldsymbol{\theta})$ where $I^{-1}(\boldsymbol{\theta}) \equiv \|I^{ij}(\boldsymbol{\theta})\|$. We claim that

$$\widehat{\beta}_1 = \frac{\sum_{i=1}^{n}(Z_{i1} - \widehat{Z}_i^{(1)})Y_i}{\sum_{i=1}^{n}(Z_{i1} - \widehat{Z}_i^{(1)})^2} \tag{6.2.26}$$

where $\widehat{Z}^{(1)}$ is the regression of $(Z_{11}, \ldots, Z_{n1})^T$ on the linear space spanned by $(Z_{1j}, \ldots, Z_{nj})^T, 2 \le j \le p$. Similarly,

$$I^{11}(\boldsymbol{\theta}) = \sigma^2/E(Z_{11} - \Pi(Z_{11} \mid Z_{12}, \ldots, Z_{1p}))^2 \tag{6.2.27}$$

where $\Pi(Z_{11} \mid Z_{12}, \ldots, Z_{1p})$ is the projection of Z_{11} on the linear span of $Z_{12}, \ldots, Z_{1p}$ (Problem 6.2.11). Thus, $\Pi(Z_{11} \mid Z_{12}, \ldots, Z_{1p}) = \sum_{j=2}^{p} a_j^* Z_{1j}$ where $(a_2^*, \ldots, a_p^*)$ minimizes $E(Z_{11} - \sum_{j=2}^{p} a_j Z_{1j})^2$ over $(a_2, \ldots, a_p) \in R^{p-1}$ (see Sections 1.4 and B.10). What (6.2.26) and (6.2.27) reveal is that there is a price paid for not knowing $\beta_2, \ldots, \beta_p$ when the variables $Z_2, \ldots, Z_p$ are in any way correlated with Z_1 and the price is measured by

$$\frac{[E(Z_{11} - \Pi(Z_{11} \mid Z_{12}, \ldots, Z_{1p})^2]^{-1}}{E(Z_{11}^2)} = \left(1 - \frac{E(\Pi(Z_{11} \mid Z_{12}, \ldots, Z_{1p}))^2}{E(Z_{11}^2)}\right)^{-1}. \tag{6.2.28}$$

In the extreme case of perfect collinearity the price is ∞ as it should be because β_1 then becomes unidentifiable. Thus, adaptation corresponds to the case where $(Z_2, \ldots, Z_p)$ have no value in predicting Z_1 linearly (see Section 1.4). Correspondingly in the Gaussian linear model (6.1.3) conditional on the $\mathbf{Z}_i$, $i = 1, \ldots, n$, $\widehat{\beta}_1$ is undefined if the denominator in (6.2.26) is 0, which corresponds to the case of collinearity and occurs with probability 1 if $E(Z_{11} - \Pi(Z_{11} \mid Z_{12}, \ldots, Z_{1p}))^2 = 0$.   □

Example 6.2.2. *M Estimates Generated by Linear Models with General Error Structure.* Suppose that the ϵ_i in (6.2.19) are i.i.d. with density $\frac{1}{\sigma}f_0\left(\frac{\cdot}{\sigma}\right)$, where f_0 is not necessarily Gaussian, for instance,

$$f_0(x) = \frac{e^{-x}}{(1 + e^{-x})^2},$$

the logistic density. Such error densities have the often more realistic, heavier tails[1] than the Gaussian density. The estimates $\widehat{\beta}_0, \widehat{\sigma}_0$ now solve

$$\sum_{i=1}^{n} Z_{ij}\psi\left(\widehat{\sigma}^{-1}\left(Y_i - \sum_{k=1}^{p}\widehat{\beta}_{k0}Z_{ik}\right)\right) = 0$$

and

$$\sum_{i=1}^{n} \frac{1}{\widehat{\sigma}}\chi\left(\widehat{\sigma}^{-1}\left(Y_i - \sum_{k=1}^{p}\widehat{\beta}_{k0}Z_{ik}\right)\right) = 0$$

where $\psi = -\frac{f_0'}{f_0}$, $\chi(y) = -\left(y\frac{f_0'}{f_0}(y) + 1\right)$, $\hat{\boldsymbol{\beta}}_0 \equiv (\hat{\beta}_{10}, \ldots, \hat{\beta}_{p0})^T$. The assumptions of Theorem 6.2.2 may be shown to hold (Problem 6.2.9) if

(i) $\log f_0$ is strictly concave, i.e., $\frac{f_0'}{f_0}$ is strictly decreasing.

(ii) $(\log f_0)''$ exists and is bounded.

Then, if further f_0 is symmetric about 0,

$$
\begin{aligned}
I(\boldsymbol{\theta}) &= \sigma^{-2} I(\boldsymbol{\beta}^T, 1) \\
&= \sigma^{-2} \begin{pmatrix} c_1 E(\mathbf{Z}^T \mathbf{Z}) & \mathbf{0} \\ \mathbf{0} & c_2 \end{pmatrix}
\end{aligned}
\tag{6.2.29}
$$

where $c_1 = \int \left(\frac{f_0'}{f_0}(x)\right)^2 f_0(x)dx$, $c_2 = \int \left(x\frac{f_0'}{f_0}(x) + 1\right)^2 f_0(x)dx$. Thus, $\hat{\boldsymbol{\beta}}_0, \hat{\sigma}_0$ are optimal estimates of $\boldsymbol{\beta}$ and σ in the sense of Theorem 6.2.2 if f_0 is true.

Now suppose f_0 generating the estimates $\hat{\boldsymbol{\beta}}_0$ and $\hat{\sigma}_0^2$ is symmetric and satisfies (i) and (ii) but the true error distribution has density f possibly different from f_0. Under suitable conditions we can apply Theorem 6.2.1 with

$$
\boldsymbol{\Psi}(\mathbf{Z}, Y, \boldsymbol{\beta}, \sigma) = (\psi_1, \ldots, \psi_p, \psi_{p+1})^T (\mathbf{Z}, Y, \boldsymbol{\beta}, \sigma)
$$

where

$$
\begin{aligned}
\psi_j(\mathbf{z}, y, \boldsymbol{\beta}, \sigma) &= \frac{z_j}{\sigma} \psi \left(\frac{y - \sum_{k=1}^p z_k \beta_k}{\sigma}\right), \quad 1 \le j \le p \\
\psi_{p+1}(\mathbf{z}, y, \boldsymbol{\beta}, \sigma) &= \frac{1}{\sigma} \chi \left(\frac{y - \sum_{k=1}^p z_k \beta_k}{\sigma}\right)
\end{aligned}
\tag{6.2.30}
$$

to conclude that

$$
\begin{aligned}
\mathcal{L}_0(\sqrt{n}(\hat{\boldsymbol{\beta}}_0 - \boldsymbol{\beta}_0)) &\to \mathcal{N}(0, \boldsymbol{\Sigma}(\boldsymbol{\Psi}, P)) \\
\mathcal{L}(\sqrt{n}(\hat{\sigma} - \sigma_0) &\to \mathcal{N}(0, \sigma^2(P))
\end{aligned}
$$

where $\boldsymbol{\beta}_0, \sigma_0$ solve

$$
\int \boldsymbol{\Psi}(\mathbf{z}, y, \boldsymbol{\beta}, \sigma) dP = \mathbf{0}
\tag{6.2.31}
$$

and $\boldsymbol{\Sigma}(\boldsymbol{\Psi}, P)$ is as in (6.2.6). What is the relation between $\boldsymbol{\beta}_0$, σ_0 and $\boldsymbol{\beta}, \sigma$ given in the Gaussian model (6.2.19)? If f_0 is symmetric about 0 and the solution of (6.2.31) is unique, then $\boldsymbol{\beta}_0 = \boldsymbol{\beta}$. But $\sigma_0 = c(f_0, \sigma)\sigma$ for some $c(f_0, \sigma)$ typically different from one. Thus, $\hat{\boldsymbol{\beta}}_0$ can be used for estimating $\boldsymbol{\beta}$ although if the true distribution of the ϵ_i is $\mathcal{N}(0, \sigma^2)$ it should perform less well than $\hat{\boldsymbol{\beta}}$. On the other hand, $\hat{\sigma}_0$ is an estimate of σ only if normalized by a constant depending on f_0. (See Problem 6.2.5.) These are issues of robustness, that is, to have a bounded sensitivity curve (Section 3.5, Problem 3.5.8), we may well wish to use a nonlinear bounded $\boldsymbol{\Psi} = (\psi_1, \ldots, \psi_p)^T$ to estimate $\boldsymbol{\beta}$ even though it is suboptimal when $\epsilon \sim \mathcal{N}(0, \sigma^2)$, and to use a suitably normalized version of $\hat{\sigma}_0$ for the same purpose. One effective choice of ψ_j is the Huber function defined in Problem 3.5.8. We will discuss these issues further in Section 6.6 and Volume II.   □

Testing and Confidence Bounds

There are three principal approaches to testing hypotheses in multiparameter models, the likelihood ratio principle, Wald tests (a generalization of pivots), and Rao's tests. All of these will be developed in Section 6.3. The three approaches coincide asymptotically but differ substantially, in performance and computationally, for fixed n. Confidence regions that parallel the tests will also be developed in Section 6.3.

Optimality criteria are not easily stated even in the fixed sample case and not very persuasive except perhaps in the case of testing hypotheses about a real parameter in the presence of other nuisance parameters such as $H : \theta_1 \leq 0$ versus $K : \theta_1 > 0$ where $\theta_2, \ldots, \theta_p$ vary freely.

6.2.3 The Posterior Distribution in the Multiparameter Case

The asymptotic theory of the posterior distribution parallels that in the one-dimensional case exactly. We simply make $\boldsymbol{\theta}$ a vector, and interpret $|\cdot|$ as the Euclidean norm in conditions A7 and A8. Using multivariate expansions as in B.8 we obtain

Theorem 6.2.3. *If the multivariate versions of* A0–A3, A4(a.s.), A5(a.s.) *and* A6–A8 *hold then, if* $\widehat{\theta}$ *denotes the MLE,*

$$\mathcal{L}(\sqrt{n}(\boldsymbol{\theta} - \widehat{\boldsymbol{\theta}}) \mid X_1, \ldots, X_n) \to \mathcal{N}(\mathbf{0}, I^{-1}(\boldsymbol{\theta})) \tag{6.2.32}$$

a.s. under $P_{\boldsymbol{\theta}}$ *for all* $\boldsymbol{\theta}$.

The consequences of Theorem 6.2.3 are the same as those of Theorem 5.5.2, the equivalence of Bayesian and frequentist optimality asymptotically.

Again the two approaches differ at the second order when the prior begins to make a difference. See Schervish (1995) for some of the relevant calculations.

A new major issue that arises is computation. Although it is easy to write down the posterior density of $\boldsymbol{\theta}$, $\pi(\boldsymbol{\theta}) \prod_{i=1}^{n} p(X_i, \boldsymbol{\theta})$, up to the proportionality constant $\int_\Theta \pi(\mathbf{t}) \prod_{i=1}^{n} p(X_i, \mathbf{t}) d\mathbf{t}$, the latter can pose a formidable problem if $p > 2$, say. The problem arises also when, as is usually the case, we are interested in the posterior distribution of some of the parameters, say (θ_1, θ_2), because we then need to integrate out $(\theta_3, \ldots, \theta_p)$. The asymptotic theory we have developed permits approximation to these constants by the procedure used in deriving (5.5.19) (Laplace's method). We have implicitly done this in the calculations leading up to (5.5.19). This approach is refined in Kass, Kadane, and Tierney (1989). However, typically there is an attempt at "exact" calculation. A class of Monte Carlo based methods derived from statistical physics loosely called Markov chain Monte Carlo has been developed in recent years to help with these problems. These methods are beyond the scope of this volume but will be discussed briefly in Volume II.

Summary. We defined minimum contrast (MC) and M-estimates in the case of p-dimensional parameters and established their convergence in law to a normal distribution. When the estimating equations defining the M-estimates coincide with the likelihood equations, this result gives the asymptotic distribution of the MLE. We find that the MLE is asymptotically efficient in the sense that it has "smaller" asymptotic covariance matrix than that of any MD or M-estimate if we know the correct model $\mathcal{P} = \{P_\theta : \theta \in \Theta\}$ and use the MLE for this model. We use an example to introduce the concept of adaptation in which an estimate $\widehat{\theta}$ is called adaptive for a model $\{P_{\theta,\eta} : \theta \in \Theta, \eta \in \mathcal{E}\}$ if the asymptotic distribution of $\sqrt{n}(\widehat{\theta} - \theta)$ has mean zero and variance matrix equal to the smallest possible for a general class of regular estimates of θ in the family of models $\{P_{\theta,\eta_0} : \theta \in \Theta\}$, η_0 specified. In linear regression, adaptive estimation of β_1 is possible iff Z_1 is uncorrelated with every linear function of $Z_2, \ldots, Z_p$. Another example deals with M-estimates based on estimating equations generated by linear models with non-Gaussian error distribution. Finally we show that in the Bayesian framework where given $\boldsymbol{\theta}$, $X_1, \ldots, X_n$ are i.i.d. $P_{\boldsymbol{\theta}}$, if $\widehat{\boldsymbol{\theta}}$ denotes the MLE for $P_{\boldsymbol{\theta}}$, then the posterior distribution of $\sqrt{n}(\boldsymbol{\theta} - \widehat{\boldsymbol{\theta}})$ converges a.s. under P_θ to the $\mathcal{N}(0, I^{-1}(\boldsymbol{\theta}))$ distribution.

6.3   LARGE SAMPLE TESTS AND CONFIDENCE REGIONS

In Section 6.1 we developed exact tests and confidence regions that are appropriate in regression and anaysis of variance (ANOVA) situations when the responses are normally distributed. We shall show (see Section 6.6) that these methods in many respects are also approximately correct when the distribution of the error in the model fitted is not assumed to be normal. However, we need methods for situations in which, as in the linear model, covariates can be arbitrary but responses are necessarily discrete (qualitative) or nonnegative and Gaussian models do not seem to be appropriate approximations. In these cases exact methods are typically not available, and we turn to asymptotic approximations to construct tests, confidence regions, and other methods of inference. We present three procedures that are used frequently: likelihood ratio, Wald and Rao large sample tests, and confidence procedures. These were treated for θ real in Section 5.4.4. In this section we will use the results of Section 6.2 to extend some of the results of Section 5.4.4 to vector-valued parameters.

6.3.1   Asymptotic Approximation to the Distribution of the Likelihood Ratio Statistic

In Sections 4.9 and 6.1 we considered the likelihood ratio test statistic,

$$\lambda(\mathbf{x}) = \frac{\sup\{p(\mathbf{x}, \boldsymbol{\theta}) : \boldsymbol{\theta} \in \Theta\}}{\sup\{p(\mathbf{x}, \boldsymbol{\theta}) : \boldsymbol{\theta} \in \Theta_0\}}$$

for testing $H : \boldsymbol{\theta} \in \Theta_0$ versus $K : \boldsymbol{\theta} \in \Theta_1$, $\Theta_1 = \Theta - \Theta_0$, and showed that in several statistical models involving normal distributions, $\lambda(\mathbf{x})$ simplified and produced intuitive tests whose critical values can be obtained from the Student t and $\mathcal{F}$ distributions.

However, in many experimental situations in which the likelihood ratio test can be used to address important questions, the exact critical value is not available analytically. In such cases we can turn to an approximation to the distribution of $\lambda(\mathbf{X})$ based on asymptotic theory, which is usually referred to as *Wilks's theorem* or *approximation*. Other approximations that will be explored in Volume II are based on Monte Carlo and bootstrap simulations. Here is an example in which Wilks's approximation to $\mathcal{L}(\lambda(\mathbf{X}))$ is useful:

Example 6.3.1. Suppose $X_1, X_2, \ldots, X_n$ are i.i.d. as X where X has the gamma, $\Gamma(\alpha, \beta)$, distribution with density

$$p(x; \theta) = \beta^\alpha x^{\alpha-1} \exp\{-\beta x\}/\Gamma(\alpha); \ x > 0; \ \alpha > 0, \beta > 0.$$

In Example 2.3.2 we showed that the MLE, $\widehat{\theta} = (\widehat{\alpha}, \widehat{\beta})$, exists and in Example 2.4.2 we showed how to find $\widehat{\theta}$ as a nonexplicit solution of likelihood equations. Thus, the numerator of $\lambda(\mathbf{x})$ is available as $p(\mathbf{x}, \widehat{\theta}) = \prod_{i=1}^{n} p(\mathbf{x}_i, \widehat{\theta})$. Suppose we want to test $H : \alpha = 1$ (exponential distribution) versus $K : \alpha \neq 1$. The MLE of β under H is readily seen from (2.3.5) to be $\widehat{\beta}_0 = 1/\bar{\mathbf{x}}$ and $p(\mathbf{x}; 1, \widehat{\beta}_0)$ is the denominator of the likelihood ratio statistic. It remains to find the critical value. This is not available analytically.   □

The approximation we shall give is based on the result "$2 \log \lambda(\mathbf{X}) \overset{\mathcal{L}}{\to} \chi_d^2$" for degrees of freedom d to be specified later. We next give an example that can be viewed as the limiting situation for which the approximation is exact:

Example 6.3.2. *The Gaussian Linear Model with Known Variance.* Let $Y_1, \ldots, Y_n$ be independent with $Y_i \sim \mathcal{N}(\mu_i, \sigma_0^2)$ where σ_0 is known. As in Section 6.1.3 we test whether $\boldsymbol{\mu} = (\mu_1, \ldots, \mu_n)^T$ is a member of a q-dimensional linear subspace of R^n, ω_0, versus the alternative that $\boldsymbol{\mu} \in \omega - \omega_0$ where ω is an r-dimensional linear subspace of R^n and $\omega \supset \omega_0$; and we transform to canonical form by setting

$$\boldsymbol{\eta} = \mathbf{A}\boldsymbol{\mu}, \ \mathbf{U} = \mathbf{A}\mathbf{Y}$$

where $\mathbf{A}_{n \times n}$ is an orthogonal matrix with rows $\mathbf{v}_1^T, \ldots, \mathbf{v}_n^T$ such that $\mathbf{v}_1, \ldots, \mathbf{v}_q$ span ω_0 and $\mathbf{v}_1, \ldots, \mathbf{v}_r$ span ω.

Set $\theta_i = \eta_i/\sigma_0$, $i = 1, \ldots, r$ and $X_i = U_i/\sigma_0$, $i = 1, \ldots, n$. Then $X_i \sim \mathcal{N}(\theta_i, 1)$, $i = 1, \ldots, r$ and $X_i \sim \mathcal{N}(0, 1)$, $i = r+1, \ldots, n$. Moreover, the hypothesis H is equivalent to $H : \theta_{q+1} = \cdots = \theta_r = 0$. Using Section 6.1.3, we conclude that under H,

$$2 \log \lambda(\mathbf{Y}) = \sum_{i=q+1}^{r} X_i^2 \sim \chi_{r-q}^2.$$

Wilks's theorem states that, under regularity conditions, when testing whether a parameter vector is restricted to an open subset of R^q or R^r, $q < r$, the χ_{r-q}^2 distribution is an approximation to $\mathcal{L}(2 \log \lambda(\mathbf{Y}))$. In this σ^2 known example, Wilks's approximation is exact.   □

We illustrate the remarkable fact that χ_{r-q}^2 holds as an approximation to the null distribution of $2 \log \lambda$ quite generally when the hypothesis is a nice q-dimensional submanifold of an r-dimensional parameter space with the following.

Example 6.3.3. *The Gaussian Linear Model with Unknown Variance.* If Y_i are as in Example 6.3.2 but σ^2 is unknown then $\boldsymbol{\theta} = (\boldsymbol{\mu}, \sigma^2)$ ranges over an $r + 1$-dimensional manifold whereas under H, $\boldsymbol{\theta}$ ranges over a $q + 1$-dimensional manifold. In Section 6.1.3, we derived

$$2 \log \lambda(\mathbf{Y}) = n \log \left(1 + \frac{\sum_{i=q+1}^{r} X_i^2}{\sum_{i=r+1}^{n} X_i^2} \right).$$

Apply Example 5.3.7 to $V_n = \sum_{i=q+1}^{r} X_i^2 / n^{-1} \sum_{i=r+1}^{n} X_i^2$ and conclude that $V_n \overset{\mathcal{L}}{\to} \chi_{r-q}^2$. Finally apply Lemma 5.3.2 with $g(t) = \log(1+t)$, $a_n = n$, $c = 0$ and conclude that $2 \log \lambda(\mathbf{Y}) \overset{\mathcal{L}}{\to} \chi_{r-q}^2$ also in the σ^2 unknown case. Note that for $\widetilde{\lambda}(\mathbf{Y})$ defined in Remark 6.1.2, $2 \log \widetilde{\lambda}(\mathbf{Y}) = V_n \overset{\mathcal{L}}{\to} \chi_{r-q}^2$ as well. $\square$

Consider the general i.i.d. case with $X_1, \ldots, X_n$ a sample from $p(x, \theta)$, where $x \in \mathcal{X} \subset R^s$, and $\boldsymbol{\theta} \in \Theta \subset R^r$. Write the log likelihood as

$$l_n(\boldsymbol{\theta}) = \sum_{i=1}^{n} \log p(X_i, \boldsymbol{\theta}).$$

We first consider the simple hypothesis $H : \boldsymbol{\theta} = \boldsymbol{\theta}_0$.

Theorem 6.3.1. *Suppose the assumptions of Theorem 6.2.2 are satisfied. Then, under $H : \boldsymbol{\theta} = \boldsymbol{\theta}_0$,*

$$2 \log \lambda(\mathbf{X}) = 2[l_n(\widehat{\boldsymbol{\theta}}_n) - l_n(\boldsymbol{\theta}_0)] \overset{\mathcal{L}}{\to} \chi_r^2.$$

Proof. Because $\widehat{\boldsymbol{\theta}}_n$ solves the likelihood equation $D_{\boldsymbol{\theta}} l_n(\boldsymbol{\theta}) = 0$, where $D_{\boldsymbol{\theta}}$ is the derivative with respect to $\boldsymbol{\theta}$, an expansion of $l_n(\boldsymbol{\theta})$ about $\widehat{\boldsymbol{\theta}}_n$ evaluated at $\boldsymbol{\theta} = \boldsymbol{\theta}_0$ gives

$$2[l_n(\widehat{\boldsymbol{\theta}}_n) - l_n(\boldsymbol{\theta}_0)] = n(\widehat{\boldsymbol{\theta}}_n - \boldsymbol{\theta}_0)^T \mathbf{I}_n(\boldsymbol{\theta}_n^*)(\widehat{\boldsymbol{\theta}}_n - \boldsymbol{\theta}_0) \qquad (6.3.1)$$

for some $\boldsymbol{\theta}_n^*$ with $|\boldsymbol{\theta}_n^* - \widehat{\boldsymbol{\theta}}_n| \leq |\widehat{\boldsymbol{\theta}}_n - \boldsymbol{\theta}_0|$. Here

$$\mathbf{I}_n(\boldsymbol{\theta}) = \left\| -\frac{1}{n} \sum_{i=1}^{n} \frac{\partial}{\partial \theta_k} \frac{\partial}{\partial \theta_j} \log p(X_i, \boldsymbol{\theta}) \right\|_{r \times r}.$$

By Theorem 6.2.2, $\sqrt{n}(\widehat{\boldsymbol{\theta}}_n - \boldsymbol{\theta}_0) \overset{\mathcal{L}}{\to} \mathcal{N}(0, I^{-1}(\boldsymbol{\theta}_0))$, where $I_{r \times r}(\boldsymbol{\theta})$ is the Fisher information matrix.

Because

$$|\boldsymbol{\theta}_n^* - \boldsymbol{\theta}_0| \leq |\boldsymbol{\theta}_n^* - \widehat{\boldsymbol{\theta}}_n| + |\widehat{\boldsymbol{\theta}}_n - \boldsymbol{\theta}_0| \leq 2|\widehat{\boldsymbol{\theta}}_n - \boldsymbol{\theta}_0|,$$

we can conclude arguing from A.3 and A.4 that that $\mathbf{I}_n(\boldsymbol{\theta}_n^*) \overset{P}{\to} E\mathbf{I}_n(\boldsymbol{\theta}_0) = I(\boldsymbol{\theta}_0)$. Hence,

$$2[l_n(\widehat{\boldsymbol{\theta}}_n) - l_n(\boldsymbol{\theta}_0)] \overset{\mathcal{L}}{\to} \mathbf{V}^T I(\boldsymbol{\theta}_0) \mathbf{V}, \mathbf{V} \sim \mathcal{N}(0, I^{-1}(\boldsymbol{\theta}_0)). \qquad (6.3.2)$$

The result follows because, by Corollary B.6.2, $\mathbf{V}^T I(\boldsymbol{\theta}_0) \mathbf{V} \sim \chi_r^2$. $\square$

As a consequence of the theorem, the test that rejects $H : \boldsymbol{\theta} = \boldsymbol{\theta}_0$ when

$$2 \log \lambda(\mathbf{X}) \geq x_r(1-\alpha),$$

where $x_r(1-\alpha)$ is the $1-\alpha$ quantile of the χ_r^2 distribution, has approximately level $1-\alpha$, and

$$\{\boldsymbol{\theta}_0 : 2[l_n(\widehat{\boldsymbol{\theta}}_n) - l_n(\boldsymbol{\theta}_0)] \leq x_r(1-\alpha)\} \tag{6.3.3}$$

is a confidence region for $\boldsymbol{\theta}$ with approximate coverage probability $1-\alpha$.

Next we turn to the more general hypothesis $H : \boldsymbol{\theta} \in \Theta_0$, where Θ is open and Θ_0 is the set of $\boldsymbol{\theta} \in \Theta$ with $\theta_j = \theta_{0,j}$, $j = q+1, \ldots, r$, and $\{\theta_{0,j}\}$ are specified values. Examples 6.3.1 and 6.3.2 illustrate such Θ_0. We set $d = r - q$, $\boldsymbol{\theta}^T = (\boldsymbol{\theta}^{(1)}, \boldsymbol{\theta}^{(2)})$, $\boldsymbol{\theta}^{(1)} = (\theta_1, \ldots, \theta_q)^T$, $\boldsymbol{\theta}^{(2)} = (\theta_{q+1}, \ldots, \theta_r)^T$, $\boldsymbol{\theta}_0^{(2)} = (\theta_{0,q+1}, \ldots, \theta_{0,r})^T$.

Theorem 6.3.2. *Suppose that the assumptions of Theorem 6.2.2 hold for $p(x, \boldsymbol{\theta})$, $\boldsymbol{\theta} \in \Theta$. Let $\mathcal{P}_0$ be the model $\{P_{\boldsymbol{\theta}} : \boldsymbol{\theta} \in \Theta_0\}$ with corresponding parametrization $\boldsymbol{\theta}^{(1)} = (\theta_1, \ldots, \theta_q)$. Suppose that $\widehat{\boldsymbol{\theta}}_0^{(1)}$ is the MLE of $\boldsymbol{\theta}^{(1)}$ under H and that $\widehat{\boldsymbol{\theta}}_0^{(1)}$ satisfies A6 for $\mathcal{P}_0$. Let $\widehat{\boldsymbol{\theta}}_{0,n}^T = (\widehat{\boldsymbol{\theta}}_0^{(1)}, \boldsymbol{\theta}_0^{(2)})$. Then under $H : \boldsymbol{\theta} \in \Theta_0$,*

$$2 \log \lambda(\mathbf{X}) \equiv 2[l_n(\widehat{\boldsymbol{\theta}}_n) - l_n(\widehat{\boldsymbol{\theta}}_{0,n})] \xrightarrow{\mathcal{L}} \chi_{r-q}^2.$$

Proof. Let $\boldsymbol{\theta}_0 \in \Theta_0$ and write

$$2 \log \lambda(\mathbf{X}) = 2[l_n(\widehat{\boldsymbol{\theta}}_n) - l_n(\boldsymbol{\theta}_0)] - 2[l_n(\widehat{\boldsymbol{\theta}}_{0,n}) - l_n(\boldsymbol{\theta}_0)]. \tag{6.3.4}$$

It is easy to see that A0–A6 for $\mathcal{P}$ imply A0–A5 for $\mathcal{P}_0$. By (6.2.10) and (6.3.1) applied to $\widehat{\boldsymbol{\theta}}_n$ and the corresponding argument applied to $\widehat{\boldsymbol{\theta}}_0^{(1)}$, $\widehat{\boldsymbol{\theta}}_{0,n}$ and (6.3.4),

$$2 \log \lambda(\mathbf{X}) = \mathbf{S}^T(\boldsymbol{\theta}_0)I^{-1}(\boldsymbol{\theta}_0)\mathbf{S}(\boldsymbol{\theta}_0) - \mathbf{S}_1^T(\boldsymbol{\theta}_0)I_0^{-1}(\boldsymbol{\theta}_0)\mathbf{S}_1(\boldsymbol{\theta}_0) + o_p(1) \tag{6.3.5}$$

where

$$\mathbf{S}(\boldsymbol{\theta}_0) = n^{-1/2} \sum_{i=1}^{n} Dl(\mathbf{X}_i, \boldsymbol{\theta})$$

and $\mathbf{S} = (\mathbf{S}_1, \mathbf{S}_2)^T$ where $\mathbf{S}_1$ is the first q coordinates of $\mathbf{S}$. Furthermore,

$$I_0(\boldsymbol{\theta}_0) = \mathrm{Var}_{\boldsymbol{\theta}_0} \mathbf{S}_1(\boldsymbol{\theta}_0).$$

Make a change of parameter, for given true $\boldsymbol{\theta}_0$ in Θ_0,

$$\boldsymbol{\eta} = M(\boldsymbol{\theta} - \boldsymbol{\theta}_0)$$

where, dropping the dependence on $\boldsymbol{\theta}_0$,

$$M = PI^{1/2} \tag{6.3.6}$$

and P is an orthogonal matrix such that, if $\boldsymbol{\Delta}_0 \equiv \{\boldsymbol{\theta} - \boldsymbol{\theta}_0 : \boldsymbol{\theta} \in \Theta_0\}$

$$M\boldsymbol{\Delta}_0 = \{\boldsymbol{\eta} : \eta_{q+1} = \cdots = \eta_r = 0, \; \boldsymbol{\eta} \in M\Theta\}.$$

Such P exists by the argument given in Example 6.2.1 because $I^{1/2}\boldsymbol{\Delta}_0$ is the intersection of a q dimensional linear subspace of R^r with $I^{1/2}\{\boldsymbol{\theta} - \boldsymbol{\theta}_0 : \boldsymbol{\theta} \in \Theta\}$. Now write $D_{\boldsymbol{\theta}}$ for differentiation with respect to $\boldsymbol{\theta}$ and $D_{\boldsymbol{\eta}}$ for differentiation with respect to $\boldsymbol{\eta}$. Note that, by definition, λ is invariant under reparametrization

$$\lambda(\mathbf{X}) = \gamma(\mathbf{X}) \tag{6.3.7}$$

where

$$\gamma(\mathbf{X}) = \sup_{\boldsymbol{\eta}}\{p(x, \boldsymbol{\theta}_0 + M^{-1}\boldsymbol{\eta})\}/\sup\{p(\mathbf{x}, \boldsymbol{\theta}_0 + M^{-1}\boldsymbol{\eta}) : \boldsymbol{\theta}_0 + M^{-1}\boldsymbol{\eta} \in \Theta_0\}$$

and from (B.8.13)

$$D_{\boldsymbol{\eta}}l(\mathbf{x}, \boldsymbol{\theta}_0 + M^{-1}\boldsymbol{\eta}) = [M^{-1}]^T D_{\boldsymbol{\theta}}l(\mathbf{x}, \boldsymbol{\theta}). \tag{6.3.8}$$

We deduce from (6.3.6) and (6.3.8) that if

$$\mathbf{T}(\boldsymbol{\eta}) \equiv n^{-1/2} \sum_{i=1}^{n} D_{\boldsymbol{\eta}}l(\mathbf{X}_i, \boldsymbol{\theta}_0 + M^{-1}\boldsymbol{\eta}),$$

then

$$\text{Var } \mathbf{T}(\mathbf{0}) = P^T I^{-1/2} II^{-1/2}P = J. \tag{6.3.9}$$

Moreover, because in terms of $\boldsymbol{\eta}$, H is $\{\boldsymbol{\eta} \in M\Theta : \eta_{q+1} = \cdots = \eta_r = 0\}$, then by applying (6.3.5) to $\gamma(\mathbf{X})$ we obtain,

$$\begin{aligned} 2\log\gamma(\mathbf{X}) &= \mathbf{T}^T(\mathbf{0})\mathbf{T}(\mathbf{0}) - \mathbf{T}_1^T(\mathbf{0})\mathbf{T}_1(\mathbf{0}) + o_p(1) \\ &= \sum_{i=1}^{r} T_i^2(\mathbf{0}) - \sum_{i=1}^{q} T_i^2(\mathbf{0}) + o_p(1) \\ &= \sum_{i=q+1}^{r} T_i^2(\mathbf{0}) + o_p(1), \end{aligned} \tag{6.3.10}$$

which has a limiting χ_{r-q}^2 distribution by Slutsky's theorem because $\mathbf{T}(\mathbf{0})$ has a limiting $\mathcal{N}_r(\mathbf{0}, J)$ distribution by (6.3.9). The result follows from (6.3.7). $\square$

Note that this argument is simply an asymptotic version of the one given in Example 6.3.2.

Thus, under the conditions of Theorem 6.3.2, rejecting if $\lambda(\mathbf{X}) \geq x_{r-q}(1 - \alpha)$ is an asymptotically level α test of $H : \boldsymbol{\theta} \in \Theta_0$. Of equal importance is that we obtain an asymptotic confidence region for $(\theta_{q+1}, \ldots, \theta_r)$, a piece of $\boldsymbol{\theta}$, with $\theta_1, \ldots, \theta_q$ acting as nuisance parameters. This asymptotic level $1 - \alpha$ confidence region is

$$\{(\theta_{q+1}, \ldots, \theta_r) : 2[l_n(\widehat{\boldsymbol{\theta}}_n) - l_n(\widehat{\theta}_{0,1}, \ldots, \widehat{\theta}_{0,q}, \theta_{q+1}, \ldots, \theta_r)] \leq x_{r-q}(1 - \alpha)\} \tag{6.3.11}$$

where $\widehat{\theta}_{0,1}, \ldots, \widehat{\theta}_{0,q}$ are the MLEs, themselves depending on $\theta_{q+1}, \ldots, \theta_r$, of $\theta_1, \ldots, \theta_q$ assuming that $\theta_{q+1}, \ldots, \theta_r$ are known.

More complicated linear hypotheses such as $H : \boldsymbol{\theta} - \boldsymbol{\theta}_0 \in \omega_0$ where ω_0 is a linear space of dimension q are also covered. We only need note that if ω_0 is a linear space spanned by an orthogonal basis $\mathbf{v}_1, \ldots, \mathbf{v}_q$ and $\mathbf{v}_{q+1}, \ldots, \mathbf{v}_r$ are orthogonal to ω_0 and $\mathbf{v}_1, \ldots, \mathbf{v}_r$ span R^r then,

$$\omega_0 = \{\boldsymbol{\theta} : \boldsymbol{\theta}^T \mathbf{v}_j = 0, \ q+1 \le j \le r\}. \tag{6.3.12}$$

The extension of Theorem 6.3.2 to this situation is easy and given in Problem 6.3.2.

The formulation of Theorem 6.3.2 is still inadequate for most applications. It can be extended as follows.

Suppose H is specified by:

There exist d functions, $g_j : \Theta \to R$, $q + 1 \le j \le r$ written as a vector $\mathbf{g}$, such that $D\mathbf{g}(\boldsymbol{\theta})$ exists and is of rank $r - q$ at all $\boldsymbol{\theta} \in \Theta$. Define $H : \boldsymbol{\theta} \in \Theta_0$ with

$$\Theta_0 = \{\boldsymbol{\theta} \in \Theta : \mathbf{g}(\boldsymbol{\theta}) = \mathbf{0}\}. \tag{6.3.13}$$

Evidently, Theorem 6.3.2 falls under this schema with $g_j(\boldsymbol{\theta}) = \theta_j - \theta_{0,j}, q + 1 \le j \le r$.

Examples such as testing for independence in contingency tables, which require the following general theorem, will appear in the next section.

Theorem 6.3.3. *Suppose the assumptions of Theorem 6.3.2 and the previously conditions on $\mathbf{g}$ hold. Suppose the MLE $\widehat{\boldsymbol{\theta}}_{0,n}$ under H is consistent for all $\boldsymbol{\theta} \in \Theta_0$. Then, if $\lambda(\mathbf{X})$ is the likelihood ratio statistic for $H : \boldsymbol{\theta} \in \Theta_0$ given in (6.3.13), $2 \log \lambda(\mathbf{X}) \xrightarrow{\mathcal{L}} \chi^2_{r-q}$ under H.*

The proof is sketched in Problems (6.3.2)–(6.3.3). The essential idea is that, if $\boldsymbol{\theta}_0$ is true, $\lambda(\mathbf{X})$ behaves asymptotically like a test for $H : \boldsymbol{\theta} \in \Theta_{00}$ where

$$\Theta_{00} = \{\boldsymbol{\theta} \in \Theta : D\mathbf{g}(\boldsymbol{\theta}_0)(\boldsymbol{\theta} - \boldsymbol{\theta}_0) = \mathbf{0}\} \tag{6.3.14}$$

a hypothesis of the form (6.3.13).

Wilks's theorem depends critically on the fact that not only is Θ open but that if Θ_0 given in (6.3.13) then the set $\{(\theta_1, \ldots, \theta_q)^T : \boldsymbol{\theta} \in \Theta\}$ is open in R^q. We need both properties because we need to analyze both the numerator and denominator of $\lambda(\mathbf{X})$. As an example of what can go wrong, let (X_{i1}, X_{i2}) be i.i.d. $\mathcal{N}(\theta_1, \theta_2, J)$, where J is the 2×2 identity matrix and $\Theta_0 = \{\boldsymbol{\theta} : \theta_1 + \theta_2 \le 1\}$. If $\theta_1 + \theta_2 = 1$,

$$\widehat{\boldsymbol{\theta}}_0 = \left(\frac{(X_{\cdot 1} + X_{\cdot 2})}{2} + \frac{1}{2}, \frac{1}{2} - \frac{(X_{\cdot 1} + X_{\cdot 2})}{2} \right)$$

and $2 \log \lambda(\mathbf{X}) \to \chi^2_1$ but if $\theta_1 + \theta_2 < 1$ clearly $2 \log \lambda(\mathbf{X}) = o_p(1)$. Here the dimension of Θ_0 and Θ is the same but the boundary of Θ_0 has lower dimension. More sophisticated examples are given in Problems 6.3.5 and 6.3.6.

6.3.2   Wald's and Rao's Large Sample Tests

The Wald Test

Suppose that the assumptions of Theorem 6.2.2 hold. Then

$$\sqrt{n}(\hat{\boldsymbol{\theta}} - \boldsymbol{\theta}) \overset{\mathcal{L}}{\to} \mathcal{N}(\mathbf{0}, I^{-1}(\boldsymbol{\theta})) \text{ as } n \to \infty. \tag{6.3.15}$$

Because $I(\boldsymbol{\theta})$ is continuous in $\boldsymbol{\theta}$ (Problem 6.3.10), it follows from Proposition B.7.1(a) that

$$I(\hat{\boldsymbol{\theta}}_n) \overset{P}{\to} I(\boldsymbol{\theta}) \text{ as } n \to \infty. \tag{6.3.16}$$

By Slutsky's theorem B.7.2, (6.3.15) and (6.3.16),

$$n(\hat{\boldsymbol{\theta}}_n - \boldsymbol{\theta})^T I(\hat{\boldsymbol{\theta}}_n)(\hat{\boldsymbol{\theta}}_n - \boldsymbol{\theta}) \overset{\mathcal{L}}{\to} \mathbf{V}^T I(\boldsymbol{\theta})\mathbf{V}, \ \mathbf{V} \sim \mathcal{N}_r(\mathbf{0}, I^{-1}(\boldsymbol{\theta}))$$

where, according to Corollary B.6.2, $\mathbf{V}^T I(\boldsymbol{\theta})\mathbf{V} \sim \chi_r^2$. It follows that the *Wald test* that rejects $H : \boldsymbol{\theta} = \boldsymbol{\theta}_0$ in favor of $K : \boldsymbol{\theta} \neq \boldsymbol{\theta}_0$ when

$$W_n(\boldsymbol{\theta}_0) = n(\hat{\boldsymbol{\theta}}_n - \boldsymbol{\theta}_0)^T I(\boldsymbol{\theta}_0)(\hat{\boldsymbol{\theta}}_n - \boldsymbol{\theta}_0) \geq x_r(1 - \alpha)$$

has asymptotic level α. More generally $I(\boldsymbol{\theta}_0)$ can be replaced by any consistent estimate of $I(\boldsymbol{\theta}_0)$, in particular $-\frac{1}{n}D^2 l_n(\boldsymbol{\theta}_0)$ or $I(\hat{\boldsymbol{\theta}}_n)$ or $-\frac{1}{n}D^2 l_n(\hat{\boldsymbol{\theta}}_n)$. The last Hessian choice is favored because it is usually computed automatically with the MLE. It and $I(\hat{\boldsymbol{\theta}}_n)$ also have the advantage that the confidence region one generates $\{\boldsymbol{\theta} : W_n(\boldsymbol{\theta}) \leq x_p(1 - \alpha)\}$ is an ellipsoid in R^r easily interpretable and computable—see (6.1.31).

For the more general hypothesis $H : \theta \in \Theta_0$ we write the MLE for $\theta \in \Theta$ as $\hat{\boldsymbol{\theta}}_n = (\hat{\boldsymbol{\theta}}_n^{(1)}, \hat{\boldsymbol{\theta}}_n^{(2)})$ where $\hat{\boldsymbol{\theta}}_n^{(1)} = (\hat{\theta}_1, \ldots, \hat{\theta}_q)$ and $\hat{\boldsymbol{\theta}}_n^{(2)} = (\hat{\theta}_{q+1}, \ldots, \hat{\theta}_r)$ and define the *Wald statistic* as

$$W_n(\boldsymbol{\theta}_0^{(2)}) = n(\hat{\boldsymbol{\theta}}_n^{(2)} - \boldsymbol{\theta}_0^{(2)})^T [I^{22}(\hat{\boldsymbol{\theta}}_n)]^{-1}(\hat{\boldsymbol{\theta}}_n^{(2)} - \boldsymbol{\theta}_0^{(2)}) \tag{6.3.17}$$

where $I^{22}(\boldsymbol{\theta})$ is the lower diagonal block of $I^{-1}(\boldsymbol{\theta})$ written as

$$I^{-1}(\boldsymbol{\theta}) = \begin{pmatrix} I^{11}(\boldsymbol{\theta}) & I^{12}(\boldsymbol{\theta}) \\ I^{21}(\boldsymbol{\theta}) & I^{22}(\boldsymbol{\theta}) \end{pmatrix}$$

with diagonal blocks of dimension $q \times q$ and $d \times d$, respectively. More generally, $I^{22}(\hat{\boldsymbol{\theta}}_n)$ is replaceable by any consistent estimate of $I^{22}(\boldsymbol{\theta})$, for instance, the lower diagonal block of the inverse of $-\frac{1}{n}D^2 l_n(\hat{\boldsymbol{\theta}}_n)$, the Hessian (Problem 6.3.9).

Theorem 6.3.4. *Under the conditions of Theorem 6.2.2, if H is true,*

$$W_n(\boldsymbol{\theta}_0^{(2)}) \overset{\mathcal{L}}{\to} \chi_{r-q}^2. \tag{6.3.18}$$

Proof. $I(\boldsymbol{\theta})$ continuous implies that $I^{-1}(\boldsymbol{\theta})$ is continuous and, hence, I^{22} is continuous. But by Theorem 6.2.2, $\sqrt{n}(\hat{\boldsymbol{\theta}}_n^{(2)} - \boldsymbol{\theta}_0^{(2)}) \overset{\mathcal{L}}{\to} \mathcal{N}_d(\mathbf{0}, I^{22}(\boldsymbol{\theta}_0))$ if $\boldsymbol{\theta}_0 \in \Theta_0$ holds. Slutsky's theorem completes the proof.   □

The *Wald* test, which rejects iff $W_n(\widehat{\boldsymbol{\theta}}_0^{(2)}) \geq x_{r-q}(1-\alpha)$, is, therefore, asymptotically level α. What is not as evident is that, under H,

$$W_n(\widehat{\boldsymbol{\theta}}_0^{(2)}) = 2 \log \lambda(\mathbf{X}) + o_p(1) \qquad (6.3.19)$$

where $\lambda(\mathbf{X})$ is the LR statistic for $H : \boldsymbol{\theta} \in \Theta_0$. The argument is sketched in Problem 6.3.9. Thus, the two tests are equivalent asymptotically.

The Wald test leads to the *Wald confidence regions* for $(\theta_{q+1}, \ldots, \theta_r)^T$ given by $\{\boldsymbol{\theta}^{(2)} : W_n(\boldsymbol{\theta}^{(2)}) \leq x_{r-q}(1-\alpha)\}$. These regions are ellipsoids in R^d. Although, as (6.3.19) indicates, the Wald and likelihood ratio tests and confidence regions are asymptotically equivalent in the sense that the same conclusions are reached for large n, in practice they can be very different.

The Rao Score Test

For the simple hypothesis $H : \boldsymbol{\theta} = \boldsymbol{\theta}_0$, Rao's score test is based on the observation that, by the central limit theorem,

$$\sqrt{n}\boldsymbol{\psi}_n(\boldsymbol{\theta}_0) \xrightarrow{\mathcal{L}} \mathcal{N}(\mathbf{0}, I(\boldsymbol{\theta}_0)) \qquad (6.3.20)$$

where $\boldsymbol{\psi}_n = n^{-1}Dl_n(\boldsymbol{\theta}_0)$ is the likelihood score vector.

It follows from this and Corollary B.6.2 that under H, as $n \to \infty$,

$$R_n(\boldsymbol{\theta}_0) = n\boldsymbol{\psi}_n^T(\boldsymbol{\theta}_0)I^{-1}(\boldsymbol{\theta}_0)\boldsymbol{\psi}_n(\boldsymbol{\theta}_0) \xrightarrow{\mathcal{L}} \chi_r^2.$$

The test that rejects H when $R_n(\boldsymbol{\theta}_0) \geq x_r(1-\alpha)$ is called the *Rao score test*. This test has the advantage that it can be carried out without computing the MLE, and the convergence $R_n(\boldsymbol{\theta}_0) \xrightarrow{\mathcal{L}} \chi_r^2$ requires much weaker regularity conditions than does the corresponding convergence for the likelihood ratio and Wald tests.

The extension of the Rao test to $H : \boldsymbol{\theta} \in \Theta_0$ runs as follows. Let

$$\boldsymbol{\Psi}_n(\boldsymbol{\theta}) = n^{-1/2}D_2l_n(\boldsymbol{\theta})$$

where D_1l_n represents the $q \times 1$ gradient with respect to the first q coordinates and D_2l_n the $d \times 1$ gradient with respect to the last d. The Rao test is based on the statistic

$$R_n(\boldsymbol{\theta}_0^{(2)}) \equiv n\boldsymbol{\Psi}_n^T(\widehat{\boldsymbol{\theta}}_{0,n})\widehat{\boldsymbol{\Sigma}}^{-1}\boldsymbol{\Psi}_n(\widehat{\boldsymbol{\theta}}_{0,n})$$

where $\widehat{\boldsymbol{\Sigma}}$ is a consistent estimate of $\boldsymbol{\Sigma}(\boldsymbol{\theta}_0)$, the asymptotic variance of $\sqrt{n}\boldsymbol{\Psi}_n(\widehat{\boldsymbol{\theta}}_{0,n})$ under H.

It can be shown that (Problem 6.3.8)

$$\boldsymbol{\Sigma}(\boldsymbol{\theta}_0) = I_{22}(\boldsymbol{\theta}_0) - I_{21}(\boldsymbol{\theta}_0)I_{11}^{-1}(\boldsymbol{\theta}_0)I_{12}(\boldsymbol{\theta}_0) \qquad (6.3.21)$$

where I_{11} is the upper left $q \times q$ block of the $r \times r$ information matrix $I(\boldsymbol{\theta}_0)$, I_{12} is the upper right $q \times d$ block, and so on. Furthermore, (Problem 6.3.9) under A0–A6 and consistency of $\widehat{\boldsymbol{\theta}}_{0,n}$ under H, a consistent estimate of $\boldsymbol{\Sigma}^{-1}(\boldsymbol{\theta}_0)$ is

$$n^{-1}[-D_2^2l_n(\widehat{\boldsymbol{\theta}}_{0,n}) + D_{21}l_n(\widehat{\boldsymbol{\theta}}_{0,n})[D_1^2l_n(\widehat{\boldsymbol{\theta}}_{0,n}]^{-1}D_{12}l_n(\widehat{\boldsymbol{\theta}}_{0,n})] \qquad (6.3.22)$$

where D_2^2 is the $d \times d$ matrix of second partials of l_n with respect to $\boldsymbol{\theta}^{(1)}$, D_{21} the $d \times d$ matrix of mixed second partials with respect to $\boldsymbol{\theta}^{(1)}, \boldsymbol{\theta}^{(2)}$, and so on.

Theorem 6.3.5. *Under $H : \boldsymbol{\theta} \in \Theta_0$ and the conditions A0–A5 of Theorem 6.2.2 but with A6 required only for $\mathcal{P}_0$*

$$R_n(\boldsymbol{\theta}_0^{(2)}) \xrightarrow{\mathcal{L}} \chi_d^2.$$

The Rao large sample critical and confidence regions are $\{R_n(\boldsymbol{\theta}_0^{(2)}) \geq x_d(1 - \alpha)\}$ and $\{\boldsymbol{\theta}^{(2)} : R_n(\boldsymbol{\theta}^{(2)}) < x_d(1 - \alpha)\}$.

The advantage of the Rao test over those of Wald and Wilks is that MLEs need to be computed only under H. On the other hand, it shares the disadvantage of the Wald test that matrices need to be computed and inverted.

Power Behavior of the LR, Rao, and Wald Tests

It is possible as in the one-dimensional case to derive the asymptotic power for these tests for alternatives of the form $\boldsymbol{\theta}_n = \boldsymbol{\theta}_0 + \frac{\boldsymbol{\Delta}}{\sqrt{n}}$ where $\boldsymbol{\theta}_0 \in \Theta_0$. The analysis for $\Theta_0 = \{\boldsymbol{\theta}_0\}$ is relatively easy. For instance, for the Wald test

$$n(\widehat{\boldsymbol{\theta}}_n - \boldsymbol{\theta}_0)^T I(\widehat{\boldsymbol{\theta}}_n)(\widehat{\boldsymbol{\theta}}_n - \boldsymbol{\theta}_0)$$
$$= (\sqrt{n}(\widehat{\boldsymbol{\theta}}_n - \boldsymbol{\theta}_n) + \boldsymbol{\Delta})I(\widehat{\boldsymbol{\theta}}_n)(\sqrt{n}(\widehat{\boldsymbol{\theta}}_n - \boldsymbol{\theta}_n) + \boldsymbol{\Delta}) \xrightarrow{\mathcal{L}} \chi_{r-q}^2(\boldsymbol{\Delta}^T I(\boldsymbol{\theta}_0)\boldsymbol{\Delta})$$

where $\chi_m^2(\gamma^2)$ is the noncentral chi square distribution with m degrees of freedom and noncentrality parameter γ^2.

It may be shown that the equivalence (6.3.19) holds under $\boldsymbol{\theta}_n$ and that the power behavior is unaffected and applies to all three tests.

Consistency for fixed alternatives is clear for the Wald test but requires conditions for the likelihood ratio and score tests—see Rao (1973) for more on this.

Summary. We considered the problem of testing $H : \boldsymbol{\theta} \in \Theta_0$ versus $K : \boldsymbol{\theta} \in \Theta - \Theta_0$ where Θ is an open subset of R^r and Θ_0 is the collection of $\boldsymbol{\theta} \in \Theta$ with the last $r - q$ coordinates $\boldsymbol{\theta}^{(2)}$ specified. We established Wilks's theorem, which states that if $\lambda(\mathbf{X})$ is the LR statistic, then, under regularity conditions, $2 \log \lambda(\mathbf{X})$ has an asymptotic χ_{r-q}^2 distribution under H. We also considered a quadratic form, called the Wald statistic, which measures the distance between the hypothesized value of $\boldsymbol{\theta}^{(2)}$ and its MLE, and showed that this quadratic form has limiting distribution χ_{r-q}^2. Finally, we introduced the Rao score test, which is based on a quadratic form in the gradient of the log likelihood. The asymptotic distribution of this quadratic form is also χ_{r-q}^2.

6.4   LARGE SAMPLE METHODS FOR DISCRETE DATA

In this section we give a number of important applications of the general methods we have developed to inference for discrete data. In particular we shall discuss problems of goodness-of-fit and special cases of log linear and generalized linear models (GLM), treated in more detail in Section 6.5.

6.4.1 Goodness-of-Fit in a Multinomial Model. Pearson's χ^2 Test

As in Examples 1.6.7, 2.2.8, and 2.3.3, consider i.i.d. trials in which $X_i = j$ if the ith trial produces a result in the jth category, $j = 1, \ldots, k$. Let $\theta_j = P(X_i = j)$ be the probability of the jth category. Because $\theta_k = 1 - \sum_{j=1}^{k-1} \theta_j$, we consider the parameter $\boldsymbol{\theta} = (\theta_1, \ldots, \theta_{k-1})^T$ and test the hypothesis $H : \theta_j = \theta_{0j}$ for specified θ_{0j}, $j = 1, \ldots, k-1$. Thus, we may be testing whether a random number generator used in simulation experiments is producing values according to a given distribution, or we may be testing whether the phenotypes in a genetic experiment follow the frequencies predicted by theory. In Example 2.2.8 we found the MLE $\hat{\theta}_j = N_j/n$, where $N_j = \sum_{i=1}^{n} 1\{X_i = j\}$. It follows that the large sample LR rejection region is

$$2\log \lambda(\mathbf{X}) = 2 \sum_{j=1}^{k} N_j \log(N_j/n\theta_{0j}) \geq x_{k-1}(1-\alpha).$$

To find the Wald test, we need the information matrix $I = \|I_{ij}\|$. For $i, j = 1, \ldots, k-1$, we find using (2.2.33) and (3.4.32) that

$$
\begin{aligned}
I_{ij} &= \left[\frac{1}{\theta_j} + \frac{1}{\theta_k} \right] && \text{if } i = j, \\
&= \frac{1}{\theta_k} && \text{if } i \neq j.
\end{aligned}
$$

Thus, with $\theta_{0k} = 1 - \sum_{j=1}^{k-1} \theta_{0j}$, the Wald statistic is

$$W_n(\boldsymbol{\theta}_0) = n \sum_{j=1}^{k-1} [\hat{\theta}_j - \theta_{0j}]^2/\theta_{0j} + n \sum_{j=1}^{k-1} \sum_{i=1}^{k-1} (\hat{\theta}_i - \theta_{0i})(\hat{\theta}_j - \theta_{0j})/\theta_{0k}.$$

The second term on the right is

$$n \left[\sum_{j=1}^{k-1} (\hat{\theta}_j - \theta_{0j}) \right]^2 \Big/ \theta_{0k} = n(\hat{\theta}_k - \theta_{0k})^2/\theta_{0k}.$$

Thus,

$$W_n(\boldsymbol{\theta}_0) = \sum_{j=1}^{k} (N_j - n\theta_{0j})^2/n\theta_{0j}.$$

The term on the right is called *Pearson's chi-square* (χ^2) *statistic* and is the statistic that is typically used for this multinomial testing problem. It is easily remembered as

$$\chi^2 = \text{SUM} \frac{(\text{Observed} - \text{Expected})^2}{\text{Expected}} \tag{6.4.1}$$

where the sum is over categories and "expected" refers to the expected frequency $E_H(N_j)$. The general form (6.4.1) of Pearson's χ^2 will reappear in other multinomial applications in this section.

To derive the Rao test, note that from Example 2.2.8,

$$\boldsymbol{\psi}_n(\boldsymbol{\theta}) = n^{-1}(\psi_1(\boldsymbol{\theta}), \ldots, \psi_{k-1}(\boldsymbol{\theta}))^T,$$

with

$$\psi_j(\boldsymbol{\theta}) = \frac{\partial}{\partial \theta_j} l_n(\boldsymbol{\theta}) = \frac{N_j}{\theta_j} - \frac{N_k}{\theta_k}, \; j = 1, \ldots, k-1.$$

To find I^{-1}, we could invert I or note that by (6.2.11), $I^{-1}(\boldsymbol{\theta}) = \Sigma = \text{Var}(\mathbf{N})$, where $\mathbf{N} = (N_1, \ldots, N_{k-1})^T$ and, by A.13.15, $\Sigma = \|\sigma_{ij}\|_{(k-1)\times(k-1)}$ with

$$\sigma_{ii} = \text{Var}(N_i) = n\theta_i(1 - \theta_i), \; \sigma_{ij} = -n\theta_i\theta_j, \; i \neq j.$$

Thus, the Rao statistic is

$$R_n(\boldsymbol{\theta}_0) = \left(n \sum_{j=1}^{k-1} \left(\frac{\widehat{\theta}_j}{\theta_{0j}} - \frac{\widehat{\theta}_k}{\theta_{0k}} \right)^2 \theta_{0j} \right)$$

$$- \left(n \sum_{j=1}^{k-1} \sum_{i=1}^{k-1} \left(\frac{\widehat{\theta}_i}{\theta_{0i}} - \frac{\widehat{\theta}_k}{\theta_{0k}} \right) \left(\frac{\widehat{\theta}_j}{\theta_{0j}} - \frac{\widehat{\theta}_k}{\theta_{0k}} \right) \theta_{0i}\theta_{0j} \right). \tag{6.4.2}$$

The second term on the right is

$$-n \left[\sum_{j=1}^{k-1} \left(\frac{\widehat{\theta}_j}{\theta_{0j}} - \frac{\widehat{\theta}_k}{\theta_{0k}} \right) \theta_{0j} \right]^2 = -n \left[\frac{\widehat{\theta}_k}{\theta_{0k}} - 1 \right]^2.$$

To simplify the first term on the right of (6.4.2), we write

$$\frac{\widehat{\theta}_j}{\theta_{0j}} - \frac{\widehat{\theta}_k}{\theta_{0k}} = \{[\theta_{0k}(\widehat{\theta}_j - \theta_{0j})] - [\theta_{0j}(\widehat{\theta}_k - \theta_{0k})]\} \frac{1}{\theta_{0j}\theta_{0k}},$$

and expand the square keeping the square brackets intact. Then, because

$$\sum_{j=1}^{k-1} (\widehat{\theta}_j - \theta_{0j}) = -(\widehat{\theta}_k - \theta_{0k}),$$

the first term on the right of (6.4.2) becomes

$$
= n \left\{ \sum_{j=1}^{k-1} \frac{1}{\theta_{0j}} (\widehat{\theta}_j - \theta_{0j})^2 + \frac{2}{\theta_{0k}} (\widehat{\theta}_k - \theta_{0k})^2 + \frac{1}{\theta_{0k}^2} (1 - \theta_{0k})(\widehat{\theta}_k - \theta_{0k})^2 \right\}
$$

$$
= n \left\{ \sum_{j=1}^{k} \frac{1}{\theta_{0j}} (\widehat{\theta}_j - \theta_{0j})^2 + \frac{1}{\theta_{0k}^2} (\widehat{\theta}_k - \theta_{0k})^2 \right\}.
$$

It follows that the Rao statistic equals Pearson's χ^2.

Example 6.4.1. *Testing a Genetic Theory.* In experiments on pea breeding, Mendel observed the different kinds of seeds obtained by crosses from peas with round yellow seeds and peas with wrinkled green seeds. Possible types of progeny were: (1) round yellow; (2) wrinkled yellow; (3) round green; and (4) wrinkled green. If we assume the seeds are produced independently, we can think of each seed as being the outcome of a multinomial trial with possible outcomes numbered 1, 2, 3, 4 as above and associated probabilities of occurrence $\theta_1, \theta_2, \theta_3, \theta_4$. Mendel's theory predicted that $\theta_1 = 9/16$, $\theta_2 = \theta_3 = 3/16$, $\theta_4 = 1/16$, and we want to test whether the distribution of types in the $n = 556$ trials he performed (seeds he observed) is consistent with his theory. Mendel observed $n_1 = 315$, $n_2 = 101$, $n_3 = 108$, $n_4 = 32$. Then, $n\theta_{10} = 312.75$, $n\theta_{20} = n\theta_{30} = 104.25$, $n\theta_{40} = 34.75$, $k = 4$

$$
\chi^2 = \frac{(2.25)^2}{312.75} + \frac{(3.25)^2}{104.25} + \frac{(3.75)^2}{104.25} + \frac{(2.75)^2}{34.75} = 0.47,
$$

which has a p-value of 0.9 when referred to a χ_3^2 table. There is insufficient evidence to reject Mendel's hypothesis. For comparison $2 \log \lambda = 0.48$ in this case. However, this value may be too small! See Note 1.

6.4.2   Goodness-of-Fit to Composite Multinomial Models. Contingency Tables

Suppose $\mathbf{N} = (N_1, \ldots, N_k)^T$ has a multinomial, $\mathcal{M}(n, \boldsymbol{\theta})$, distribution. We will investigate how to test $H : \boldsymbol{\theta} \in \Theta_0$ versus $K : \boldsymbol{\theta} \notin \Theta_0$, where Θ_0 is a composite "smooth" subset of the $(k-1)$-dimensional parameter space

$$
\Theta = \{ \boldsymbol{\theta} : \theta_i \geq 0, \ 1 \leq i \leq k, \ \sum_{i=1}^{k} \theta_i = 1 \}.
$$

For example, in the Hardy–Weinberg model (Example 2.1.4),

$$
\Theta_0 = \{ (\eta^2, 2\eta(1-\eta), (1-\eta)^2), \ 0 \leq \eta \leq 1 \},
$$

which is a one-dimensional curve in the two-dimensional parameter space Θ. Here testing the adequacy of the Hardy–Weinberg model means testing $H : \boldsymbol{\theta} \in \Theta_0$ versus $K : \boldsymbol{\theta} \in$

Θ_1 where $\Theta_1 = \Theta - \Theta_0$. Other examples, which will be pursued further later in this section, involve restrictions on the θ_i obtained by specifying independence assumptions on classifications of cases into different categories.

We suppose that we can describe Θ_0 parametrically as

$$\Theta_0 = \{(\theta_1(\boldsymbol{\eta}), \ldots, \theta_k(\boldsymbol{\eta}) : \boldsymbol{\eta} \in \mathcal{E}\},$$

where $\boldsymbol{\eta} = (\eta_1, \ldots, \eta_q)^T$, is a subset of q-dimensional space, and the map $\boldsymbol{\eta} \to (\theta_1(\boldsymbol{\eta}), \ldots, \theta_k(\boldsymbol{\eta}))^T$ takes $\mathcal{E}$ into Θ_0. To avoid trivialities we assume $q < k - 1$.

Consider the likelihood ratio test for $H : \theta \in \Theta_0$ versus $K : \theta \notin \Theta_0$. Let $p(n_1, \ldots, n_k, \boldsymbol{\theta})$ denote the frequency function of $\mathbf{N}$. Maximizing $p(n_1, \ldots, n_k, \boldsymbol{\theta})$ for $\boldsymbol{\theta} \in \Theta_0$ is the same as maximizing $p(n_1, \ldots, n_k, \boldsymbol{\theta}(\boldsymbol{\eta}))$ for $\boldsymbol{\eta} \in \mathcal{E}$. If a maximizing value, $\widehat{\boldsymbol{\eta}} = (\widehat{\eta}_1, \ldots, \widehat{\eta}_q)$ exists, the log likelihood ratio is given by

$$\log \lambda(n_1, \ldots, n_k) = \sum_{i=1}^{k} n_i [\log(n_i/n) - \log \theta_i(\widehat{\boldsymbol{\eta}})].$$

If $\boldsymbol{\eta} \to \theta(\boldsymbol{\eta})$ is differentiable in each coordinate, $\mathcal{E}$ is open, and $\widehat{\boldsymbol{\eta}}$ exists, then it must solve the likelihood equation for the model, $\{p(\cdot, \boldsymbol{\theta}(\boldsymbol{\eta})) : \boldsymbol{\eta} \in \mathcal{E}\}$. That is, $\widehat{\boldsymbol{\eta}}$ satisfies

$$\frac{\partial}{\partial \eta_j} \log p(n_1, \ldots, n_k, \boldsymbol{\theta}(\boldsymbol{\eta})) = 0, \ 1 \le j \le q \qquad (6.4.3)$$

or

$$\sum_{i=1}^{k} \frac{n_i}{\theta_i(\boldsymbol{\eta})} \frac{\partial}{\partial \eta_j} \theta_i(\boldsymbol{\eta}) = 0, \ 1 \le j \le q.$$

If $\mathcal{E}$ is not open sometimes the closure of $\mathcal{E}$ will contain a solution of (6.4.3).

To apply the results of Section 6.3 and conclude that, under H, $2 \log \lambda$ approximately has a χ^2_{r-q} distribution for large n, we define $\theta'_j = g_j(\boldsymbol{\theta})$, $j = 1, \ldots, r$, where g_j is chosen so that H becomes equivalent to "$(\theta'_1, \ldots, \theta'_q)^T$ ranges over an open subset of R^q and $\theta_j = \theta_{0j}$, $j = q + 1, \ldots, r$ for specified θ_{0j}." For instance, to test the Hardy–Weinberg model we set $\theta'_1 = \theta_1$, $\theta'_2 = \theta_2 - 2\sqrt{\theta_1}(1 - \sqrt{\theta_1})$ and test $H : \theta'_2 = 0$. Then we can conclude from Theorem 6.3.3 that $2 \log \lambda$ approximately has a χ^2_1 distribution under H.

The Rao statistic is also invariant under reparametrization and, thus, approximately χ^2_{r-q}. Moreover, we obtain the Rao statistic for the composite multinomial hypothesis by replacing θ_{0j} in (6.4.2) by $\theta_j(\widehat{\boldsymbol{\eta}})$. The algebra showing $R_n(\boldsymbol{\theta}_0) = \chi^2$ in Section 6.4.1 now leads to the Rao statistic

$$R_n(\boldsymbol{\theta}(\widehat{\boldsymbol{\eta}})) = \sum_{j=1}^{k} \frac{[N_i - n\theta_j(\widehat{\boldsymbol{\eta}})]^2}{n\theta_j(\widehat{\boldsymbol{\eta}})} = \chi^2$$

where the right-hand side is Pearson's χ^2 as defined in general by (6.4.1).

The Wald statistic is only asymptotically invariant under reparametrization. However, the Wald statistic based on the parametrization $\boldsymbol{\theta}(\boldsymbol{\eta})$ obtained by replacing θ_{0j} by $\theta_j(\widehat{\boldsymbol{\eta}})$, is, by the algebra of Section 6.4.1, also equal to Pearson's χ^2.

Example 6.4.4. *Hardy–Weinberg.* We found in Example 2.2.6 that $\widehat{\eta} = (2n_1 + n_2)/2n$. Thus, H is rejected if $\chi^2 \geq x_1(1 - \alpha)$ with

$$\boldsymbol{\theta}(\widehat{\eta}) = \left(\left(\frac{2n_1 + n_2}{2n} \right)^2, \frac{(2n_1 + n_2)(2n_3 + n_2)}{2n^2}, \left(\frac{2n_3 + n_2}{2n} \right)^2 \right)^T.$$

$\square$

Example 6.4.5. *The Fisher Linkage Model.* A self-crossing of maize heterozygous on two characteristics (starchy versus sugary; green base leaf versus white base leaf) leads to four possible offspring types: (1) sugary-white; (2) sugary-green; (3) starchy-white; (4) starchy-green. If N_i is the number of offspring of type i among a total of n offspring, then $(N_1, \ldots, N_4)$ has a $\mathcal{M}(n, \theta_1, \ldots, \theta_4)$ distribution. A linkage model (Fisher, 1958, p. 301), specifies that

$$\theta_1 = \frac{1}{4}(2 + \eta), \ \theta_2 = \theta_3 = \frac{1}{4}(1 - \eta), \ \theta_4 = \frac{1}{4}\eta$$

where η is an unknown number between 0 and 1. To test the validity of the linkage model we would take $\Theta_0 = \left\{ \left(\frac{1}{4}(2 + \eta), \frac{1}{4}(1 - \eta), \frac{1}{4}(1 - \eta), \frac{1}{4}\eta \right) : 0 \leq \eta \leq 1 \right\}$ a "one-dimensional curve" of the three-dimensional parameter space Θ.

The likelihood equation (6.4.3) becomes

$$\frac{n_1}{(2 + \eta)} - \frac{(n_2 + n_3)}{(1 - \eta)} + \frac{n_4}{\eta} = 0, \tag{6.4.4}$$

which reduces to a quadratic equation in $\widehat{\eta}$. The only root of this equation in $[0, 1]$ is the desired estimate (see Problem 6.4.1). Because $q = 1, k = 4$, we obtain critical values from the χ_2^2 tables.

$\square$

Testing Independence of Classifications in Contingency Tables

Many important characteristics have only two categories. An individual either is or is not inoculated against a disease; is or is not a smoker; is male or female; and so on. We often want to know whether such characteristics are linked or are independent. For instance, do smoking and lung cancer have any relation to each other? Are sex and admission to a university department independent classifications? Let us call the possible categories or states of the first characteristic A and $\bar{A}$ and of the second B and $\bar{B}$. Then a randomly selected individual from the population can be one of four types $AB, A\bar{B}, \bar{A}B, \bar{A}\bar{B}$. Denote the probabilities of these types by $\theta_{11}, \theta_{12}, \theta_{21}, \theta_{22}$, respectively. Independent classification then means that the events [being an A] and [being a B] are independent or in terms of the θ_{ij},

$$\theta_{ij} = (\theta_{i1} + \theta_{i2})(\theta_{1j} + \theta_{2j}).$$

To study the relation between the two characteristics we take a random sample of size n from the population. The results are assembled in what is called a 2×2 *contingency table* such as the one shown.

	B	$\bar{B}$
A	N_{11}	N_{12}
$\bar{A}$	N_{21}	N_{22}

The entries in the boxes of the table indicate the number of individuals in the sample who belong to the categories of the appropriate row and column. Thus, for example N_{12} is the number of sampled individuals who fall in category A of the first characteristic and category $\bar{B}$ of the second characteristic. Then, if $\mathbf{N} = (N_{11}, N_{12}, N_{21}, N_{22})^T$, we have $\mathbf{N} \sim \mathcal{M}(n, \theta_{11}, \theta_{12}, \theta_{21}, \theta_{22})$. We test the hypothesis $H : \boldsymbol{\theta} \in \Theta_0$ versus $K : \theta \notin \Theta_0$, where Θ_0 is a two-dimensional subset of Θ given by

$$\Theta_0 = \{(\eta_1 \eta_2, \eta_1(1 - \eta_2), \eta_2(1 - \eta_1), (1 - \eta_1)(1 - \eta_2)) : 0 \le \eta_1 \le 1, \ 0 \le \eta_2 \le 1\}.$$

Here we have relabeled $\theta_{11} + \theta_{12}$, $\theta_{11} + \theta_{21}$ as η_1, η_2 to indicate that these are parameters, which vary freely.

For $\boldsymbol{\theta} \in \Theta_0$, the likelihood equations (6.4.3) become

$$\begin{aligned}
\frac{(n_{11} + n_{12})}{\widehat{\eta}_1} &= \frac{(n_{21} + n_{22})}{(1 - \widehat{\eta}_1)} \\
\frac{(n_{11} + n_{21})}{\widehat{\eta}_2} &= \frac{(n_{12} + n_{22})}{(1 - \widehat{\eta}_2)}
\end{aligned} \tag{6.4.5}$$

whose solutions are

$$\begin{aligned}
\widehat{\eta}_1 &= (n_{11} + n_{12})/n \\
\widehat{\eta}_2 &= (n_{11} + n_{21})/n,
\end{aligned} \tag{6.4.6}$$

the proportions of individuals of type A and type B, respectively. These solutions are the maximum likelihood estimates. Pearson's statistic is then easily seen to be

$$\chi^2 = n \sum_{i=1}^{2} \sum_{j=1}^{2} \frac{(N_{ij} - R_i C_j/n)^2}{R_i C_j}, \tag{6.4.7}$$

where $R_i = N_{i1} + N_{i2}$ is the ith row sum, $C_j = N_{1j} + N_{2j}$ is the jth column sum.

By our theory if H is true, because $k = 4$, $q = 2$, χ^2 has approximately a χ_1^2 distribution. This suggests that χ^2 may be written as the square of a single (approximately) standard normal variable. In fact (Problem 6.4.2), the $(N_{ij} - R_i C_j/n)$ are all the same in absolute value and,

$$\chi^2 = Z^2$$

where

$$\begin{aligned}
Z &= \left(N_{11} - \frac{R_1 C_1}{n}\right) \sqrt{\sum_{i=1}^{2} \sum_{j=1}^{2} \left[\frac{R_i C_j}{n}\right]^{-1}} \\
&= \left(N_{11} - \frac{R_1 C_1}{n}\right) \left[\frac{R_1 R_2 C_1 C_2}{n^3}\right]^{-1/2}.
\end{aligned}$$

An important alternative form for Z is given by

$$Z = \left(\frac{N_{11}}{C_1} - \frac{N_{12}}{C_2} \right) \sqrt{\frac{C_1 C_2 n}{R_1 R_2}}. \tag{6.4.8}$$

Thus,

$$Z = \sqrt{n}[\widehat{P}(A \mid B) - \widehat{P}(A \mid \bar{B})] \left[\frac{\widehat{P}(B)}{\widehat{P}(A)} \frac{\widehat{P}(\bar{B})}{\widehat{P}(\bar{A})} \right]^{1/2}$$

where $\widehat{P}$ is the empirical distribution and where we use $A, B, \bar{A}, \bar{B}$ to denote the event that a randomly selected individual has characteristic $A, B, \bar{A}, \bar{B}$. Thus, if χ^2 measures deviations from independence, Z indicates what directions these deviations take. Positive values of Z indicate that A and B are positively associated (i.e., that A is more likely to occur in the presence of B than it would in the presence of $\bar{B}$). It may be shown (Problem 6.4.3) that if A and B are independent, that is, $P(A \mid B) = P(A \mid \bar{B})$, then Z is approximately distributed as $\mathcal{N}(0, 1)$. Therefore, it is reasonable to use the test that rejects, if and only if,

$$Z \geq z(1 - \alpha)$$

as a level α one-sided test of $H : P(A \mid B) = P(A \mid \bar{B})$ (or $P(A \mid B) \leq P(A \mid \bar{B})$) versus $K : P(A \mid B) > P(A \mid \bar{B})$. The χ^2 test is equivalent to rejecting (two-sidedly) if, and only if,

$$|Z| \geq z \left(1 - \frac{\alpha}{2} \right).$$

Next we consider contingency tables for two nonnumerical characteristics having a and b states, respectively, $a, b \geq 2$ (e.g., eye color, hair color). If we take a sample of size n from a population and classify them according to each characteristic we obtain a vector N_{ij}, $i = 1, \dots, a$, $j = 1, \dots, b$ where N_{ij} is the number of individuals of type i for characteristic 1 and j for characteristic 2. If $\theta_{ij} = P[A$ randomly selected individual is of type i for 1 and j for 2], then

$$\{N_{ij} : 1 \leq i \leq a, \ 1 \leq j \leq b\} \sim \mathcal{M}(n, \theta_{ij} : 1 \leq i \leq a, \ 1 \leq j \leq b).$$

The hypothesis that the characteristics are assigned independently becomes $H : \theta_{ij} = \eta_{i1} \eta_{j2}$ for $1 \leq i \leq a, \ 1 \leq j \leq b$ where the η_{i1}, η_{j2} are nonnegative and $\sum_{i=1}^{a} \eta_{i1} = \sum_{j=1}^{b} \eta_{j2} = 1$.

The N_{ij} can be arranged in a $a \times b$ contingency table,

	1	2	$\cdots$	b	
1	N_{11}	N_{12}	$\cdots$	N_{1b}	R_1
$\vdots$	$\vdots$	$\vdots$	$\cdots$	$\vdots$	$\vdots$
a	N_{a1}	$\cdot$	$\cdots$	N_{ab}	R_a
	C_1	C_2	$\cdots$	C_b	n

with row and column sums as indicated. Maximum likelihood and dimensionality calculations similar to those for the 2×2 table show that Pearson's χ^2 for the hypothesis of independence is given by

$$\chi^2 = n \sum_{i=1}^{a} \sum_{j=1}^{b} \frac{\left(N_{ij} - \frac{R_i C_j}{n}\right)^2}{R_i C_j}, \tag{6.4.9}$$

which has approximately a $\chi^2_{(a-1)(b-1)}$ distribution under H. The argument is left to the problems as are some numerical applications.

6.4.3   Logistic Regression for Binary Responses

In Section 6.1 we considered linear models that are appropriate for analyzing continuous responses $\{Y_i\}$ that are, perhaps after a transformation, approximately normally distributed and whose means are modeled as $\mu_i = \sum_{j=1}^{p} z_{ij}\beta_j = \mathbf{z}_i^T \boldsymbol{\beta}$ for known constants $\{z_{ij}\}$ and unknown parameters $\beta_1, \ldots, \beta_p$. In this section we will consider Bernoulli responses Y that can only take on the values 0 and 1. Examples are (1) medical trials where at the end of the trial the patient has either recovered ($Y = 1$) or has not recovered ($Y = 0$), (2) election polls where a voter either supports a proposition ($Y = 1$) or does not ($Y = 0$), or (3) market research where a potential customer either desires a new product ($Y = 1$) or does not ($Y = 0$). As is typical, we call $Y = 1$ a "success" and $Y = 0$ a "failure."

We assume that the distribution of the response Y depends on the known covariate vector $\mathbf{z}^T$. In this section we assume that the data are grouped or replicated so that for each fixed i, we observe the number of successes $X_i = \sum_{j=1}^{m_i} Y_{ij}$ where Y_{ij} is the response on the jth of the m_i trials in block i, $1 \le i \le k$. Thus, we observe independent $X_1, \ldots, X_k$ with X_i binomial, $\mathcal{B}(m_i, \pi_i)$, where $\pi_i = \pi(\mathbf{z}_i)$ is the probability of success for a case with covariate vector $\mathbf{z}_i$. Next we choose a parametric model for $\pi(\mathbf{z})$ that will generate useful procedures for analyzing experiments with binary responses. Because $\pi(\mathbf{z})$ varies between 0 and 1, a simple linear representation $\mathbf{z}^T \boldsymbol{\beta}$ for $\pi(\cdot)$ over the whole range of $\mathbf{z}$ is impossible. Instead we turn to the *logistic transform* $g(\pi)$, usually called the *logit*, which we introduced in Example 1.6.8 as the canonical parameter

$$\eta = g(\pi) = \log[\pi/(1-\pi)]. \tag{6.4.10}$$

Other transforms, such as the *probit* $g_1(\pi) = \Phi^{-1}(\pi)$ where Φ is the $\mathcal{N}(0,1)$ d.f. and the *log-log transform* $g_2(\pi) = \log[-\log(1-\pi)]$ are also used in practice.

The log likelihood of $\boldsymbol{\pi} = (\pi_1, \ldots, \pi_k)^T$ based on $\mathbf{X} = (X_1, \ldots, X_k)^T$ is

$$\sum_{i=1}^{k} \left[X_i \log\left(\frac{\pi_i}{1-\pi_i}\right) + m_i \log(1-\pi_i) \right] + \sum_{i=1}^{k} \log\binom{m_i}{X_i}. \tag{6.4.11}$$

When we use the logit transform $g(\pi)$, we obtain what is called the *logistic linear regression model* where

$$\eta_i = \log[\pi_i/(1-\pi_i)] = \mathbf{z}_i^T \boldsymbol{\beta}.$$

The special case $p = 2$, $\mathbf{z}_i = (1, z_i)^T$ is the *logistic regression* model of Problem 2.3.1. The log likelihood $l(\boldsymbol{\pi}(\boldsymbol{\beta})) \equiv l_N(\boldsymbol{\beta})$ of $\boldsymbol{\beta} = (\beta_1, \ldots, \beta_p)^T$ is, if $N = \sum_{i=1}^{k} m_i$,

$$l_N(\boldsymbol{\beta}) = \sum_{j=1}^{p} \beta_j T_j - \sum_{i=1}^{k} m_i \log(1 + \exp\{\mathbf{z}_i \boldsymbol{\beta}\}) + \sum_{i=1}^{k} \log \binom{m_i}{X_i} \qquad (6.4.12)$$

where $T_j = \sum_{i=1}^{k} z_{ij} X_i$ and we make the dependence on N explicit. Note that $l_N(\boldsymbol{\beta})$ is the log likelihood of a p-parameter canonical exponential model with parameter vector $\boldsymbol{\beta}$ and sufficient statistic $\mathbf{T} = (T_1, \ldots, T_p)^T$. It follows that the MLE of $\boldsymbol{\beta}$ solves $E_{\boldsymbol{\beta}}(T_j) = T_j$, $j = 1, \ldots, p$, or $E_{\boldsymbol{\beta}}(\mathbf{Z}^T \mathbf{X}) = \mathbf{Z}^T \mathbf{X}$, where $\mathbf{Z} = \|z_{ij}\|_{m \times p}$ is the design matrix. Thus, Theorem 2.3.1 applies and we can conclude that if $0 < X_i < m_i$ and $\mathbf{Z}$ has rank p, the solution to this equation exists and gives the unique MLE $\widehat{\boldsymbol{\beta}}$ of $\boldsymbol{\beta}$. The condition is sufficient but not necessary for existence—see Problem 2.3.1. We let $\mu_i = E(X_i) = m_i \pi_i$. Then $E(T_j) = \sum_{i=1}^{k} z_{ij} \mu_i$, the likelihood equations are just

$$\mathbf{Z}^T (\mathbf{X} - \boldsymbol{\mu}) = 0. \qquad (6.4.13)$$

By Theorem 2.3.1 and by Proposition 3.4.4 the Fisher information matrix is

$$I(\boldsymbol{\beta}) = \mathbf{Z}^T \mathbf{W} \mathbf{Z} \qquad (6.4.14)$$

where $\mathbf{W} = \mathrm{diag}\{m_i \pi_i (1 - \pi_i)\}_{k \times k}$. The coordinate ascent iterative procedure of Section 2.4.2 can be used to compute the MLE of $\boldsymbol{\beta}$.

Alternatively with a good initial value the Newton–Raphson algorithm can be employed. Although unlike coordinate ascent Newton–Raphson need not converge, we can guarantee convergence with probability tending to 1 as $N \to \infty$ as follows. As the initial estimate use

$$\widehat{\boldsymbol{\beta}}_0 = (\mathbf{Z}^T \mathbf{Z})^{-1} \mathbf{Z}^T \mathbf{V} \qquad (6.4.15)$$

where $\mathbf{V} = (V_1, \ldots, V_k)^T$ with

$$V_i = \log \left(\frac{X_i + \frac{1}{2}}{m_i - X_i + \frac{1}{2}} \right), \qquad (6.4.16)$$

the *empirical logistic transform*. Because $\boldsymbol{\beta} = (\mathbf{Z}^T \mathbf{Z})^{-1} \mathbf{Z}^T \boldsymbol{\eta}$ and $\eta_i = \log[\pi_i (1 - \pi_i)^{-1}]$, $\widehat{\boldsymbol{\beta}}_0$ is a plug-in estimate of $\boldsymbol{\beta}$ where π_i and $(1 - \pi_i)$ in η_i has been replaced by

$$\pi_i^* = \frac{X_i}{m_i} + \frac{1}{2m_i}, \quad (1 - \pi_i)^* = 1 - \frac{X_i}{m_i} + \frac{1}{2m_i}.$$

Here the adjustment $1/2m_i$ is used to avoid $\log 0$ and $\log \infty$. Similarly, in (6.4.14), $\mathbf{W}$ is estimated using

$$\widehat{\mathbf{W}}_0 = \mathrm{diag}\{m_i \pi_i^* (1 - \pi_i)^*\}.$$

Using the δ-method, it follows by Theorem 5.3.3, that if $m \to \infty$, $\pi_i > 0$ for $1 \le i \le k$

$$m_i^{\frac{1}{2}} \left(V_i - \log \left(\frac{\pi_i}{1 - \pi_i} \right) \right) \xrightarrow{\mathcal{L}} \mathcal{N}(0, [\pi_i (1 - \pi_i)]^{-1}).$$

Because **Z** has rank p, it follows (Problem 6.4.14) that $\widehat{\boldsymbol{\beta}}_0$ is consistent.

To get expressions for the MLEs of $\boldsymbol{\pi}$ and $\boldsymbol{\mu}$, recall from Example 1.6.8 that the inverse of the logit transform g is the logistic distribution function

$$g^{-1}(y) = (1 + e^{-y})^{-1}.$$

Thus, the MLE of π_i is $\widehat{\pi}_i = g^{-1}\left(\sum_{j=1}^{p} x_{ij}\widehat{\beta}_j\right).$

Testing

In analogy with Section 6.1, we let $\omega = \{\boldsymbol{\eta} : \eta_i = \mathbf{z}_i^T \boldsymbol{\beta}, \boldsymbol{\beta} \in R^p\}$ and let r be the dimension of ω. We want to contrast ω to the case where there are no restrictions on $\boldsymbol{\eta}$; that is, we set $\Omega = R^k$ and consider $\boldsymbol{\eta} \in \Omega$. In this case the likelihood is a product of independent binomial densities, and the MLEs of π_i and μ_i are X_i/m_i and X_i. The LR statistic $2\log\lambda$ for testing $H : \boldsymbol{\eta} \in \omega$ versus $K : \boldsymbol{\eta} \in \Omega - \omega$ is denoted by $D(\mathbf{y}, \widehat{\boldsymbol{\mu}})$, where $\widehat{\boldsymbol{\mu}}$ is the MLE of $\boldsymbol{\mu}$ for $\boldsymbol{\eta} \in \omega$. Thus, from (6.4.11) and (6.4.12)

$$D(\mathbf{X}, \widehat{\boldsymbol{\mu}}) = 2\sum_{i=1}^{k}[X_i \log(X_i/\widehat{\mu}_i) + X_i' \log(X_i'/\widehat{\mu}_i')] \tag{6.4.17}$$

where $X_i' = m_i - X_i$ and $\widehat{\mu}_i' = m_i - \widehat{\mu}_i$. $D(\mathbf{X}, \widehat{\boldsymbol{\mu}})$ measures the distance between the fit $\widehat{\boldsymbol{\mu}}$ based on the model ω and the data $\mathbf{X}$. By the multivariate delta method, Theorem 5.3.4, $D(\mathbf{X}, \widehat{\boldsymbol{\mu}})$ has asymptotically a χ^2_{k-r} distribution for $\boldsymbol{\eta} \in \omega$ as $m_i \to \infty$, $i = 1, \ldots, k < \infty$—see Problem 6.4.13.

As in Section 6.1 linear subhypotheses are important. If ω_0 is a q-dimensional linear subspace of ω with $q < r$, then we can form the LR statistic for $H : \boldsymbol{\eta} \in \omega_0$ versus $K : \boldsymbol{\eta} \in \omega - \omega_0$

$$2\log\lambda = 2\sum_{i=1}^{k}\left[X_i \log\left(\frac{\widehat{\mu}_i}{\widehat{\mu}_{0i}}\right) + X_i' \log\left(\frac{\widehat{\mu}_i'}{\widehat{\mu}_{0i}'}\right)\right] \tag{6.4.18}$$

where $\widehat{\mu}_0$ is the MLE of $\boldsymbol{\mu}$ under H and $\widehat{\mu}_{0i}' = m_i - \widehat{\mu}_{0i}$. In the present case, by Problem 6.4.13, $2\log\lambda$ has an asymptotic χ^2_{r-q} distribution as $m_i \to \infty$, $i = 1, \ldots, k$. Here is a special case.

Example 6.4.1. *The Binomial One-Way Layout.* Suppose that k treatments are to be tested for their effectiveness by assigning the ith treatment to a sample of m_i patients and recording the number X_i of patients that recover, $i = 1, \ldots, k$. The samples are collected independently and we observe $X_1, \ldots, X_k$ independent with $X_i \sim B(\pi_i, m_i)$. For a second example, suppose we want to compare k different locations with respect to the percentage that have a certain attribute such as the intention to vote for or against a certain proposition. We obtain k independent samples, one from each location, and for the ith location count the number X_i among m_i that has the given attribute.

This model corresponds to the one-way layout of Section 6.1, and as in that section, an important hypothesis is that the populations are homogenous. Thus, we test $H : \pi_1 = \pi_2 = \cdots = \pi_k = \pi$, $\pi \in (0,1)$, versus the alternative that the π's are not all equal. Under H the log likelihood in canonical exponential form is

$$\beta T - N \log(1 + \exp\{\beta\}) + \sum_{i=1}^{k} \log \binom{m_i}{X_i}$$

where $T = \sum_{i=1}^{k} X_i$, $N = \sum_{i=1}^{k} m_i$, and $\beta = \log[\pi/(1-\pi)]$. It follows from Theorem 2.3.1 that if $0 < T < N$ the MLE exists and is the solution of (6.4.13), where $\boldsymbol{\mu} = (m_1\pi, \ldots, m_k\pi)^T$. Using $\mathbf{Z}$ as given in the one-way layout in Example 6.1.3, we find that the MLE of π under H is $\widehat{\pi} = T/N$. The LR statistic is given by (6.4.18) with $\widehat{\mu}_{0i} = m_i\widehat{\pi}$. The Pearson statistic

$$\chi^2 = \sum_{i=1}^{k-1} \frac{(X_i - m_i\widehat{\pi})^2}{m_i\widehat{\pi}(1-\widehat{\pi})}$$

is a Wald statistic and the χ^2 test is equivalent asymptotically to the LR test (Problem 6.4.15).

Summary. We used the large sample testing results of Section 6.3 to find tests for important statistical problems involving discrete data. We found that for testing the hypothesis that a multinomial parameter equals a specified value, the Wald and Rao statistics take a form called "Pearson's χ^2," which equals the sum of standardized squared distances between observed frequencies and expected frequencies under H. When the hypothesis is that the multinomial parameter is in a q-dimensional subset of the $k-1$-dimensional parameter space Θ, the Rao statistic is again of the Pearson χ^2 form. In the special case of testing independence of multinomial frequencies representing classifications in a two-way contingency table, the Pearson statistic is shown to have a simple intuitive form. Finally, we considered logistic regression for binary responses in which the logit transformation of the probability of success is modeled to be a linear function of covariates. We derive the likelihood equations, discuss algorithms for computing MLEs, and give the LR test. In the special case of testing equality of k binomial parameters, we give explicitly the MLEs and χ^2 test.

6.5   GENERALIZED LINEAR MODELS

In Sections 6.1 and 6.4.3 we considered experiments in which the mean μ_i of a response Y_i is expressed as a function of a linear combination

$$\xi_i \equiv \mathbf{z}_i^T \boldsymbol{\beta} = \sum_{j=1}^{p} z_{ij}\beta_j$$

of covariate values. In particular, in the case of a Gaussian response, $\mu_i = \xi_i$. In Section 6.4.3, if $\mu_i = EY_{ij}$, then $\mu_i = \pi_i = g^{-1}(\xi_i)$, where $g^{-1}(y)$ is the logistic distribution function. More generally, McCullagh and Nelder (1983, 1989) synthesized a number of previous generalizations of the linear model, most importantly the log linear model developed by Goodman and Haberman. See Haberman (1974).

The generalized linear model with dispersion depending only on the mean

The data consist of an observation $(\mathbf{Z}, \mathbf{Y})$ where $\mathbf{Y} = (Y_1, \ldots, Y_n)^T$ is $n \times 1$ and $\mathbf{Z}_{p \times n}^T = (\mathbf{z}_1, \ldots, \mathbf{z}_n)$ with $\mathbf{z}_i = (z_{i1}, \ldots, z_{ip})^T$ nonrandom and $\mathbf{Y}$ has density $p(\mathbf{y}, \boldsymbol{\eta})$ given by

$$p(\mathbf{y}, \boldsymbol{\eta}) = \exp\{\boldsymbol{\eta}^T \mathbf{y} - A(\boldsymbol{\eta})\} h(\mathbf{y}) \tag{6.5.1}$$

where $\boldsymbol{\eta}$ is not in $\mathcal{E}$, the natural parameter space of the n-parameter canonical exponential family (6.5.1), but in a subset of $\mathcal{E}$ obtained by restricting η_i to be of the form

$$\eta_i = h(\mathbf{z}_i, \boldsymbol{\beta}), \ \boldsymbol{\beta} \in \mathcal{B} \subset R^p, \ p \leq n$$

where h is a known function. As we know from Corollary 1.6.1, the mean $\boldsymbol{\mu}$ of $\mathbf{Y}$ is related to $\boldsymbol{\eta}$ via $\boldsymbol{\mu} = \dot{A}(\boldsymbol{\eta})$. Typically, $A(\boldsymbol{\eta}) = \sum_{i=1}^{n} A_0(\eta_i)$ for some A_0, in which case $\mu_i = A_0'(\eta_i)$. We assume that there is a one-to-one transform $\mathbf{g}(\boldsymbol{\mu})$ of $\boldsymbol{\mu}$, called the *link function*, such that

$$\mathbf{g}(\boldsymbol{\mu}) = \sum_{j=1}^{p} \beta_j \mathbf{Z}^{(j)} = \mathbf{Z}\boldsymbol{\beta}$$

where $\mathbf{Z}^{(j)} = (z_{1j}, \ldots, z_{nj})^T$ is the jth column vector of $\mathbf{Z}$. Note that if $\dot{A}$ is one-one, $\boldsymbol{\mu}$ determines $\boldsymbol{\eta}$ and thereby $\mathrm{Var}(\mathbf{Y}) = \ddot{A}(\boldsymbol{\eta})$. Typically, $\mathbf{g}(\boldsymbol{\mu})$ is of the form $(g(\mu_1), \ldots, g(\mu_n))^T$, in which case g is also called the link function.

Canonical links

The most important case corresponds to the link being *canonical*; that is, $\mathbf{g} = \dot{A}^{-1}$ or

$$\boldsymbol{\eta} = \sum_{j=1}^{p} \beta_j \mathbf{Z}^{(j)} = \mathbf{Z}\boldsymbol{\beta}.$$

In this case, the GLM is the canonical subfamily of the original exponential family generated by $\mathbf{Z}^T \mathbf{Y}$, which is $p \times 1$.

Special cases are:

(i) *The linear model with known variance.*

These Y_i are independent Gaussian with known variance 1 and means $\mu_i = \sum_{j=1}^{p} Z_{ij}\beta_j$. The model is GLM with canonical $\mathbf{g}(\boldsymbol{\mu}) = \boldsymbol{\mu}$, the identity.

(ii) *Log linear models.*

Suppose $(Y_1, \ldots, Y_p)^T$ is $\mathcal{M}(n, \theta_1, \ldots, \theta_p)$, $\theta_j > 0$, $1 \leq j \leq p$, $\sum_{j=1}^{p} \theta = 1$. Then, as seen in Example 1.6.7,

$$\eta_j = \log \theta_j, \ 1 \leq j \leq p$$

are canonical parameters. If we take $\mathbf{g}(\boldsymbol{\mu}) = (\log \mu_1, \ldots, \log \mu_p)^T$, the link is canonical. The models we obtain are called *log linear*—see Haberman (1974) for an extensive treatment. Suppose, for example, that $\mathbf{Y} = \|Y_{ij}\|_{1 \leq i \leq a}$, $1 \leq j \leq b$, so that Y_{ij} is the indicator of, say, classification i on characteristic 1 and j on characteristic 2. Then

$$\boldsymbol{\theta} = \|\theta_{ij}\|, \ 1 \leq i \leq a, \ 1 \leq j \leq b,$$

and the log linear model corresponding to

$$\log \theta_{ij} = \beta_i + \beta_j,$$

where β_i, β_j are free (unidentifiable parameters), is that of independence

$$\theta_{ij} = \theta_{i+}\theta_{+j}$$

where $\theta_{i+} = \sum_{j=1}^{b} \theta_{ij}$, $\theta_{+j} = \sum_{i=1}^{b} \theta_{ij}$.

The log linear label is also attached to models obtained by taking the Y_i independent Bernoulli (θ_i), $0 < \theta_i < 1$ with canonical link $g(\theta) = \log[\theta(1-\theta)]$. This is just the logistic linear model of Section 6.4.3. See Haberman (1974) for a further discussion.

Algorithms

If the link is canonical, by Theorem 2.3.1, if maximum likelihood estimates $\widehat{\boldsymbol{\beta}}$ exist, they necessarily uniquely satisfy the equation

$$\mathbf{Z}^T \mathbf{Y} = \mathbf{Z}^T E_{\boldsymbol{\beta}} \mathbf{Y} = \mathbf{Z}^T \dot{A}(\mathbf{Z}\widehat{\boldsymbol{\beta}})$$

or

$$\mathbf{Z}^T (\mathbf{Y} - \boldsymbol{\mu}(\widehat{\boldsymbol{\beta}})) = 0. \tag{6.5.2}$$

It's interesting to note that (6.5.2) can be interpreted geometrically in somewhat the same way as in the Gaussian linear model—the "residual" vector $\mathbf{Y} - \boldsymbol{\mu}(\widehat{\boldsymbol{\beta}})$ is orthogonal to the column space of $\mathbf{Z}$. But, in general, $\boldsymbol{\mu}(\widehat{\boldsymbol{\beta}})$ is not a member of that space.

The coordinate ascent algorithm can be used to solve (6.5.2) (or ascertain that no solution exists). With a good starting point $\boldsymbol{\beta}_0$ one can achieve faster convergence with the Newton–Raphson algorithm of Section 2.4. In this case, that procedure is just

$$\widehat{\boldsymbol{\beta}}_{m+1} = \widehat{\boldsymbol{\beta}}_m + (\mathbf{Z}^T \mathbf{W}_m \mathbf{Z})^{-1} \mathbf{Z}^T (\mathbf{Y} - \widehat{\boldsymbol{\mu}}(\widehat{\boldsymbol{\beta}}_m)) \tag{6.5.3}$$

where

$$\mathbf{W}_m = \ddot{A}(\mathbf{Z}\widehat{\boldsymbol{\beta}}_m).$$

In this situation and more generally even for noncanonical links, Newton–Raphson coincides with Fisher's method of scoring described in Problem 6.5.1. If $\widehat{\boldsymbol{\beta}}_0 \xrightarrow{P} \boldsymbol{\beta}_0$, the true value of β, as $n \to \infty$, then with probability tending to 1 the algorithm converges to the MLE if it exists.

In this context, the algorithm is also called iterated weighted least squares. This name stems from the following interpretation. Let $\widehat{\Delta}_{m+1} \equiv \widehat{\beta}_{m+1} - \widehat{\beta}_m$, which satisfies the equation

$$\widehat{\Delta}_{m+1} = (\mathbf{Z}^T \mathbf{W}_m \mathbf{Z})^{-1} \mathbf{Z}^T \mathbf{W}_m (\mathbf{Y} - \boldsymbol{\mu}(\widehat{\beta}_m)). \tag{6.5.4}$$

That is, the correction $\widehat{\Delta}_{m+1}$ is given by the weighted least squares formula (2.2.20) when the data are the residuals from the fit at stage m, the variance covariance matrix is $\mathbf{W}_m$ and the regression is on the columns of $\mathbf{W}_m \mathbf{Z}$—Problem 6.5.2.

Testing in GLM

Testing hypotheses in GLM is done via the LR statistic. As in the linear model we can define the biggest possible GLM $\mathcal{M}$ of the form (6.5.1) for which $p = n$. In that case the MLE of μ is $\widehat{\mu}_{\mathcal{M}} = (Y_1, \ldots, Y_n)^T$ (assume that $\mathbf{Y}$ is in the interior of the convex support of $\{y : p(\mathbf{y}, \boldsymbol{\eta}) > 0\}$). Write $\boldsymbol{\eta}(\cdot)$ for $\dot{A}^{-1}$.

We can think of the test statistic

$$2 \log \lambda = 2[l(\mathbf{Y}, \boldsymbol{\eta}(\mathbf{Y})) - l(\mathbf{Y}, \boldsymbol{\eta}(\boldsymbol{\mu}_0)]$$

for the hypothesis that $\boldsymbol{\mu} = \boldsymbol{\mu}_0$ within $\mathcal{M}$ as a "measure" of (squared) distance between $\mathbf{Y}$ and $\boldsymbol{\mu}_0$. This quantity, called the *deviance* between $\mathbf{Y}$ and $\boldsymbol{\mu}_0$,

$$D(\mathbf{Y}, \boldsymbol{\mu}_0) = 2\{[\boldsymbol{\eta}^T(\mathbf{Y}) - \boldsymbol{\eta}^T(\boldsymbol{\mu}_0)]\mathbf{Y} - [A(\boldsymbol{\eta}(\mathbf{Y})) - A(\boldsymbol{\eta}(\boldsymbol{\mu}_0))]\} \tag{6.5.5}$$

is always ≥ 0. For the Gaussian linear model with known variance σ_0^2

$$D(\mathbf{y}, \boldsymbol{\mu}_0) = |\mathbf{y} - \boldsymbol{\mu}_0|^2 / \sigma_0^2$$

(Problem 6.5.4). The LR statistic for $H : \boldsymbol{\mu} \in \omega_0$ is just

$$D(\mathbf{Y}, \widehat{\boldsymbol{\mu}}_0) \equiv \inf\{D(\mathbf{Y}, \boldsymbol{\mu}) : \boldsymbol{\mu} \in \omega_0\}$$

where $\widehat{\boldsymbol{\mu}}_0$ is the MLE of $\boldsymbol{\mu}$ in ω_0. The LR statistic for $H : \boldsymbol{\mu} \in \omega_0$ versus $K : \boldsymbol{\mu} \in \omega_1 - \omega_0$ with $\omega_1 \supset \omega_0$ is

$$D(\mathbf{Y}, \widehat{\boldsymbol{\mu}}_0) - D(\mathbf{Y}, \widehat{\boldsymbol{\mu}}_1)$$

where $\widehat{\boldsymbol{\mu}}_1$ is the MLE under ω_1. We can then formally write an analysis of deviance analogous to the analysis of variance of Section 6.1. If $\omega_0 \subset \omega_1$ we can write

$$D(\mathbf{Y}, \widehat{\boldsymbol{\mu}}_0) = D(\mathbf{Y}, \widehat{\boldsymbol{\mu}}_1) + \Delta(\widehat{\boldsymbol{\mu}}_0, \widehat{\boldsymbol{\mu}}_1) \tag{6.5.6}$$

a decomposition of the deviance between $\mathbf{Y}$ and $\widehat{\boldsymbol{\mu}}_0$ as the sum of two nonnegative components, the deviance of $\mathbf{Y}$ to $\widehat{\boldsymbol{\mu}}_1$ and $\Delta(\widehat{\boldsymbol{\mu}}_0, \widehat{\boldsymbol{\mu}}_1) \equiv D(\mathbf{Y}, \widehat{\boldsymbol{\mu}}_0) - D(\mathbf{Y}, \widehat{\boldsymbol{\mu}}_1)$, each of which can be thought of as a squared distance between their arguments. Unfortunately $\Delta \neq D$ generally except in the Gaussian case.

Formally if ω_0 is a GLM of dimension p and ω_1 of dimension q with canonical links, then $\Delta(\widehat{\mu}_0, \widehat{\mu}_1)$ is thought of as being asymptotically χ^2_{p-q}. This can be made precise for stochastic GLMs obtained by conditioning on $\mathbf{Z}_1, \ldots, \mathbf{Z}_n$ in the sample $(\mathbf{Z}_1, Y_1), \ldots, (\mathbf{Z}_n, Y_n)$ from the family with density

$$p(\mathbf{z}, y, \boldsymbol{\beta}) = h(y)q_0(\mathbf{z}) \exp\{(\mathbf{z}^T \boldsymbol{\beta})y - A_0(\mathbf{z}^T \boldsymbol{\beta})\}. \tag{6.5.7}$$

More details are discussed in what follows.

Asymptotic theory for estimates and tests

If $(\mathbf{Z}_1, Y_1), \ldots, (\mathbf{Z}_n, Y_n)$ can be viewed as a sample from a population and the link is canonical, the theory of Sections 6.2 and 6.3 applies straightforwardly in view of the general smoothness properties of canonical exponential families. Thus, if we take $\mathbf{Z}_i$ as having marginal density q_0, which we temporarily assume known, then $(\mathbf{Z}_1, Y_1), \ldots, (\mathbf{Z}_n, Y_n)$ has density

$$p(\mathbf{z}, \mathbf{y}, \boldsymbol{\beta}) = \prod_{i=1}^{n} h(y_i)q_0(\mathbf{z}_i) \exp \left\{ \sum_{i=1}^{n} [(\mathbf{z}_i^T \boldsymbol{\beta})y_i - A_0(\mathbf{z}_i^T \boldsymbol{\beta})] \right\}. \tag{6.5.8}$$

This is *not* unconditionally an exponential family in view of the $A_0(\mathbf{z}_i^T \boldsymbol{\beta})$ term. However, there are easy conditions under which conditions of Theorems 6.2.2, 6.3.2, and 6.3.3 hold (Problem 6.5.3), so that the MLE $\widehat{\boldsymbol{\beta}}$ is unique, asymptotically exists, is consistent with probability 1, and

$$\sqrt{n}(\widehat{\boldsymbol{\beta}} - \boldsymbol{\beta}) \xrightarrow{\mathcal{L}} \mathcal{N}(0, I^{-1}(\boldsymbol{\beta})). \tag{6.5.9}$$

What is $I^{-1}(\boldsymbol{\beta})$? The efficient score function $\frac{\partial}{\partial \beta_i} \log p(\mathbf{z}, y, \boldsymbol{\beta})$ is

$$(Y_i - \dot{A}_0(\mathbf{Z}_i^T \boldsymbol{\beta}))\mathbf{Z}_i^T$$

and so,

$$I(\boldsymbol{\beta}) = E(\mathbf{Z}_1^T \mathbf{Z}_1 \ddot{A}_0(\mathbf{Z}_1^T \boldsymbol{\beta})),$$

which, in order to obtain approximate confidence procedures, can be estimated by $\widehat{I} = \widehat{\Sigma} \ddot{A}(\mathbf{Z}\widehat{\boldsymbol{\beta}})$ where $\widehat{\Sigma}$ is the sample variance matrix of the covariates. For instance, if we assume the covariates in logistic regression with canonical link to be stochastic, we obtain

$$I(\boldsymbol{\beta}) = E(\mathbf{Z}_1^T \mathbf{Z}_1 \pi(\mathbf{Z}_1^T \boldsymbol{\beta}_1)(1 - \pi(\mathbf{Z}_1^T \boldsymbol{\beta}))).$$

If we wish to test hypotheses such as $H : \beta_1 = \cdots = \beta_d = 0, d < p$, we can calculate

$$2 \log \lambda(\mathbf{Z}_i, Y_i : 1 \leq i \leq n) = 2 \sum_{i=1}^{n} [\mathbf{Z}_i^T (\widehat{\boldsymbol{\beta}}_H - \widehat{\boldsymbol{\beta}})Y_i + A_0(\mathbf{Z}_i^T \widehat{\boldsymbol{\beta}}_H) - A(\mathbf{Z}_i^T \widehat{\boldsymbol{\beta}})] \tag{6.5.10}$$

where $\widehat{\boldsymbol{\beta}}_H$ is the $(p \times 1)$ MLE for the GLM with $\boldsymbol{\beta}_{p \times 1}^T = (0, \ldots, 0, \beta_{d+1}, \ldots, \beta_p)$, and can conclude that the statistic of (6.5.10) is asymptotically χ^2_d under H. Similar conclusions follow for the Wald and Rao statistics. Note that these tests can be carried out without knowing the density q_0 of $\mathbf{Z}_1$.

These conclusions remain valid for the usual situation in which the $\mathbf{Z}_i$ are not random but their proof depends on asymptotic theory for independent nonidentically distributed variables, which we postpone to Volume II.

The generalized linear model

The GLMs considered so far force the variance of the response to be a function of its mean. An additional "dispersion" parameter can be introduced in some exponential family models by making the function h in (6.5.1) depend on an additional scalar parameter τ. It is customary to write the model as, for $c(\tau) > 0$,

$$p(\mathbf{y}, \boldsymbol{\eta}, \tau) = \exp\{c^{-1}(\tau)(\boldsymbol{\eta}^T \mathbf{y} - A(\boldsymbol{\eta}))\} h(\mathbf{y}, \tau). \tag{6.5.11}$$

Because $\int p(\mathbf{y}, \boldsymbol{\eta}, \tau) d\mathbf{y} = 1$, then

$$A(\boldsymbol{\eta})/c(\tau) = \log \int \exp\{c^{-1}(\tau)\boldsymbol{\eta}^T \mathbf{y}\} h(\mathbf{y}, \tau) d\mathbf{y}. \tag{6.5.12}$$

The left-hand side of (6.5.12) is of product form $A(\boldsymbol{\eta})[1/c(\tau)]$ whereas the right-hand side cannot always be put in this form. However, when it can, then it is easy to see that

$$E(\mathbf{Y}) = \dot{A}(\boldsymbol{\eta}) \tag{6.5.13}$$

$$\text{Var}(\mathbf{Y}) = c(\tau)\ddot{A}(\boldsymbol{\eta}) \tag{6.5.14}$$

so that the variance can be written as the product of a function of the mean and a general dispersion parameter.

Important special cases are the $\mathcal{N}(\mu, \sigma^2)$ and gamma (p, λ) families. For further discussion of this generalization see McCullagh and Nelder (1983, 1989).

General link functions

Links other than the canonical one can be of interest. For instance, if in the binary data regression model of Section 6.4.3, we take $g(\mu) = \Phi^{-1}(\mu)$ so that

$$\pi_i = \Phi(\mathbf{z}_i^T \boldsymbol{\beta})$$

we obtain the so-called *probit* model. Cox (1970) considers the variance stabilizing transformation

$$g(\mu) = \sin^{-1}(\sqrt{\mu}), \ 0 \leq \mu \leq 1,$$

which makes asymptotic analysis equivalent to that in the standard Gaussian linear model. As he points out, the results of analyses with these various transformations over the range $.1 \leq \mu \leq .9$ are rather similar. From the point of analysis for fixed n, noncanonical links can cause numerical problems because the models are now curved rather than canonical exponential families. Existence of MLEs and convergence of algorithm questions all become more difficult and so canonical links tend to be preferred.

Summary. We considered generalized linear models defined as a canonical exponential model where the mean vector of the vector $\mathbf{Y}$ of responses can be written as a function, called the link function, of a linear predictor of the form $\Sigma \beta_j \mathbf{Z}^{(j)}$, where the $\mathbf{Z}^{(j)}$ are observable covariate vectors and β is a vector of regression coefficients. We considered the canonical link function that corresponds to the model in which the canonical exponential model parameter equals the linear predictor. We discussed algorithms for computing MLEs of $\widehat{\beta}$. In the random design case, we use the asymptotic results of the previous sections to develop large sample estimation results, confidence procedures, and tests.

6.6 ROBUSTNESS PROPERTIES AND SEMIPARAMETRIC MODELS

As we most recently indicated in Example 6.2.2, the distributional and implicit structural assumptions of parametric models are often suspect. In Example 6.2.2, we studied what procedures would be appropriate if the linearity of the linear model held but the error distribution failed to be Gaussian. We found that if we assume the error distribution f_0, which is symmetric about 0, the resulting MLEs for β optimal under f_0 continue to estimate β as defined by (6.2.19) even if the true errors are $\mathcal{N}(0, \sigma^2)$ or, in fact, had any distribution symmetric around 0 (Problem 6.2.5). That is, roughly speaking, if we consider the semi-parametric model, $\mathcal{P}_1 = \{P : (\mathbf{Z}^T, Y) \sim P$ given by (6.2.19) with ϵ_i i.i.d. with density f for some f symmetric about $0\}$, then the LSE $\widehat{\beta}$ of β is, under further mild conditions on P, still a consistent asymptotically normal estimate of β and, in fact, so is any estimate solving the equations based on (6.2.31) with f_0 symmetric about 0. There is another semi-parametric model $\mathcal{P}_2 = \{P : (\mathbf{Z}^T, Y)^T \sim P$ that satisfies $E_P(Y \mid \mathbf{Z}) = \mathbf{Z}^T\beta$, $E_P \mathbf{Z}^T \mathbf{Z}$ nonsingular, $E_P Y^2 < \infty\}$. For this model it turns out that the LSE is optimal in a sense to be discussed in Volume II. Furthermore, if $\mathcal{P}_3$ is the nonparametric model where we assume only that $(\mathbf{Z}, Y)$ has a joint distribution and if we are interested in estimating the best linear predictor $\mu_L(\mathbf{Z})$ of Y given $\mathbf{Z}$, the right thing to do if one assumes the $(\mathbf{Z}_i, Y_i)$ are i.i.d., is "act as if the model were the one given in Example 6.1.2." Of course, for estimating $\mu_L(\mathbf{Z})$ in a submodel of $\mathcal{P}_3$ with

$$Y = \mathbf{Z}^T\beta + \sigma_0(\mathbf{Z})\epsilon \tag{6.6.1}$$

where ϵ is independent of $\mathbf{Z}$ but $\sigma_0(\mathbf{Z})$ is not constant and σ_0 is assumed known, the LSE is not the best estimate of β. These issues, whose further discussion we postpone to Volume II, have a fixed covariate exact counterpart, the Gauss–Markov theorem, as discussed below. Another even more important set of questions having to do with selection between nested models of different dimension are touched on in Problem 6.6.8 but otherwise also postponed to Volume II.

Robustness in Estimation

We drop the assumption that the errors $\epsilon_1, \ldots, \epsilon_n$ in the linear model $(6.1.3)$ are normal. Instead we assume the *Gauss–Markov linear model* where

$$\mathbf{Y} = \mathbf{Z}\boldsymbol{\beta} + \boldsymbol{\epsilon}, \ E(\boldsymbol{\epsilon}) = \mathbf{0}, \ \mathrm{Var}(\boldsymbol{\epsilon}) = \sigma^2 \mathbf{J} \tag{6.6.2}$$

where $\mathbf{Z}$ is an $n \times p$ matrix of constants, $\boldsymbol{\beta}$ is $p \times 1$, and $\mathbf{Y}, \boldsymbol{\epsilon}$ are $n \times 1$. The optimality of the estimates $\widehat{\mu}_i$, $\widehat{\beta}_i$ of Section 6.1.1 and the LSE in general still holds *when they are compared to other linear estimates*.

Theorem 6.6.1. *Suppose the Gauss–Markov linear model* $(6.6.2)$ *holds. Then, for any parameter of the form* $\alpha = \sum_{i=1}^n a_i \mu_i$ *for some constants* $a_1, \ldots, a_n$, *the estimate* $\widehat{\alpha} = \sum_{i=1}^n a_i \widehat{\mu}_i$ *has uniformly minimum variance among all unbiased estimates linear in* Y_1, $\ldots, Y_n$.

Proof. Because $E(\epsilon_i) = 0$, $E(\widehat{\alpha}) = \sum_{i=1}^n a_i \mu_i = \alpha$, and $\widehat{\alpha}$ is unbiased. Moreover, $\mathrm{Cov}(Y_i, Y_j) = \mathrm{Cov}(\epsilon_i, \epsilon_j)$, and by (B.5.6),

$$\mathrm{Var}_{GM}(\widehat{\alpha}) = \sum_{i=1}^n a_i^2 \, \mathrm{Var}(\epsilon_i) + 2 \sum_{i<j} a_i a_j \, \mathrm{Cov}(\epsilon_i, \epsilon_j) = \sigma^2 \sum_{i=1}^n a_i^2$$

where Var_{GM} refers to the variance computed under the Gauss–Markov assumptions. Let $\widetilde{\alpha}$ stand for any estimate linear in $Y_1, \ldots, Y_n$. The preceding computation shows that $\mathrm{Var}_{GM}(\widetilde{\alpha}) = \mathrm{Var}_G(\widetilde{\alpha})$, where $\mathrm{Var}_G(\widetilde{\alpha})$ stands for the variance computed under the Gaussian assumption that $\epsilon_1, \ldots, \epsilon_n$ are i.i.d. $\mathcal{N}(0, \sigma^2)$. By Theorems 6.1.2(iv) and 6.1.3(iv), $\mathrm{Var}_G(\widehat{\alpha}) \leq \mathrm{Var}_G(\widetilde{\alpha})$ for all unbiased $\widetilde{\alpha}$. Because $\mathrm{Var}_G = \mathrm{Var}_{GM}$ for all linear estimators, the result follows. $\qquad \square$

Note that the preceding result and proof are similar to Theorem 1.4.4 where it was shown that the optimal linear predictor in the random design case is the same as the optimal predictor in the multivariate normal case. In fact, in Example 6.1.2, our current $\widehat{\mu}$ coincides with the empirical plug-in estimate of the optimal linear predictor $(1.4.14)$. See Problem 6.1.3.

Many of the properties stated for the Gaussian case carry over to the Gauss–Markov case:

Proposition 6.6.1. *If we replace the Gaussian assumptions on the errors* $\epsilon_1, \ldots, \epsilon_n$ *with the Gauss–Markov assumptions, the conclusions* (1) *and* (2) *of Theorem 6.1.4 are still valid; moreover,* $\widehat{\boldsymbol{\beta}}$ *and* $\widehat{\boldsymbol{\mu}}$ *are still unbiased and* $\mathrm{Var}(\widehat{\boldsymbol{\mu}}) = \sigma^2 \mathbf{H}$, $\mathrm{Var}(\widehat{\boldsymbol{\epsilon}}) = \sigma^2 (\mathbf{I} - \mathbf{H})$, *and if* $p = r$, $\mathrm{Var}(\widehat{\boldsymbol{\beta}}) = \sigma^2 (\mathbf{Z}^T \mathbf{Z})^{-1}$.

Example 6.6.1. *One Sample (continued).* In Example 6.1.1, $\mu = \beta_1$, and $\widehat{\mu} = \widehat{\beta}_1 = \bar{Y}$. The Gauss–Markov theorem shows that $\bar{Y}$, in addition to being UMVU in the normal case, is UMVU in the class of linear estimates for all models with $EY_i^2 < \infty$. However, for

n large, $\bar{Y}$ has a larger variance (Problem 6.6.5) than the nonlinear estimate $\widehat{Y}$ = sample median when the density of Y is the Laplace density

$$\frac{1}{2\lambda} \exp\{-\lambda|y - \mu|\}, \ y \in R, \ \mu \in R, \ \lambda > 0.$$

For this density and all symmetric densities, $\widehat{Y}$ is unbiased (Problem 3.4.12).

Remark 6.6.1. As seen in Example 6.6.1, a major weakness of the Gauss–Markov theorem is that it only applies to linear estimates. Another weakness is that it only applies to the homoscedastic case where $\mathrm{Var}(Y_i)$ is the same for all i. Suppose we have a heteroscedastic version of the linear model where $E(\boldsymbol{\epsilon}) = \mathbf{0}$, but $\mathrm{Var}(\epsilon_i) = \sigma_i^2$ depends on i. If the σ_i^2 are known, we can use the Gauss–Markov theorem to conclude that the weighted least squares estimates of Section 2.2 are UMVU. However, when the σ_i^2 are unknown, our estimates are not optimal even in the class of linear estimates.

Robustness of Tests

In Section 5.3 we investigated the robustness of the significance levels of t tests for the one- and two-sample problems using asymptotic and Monte Carlo methods. Now we will use asymptotic methods to investigate the robustness of levels more generally. The δ-method implies that if the observations $\mathbf{X}_i$ come from a distribution P, which does not belong to the model, the asymptotic behavior of the LR, Wald and Rao tests depends critically on the asymptotic behavior of the underlying MLEs $\widehat{\boldsymbol{\theta}}$ and $\widehat{\boldsymbol{\theta}}_0$.

From the theory developed in Section 6.2 we know that if $\Psi(\cdot, \boldsymbol{\theta}) = Dl(\cdot, \boldsymbol{\theta})$ and P is true we expect that $\widehat{\boldsymbol{\theta}}_n \overset{P}{\to} \boldsymbol{\theta}(P)$, which is the unique solution of

$$\int \Psi(\mathbf{x}, \boldsymbol{\theta}) dP(\mathbf{x}) = \mathbf{0} \tag{6.6.3}$$

and

$$\sqrt{n}(\widehat{\boldsymbol{\theta}} - \boldsymbol{\theta}(P)) \overset{\mathcal{L}}{\to} \mathcal{N}(\mathbf{0}, \Sigma(\Psi, P)) \tag{6.6.4}$$

with Σ given by (6.2.6). Thus, for instance, consider $H : \boldsymbol{\theta} = \boldsymbol{\theta}_0$ and the Wald test statistics $T_W = n(\widehat{\boldsymbol{\theta}} - \boldsymbol{\theta}_0)^T I(\boldsymbol{\theta}_0)(\widehat{\boldsymbol{\theta}} - \boldsymbol{\theta}_0)$. If $\boldsymbol{\theta}(P) \neq \boldsymbol{\theta}_0$ evidently we have $T_W \overset{P}{\to} \infty$. But if $\boldsymbol{\theta}(P) = \boldsymbol{\theta}_0$, then $T_W \overset{\mathcal{L}}{\to} \mathbf{V}^T I(\boldsymbol{\theta}_0) \mathbf{V}$ where $\mathbf{V} \sim \mathcal{N}(\mathbf{0}, \Sigma(\Psi, P))$. Because $\Sigma(\Psi, P) \neq I^{-1}(\boldsymbol{\theta}_0)$ in general, the asymptotic distribution of T_W is not χ_r^2. This observation holds for the LR and Rao tests as well—but see Problem 6.6.4. There is an important special case where all is well: the linear model we have discussed in Example 6.2.2.

Example 6.6.2. *The Linear Model with Stochastic Covariates.* Suppose $(\mathbf{Z}_1, Y_1), \ldots,$ $(\mathbf{Z}_n, Y_n)$ are i.i.d. as $(\mathbf{Z}, Y)$ where $\mathbf{Z}$ is a $(p \times 1)$ vector of random covariates and we model the relationship between $\mathbf{Z}$ and Y as

$$Y = \mathbf{Z}^T \boldsymbol{\beta} + \epsilon \tag{6.6.5}$$

with the distribution P of $(\mathbf{Z}, Y)$ such that ϵ and Z are independent, $E_P \epsilon = 0$, $E_P \epsilon^2 < \infty$, and we consider, say, $H : \boldsymbol{\beta} = \mathbf{0}$. Then, when ϵ is $\mathcal{N}(0, \sigma^2)$, (see Examples 6.2.1 and 6.2.2) the LR, Wald, and Rao tests all are equivalent to the F test: Reject if

$$T_n \equiv \widehat{\boldsymbol{\beta}}^T \mathbf{Z}_{(n)}^T \mathbf{Z}_{(n)} \widehat{\boldsymbol{\beta}} / s^2 \geq f_{p, n-p}(1 - \alpha) \qquad (6.6.6)$$

where $\widehat{\boldsymbol{\beta}}$ is the LSE, $\mathbf{Z}_{(n)} = (\mathbf{Z}_1, \ldots, \mathbf{Z}_n)^T$ and

$$s^2 = \frac{1}{n - p} \sum_{i=1}^n (Y_i - \widehat{Y}_i)^2 = \frac{1}{n - p} |\mathbf{Y} - \mathbf{Z}_{(n)} \widehat{\boldsymbol{\beta}}|^2.$$

Now, even if ϵ is not Gaussian, it is still true that if $\boldsymbol{\Psi}$ is given by (6.2.30) then $\boldsymbol{\beta}(P)$ specified by (6.6.3) equals $\boldsymbol{\beta}$ in (6.6.5) and $\sigma^2(P) = \mathrm{Var}_P(\epsilon)$. For instance,

$$\int \left(\frac{(y - \mathbf{z}^T \boldsymbol{\beta})^2}{\sigma^2} - 1 \right) dP = 0 \text{ iff } \sigma^2 = \mathrm{Var}_P(\epsilon).$$

Thus, it is still true by Theorem 6.2.1 that

$$\sqrt{n}(\widehat{\boldsymbol{\beta}} - \boldsymbol{\beta}) \xrightarrow{\mathcal{L}} \mathcal{N}(\mathbf{0}, E^{-1}(\mathbf{Z}\mathbf{Z}^T)\sigma^2(P)) \qquad (6.6.7)$$

and

$$s^2 \xrightarrow{P} \sigma^2(P). \qquad (6.6.8)$$

Moreover, by the law of large numbers,

$$n^{-1} \mathbf{Z}_{(n)}^T \mathbf{Z}_{(n)} = \frac{1}{n} \sum_{i=1}^n \mathbf{Z}_i \mathbf{Z}_i^T \xrightarrow{P} E(\mathbf{Z}\mathbf{Z}^T) \qquad (6.6.9)$$

so that the confidence procedures based on approximating $\mathrm{Var}(\widehat{\boldsymbol{\beta}})$ by $n^{-1} \mathbf{Z}_{(n)}^T \mathbf{Z}_{(n)} s^2 / \sqrt{n}$ are still asymptotically of correct level. It follows by Slutsky's theorem that, under $H : \boldsymbol{\beta} = \mathbf{0}$,

$$T_n \xrightarrow{\mathcal{L}} \mathbf{W}^T [\mathrm{Var}(\mathbf{W})]^{-1} \mathbf{W}$$

where $\mathbf{W}_{p \times p}$ is Gaussian with mean $\mathbf{0}$ and, hence, the limiting distribution of T_n is χ_p^2. Because $f_{p, n-p}(1 - \alpha) \to x_p(1 - \alpha)$ by Example 5.3.7, the test (6.6.5) has asymptotic level α even if the errors are not Gaussian.

This kind of robustness holds for $H : \beta_{q+1} = \beta_{0, q+1}, \ldots, \beta_p = \beta_{0, p}$ or more generally $\boldsymbol{\beta} \in \mathcal{L}_0 + \boldsymbol{\beta}_0$, a q-dimensional affine subspace of R^p. It is intimately linked to the fact that even though the parametric model on which the test was based is false,

(i) the set of P satisfying the hypothesis remains the same and

(ii) the (asymptotic) variance of $\widehat{\boldsymbol{\beta}}$ is the same as under the Gaussian model.

If the first of these conditions fails, then it's not clear what $H : \boldsymbol{\theta} = \boldsymbol{\theta}_0$ means anymore. If it holds but the second fails, the theory goes wrong. We illustrate with two final examples.

Example 6.6.3. *The Two-Sample Scale Problem.* Suppose our model is that $X = (X^{(1)}, X^{(2)})$ where $X^{(1)}, X^{(2)}$ are independent $\mathcal{E}\left(\frac{1}{\theta_1}\right), \mathcal{E}\left(\frac{1}{\theta_2}\right)$ respectively, the lifetimes of paired pieces of equipment. If we take $H : \theta_1 = \theta_2$ the standard Wald test (6.3.17) is

$$\text{``Reject iff } n\frac{(\widehat{\theta}_1 - \widehat{\theta}_2)^2}{\widehat{\sigma}^2} \geq x_1(1 - \alpha)\text{''} \tag{6.6.10}$$

where

$$\widehat{\theta}_j = \frac{1}{n} \sum_{i=1}^n X_i^{(j)} \tag{6.6.11}$$

and

$$\widehat{\sigma} = \frac{1}{\sqrt{2n}} \sum_{i=1}^n (X_i^{(1)} + X_i^{(2)}). \tag{6.6.12}$$

The conditions of Theorem 6.3.2 clearly hold. However suppose now that $X^{(1)}/\theta_1$ and $X^{(2)}/\theta_2$ are identically distributed but not exponential. Then H is still meaningful, $X^{(1)}$ and $X^{(2)}$ are identically distributed. But the test (6.6.10) does not have asymptotic level α in general. To see this, note that, under H,

$$\sqrt{n}(\widehat{\theta}_1 - \widehat{\theta}_2) \xrightarrow{\mathcal{L}} \mathcal{N}(0, 2\,\mathrm{Var}_P X^{(1)}) \tag{6.6.13}$$

but

$$\widehat{\sigma} \xrightarrow{P} \sqrt{2}E_P(X^{(1)}) \neq \sqrt{2\,\mathrm{Var}_P X^{(1)}}$$

in general. It is possible to construct a test equivalent to the Wald test under the parametric model and valid in general. Simply replace $\widehat{\sigma}^2$ by

$$\widetilde{\sigma}^2 = \frac{1}{n} \sum_{i=1}^n \left\{ \left(X_i^{(1)} - \frac{(\bar{X}^{(1)} + \bar{X}^{(2)})}{2} \right)^2 + \left(X_i^{(2)} - \frac{(\bar{X}^{(1)} + \bar{X}^{(2)})}{2} \right)^2 \right\}. \tag{6.6.14}$$

$\square$

Example 6.6.4. *The Linear Model with Stochastic Covariates with ϵ and $\mathbf{Z}$ Dependent.* Suppose $E(\epsilon \mid \mathbf{Z}) = 0$ so that the parameters $\boldsymbol{\beta}$ of (6.6.5) are still meaningful but ϵ and $\mathbf{Z}$ are dependent. For simplicity, let the distribution of ϵ given $\mathbf{Z} = \mathbf{z}$ be that of $\sigma(\mathbf{z})\epsilon'$ where ϵ' is independent of $\mathbf{Z}$. That is, we assume variances are heteroscedastic. Suppose without loss of generality that $\mathrm{Var}(\epsilon') = 1$. Then (6.6.7) fails and in fact by Theorem 6.2.1

$$\sqrt{n}(\widehat{\boldsymbol{\beta}} - \boldsymbol{\beta}) \rightarrow \mathcal{N}(\mathbf{0}, \mathbf{Q}) \tag{6.6.15}$$

where

$$\mathbf{Q} = E^{-1}(\mathbf{ZZ}^T)E(\sigma^2(\mathbf{Z})\mathbf{ZZ}^T)E^{-1}(\mathbf{ZZ}^T)$$

(Problem 6.6.6) and, in general, the test (6.6.6) does not have the correct level. Special-ize to the case $\mathbf{Z} = (1, I_1 - \lambda_1, \ldots, I_d - \lambda_d)^T$ where $(I_1, \ldots, I_d)$ has a multinomial $(\lambda_1, \ldots, \lambda_d)$ distribution, $0 < \lambda_j < 1$, $1 \leq j \leq d$. This is the stochastic version of the d-sample model of Example 6.1.3. It is easy to see that our tests of $H : \beta_2 = \cdots = \beta_d = 0$ fail to have correct asymptotic levels in general unless $\sigma^2(\mathbf{Z})$ is constant or $\lambda_1 = \cdots = \lambda_d = 1/d$ (Problem 6.6.6). A solution is to replace $\mathbf{Z}_{(n)}^T \mathbf{Z}_{(n)}/s^2$ in (6.6.6) by $\widehat{\mathbf{Q}}^{-1}$, where $\widehat{\mathbf{Q}}$ is a consistent estimate of $\mathbf{Q}$. For $d = 2$ above this is just the asymptotic solution of the Behrens–Fisher problem discussed in Section 4.9.4, the two-sample problem with unequal sample sizes and variances.   □

To summarize: If hypotheses remain meaningful when the model is false then, in gen-eral, LR, Wald, and Rao tests need to be modified to continue to be valid asymptotically. The simplest solution at least for Wald tests is to use as an estimate of $[\text{Var } \sqrt{n}\widehat{\boldsymbol{\theta}}]^{-1}$ not $I(\widehat{\boldsymbol{\theta}})$ or $-\frac{1}{n}D^2 l_n(\widehat{\boldsymbol{\theta}})$ but the so-called "sandwich estimate" (Huber, 1967),

$$\left[\frac{1}{n}D^2 l_n(\widehat{\boldsymbol{\theta}}_n)\right]^{-1} \frac{1}{n}\sum_{i=1}^{n}[Dl_1][Dl_1]^T(\mathbf{X}_i, \widehat{\boldsymbol{\theta}}_n)\left[\frac{1}{n}D^2 l_n(\widehat{\boldsymbol{\theta}}_n)\right]^{-1}. \qquad (6.6.16)$$

Summary. We considered the behavior of estimates and tests when the model that gen-erated them does not hold. The Gauss–Markov theorem states that the linear estimates that are optimal in the linear model continue to be so if the i.i.d. $\mathcal{N}(0, \sigma^2)$ assumption on the errors is replaced by the assumption that the errors have mean zero, identical vari-ances, and are uncorrelated, provided we restrict the class of estimates to linear functions of $Y_1, \ldots, Y_n$. In the linear model with a random design matrix, we showed that the MLEs and tests generated by the model where the errors are i.i.d. $\mathcal{N}(0, \sigma^2)$ are still reasonable when the true error distribution is not Gaussian. In particular, the confidence procedures derived from the normal model of Section 6.1 are still approximately valid as are the LR, Wald, and Rao tests. We also demonstrated that when either the hypothesis H or the vari-ance of the MLE is not preserved when going to the wider model, then the MLE and LR procedures for a specific model will fail asymptotically in the wider model. In this case, the methods need adjustment, and we gave the sandwich estimate as one possible adjustment to the variance of the MLE for the smaller model.

6.7 PROBLEMS AND COMPLEMENTS

Problems for Section 6.1

1. Show that in the canonical exponential model (6.1.11) with both $\boldsymbol{\eta}$ and σ^2 unknown, (i) the MLE does not exist if $n = r$, (ii) if $n \geq r + 1$, then the MLE $(\widehat{\eta}_1, \ldots, \widehat{\eta}_r, \widehat{\sigma}^2)^T$ of $(\eta_1, \ldots, \eta_r, \sigma^2)^T$ is $(U_1, \ldots, U_r, n^{-1}\sum_{i=r+1}^{n} U_i^2)^T$. In particular, $\widehat{\sigma}^2 = n^{-1}|\mathbf{Y} - \widehat{\boldsymbol{\mu}}|^2$.

2. For the canonical linear Gaussian model with σ^2 unknown, use Theorem 3.4.3 to com-pute the information lower bound on the variance of an unbiased estimator of σ^2. Compare this bound to the variance of s^2.

Hint: By A.13.22, $\text{Var}(U_i^2) = 2\sigma^4$.

3. Show that $\hat{\mu}$ of Example 6.1.2 with $p = r$ coincides with the empirical plug-in estimate of $\mu_L = (\mu_{L1}, \ldots, \mu_{Ln})^T$, where $\mu_{Li} = \mu_Y + (\mathbf{z}_i^* - \mu_{\mathbf{z}})\boldsymbol{\beta}$, $\mathbf{z}_i^* = (z_{i2}, \ldots, z_{ip})^T$, $\mu_Y = \beta_1$; and $\boldsymbol{\beta}$ and $\mu_{\mathbf{z}}$ are as defined in (1.4.14). Here the empirical plug-in estimate is based on i.i.d. $(\mathbf{Z}_1^*, Y_1), \ldots, (\mathbf{Z}_n^*, Y_n)$ where $\mathbf{Z}_i^* = (Z_{i2}, \ldots, Z_{ip})^T$.

4. Let Y_i denote the response of a subject at time i, $i = 1, \ldots, n$. Suppose that Y_i satisfies the following model

$$Y_i = \theta + \epsilon_i, \; i = 1, \ldots, n$$

where ϵ_i can be written as $\epsilon_i = c e_{i-1} + e_i$ for given constant c satisfying $0 \le c \le 1$, and the e_i are independent identically distributed with mean zero and variance σ^2, $i = 1, \ldots, n$; $e_0 = 0$ (the ϵ_i are called moving average errors, see Problem 2.2.29). Let

$$\bar{Y} = \frac{1}{n}\sum_{i=1}^{n} Y_i, \; \hat{\theta} = \sum_{j=1}^{n} a_j Y_j$$

where

$$a_j = \sum_{i=0}^{n-j} (-c)^i \left(\frac{1 - (-c)^{j+1}}{1+c}\right) \Big/ \sum_{i=1}^{n} \left(\frac{1-(-c)^i}{1+c}\right)^2.$$

(a) Show that $\hat{\theta}$ is the weighted least squares estimate of θ.

(b) Show that if $e_i \sim \mathcal{N}(0, \sigma^2)$, then $\hat{\theta}$ is the MLE of θ.

(c) Show that $\bar{Y}$ and $\hat{\theta}$ are unbiased.

(d) Show that $\text{Var}(\hat{\theta}) \le \text{Var}(\bar{Y})$.

(e) Show that $\text{Var}(\hat{\theta}) < \text{Var}(\bar{Y})$ unless $c = 0$.

5. Show that $\widetilde{\lambda}(\mathbf{Y})$ defined in Remark 6.1.2 coincides with the likelihood ratio statistic $\lambda(\mathbf{Y})$ for the σ^2 known case with σ^2 replaced by $\hat{\sigma}^2$.

6. Consider the model (see Example 1.1.5)

$$Y_i = \theta + e_i, \; i = 1, \ldots, n$$

where $e_i = c e_{i-1} + \epsilon_i$, $i = 1, \ldots, n$, $\epsilon_0 = 0$, $c \in [0, 1]$ is a known constant and $\epsilon_1, \ldots, \epsilon_n$ are i.i.d. $\mathcal{N}(0, \sigma^2)$. Find the MLE $\hat{\theta}$ of θ.

7. Derive the formula (6.1.27) for the noncentrality parameter θ^2 in the regression example.

8. Derive the formula (6.1.29) for the noncentrality parameter δ^2 in the one-way layout.

9. Show that in the regression example with $p = r = 2$, the $100(1 - \alpha)\%$ confidence interval for β_1 in the Gaussian linear model is

$$\beta_1 = \hat{\beta}_1 \pm t_{n-2}\left(1 - \tfrac{1}{2}\alpha\right) s \left[\frac{1}{n} + \frac{z_{.2}^2}{\sum_{i=1}^{n}(z_{i2} - z_{.2})^2}\right].$$

10. Show that if $p = r = 2$ in Example 6.1.2, then the hat matrix $\mathbf{H} = (h_{ij})$ is given by

$$h_{ij} = \frac{1}{n} + \frac{(z_{i2} - z_{\cdot 2})(z_{j2} - z_{\cdot 2})}{\sum (z_{i2} - z_{\cdot 2})^2}.$$

11. Show that for the estimates $\widehat{\alpha}$ and $\widehat{\delta}_k$ in the one-way layout

(a) $\text{Var}(\widehat{\alpha}) = \frac{\sigma^2}{p^2} \sum_{k=1}^{p} \frac{1}{n_k}$, $\text{Var}(\widehat{\delta}_k) = \frac{\sigma^2}{p^2} \left(\frac{(p-1)^2}{n_i} + \sum_{k \neq i} \frac{1}{n_k} \right)$.

(b) If n is fixed and divisible by p, then $\text{Var}(\widehat{\alpha})$ is minimized by choosing $n_i = c = n/p$.

(c) If n is fixed and divisible by $2(p-1)$, then $\text{Var}(\widehat{\delta}_k)$ is minimized by choosing $n_i = n/2, n_2 = \cdots = n_p = n/2(p-1)$.

(d) Give the $100(1-\alpha)\%$ confidence intervals for α and δ_k.

12. Let $Y_1, \ldots, Y_n$ be a sample from a population with mean μ and variance σ^2, where n is even. Consider the three estimates $T_1 = \bar{Y}, T_2 = (1/2n) \sum_{i=1}^{\frac{1}{2}n} Y_i + (3/2n) \sum_{i=\frac{1}{2}n+1}^{n} Y_i$, and $T_3 = \frac{1}{2}(\bar{Y} - 2)$.

(a) Why can you conclude that T_1 has a smaller MSE (mean square error) than T_2?

(b) Which estimate has the smallest MSE for estimating $\theta = \frac{1}{2}(\mu - 2)$?

13. In the one-way layout,

(a) Show that level $(1-\alpha)$ confidence intervals for linear functions of the form $\beta_j - \beta_i$ are given by

$$\beta_j - \beta_i = Y_{j\cdot} - Y_{i\cdot} \pm st_{n-p}\left(1 - \tfrac{1}{2}\alpha\right) \sqrt{\frac{n_i + n_j}{n_i n_j}}$$

and that a level $(1-\alpha)$ confidence interval for σ^2 is given by

$$(n-p)s^2/x_{n-p}\left(1 - \tfrac{1}{2}\alpha\right) \leq \sigma^2 \leq (n-p)s^2/x_{n-p}\left(\tfrac{1}{2}\alpha\right).$$

(b) Find confidence intervals for $\psi = \frac{1}{2}(\beta_2 + \beta_3) - \beta_1$ and $\sigma_\psi^2 = \text{Var}(\widehat{\psi})$ where $\widehat{\psi} = \frac{1}{2}(\widehat{\beta}_2 + \widehat{\beta}_3) - \widehat{\beta}_1$.

14. Assume the linear regression model with $p = r = 2$. We want to predict the value of a future observation Y to be taken at the pont z.

(a) Find a level $(1-\alpha)$ confidence interval for the best MSPE predictor $E(Y) = \beta_1 + \beta_2 z$.

(b) Find a level $(1-\alpha)$ *prediction interval* for Y (i.e., statistics $\underline{t}(Y_1, \ldots, Y_n), \bar{t}(Y_1, \ldots, Y_n)$ such that $P[\underline{t} \leq Y \leq \bar{t}] = 1 - \alpha$). Note that Y is independent of $Y_1, \ldots, Y_n$.

15. Often a treatment that is beneficial in small doses is harmful in large doses. The following model is useful in such situations. Consider a covariate x, which is the amount or dose of a treatment, and a response variable Y, which is yield or production. Suppose a good fit is obtained by the equation

$$y_i = e^{\beta_1} e^{\beta_2 x_i} x_i^{\beta_3}$$

where y_i is observed yield for dose x_i. Assume the model

$$\log Y_i = \beta_1 + \beta_2 x_i + \beta_3 \log x_i + \epsilon_i, \quad i = 1, \ldots, n$$

where $\epsilon_1, \ldots, \epsilon_n$ are independent $N(0, \sigma^2)$.

(a) For the following data (from Hald, 1952, p. 653), compute $\widehat{\beta}_1, \widehat{\beta}_2, \widehat{\beta}_3$ and level 0.95 confidence intervals for $\beta_1, \beta_2, \beta_3$.

(b) Plot (x_i, y_i) and $(x_i, \widehat{y}_i)$ where $\widehat{y}_i = e^{\beta_1} e^{\beta_2 x_i} x_i^{\beta_3}$. Find the value of x that maximizes the estimated yield $\widehat{y} = e^{\beta_1} e^{\beta_2 x} x^{\beta_3}$.

x (nitrogen)	0.09	0.32	0.69	1.51	2.29	3.06	3.39	3.63	3.77
y (yield)	15.1	57.3	103.3	174.6	191.5	193.2	178.7	172.3	167.5

Hint: Do the regression for $\mu_i = \beta_1 + \beta_2 z_{i1} + \beta_2 z_{i1} + \beta_3 z_{i2}$ where $z_{i1} = x_i - \bar{x}$, $z_{i2} = \log x_i - \frac{1}{n} \sum_{i=1}^{n} \log x_i$. You may use $s = 0.0289$.

16. Show that if C is an $n \times r$ matrix of rank r, $r \leq n$, then the $r \times r$ matrix $C'C$ is of rank r and, hence, nonsingular.

Hint: Because C' is of rank r, $xC' = 0 \Rightarrow x = 0$ for any r-vector x. But $xC'C = 0 \Rightarrow \|xC'\|^2 = xC'Cx' = 0 \Rightarrow xC' = 0$.

17. In the Gaussian linear model with $n \geq p+1$ show that the parametrization $(\beta, \sigma^2)^T$ is identifiable if and only if $r = p$.

Problems for Section 6.2

1. (a) Check A0,...,A6 when $\theta = (\mu, \sigma^2)$, $\mathcal{P} = \{N(\mu, \sigma^2) : \mu \in R, \sigma^2 > 0\}$, $\psi(x, \theta) = \nabla_\theta \rho(x, \theta)$, $\rho(x, \theta) = -\log p(x, \theta)$, and $Q = \mathcal{P}$.

(b) Let θ, $\mathcal{P}$, and $\rho = -\log p$, $p \in \mathcal{P}$ be as in (a) but let Q be the class of distributions with densities of the form

$$(1 - \epsilon)\varphi_{(\mu, \sigma^2)}(x) + \epsilon\varphi_{(\mu, \tau^2)}(x), \quad \sigma^2 < \tau^2, \quad \epsilon \leq \frac{1}{2}$$

where $\varphi_{(\mu, \sigma^2)}$ is the $N(\mu, \sigma^2)$ density. For fixed ϵ and τ^2, check A1–A4.

2. (a) Show that A0–A4 and A6 hold for $\mathcal{P} = Q$ where $\mathcal{P} = \{P_\theta\}$, $\theta = (\mu, \tau)^T$, with densities of the form

$$\frac{1}{2}\varphi_{(\mu, 1)} + \frac{1}{2}\varphi_{(\mu, \tau^2)}, \quad \tau^2 > 0.$$

(b) Show that the MLE of (μ, τ^2) does not exist so that A5 doesn't hold.

(c) Construct a method of moment estimate $\bar{\boldsymbol{\theta}}_n$ of $\boldsymbol{\theta} = (\mu, \tau^2)$ based on the first two moments which are $\sqrt{n}$ consistent.

(d) Deduce from Problem 6.2.10 that the estimate $\widehat{\boldsymbol{\theta}}$ derived as the limit of Newton–Raphson estimates from $\bar{\boldsymbol{\theta}}_n$ is efficient.

3. In Example 6.2.1, show that $([E\mathbf{Z}\mathbf{Z}^T]^{-1})_{(1,1)} \geq [EZ_1^2]^{-1}$ with equality if and only if $EZ_1 Z_i = 0, i > 1$.

4. In Example 6.2.2, show that the assumptions of Theorem 6.2.2 hold if (i) and (ii) hold.

5. In Example 6.2.2, show that $c(f_0, \sigma) = \sigma_0/\sigma$ is 1 if f_0 is normal and is different from 1 if f_0 is logistic. You may assume that $E(\epsilon F_0(\epsilon)) \neq \frac{1}{2}$, where $F_0 = $ logistic df and $\epsilon \sim \mathcal{N}(0, 1)$.

6. (a) In Example 6.2.1 show that MLEs of β, μ, and σ^2 are as given in (6.2.20), (6.2.21).
    Hint: $f_X(x) = f_{Y|\mathbf{Z}}(y) f_{\mathbf{Z}}(\mathbf{z})$.

(b) Suppose that the distribution of $\mathbf{Z}$ is not known so that the model is semiparametric, $X \sim P_{(\theta,H)}, \{P_{(\theta,H)} : \theta \in \Theta, H \in \mathcal{H}\}, \theta$ Euclidean, $\mathcal{H}$ abstract. In some cases it is possible to find $T(X)$ such that the distribution of X given $T(X) = t$ is Q_θ, which doesn't depend on $H \in \mathcal{H}$. The MLE of θ based on (X, t) is then called a *conditional* MLE. Show that if we identify $X = (\mathbf{Z}^{(n)}, Y), T(X) = \mathbf{Z}^{(n)}$, then $(\widehat{\beta}, \widehat{\mu}, \widehat{\sigma}^2)$ are conditional MLEs.
    Hint: (a),(b) The MLE minimizes $\frac{1}{\sigma^2}|\mathbf{Y} - \mathbf{Z}^{(n)}\beta|^2$.

7. Fill in the details of the proof of Theorem 6.2.1.

8. Establish (6.2.24) directly as follows:

(a) Show that if $\bar{\mathbf{Z}}_n = \frac{1}{n}\sum_{i=1}^n \mathbf{Z}_i$ then, given $\mathbf{Z}_{(n)}, \sqrt{n}(\widehat{\mu} - \mu, (\widehat{\beta} - \beta)^T)^T$ has a multivariate normal distribution with mean $\mathbf{0}$ and variance,

$$\begin{pmatrix} \sigma^2 & \mathbf{0} \\ \mathbf{0} & n[\mathbf{Z}_{(n)}^T \mathbf{Z}_{(n)}]^{-1} \end{pmatrix},$$

and that $\widehat{\sigma}^2$ is independent of the preceding vector with $n\widehat{\sigma}^2/\sigma^2$ having a χ_{r-p}^2 distribution.

(b) Apply the law of large numbers to conclude that

$$n^{-1}\mathbf{Z}_{(n)}^T \mathbf{Z}_{(n)} \xrightarrow{P} E(\mathbf{Z}\mathbf{Z}^T).$$

(c) Apply Slutsky's theorem to conclude that

$$\mathcal{L}(\sqrt{n}[E\mathbf{Z}\mathbf{Z}^T]^{-1/2}(\widehat{\beta} - \beta)) \to \mathcal{N}(0, \sigma^2 J)$$

and, hence, that
(d) $(\widehat{\beta} - \beta)^T \mathbf{Z}_{(n)}^T \mathbf{Z}_{(n)}(\widehat{\beta} - \beta) = o_p(n^{-1/2})$.
(e) Show that $\widehat{\sigma}^2$ is unconditionally independent of $(\widehat{\mu}, \widehat{\beta})$.
(f) Combine (a)–(e) to establish (6.2.24).

9. Let $Y_1, \ldots, Y_n$ real be independent identically distributed

$$Y_i = \mu + \sigma \epsilon_i$$

where $\mu \in R$, $\sigma > 0$ are unknown and ϵ has known density $f > 0$ such that if $\rho(x) \equiv -\log f(x)$ then $\rho'' > 0$ and, hence, ρ is strictly convex. Examples are f Gaussian, and $f(x) = e^{-x}(1 + e^{-x})^{-2}$, (logistic).

(a) Show that if $\sigma = \sigma_0$ is assumed known a unique MLE for μ exists and uniquely solves

$$\sum_{i=1}^{n} \rho'\left(\frac{x_i - \mu}{\sigma_0}\right) = 0.$$

(b) Write $\theta_1 = \frac{1}{\sigma}$, $\theta_2 = \frac{\mu}{\sigma}$. Show that if $\theta_2 = \theta_2^0$ a unique MLE for θ_1 exists and uniquely solves

$$\frac{1}{n} \sum_{i=1}^{n} X_i \rho'(\theta_1 X_i - \theta_2^0) = \frac{1}{\theta_1}.$$

10. Suppose A0–A4 hold and $\boldsymbol{\theta}_n^*$ is $\sqrt{n}$ consistent; that is, $\boldsymbol{\theta}_n^* = \boldsymbol{\theta}_0 + O_p(n^{-1/2})$.

(a) Let $\bar{\boldsymbol{\theta}}_n$ be the first iterate of the Newton–Raphson algorithm for solving (6.2.1) starting at $\boldsymbol{\theta}_n^*$,

$$\bar{\boldsymbol{\theta}}_n = \boldsymbol{\theta}_n^* - \left[\frac{1}{n}\sum_{i=1}^{n} D\psi(X_i, \boldsymbol{\theta}_n^*)\right]^{-1} \frac{1}{n}\sum_{i=1}^{n} \boldsymbol{\Psi}(X_i, \boldsymbol{\theta}_n^*).$$

Show that $\bar{\boldsymbol{\theta}}_n$ satisfies (6.2.3).
    Hint:

$$\frac{1}{n}\sum_{i=1}^{n} \boldsymbol{\Psi}(X_i, \boldsymbol{\theta}_n^*) = \frac{1}{n}\sum_{i=1}^{n} \boldsymbol{\Psi}(X_i, \boldsymbol{\theta}_0) - \left(\frac{1}{n}\sum_{i=1}^{n} D\psi(X_i, \boldsymbol{\theta}_n^*) + o_p(1)\right)(\boldsymbol{\theta}_n^* - \boldsymbol{\theta}_0).$$

(b) Show that under A0–A4 there exists $\epsilon > 0$ such that with probability tending to 1, $\frac{1}{n}\sum_{i=1}^{n} \boldsymbol{\Psi}(X_i, \boldsymbol{\theta})$ has a unique 0 in $S(\boldsymbol{\theta}_0, \epsilon)$, the ϵ ball about $\boldsymbol{\theta}_0$.
    Hint: You may use a uniform version of the inverse function theorem: If $g_n : R^d \to R^d$ are such that:

(i) $\sup\{|D\mathbf{g}_n(\boldsymbol{\theta}) - D\mathbf{g}(\boldsymbol{\theta})| : |\boldsymbol{\theta} - \boldsymbol{\theta}_0| \le \epsilon\} \to 0$,

(ii) $\mathbf{g}_n(\boldsymbol{\theta}_0) \to \mathbf{g}(\boldsymbol{\theta}_0)$,

(iii) $D\mathbf{g}(\boldsymbol{\theta}_0)$ is nonsingular,

(iv) $D\mathbf{g}(\boldsymbol{\theta})$ is continuous at $\boldsymbol{\theta}_0$, then, for n sufficiently large, there exists a $\delta > 0$, $\epsilon > 0$ such that $\mathbf{g}_n$ are $1 - 1$ on $\delta(\boldsymbol{\theta}_0, \delta)$ and their image contains a ball $S(\mathbf{g}(\boldsymbol{\theta}_0), \delta)$.

(c) Conclude that with probability tending to 1, iteration of the Newton–Raphson algorithm starting at $\boldsymbol{\theta}_n^*$ converges to the unique root $\bar{\boldsymbol{\theta}}_n$ described in (b) and that $\bar{\boldsymbol{\theta}}_n$ satisfies (6.2.3).

Hint: You may use the fact that if the initial value of Newton–Raphson is close enough to a unique solution, then it converges to that solution.

11. Establish (6.2.26) and (6.2.27).

Hint: Write

$$\sum_{i=1}^{n}(Y_i - \mathbf{Z}_i^T \boldsymbol{\beta})^2 = \sum_{i=1}^{n}\left(Y_i - (Z_{i1} - \widehat{Z}_i^{(1)})\beta_1 - \sum_{j=2}^{p}(\beta_j + c_j\beta_1)Z_{ij} \right)^2$$

where $\widehat{Z}_i^{(1)} = \sum_{j=2}^{p} c_j Z_{ij}$ and the c_j do not depend on $\boldsymbol{\beta}$. Thus, minimizing $\sum_{i=1}^{n}(Y_i - \mathbf{Z}_i^T \boldsymbol{\beta})^2$ over all $\boldsymbol{\beta}$ is the same as minimizing

$$\sum_{i=1}^{n}\left(Y_i - (Z_{i1} - \widehat{Z}_i^{(1)})\beta_1 - \sum_{j=2}^{p}\gamma_j Z_{ij} \right)^2,$$

where $\gamma_j = \beta_j + c_j\beta_1$. Differentiate with respect to β_1. Similarly compute the information matrix when the model is written as

$$Y_i = \beta_1(Z_{i1} - \Pi(Z_{i1} \mid Z_{i2}, \ldots, Z_{ip})) + \sum_{j=2}^{p}\gamma_j Z_{ij} + \epsilon_i$$

where $\beta_1, \gamma_2, \ldots, \gamma_p$ range freely and ϵ_i are i.i.d. $\mathcal{N}(0, \sigma^2)$.

Problems for Section 6.3

1. Suppose responses $Y_1, \ldots, Y_n$ are independent Poisson variables with $Y_i \sim \mathcal{P}(\lambda_i)$, and

$$\log \lambda_i = \theta_1 + \theta_2 z_i, \; 0 < z_1 < \cdots < z_n$$

for given covariate values $z_1, \ldots, z_n$. Find the asymptotic likelihood ratio, Wald, and Rao tests for testing $H : \theta_2 = 0$ versus $K : \theta_2 \neq 0$.

2. Suppose that ω_0 is given by (6.3.12) and the assumptions of Theorem (6.3.2) hold for $p(x, \boldsymbol{\theta}), \boldsymbol{\theta} \in \Theta$. Reparametrize $\mathcal{P}$ by $\boldsymbol{\eta}(\boldsymbol{\theta}) = \sum_{j=1}^{q} \eta_j(\boldsymbol{\theta})\mathbf{v}_j$ where $\eta_j(\boldsymbol{\theta}) \equiv \boldsymbol{\theta}^T \mathbf{v}_j$, $\{\mathbf{v}_j\}$ are orthonormal, $q(\cdot, \boldsymbol{\eta}) \equiv p(\cdot, \boldsymbol{\theta})$ for $\boldsymbol{\eta} \in \Xi$ and $\Xi \equiv \{\boldsymbol{\eta}(\boldsymbol{\theta}) : \boldsymbol{\theta} \in \Theta\}$. Show that if $\Xi_0 = \{\boldsymbol{\eta} \in \Xi : \eta_j = 0, q + 1 < j \leq r\}$ then $\lambda(\mathbf{X})$ for the original testing problem is given by

$$\lambda(\mathbf{X}) = \sup\{q(\mathbf{X} \cdot \boldsymbol{\eta}) : \boldsymbol{\eta} \in \Xi\} / \sup\{q(\mathbf{X}, \boldsymbol{\eta}) : \boldsymbol{\eta} \in \Xi_0\}$$

and, hence, that Theorem 6.3.2 holds for Θ_0 as given in (6.3.12).

3. Suppose that $\boldsymbol{\theta}_0 \in \Theta_0$ and the conditions of Theorem 6.3.3 hold. There exists an open ball about $\boldsymbol{\theta}_0$, $S(\boldsymbol{\theta}_0) \subset \Theta$ and a map

$$\boldsymbol{\eta} : R^r \rightarrow R^r,$$

which is continuously differentiable such that

(i) $\eta_j(\boldsymbol{\theta}) = g_j(\boldsymbol{\theta})$ on $S(\boldsymbol{\theta}_0)$, $q + 1 \le j \le r$ and, hence,

$$S(\boldsymbol{\theta}_0) \cap \Theta_0 = \{\boldsymbol{\theta} \in S(\boldsymbol{\theta}_0) : \eta_j(\boldsymbol{\theta}) = 0, \ q + 1 \le j \le r\}.$$

(ii) $\boldsymbol{\eta}$ is $1-1$ on $S(\boldsymbol{\theta}_0)$ and $D\boldsymbol{\eta}(\boldsymbol{\theta})$ is a nonsingular $r \times r$ matrix for all $\boldsymbol{\theta} \in S(\boldsymbol{\theta}_0)$.

(Adjoin to $g_{q+1}, \ldots, g_r, \mathbf{a}_1^T \boldsymbol{\theta}, \ldots, \mathbf{a}_q^T \boldsymbol{\theta}$ where $\mathbf{a}_1, \ldots, \mathbf{a}_q$ are orthogonal to the linear span of $\left\| \frac{\partial g_j}{\partial \theta_i}(\boldsymbol{\theta}_0) \right\|_{p \times d}$.)

Show that if we reparametrize $\{P_{\boldsymbol{\theta}} : \boldsymbol{\theta} \in S(\boldsymbol{\theta}_0)\}$ by $q(\cdot, \boldsymbol{\eta}(\boldsymbol{\theta})) \equiv p(\cdot, \boldsymbol{\theta})$ where $q(\cdot, \boldsymbol{\eta})$ is uniquely defined on $\Xi = \{\boldsymbol{\eta}(\boldsymbol{\theta}) : \boldsymbol{\theta} \in S(\boldsymbol{\theta}_0)\}$ then, $q(\cdot, \boldsymbol{\eta})$ and $\widehat{\boldsymbol{\eta}}_n \equiv \boldsymbol{\eta}(\widehat{\boldsymbol{\theta}}_n)$ and, $\widehat{\boldsymbol{\eta}}_{0,n} \equiv \boldsymbol{\eta}(\widehat{\boldsymbol{\theta}}_{0,n})$ satisfy the conditions of Problem 6.3.2. Deduce that Theorem 6.3.3 is valid.

4. *Testing Simple versus Simple.* Let $\Theta = \{\theta_0, \theta_1\}$, $X_i, \ldots, X_n$ i.i.d. with density $p(\cdot, \theta)$. Consider testing $H : \theta = \theta_0$ versus $K : \theta = \theta_1$. Assume that $P_{\theta_1} \ne P_{\theta_0}$, and that for some $\delta \ge 0$,

(i) $E_{\theta_0} \left| \log \frac{p(X_1, \theta_1)}{p(X_1, \theta_0)} \right|^\delta < \infty$,

(ii) $E_{\theta_1} \left| \log \frac{p(X_1, \theta_1)}{p(X_1, \theta_0)} \right|^\delta < \infty$.

(a) Let $\lambda(X_1, \ldots, X_n)$ be the likelihood ratio statistic. Show that under H, even if $\delta = 0$, $2 \log \lambda(X_1, \ldots, X_n) \xrightarrow{P} 0$.

(b) If $\delta = 2$ show that asymptotically the critical value of the most powerful (Neyman–Pearson) test with $T_n = \sum_{i=1}^n (l(X_i, \theta_1) - l(X_i, \theta_0))$ is $-nK(\theta_0, \theta_1) + z_{1-\alpha} \sqrt{n} \sigma(\theta_0, \theta_1)$ where $K(\theta_0, \theta_1)$ is the Kullback–Leibler divergence

$$K(\theta_0, \theta_1) = E_{\theta_0} \log \frac{p(X_1, \theta_0)}{p(X_1, \theta_1)}$$

and

$$\sigma^2(\theta_0, \theta_1) = \text{Var}_{\theta_0} \left(\log \frac{p(X_1, \theta_1)}{p(X_1, \theta_0)} \right).$$

5. Let (X_i, Y_i), $1 \le i \le n$, be i.i.d. with X_i and Y_i independent, $\mathcal{N}(\theta_1, 1)$, $\mathcal{N}(\theta_2, 1)$, respectively. Suppose $\theta_j \ge 0$, $j = 1, 2$. Consider testing $H : \theta_1 = \theta_2 = 0$ versus $K : \theta_1 > 0$ or $\theta_2 > 0$.

(a) Show that whatever be n, under H, $2 \log \lambda(X_i, Y_i : 1 \leq i \leq n)$ is distributed as a mixture of point mass at 0, χ_1^2 and χ_2^2 with probabilities $\frac{1}{4}, \frac{1}{2}, \frac{1}{4}$, respectively.

Hint: By sufficiency reduce to $n = 1$. Then

$$2 \log \lambda(X_1, Y_1) = X_1^2 1(X_1 > 0) + Y_1^2 1(Y_1 > 0).$$

(b) Suppose X_i, Y_i are as above with the same hypothesis but $\Theta = \{(\theta_1, \theta_2) : 0 \leq \theta_2 \leq c\theta_1, \theta_1 \geq 0\}$. Show that $2 \log \lambda(X_i, Y_i : 1 \leq i \leq n)$ has a null distribution, which is a mixture of point mass at 0, χ_1^2 and χ_2^2 but with probabilities $\frac{1}{2} - \frac{\Delta}{2\pi}, \frac{1}{2}$ and $\frac{\Delta}{2\pi}$ where $\sin \Delta = \frac{c}{\sqrt{1+c^2}}, 0 \leq \Delta \leq \frac{\pi}{2}$.

(c) Let (X_1, Y_1) have an $\mathcal{N}_2(\theta_1, \theta_2, \sigma_{10}^2, \sigma_{20}^2, \rho_0)$ distribution and $(X_i, Y_i), 1 \leq i \leq n$, be i.i.d. Let $\theta_1, \theta_2 \geq 0$ and H be as above. Exhibit the null distribution of $2 \log \lambda(X_i, Y_i : 1 \leq i \leq n)$.

Hint: Consider $\sigma_{10}^2 = \sigma_{20}^2 = 1$ and $Z_1 = X_1, Z_2 = \frac{\rho_0 X_1 - Y_1}{\sqrt{1 - \rho_0^2}}$.

6. In the model of Problem 5(a) compute the MLE $(\widehat{\theta}_1, \widehat{\theta}_2)$ under the model and show that

(a) If $\theta_1 > 0, \theta_2 > 0$,

$$\mathcal{L}(\sqrt{n}(\widehat{\theta}_1 - \theta_1, \widehat{\theta}_2 - \theta_2)) \to \mathcal{N}(0, 0, 1, 1, 0).$$

(b) If $\theta_1 = \theta_2 = 0$

$$\mathcal{L}(\sqrt{n}(\widehat{\theta}_1, \widehat{\theta}_2)) \to \mathcal{L}(|U|, |V|)$$

where $U \sim \mathcal{N}(0, 1)$ with probability $\frac{1}{2}$ and 0 with probability $\frac{1}{2}$ and V is independent of U with the same distribution.

(c) Obtain the limit distribution of $\sqrt{n}(\widehat{\theta}_1 - \theta_1, \widehat{\theta}_2 - \theta_2)$ if $\theta_1 = 0, \theta_2 > 0$.

(d) Relate the result of (b) to the result of Problem 4(a).

Note: The results of Problems 4 and 5 apply generally to models obeying A0–A6 when we restrict the parameter space to a cone (Robertson, Wright, and Dykstra, 1988). Such restrictions are natural if, for instance, we test the efficacy of a treatment on the basis of two correlated responses per individual.

7. Show that $(6.3.19)$ holds.

Hint:

(i) Show that $I(\widehat{\theta}_n)$ can be replaced by $I(\theta)$.

(ii) Show that $W_n(\theta_0^{(2)})$ is invariant under affine reparametrizations $\boldsymbol{\eta} = \mathbf{a} + B\boldsymbol{\theta}$ where B is nonsingular.

(iii) Reparametrize as in Theorem 6.3.2 and compute $W_n(\theta_0^{(2)})$ showing that its leading term is the same as that obtained in the proof of Theorem 6.3.2 for $2 \log \lambda(\mathbf{X})$.

8. Show that under A0–A5 and A6 for $\widehat{\boldsymbol{\theta}}_n^{(1)}$

$$\sqrt{n}\boldsymbol{\Psi}_n(\widehat{\boldsymbol{\theta}}_{n,0}) \xrightarrow{\mathcal{L}} \mathcal{N}(0, \boldsymbol{\Sigma}(\boldsymbol{\theta}_0))$$

where $\boldsymbol{\Sigma}(\boldsymbol{\theta}_0)$ is given by (6.3.21).
    Hint: Write

$$\boldsymbol{\Psi}_n(\widehat{\boldsymbol{\theta}}_{n0}) = \boldsymbol{\Psi}_n(\boldsymbol{\theta}_0) + \frac{1}{n}D_{21}l_n(\widehat{\boldsymbol{\theta}}_n^*)(\sqrt{n}(\widehat{\boldsymbol{\theta}}_n^{(1)} - \boldsymbol{\theta}_0^{(1)}))$$

and apply Theorem 6.2.2 to $\widehat{\boldsymbol{\theta}}_n^{(1)}$.

9. Under conditions A0–A6 for (a) and A0–A6 with A6 for $\widehat{\boldsymbol{\theta}}_n^{(1)}$ for (b) establish that

(a) $\left[-\frac{1}{n}D^2 l_n(\widehat{\boldsymbol{\theta}}_n)\right]^{-1}$ is a consistent estimate of $I^{-1}(\boldsymbol{\theta}_0)$.

(b) (6.3.22) is a consistent estimate of $\boldsymbol{\Sigma}^{-1}(\boldsymbol{\theta}_0)$.
    Hint: Argue as in Problem 5.3.10.

10. Show that under A2, A3, A6 $\boldsymbol{\theta} \to I(\boldsymbol{\theta})$ is continuous.

Problems for Section 6.4

1. Exhibit the two solutions of (6.4.4) explicitly and find the one that corresponds to the maximizer of the likelihood.

2. (a) Show that for any 2×2 contingency table the table obtained by subtracting (estimated) expectations from each entry has all rows and columns summing to zero, hence, is of the form

Δ	$-\Delta$
$-\Delta$	Δ

(b) Deduce that $\chi^2 = Z^2$ where Z is given by (6.4.8)

(c) Derive the alternative form (6.4.8) for Z.

3. In the 2×2 contingency table model let $X_i = 1$ or 0 according as the ith individual sampled is an A or $\bar{A}$ and $Y_i = 1$ or 0 according as the ith individual sampled is a B or $\bar{B}$.

(a) Show that the correlation of X_1 and Y_1 is

$$\rho = \frac{P(A \cap B) - P(A)P(B)}{\sqrt{P(A)(1 - P(A))P(B)(1 - P(B))}}.$$

(b) Show that the sample correlation coefficient r studied in Example 5.3.6 is related to Z of (6.4.8) by $Z = \sqrt{n}r$.

(c) Conclude that if A and B are independent, $0 < P(A) < 1$, $0 < P(B) < 1$, then Z has a limiting $\mathcal{N}(0, 1)$ distribution.

4. (a) Let $(N_{11}, N_{12}, N_{21}, N_{22}) \sim \mathcal{M}(n, \theta_{11}, \theta_{12}, \theta_{21}, \theta_{22})$ as in the contingency table. Let $R_i = N_{i1} + N_{i2}$, $C_i = N_{1i} + N_{2i}$. Show that given $R_1 = r_1$, $R_2 = r_2 = n - r_1$, N_{11} and N_{21} are independent $\mathcal{B}(r_1, \theta_{11}/(\theta_{11} + \theta_{12}))$, $\mathcal{B}(r_2, \theta_{21}/(\theta_{21} + \theta_{22}))$.

(b) Show that $\theta_{11}/(\theta_{11} + \theta_{12}) = \theta_{21}/(\theta_{21} + \theta_{22})$ iff R_1 and C_1 are independent.

(c) Show that under independence the conditional distribution of N_{ii} given $R_i = r_i$, $C_i = c_i$, $i = 1, 2$ is $\mathcal{H}(c_i, n, r_i)$ (the hypergeometric distribution).

5. *Fisher's Exact Test*
From the result of Problem 6.2.4 deduce that if $j(\alpha)$ (depending on r_1, c_1, n) can be chosen so that

$$P[\mathcal{H}(c_1, n, r_1) \geq j(\alpha)] \leq \alpha, \ P[\mathcal{H}(c_1, n, r_1) \geq j(\alpha) - 1] \geq \alpha$$

then the test that rejects (conditionally on $R_1 = r_1$, $C_1 = c_1$) if $N_{11} \geq j(\alpha)$ is exact level α. This is known as Fisher's exact test. It may be shown (see Volume II) that the (approximate) tests based on Z and Fisher's test are asymptotically equivalent in the sense of $(5.4.54)$.

6. Let N_{ij} be the entries of an $a \times b$ contingency table with associated probabilities θ_{ij} and let $\eta_{i1} = \sum_{j=1}^{b} \theta_{ij}$, $\eta_{j2} = \sum_{i=1}^{a} \theta_{ij}$. Consider the hypothesis $H : \theta_{ij} = \eta_{i1}\eta_{j2}$ for all i, j.

(a) Show that the maximum likelihood estimates of η_{i1}, η_{j2} are given by

$$\widehat{\eta}_{i1} = \frac{R_i}{n}, \ \widehat{\eta}_{j2} = \frac{C_j}{n}$$

where $R_i = \sum_j N_{ij}$, $C_j = \sum_i N_{ij}$.

(b) Deduce that Pearson's χ^2 is given by $(6.4.9)$ and has approximately a $\chi^2_{(a-1)(b-1)}$ distribution under H.
Hint: (a) Consider the likelihood as a function of η_{i1}, $i = 1, \ldots, a - 1, \eta_{j2}$, $j = 1, \ldots, b - 1$ only.

7. Suppose in Problem 6.4.6 that H is true.

(a) Show that then

$$P[N_{ij} = n_{ij}; i = 1, \ldots, a, \ j = 1, \ldots, b \mid R_i = r_i, \ C_j = c_j]$$

$$= \frac{\begin{pmatrix} c_1 \\ n_{11}, \ldots, n_{a1} \end{pmatrix} \begin{pmatrix} c_2 \\ n_{12}, \ldots, n_{a2} \end{pmatrix} \cdots \begin{pmatrix} c_a \\ n_{a1}, \ldots, n_{ab} \end{pmatrix}}{\begin{pmatrix} n \\ r_1, \ldots, r_a \end{pmatrix}}$$

where $\begin{pmatrix} A \\ B, C, D, \ldots \end{pmatrix} = \frac{A!}{B!C!D!\ldots}$ are the multinomial coefficients.

(b) How would you, in principle, use this result to construct a test of H similar to the χ^2 test with probability of type I error independent of η_{i1}, η_{j2}?

8. The following table gives the number of applicants to the graduate program of a small department of the University of California, classified by sex and admission status. Would you accept or reject the hypothesis of independence at the 0.05 level

(a) using the χ^2 test with approximate critical value?

(b) using Fisher's exact test of Problem 6.4.5?

	Admit	Deny
Men	19	12
Women	5	0

Hint: (b) It is easier to work with N_{22}. Argue that the Fisher test is equivalent to rejecting H if $N_{22} \geq q_2 + n - (r_1 + c_1)$ or $N_{22} \leq q_1 + n - (r_1 + c_1)$, and that under H, N_{22} is conditionally distributed $\mathcal{H}(r_2, n, c_2)$.

9. (a) If A, B, C are three events, consider the assertions,

(i) $P(A \cap B \mid C) = P(A \mid C)P(B \mid C)$ $(A, B$ independent given $C)$

(ii) $P(A \cap B \mid \bar{C}) = P(A \mid \bar{C})P(B \mid \bar{C})$ $(A, B$ independent given $\bar{C})$

(iii) $P(A \cap B) = P(A)P(B)$ $(A, B$ independent$)$

($\bar{C}$ is the complement of C.) Show that (i) and (ii) imply (iii), if A and C are independent or B and C are independent.

(b) Construct an experiment and three events for which (i) and (ii) hold, but (iii) does not.

(c) The following 2×2 tables classify applicants for graduate study in different departments of the university according to admission status and sex. Test in both cases whether the events [being a man] and [being admitted] are independent. Then combine the two tables into one, and perform the same test on the resulting table. Give p-values for the three cases.

	Admit	Deny	
Men	235	35	270
Women	38	7	45
	273	42	
	$n = 315$		

	Admit	Deny	
Men	122	93	215
Women	103	69	172
	225	162	
	$n = 387$		

(d) Relate your results to the phenomenon discussed in (a), (b).

10. Establish (6.4.14).

11. Suppose that we know that $\beta_1 = 0$ in the logistic model, $\eta_i = \beta_1 + \beta_2 z_i$, z_i not all equal, and that we wish to test $H : \beta_2 \leq \beta_2^0$ versus $K : \beta_2 > \beta_2^0$.

Show that, for suitable α, there is a UMP level α test, which rejects, if and only if, $\sum_{i=1}^{p} z_i N_i \geq k$, where $P_{\beta_2^0}[\sum_{i=1}^{p} z_i N_i \geq k] = \alpha$.

12. Suppose the z_i in Problem 6.4.11 are obtained as realization of i.i.d. Z_i and $m_i \equiv m$ so that (Z_i, X_i) are i.i.d. with $(X_i \mid Z_i) \sim \mathcal{B}(m, \pi(\beta_2 Z_i))$.

(a) Compute the Rao test for $H : \beta_2 \leq \beta_2^0$ and show that it agrees with the test of Problem 6.4.11.

(b) Suppose that β_1 is unknown. Compute the Rao test statistic for $H : \beta_2 \leq \beta_2^0$ in this case.

(c) By conditioning on $\sum_{i=1}^{k} X_i$ and using the approach of Problem 6.4.5 construct an exact test (level independent of β_1).

13. Show that if $\omega_0 \subset \omega_1$ are nested logistic regression models of dimension $q < r \leq k$ and $m_1, \ldots, m_k \to \infty$ and $H : \eta \in \omega_0$ is true then the law of the statistic of (6.4.18) tends to χ_{r-q}^2.

Hint: $(X_i - \mu_i)/\sqrt{m_i \pi_i(1 - \pi_i)}, 1 \leq i \leq k$ are independent, asymptotically $\mathcal{N}(0, 1)$. Use this to imitate the argument of Theorem 6.3.3, which is valid for the i.i.d. case.

14. Show that, in the logistic regression model, if the design matrix has rank p, then $\widehat{\beta}_0$ as defined by (6.4.15) is consistent.

15. In the binomial one-way layout show that the LR test is asymptotically equivalent to Pearson's χ^2 test in the sense that $2 \log \lambda - \chi^2 \xrightarrow{P} 0$ under H.

16. Let $X_1, \ldots, X_k$ be independent $X_i \sim \mathcal{N}(\theta_i, \sigma^2)$ where either $\sigma^2 = \sigma_0^2$ (known) and $\theta_1, \ldots, \theta_k$ vary freely, or $\theta_i = \theta_{i0}$ (known) $i = 1, \ldots, k$ and σ^2 is unknown.

Show that the likelihood ratio test of $H : \theta_1 = \theta_{10}, \ldots, \theta_k = \theta_{k0}, \sigma^2 = \sigma_0^2$ is of the form: Reject if $(1/\sigma_0^2) \sum_{i=1}^{k} (X_i - \theta_{i0})^2 \geq k_2$ or $\leq k_1$. This is an approximation (for large k, n) and simplification of a model under which $(N_1, \ldots, N_k) \sim \mathcal{M}(n, \theta_{10}, \ldots, \theta_{k0})$ under H, but under K may be either multinomial with $\theta \neq \theta_0$ or have $E_\theta(N_i) = n\theta_{i0}$, but $\text{Var}_\theta(N_i) < n\theta_{i0}(1 - \theta_{i0})$("Cooked data").

Problems for Section 6.5

1. *Fisher's Method of Scoring*
The following algorithm for solving likelihood equations was proosed by Fisher—see Rao (1973), for example. Given an initial value $\widehat{\theta}_0$ define iterates

$$\widehat{\theta}_{m+1} = \widehat{\theta}_m + I^{-1}(\widehat{\theta}_m) Dl(\widehat{\theta}_m).$$

Show that for GLM this method coincides with the Newton–Raphson method of Section 2.4.

2. Verify that (6.5.4) is as claimed formula (2.2.20) for the regression described after (6.5.4).

3. Suppose that $(\mathbf{Z}_1, Y_1), \ldots, (\mathbf{Z}_n, Y_n)$ have density as in (6.5.8) and,

(a) $P[\mathbf{Z}_1 \in \{\mathbf{z}^{(1)}, \ldots, \mathbf{z}^{(k)}\}] = 1$

(b) The linear span of $\{z^{(1)}, \ldots, z^{(k)}\}$ is R^p.

(c) $P[\mathbf{Z}_1 = \mathbf{z}^{(j)}] > 0$ for all j. Show that the conditions A0–A6 hold for $P = P_{\boldsymbol{\beta}_0} \in \mathcal{P}$ (where q_0 is assumed known).

Hint: Show that if the convex support of the conditional distribution of Y_1 given $\mathbf{Z}_1 = \mathbf{z}^{(j)}$ contains an open interval about μ_j for $j = 1, \ldots, k$, then the convex support of the conditional distribution of $\sum_{j=1}^k \lambda_j Y_j \mathbf{z}^{(j)}$ given $\mathbf{Z}_j = \mathbf{z}^{(j)}, j = 1, \ldots, k$, contains an open ball about $\sum_{j=1}^k \lambda_j \mu_j \mathbf{z}^{(j)}$ in R^p.

4. Show that for the Gaussian linear model with known variance σ_0^2, the deviance is $D(\mathbf{y}, \boldsymbol{\mu}_0) = |\mathbf{y} - \boldsymbol{\mu}_0|^2/\sigma_0^2$.

5. Let $Y_1, \ldots, Y_n$ be independent responses and suppose the distribution of Y_i depends on a covariate vector $\mathbf{z}_i$. Assume that there exist functions $h(y, \tau)$, $b(\theta)$, $g(\mu)$ and $c(\tau)$ such that the model for Y_i can be written as

$$p(y, \theta_i) = h(y, \tau) \exp \left\{ \frac{\theta_i y - b(\theta_i)}{c(\tau)} \right\}$$

where τ is known, $g(\mu_i) = \mathbf{z}_i^T \boldsymbol{\beta}$, and b' and g are monotone. Set $\xi = g(\mu)$ and $v(\mu) = \text{Var}(Y)/c(\tau) = b''(\theta)$.

(a) Show that the likelihood equations are

$$\sum_{i=1}^n \frac{d\mu_i}{d\xi_i} \frac{(y_i - \mu_i)z_{ij}}{v(\mu_i)} = 0, \; j = 1, \ldots, p.$$

Hint: By the chain rule

$$\frac{\partial}{\partial \beta_j} l(y, \theta) = \frac{\partial l}{\partial \theta} \frac{d\theta}{d\mu} \frac{d\mu}{d\xi} \frac{\partial \xi}{\partial \beta_j}.$$

(b) Show that the Fisher information is $\mathbf{Z}_D^T \mathbf{W} \mathbf{Z}_D$ where $\mathbf{Z}_D = \|z_{ij}\|$ is the design matrix and $\mathbf{W} = \text{diag}(w_1, \ldots, w_n)$, $w_i = w(\mu_i) = 1/v(\mu_i)(d\xi_i/d\mu_i)^2$.

(c) Suppose $(\mathbf{Z}_1, Y_1), \ldots, (\mathbf{Z}_n, Y_n)$ are i.i.d. as $(\mathbf{Z}, Y)$ and that given $\mathbf{Z} = \mathbf{z}$, Y follow the model $p(y, \theta(\mathbf{z}))$ where $\theta(\mathbf{z})$ solves $b'(\theta) = g^{-1}(\mathbf{z}^T \boldsymbol{\beta})$. Show that, under appropriate conditions,

$$\sqrt{n}(\hat{\boldsymbol{\beta}} - \boldsymbol{\beta}) \xrightarrow{\mathcal{L}} \mathcal{N}(0, w(\mathbf{Z}^T \boldsymbol{\beta})\mathbf{Z}\mathbf{Z}^T).$$

(d) *Gaussian GLM.* Suppose $Y_i \sim \mathcal{N}(\mu_i, \sigma_0^2)$. Give $\theta, \tau, h(y, \tau), b(\theta), c(\tau)$, and $v(\mu)$. Show that when g is the canonical link, $g = (b')^{-1}$, the result of (c) coincides with (6.5.9).

(e) Suppose that Y_i has the Poisson, $\mathcal{P}(\mu_i)$, distribution. Give $\theta, \tau, h(y, \tau), b(\theta), c(\tau)$, and $v(\mu)$. In the random design case, give the asymptotic distribution of $\sqrt{n}(\hat{\boldsymbol{\beta}} - \boldsymbol{\beta})$. Find the canonical link function and show that when g is the canonical link, your result coincides with (6.5.9).

Problems for Section 6.6

1. Consider the linear model of Example 6.6.2 and the hypothesis

$$\beta_{q+1} = \beta_{0,q+1}, \ldots, \beta_p = \beta_{0,p}$$

under the sole assumption that $E\epsilon = 0, 0 < \mathrm{Var}\,\epsilon < \infty$. Show that the LR, Wald, and Rao tests are still asymptotically equivalent in the sense that if $2\log\lambda_n$, W_n, and R_n are the corresponding test statistics, then under H,

$$
\begin{aligned}
2\log\lambda_n &= W_n + o_p(1) \\
R_n &= W_n + o_p(1).
\end{aligned}
$$

Note: $2\log\lambda_n$, W_n and R_n are computed under the assumption of the *Gaussian* linear model with σ^2 known.

Hint: Retrace the arguments given for the asymptotic equivalence of these statistics under parametric models and note that the only essential property used is that the MLEs under the model satisfy an appropriate estimating equation. Apply Theorem 6.2.1.

2. Show that the standard Wald test for the problem of Example 6.6.3 is as given in (6.6.10).

3. Show that $\tilde{\sigma}^2$ given in (6.6.14) is a consistent estimate of $2\,\mathrm{Var}_P X^{(1)}$ in Example 6.6.3 and, hence, replacing $\hat{\sigma}^2$ by $\tilde{\sigma}^2$ in (6.6.10) creates a valid level α test.

4. Consider the Rao test for $H : \boldsymbol{\theta} = \boldsymbol{\theta}_0$ for the model $\mathcal{P} = \{P_{\boldsymbol{\theta}} : \boldsymbol{\theta} \in \Theta\}$ and assume that A0–A6 hold. Suppose that the true P does not belong to $\mathcal{P}$ but if $\boldsymbol{\theta}(P)$ is defined by (6.6.3) then $\boldsymbol{\theta}(P) = \boldsymbol{\theta}_0$. Show that, if $\mathrm{Var}_P Dl(X, \boldsymbol{\theta}_0)$ is estimated by $I(\boldsymbol{\theta}_0)$, then the Rao test does not in general have the correct asymptotic level, but that if the estimate $\frac{1}{n}\sum_{i=1}^n [Dl][Dl]^T(X_i, \boldsymbol{\theta}_0)$ is used, then it does.

5. Suppose $X_1, \ldots, X_n$ are i.i.d. P. By Problem 5.4.1, if P has a positive density f at $\nu(P)$, the unique median of P, then the sample median $\hat{X}$ satisfies

$$\sqrt{n}(\hat{X} - \nu(P)) \to \mathcal{N}(0, \sigma^2(P))$$

where $\sigma^2(P) = 1/4f(\nu(p))$.

(a) Show that if f is symmetric about μ, then $\nu(P) = \mu$.

(b) Show that if f is $\mathcal{N}(\mu, \sigma^2)$, then $\sigma^2(P) > \sigma^2 = \mathrm{Var}_P(X_1)$, the information bound and asymptotic variance of $\sqrt{n}(\bar{X} - \mu)$, but if $f_\mu(x) = \frac{1}{2}\exp{-|x - \mu|}$, then $\sigma^2(P) < \sigma^2$, in fact, $\sigma^2(P)/\sigma^2 = 2/\pi$.

6. Establish (6.6.15) by verifying the condition of Theorem 6.2.1 under this model and verifying the formula given.

7. In the binary data regression model of Section 6.4.3, let $\pi = s(\mathbf{z}_i^T \boldsymbol{\beta})$ where $s(t)$ is the continuous distribution function of a random variable symmetric about 0; that is,

$$s(t) = 1 - s(-t), \ t \in R. \tag{6.7.1}$$

(a) Show that π can be written in this form for both the probit and logit models.

(b) Suppose that $\mathbf{z}_i$ are realizations of i.i.d. $\mathbf{Z}_i$, that $\mathbf{Z}_1$ is bounded with probability 1 and let $\widehat{\beta}_L(\mathbf{X}^{(n)})$, where $\mathbf{X}^{(n)} = \{(Y_i, \mathbf{Z}_i) : 1 \leq i \leq n\}$, be the MLE for the logit model. Show that if the correct model has π_i given by s as above and $\beta = \beta_0$, then $\widehat{\beta}_L$ is not a consistent estimate of β_0 unless $s(t)$ is the logistic distribution. But if β_L is defined as the solution of $E\mathbf{Z}_1 s(\mathbf{Z}_1^T \beta_0) = \mathbf{Q}(\beta)$ where $\mathbf{Q}(\beta) = E(\mathbf{Z}_1^T \dot{A}(\mathbf{Z}_1^T \beta))$ is $p \times 1$, then

$$\sqrt{n}(\widehat{\beta}_L - \beta_L)$$

has a limiting normal distribution with mean 0 and variance

$$\dot{\mathbf{Q}}^{-1}(\beta) \text{Var}(\mathbf{Z}_1(Y_1 - \dot{A}(\mathbf{Z}_1^T \beta)))[\dot{Q}^{-1}(\beta)]$$

where $\dot{\mathbf{Q}}(\beta) = E(\mathbf{Z}_1^T \ddot{A}(\mathbf{Z}_1^T \beta_0)\mathbf{Z}_1)$ is $p \times p$ and necessarily nonsingular.
   Hint: Apply Theorem 6.2.1.

8. (Model Selection) Consider the classical Gaussian linear model (6.1.1) $Y_i = \mu_i + \epsilon_i$, $i = 1, \ldots, n$, where ϵ_i are i.i.d. Gaussian with mean zero $\mu_i = \mathbf{z}_i^T \beta$ and variance σ^2, $\mathbf{z}_i$ are d-dimensional vectors of covariate (factor) values. Suppose that the covariates are ranked in order of importance and that we entertain the possibility that the last $d - p$ don't matter for prediction, that is, the model with $\beta_{p+1} = \cdots = \beta_d = 0$ may produce better predictors.

Let $\widehat{\beta}(p)$ be the LSE using only the first p covariates and $\widehat{Y}_i^{(p)}$ the corresponding fitted value.

A natural goal to entertain is to predict future values $Y_1^*, \ldots, Y_n^*$ at $\mathbf{z}_1, \ldots, \mathbf{z}_n$ and evaluate the performance of $\widehat{Y}_1^{(p)}, \ldots, \widehat{Y}_n^{(p)}$ as predictors of $Y_1^*, \ldots, Y_n^*$ and, hence, of the model with $\beta_{d+1} = \cdots = \beta_p = 0$, by the (average) expected prediction error

$$EPE(p) = n^{-1} E \sum_{i=1}^{n} (Y_i^* - \widehat{Y}_i^{(p)})^2.$$

Here $Y_1^*, \ldots, Y_n^*$ are independent of $Y_1, \ldots, Y_n$ and Y_i^* is distributed as Y_i, $i = 1, \ldots, n$. Let $RSS(p) = \sum (Y_i - \widehat{Y}_i^{(p)})^2$ be the residual sum of squares. Suppose that σ^2 is known.

(a) Show that $EPE(p) = \sigma^2 \left(1 + \frac{p}{n}\right) + \frac{1}{n}\sum_{i=1}^{n}(\mu_i - \mu_i^{(p)})^2$ where $\mu_i^{(p)} = \mathbf{z}_i^T \beta^{(p)}$ and $\beta^{(p)} = (\beta_1, \ldots, \beta_p, 0, \ldots, 0)^T$.

(b) Show that

$$E\left(\frac{1}{n}RSS(p)\right) = \sigma^2 \left(1 - \frac{p}{n}\right) + \frac{1}{n}\sum_{i=1}^{n}(\mu_i - \mu_i^{(p)})^2.$$

(c) Show that $\widehat{EPE}(p) \equiv \frac{1}{n}RSS(p) + \frac{2p}{n}\sigma^2$ is an unbiased estimate of $EPE(p)$.

(d) Show that (a),(b) and (c) continue to hold if we assume the Gauss–Markov linear model (6.6.2).

Model selection consists in selecting $p = \widehat{p}$ to minimize $\widehat{EPE}(p)$ and then using $\widehat{\mathbf{Y}}(\widehat{p})$ as a predictor (Mallows, 1973, for instance).

(e) Suppose $d = 2$ and $\mu_i = \beta_1 z_{i1} + \beta_2 z_{i2}$. Evaluate $EPE(p)$ for (i) $p = 1$ and (ii) $p = 2$. Give values of β_1, β_2 and $\{z_{i1}, z_{i2}\}$ such that the EPE in case (i) is smaller than in case (ii) and vice versa. Use $\sigma^2 = 1$ and $n = 10$.

Hint: (a), (b) and (c). Let $\widehat{\mu}_i^{(p)} = \widehat{Y}_i^{(p)}$ and note that

$$EPE(p) = \sigma^2 + \frac{1}{n}\sum_{i=1}^{n}E(\widehat{\mu}_i^{(p)} - \mu_i^{(p)})^2$$

$$+\frac{1}{n}\sum_{i=1}^{n}(\mu_i^{(p)} - \mu_i)^2$$

$$\frac{1}{n}RSS(p) = \frac{1}{n}\sum_{i=1}^{n}(Y_i - \mu_i)^2 - \frac{1}{n}\sum_{i=1}^{n}(\widehat{\mu}_i^{(p)}) - \mu_i)^2.$$

Derive the result for the canonical model.

(d) The result depends only on the mean and covariance structure of the $Y_i, Y_i^*, \widehat{\mu}_i^{(p)}$, $i = 1, \ldots, n$. See the proof of Theorems 1.4.4 and 6.6.1.

6.8 NOTES

Note for Section 6.1

(1) From the L. A. Heart Study after Dixon and Massey (1969).

Note for Section 6.2

(1) See Problem 3.5.9 for a discussion of densities with heavy tails.

Note for Section 6.4

(1) R. A. Fisher pointed out that the agreement of this and other data of Mendel's with his hypotheses is *too* good. To guard against such situations he argued that the test should be used in a two-tailed fashion and that we should reject H both for large and for *small* values of χ^2. Of course, this makes no sense for the model we discussed in this section, but it is reasonable, if we consider alternatives to H, which are not multinomial. For instance, we might envision the possibility that an overzealous assistant of Mendel "cooked" the data. LR test statistics for enlarged models of this type do indeed reject H for data corresponding to small values of χ^2 as well as large ones (Problem 6.4.16). The moral of the story is that the practicing statisticians should be on their guard! For more on this theme see Section 6.6.

6.9 REFERENCES

Cox, D. R., *The Analysis of Binary Data* London: Methuen, 1970.

DIXON, W. AND F. MASSEY, *Introduction to Statistical Analysis*, 3rd ed. New York: McGraw–Hill, 1969.

FISHER, R. A., *Statistical Methods for Research Workers*, 13th ed. New York: Hafner, 1958.

GRAYBILL, F., *An Introduction to Linear Statistical Models*, Vol. I New York: McGraw–Hill, 1961.

HABERMAN, S., *The Analysis of Frequency Data* Chicago: University of Chicago Press, 1974.

HALD, A., *Statistical Theory with Engineering Applications* New York: Wiley, 1952.

HUBER, P. J., "The behavior of the maximum likelihood estimator under nonstandard conditions," *Proc. Fifth Berkeley Symp. Math. Statist. Prob. 1, Univ. of California Press*, 221–233 (1967).

KASS, R., J. KADANE AND L. TIERNEY, "Approximate marginal densities of nonlinear functions," *Biometrika, 76*, 425–433 (1989).

KOENKER, R. AND V. D'OREY, "Computing regression quantiles," *J. Roy. Statist. Soc. Ser. C, 36*, 383–393 (1987).

LAPLACE, P.-S., "Sur quelques points du système du monde," *Memoires de l'Académie des Sciences de Paris* (Reprinted in *Oevres Complétes, 11*, 475–558. Gauthier–Villars, Paris) (1789).

MALLOWS, C., "Some comments on C_p," *Technometrics, 15*, 661–675 (1973).

McCULLAGH, P. AND J. A. NELDER, *Generalized Linear Models* London: Chapman and Hall, New York, 1983; second edition, 1989.

PORTNOY, S. AND R. KOENKER, "The Gaussian Hare and the Laplacian Tortoise: Computability of squared-error versus absolute-error estimators," *Statistical Science, 12*, 279–300 (1997).

RAO, C. R., *Linear Statistical Inference and Its Applications*, 2nd ed. New York: J. Wiley & Sons, 1973.

ROBERTSON, T., F. T. WRIGHT, AND R. L. DYKSTRA, *Order Restricted Statistical Inference* New York: Wiley, 1988.

SCHEFFÉ, H., *The Analysis of Variance* New York: Wiley, 1959.

SCHERVISCH, M., *Theory of Statistics* New York: Springer, 1995.

STIGLER, S., *The History of Statistics: The Measurement of Uncertainty Before 1900* Cambridge, MA: Harvard University Press, 1986.

WEISBERG, S., *Applied Linear Regression*, 2nd ed. New York: Wiley, 1985.

Appendix A

A REVIEW OF BASIC PROBABILITY THEORY

In statistics we study techniques for obtaining and using information in the presence of uncertainty. A prerequisite for such a study is a mathematical model for randomness and some knowledge of its properties. The Kolmogorov model and the modern theory of probability based on it are what we need. The reader is expected to have had a basic course in probability theory. The purpose of this appendix is to indicate what results we consider basic and to introduce some of the notation that will be used in the rest of the book. Because the notation and the level of generality differ somewhat from that found in the standard textbooks in probability at this level, we include some commentary. Sections A.14 and A.15 contain some results that the student may not know, which are relevant to our study of statistics. Therefore, we include some proofs as well in these sections.

In Appendix B we will give additional probability theory results that are of special interest in statistics and may not be treated in enough detail in some probability texts.

A.1 THE BASIC MODEL

Classical mechanics is built around the principle that like causes produce like effects. Probability theory provides a model for situations in which like or similar causes can produce one of a number of unlike effects. A coin that is tossed can land heads or tails. A group of ten individuals selected from the population of the United States can have a majority for or against legalized abortion. The intensity of solar flares in the same month of two different years can vary sharply.

The situations we are going to model can all be thought of as *random experiments*. Viewed naively, an experiment is an action that consists of observing or preparing a set of circumstances and then observing the outcome of this situation. We add to this notion the requirement that to be called an experiment such an action must be repeatable, at least conceptually. The adjective *random* is used only to indicate that we do not, in addition, *require* that every repetition yield the same outcome, although we do not exclude this case. What we expect and observe in practice when we repeat a random experiment many times is that the relative frequency of each of the possible outcomes will tend to stabilize. This long-term relative frequency n_A/n, where n_A is the number of times the possible outcome A occurs in n repetitions, is to many statisticians, including the authors, the operational interpretation of the mathematical concept of probability. In this sense, almost any kind of activity involving uncertainty, from horse races to genetic experiments, falls under the vague heading of "random experiment."

Another school of statisticians finds this formulation too restrictive. By interpreting probability as a subjective measure, they are willing to assign probabilities in any situation involving uncertainty, whether it is conceptually repeatable or not. For a discussion of this approach and further references the reader may wish to consult Savage (1954), Raiffa and Schlaiffer (1961), Savage (1962), Lindley (1965), de Groot (1970), and Berger (1985). We now turn to the mathematical abstraction of a random experiment, the probability model.

In this section and throughout the book, we presume the reader to be familiar with elementary set theory and its notation at the level of Chapter 1 of Feller (1968) or Chapter 1 of Parzen (1960). We shall use the symbols $\cup$, $\cap$, ^{c}, $-$, $\subset$ for union, intersection, complementation, set theoretic difference, and inclusion as is usual in elementary set theory.

A random experiment is described mathematically in terms of the following quantities.

A.1.1 The *sample space* is the set of all possible outcomes of a random experiment. We denote it by Ω. Its complement, the *null set* or *impossible event*, is denoted by $\emptyset$.

A.1.2 A *sample point* is any member of Ω and is typically denoted by ω.

A.1.3 Subsets of Ω are called *events*. We denote events by A, B, and so on or by a description of their members, as we shall see subsequently. The relation between the experiment and the model is given by the correspondence "A occurs if and only if the actual outcome of the experiment is a member of A." The set operations we have mentioned have interpretations also. For example, the relation $A \subset B$ between sets considered as events means that the occurrence of A implies the occurrence of B. If $\omega \in \Omega$, $\{\omega\}$ is called an *elementary* event. If A contains more than one point, it is called a *composite* event.

A.1.4 We will let $\mathcal{A}$ denote a class of subsets of Ω to which we can assign probabilities. For technical mathematical reasons it may not be possible to assign a probability P to every subset of Ω. However, $\mathcal{A}$ is always taken to be a sigma field, which by definition is a nonempty class of events closed under countable unions, intersections, and complementation (cf. Chung, 1974; Grimmett and Stirzaker, 1992; and Loeve, 1977). A *probability distribution* or *measure* is a nonnegative function P on $\mathcal{A}$ having the following properties:

(i) $P(\Omega) = 1$.

(ii) If $A_1, A_2, \ldots$ are pairwise disjoint sets in $\mathcal{A}$, then

$$P\left(\bigcup_{i=1}^{\infty} A_i\right) = \sum_{i=1}^{\infty} P(A_i).$$

Recall that $\cup_{i=1}^{\infty} A_i$ is just the collection of points that are in any one of the sets A_i and that two sets are disjoint if they have no points in common.

A.1.5 The three objects Ω, $\mathcal{A}$, and P together describe a random experiment mathematically. We shall refer to the triple $(\Omega, \mathcal{A}, P)$ either as a *probability model* or identify the model with what it represents as a (random) *experiment*. For convenience, when we refer to *events* we shall automatically exclude those that are not members of $\mathcal{A}$.

References

> Gnedenko (1967) Chapter 1, Sections 1–3, 6–8
> Grimmett and Stirzaker (1992) Sections 1.1–1.3
> Hoel, Port, and Stone (1971) Sections 1.1, 1.2
> Parzen (1960) Chapter 1, Sections 1–5
> Pitman (1993) Sections 1.2 and 1.3

A.2   ELEMENTARY PROPERTIES OF PROBABILITY MODELS

The following are consequences of the definition of P.

A.2.1 If $A \subset B$, then $P(B - A) = P(B) - P(A)$.

A.2.2 $P(A^c) = 1 - P(A)$, $P(\emptyset) = 0$.

A.2.3 If $A \subset B$, $P(B) \geq P(A)$.

A.2.4 $0 \leq P(A) \leq 1$.

A.2.5 $P\left(\bigcup_{n=1}^{\infty} A_n\right) \leq \sum_{n=1}^{\infty} P(A_n)$.

A.2.6 If $A_1 \subset A_2 \subset \cdots \subset A_n \ldots$, then $P\left(\bigcup_{n=1}^{\infty} A_n\right) = \lim_{n \to \infty} P(A_n)$.

A.2.7 $P\left(\bigcap_{i=1}^{k} A_i\right) \geq 1 - \sum_{i=1}^{k} P(A_i^c)$ (Bonferroni's inequality).

References

> Gnedenko, (1967) Chapter 1, Section 8
> Grimmett and Stirzaker (1992) Section 1.3
> Hoel, Port, and Stone (1992) Section 1.3
> Parzen (1960) Chapter 1, Sections 4–5
> Pitman (1993) Section 1.3

A.3   DISCRETE PROBABILITY MODELS

A.3.1 A probability model is called *discrete* if Ω is finite or countably infinite and every subset of Ω is assigned a probability. That is, we can write $\Omega = \{\omega_1, \omega_2, \ldots\}$ and $\mathcal{A}$ is the collection of subsets of Ω. In this case, by axiom (ii) of (A.1.4), we have for any event A,

$$P(A) = \Sigma_{\omega_i \in A} P(\{\omega_i\}). \tag{A.3.2}$$

An important special case arises when Ω has a finite number of elements, say N, all of which are equally likely. Then $P(\{\omega\}) = 1/N$ for every $\omega \in \Omega$, and

$$P(A) = \frac{\text{Number of elements in } A}{N}. \qquad (A.3.3)$$

A.3.4 Suppose that $\omega_1, \ldots, \omega_N$ are the members of some population (humans, guinea pigs, flowers, machines, etc.). Then selecting an individual from this population in such a way that no one member is more likely to be drawn than another, *selecting at random*, is an experiment leading to the model of (A.3.3). Such selection can be carried out if N is small by putting the "names" of the ω_i in a hopper, shaking well, and drawing. For large N, a random number table or computer can be used.

References

Gnedenko (1967) Chapter 1, Sections 4–5
Parzen (1960) Chapter 1, Sections 6–7
Pitman (1993) Section 1.1

A.4   CONDITIONAL PROBABILITY AND INDEPENDENCE

Given an event B such that $P(B) > 0$ and any other event A, we define the *conditional probability* of A *given* B, which we write $P(A \mid B)$, by

$$P(A \mid B) = \frac{P(A \cap B)}{P(B)}. \qquad (A.4.1)$$

If $P(A)$ corresponds to the frequency with which A occurs in a large number of repetitions of the experiment, then $P(A \mid B)$ corresponds to the frequency of occurrence of A relative to the class of trials in which B does occur. From a heuristic point of view $P(A \mid B)$ is the chance we would assign to the event A if we were told that B has occurred.

If $A_1, A_2, \ldots$ are (pairwise) disjoint events and $P(B) > 0$, then

$$P\left(\bigcup_{i=1}^{\infty} A_i \mid B\right) = \sum_{i=1}^{\infty} P(A_i \mid B). \qquad (A.4.2)$$

In fact, for fixed B as before, the function $P(\cdot \mid B)$ is a probability measure on $(\Omega, \mathcal{A})$ which is referred to as the *conditional probability measure* given B.

Transposition of the denominator in (A.4.1) gives the *multiplication rule*,

$$P(A \cap B) = P(B)P(A \mid B). \qquad (A.4.3)$$

If $B_1, B_2, \ldots, B_n$ are (pairwise) disjoint events of positive probability whose union is Ω, the identity $A = \bigcup_{j=1}^{n}(A \cap B_j)$, (A.1.4)(ii) and (A.4.3) yield

$$P(A) = \sum_{j=1}^{n} P(A \mid B_j)P(B_j). \qquad (A.4.4)$$

If $P(A)$ is positive, we can combine (A.4.1), (A.4.3), and (A.4.4) and obtain *Bayes rule*

$$P(B_i \mid A) = \frac{P(A \mid B_i)P(B_i)}{\sum_{j=1}^{n} P(A \mid B_j)P(B_j)}. \qquad (A.4.5)$$

The *conditional probability* of A given $B_1, \ldots, B_n$ is written $P(A \mid B_1, \ldots, B_n)$ and defined by

$$P(A \mid B_1, \ldots, B_n) = P(A \mid B_1 \cap \cdots \cap B_n) \qquad (A.4.6)$$

for any events $A, B_1, \ldots, B_n$ such that $P(B_1 \cap \cdots \cap B_n) > 0$.

Simple algebra leads to the *multiplication rule*,

$$P(B_1 \cap \cdots \cap B_n) = P(B_1)P(B_2 \mid B_1)P(B_3 \mid B_1, B_2) \ldots P(B_n \mid B_1, \ldots, B_{n-1}) \qquad (A.4.7)$$

whenever $P(B_1 \cap \cdots \cap B_{n-1}) > 0$.

Two events A and B are said to be *independent* if

$$P(A \cap B) = P(A)P(B). \qquad (A.4.8)$$

If $P(B) > 0$, the relation (A.4.8) may be written

$$P(A \mid B) = P(A). \qquad (A.4.9)$$

In other words, A and B are independent if knowledge of B does not affect the probability of A.

The events $A_1, \ldots, A_n$ are said to be *independent* if

$$P(A_{i_1} \cap \cdots \cap A_{i_k}) = \prod_{j=1}^{k} P(A_{i_j}) \qquad (A.4.10)$$

for any subset $\{i_1, \ldots, i_k\}$ of the integers $\{1, \ldots, n\}$. If all the $P(A_i)$ are positive, relation (A.4.10) is equivalent to requiring that

$$P(A_j \mid A_{i_1}, \ldots, A_{i_k}) = P(A_j) \qquad (A.4.11)$$

for any j and $\{i_1, \ldots, i_k\}$ such that $j \notin \{i_1, \ldots, i_k\}$.

References

Gnedenko (1967) Chapter 1, Sections 9

Grimmett and Stirzaker (1992) Section 1.4

Hoel, Port, and Stone (1971) Sections 1.4, 1.5

Parzen (1960) Chapter 2, Section 4; Chapter 3, Sections 1,4

Pitman (1993) Section 1.4

A.5   COMPOUND EXPERIMENTS

There is an intuitive notion of independent experiments. For example, if we toss a coin twice, the outcome of the first experiment (toss) reasonably has nothing to do with the outcome of the second. On the other hand, it is easy to give examples of dependent experiments: If we draw twice at random from a hat containing two green chips and one red chip, and if we do not replace the first chip drawn before the second draw, then the probability of a given chip in the second draw will depend on the outcome of the first draw. To be able to talk about independence and dependence of experiments, we introduce the notion of a compound experiment.

Informally, a compound experiment is one made up of two or more component experiments. There are certain natural ways of defining sigma fields and probabilities for these experiments. These will be discussed in this section. The reader not interested in the formalities may skip to Section A.6 where examples of compound experiments are given.

A.5.1 Recall that if $A_1, \ldots, A_n$ are events, the *Cartesian product* $A_1 \times \cdots \times A_n$ of $A_1, \ldots, A_n$ is by definition $\{(\omega_1, \ldots, \omega_n) : \omega_i \in A_i, \ 1 \leq i \leq n\}$. If we are given n experiments (probability models) $\mathcal{E}_1, \ldots, \mathcal{E}_n$ with respective sample spaces $\Omega_1, \ldots, \Omega_n$, then the *sample space Ω of the n stage compound experiment* is by definition $\Omega_1 \times \cdots \times \Omega_n$. The $(n$ stage) compound experiment consists in performing *component experiments* $\mathcal{E}_1, \ldots, \mathcal{E}_n$ and recording all n outcomes. The interpretation of the sample space Ω is that $(\omega_1, \ldots, \omega_n)$ is a sample point in Ω if and only if ω_1 is the outcome of $\mathcal{E}_1$, ω_2 is the outcome of $\mathcal{E}_2$ and so on. To say that $\mathcal{E}_i$ has had outcome $\omega_i^0 \in \Omega_i$ corresponds to the occurrence of the compound event (in Ω) given by $\Omega_1 \times \cdots \times \Omega_{i-1} \times \{\omega_i^0\} \times \Omega_{i+1} \times \cdots \times \Omega_n = \{(\omega_1, \ldots, \omega_n) \in \Omega : \omega_i = \omega_i^0\}$. More generally, if $A_i \in \mathcal{A}_i$, the sigma field corresponding to $\mathcal{E}_i$, then A_i corresponds to $\Omega_1 \times \cdots \times \Omega_{i-1} \times A_i \times \Omega_{i+1} \times \cdots \times \Omega_n$ in the compound experiment. If we want to make the $\mathcal{E}_i$ independent, then intuitively we should have all classes of events $A_1, \ldots, A_n$ with $A_i \in \mathcal{A}_i$, independent. This makes sense in the compound experiment. If P is the probability measure defined on the sigma field $\mathcal{A}$ of the compound experiment, that is, the subsets A of Ω to which we can assign probability[1], we should have

$$P([A_1 \times \Omega_2 \times \cdots \times \Omega_n] \cap [\Omega_1 \times A_2 \times \cdots \times \Omega_n] \cap \ldots)$$
$$= P(A_1 \times \cdots \times A_n)$$
$$= P(A_1 \times \Omega_2 \times \cdots \times \Omega_n) P(\Omega_1 \times A_2 \times \cdots \times \Omega_n) \ldots P(\Omega_1 \times \cdots \times \Omega_{n-1} \times A_n).$$
$$\text{(A.5.2)}$$

If we are given probabilities P_1 on $(\Omega_1, \mathcal{A}_1)$, P_2 on $(\Omega_2, \mathcal{A}_2), \ldots, P_n$ on $(\Omega_n, \mathcal{A}_n)$, then (A.5.2) *defines* P for $A_1 \times \cdots \times A_n$ by

$$P(A_1 \times \cdots \times A_n) = P_1(A_1) \ldots P_n(A_n). \qquad \text{(A.5.3)}$$

It may be shown (Billingsley, 1995; Chung, 1974; Loeve, 1977) that if P is defined by (A.5.3) for events $A_1 \times \cdots \times A_n$, it can be uniquely extended to the sigma field $\mathcal{A}$ specified in note (1) at the end of this appendix. We shall speak of *independent experiments* $\mathcal{E}_1, \ldots, \mathcal{E}_n$ if the n stage compound experiment has its probability structure specified by (A.5.3). In the discrete case (A.5.3) holds provided that

$$P(\{(\omega_1, \ldots, \omega_n)\}) = P_1(\{\omega_1\}) \ldots P_n(\{\omega_n\}) \text{ for all } \omega_i \in \Omega_i, \ 1 \leq i \leq n. \qquad \text{(A.5.4)}$$

Specifying P when the $\mathcal{E}_i$ are dependent is more complicated. In the discrete case we know P once we have specified $P(\{(\omega_1 \ldots, \omega_n)\})$ for each $(\omega_1, \ldots, \omega_n)$ with $\omega_i \in \Omega_i$, $i = 1, \ldots, n$. By the multiplication rule (A.4.7) we have, in the discrete case, the following.

A.5.5 $P(\{(\omega_1, \ldots, \omega_n)\}) = P(\mathcal{E}_1 \text{ has outcome } \omega_1) \, P(\mathcal{E}_2 \text{ has outcome } \omega_2 \mid \mathcal{E}_1 \text{ has outcome } \omega_i) \ldots P(\mathcal{E}_n \text{ has outcome } \omega_n \mid \mathcal{E}_1 \text{ has outcome } \omega_1, \ldots, \mathcal{E}_{n-1} \text{ has outcome } \omega_{n-1})$. The probability structure is determined by these conditional probabilities and conversely.

References

Grimmett and Stirzaker (1992) Sections 1.5, 1.6
Hoel, Port, and Stone (1971) Section 1.5
Parzen (1960) Chapter 3

A.6   BERNOULLI AND MULTINOMIAL TRIALS, SAMPLING WITH AND WITHOUT REPLACEMENT

A.6.1 Suppose that we have an experiment with only two possible outcomes, which we shall denote by S (success) and F (failure). If we assign $P(\{S\}) = p$, we shall refer to such an experiment as a *Bernoulli trial* with probability of success p. The simplest example of such a Bernoulli trial is tossing a coin with probability p of landing heads (success). Other examples will appear naturally in what follows. If we repeat such an experiment n times independently, we say we have performed n *Bernoulli trials with success probability p*. If Ω is the sample space of the compound experiment, any point $\omega \in \Omega$ is an n-dimensional vector of S's and F's and,

$$P(\{\omega\}) = p^{k(\omega)}(1 - p)^{n-k(\omega)} \tag{A.6.2}$$

where $k(\omega)$ is the number of S's appearing in ω. If A_k is the event [exactly k S's occur], then

$$P(A_k) = \binom{n}{k} p^k (1 - p)^{n-k}, \; k = 0, 1, \ldots, n, \tag{A.6.3}$$

where

$$\binom{n}{k} = \frac{n!}{k!(n - k)!}.$$

The formula (A.6.3) is known as the *binomial probability*.

A.6.4 More generally, if an experiment has q possible outcomes $\omega_1, \ldots, \omega_q$ and $P(\{\omega_i\}) = p_i$, we refer to such an experiment as a *multinomial trial with probabilities $p_1, \ldots, p_q$*. If the experiment is performed n times independently, the compound experiment is called n *multinomial trials with probabilities $p_1, \ldots, p_q$*. If Ω is the sample space of this experiment and $\omega \in \Omega$, then

$$P(\{\omega\}) = p_1^{k_1(\omega)} \ldots p_q^{k_q(\omega)} \tag{A.6.5}$$

where $k_i(\omega) =$ number of times ω_i appears in the sequence ω. If $A_{k_1,\ldots,k_q}$ is the event (exactly $k_1\omega_1$'s are observed, exactly $k_2\omega_2$'s are observed, $\ldots$, exactly $k_q\omega_q$'s are observed), then

$$P(A_{k_1,\ldots,k_q}) = \frac{n!}{k_1!\ldots k_q!}p_1^{k_1}\ldots p_q^{k_q} \tag{A.6.6}$$

where the k_i are natural numbers adding up to n.

A.6.7 If we perform an experiment given by $(\Omega, \mathcal{A}, P)$ independently n times, we shall sometimes refer to the outcome of the compound experiment as a *sample of size n from the population given by* $(\Omega, \mathcal{A}, P)$. When Ω is finite the term, *with replacement* is added to distinguish this situation from that described in (A.6.8) as follows.

A.6.8 If we have a finite population of cases $\Omega = \{\omega_1 \ldots, \omega_N\}$ and we select cases ω_i successively at random n times *without replacement*, the component experiments are not independent and, for any outcome $a = (\omega_{i_1}, \ldots, \omega_{i_n})$ of the compound experiment,

$$P(\{a\}) = \frac{1}{(N)_n} \tag{A.6.9}$$

where

$$(N)_n = \frac{N!}{(N-n)!}.$$

If the case drawn is replaced before the next drawing, we are sampling *with replacement*, and the component experiments are independent and $P(\{a\}) = 1/N^n$. If Np of the members of Ω have a "special" characteristic S and $N(1-p)$ have the opposite characteristic F and $A_k = $ (exactly k "special" individuals are obtained in the sample), then

$$P(A_k) = \binom{n}{k}\frac{(Np)_k(N(1-p))_{n-k}}{(N)_n} = \frac{\binom{Np}{k}\binom{N(1-p)}{n-k}}{\binom{N}{n}} \tag{A.6.10}$$

for $\max(0, n - N(1-p)) \le k \le \min(n, Np)$, and $P(A_k) = 0$ otherwise. The formula (A.6.10) is known as the *hypergeometric* probability.

References

Gnedenko (1967) Chapter 2, Section 11
Hoel, Port, and Stone (1971) Section 2.4
Parzen (1960) Chapter 3, Sections 1–4
Pitman (1993) Section 2.1

A.7 PROBABILITIES ON EUCLIDEAN SPACE

Random experiments whose outcomes are real numbers play a central role in theory and practice. The probability models corresponding to such experiments can all be thought of as having a Euclidean space for sample space.

We shall use the notation R^k of k-dimensional Euclidean space and denote members of R^k by symbols such as $\mathbf{x}$ or $(x_1, \ldots, x_k)'$, where $(\ )'$ denotes transpose.

A.7.1 If $(a_1, b_1), \ldots, (a_k, b_k)$ are k open intervals, we shall call the set $(a_1, b_1) \times \cdots \times (a_k, b_k) = \{(x_1, \ldots, x_k) : a_i < x_i < b_i, \ 1 \le i \le k\}$ *an open k rectangle*.

A.7.2 The *Borel field* in R^k, which we denote by $\mathcal{B}^k$, is defined to be the smallest sigma field having all open k rectangles as members. Any subset of R^k we might conceivably be interested in turns out to be a member of $\mathcal{B}^k$. We will write R for R^1 and $\mathcal{B}$ for $\mathcal{B}^1$.

A.7.3 A *discrete (probability) distribution* on R^k is a probability measure P such that $\sum_{i=1}^{\infty} P(\{\mathbf{x}_i\}) = 1$ for some sequence of points $\{\mathbf{x}_i\}$ in R^k. That is, only an $\mathbf{x}_i$ can occur as an outcome of the experiment. This definition is consistent with (A.3.1) because the study of this model and that of the model that has $\Omega = \{\mathbf{x}_1, \ldots, \mathbf{x}_n, \ldots\}$ are equivalent.

The *frequency function* p of a discrete distribution is defined on R^k by

$$p(\mathbf{x}) = P(\{\mathbf{x}\}). \tag{A.7.4}$$

Conversely, any nonnegative function p on R^k vanishing except on a sequence $\{\mathbf{x}_1, \ldots, \mathbf{x}_n, \ldots\}$ of vectors and that satisfies $\sum_{i=1}^{\infty} p(\mathbf{x}_i) = 1$ defines a unique discrete probability distribution by the relation

$$P(A) = \sum_{\mathbf{x}_i \in A} p(\mathbf{x}_i). \tag{A.7.5}$$

A.7.6 A nonnegative function p on R^k, which is integrable and which has

$$\int_{R^k} p(\mathbf{x})d\mathbf{x} = 1,$$

where $d\mathbf{x}$ denotes $dx_1 \ldots dx_n$, is called a *density function*. Integrals should be interpreted in the sense of Lebesgue. However, for practical purposes, Riemann integrals are adequate.

A.7.7 A *continuous probability distribution* on R^k is a probability P that is defined by the relation

$$P(A) = \int_A p(\mathbf{x})d\mathbf{x} \tag{A.7.8}$$

for some density function p and all events A. P defined by A.7.8 are usually called absolutely continuous. We will only consider continuous probability distributions that are also absolutely continuous and drop the term absolutely. It may be shown that a function P so defined satisfies (A.1.4). Recall that the integral on the right of (A.7.8) is by definition

$$\int_{R^k} 1_A(\mathbf{x})p(\mathbf{x})d\mathbf{x}$$

where $1_A(\mathbf{x}) = 1$ if $\mathbf{x} \in A$, and 0 otherwise. Geometrically, $P(A)$ is the volume of the "cylinder" with base A and height $p(\mathbf{x})$ at $\mathbf{x}$. An important special case of (A.7.8) is given by

$$P((a_1, b_1) \times \cdots \times (a_k, b_k)) = \int_{a_k}^{b_k} \cdots \int_{a_1}^{b_1} p(\mathbf{x})d\mathbf{x}. \tag{A.7.9}$$

It turns out that a continuous probability distribution determines the density that generates it "uniquely."[1]

Although in a continuous model $P(\{\mathbf{x}\}) = 0$ for every $\mathbf{x}$, the density function has an operational interpretation close to that of the frequency function. For instance, if p is a continuous density on R, x_0 and x_1 are in R, and h is close to 0, then by the mean value theorem

$$P([x_0 - h, x_0 + h]) \approx 2hp(x_0) \text{ and } \frac{P([x_0 - h, x_0 + h])}{P([x_1 - h, x_1 + h])} \approx \frac{p(x_0)}{p(x_1)}. \qquad \text{(A.7.10)}$$

The ratio $p(x_0)/p(x_1)$ can, thus, be thought of as measuring approximately how much more or less likely we are to obtain an outcome in a neighborhood of x_0 then one in a neighborhood of x_1.

A.7.11 The *distribution function* (d.f.) F is defined by

$$F(x_1, \ldots, x_k) = P((-\infty, x_1] \times \cdots \times (-\infty, x_k]). \qquad \text{(A.7.12)}$$

The d.f. defines P in the sense that if P and Q are two probabilities with the same d.f., then $P = Q$. When $k = 1$, F is a function of a real variable characterized by the following properties:

$$0 \le F \le 1 \qquad \text{(A.7.13)}$$

$$x \le y \Rightarrow F(x) \le F(y) \text{ (Monotone)} \qquad \text{(A.7.14)}$$

$$x_n \downarrow x \Rightarrow F(x_n) \to F(x) \text{ (Continuous from the right)} \qquad \text{(A.7.15)}$$

$$\lim_{x \to \infty} F(x) = 1$$
$$\lim_{x \to -\infty} F(x) = 0. \qquad \text{(A.7.16)}$$

It may be shown that any function F satisfying (A.7.13)–(A.7.16) defines a unique P on the real line. We always have

$$F(x) - F(x - 0)^{(2)} = P(\{x\}). \qquad \text{(A.7.17)}$$

Thus, F is continuous at x if and only if $P(\{x\}) = 0$.

References

Gnedenko (1967) Chapter 4, Sections 21, 22
Hoel, Port, and Stone (1971) Sections 3.1, 3.2, 5.1, 5.2
Parzen (1960) Chapter 4, Sections 1–4, 7
Pitman (1993) Sections 3.4, 4.1 and 4.5

A.8   RANDOM VARIABLES AND VECTORS: TRANSFORMATIONS

Although sample spaces can be very diverse, the statistician is usually interested primarily in one or more numerical characteristics of the sample point that has occurred. For example, we measure the weight of pigs drawn at random from a population, the time to breakdown and length of repair time for a randomly chosen machine, the yield per acre of a field of wheat in a given year, the concentration of a certain pollutant in the atmosphere, and so on. In the probability model, these quantities will correspond to random variables and vectors.

A.8.1 A *random variable* X is a function from Ω to R such that the set $\{\omega : X(\omega) \in B\} = X^{-1}(B)$ is in $\mathcal{A}$ for every $B \in \mathcal{B}$.[1]

A.8.2 A *random vector* $\mathbf{X} = (X_1, \ldots, X_k)^T$ is k-tuple of random variables, or equivalently a function from Ω to R^k such that the set $\{\omega : \mathbf{X}(\omega) \in B\} = \mathbf{X}^{-1}(B)$ is in $\mathcal{A}$ for every $B \in \mathcal{B}^k$.[1] For $k = 1$ random vectors are just random variables. The event $\mathbf{X}^{-1}(B)$ will usually be written $[\mathbf{X} \in B]$ and $P([\mathbf{X} \in B])$ will be written $P[\mathbf{X} \in B]$.

The *probability distribution* of a random vector $\mathbf{X}$ is, by definition, the probability measure $P_{\mathbf{X}}$ in the model $(R^k, \mathcal{B}^k, P_{\mathbf{X}})$ given by

$$P_{\mathbf{X}}(B) = P[\mathbf{X} \in B]. \tag{A.8.3}$$

A.8.4 A random vector is said to have a *continuous* or *discrete distribution* (or to be *continuous* or *discrete*) according to whether its probability distribution is continuous or discrete. Similarly, we will refer to the *frequency function, density, d.f.*, and so on of a random vector when we are, in fact, referring to those features of its probability distribution. The subscript $\mathbf{X}$ or X will be used for densities, d.f.'s, and so on to indicate which vector or variable they correspond to unless the reference is clear from the context in which case they will be omitted.

The probability of any event that is expressible purely in terms of $\mathbf{X}$ can be calculated if we know only the probability distribution of $\mathbf{X}$. In the discrete case this means we need only know the frequency function and in the continuous case the density. Thus, from (A.7.5) and (A.7.8)

$$P[\mathbf{X} \in A] = \sum_{\mathbf{x} \in A} p(\mathbf{x}), \text{ if } \mathbf{X} \text{ is discrete}$$

$$= \int_A p(\mathbf{x}) d\mathbf{x}, \text{ if } \mathbf{X} \text{ is continuous.} \tag{A.8.5}$$

When we are interested in particular random variables or vectors, we will describe them purely in terms of their probability distributions without any further specification of the underlying sample space on which they are defined.

The study of real- or vector-valued functions of a random vector $\mathbf{X}$ is central in the theory of probability and of statistics. Here is the formal definition of such transformations. Let $\mathbf{g}$ be any function from R^k to R^m, $k, m \geq 1$, such that[2] $\mathbf{g}^{-1}(B) = \{\mathbf{y} \in R^k : \mathbf{g}(\mathbf{y}) \in$

$B\} \in \mathcal{B}^k$ for every $B \in \mathcal{B}^m$. Then the *random transformation* $\mathbf{g}(\mathbf{X})$ is defined by

$$\mathbf{g}(\mathbf{X})(\omega) = \mathbf{g}(\mathbf{X}(\omega)). \tag{A.8.6}$$

An example of a transformation often used in statistics is $\mathbf{g} = (g_1, g_2)'$ with $g_1(\mathbf{X}) = k^{-1}\sum_{i=1}^{k} X_i = \bar{X}$ and $g_2(\mathbf{X}) = k^{-1}\sum_{i=1}^{k}(X_i - \bar{X})^2$. Another common example is $\mathbf{g}(\mathbf{X}) = (\min\{X_i\}, \max\{X_i\})'$.

The probability distribution of $\mathbf{g}(\mathbf{X})$ is completely determined by that of $\mathbf{X}$ through

$$P[\mathbf{g}(\mathbf{X}) \in B] = P[\mathbf{X} \in \mathbf{g}^{-1}(B)]. \tag{A.8.7}$$

If $\mathbf{X}$ is discrete with frequency function $p_{\mathbf{X}}$, then $\mathbf{g}(\mathbf{X})$ is discrete and has frequency function

$$p_{\mathbf{g}(\mathbf{X})}(\mathbf{t}) = \sum_{\{\mathbf{x}:\mathbf{g}(\mathbf{x})=\mathbf{t}\}} p_{\mathbf{X}}(\mathbf{x}). \tag{A.8.8}$$

Suppose that X is continuous with density p_X and g is real-valued and one-to-one[3] on an open set S such that $P[X \in S] = 1$. Furthermore, assume that the derivative g' of g exists and does not vanish on S. Then $g(X)$ is continuous with density given by

$$p_{g(X)}(t) = \frac{p_X(g^{-1}(t))}{|g'(g^{-1}(t))|} \tag{A.8.9}$$

for $t \in g(S)$, and 0 otherwise. This is called the *change of variable formula*.

If $g(X) = \sigma X + \mu$, $\sigma \neq 0$, and X is continuous, then

$$p_{g(X)}(t) = \frac{1}{|\sigma|}p_X\left(\frac{t-\mu}{\sigma}\right). \tag{A.8.10}$$

From (A.8.8) it follows that if $(X, Y)^T$ is a discrete random vector with frequency function $p_{(X,Y)}$, then the frequency function of X, known as the *marginal* frequency function, is given by[4]

$$p_X(x) = \sum_y p_{(X,Y)}(x, y). \tag{A.8.11}$$

Similarly, if $(X, Y)^T$ is continuous with density $p_{(X,Y)}$, it may be shown (as a consequence of (A.8.7) and (A.7.8)) that X has a *marginal* density function given by

$$p_X(x) = \int_{-\infty}^{\infty} p_{(X,Y)}(x, y)dy.\text{[5]} \tag{A.8.12}$$

These notions generalize to the case $\mathbf{Z} = (\mathbf{X}, \mathbf{Y})$, a random vector obtained by putting two random vectors together. The (marginal) frequency or density of $\mathbf{X}$ is found as in (A.8.11) and (A.8.12) by summing or integrating out over $\mathbf{y}$ in $p_{(\mathbf{X},\mathbf{Y})}(\mathbf{x}, \mathbf{y})$.

Discrete random variables may be used to approximate continuous ones arbitrarily closely and vice versa.

In practice, all random variables are discrete because there is no instrument that can measure with perfect accuracy. Nevertheless, it is common in statistics to work with continuous distributions, which may be easier to deal with. The justification for this may be theoretical or pragmatic. One possibility is that the observed random variable or vector is obtained by rounding off to a large number of places the true unobservable continuous random variable specified by some idealized physical model. Or else, the approximation of a discrete distribution by a continuous one is made reasonable by one of the limit theorems of Sections A.15 and B.7.

A.8.13 A *convention*: We shall write $X = Y$ if the probability of the event $[X \neq Y]$ is 0.

References

Gnedenko (1967) Chapter 4, Sections 21–24
Grimmett and Stirzaker (1992) Section 4.7
Hoel, Port, and Stone (1971) Sections 3.3, 5.2, 6.1, 6.4
Parzen (1960) Chapter 7, Sections 1–5, 8, 9
Pitman (1993) Section 4.4

A.9   INDEPENDENCE OF RANDOM VARIABLES AND VECTORS

A.9.1 Two random variables X_1 and X_2 are said to be *independent* if and only if for sets A and B in $\mathcal{B}$, the events $[X_1 \in A]$ and $[X_2 \in B]$ are independent.

A.9.2 The random variables $X_1, \ldots, X_n$ are said to be *(mutually) independent* if and only if for any sets $A_1, \ldots, A_n$ in $\mathcal{B}$, the events $[X_1 \in A_1], \ldots, [X_n \in A_n]$ are independent. To generalize these definitions to random vectors $\mathbf{X}_1, \ldots, \mathbf{X}_n$ (not necessarily of the same dimensionality) we need only use the events $[\mathbf{X}_i \in A_i]$ where A_i is a set in the range of $\mathbf{X}_i$.

A.9.3 By (A.8.7), if $\mathbf{X}$ and $\mathbf{Y}$ are independent, so are $\mathbf{g}(\mathbf{X})$ and $\mathbf{h}(\mathbf{Y})$, whatever be $\mathbf{g}$ and $\mathbf{h}$. For example, if (X_1, X_2) and (Y_1, Y_2) are independent, so are $X_1 + X_2$ and $Y_1 Y_2$, $(X_1, X_1 X_2)$ and Y_2, and so on.

Theorem A.9.1. *Suppose* $\mathbf{X} = (X_1, \ldots, X_n)$ *is either a discrete or continuous random vector. Then the random variables* $X_1, \ldots, X_n$ *are independent if, and only if, either of the following two conditions hold:*

$$F_{\mathbf{X}}(x_1, \ldots, x_n) = F_{X_1}(x_1) \ldots F_{X_n}(x_n) \text{ for all } x_1, \ldots, x_n \tag{A.9.4}$$

$$p_{\mathbf{X}}(x_1, \ldots, x_n) = p_{X_1}(x_1) \ldots p_{X_n}(x_n) \text{ for all } x_1, \ldots, x_n. \tag{A.9.5}$$

A.9.6 If the X_i are all continuous and independent, then $\mathbf{X} = (X_1, \ldots, , X_n)$ is continuous.

A.9.7 The preceding equivalences are valid for random vectors $\mathbf{X}_1, \ldots, \mathbf{X}_n$ with $\mathbf{X} = (\mathbf{X}_1, \ldots, \mathbf{X}_n)$.

A.9.8 If $\mathbf{X}_1, \ldots, \mathbf{X}_n$ are independent identically distributed k-dimensional random vectors with d.f. $F_{\mathbf{X}}$ or density (frequency function) $p_{\mathbf{X}}$, then $\mathbf{X}_1, \ldots, \mathbf{X}_n$ is called a *random sample of size* n from a population with d.f. $F_{\mathbf{X}}$ or density (frequency function) $p_{\mathbf{X}}$. In statistics, such a random sample is often obtained by selecting n members at random in the sense of (A.3.4) from a population and measuring k characteristics on each member.

If A is any event, we define the random variable $1(A)$, *the indicator of the event* A, by

$$
\begin{aligned}
1(A)(\omega) &= 1 \text{ if } \omega \in A \\
&= 0 \text{ otherwise.}
\end{aligned}
\tag{A.9.9}
$$

If we perform n Bernoulli trials with probability of success p and we let X_i be the indicator of the event (success on the ith trial), then the X_i form a sample from a distribution that assigns probability p to 1 and $(1-p)$ to 0. Such samples will be referred to as the *indicators of* n *Bernoulli trials with probability of success* p.

References

Gnedenko (1967) Chapter 4, Sections 23, 24
Grimmett and Stirzaker (2001) Sections 3.2, 4.2
Hoel, Port, and Stone (1971) Section 3.4
Parzen (1960) Chapter 7, Sections 6, 7
Pitman (1993) Sections 2.5, 5.3

A.10   THE EXPECTATION OF A RANDOM VARIABLE

Let X be the height of an individual sampled at random from a finite population. Then a reasonable measure of the center of the distribution of X is the average height of an individual in the given population. If $x_1, \ldots, x_q$ are the only heights present in the population, it follows that this average is given by $\sum_{i=1}^{q} x_i P[X = x_i]$ where $P[X = x_i]$ is just the proportion of individuals of height x_i in the population. The same quantity arises (approximately) if we use the long-run frequency interpretation of probability and calculate the average height of the individuals in a large sample from the population in question. In line with these ideas we develop the general concept of expectation as follows.

If X is a nonnegative, discrete random variable with possible values $\{x_1, x_2, \ldots\}$, we define the *expectation* or *mean* of X, written $E(X)$, by

$$
E(X) = \sum_{i=1}^{\infty} x_i p_X(x_i).
\tag{A.10.1}
$$

(Infinity is a possible value of $E(X)$. Take

$$
x_i = i, \ p_X(i) = \frac{1}{i(i+1)}, \ i = 1, 2, \ldots.)
$$

A.10.2 More generally, if X is discrete, decompose $\{x_1, x_2, \ldots\}$ into two sets A and B where A consists of all nonnegative x_i and B of all negative x_i. If either $\sum_{x_i \in A} x_i p_X(x_i) <$

∞ or $\sum_{x_i \in B}(-x_i)p_X(x_i) < \infty$, we define $E(X)$ unambiguously by (A.10.1). Otherwise, we leave $E(X)$ undefined.

Here are some properties of the expectation that hold when X is discrete. If X is a constant, $X(\omega) = c$ for all ω, then

$$E(X) = c. \tag{A.10.3}$$

If $X = 1(A)$ (cf. (A.9.9)), then

$$E(X) = P(A). \tag{A.10.4}$$

If $\mathbf{X}$ is an n-dimensional random vector, if g is a real-valued function on R^n, and if $E(|g(\mathbf{X})|) < \infty$, then it may be shown that

$$E(g(\mathbf{X})) = \sum_{i=1}^{\infty} g(\mathbf{x}_i)p_{\mathbf{X}}(x_i). \tag{A.10.5}$$

As a consequence of this result, we have

$$E(|X|) = \sum_{i=1}^{\infty} |x_i|p_X(x_i). \tag{A.10.6}$$

Taking $g(\mathbf{x}) = \sum_{i=1}^{n}\alpha_i x_i$ we obtain the fundamental relationship

$$E\left(\sum_{i=1}^{n}\alpha_i X_i\right) = \sum_{i=1}^{n}\alpha_i E(X_i) \tag{A.10.7}$$

if $\alpha_1, \ldots, \alpha_n$ are constants and $E(|X_i|) < \infty$, $i = 1, \ldots, n$.

A.10.8 From (A.10.7) it follows that if $X \leq Y$ and $E(X), E(Y)$ are defined, then $E(X) \leq E(Y)$.

If X is a continuous random variable, it is natural to attempt a definition of the expectation via approximation from the discrete case. Those familiar with Lebesgue integration will realize that this leads to

$$E(X) = \int_{-\infty}^{\infty} x p_X(x)dx \tag{A.10.9}$$

as the definition of the *expectation* or *mean* of X whenever $\int_0^{\infty} x p_X(x)dx$ or $\int_{-\infty}^{0} x p_X(x)dx$ is finite. Otherwise, $E(X)$ is left undefined.

A.10.10 A random variable X is said to be *integrable* if $E(|X|) < \infty$.

It may be shown that if $\mathbf{X}$ is a continuous k-dimensional random vector and $g(\mathbf{X})$ is any random variable such that

$$\int_{R^k} |g(\mathbf{x})|p_{\mathbf{X}}(\mathbf{x})d\mathbf{x} < \infty,$$

then $E(g(\mathbf{X}))$ exists and

$$E(g(\mathbf{X})) = \int_{R^k} g(\mathbf{x})p_{\mathbf{X}}(\mathbf{x})d\mathbf{x}. \qquad (A.10.11)$$

In the continuous case expectation properties (A.10.3), (A.10.4), (A.10.7), and (A.10.8) as well as continuous analogues of (A.10.5) and (A.10.6) hold. It is possible to define the expectation of a random variable in general using discrete approximations. The interested reader may consult an advanced text such as Chung (1974), Chapter 3.

The formulae (A.10.5) and (A.10.11) are both sometimes written as

$$E(g(\mathbf{X})) = \int_{R^k} g(\mathbf{x})dF(\mathbf{x}) \text{ or } \int_{R^k} g(\mathbf{x})dP(\mathbf{x}) \qquad (A.10.12)$$

where F denotes the distribution function of $\mathbf{X}$ and P is the probability function of $\mathbf{X}$ defined by (A.8.5).

A convenient notation is $dP(x) = p(x)d\mu(x)$, which means

$$
\begin{aligned}
\int g(\mathbf{x})dP(\mathbf{x}) &= \sum_{i=1}^{\infty} \mathbf{g}(\mathbf{x}_i)p(\mathbf{x}_i), \text{ discrete case} \\
&= \int_{-\infty}^{\infty} g(\mathbf{x})p(\mathbf{x})d\mathbf{x}, \text{ continuous case.}
\end{aligned}
\qquad (A.10.13)
$$

We refer to $\mu = \mu_P$ as the *dominating measure* for P. In the discrete case μ assigns weight one to each of the points in $\{\mathbf{x} : p(\mathbf{x}) > 0\}$ and it is called *counting measure*. In the continuous case $d\mu(\mathbf{x}) = d\mathbf{x}$ and $\mu(\mathbf{x})$ is called *Lebesgue measure*. We will often refer to $p(\mathbf{x})$ as the *density* of $\mathbf{X}$ in the discrete case as well as the continuous case.

References

Chung (1974) Chapter 3
Gnedenko (1967) Chapter 5, Section 26
Grimmett and Stirzaker (1992) Sections 3.3, 4.3
Hoel, Port, and Stone (1971) Sections 4.1, 7.1
Parzen (1960) Chapter 5; Chapter 8, Sections 1–4
Pitman (1993) Sections 3.3, 3.4, 4.1

A.11 MOMENTS

A.11.1 If k is any natural number and X is a random variable, *the kth moment of X is* defined to be the expectation of X^k. We assume that all moments written here exist.

By (A.10.5) and (A.10.11),

$$
\begin{aligned}
E(X^k) &= \sum_x x^k p_X(x) \text{ if } X \text{ is discrete} \\
&= \int_{-\infty}^{\infty} x^k p_X(x)dx \text{ if } X \text{ is continuous.}
\end{aligned}
\qquad (A.11.2)
$$

In general, the moments depend on the distribution of X only.

A.11.3 The distribution of a random variable is typically uniquely specified by its moments. This is the case, for example, if the random variable possesses a moment generating function (cf. (A.12.1)).

A.11.4 The *kth central moment* of X is by definition $E[(X - E(X))^k]$, the kth moment of $(X - E(X))$, and is denoted by μ_k.

A.11.5 The second central moment is called the *variance* of X and will be written Var X. The nonnegative square root of Var X is called the *standard deviation* of X. The standard deviation measures the spread of the distribution of X about its expectation. It is also called a measure of *scale*. Another measure of the same type is $E(|X - E(X)|)$, which is often referred to as the *mean deviation*.

The variance of X is finite if and only if the second moment of X is finite (cf. (A.11.15)). If a and b are constants, then by (A.10.7)

$$\text{Var}(aX + b) = a^2 \text{ Var } X. \tag{A.11.6}$$

(One side of the equation exists if and only if the other does.)

A.11.7 If X is any random variable with well-defined (finite) mean and variance, the *standardized version* or *Z-score* of X is the random variable $Z = (X - E(X))/\sqrt{\text{Var } X}$. By (A.10.7) and (A.11.6) it follows then that

$$E(Z) = 0 \text{ and Var } Z = 1. \tag{A.11.8}$$

A.11.9 If $E(X^2) = 0$, then $X = 0$. If Var $X = 0$, $X = E(X)$ (a constant). These results follow, for instance, from (A.15.2).

A.11.10 The third and fourth central moments are used in the *coefficient of skewness* γ_1 and the *kurtosis* γ_2, which are defined by

$$\gamma_1 = \mu_3/\sigma^3, \ \gamma_2 = (\mu_4/\sigma^4) - 3$$

where $\sigma^2 = \text{Var } X$. See also Section A.12 where γ_1 and γ_2 are expressed in terms of cumulants. These descriptive measures are useful in comparing the shapes of various frequently used densities.

A.11.11 If $Y = a + bX$ with $b > 0$, then the coefficient of skewness and the kurtosis of Y are the same as those of X. If $X \sim \mathcal{N}(\mu, \sigma^2)$, then $\gamma_1 = \gamma_2 = 0$.

A.11.12 It is possible to generalize the notion of moments to random vectors. For simplicity we consider the case $k = 2$. If X_1 and X_2 are random variables and i, j are natural numbers, then the *product moment* of order (i, j) of X_1 and X_2 is, by definition, $E(X_1^i X_2^j)$. The *central product moment* of order (i, j) of X_1 and X_2 is again by definition $E[(X_1 - E(X_1))^i (X_2 - E(X_2))^j]$. The central product moment of order $(1, 1)$ is called the *covariance* of X_1 and X_2 and is written $\text{Cov}(X_1, X_2)$. By expanding the product $(X_1 - E(X_1))(X_2 - E(X_2))$ and using (A.10.3) and (A.10.7), we obtain the relations,

$$\text{Cov}(aX_1 + bX_2, cX_3 + dX_4)$$
$$= ac\,\text{Cov}(X_1, X_3) + bc\,\text{Cov}(X_2, X_3) + ad\,\text{Cov}(X_1, X_4) + bd\,\text{Cov}(X_2, X_4)$$

$$(A.11.13)$$

and

$$\text{Cov}(X_1, X_2) = E(X_1 X_2) - E(X_1)E(X_2). \qquad (A.11.14)$$

If X_1' and X_2' are distributed as X_1 and X_2 and are independent of X_1 and X_2, then

$$\text{Cov}(X_1, X_2) = \frac{1}{2}E(X_1 - X_1')(X_2 - X_2').$$

If we put $X_1 = X_2 = X$ in (A.11.14), we get the formula

$$\text{Var}\, X = E(X^2) - [E(X)]^2. \qquad (A.11.15)$$

The covariance is defined whenever X_1 and X_2 have finite variances and in that case

$$|\text{Cov}(X_1, X_2)| \leq \sqrt{(\text{Var}\, X_1)(\text{Var}\, X_2)} \qquad (A.11.16)$$

with equality holding if and only if

(1) X_1 or X_2 is a constant

or

(2) $(X_1 - E(X_1)) = \dfrac{\text{Cov}(X_1, X_2)}{\text{Var}\, X_2}(X_2 - E(X_2)).$

This is the *correlation inequality*. It may be obtained from the *Cauchy–Schwarz inequality*,

$$|E(Z_1 Z_2)| \leq \sqrt{E(Z_1^2)E(Z_2^2)} \qquad (A.11.17)$$

for any two random variables Z_1, Z_2 such that $E(Z_1^2) < \infty$, $E(Z_2^2) < \infty$. Equality holds if and only if one of Z_1, Z_2 equals 0 or $Z_1 = aZ_2$ for some constant a. The correlation inequality corresponds to the special case $Z_1 = X_1 - E(X_1)$, $Z_2 = X_2 - E(X_2)$. A proof of the Cauchy–Schwarz inequality is given in Remark 1.4.1.

The *correlation* of X_1 and X_2, denoted by $\text{Corr}(X_1, X_2)$, is defined whenever X_1 and X_2 are not constant and the variances of X_1 and X_2 are finite by

$$\text{Corr}(X_1, X_2) = \frac{\text{Cov}(X_1, X_2)}{\sqrt{(\text{Var}\, X_1)(\text{Var}\, X_2)}}. \qquad (A.11.18)$$

The correlation of X_1 and X_2 is the covariance of the standardized versions of X_1 and X_2. The correlation inequality is equivalent to the statement

$$|\text{Corr}(X_1, X_2)| \leq 1. \qquad (A.11.19)$$

Equality holds if and only if X_2 is linear function $(X_2 = a + bX_1, b \neq 0)$ of X_1.

If $X_1, \ldots, X_n$ have finite variances, we obtain as a consequence of (A.11.13) the relation

$$\text{Var}(X_1 + \cdots + X_n) = \sum_{i=1}^{n} \text{Var } X_i + 2 \sum_{i<j} \text{Cov}(X_i, X_j). \qquad (A.11.20)$$

If X_1 and X_2 are independent and X_1 and X_2 are integrable, then

$$E(X_1 X_2) = E(X_1)E(X_2) \qquad (A.11.21)$$

or in view of (A.11.14),

$$\text{Cov}(X_1, X_2) = \text{Corr}(X_1, X_2) = 0 \text{ when } \text{Var}(X_i) > 0, \ i = 1, 2. \qquad (A.11.22)$$

This may be checked directly. It is not true in general that X_1 and X_2 that satisfy (A.11.22) (i.e., are *uncorrelated*) need be independent.

The correlation coefficient roughly measures the amount and sign of linear relationship between X_1 and X_2. It is -1 or 1 in the case of perfect relationship ($X_2 = a + bX_1, b < 0$ or $b > 0$, respectively). See also Section 1.4.

As a consequence of (A.11.22) and (A.11.20), we see that if $X_1, \ldots, X_n$ are independent with finite variances, then

$$\text{Var}(X_1 + \cdots + X_n) = \sum_{i=1}^{n} \text{Var } X_i. \qquad (A.11.23)$$

References

Gnedenko (1967) Chapter 5, Sections 27, 28, 30
Hoel, Port, and Stone (1971) Sections 4.2–4.5, 7.3
Parzen (1960) Chapter 5; Chapter 8, Sections 1–4
Pitman (1993) Section 6.4

A.12   MOMENT AND CUMULANT GENERATING FUNCTIONS

A.12.1 If $E(e^{s_0|X|}) < \infty$ for some $s_0 > 0$, $M_X(s) = E(e^{sX})$ is well defined for $|s| \leq s_0$ and is called the *moment generating function of X*. By (A.10.5) and (A.10.11),

$$
\begin{aligned}
M_X(s) &= \sum_{i=1}^{\infty} e^{sx_i} p_X(x_i) && \text{if } X \text{ is discrete} \\
&= \int_{-\infty}^{\infty} e^{sx} p_X(x)dx && \text{if } X \text{ is continuous.}
\end{aligned}
\qquad (A.12.2)
$$

If M_X is well defined in a neighborhood $\{s : |s| \leq s_0\}$ of zero, all moments of X are finite and

$$M_X(s) = \sum_{k=0}^{\infty} \frac{E(X^k)}{k!} s^k, \ |s| \leq s_0. \qquad (A.12.3)$$

A.12.4 The moment generating function M_X has derivatives of all orders at $s = 0$ and

$$\frac{d^k}{ds^k} M_X(s)\bigg|_{s=0} = E(X^k).$$

A.12.5 If defined, M_X determines the distribution of X uniquely and is itself uniquely determined by the distribution of X.

If $X_1, \ldots, X_n$ are independent random variables with moment generating functions $M_{X_1}, \ldots, M_{X_n}$, then $X_1 + \cdots + X_n$ has moment generating function given by

$$M_{(X_1 + \cdots + X_n)}(s) = \prod_{i=1}^{n} M_{X_i}(s). \tag{A.12.6}$$

This follows by induction from the definition and (A.11.21). For a generalization of the notion of moment generating function to random vectors, see Section B.5.

The function

$$K_X(s) = \log M_X(s) \tag{A.12.7}$$

is called the *cumulant generating function* of X. If M_X is well defined in some neighborhood of zero, K_X can be represented by the convergent Taylor expansion

$$K_X(s) = \sum_{j=0}^{\infty} \frac{c_j}{j!} s^j \tag{A.12.8}$$

where

$$c_j = c_j(X) = \frac{d^j}{ds^j} K_X(s)|_{s=0} \tag{A.12.9}$$

is called the jth *cumulant* of X, $j \geq 1$. For $j \geq 2$ and any constant a, $c_j(X + a) = c_j(X)$. If X and Y are independent, then $c_j(X + Y) = c_j(X) + c_j(Y)$. The first cumulant c_1 is the mean μ of X, c_2 and c_3 equal the second and third central moments μ_2 and μ_3 of X, and $c_4 = \mu_4 - 3\mu_2^2$. The coefficients of skewness and kurtosis (see (A.11.10)) can be written as $\gamma_1 = c_3/c_2^{\frac{3}{2}}$ and $\gamma_2 = c_4/c_2^2$. If X is normally distributed, $c_j = 0$ for $j \geq 3$. See Problem B.3.8.

References

Hoel, Port, and Stone (1971) Chapter 8, Section 8.1
Parzen (1960) Chapter 5, Section 3; Chapter 8, Sections 2–3
Rao (1973) Section 2b.4

A.13 SOME CLASSICAL DISCRETE AND CONTINUOUS DISTRIBUTIONS

By definition, the probability distribution of a random variable or vector is just a probability measure on a suitable Euclidean space. In this section we introduce certain families of distributions, which arise frequently in probability and statistics, and list some of their properties. Following the name of each distribution we give a shorthand notation that will sometimes be used as will obvious abbreviations such as "binomial (n, θ)" for "the binomial distribution with parameter (n, θ)". The symbol p as usual stands for a frequency or density function. *If anywhere below p is not specified explicitly for some value of x it shall be assumed that p vanishes at that point. Similarly, if the value of the distribution function F is not specified outside some set, it is assumed to be zero to the "left" of the set and one to the "right" of the set.*

I. Discrete Distributions

The binomial distribution with parameters n and θ : $\mathcal{B}(n, \theta)$.

$$p(k) = \left(\begin{array}{c} n \\ k \end{array} \right) \theta^k (1 - \theta)^{n-k}, \ k = 0, 1, \ldots, n. \tag{A.13.1}$$

The parameter n can be any integer ≥ 0 whereas θ may be any number in $[0, 1]$.

A.13.2 If X is the total number of successes obtained in n Bernoulli trials with probability of success θ, then X has a $\mathcal{B}(n, \theta)$ distribution (see (A.6.3)).

If X has a $\mathcal{B}(n, \theta)$ distribution, then

$$E(X) = n\theta, \ \text{Var} \ X = n\theta(1 - \theta). \tag{A.13.3}$$

Higher-order moments may be computed from the moment generating function

$$M_X(t) = [\theta e^t + (1 - \theta)]^n. \tag{A.13.4}$$

A.13.5 If $X_1, X_2, \ldots, X_k$ are independent random variables distributed as $\mathcal{B}(n_1, \theta)$, $\mathcal{B}(n_2, \theta), \ldots, \mathcal{B}(n_k, \theta)$, respectively, then $X_1 + X_2 + \cdots + X_k$ has a $\mathcal{B}(n_1 + \cdots + n_k, \theta)$ distribution. This result may be derived by using (A.12.5) and (A.12.6) in conjunction with (A.13.4).

The hypergeometric distribution with parameters D, N, and n : $\mathcal{H}(D, N, n)$.

$$p(k) = \frac{\left(\begin{array}{c} D \\ k \end{array} \right) \left(\begin{array}{c} N - D \\ n - k \end{array} \right)}{\left(\begin{array}{c} N \\ n \end{array} \right)} \tag{A.13.6}$$

for k a natural number with $\max(0, n - (N - D)) \leq k \leq \min(n, D)$. The parameters D and n may be any natural numbers that are less than or equal to the natural number N.

A.13.7 If X is the number of defectives (special objects) in a sample of size n taken without replacement from a population with D defectives and $N - D$ nondefectives, then X has an $\mathcal{H}(D, N, n)$ distribution (see (A.6.10)). If the sample is taken with replacement, X has a $\mathcal{B}(n, D/N)$ distribution.

If X has an $\mathcal{H}(D, N, n)$ distribution, then

$$E(X) = n\frac{D}{N}, \ \text{Var} \ X = n\frac{D}{N}\left(1 - \frac{D}{N}\right)\frac{N-n}{N-1}. \tag{A.13.8}$$

Formulae (A.13.8) may be obtained directly from the definition (A.13.6). An easier way is to use the interpretation (A.13.7) by writing $X = \sum_{j=1}^{n} I_j$ where $I_j = 1$ if the jth object sampled is defective and 0 otherwise, and then applying formulae (A.10.4), (A.10.7), and (A.11.20).

The Poisson distribution with parameter $\lambda : \mathcal{P}(\lambda)$.

$$p(k) = \frac{e^{-\lambda}\lambda^k}{k!} \tag{A.13.9}$$

for $k = 0, 1, 2, \ldots$. The parameter λ can be any positive number.

If X has a $\mathcal{P}(\lambda)$ distribution, then

$$E(X) = \text{Var} \ X = \lambda. \tag{A.13.10}$$

The moment generating function of X is given by

$$M_X(t) = e^{\lambda(e^t - 1)}. \tag{A.13.11}$$

A.13.12 If $X_1, X_2, \ldots, X_n$ are independent random variables with $\mathcal{P}(\lambda_1), \mathcal{P}(\lambda_2), \ldots,$ $\mathcal{P}(\lambda_n)$ distributions, respectively, then $X_1 + X_2 + \ + X_n$ has the $\mathcal{P}(\lambda_1 + \lambda_2 + \cdots + \lambda_n)$ distribution. This result may be derived in th same manner as the corresponding fact for the binomial distribution.

The multinomial distribution with parameters $n, \theta_1, \ldots, \theta_q : \mathcal{M}(n, \theta_1, \ldots, \theta_q)$.

$$p(k_1, \ldots, k_q) = \frac{n!}{k_1!\ldots k_q!}\theta_1^{k_1}\ldots\theta_q^{k_q} \tag{A.13.13}$$

whenever k_i are nonnegative integers such that $\sum_{i=1}^{q} k_i = n$. The parameter n is any natural number while $(\theta_1, \ldots, \theta_q)$ is any vector in

$$\Theta = \left\{(\theta_1, \ldots, \theta_q) : \theta_i \geq 0, \ 1 \leq i \leq q, \ \sum_{i=1}^{q}\theta_i = 1\right\}.$$

A.13.14 If $\mathbf{X} = (X_1, \ldots, X_q)'$, where X_i is the number of times outcome ω_i occurs in n multinomial trials with probabilities $(\theta_1, \ldots, \theta_q)$, then $\mathbf{X}$ has a $\mathcal{M}(n, \theta_1, \ldots, \theta_q)$ distribution (see (A.6.6)).

If $\mathbf{X}$ has a $\mathcal{M}(n, \theta_1, \ldots, \theta_q)$ distribution,

$$
\begin{aligned}
E(X_i) &= n\theta_i, \ \text{Var} \ X_i = n\theta_i(1 - \theta_i) \\
\text{Cov}(X_i, X_j) &= -n\theta_i\theta_j, \ i \neq j, \ i, j = 1, \ldots, q.
\end{aligned}
\tag{A.13.15}
$$

These results may either be derived directly or by a representation such as that discussed in
(A.13.8) and an application of formulas (A.10.4), (A.10.7), (A.13.13), and (A.11.20).

A.13.16 If $\mathbf{X}$ has a $\mathcal{M}(n, \theta_1, \ldots, \theta_q)$ distribution, then $(X_{i_1}, \ldots, X_{i_s}, n - \sum_{j=1}^{s} X_{i_j})'$
has a $\mathcal{M}(n, \theta_{i_1}, \ldots, \theta_{i_s}, 1 - \sum_{j=1}^{s} \theta_{i_j})$ distribution for any set $\{i_1, \ldots, i_s\} \subset \{1, \ldots, q\}$.
Therefore, X_j has $\mathcal{B}(n, \theta_j)$ distributions for each j and more generally $\sum_{j=1}^{s} X_{i_j}$ has
a $\mathcal{B}(n, \sum_{j=1}^{s} \theta_{i_j})$ distribution if $s < q$. These remarks follow from the interpretation
(A.13.14).

II. Continuous Distributions

Before beginning our listing we introduce some convenient notations: $X \sim F$ will
mean that X is a random variable with d.f. F, and $X \sim p$ will similarly mean that X has
density or frequency function p.

Let Y be a random variable with d.f. F. Let F_μ be the d.f. of $Y + \mu$. The family
$\mathcal{F}_L = \{F_\mu : -\infty < \mu < \infty\}$ is called a *location parameter family*, μ is called a *location
parameter*, and we say that Y *generates* $\mathcal{F}_L$. By definition, for any μ, $X \sim F_\mu \Leftrightarrow X - \mu \sim$
F. Therefore, for any μ, γ,

$$F_\mu(x) = F(x - \mu) = F_0(x - \mu) = F_\gamma(x + (\gamma - \mu))$$

and all calculations involving F_μ can be referred back to F or any other member of the
family. Similarly, if Y generates $\mathcal{F}_L$ so does $Y + \gamma$ for any fixed γ. If Y has a first
moment, it follows that we may without loss of generality (as far as generating $\mathcal{F}_L$ goes)
assume that $E(Y) = 0$. Then if $X \sim F_\mu$, $E(X) = \mu$.

Similarly let F_σ^* be the d.f. of σY, $\sigma > 0$. The family $\mathcal{F}_S = \{F_\sigma^* : \sigma > 0\}$ is called a
scale parameter family, σ is a *scale parameter*, and Y is said to *generate* $\mathcal{F}_S$. By definition,
for any $\sigma > 0$, $X \sim F_\sigma^* \Leftrightarrow X/\sigma \sim F$. Again all calculations involving one member of the
family can be referred back to any other because for any $\sigma, \tau > 0$,

$$F_\sigma^*(x) = F_\tau^* \left(\frac{\tau x}{\sigma} \right).$$

If Y generates $\mathcal{F}_S$ and Y has a first moment different from 0, we may without loss of
generality take $E(Y) = 1$ and, hence, if $X \sim F_\sigma^*$, then $E(X) = \sigma$. Alternatively, if Y has
a second moment, we may select F as being the unique member of the family $\mathcal{F}_S$ having
$\mathrm{Var}\, Y = 1$ and then $X \sim F_\sigma^* \Rightarrow \mathrm{Var}\, X = \sigma^2$. Finally, define $F_{\mu,\sigma}$ as the d.f. of $\sigma Y + \mu$.
The family $\mathcal{F}_{L,S} = \{F_{\mu,\sigma} : -\infty < \mu < \infty, \sigma > 0\}$ is called a *location-scale parameter
family*, μ is called a *location parameter*, and σ a *scale parameter*, and Y is said to generate
$\mathcal{F}_{L,S}$. From

$$F_{\mu,\sigma}(x) = F \left(\frac{x - \mu}{\sigma} \right) = F_{\gamma,\tau} \left(\frac{\tau(x - \mu)}{\sigma} + \gamma \right),$$

we see as before how to refer calculations involving one member of the family back to any
other. Without loss of generality, if Y has a second moment, we may take

$$E(Y) = 0, \ \mathrm{Var}\, Y = 1.$$

Then if $X \sim F_{\mu,\sigma}$, we obtain

$$E(X) = \mu, \ \text{Var} \ X = \sigma^2.$$

Clearly $F_\mu = F_{\mu,1}$, $F_\sigma^* = F_{0,\sigma}$.

The relation between the density of $F_{\mu,\sigma}$ and that of F is given by (A.8.10). All the families of densities we now define are location-scale or scale families.

The normal (Gaussian) distribution with parameters μ and σ^2 : $\mathcal{N}(\mu, \sigma^2)$.

$$p(x) = \frac{1}{\sqrt{2\pi} \ \sigma} \exp\left\{ -\frac{1}{2\sigma^2}(x - \mu)^2 \right\}. \qquad (A.13.17)$$

The parameter μ can be any real number while σ is positive. The normal distribution with $\mu = 0$ and $\sigma = 1$ is known as the *standard normal distribution*. Its density will be denoted by $\varphi(z)$ and its d.f. by $\Phi(z)$.

A.13.18 The family of $\mathcal{N}(\mu, \sigma^2)$ distributions is a location-scale family. If Z has a $\mathcal{N}(0, 1)$ distribution, then $\sigma Z + \mu$ has a $\mathcal{N}(\mu, \sigma^2)$ distribution, and conversely if X has a $\mathcal{N}(\mu, \sigma^2)$ distribution, then $(X - \mu)/\sigma$ has a standard normal distribution.

If X has a $\mathcal{N}(\mu, \sigma^2)$ distribution, then

$$E(X) = \mu, \ \text{Var} \ X = \sigma^2. \qquad (A.13.19)$$

More generally, all moments may be obtained from

$$M_X(t) = \exp\left\{ \mu t + \frac{\sigma^2 t^2}{2} \right\} \qquad (A.13.20)$$

for $-\infty < t < \infty$. In particular if $\mu = 0$, $\sigma^2 = 1$, then

$$M_X(t) = \sum_{k=0}^{\infty} \left[\frac{(2k)!}{2^k k!} \right] \frac{t^{2k}}{(2k)!} \qquad (A.13.21)$$

and, hence, in this case we can conclude from (A.12.4) that

$$\begin{aligned} E(X^k) &= 0 && \text{if } k \geq 0 \text{ is odd} \\ E(X^k) &= \frac{k!}{2^{k/2}(k/2)!} && \text{if } k \geq 0 \text{ is even.} \end{aligned} \qquad (A.13.22)$$

A.13.23 If $X_1, \ldots, X_n$ are independent normal random variables such that $E(X_i) = \mu_i$, $\text{Var} \ X_i = \sigma_i^2$, and $c_1, \ldots, c_n$ are any constants that are not all 0, then $\sum_{i=1}^{n} c_i X_i$ has a $\mathcal{N}(c_1\mu_1 + \cdots + c_n\mu_n, c_1^2\sigma_1^2 + \cdots + c_n^2\sigma_n^2)$ distribution. This follows from (A.13.20), (A.12.5), and (A.12.6).

Further information about the normal distribution may be found in Section A.15 and Appendix B.

The exponential distribution with parameter λ : $\mathcal{E}(\lambda)$.

$$p(x) = \lambda e^{-\lambda x}, \ x > 0. \tag{A.13.24}$$

The range of λ is $(0, \infty)$. The distribution function corresponding to this p is given by

$$F(x) = 1 - e^{-\lambda x} \text{ for } x > 0. \tag{A.13.25}$$

A.13.26 If $\sigma = 1/\lambda$, then σ is a scale parameter. $\mathcal{E}(1)$ is called the *standard exponential distribution*.
    If X has an $\mathcal{E}(\lambda)$ distribution,

$$E(X) = \frac{1}{\lambda}, \ \text{Var } X = \frac{1}{\lambda^2}. \tag{A.13.27}$$

More generally, all moments may be obtained from

$$M_X(t) = \frac{1}{1 - (t/\lambda)} = \sum_{k=0}^{\infty} \left[\frac{k!}{\lambda^k} \right] \frac{t^k}{k!} \tag{A.13.28}$$

which is well defined for $t < \lambda$.
    Further information about the exponential distribution may be found in Appendix B.

The uniform distribution on $(a, b) : \mathcal{U}(a, b)$**.**

$$p(x) = \frac{1}{(b - a)}, \ a < x < b \tag{A.13.29}$$

where (a, b) is any pair of real numbers such that $a < b$. The corresponding distribution function is given by

$$F(x) = \frac{(x - a)}{(b - a)} \text{ for } a < x < b. \tag{A.13.30}$$

If X has a $\mathcal{U}(a, b)$ distribution, then

$$E(X) = \frac{a + b}{2}, \ \text{Var } X = \frac{(b - a)^2}{12}. \tag{A.13.31}$$

A.13.32 If we set $\mu = a$, $\sigma = (b - a)$, then we can check that the $\mathcal{U}(a, b)$ family is a location-scale family generated by Y, where $Y \sim \mathcal{U}(0, 1)$.

References

Gnedenko (1967) Chapter 4, Sections 21–24; Chapter 5, Sections 26–28, 30
Hoel, Port, and Stone (1971) Sections 3.4.1, 5.3.1, 5.3.2
Parzen (1962) Chapter 4, Sections 4–6; Chapter 5; Chapter 6
Pitman (1993) pages 475–487

A.14 MODES OF CONVERGENCE OF RANDOM VARIABLES AND LIMIT THEOREMS

Much of probability theory can be viewed as variations, extensions, and generalizations of two basic results, the central limit theorem and the law of large numbers. Both of these theorems deal with the limiting behavior of sequences of random variables. The notions of limit that are involved are the subject of this section. All limits in the section are as $n \to \infty$.

A.14.1 We say that the sequence of random variables $\{Z_n\}$ *converges* to the random variable Z in *probability* and write $Z_n \overset{P}{\to} Z$ if $P[|Z_n - Z| \geq \epsilon] \to 0$ as $n \to \infty$ for every $\epsilon > 0$. That is, $Z_n \overset{P}{\to} Z$ if the chance that Z_n and Z differ by any given amount is negligible for n large enough.

A.14.2 We say that the sequence $\{Z_n\}$ *converges in law (in distribution)* to Z and write $Z_n \overset{\mathcal{L}}{\to} Z$ if $F_{Z_n}(t) \to F_Z(t)$ *for every point t such that F_Z is continuous at t.* (Recall that F_Z is continuous at t if and only if $P[Z = t] = 0$ (A.7.17).) This is the mode of convergence needed for approximation of one distribution by another.

$$\text{If } Z_n \overset{P}{\to} Z, \text{ then } Z_n \overset{\mathcal{L}}{\to} Z. \tag{A.14.3}$$

Because convergence in law requires nothing of the joint distribution of the Z_n and Z whereas convergence in probability does, it is not surprising and easy to show that, in general, convergence in law does not imply convergence in probability (e.g., Chung, 1974), but consider the following.

A.14.4 If $Z = z_0$ (a constant), convergence in law of $\{Z_n\}$ to Z implies convergence in probability.

Proof. Note that $z_0 \pm \epsilon$ are points of continuity of F_Z for every $\epsilon > 0$. Then

$$
\begin{aligned}
P[|Z_n - z_0| \geq \epsilon] &= 1 - P(Z_n < z_0 + \epsilon) + P(Z_n \leq z_0 - \epsilon) \\
&\leq 1 - F_{Z_n}\left(z_0 + \frac{\epsilon}{2}\right) + F_{Z_n}(z_0 - \epsilon).
\end{aligned}
\tag{A.14.5}
$$

By assumption the right-hand side of (A.14.5) converges to $(1 - F_Z(z_0 + \epsilon/2)) + F_Z(z_0 - \epsilon) = 0$. □

A.14.6 If $Z_n \overset{P}{\to} z_0$ (a constant) and g is continuous at z_0, then $g(Z_n) \overset{P}{\to} g(z_0)$.

Proof. If ϵ is positive, there exists a δ such that $|z - z_0| < \delta$ implies $|g(z) - g(z_0)| < \epsilon$. Therefore,

$$P[|g(Z_n) - g(z_0)| < \epsilon] \geq P[|Z_n - z_0| < \delta] = 1 - P[|Z_n - z_0| \geq \delta]. \tag{A.14.7}$$

Because the right-hand side of (A.14.7) converges to 1, by the definition (A.14.1) the result follows. □

A more general result is given by the following.

A.14.8 If $Z_n \overset{\mathcal{L}}{\to} Z$ and g is continuous, then $g(Z_n) \overset{\mathcal{L}}{\to} g(Z)$.
The following theorem due to Slutsky will be used repeatedly.

Theorem A.14.9 *If $Z_n \overset{\mathcal{L}}{\to} Z$ and $U_n \overset{P}{\to} u_0$ (a constant), then*

(a) $Z_n + U_n \overset{\mathcal{L}}{\to} Z + u_0$,

(b) $U_n Z_n \overset{\mathcal{L}}{\to} u_0 Z$.

Proof. We prove (a). The other claim follows similarly. Begin by writing

$$F_{(Z_n+U_n)}(t) \;=\; P[Z_n + U_n \leq t,\, U_n \geq u_0 - \epsilon]$$
$$+ P[Z_n + U_n \leq t,\, U_n < u_0 - \epsilon]. \tag{A.14.10}$$

Let t be a point of continuity of $F_{(Z+u_0)}$. Because a distribution function has at most countably many points of discontinuity, we may for any t choose ϵ positive and arbitrarily small such that $t \pm \epsilon$ are both points of continuity of $F_{(Z+u_0)}$. Now by (A.14.10),

$$F_{(Z_n+U_n)}(t) \leq P[Z_n \leq t - u_0 + \epsilon] + P[|U_n - u_0| > \epsilon]. \tag{A.14.11}$$

Moreover,

$$P[Z_n \leq t - u_0 + \epsilon] = F_{(Z_n+u_0)}(t + \epsilon). \tag{A.14.12}$$

Because $F_{Z_n+u_0}(t) = P[Z_n \leq t - u_0] = F_{Z_n}(t - u_0)$, we must have $Z_n + u_0 \overset{\mathcal{L}}{\to} Z + u_0$. Thus,

$$\limsup_n F_{(Z_n+U_n)}(t) \leq \lim_n F_{(Z_n+u_0)}(t + \epsilon) + \lim_n P[|U_n - u_0| \geq \epsilon] = F_{(Z+u_0)}(t + \epsilon). \tag{A.14.13}$$

Similarly,

$$1 - F_{(Z_n+U_n)}(t) = P[Z_n + U_n > t] \leq P[Z_n > t - u_0 - \epsilon] + P[|U_n - u_0| > \epsilon]. \tag{A.14.14}$$

and, hence,

$$\liminf_n F_{(Z_n+U_n)}(t) \geq \lim_n F_{(Z_n+u_0)}(t - \epsilon) = F_{(Z+u_0)}(t - \epsilon). \tag{A.14.15}$$

Therefore,

$$F_{(Z+u_0)}(t - \epsilon) \leq \liminf_n F_{(Z_n+U_n)}(t) \leq \limsup_n F_{(Z_n+U_n)}(t) \leq F_{(Z+u_0)}(t + \epsilon). \tag{A.14.16}$$

Because ϵ may be permitted to tend to 0 and $F_{(Z+u_0)}$ is continuous at t, the result follows. □

A.14.17 Corollary. *Suppose that a_n is a sequence of constants tending to ∞, b is a fixed number, and $a_n(Z_n - b) \overset{\mathcal{L}}{\to} X$. Let g be a function of a real variable that is differentiable and whose derivative g' is continuous at b.*[1] *Then*

$$a_n[g(Z_n) - g(b)] \overset{\mathcal{L}}{\to} g'(b)X. \tag{A.14.18}$$

Proof. By Slutsky's theorem

$$Z_n - b = \frac{1}{a_n}[a_n(Z_n - b)] \overset{\mathcal{L}}{\to} 0 \cdot X = 0. \qquad (A.14.19)$$

By (A.14.4), $|Z_n - b| \overset{P}{\to} 0$. Now apply the mean value theorem to $g(Z_n) - g(b)$ getting

$$a_n[g(Z_n) - g(b)] = a_n[g'(Z_n^*)(Z_n - b)]$$

where $|Z_n^* - b| \leq |Z_n - b|$. Because $|Z_n - b| \overset{P}{\to} 0$, so does $|Z_n^* - b|$ and, hence, $Z_n^* \overset{P}{\to} b$. By the continuity of g' and (A.14.6), $g'(Z_n^*) \overset{P}{\to} g'(b)$. Therefore, applying (A.14.9) again, $g'(Z_n^*)[a_n(Z_n - b)] \overset{\mathcal{L}}{\to} g'(b)X$. $\qquad \square$

A.14.20 Suppose that $\{Z_n\}$ takes on only natural number values and $p_{Z_n}(z) \to p_Z(z)$ for all z. Then $Z_n \overset{\mathcal{L}}{\to} Z$.

This is immediate because whatever be z, $F_{Z_n}(z) = \sum_{k=0}^{[z]} p_{Z_n}(k) \to \sum_{k=0}^{[z]} p_Z(k) = F_Z(z)$ where $[z] = $ greatest integer $\leq z$. The converse is also true and easy to establish.

A.14.21 (Scheffé) Suppose that $\{Z_n\}$, Z are continuous and $p_{Z_n}(z) \to p_Z(z)$ for (almost) all z. Then $Z_n \overset{\mathcal{L}}{\to} Z$ (Hajek and Sidak, 1967, p. 64).

References

Grimmett and Stirzaker (1992) Sections 7.1–7.4

A.15 FURTHER LIMIT THEOREMS AND INEQUALITIES

The Bernoulli law of large numbers, which we give next, brings us back to the motivation of our definition of probability. There we noted that in practice the relative frequency of occurrence of an event A in many repetitions of an experiment tends to stabilize and that this "limit" corresponds to what we mean by the probability of A. Now having defined probability and independence of experiments abstractly we can prove that, in fact, the relative frequency of occurrence of an event A approaches its probability as the number of repetitions of the experiment increases. Such results and their generalizations are known as laws of large numbers. The first was discovered by Bernoulli in 1713. Its statement is as follows.

Bernoulli's (Weak) Law of Large Numbers

If $\{S_n\}$ is a sequence of random variables such that S_n has a $\mathcal{B}(n, p)$ distribution for $n \geq 1$, then

$$\frac{S_n}{n} \overset{P}{\to} p. \qquad (A.15.1)$$

As in (A.13.2), think of S_n as the number of successes in n binomial trials in which we identify success with occurrence of A and failure with occurrence of A^c. Then S_n/n can be interpreted as the relative frequency of occurrence of A in n independent repetitions of the experiment in which A is an event and the Bernoulli law is now evidently a statement of the type we wanted.

Bernoulli's proof of this result was rather complicated and it remained for the Russian mathematician Chebychev to give a two-line argument. His generalization of Bernoulli's result is based on an inequality that has proved to be of the greatest importance in probability and statistics.

Chebychev's Inequality

If X is any random variable, then

$$P[|X| \geq a] \leq \frac{E(X^2)}{a^2}. \tag{A.15.2}$$

The Bernoulli law follows readily from (A.15.2) and (A.13.3) via the calculation

$$P\left[\left|\frac{S_n}{n} - p\right| \geq \epsilon\right] \leq \frac{E(S_n/n - p)^2}{\epsilon^2} = \frac{\text{Var } S_n}{n^2\epsilon^2} = \frac{p(1-p)}{n\epsilon^2} \to 0 \text{ as } n \to \infty. \tag{A.15.3}$$

A generalization of (A.15.2), which contains various important and useful inequalities, is the following. Let g be a nonnegative function on R such that g is nondecreasing on the range of a random variable Z. Then

$$P[Z \geq a] \leq \frac{E(g(Z))}{g(a)}. \tag{A.15.4}$$

If we put $Z = |X|$, $g(t) = t^2$ if $t \geq 0$ and 0 otherwise, we get (A.15.2). Other important cases are obtained by taking $Z = |X|$ and $g(t) = t$ if $t \geq 0$ and 0 otherwise (*Markov's inequality*), and $Z = X$ and $g(t) = e^{st}$ for $s > 0$ and all real t (*Bernstein's inequality*, see B.8.1 for the binomial case Bernstein's inequality).

Proof of (A.15.4). Note that by the properties of g,

$$g(a)I_{[Z \geq a]} \leq g(Z)I_{[Z \geq a]} \leq g(Z). \tag{A.15.5}$$

Therefore, by (A.10.8)

$$g(a)P[Z \geq a] = E(g(a)I_{[Z \geq a]}) \leq E(g(Z)), \tag{A.15.6}$$

which is equivalent to (A.15.4).   □

The following result, which follows from Chebychev's inequality, is a useful generalization of Bernoulli's law.

Khintchin's (Weak) Law of Large Numbers

Let $\{X_i\}$, $i \geq 1$, be a sequence of independent identically distributed random variables with finite mean μ and define $S_n = \sum_{i=1}^{n} X_i$. Then

$$\frac{S_n}{n} \xrightarrow{P} \mu. \tag{A.15.7}$$

Upon taking the X_i to be indicators of binomial trials, we obtain (A.15.1).

De Moivre–Laplace Theorem

Suppose that $\{S_n\}$ is a sequence of random variables such that for each n, S_n has a $\mathcal{B}(n, p)$ distribution where $0 < p < 1$. Then

$$\frac{S_n - np}{\sqrt{np(1-p)}} \xrightarrow{\mathcal{L}} Z, \tag{A.15.8}$$

where Z has a standard normal distribution. That is, the standardized versions of S_n converge in law to a standard normal random variable. If we write

$$\frac{S_n - np}{\sqrt{np(1-p)}} = \frac{\sqrt{n}}{\sqrt{p(1-p)}} \left(\frac{S_n}{n} - p \right)$$

and use (A.14.9), it is easy to see that (A.15.8) implies (A.15.1).

The De Moivre–Laplace theorem is generalized by the following.

Central Limit Theorem

Let $\{X_i\}$ be a sequence of independent identically distributed random variables with (common) expectation μ and variance σ^2 such that $0 < \sigma^2 < \infty$. Then, if $S_n = \sum_{i=1}^{n} X_i$

$$\frac{S_n - n\mu}{\sigma\sqrt{n}} \xrightarrow{\mathcal{L}} Z, \tag{A.15.9}$$

where Z has the standard normal distribution.

The last two results are most commonly used in statistics as approximation theorems. Let k and l be nonnegative integers. The De Moivre–Laplace theorem is used as

$$
\begin{aligned}
P[k \le S_n \le l] &= P\left[k - \frac{1}{2} \le S_n \le l + \frac{1}{2} \right] \\
&= P\left[\frac{k - np - \frac{1}{2}}{\sqrt{npq}} \le \frac{S_n - np}{\sqrt{npq}} \le \frac{l - np + \frac{1}{2}}{\sqrt{npq}} \right] \\
&\approx \Phi\left(\frac{l - np + \frac{1}{2}}{\sqrt{npq}} \right) - \Phi\left(\frac{k - np - \frac{1}{2}}{\sqrt{npq}} \right)
\end{aligned}
\tag{A.15.10}
$$

where $q = 1 - p$. The $\frac{1}{2}$ appearing in $k - \frac{1}{2}$ and $l + \frac{1}{2}$ is called the *continuity correction*. We have an excellent idea of how good this approximation is. An illustrative discussion is given in Feller (1968, pp. 187–188). A rule of thumb is that for most purposes the approximation can be used when np and $n(1 - p)$ are both larger than 5.

Only when the X_i are integer-valued is the first step of (A.15.10) followed. Otherwise (A.15.9) is applied in the form

$$P[a \le S_n \le b] \approx \Phi\left(\frac{b - n\mu}{\sqrt{n}\sigma} \right) - \Phi\left(\frac{a - n\mu}{\sqrt{n}\sigma} \right). \tag{A.15.11}$$

The central limit theorem (and some of its generalizations) are also used to justify the assumption that "most" random variables that are measures of numerical characteristics of real populations, such as intelligence, height, weight, and blood pressure, are approximately normally distributed. The argument is that the observed numbers are sums of a large number of small (unobserved) independent factors. That is, each of the characteristic variables is expressible as a sum of a large number of small variables such as influences of particular genes, elements in the diet, and so on. For example, height is a sum of factors corresponding to heredity and environment.

If a bound for $E|X_i - \mu|^3$ is known, it is possible to give a theoretical estimate of the error involved in replacing $P(S_n \leq b)$ by its normal approximation:

Berry–Esséen Theorem

Suppose that $X_1, \ldots, X_n$ are i.i.d. with mean μ and variance $\sigma^2 > 0$. Then, for all n,

$$\sup_t \left| P\left(\frac{S_n - n\mu}{\sqrt{n}\sigma} \leq t \right) - \Phi(t) \right| \leq \frac{33}{4} \frac{E|X_1 - \mu|^3}{\sqrt{n}\sigma^3}. \qquad (A.15.12)$$

For a proof, see Chung (1974, p. 224).

In practice, if we need the distribution of S_n we try to calculate it exactly for small values of n and then observe empirically when the approximation can be used with safety. This process of combining a limit theorem with empirical investigations is applicable in many statistical situations where the distributions of transformations $g(\mathbf{x})$ (see A.8.6) of interest become progressively more difficult to compute as the sample size increases and yet tend to stabilize. Examples of this process may be found in Chapter 5.

We conclude this section with two simple limit theorems that lead to approximations of one classical distribution by another. The very simple proofs of these results may, for instance, be found in Gnedenko (1967, p. 53 and p. 105).

A.15.12 The first of these results reflects the intuitively obvious fact that if the populations sampled are large and the samples are comparatively small, sampling with and without replacement leads to approximately the same probability distribution. Specifically, suppose that $\{X_N\}$ is a sequence of random variables such that X_N has a hypergeometric $\mathcal{H}(D_N, N, n)$, distribution where $D_N/N \to p$ as $N \to \infty$ and n is fixed. Then

$$p_{X_N}(k) \to \binom{n}{k} p^k (1 - p)^{n-k} \qquad (A.15.13)$$

as $N \to \infty$ for $k = 0, 1, \ldots, n$. By (A.14.20) we conclude that

$$X_n \xrightarrow{\mathcal{L}} X, \qquad (A.15.14)$$

where X has a $\mathcal{B}(n, p)$ distribution. The approximation of the hypergeometric distribution by the binomial distribution indicated by this theorem is rather good. For instance, if $N = 50$, $n = 5$, and $D = 20$, the approximating binomial distribution to $\mathcal{H}(D, N, n)$ is $\mathcal{B}(5, 0.4)$. If $\mathcal{H}$ holds, $P[X \leq 2] = 0.690$ while under the approximation,

$P[X \leq 2] = 0.683$. As indicated in this example, the approximation is reasonable when $(n/N) \leq 0.1$.

The next elementary result, due to Poisson, plays an important role in advanced probability theory.

Poisson's Theorem

Suppose that $\{X_n\}$ is a sequence of random variables such that X_n has a $\mathcal{B}(n, p_n)$ distribution and $np_n \to \lambda$ as $n \to \infty$, where $0 \leq \lambda < \infty$. Then

$$p_{X_n}(k) \to \frac{e^{-\lambda}\lambda^k}{k!} \tag{A.15.15}$$

for $k = 0, 1, 2, \ldots$ as $n \to \infty$. By (A.14.20) it follows that $X_n \overset{\mathcal{L}}{\to} X$ where X has a $\mathcal{P}(\lambda)$ distribution. This theorem suggests that we approximate the $\mathcal{B}(n, p)$ distribution by the $\mathcal{P}(np)$ distribution. Tables 3 on p. 108 and 2 on p. 154 of Feller (1968) indicate the excellence of the approximation when p is small and np is moderate. It may be shown that the error committed is always bounded by np^2.

References

Gnedenko (1967) Chapter 2, Section 13; Chapter 6, Section 32; Chapter 8, Section 42
Hoel, Port, and Stone (1971) Chapter 3, Section 3.4.2
Parzen (1960) Chapter 5, Sections 4, 5; Chapter 6, Section 2; Chapter 10, Section 2

A.16 POISSON PROCESS

A.16.1 A *Poisson process with parameter* λ is a collection of random variables $\{N(t)\}$, $t > 0$, such that

(i) $N(t)$ has a $\mathcal{P}(\lambda t)$ distribution for each t.

(ii) $N(t + h) - N(t)$ is independent of $N(s)$ for all $s \leq t$, $h > 0$, and has a $\mathcal{P}(\lambda h)$ distribution.

Poisson processes are frequently applicable when we study phenomena involving events that occur "rarely" in small time intervals. For example, if $N(t)$ is the number of disintegrations of a fixed amount of some radioactive substance in the period from time 0 to time t, then $\{N(t)\}$ is a Poisson process. The numbers $N(t)$ of "customers" (people, machines, etc.) arriving at a service counter from time 0 to time t are sometimes well approximated by a Poisson process as is the number of people who visit a WEB site from time 0 to t. Many interesting examples are discussed in the books of Feller (1968), Parzen (1962), Karlin (1969). In each of the preceding examples of a Poisson process $N(t)$ represents the number of times an "event" (radioactive disintegration, arrival of a customer) has occurred in the time from 0 to t. We use the word *event* here for lack of a better one because these are not events in terms of the probability model on which the $N(t)$ are defined. If we keep temporarily to this notion of event as a recurrent phenomenon that is randomly determined in some fashion and define $N(t)$ as the number of events occurring between time 0 and time t, we can ask under what circumstances $\{N(t)\}$ will form a Poisson process.

A.16.2 Formally, let $\{N(t)\}$, $t > 0$ be a collection of natural number valued random variables. It turns out that, $\{N(t)\}$ is a Poisson process with parameter λ if and only if the following conditions hold:

  (a) $N(t+h) - N(t)$ is independent of $N(s)$, $s \leq t$, for $h > 0$,

  (b) $N(t+h) - N(t)$ has the same distribution as $N(h)$ for $h > 0$,

  (c) $P[N(h) = 1] = \lambda h + o(h)$, and

  (d) $P[N(h) > 1] = o(h)$.

(The quantity $o(h)$ is such that $o(h)/h \to 0$ as $h \to 0$.) Physically, these assumptions may be interpreted as follows.

  (i) The time of recurrence of the "event" is unaffected by past occurrences.

  (ii) The distribution of the number of occurrences of the "event" depends only on the length of the time for which we observe the process.

  (iii) and (iv) The chance of any occurrence in a given time period goes to 0 as the period shrinks and having only one occurrence becomes far more likely than multiple occurrences.

This assertion may be proved as follows. Fix t and divide $[0, t]$ into n intervals $[0, t/n]$, $(t/n, 2t/n], \ldots, ((n-1)t/n, t]$. Let I_{jn} be the indicator of the event $[N(jt/n) - N((j-1)t/n) \geq 1]$ and define $N_n(t) = \sum_{j=1}^{n} I_{jn}$. Then $N_n(t)$ differs from $N(t)$ only insofar as multiple occurrences in one of the small subintervals are only counted as one occurrence. By (a) and (b), $N_n(t)$ has a $\mathcal{B}(n, P[N(t/n) \geq 1])$ distribution. From (c) and (d) and Theorem (A.15.15) we see that $N_n(t) \overset{\mathcal{L}}{\to} Z$, where Z has a $\mathcal{P}(\lambda t)$ distribution. On the other hand,

$$
\begin{aligned}
P[|N_n(t) - N(t)| \geq \epsilon] &\leq P[N_n(t) \neq N(t)] \\
&\leq P\left[\bigcup_{j=1}^{n}\left[\left(N\left(\frac{jt}{n}\right) - N\left(\frac{(j-1)t}{n}\right)\right) > 1\right]\right] \\
&\leq \sum_{j=1}^{n} P\left[\left(N\left(\frac{jt}{n}\right) - N\left(\frac{(j-1)t}{n}\right)\right) > 1\right] \\
&= nP\left[N\left(\frac{t}{n}\right) > 1\right] \\
&= no\left(\frac{t}{n}\right) \to 0 \text{ as } n \to \infty.
\end{aligned}
$$

$$(A.16.3)$$

The first of the inequalities in (A.16.3) is obvious, the second says that if $N_n(t) \neq N(t)$ there must have been a multiple occurrence in a small subinterval, the third is just (A.2.5), and the remaining identities follow from (b) and (d). The claim (A.16.3) now follows from Slutsky's theorem (A.14.9) upon writing $N(t) = N_n(t) + (N(t) - N_n(t))$.

A.16.4 Let T_1 be the time at which the "event" first occurs in a Poisson process (the first t such that $N(t) = 1$), T_2 be the time at which the "event" occurs for the second time, and so on. Then $T_1, T_2 - T_1, \ldots, T_n - T_{n-1}, \ldots$ are independent, identically distributed $\mathcal{E}(\lambda)$ random variables.

References

Gnedenko (1967) Chapter 10, Section 51
Grimmett and Stirzaker (1992) Section 6.8
Hoel, Port, and Stone (1971) Section 9.3
Parzen (1962) Chapter 6, Section 5
Pitman (1993) Sections 3.5, 4.2

A.17   NOTES

Notes for Section A.5

(1) We define $\mathcal{A}$ to be the smallest sigma field that has every set of the form $A_1 \times \cdots \times A_n$ with $A_i \in \mathcal{A}_i$, $1 \leq i \leq n$, as a member.

Notes for Section A.7

(1) Strictly speaking, the density is only defined up to a set of Lebesgue measure 0.

(2) We shall use the notation $g(x+0)$ for $\lim_{x_n \downarrow x} g(x_n)$ and $g(x-0)$ for $\lim_{x_n \uparrow x} g(x_n)$ for a function g of a real variable that possesses such limits.

Notes for Section A.8

(1) The requirement on the sets $X^{-1}(B)$ is purely technical. It is no restriction in the discrete case and is satisfied by any function of interest when Ω is R^k or a subset of R^k. Sets B that are members of $\mathcal{B}^k$ are called *measurable*. When considering subsets of R^k, we will assume automatically that they are measurable.

(2) Such functions $\mathbf{g}$ are called *measurable*. This condition ensures that $\mathbf{g}(\mathbf{X})$ satisfies definitions (A.8.1) and (A.8.2). For convenience, when we refer to functions we shall assume automatically that this condition is satisfied.

(3) A function $\mathbf{g}$ is said to be one to one if $\mathbf{g}(\mathbf{x}) = \mathbf{g}(\mathbf{y})$ implies $\mathbf{x} = \mathbf{y}$.

(4) Strictly speaking, (X, Y) and (x, y) in (A.8.11) and (A.8.12) should be transposed. However, we avoid this awkward notation when the meaning is clear.

(5) The integral in (A.8.12) may only be finite for "almost all" x. In the regular cases we study this will not be a problem.

Notes for Section A.14

(1) It may be shown that one only needs the existence of the derivative g' at b for (A.14.17) to hold. See Theorem 5.3.3.

A.18   REFERENCES

BERGER, J. O., *Statistical Decision Theory and Bayesian Analysis* New York: Springer, 1985.

BILLINGSLEY, P., *Probability and Measure*, 3rd ed. New York: J. Wiley & Sons, 1995.

CHUNG, K. L., *A Course in Probability Theory* New York: Academic Press, 1974.

DEGROOT, M. H., *Optimal Statistical Decisions* New York: McGraw Hill, 1970.

FELLER, W., *An Introduction to Probability Theory and Its Applications*, Vol. I, 3rd ed. New York: J. Wiley & Sons, 1968.

GNEDENKO, B. V., *The Theory of Probability*, 4th ed. New York: Chelsea, 1967.

GRIMMETT, G. R., AND D. R. STIRZAKER, *Probability and Random Processes* Oxford: Clarendon Press, 1992.

HÁJEK, J. AND Z. SIDÁK, *Theory of Rank Tests* New York: Academic Press, 1967.

HOEL, P. G., S. C. PORT, AND C. J. STONE, *Introduction to Probability Theory* Boston: Houghton Mifflin, 1971.

KARLIN, S., *A First Course in Stochastic Processes* New York: Academic Press, 1969.

LINDLEY, D. V., *Introduction to Probability and Statistics from a Bayesian Point of View*, Part I: *Probability*; Part II: *Inference* London: Cambridge University Press, 1965.

LOÉVE, M., *Probability Theory*, Vol. I, 4th ed. Berlin: Springer, 1977.

PARZEN, E., *Modern Probability Theory and Its Application* New York: J. Wiley & Sons, 1960.

PARZEN, E., *Stochastic Processes* San Francisco: Holden–Day, 1962.

PITMAN, J., *Probability* New York: Springer, 1993.

RAIFFA, H., AND R. SCHLAIFFER, *Applied Statistical Decision Theory*, Division of Research, Graduate School of Business Administration, Boston: Harvard University, 1961.

RAO, C. R., *Linear Statistical Inference and Its Applications*, 2nd ed. New York: J. Wiley & Sons, 1973.

SAVAGE, L. J., *The Foundations of Statistics* New York: J. Wiley & Sons, 1954.

SAVAGE, L. J., *The Foundation of Statistical Inference* London: Methuen & Co., 1962.

Appendix B

ADDITIONAL TOPICS IN PROBABILITY AND ANALYSIS

In this appendix we give some results in probability theory, matrix algebra, and analysis that are essential in our treatment of statistics and that may not be treated in enough detail in more specialized texts. Some of the material in this appendix, as well as extensions, can be found in Anderson (1958), Billingsley (1995), Breiman (1968), Chung (1978), Dempster (1969), Feller (1971), Loeve (1977), and Rao (1973).

Measure theory will not be used. We make the blanket assumption that all sets and functions considered are measurable.

B.1 CONDITIONING BY A RANDOM VARIABLE OR VECTOR

The concept of conditioning is important in studying associations between random variables or vectors. In this section we present some results useful for prediction theory, estimation theory, and regression.

B.1.1 The Discrete Case

The reader is already familiar with the notion of the conditional probability of an event A given that another event B has occurred. If $\mathbf{Y}$ and $\mathbf{Z}$ are discrete random vectors possibly of different dimensions, we want to study the conditional probability structure of $\mathbf{Y}$ given that $\mathbf{Z}$ has taken on a particular value $\mathbf{z}$.

Define the *conditional frequency function* $p(\cdot \mid \mathbf{z})$ *of* $\mathbf{Y}$ *given* $\mathbf{Z} = \mathbf{z}$ by

$$p(\mathbf{y} \mid \mathbf{z}) = P[\mathbf{Y} = \mathbf{y} \mid \mathbf{Z} = \mathbf{z}] = \frac{p(\mathbf{y}, \mathbf{z})}{p_{\mathbf{Z}}(\mathbf{z})} \qquad (\text{B.1.1})$$

where p and $p_{\mathbf{Z}}$ are the frequency functions of $(\mathbf{Y}, \mathbf{Z})$ and $\mathbf{Z}$. The conditional frequency function p is defined only for values of $\mathbf{z}$ such that $p_{\mathbf{Z}}(\mathbf{z}) > 0$. With this definition it is

TABLE B.1

y \ z	0	10	20	$p_Y(y)$
0	0.25	0.05	0.05	0.35
1	0.05	0.15	0.05	0.25
2	0.05	0.10	0.25	0.40
$p_Z(z)$	0.35	0.30	0.35	1

clear that $p(\cdot \mid \mathbf{z})$ is the frequency of a probability distribution because

$$\Sigma_{\mathbf{y}} p(\mathbf{y} \mid \mathbf{z}) = \frac{\Sigma_{\mathbf{y}} p(\mathbf{y}, \mathbf{z})}{p_{\mathbf{Z}}(\mathbf{z})} = \frac{p_{\mathbf{Z}}(\mathbf{z})}{p_{\mathbf{Z}}(\mathbf{z})} = 1$$

by $(A.8.11)$. This probability distribution is called the *conditional distribution of* $\mathbf{Y}$ *given that* $\mathbf{Z} = \mathbf{z}$.

Example B.1.1 Let $\mathbf{Y} = (Y_1, \ldots, Y_n)$, where the Y_i are the indicators of a set of n Bernoulli trials with success probability p. Let $Z = \Sigma_{i=1}^{n} Y_i$, the total number of successes. Then Z has a binomial, $\mathcal{B}(n, p)$, distribution and

$$p(\mathbf{y} \mid z) = \frac{P[\mathbf{Y} = \mathbf{y}, Z = z]}{\binom{n}{z} p^z (1-p)^{n-z}} = \frac{p^z (1-p)^{n-z}}{\binom{n}{z} p^z (1-p)^{n-z}} = \frac{1}{\binom{n}{z}} \quad \text{(B.1.2)}$$

if the y_i are all 0 or 1 and $\Sigma y_i = z$.

Thus, if we are told we obtained k successes in n binomial trials, then these successes are as likely to occur on one set of trials as on any other. □

Example B.1.2 Let Y and Z have the joint frequency function given by the table
For instance, suppose Z is the number of cigarettes that a person picked at random from a certain population smokes per day (to the nearest 10), and Y is a general health rating for the same person with 0 corresponding to good, 2 to poor, and 1 to neither. We find for $z = 20$

y	0	1	2
$p(y \mid 20)$	$\frac{1}{7}$	$\frac{1}{7}$	$\frac{5}{7}$

These figures would indicate an association between heavy smoking and poor health because $p(2 \mid 20)$ is almost twice as large as $p_Y(2)$. □

The conditional distribution of $\mathbf{Y}$ given $\mathbf{Z} = \mathbf{z}$ is easy to calculate in two special cases.

(i) If $\mathbf{Y}$ and $\mathbf{Z}$ are independent, then $p(\mathbf{y} \mid \mathbf{z}) = p_{\mathbf{Y}}(\mathbf{y})$ and the conditional distribution coincides with the marginal distribution.

(ii) If $\mathbf{Y}$ is a function of $\mathbf{Z}$, $h(\mathbf{Z})$, then the conditional distribution of $\mathbf{Y}$ is degenerate, $\mathbf{Y} = h(\mathbf{Z})$ with probability 1.

Both of these assertions follow immediately from Definition(B.1.1).

Two important formulae follow from (B.1.1) and (A.4.5). Let $q(\mathbf{z} \mid \mathbf{y})$ denote the conditional frequency function of $\mathbf{Z}$ given $\mathbf{Y} = \mathbf{y}$. Then

$$p(\mathbf{y}, \mathbf{z}) = p(\mathbf{y} \mid \mathbf{z})p_{\mathbf{Z}}(\mathbf{z}) \tag{B.1.3}$$

$$p(\mathbf{y} \mid \mathbf{z}) = \frac{q(\mathbf{z} \mid \mathbf{y})p_{\mathbf{Y}}(\mathbf{y})}{\Sigma_{\mathbf{y}}q(\mathbf{z} \mid \mathbf{y})p_{\mathbf{Y}}(\mathbf{y})} \qquad \text{Bayes' Theorem} \tag{B.1.4}$$

whenever the denominator of the right-hand side is positive.

Equation (B.1.3) can be used for model construction. For instance, suppose that the number Z of defectives in a lot of N produced by a manufacturing process has a $\mathcal{B}(N, \theta)$ distribution. Suppose the lot is sampled n times without replacement and let Y be the number of defectives found in the sample. We know that given $Z = z$, Y has a hypergeometric, $\mathcal{H}(z, N, n)$, distribution. We can now use (B.1.3) to write down the joint distribution of Y and Z

$$P[Y = y, Z = z] = \binom{N}{z} \theta^z(1 - \theta)^{N-z}\frac{\binom{z}{y}\binom{N-z}{n-y}}{\binom{N}{n}}$$

where the combinatorial coefficients $\binom{a}{b}$ vanish unless a, b are integers with $b \le a$.

We can also use this model to illustrate (B.1.4). Because we would usually only observe Y, we may want to know what the conditional distribution of Z given $Y = y$ is. By (B.1.4) this is

$$P[Z = z \mid Y = y] = \binom{N}{z} \theta^z(1 - \theta)^{N-z} \binom{z}{y}\binom{N-z}{n-y} \Big/ c(y) \tag{B.1.5}$$

where

$$c(y) = \Sigma_z \binom{N}{z} \theta^z(1 - \theta)^{N-z} \binom{z}{y}\binom{N-z}{n-y}.$$

This formula simplifies to (see Problem B.1.11) the binomial probability,

$$P[Z = z \mid Y = y] = \binom{N - n}{z - y} \theta^{z-y}(1 - \theta)^{N-n-(z-y)}. \tag{B.1.6}$$

B.1.2   Conditional Expectation for Discrete Variables

Suppose that Y is a random variable with $E(|Y|) < \infty$. Define the *conditional expectation of Y given $\mathbf{Z} = \mathbf{z}$*, written $E(Y \mid \mathbf{Z} = \mathbf{z})$, by

$$E(Y \mid \mathbf{Z} = \mathbf{z}) = \Sigma_y yp(y \mid \mathbf{z}). \tag{B.1.7}$$

Note that by (B.1.1), if $p_\mathbf{Z}(\mathbf{z}) > 0$,

$$E(|Y| \mid \mathbf{Z} = \mathbf{z}) = \Sigma_y |y| p(y \mid \mathbf{z}) \leq \Sigma_y |y| \frac{p_Y(y)}{p_\mathbf{Z}(\mathbf{z})} = \frac{E(|Y|)}{p_\mathbf{Z}(\mathbf{z})}. \tag{B.1.8}$$

Thus, when $p_\mathbf{Z}(\mathbf{z}) > 0$, the conditional expected value of Y is finite whenever the expected value is finite.

Example B.1.3 Suppose Y and Z have the joint frequency function of Table B.1. We find

$$E(Y \mid Z = 20) = 0 \cdot \frac{1}{7} + 1 \cdot \frac{1}{7} + 2 \cdot \frac{5}{7} = \frac{11}{7} = 1.57.$$

Similarly, $E(Y \mid Z = 10) = \frac{7}{6} = 1.17$ and $E(Y \mid Z = 0) = \frac{3}{7} = 0.43$. Note that in the health versus smoking context, we can think of $E(Y \mid Z = z)$ as the mean health rating for people who smoke z cigarettes a day. $\qquad \square$

Let $g(\mathbf{z}) = E(Y \mid \mathbf{Z} = \mathbf{z})$. The random variable $g(\mathbf{Z})$ is written $E(Y \mid \mathbf{Z})$ and is called the *conditional expectation of Y given* $\mathbf{Z}$.[1]

As an example we calculate $E(Y_1 \mid Z)$ where Y_1 and Z are given in Example B.1.1. We have

$$E(Y_1 \mid Z = i) = P[Y_1 = 1 \mid Z = i] = \frac{\binom{n-1}{i-1}}{\binom{n}{i}} = \frac{i}{n}. \tag{B.1.9}$$

The first of these equalities holds because Y_1 is an indicator. The second follows from (B.1.2) because $\binom{n-1}{i-1}$ is just the number of ways i successes can occur in n Bernoulli trials with the first trial being a success. Therefore,

$$E(Y_1 \mid Z) = \frac{Z}{n}. \tag{B.1.10}$$

B.1.3   Properties of Conditional Expected Values

In the context of Section A.4, the conditional distribution of a random vector $\mathbf{Y}$ given $\mathbf{Z} = \mathbf{z}$ corresponds to a single probability measure $P_\mathbf{z}$ on $(\Omega, \mathcal{A})$. Specifically, define for $A \in \mathcal{A}$,

$$P_\mathbf{z}(A) = P(A \mid [\mathbf{Z} = \mathbf{z}]) \text{ if } p_\mathbf{Z}(\mathbf{z}) > 0. \tag{B.1.11}$$

This $P_\mathbf{z}$ is just the conditional probability measure on $(\Omega, \mathcal{A})$ mentioned in (A.4.2). Now the conditional distribution of $\mathbf{Y}$ given $\mathbf{Z} = \mathbf{z}$ is the same as the distribution of $\mathbf{Y}$ if $P_\mathbf{z}$ is the probability measure on $(\Omega, \mathcal{A})$. Therefore, the conditional expectation *is* an ordinary expectation with respect to the probability measure $P_\mathbf{z}$. It follows that all the properties of the expectation given in (A.10.3)–(A.10.8) hold for the conditional expectation given $\mathbf{Z} = \mathbf{z}$. Thus, for any real-valued function $r(\mathbf{Y})$ with $E|r(\mathbf{Y})| < \infty$,

$$E(r(\mathbf{Y}) \mid \mathbf{Z} = \mathbf{z}) = \Sigma_\mathbf{y} r(\mathbf{y}) p(\mathbf{y} \mid \mathbf{z})$$

and

$$E(\alpha Y_1 + \beta Y_2 \mid \mathbf{Z} = \mathbf{z}) = \alpha E(Y_1 \mid \mathbf{Z} = \mathbf{z}) + \beta E(Y_2 \mid \mathbf{Z} = \mathbf{z}) \tag{B.1.12}$$

identically in $\mathbf{z}$ for any Y_1, Y_2 such that $E(|Y_1|)$, $E(|Y_2|)$ are finite. Because the identity holds for all $\mathbf{z}$, we have

$$E(\alpha Y_1 + \beta Y_2 \mid \mathbf{Z}) = \alpha E(Y_1 \mid \mathbf{Z}) + \beta E(Y_2 \mid \mathbf{Z}). \tag{B.1.13}$$

This process can be repeated for each of (A.10.3)–(A.10.8) to obtain analogous properties of the conditional expectations.

In two special cases we can calculate conditional expectations immediately. If Y and $\mathbf{Z}$ are independent and $E(|Y|) < \infty$, then

$$E(Y \mid \mathbf{Z}) = E(Y). \tag{B.1.14}$$

This is clear by (i).

On the other hand, by (ii)

$$E(h(\mathbf{Z}) \mid \mathbf{Z}) = h(\mathbf{Z}). \tag{B.1.15}$$

The notion implicit in (B.1.15) is that given $\mathbf{Z} = \mathbf{z}$, $\mathbf{Z}$ acts as a constant. If we carry this further, we have a relation that we shall call the *substitution theorem for conditional expectations*:

$$E(q(\mathbf{Y}, \mathbf{Z}) \mid \mathbf{Z} = \mathbf{z}) = E(q(\mathbf{Y}, \mathbf{z}) \mid \mathbf{Z} = \mathbf{z}). \tag{B.1.16}$$

This is valid for all $\mathbf{z}$ such that $p_{\mathbf{Z}}(\mathbf{z}) > 0$ if $E|q(\mathbf{Y}, \mathbf{Z})| < \infty$. This follows from definitions (B.1.11) and (B.1.7) because

$$P[q(\mathbf{Y}, \mathbf{Z}) = a \mid \mathbf{Z} = \mathbf{z}] = P[q(\mathbf{Y}, \mathbf{Z}) = a, \ \mathbf{Z} = \mathbf{z} \mid \mathbf{Z} = \mathbf{z}] = P[q(\mathbf{Y}, \mathbf{z}) = a \mid \mathbf{Z} = \mathbf{z}] \tag{B.1.17}$$

for any a.

If we put $q(\mathbf{Y}, \mathbf{Z}) = r(\mathbf{Y})h(\mathbf{Z})$, where $E|r(\mathbf{Y})h(\mathbf{Z})| < \infty$, we obtain by (B.1.16),

$$E(r(\mathbf{Y})h(\mathbf{Z}) \mid \mathbf{Z} = \mathbf{z}) = E(r(\mathbf{Y})h(\mathbf{z}) \mid \mathbf{Z} = \mathbf{z}) = h(\mathbf{z})E(r(\mathbf{Y}) \mid \mathbf{Z} = \mathbf{z}). \tag{B.1.18}$$

Therefore,

$$E(r(\mathbf{Y})h(\mathbf{Z}) \mid \mathbf{Z}) = h(\mathbf{Z})E(r(\mathbf{Y}) \mid \mathbf{Z}). \tag{B.1.19}$$

Another intuitively reasonable result is that the mean of the conditional means is the mean:

$$E(E(Y \mid \mathbf{Z})) = E(Y), \tag{B.1.20}$$

whenever Y has a finite expectation. We refer to this as the *double* or *iterated expectation theorem*.

To prove (B.1.20) we write, in view of (B.1.7) and (A.10.5),

$$E(E(Y \mid \mathbf{Z})) = \Sigma_{\mathbf{z}} p_{\mathbf{Z}}(\mathbf{z})[\Sigma_y y p(y \mid \mathbf{z})] = \Sigma_{y,\mathbf{z}} y p(y \mid \mathbf{z}) p_{\mathbf{Z}}(\mathbf{z}) = \Sigma_{y,\mathbf{z}} y p(y, \mathbf{z}) = E(Y). \tag{B.1.21}$$

The interchange of summation used is valid because the finiteness of $E(|Y|)$ implies that all sums converge absolutely.

As an illustration, we check (B.1.20) for $E(Y_1 \mid Z)$ given by (B.1.10). In this case,

$$E(E(Y_1 \mid Z)) = E\left(\frac{Z}{n}\right) = \frac{np}{n} = p = E(Y_1). \tag{B.1.22}$$

If we apply (B.1.20) to $Y = r(\mathbf{Y})h(\mathbf{Z})$ and use (B.1.19), we obtain the *product expectation formula*:

Theorem B.1.1 *If $E|r(\mathbf{Y})h(\mathbf{Z})| < \infty$, then*

$$E(r(\mathbf{Y})h(\mathbf{Z})) = E(h(\mathbf{Z})E(r(\mathbf{Y}) \mid \mathbf{Z})). \tag{B.1.23}$$

Note that we can express the conditional probability that $\mathbf{Y} \in A$ given $\mathbf{Z} = \mathbf{z}$ as

$$P[\mathbf{Y} \in A \mid \mathbf{Z} = \mathbf{z}] = E(1[\mathbf{Y} \in A] \mid \mathbf{Z} = \mathbf{z}) = \Sigma_{\mathbf{y} \in A} p(\mathbf{y} \mid \mathbf{z}).$$

Then by taking $r(\mathbf{Y}) = 1[\mathbf{Y} \in A]$, $h = 1$ in Theorem B.1.1 we can express the (unconditional) probability that $\mathbf{Y} \in A$ as

$$P[\mathbf{Y} \in A] = E(E(r(\mathbf{Y}) \mid \mathbf{Z})) = \Sigma_{\mathbf{z}} P[\mathbf{Y} \in A \mid \mathbf{Z} = \mathbf{z}] p_{\mathbf{Z}}(\mathbf{z}) = E[P(Y \in A \mid \mathbf{Z})]. \tag{B.1.24}$$

For example, if Y and Z are as in (B.1.5),

$$P[Y \le y] = \Sigma_z \binom{N}{z} \theta^z (1 - \theta)^{n-z} H_z(y)$$

where H_z is the distribution function of a hypergeometric distribution with parameters (z, N, n).

B.1.4 Continuous Variables

Suppose now that $(\mathbf{Y}, \mathbf{Z})$ is a continuous random vector having coordinates that are themselves vectors and having density function $p(\mathbf{y}, \mathbf{z})$. We define, following the analogy between frequency and density functions, the *conditional density* function of $\mathbf{Y}$ given $\mathbf{Z} = \mathbf{z}$ by

$$p(\mathbf{y} \mid \mathbf{z}) = \frac{p(\mathbf{y}, \mathbf{z})}{p_{\mathbf{Z}}(\mathbf{z})} \tag{B.1.25}$$

if $p_{\mathbf{Z}}(\mathbf{z}) > 0$.

Because the marginal density of $\mathbf{Z}$, $p_{\mathbf{Z}}(\mathbf{z})$, is given by (A.8.12), it is clear that $p(\cdot \mid \mathbf{z})$ is a density. Because (B.1.25) does not differ formally from (B.1.1), equations (B.1.3) and (B.1.6) go over verbatim. Expression (B.1.4) becomes

$$p(\mathbf{y} \mid \mathbf{z}) = \frac{p_{\mathbf{Y}}(\mathbf{y}) q(\mathbf{z} \mid \mathbf{y})}{\int_{-\infty}^{\infty} \cdots \int_{-\infty}^{\infty} p_{\mathbf{Y}}(\mathbf{t}) q(\mathbf{z} \mid \mathbf{t}) dt_1 \ldots dt_n}, \tag{B.1.26}$$

where q is the conditional density of $\mathbf{Z}$ given $\mathbf{Y} = \mathbf{y}$. This is also called *Bayes' Theorem*.

If **Y** and **Z** are independent, the conditional distributions equal the marginals as in the discrete case.

Example B.1.4 Let Y_1 and Y_2 be independent and uniformly, $\mathcal{U}(0,1)$, distributed. Let $Z = \min(Y_1, Y_2)$, $Y = \max(Y_1, Y_2)$. The joint distribution of Z and Y is given by

$$
\begin{aligned}
F(z,y) &= 2P[Y_1 < Y_2, Y_1 < z, Y_2 < y] \\
&= 2 \int_0^y \int_0^{\min(y_2,z)} dy_1\, dy_2 = 2 \int_0^y \min(y_2, z) dy_2
\end{aligned}
\tag{B.1.27}
$$

if $0 \leq z, y \leq 1$.

The joint density is, therefore,

$$
\begin{aligned}
p(z,y) &= 2 \text{ if } 0 < z \leq y < 1 \\
&= 0 \text{ otherwise.}
\end{aligned}
\tag{B.1.28}
$$

The marginal density of Z is given by

$$
\begin{aligned}
p_Z(z) &= \int_z^1 2dy = 2(1-z),\ 0 < z < 1 \\
&= 0 \text{ otherwise.}
\end{aligned}
\tag{B.1.29}
$$

We conclude that the conditional density of Y given $Z = z$ is uniform on the interval $(z,1)$.   □

If $E(|Y|) < \infty$, we denote the *conditional expectation of Y given $\mathbf{Z} = \mathbf{z}$* in analogy to the discrete case as the expected value of a random variable with density $p(y \mid \mathbf{z})$. More generally, if $E(|r(\mathbf{Y})|) < \infty$, (A.10.11) shows that the conditional expectation of $r(\mathbf{Y})$ given $\mathbf{Z} = \mathbf{z}$ can be obtained from

$$
E(r(\mathbf{Y}) \mid \mathbf{Z} = \mathbf{z}) = \int_{-\infty}^{\infty} r(\mathbf{y})p(\mathbf{y} \mid \mathbf{z})d\mathbf{y}.
\tag{B.1.30}
$$

As before, if $g(\mathbf{z}) = E(r(\mathbf{Y}) \mid \mathbf{Z} = \mathbf{z})$, we write $g(\mathbf{Z})$ as $E(r(\mathbf{Y}) \mid \mathbf{Z})$, the conditional expectation of $r(\mathbf{Y})$ given $\mathbf{Z}$. With this definition we can show that formulas 12, 13, 14, 19, 20, 23, and 24 of this section hold in the continuous case also. As an illustration, we next derive B.1.23:

Let $g(\mathbf{z}) = E[r(\mathbf{Y}) \mid \mathbf{Z}]$, then, by (A.10.11),

$$
\begin{aligned}
E(h(\mathbf{Z})E(r(\mathbf{Y}) \mid \mathbf{Z})) &= E(h(\mathbf{Z})g(\mathbf{Z})) = \int_{-\infty}^{\infty} h(\mathbf{z})g(\mathbf{z})p_Z(\mathbf{z})d\mathbf{z} \\
&= \int_{-\infty}^{\infty} h(\mathbf{z})p_Z(\mathbf{z}) \left[\int_{-\infty}^{\infty} r(\mathbf{y})p(\mathbf{y} \mid \mathbf{z})d\mathbf{y} \right] d\mathbf{z}.
\end{aligned}
\tag{B.1.31}
$$

By a standard theorem on double integrals, we conclude that the right-hand side of (B.1.31) equals

$$\int_{-\infty}^{\infty}\int_{-\infty}^{\infty} r(\mathbf{y})h(\mathbf{z})p_Z(\mathbf{z})p(\mathbf{y} \mid \mathbf{z})d\mathbf{y}d\mathbf{z}$$

$$= \int_{-\infty}^{\infty}\int_{-\infty}^{\infty} r(\mathbf{y})h(\mathbf{z})p(\mathbf{y},\mathbf{z})d\mathbf{y}d\mathbf{z} = E(r(\mathbf{Y})h(\mathbf{Z})) \tag{B.1.32}$$

by (A.10.11), and we have established B.1.23.

To illustrate these formulae, we calculate $E(Y \mid Z)$ in Example B.1.4. Here,

$$E(Y \mid Z = z) = \int_0^1 yp(y \mid z)dy = \frac{1}{(1-z)}\int_z^1 ydy = \frac{1+z}{2}, \ 0 < z < 1,$$

and, hence,

$$E(Y \mid Z) = \frac{1+Z}{2}.$$

B.1.5 Comments on the General Case

Clearly the cases $(\mathbf{Y},\mathbf{Z})$ discrete and $(\mathbf{Y},\mathbf{Z})$ continuous do not cover the field. For example, if Y is uniform on $(0,1)$ and $Z = Y^2$, then (Y,Z) neither has a joint frequency function nor a joint density. (The density would have to concentrate on $z = y^2$, but then it cannot satisfy $\int_0^1\int_0^1 f(y,z)dydz = 1$.) Thus, (Y,Z) is neither discrete nor continuous in our sense. On the other hand, we should have a concept of conditional probability for which $P[Y = u \mid Z = \sqrt{u}] = 1$. To cover the general theory of conditioning is beyond the scope of this book. The interested student should refer to the books by Breiman (1968), Loève (1977), Chung (1974), or Billingsley (1995). We merely note that it is possible to define $E(\mathbf{Y} \mid \mathbf{Z} = \mathbf{z})$ and $E(\mathbf{Y} \mid \mathbf{Z})$ in such a way that they coincide with (B.1.7) and (B.1.30) in the discrete and continuous cases and moreover so that equations 15, 16, 20, and 23 of this section hold.

As an illustration, suppose that in Example B.1.4 we want to find the conditional expectation of $\sin(ZY)$ given $Z = z$. By our discussion we can calculate $E(\sin(ZY) \mid Z = z)$ as follows: First, apply (B.1.16) to get

$$E(\sin(ZY) \mid Z = z) = E(\sin(zY) \mid Z = z).$$

Because, given $Z = z$, Y has a $\mathcal{U}(z,1)$ distribution, we can complete the computation by applying (A.10.11) to get

$$E(\sin(zY) \mid Z = z) = \frac{1}{(1-z)}\int_z^1 \sin(zy)dy = \frac{1}{z(1-z)}[\cos z^2 - \cos z].$$

B.2   DISTRIBUTION THEORY FOR TRANSFORMATIONS OF RANDOM VECTORS

B.2.1   The Basic Framework

In statistics we will need the distributions of functions of the random variables appearing in an experiment. Examples of such functions are sums, averages, differences, sums of squares, and so on. In this section we will develop a result that often is useful in finding the joint distribution of several functions of a continuous random vector. The result will generalize (A.8.9), which gives the density of a real-valued function of a continuous random variable.

Let $\mathbf{h} = (h_1, \ldots, h_k)^T$, where each h_i is a real-valued function on R^k. Thus, $\mathbf{h}$ is a transformation from R^k to R^k. Recall that the *Jacobian* $J_{\mathbf{h}}(\mathbf{t})$ of $\mathbf{h}$ evaluated at $\mathbf{t} = (t_1, \ldots, t_k)^T$ is by definition the determinant

$$
J_{\mathbf{h}}(\mathbf{t}) = \begin{vmatrix} \dfrac{\partial}{\partial t_1} h_1(\mathbf{t}) & \cdots & \dfrac{\partial}{\partial t_1} h_k(\mathbf{t}) \\ \vdots & & \vdots \\ \dfrac{\partial}{\partial t_k} h_1(\mathbf{t}) & \cdots & \dfrac{\partial}{\partial t_k} h_k(\mathbf{t}) \end{vmatrix}.
$$

The principal result of this section, Theorem B.2.2, rests on the change of variable theorem for multiple integrals from calculus. We now state this theorem without proof (see Apostol, 1974, p. 421).

Theorem B.2.1 *Let* $\mathbf{h} = (h_1, \ldots, h_k)^T$ *be a transformation defined on an open subset B of R^k. Suppose that:*[1]

(i) $\mathbf{h}$ *has continuous first partial derivatives in B.*

(ii) $\mathbf{h}$ *is one-to-one on B.*

(iii) *The Jacobian of* $\mathbf{h}$ *does not vanish on B.*

Let f be a real-valued function (defined and measurable) on the range $\mathbf{h}(B) = \{(h_1(\mathbf{t}), \ldots, h_k(\mathbf{t})) : \mathbf{t} \in B\}$ *of* $\mathbf{h}$ *and suppose f satisfies*

$$
\int_{\mathbf{h}(B)} |f(\mathbf{x})| d\mathbf{x} < \infty.
$$

Then for every (measurable) subset K of $\mathbf{h}(B)$ *we have*

$$
\int_K f(\mathbf{x}) d\mathbf{x} = \int_{\mathbf{h}^{-1}(K)} f(\mathbf{h}(\mathbf{t})) |J_{\mathbf{h}}(\mathbf{t})| d\mathbf{t}. \tag{B.2.1}
$$

In these expressions we write $d\mathbf{x}$ for $dx_1 \dots dx_k$. Moreover, $\mathbf{h}^{-1}$ denotes the inverse of the transformation $\mathbf{h}$; that is, $\mathbf{h}^{-1}(\mathbf{x}) = \mathbf{t}$ if, and only if, $\mathbf{x} = \mathbf{h}(\mathbf{t})$. We also need another result from the calculus (see Apostol, 1974, p. 417),

$$J_{\mathbf{h}^{-1}}(\mathbf{t}) = \frac{1}{J_{\mathbf{h}}(\mathbf{h}^{-1}(\mathbf{t}))}. \tag{B.2.2}$$

It follows that a transformation $\mathbf{h}$ satisfies the conditions of Theorem B.2.1 if, and only if, $\mathbf{h}^{-1}$ does.

We can now derive the density of $\mathbf{Y} = \mathbf{g}(\mathbf{X}) = (g_1(\mathbf{X}), \dots, g_k(\mathbf{X}))^T$ when $\mathbf{g}$ satisfies the conditions of Theorem B.2.1 and $\mathbf{X} = (X_1, \dots, X_k)^T$ is a continuous random vector.

Theorem B.2.2 *Let $\mathbf{X}$ be continuous and let S be an open subset of R^k such that $P(\mathbf{X} \in S) = 1$. If $\mathbf{g} = (g_1, \dots, g_k)^T$ is a transformation from S to R^k such that $\mathbf{g}$ and S satisfy the conditions of Theorem B.2.1, then the density of $\mathbf{Y} = \mathbf{g}(\mathbf{X})$ is given by*

$$p_{\mathbf{Y}}(\mathbf{y}) = p_{\mathbf{X}}(\mathbf{g}^{-1}(\mathbf{y}))|J_{\mathbf{g}^{-1}}(\mathbf{y})| \tag{B.2.3}$$

for $\mathbf{y} \in \mathbf{g}(S)$.

Proof. The distribution function of $\mathbf{Y}$ is (see (A.7.8))

$$F_{\mathbf{Y}}(\mathbf{y}) = \int \cdots \int_{A_k} p_{\mathbf{X}}(x_1, \dots, x_k) dx_1 \dots dx_k$$

where $A_k = \{\mathbf{x} \in R^k : g_i(\mathbf{x}) \le y_i, \ i = 1, \dots, k\}$. Next we apply Theorem B.2.1 with $\mathbf{h} = \mathbf{g}^{-1}$ and $f = p_{\mathbf{X}}$. Because $\mathbf{h}^{-1}(A_k) = \mathbf{g}(A_k) = \{\mathbf{g}(\mathbf{x}) : g_i(\mathbf{x}) \le y_i, \ i = 1, \dots, k\} = \{\mathbf{t} : \mathbf{t} \le y_i, \ i = 1, \dots, k\}$, we obtain

$$F_{\mathbf{Y}}(\mathbf{y}) = \int_{-\infty}^{y_k} \cdots \int_{-\infty}^{y_1} p_{\mathbf{X}}(\mathbf{g}^{-1}(\mathbf{t}))|J_{\mathbf{g}^{-1}}(\mathbf{t})| dt_1 \dots dt_k.$$

The result now follows if we recall from Section A.7 that whenever $F_{\mathbf{Y}}(\mathbf{y}) = \int_{-\infty}^{y_k} \cdots \int_{-\infty}^{y_1} q(t_1, \dots, t_k) dt_1 \dots dt_k$ for some nonnegative function q, then q must be the density of $\mathbf{Y}$. $\qquad\square$

Example B.2.1 Suppose $\mathbf{X} = (X_1, X_2)^T$ where X_1 and X_2 are independent with $\mathcal{N}(0, 1)$ and $\mathcal{N}(0, 4)$ distributions, respectively. What is the joint distribution of $Y_1 = X_1 + X_2$ and $Y_2 = X_1 - X_2$? Here (see (A.13.17)),

$$p_{\mathbf{X}}(x_1, x_2) = \frac{1}{4\pi} \exp -\frac{1}{2}\left[x_1^2 + \frac{1}{4}x_2^2\right].$$

In this case, $S = R^2$. Also note that $g_1(\mathbf{x}) = x_1 + x_2$, $g_2(\mathbf{x}) = x_1 - x_2$, $g_1^{-1}(\mathbf{y}) = \frac{1}{2}(y_1 + y_2)$, $g_2^{-1}(\mathbf{y}) = \frac{1}{2}(y_1 - y_2)$, that the range $\mathbf{g}(S)$ is R^2 and that

$$J_{\mathbf{g}^{-1}}(\mathbf{y}) = \begin{vmatrix} \frac{1}{2} & \frac{1}{2} \\ \frac{1}{2} & -\frac{1}{2} \end{vmatrix} = -\frac{1}{2}.$$

Upon substituting these quantities in (B.2.3), we obtain

$$
\begin{aligned}
p_{\mathbf{Y}}(y_1, y_2) &= \frac{1}{2} p_{\mathbf{X}}\left(\frac{1}{2}(y_1 + y_2), \frac{1}{2}(y_1 - y_2)\right) \\
&= \frac{1}{8\pi} \exp -\frac{1}{2}\left[\frac{1}{4}(y_1 + y_2)^2 + \frac{1}{16}(y_1 - y_2)^2\right] \\
&= \frac{1}{8\pi} \exp -\frac{1}{32}[5y_1^2 + 5y_2^2 + 6y_1 y_2].
\end{aligned}
$$

This is an example of bivariate normal density. Such densities will be considered further in Section B.4.   □

Upon combining (B.2.2) and (B.2.3) we see that for $\mathbf{y} \in \mathbf{g}(S)$,

$$
p_{\mathbf{Y}}(\mathbf{y}) = \frac{p_{\mathbf{X}}(\mathbf{g}^{-1}(\mathbf{y}))}{|J_{\mathbf{g}}(\mathbf{g}^{-1}(\mathbf{y}))|}. \tag{B.2.4}
$$

If X is a random variable ($k = 1$), the Jacobian of g is just its derivative and the requirements (i) and (iii) that g' be continuous and nonvanishing imply that g is strictly monotone and, hence, satisfies (ii). In this case (B.2.4) reduces to the familiar formula (A.8.9).

It is possible to give useful generalizations of Theorem B.2.2 to situations where $\mathbf{g}$ is not one-to-one (Problem B.2.7).

Theorem B.2.2 provides one of the instances in which frequency and density functions differ. If $\mathbf{X}$ is discrete, $\mathbf{g}$ is one-to-one, and $\mathbf{Y} = \mathbf{g}(\mathbf{X})$, then $p_{\mathbf{Y}}(\mathbf{y}) = p_{\mathbf{X}}(\mathbf{g}^{-1}(\mathbf{y}))$. The extra factor in the continuous case appears roughly as follows. If $A(\mathbf{y})$ is a "small" cube surrounding $\mathbf{y}$ and we let $V(B)$ denote the volume of a set B, then

$$
\begin{aligned}
p_{\mathbf{Y}}(\mathbf{y}) &\approx \frac{P[\mathbf{g}(\mathbf{X}) \in A(\mathbf{y})]}{V(A(\mathbf{y}))} = \frac{P[\mathbf{X} \in \mathbf{g}^{-1}(A(\mathbf{y}))]}{V(\mathbf{g}^{-1}(A(\mathbf{y})))} \cdot \frac{V(\mathbf{g}^{-1}(A(\mathbf{y})))}{V(A(\mathbf{y}))} \\
&\approx p_{\mathbf{X}}(\mathbf{g}^{-1}(\mathbf{y})) \cdot \frac{V(\mathbf{g}^{-1}(A(\mathbf{y})))}{V(A(\mathbf{y}))}.
\end{aligned}
$$

Using the fact that $\mathbf{g}^{-1}$ is approximately linear on $A(\mathbf{y})$, it is not hard to show that

$$
\frac{V(g^{-1}(A(\mathbf{y})))}{V(A(\mathbf{y}))} \approx |J_{\mathbf{g}^{-1}}(\mathbf{y})|.
$$

The justification of these approximations is the content of Theorem B.2.2.

The following generalization of (A.8.10) is very important. For a review of the elementary properties of matrices needed in its formulation, we refer the reader to Section B.10.

Recall that $\mathbf{g}$ is called an *affine transformation* of R^k if there exists a $k \times k$ matrix $\mathbf{A}$ and a $k \times 1$ vector $\mathbf{c}$ such that $\mathbf{g}(\mathbf{x}) = \mathbf{A}\mathbf{x} + \mathbf{c}$. If $\mathbf{c} = \mathbf{0}$, $\mathbf{g}$ is called a *linear transformation*. The function $\mathbf{g}$ is one-to-one if, and only if, $\mathbf{A}$ is nonsingular and then

$$
\mathbf{g}^{-1}(\mathbf{y}) = \mathbf{A}^{-1}(\mathbf{y} - \mathbf{c}), \tag{B.2.5}
$$

$\mathbf{y} \in R^k$, where $\mathbf{A}^{-1}$ is the inverse of $\mathbf{A}$.

Corollary B.2.1 *Suppose* $\mathbf{X}$ *is continuous and* S *is such that* $P(\mathbf{X} \in S) = 1$. *If* $\mathbf{g}$ *is a one-to-one affine transformation as defined earlier, then* $\mathbf{Y} = \mathbf{g}(\mathbf{X})$ *has density*

$$p_{\mathbf{Y}}(\mathbf{y}) = |\det \mathbf{A}|^{-1} p_{\mathbf{X}}(\mathbf{A}^{-1}(\mathbf{y} - \mathbf{c})) \tag{B.2.6}$$

for $\mathbf{y} \in \mathbf{g}(S)$, *where* $\det \mathbf{A}$ *is the determinant of* $\mathbf{A}$.

The corollary follows from (B.2.4), (B.2.5), and the relation,

$$J_{\mathbf{g}}(\mathbf{g}^{-1}(\mathbf{y})) \equiv \det \mathbf{A}. \tag{B.2.7}$$

Example B.2.1 is a special case of the corollary. Further applications appear in the next section. $\qquad\qquad\square$

B.2.2   The Gamma and Beta Distributions

As a consequence of the transformation theorem we obtain basic properties of two important families of distributions, which will also figure in the next section. The first family has densities given by

$$g_{p,\lambda}(x) = \frac{\lambda^p x^{p-1} e^{-\lambda x}}{\Gamma(p)} \tag{B.2.8}$$

for $x > 0$, where the parameters p and λ are taken to be positive and $\Gamma(p)$ denotes the *Euler gamma function* defined by

$$\Gamma(p) = \int_0^\infty t^{p-1} e^{-t} dt. \tag{B.2.9}$$

A useful fact is that $\Gamma\left(\frac{1}{2}\right) = \sqrt{\pi}$. It follows by integration by parts that, for all $p > 0$,

$$\Gamma(p+1) = p\Gamma(p) \text{ and that } \Gamma(k) = (k-1)! \text{ for positive integers } k. \tag{B.2.10}$$

The family of distributions with densities given by (B.2.8) is referred to as the *gamma* family of distributions and we shall write $\Gamma(p, \lambda)$ for the distribution corresponding to $g_{p,\lambda}$. The special case $p = 1$ corresponds to the familiar exponential distribution $\mathcal{E}(\lambda)$ of (A.13.24). By (A.8.10), X is distributed $\Gamma(p, \lambda)$ if, and only if, λX is distributed $\Gamma(p, 1)$. Thus, $1/\lambda$ is a scale parameter for the $\Gamma(p, \lambda)$ family.

Let k be a positive integer. In statistics, the gamma density $g_{p,\lambda}$ with $p = \frac{1}{2}k$ and $\lambda = \frac{1}{2}$ is referred to as the *chi squared density with k degrees of freedom* and is denoted by χ_k^2.

The other family of distributions we wish to consider is the *beta* family, which is indexed by the positive parameters r and s. Its densities are given by

$$b_{r,s}(x) = \frac{x^{r-1}(1-x)^{s-1}}{B(r, s)} \tag{B.2.11}$$

for $0 < x < 1$, where $B(r, s) = [\Gamma(r)\Gamma(s)]/[\Gamma(r+s)]$ is the *beta function*. The distribution corresponding to $b_{r,s}$ will be written $\beta(r, s)$. Figures B.2.1 and B.2.2 show some typical members of the two families.

Theorem B.2.3 *If X_1 and X_2 are independent random variables with $\Gamma(p, \lambda)$ and $\Gamma(q, \lambda)$ distributions, respectively, then $Y_1 = X_1 + X_2$ and $Y_2 = X_1/(X_1 + X_2)$ are independent and have, respectively, $\Gamma(p + q, \lambda)$ and $\beta(p, q)$ distributions.*

Proof. If $\lambda = 1$, the joint density of X_1 and X_2 is

$$p(x_1, x_2) = [\Gamma(p)\Gamma(q)]^{-1} e^{-(x_1 + x_2)} x_1^{p-1} x_2^{q-1} \tag{B.2.12}$$

for $x_1 > 0$, $x_2 > 0$. Let

$$(y_1, y_2)^T = \mathbf{g}(x_1, x_2) = \left(x_1 + x_2, \frac{x_1}{x_1 + x_2} \right)^T.$$

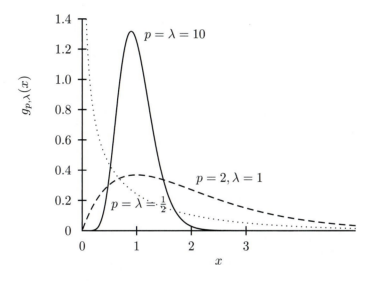

Figure B.2.1 The gamma density, $g_{p,\lambda}(x)$, for selected p, λ.

Then $\mathbf{g}$ is one-to-one on $S = \{(x_1, x_2)^T : x_1 > 0, \; x_2 > 0\}$ and its range is $S_1 = \{(y_1, y_2)^T : y_1 > 0, \; 0 < y_2 < 1\}$. We note that on S_1

$$\mathbf{g}^{-1}(y_1, y_2) = (y_1 y_2, \; y_1 - y_1 y_2)^T. \tag{B.2.13}$$

Therefore,

$$J_{\mathbf{g}^{-1}}(y_1, y_2) = \begin{vmatrix} y_2 & 1 - y_2 \\ y_1 & -y_1 \end{vmatrix} = -y_1. \tag{B.2.14}$$

If we now substitute (B.2.13) and (B.2.14) in (B.2.4) we get for the density of $(Y_1, Y_2)^T = \mathbf{g}(X_1, X_2)$,

$$p_\mathbf{Y}(y_1, y_2) = \frac{e^{-y_1}(y_1 y_2)^{p-1}(y_1 - y_1 y_2)^{q-1} y_1}{\Gamma(p)\Gamma(q)} \qquad \text{(B.2.15)}$$

for $y_1 > 0, 0 < y_2 < 1$. Simplifying (B.2.15) leads to

$$p_\mathbf{Y}(y_1, y_2) = g_{p+q,1}(y_1) b_{p,q}(y_2). \qquad \text{(B.2.16)}$$

The result is proved for $\lambda = 1$. If $\lambda \neq 1$ define $X_1' = \lambda X_1$ and $X_2' = \lambda X_2$. Now X_1' and X_2' are independent $\Gamma(p, 1)$, $\Gamma(q, 1)$ variables respectively. Because $X_1' + X_2' = \lambda(X_1 + X_2)$ and $X_1'(X_1' + X_2')^{-1} = X_1(X_1 + X_2)^{-1}$ the theorem follows. $\qquad\square$

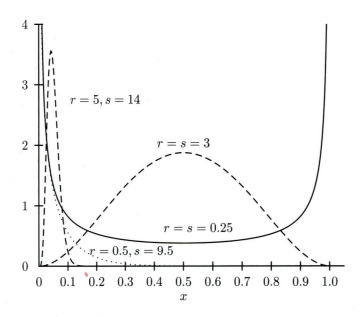

Figure B.2.2 The beta density, $b_{r,s}(x)$, for selected r, s.

By iterating the argument of Theorem B.2.3, we obtain the following general result.

Corollary B.2.2 *If $X_1, \ldots, X_n$ are independent random variables such that X_i has a $\Gamma(p_i, \lambda)$ distribution, $i = 1, \ldots, n$, then $\Sigma_{i=1}^n X_i$ has a $\Gamma(\Sigma_{i=1}^n p_i, \lambda)$ distribution.*

Some other properties of the gamma and beta families are given in the problems and in the next section.

B.3   DISTRIBUTION THEORY FOR SAMPLES FROM A NORMAL POPULATION

In this section we introduce some distributions that appear throughout modern statistics. We derive their densities as an illustration of the theory of Section B.2. However, these distributions should be remembered in terms of their definitions and qualitative properties rather than density formulas.

B.3.1   The χ^2, F, and t Distributions

Throughout this section we shall suppose that $\mathbf{X} = (X_1, \ldots, X_n)^T$ where the X_i form a sample from a $\mathcal{N}(0, \sigma^2)$ population. Some results for normal populations, whose mean differs from 0, are given in the problems. We begin by investigating the distribution of the $\Sigma_{i=1}^n X_i^2$, the squared distance of $\mathbf{X}$ from the origin.

Theorem B.3.1 *The random variable $V = \Sigma_{i=1}^n X_i^2 / \sigma^2$ has a χ_n^2 distribution. That is, V has density*

$$p_V(v) = \frac{v^{\frac{1}{2}(n-2)} e^{-\frac{1}{2}v}}{2^{n/2} \Gamma(n/2)} \tag{B.3.1}$$

for $v > 0$.

Proof. Let $Z_i = X_i / \sigma$, $i = 1, \ldots, n$. Then $Z_i \sim \mathcal{N}(0, 1)$. Because the Z_i^2 are independent, it is enough to prove the theorem for $n = 1$ and then apply Corollary B.2.2. If $T = Z_1^2$, then the distribution function of T is

$$P[Z_1^2 \le t] = P[-\sqrt{t} \le Z_1 \le \sqrt{t}] \tag{B.3.2}$$

and, thus,

$$F_T(t) = \Phi(\sqrt{t}) - \Phi(-\sqrt{t}). \tag{B.3.3}$$

Differentiating both sides we get the density of T

$$p_T(t) = t^{-\frac{1}{2}} \varphi(\sqrt{t}) = \frac{1}{\sqrt{2\pi}} t^{-\frac{1}{2}} e^{-t/2} \tag{B.3.4}$$

for $t > 0$, which agrees with $g_{\frac{1}{2}, \frac{1}{2}}$ up to a multiplicative constant. Because the constant is determined by the requirement that p_T and $g_{\frac{1}{2}, \frac{1}{2}}$ are densities, we must have $p_T = g_{\frac{1}{2}, \frac{1}{2}}$ and the result follows.   □

Let V and W be independent and have χ_k^2 and χ_m^2 distributions, respectively, and let $S = (V/k)/(W/m)$. The distribution of S is called the F *distribution with k and m degrees of freedom.* We shall denote it by $\mathcal{F}_{k,m}$.

Next, we introduce the t *distribution with k degrees of freedom,* which we shall denote by $\mathcal{T}_k$. By definition $\mathcal{T}_k$ is the distribution of $Q = Z/\sqrt{V/k}$, where Z and V are independent with $\mathcal{N}(0, 1)$ and χ_k^2 distributions, respectively. We can now state the following elementary consequence of Theorem B.3.1.

Corollary B.3.1 *The random variable* $(m/k)\Sigma_{i=1}^{k}X_i^2/\Sigma_{i=k+1}^{k+m}X_i^2$ *has an* $\mathcal{F}_{k,m}$ *distribution. The random variable* $X_1/\sqrt{(1/k)\Sigma_{i=2}^{k+1}X_i^2}$ *has a* $\mathcal{T}_k$ *distribution.*

Proof. For the first assertion we need only note that

$$\sum_{i=1}^{k}X_i^2 \Big/ \sum_{i=k+1}^{k+m}X_i^2 = \frac{1}{\sigma^2}\sum_{i=1}^{k}X_i^2 \Big/ \frac{1}{\sigma^2}\sum_{i=k+1}^{k+m}X_i^2 \tag{B.3.5}$$

and apply the theorem and the definition of $\mathcal{F}_{k,m}$. The second assertion follows in the same way.   □

To make the definitions of the $\mathcal{F}_{k,m}$ and $\mathcal{T}_k$ distributions useful for computation, we need their densities. We assume the S, Q, V, W are as in the definitions of these distributions.

To derive the density of S note that, if $U = V/(V + W)$, then

$$S = \frac{V/k}{W/m} = \frac{m}{k}\frac{U}{1-U}. \tag{B.3.6}$$

Because $V \sim \Gamma\left(\frac{1}{2}k, \frac{1}{2}\right)$, $W \sim \Gamma\left(\frac{1}{2}m, \frac{1}{2}\right)$ and V and W are independent, then by Theorem B.2.3, U has a beta distribution with parameters $\frac{1}{2}k$ and $\frac{1}{2}m$. To obtain the density of S we need only apply the change of variable formula (A.8.9) to U with $g(u) = (m/k)u/(1-u)$. After some calculation we arrive at the $\mathcal{F}_{k,m}$ density (see Figure B.3.1)

$$p_S(s) = \frac{(k/m)^{\frac{1}{2}k}s^{\frac{1}{2}(k-2)}(1+(k/m)s)^{-\frac{1}{2}(k+m)}}{B\left(\frac{1}{2}k, \frac{1}{2}m\right)} \tag{B.3.7}$$

for $s > 0$.

To get the density of Q we argue as follows. Because $-Z$ has the same distribution as Z, we may conclude that Q and $-Q$ are identically distributed. It follows that

$$
\begin{aligned}
P[0 < Q < q] &= P[0 < -Q < q] \\
&= P[-q < Q < 0] = \frac{1}{2}P[0 < Q^2 < q^2].
\end{aligned} \tag{B.3.8}
$$

Differentiating $P[0 < Q < q]$, $P[-q < Q < 0]$ and $\frac{1}{2}P[0 < Q^2 < q^2]$ we get

$$p_Q(q) = p_Q(-q) = qp_{Q^2}(q^2) \text{ if } q > 0. \tag{B.3.9}$$

Now Q^2 has by Corollary B.3.1 an $\mathcal{F}_{1,k}$ distribution. We can, therefore, use (B.3.7) and (B.3.9) to conclude

$$p_Q(q) = \frac{\Gamma\left(\frac{1}{2}(k+1)\right)(1+(q^2/k))^{-\frac{1}{2}(k+1)}}{\sqrt{\pi k}\Gamma\left(\frac{1}{2}k\right)} \tag{B.3.10}$$

for $-\infty < q < \infty$.

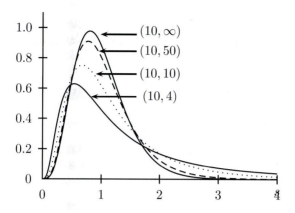

Figure B.3.1 The $\mathcal{F}_{k,m}$ density for selected (k,m).

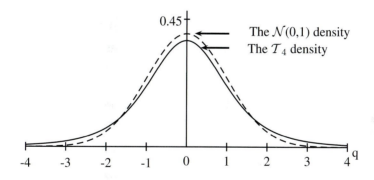

Figure B.3.2 The Student t and standard normal densities.

The χ^2, $\mathcal{T}$, and $\mathcal{F}$ cumulative distribution functions are given in Tables II, III, and IV, respectively. More precisely, these tables give the inverses or quantiles of these distributions. For $\alpha \in (0,1)$, an αth *quantile* or $100\,\alpha$th *percentile* of the continuous distribution F is by definition any number $x(\alpha)$ such that $F(x(\alpha)) = \alpha$.

Continuity of F guarantees the existence of $x(\alpha)$ for all α. If F is strictly increasing, $x(\alpha)$ is unique for each α. As an illustration, we read from Table III that the (0.95)th quantile or 95th percentile of the $\mathcal{T}_{20}$ distribution is $t(0.95) = 1.725$.

B.3.2   Orthogonal Transformations

We turn now to orthogonal transformations of normal samples. Let us begin by recalling some classical facts and definitions involving matrices, which may be found in standard texts, for example, Birkhoff and MacLane (1965).

Suppose that $\mathbf{A}$ is a $k \times m$ matrix with entry a_{ij} in the ith row and jth column, $i = 1, \ldots, k; j = 1, \ldots, m$. Then the *transpose* of $\mathbf{A}$, written $\mathbf{A}^T$, is the $m \times k$ matrix, whose entry in the ith row and jth column is a_{ji}. Thus, the transpose of a row vector is a column vector and the transpose of a square matrix is a square matrix.

An $n \times n$ matrix $\mathbf{A}$ is said to be *orthogonal* if, and only if,

$$\mathbf{A}^T = \mathbf{A}^{-1} \tag{B.3.11}$$

or equivalently if, and only if, either one of the following two matrix equations is satisfied,

$$\mathbf{A}^T \mathbf{A} = \mathbf{I} \tag{B.3.12}$$

$$\mathbf{A}\mathbf{A}^T = \mathbf{I} \tag{B.3.13}$$

where $\mathbf{I}$ is the $n \times n$ identity matrix. Equation (B.3.12) requires that the column vectors of $\mathbf{A}$ be of length 1 and mutually perpendicular, whereas (B.3.13) imposes the same requirement on the rows. Clearly, (B.3.12) and (B.3.13) are equivalent. Considered as transformations on R^n orthogonal matrices are rigid motions, which preserve the distance between points. That is, if $\mathbf{a} = (a_1, \ldots, a_n)^T$, $\mathbf{b} = (b_1, \ldots, b_n)^T$, $|\mathbf{a} - \mathbf{b}| = \sqrt{\Sigma_{i=1}^n (a_i - b_i)^2}$ is the Euclidean distance between $\mathbf{a}$ and $\mathbf{b}$, and $\mathbf{A}$ is orthogonal, then

$$|\mathbf{a} - \mathbf{b}| = |\mathbf{A}\mathbf{a} - \mathbf{A}\mathbf{b}|. \tag{B.3.14}$$

To see this, note that

$$\begin{aligned} |\mathbf{a} - \mathbf{b}|^2 &= (\mathbf{a} - \mathbf{b})^T(\mathbf{a} - \mathbf{b}) = (\mathbf{a} - \mathbf{b})^T \mathbf{A}^T \mathbf{A}(\mathbf{a} - \mathbf{b}) \\ &= [\mathbf{A}(\mathbf{a} - \mathbf{b})]^T[\mathbf{A}(\mathbf{a} - \mathbf{b})] = |\mathbf{A}\mathbf{a} - \mathbf{A}\mathbf{b}|^2. \end{aligned} \tag{B.3.15}$$

Finally, we shall use the fact that if $\mathbf{A}$ is orthogonal,

$$|\det \mathbf{A}| = 1. \tag{B.3.16}$$

This follows from[1]

$$[\det \mathbf{A}]^2 = [\det \mathbf{A}][\det \mathbf{A}^T] = \det[\mathbf{A}\mathbf{A}^T] = \det \mathbf{I} = 1. \tag{B.3.17}$$

Because $\mathbf{X} = (X_1, \ldots, X_n)^T$ is a vector of independent identically distributed $\mathcal{N}(0, \sigma^2)$ random variables we can write the density of $\mathbf{X}$ as

$$\begin{aligned} p_{\mathbf{X}}(x_1, \ldots, x_n) &= \frac{1}{[\sqrt{2\pi}\sigma]^n} \exp\left[-\frac{1}{2\sigma^2}\sum_{i=1}^n x_i^2\right] \\ &= \frac{1}{[\sqrt{2\pi}\sigma]^n} \exp\left[-\frac{1}{2\sigma^2}|\mathbf{x}|^2\right]. \end{aligned} \tag{B.3.18}$$

We seek the density of $\mathbf{Y} = g(\mathbf{X}) = \mathbf{AX} + \mathbf{c}$, where $\mathbf{A}$ is orthogonal, $n \times n$, and $\mathbf{c} = (c_1, \ldots, c_n)$. By Corollary B.2.1, $\mathbf{Y} = (Y_1, \ldots, Y_n)^T$ has density

$$p_{\mathbf{Y}}(\mathbf{y}) = \frac{1}{|\det \mathbf{A}|} p_{\mathbf{X}}(\mathbf{A}^{-1}(\mathbf{y} - \mathbf{c}))$$
$$= p_{\mathbf{X}}(\mathbf{A}^T(\mathbf{y} - \mathbf{c})) \tag{B.3.19}$$

by (B.3.11) and (B.3.16). If we substitute $\mathbf{A}^T(\mathbf{y} - \mathbf{c})$ for $\mathbf{x}$ in (B.3.18) and apply (B.3.15), we get

$$p_{\mathbf{Y}}(\mathbf{y}) = \frac{1}{[\sqrt{2\pi}\sigma]^n} \exp\left[-\frac{1}{2\sigma^2}|\mathbf{y} - \mathbf{c}|^2\right] = p_{\mathbf{X}}(\mathbf{y} - \mathbf{c}). \tag{B.3.20}$$

Because

$$p_{\mathbf{X}}(\mathbf{y} - \mathbf{c}) = \prod_{i=1}^{n}\left\{\frac{1}{\sigma}\varphi\left(\frac{y_i - c_i}{\sigma}\right)\right\} \tag{B.3.21}$$

we see that the Y_i are independent normal random variables with $E(Y_i) = c_i$ and common variance σ^2. If, in particular, $\mathbf{c} = \mathbf{0}$, then $Y_1, \ldots, Y_n$ are again a sample from a $\mathcal{N}(0, \sigma^2)$ population.

More generally it follows that if $\mathbf{Z} = \mathbf{X} + \mathbf{d}$, then $\mathbf{Y} = g(\mathbf{Z}) = \mathbf{A}(\mathbf{X} + \mathbf{d}) + \mathbf{c} = \mathbf{AX} + (\mathbf{Ad} + \mathbf{c})$ has density

$$p_{\mathbf{Y}}(\mathbf{y}) = p_{\mathbf{X}}(\mathbf{y} - (\mathbf{Ad} + \mathbf{c})). \tag{B.3.22}$$

Because $\mathbf{d} = (d_1, \ldots, d_n)^T$ is arbitrary and, by definition, $E(Z_i) = E(X_i + d_i) = d_i$, $i = 1, \ldots, n$, we see that we have proved the following theorem.

Theorem B.3.2 *If* $\mathbf{Z} = (Z_1, \ldots, Z_n)^T$ *has independent normally distributed components with the same variance* σ^2 *and* g *is an affine transformation defined by the orthogonal matrix* $\mathbf{A}$ *and vector* $\mathbf{c} = (c_1, \ldots, c_n)^T$, *then* $\mathbf{Y} = g(\mathbf{Z}) = (Y_1, \ldots, Y_n)^T$ *has independent normally distributed components with variance* σ^2. *Furthermore, if* $\mathbf{A} = (a_{ij})$

$$E(Y_i) = c_i + \sum_{j=1}^{n} a_{ij} E(Z_j) \tag{B.3.23}$$

for $i = 1, \ldots, n$.

This fundamental result will be used repeatedly in the sequel. As an application, we shall derive another classical property of samples from a normal population.

Theorem B.3.3 *Let* $(Z_1, \ldots, Z_n)^T$ *be a sample from a* $\mathcal{N}(\mu, \sigma^2)$ *population. Define*

$$\bar{Z} = \frac{1}{n}\sum_{i=1}^{n} Z_i. \tag{B.3.24}$$

Then $\bar{Z}$ *and* $\sum_{k=1}^{n}(Z_i - \bar{Z})^2$ *are independent. Furthermore,* $\bar{Z}$ *has a* $\mathcal{N}(\mu, \sigma^2/n)$ *distribution while* $(1/\sigma^2)\sum_{i=1}^{n}(Z_i - \bar{Z})^2$ *is distributed as* χ_{n-1}^2.

Proof. Construct an orthogonal matrix $\mathbf{A} = (a_{ij})$ whose first row is

$$\mathbf{a}_1 = \left(\frac{1}{\sqrt{n}}, \ldots, \frac{1}{\sqrt{n}} \right).$$

This is equivalent to finding one of the many orthogonal bases in R^n whose first member is $\mathbf{a}_1$ and may, for instance, be done by the Gram–Schmidt process (Birkhoff and MacLane, 1965, p. 180). An example of such an $\mathbf{A}$ is given in Problem B.3.15. Let $\mathbf{AZ} = (Y_1, \ldots, Y_n)^T$. By Theorem B.3.2, the Y_i are independent and normally distributed with variance σ^2 and means

$$E(Y_i) = \sum_{j=1}^{n} a_{ij} E(Z_j) = \mu \sum_{j=1}^{n} a_{ij}. \tag{B.3.25}$$

Because $a_{1j} = 1/\sqrt{n}$, $1 \le j \le n$, and $\mathbf{A}$ is orthogonal we see that

$$\sum_{j=1}^{n} a_{kj} = \sqrt{n} \sum_{j=1}^{n} a_{1j} a_{kj} = 0, \ k = 2, \ldots, n. \tag{B.3.26}$$

Therefore,

$$E(Y_1) = \mu \sqrt{n}, \ E(Y_k) = 0, \ k = 2, \ldots, n. \tag{B.3.27}$$

By Theorem B.3.1, $(1/\sigma^2)\Sigma_{k=2}^{n} Y_k^2$ has a χ_{n-1}^2 distribution. Because by the definition of $\mathbf{A}$,

$$\bar{Z} = \frac{Y_1}{\sqrt{n}}, \tag{B.3.28}$$

the theorem will be proved once we establish the identity

$$\sum_{k=2}^{n} Y_k^2 = \sum_{k=1}^{n} (Z_k - \bar{Z})^2. \tag{B.3.29}$$

Now

$$\sum_{k=1}^{n} (Z_k - \bar{Z})^2 = \sum_{k=1}^{n} Z_k^2 - 2\bar{Z} \sum_{k=1}^{n} Z_k + n\bar{Z}^2 = \sum_{k=1}^{n} Z_k^2 - n\bar{Z}^2. \tag{B.3.30}$$

Therefore, by (B.3.28),

$$\sum_{k=1}^{n} (Z_k - \bar{Z})^2 = \sum_{k=1}^{n} Z_k^2 - Y_1^2. \tag{B.3.31}$$

Finally,

$$\sum_{k=1}^{n} Y_k^2 = |\mathbf{Y}|^2 = |\mathbf{AZ} - \mathbf{A0}|^2 = |\mathbf{Z}|^2 = \sum_{k=1}^{n} Z_k^2. \tag{B.3.32}$$

Assertion (B.3.29) follows.   □

B.4   THE BIVARIATE NORMAL DISTRIBUTION

The normal distribution is the most ubiquitous object in statistics. It appears in theory as an approximation to the distribution of sums of independent random variables, of order statistics, of maximum likelihood estimates, and so on. In practice it turns out that variables arising in all sorts of situations, such as errors of measurement, height, weight, yields of chemical and biological processes, and so on, are approximately normally distributed.

In the same way, the family of k-variate normal distributions arises on theoretical grounds when we consider the limiting behavior of sums of independent k-vectors of random variables and in practice as an approximation to the joint distribution of k-variables. Examples are given in Section 6.2. In this section we focus on the important case $k = 2$ where all properties can be derived relatively easily without matrix calculus and we can draw pictures. The general k-variate distribution is presented in Section B.6 following a more thorough introduction to moments of random vectors.

Recall that if Z has a standard normal distribution, we obtain the $\mathcal{N}(\mu, \sigma^2)$ distribution as the distribution of $g(Z) = \sigma Z + \mu$. Thus, Z generates the location-scale family of $\mathcal{N}(\mu, \sigma^2)$ distributions. The analogue of the standard normal distribution in two dimensions is the distribution of a random pair with two independent standard normal components, whereas the generalization of the family of maps $g(z) = \sigma z + \mu$ is the group of affine transformations. This suggests the following definition, in which we let the independent $\mathcal{N}(0, 1)$ random variables Z_1 and Z_2 generate our family of bivariate distributions.

A planar vector (X, Y) has a *bivariate normal distribution* if, and only if, there exist constants $a_{ij}, 1 \leq i, j \leq 2, \mu_1, \mu_2$, and independent standard normal random variables Z_1, Z_2 such that

$$\begin{aligned} X &= \mu_1 + a_{11}Z_1 + a_{12}Z_2 \\ Y &= \mu_2 + a_{21}Z_1 + a_{22}Z_2. \end{aligned} \tag{B.4.1}$$

In matrix notation, if $\mathbf{A} = (a_{ij})$, $\boldsymbol{\mu} = (\mu_1, \mu_2)^T$, $\mathbf{X} = (X, Y)^T$, $\mathbf{Z} = (Z_1, Z_2)^T$, the definition is equivalent to

$$\mathbf{X} = \mathbf{AZ} + \boldsymbol{\mu}. \tag{B.4.2}$$

Two important properties follow from the Definition (B.4.1).

Proposition B.4.1 *The marginal distributions of the components of a bivariate normal random vector are (univariate) normal or degenerate (concentrate on one point).*

This is a consequence of (A.13.23). The converse is not true. See Problem B.4.10. Note that

$$E(X) = \mu_1 + a_{11}E(Z_1) + a_{12}E(Z_2) = \mu_1, \ E(Y) = \mu_2 \tag{B.4.3}$$

and define

$$\sigma_1 = \sqrt{\text{Var } X}, \ \sigma_2 = \sqrt{\text{Var } Y}. \tag{B.4.4}$$

Then X has a $\mathcal{N}(\mu_1, \sigma_1^2)$ and Y a $\mathcal{N}(\mu_2, \sigma_2^2)$ distribution.

Proposition B.4.2 *If we apply an affine transformation* $\mathbf{g}(\mathbf{x}) = \mathbf{Cx} + \mathbf{d}$ *to a vector* $\mathbf{X}$, *which has a bivariate normal distribution, then* $\mathbf{g}(\mathbf{X})$ *also has such a distribution.*

This is clear because

$$CX + d = C(AZ + \mu) + d = (CA)Z + (C\mu + d). \tag{B.4.5}$$

We now show that the bivariate normal distribution can be characterized in terms of first- and second-order moments and derive its density. As in Section A.11, let

$$\rho = \text{Corr}(X, Y) = \frac{\text{Cov}(X, Y)}{\sigma_1 \sigma_2} \tag{B.4.6}$$

if $\sigma_1 \sigma_2 \neq 0$. If $\sigma_1 \sigma_2 = 0$, it will be convenient to let $\rho = 0$. We define the *variance-covariance matrix* of (X, Y) (or of the distribution of (X, Y)) as the matrix of central second moments

$$\Sigma = \begin{pmatrix} \sigma_1^2 & \rho \sigma_1 \sigma_2 \\ \rho \sigma_1 \sigma_2 & \sigma_2^2 \end{pmatrix}. \tag{B.4.7}$$

This symmetric matrix is in many ways the right generalization of the variance to two dimensions. We see this in Theorem B.4.1 and (B.4.21). A general definition and its properties (for k dimensions) are given in Section B.5.

Theorem B.4.1 *Suppose that $\sigma_1 \sigma_2 \neq 0$ and $|\rho| < 1$. Then*

$$p_{\mathbf{X}}(\mathbf{x}) = \frac{1}{2\pi\sqrt{\det \Sigma}} \exp\left[-\frac{1}{2}((\mathbf{x} - \boldsymbol{\mu})^T \Sigma^{-1} (\mathbf{x} - \boldsymbol{\mu}))\right] \tag{B.4.8}$$

$$= \frac{1}{2\pi\sigma_1\sigma_2\sqrt{1-\rho^2}} \exp\left[-\frac{1}{2(1-\rho^2)}\left\{\left(\frac{x-\mu_1}{\sigma_1}\right)^2 \right.\right.$$
$$\left.\left. -2\rho\frac{(x-\mu_1)}{\sigma_1}\frac{(y-\mu_2)}{\sigma_2} + \left(\frac{y-\mu_2}{\sigma_2}\right)^2\right\}\right]. \tag{B.4.9}$$

Proof. Because (X, Y) is an affine transformation of (Z_1, Z_2), we can use Corollary B.2.1 to obtain the joint density of (X, Y) provided $\mathbf{A}$ is nonsingular. We start by showing that $\mathbf{A}\mathbf{A}^T = \Sigma$. Note that

$$\mathbf{A}\mathbf{A}^T = \begin{pmatrix} a_{11}^2 + a_{12}^2 & a_{11}a_{21} + a_{12}a_{22} \\ a_{11}a_{21} + a_{12}a_{22} & a_{21}^2 + a_{22}^2 \end{pmatrix}$$

while

$$\begin{aligned} \sigma_1^2 &= \text{Var}(a_{11}Z_1) + \text{Var}(a_{12}Z_2) = a_{11}^2 \text{Var } Z_1 + a_{12}^2 \text{Var } Z_2 \\ &= a_{11}^2 + a_{12}^2, \ \sigma_2^2 = a_{21}^2 + a_{22}^2 \end{aligned} \tag{B.4.10}$$

and

$$\begin{aligned} \rho\sigma_1\sigma_2 &= \text{Cov}(a_{11}Z_1 + a_{12}Z_2, a_{21}Z_1 + a_{22}Z_2) \\ &= a_{11}a_{21} \text{Cov}(Z_1, Z_1) \\ &\quad + (a_{12}a_{21} + a_{11}a_{22}) \text{Cov}(Z_1, Z_2) \\ &\quad + a_{12}a_{22} \text{Cov}(Z_2, Z_2) \\ &= a_{11}a_{21} + a_{12}a_{22}. \end{aligned} \tag{B.4.11}$$

Therefore, $\mathbf{A}\mathbf{A}^T = \mathbf{\Sigma}$ and by using elementary properties of determinants we obtain

$$
\begin{aligned}
|\det \mathbf{A}| &= \sqrt{[\det \mathbf{A}]^2} = \sqrt{\det \mathbf{A}\det \mathbf{A}^T} = \sqrt{\det \mathbf{A}\mathbf{A}^T} \\
&= \sqrt{\det \mathbf{\Sigma}} = \sigma_1\sigma_2\sqrt{1 - \rho^2}.
\end{aligned} \tag{B.4.12}
$$

Because $|\rho| < 1$ and $\sigma_1\sigma_2 \neq 0$, we see that $\mathbf{A}$ is nonsingular and can apply Corollary B.2.1 to obtain the density of $\mathbf{X}$. The density of $\mathbf{Z}$ can be written

$$
p_{\mathbf{Z}}(\mathbf{z}) = \frac{1}{2\pi} \exp\left(-\frac{1}{2}\mathbf{z}^T\mathbf{z}\right). \tag{B.4.13}
$$

As in (B.3.19),

$$
\begin{aligned}
p_{\mathbf{X}}(\mathbf{x}) &= \frac{1}{2\pi|\det \mathbf{A}|}\left\{\exp -\frac{1}{2}[\mathbf{A}^{-1}(\mathbf{x} - \boldsymbol{\mu})]^T[\mathbf{A}^{-1}(\mathbf{x} - \boldsymbol{\mu})]\right\} \\
&= \frac{1}{2\pi|\det \mathbf{A}|}\left\{\exp -\frac{1}{2}(\mathbf{x} - \boldsymbol{\mu})^T[\mathbf{A}^{-1}]^T\mathbf{A}^{-1}(\mathbf{x} - \boldsymbol{\mu})\right\}.
\end{aligned} \tag{B.4.14}
$$

Because

$$
[\mathbf{A}^{-1}]^T \cdot \mathbf{A}^{-1} = [\mathbf{A}\mathbf{A}^T]^{-1} = \mathbf{\Sigma}^{-1} \tag{B.4.15}
$$

we arrive at (B.4.8). Finally (B.4.9) follows because by the formulas for an inverse

$$
\mathbf{\Sigma}^{-1} = \begin{pmatrix} \frac{1}{\sigma_1^2(1-\rho^2)} & \frac{-\rho}{\sigma_1\sigma_2(1-\rho^2)} \\ \frac{-\rho}{\sigma_1\sigma_2(1-\rho^2)} & \frac{1}{\sigma_2^2(1-\rho^2)} \end{pmatrix}. \tag{B.4.16}
$$

$\square$

From (B.4.7) it is clear that $\mathbf{\Sigma}$ is nonsingular if, and only if, $\sigma_1\sigma_2 \neq 0$ and $|\rho| < 1$. Bivariate normal distributions with $\sigma_1\sigma_2 \neq 0$ and $|\rho| < 1$ are referred to as *nondegenerate*, whereas others are *degenerate*. If $\sigma_1^2 = a_{11}^2 + a_{12}^2 = 0$, then $X \equiv \mu_1$, Y is necessarily distributed as $\mathcal{N}(\mu_2, \sigma_2^2)$, while $\sigma_2^2 = 0$ implies that $Y \equiv \mu_2$ and X has a $\mathcal{N}(\mu_1, \sigma_1^2)$ distribution. Finally, $\sigma_1\sigma_2 \neq 0$ and $|\rho| = 1$ implies by (A.11.16) that

$$
\frac{(Y - \mu_2)}{\sigma_2} = \rho\frac{(X - \mu_1)}{\sigma_1}. \tag{B.4.17}
$$

Because the marginal distributions of X and Y are, as we have already noted, $\mathcal{N}(\mu_1, \sigma_1^2)$ and $\mathcal{N}(\mu_2, \sigma_2^2)$ respectively, relation (B.4.17) specifies the joint distribution of (X, Y) completely. Degenerate distributions do not have densities but correspond to random vectors whose marginal distributions are normal or degenerate and are such that (X, Y) falls on a fixed line or a point with probability 1.

Note that when $\rho = 0$, $p_{\mathbf{X}}(\mathbf{x})$ becomes the joint density of two independent normal variables. Thus, in the bivariate normal case, independence is equivalent to correlation zero. This is not true in general. An example is given in Problem B.4.11.

Now, suppose that we are given nonnegative constants σ_1, σ_2, a number ρ such that $|\rho| \leq 1$ and numbers μ_1, μ_2. Then we can construct a random vector (X, Y) having a

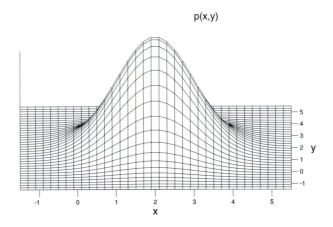

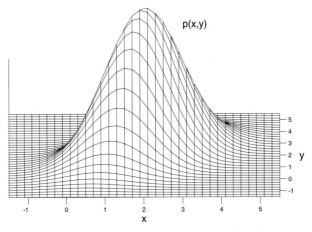

Figure B.4.1 A plot of the bivariate normal density $p(x,y)$ for $\mu_1 = \mu_2 = 2$,
$\sigma_1 = \sigma_2 = 1$, $\rho = 0$ (top) and $\rho = 0.5$ (bottom).

bivariate normal distribution with vector of means (μ_1, μ_2) and variance-covariance matrix Σ given by (B.4.7). For example, take

$$X = \mu_1 + \sigma_1 Z_1, \; Y = \mu_2 + \sigma_2(\rho Z_1 + \sqrt{1 - \rho^2} Z_2) \qquad \text{(B.4.18)}$$

and apply (B.4.10) and (B.4.11). A bivariate normal distribution with this moment structure will be referred to as $\mathcal{N}(\mu_1, \mu_2, \sigma_1^2, \sigma_2^2, \rho)$ or $\mathcal{N}(\boldsymbol{\mu}, \boldsymbol{\Sigma})$.

Now suppose that $\mathbf{U} = (U_1, U_2)^T$ is obtained by an affine transformation,

$$
\begin{aligned}
U_1 &= c_{11}X + c_{12}Y + \nu_1 \\
U_2 &= c_{21}X + c_{22}Y + \nu_2
\end{aligned}
\tag{B.4.19}
$$

from a vector $(X, Y)^T$ having a $\mathcal{N}(\mu_1, \mu_2, \sigma_1^2, \sigma_2^2, \rho)$ distribution. By Proposition B.4.2, $\mathbf{U}$ has a bivariate normal distribution. In view of our discussion, this distribution is completely determined by the means, variances, and covariances of U_1 and U_2, which in turn may be expressed in terms of the μ_i, σ_i^2, ρ, c_{ij}, and ν_i. Explicitly,

$$
\begin{aligned}
E(U_1) &= \nu_1 + c_{11}\mu_1 + c_{12}\mu_2, \quad E(U_2) = \nu_2 + c_{21}\mu_1 + c_{22}\mu_2 \\
\operatorname{Var} U_1 &= c_{11}^2\sigma_1^2 + c_{12}^2\sigma_2^2 + 2c_{12}c_{11}\rho\sigma_1\sigma_2 \\
\operatorname{Var} U_2 &= c_{21}^2\sigma_1^2 + c_{22}^2\sigma_2^2 + 2c_{21}c_{22}\rho\sigma_1\sigma_2 \\
\operatorname{Cov}(U_1, U_2) &= c_{11}c_{21}\sigma_1^2 + c_{12}c_{22}\sigma_2^2 + (c_{11}c_{22} + c_{12}c_{21})\rho\sigma_1\sigma_2.
\end{aligned}
\tag{B.4.20}
$$

In matrix notation we can write compactly

$$
\begin{aligned}
(E(U_1), E(U_2)) &= (\nu_1, \nu_2)^T + \mathbf{C}(\mu_1, \mu_2)^T = \boldsymbol{\nu} + \mathbf{C}\boldsymbol{\mu}, \\
\boldsymbol{\Sigma}(\mathbf{U}) &= \mathbf{C}\boldsymbol{\Sigma}\mathbf{C}^T
\end{aligned}
\tag{B.4.21}
$$

where $\boldsymbol{\Sigma}(\mathbf{U})$ denotes the covariance matrix of $\mathbf{U}$.

If the distribution of (X, Y) is nondegenerate and we take

$$
\mathbf{C}^T = \begin{pmatrix} \frac{1}{\sigma_1} & \frac{-\rho}{\sigma_1\sqrt{1-\rho^2}} \\ 0 & \frac{1}{\sigma_2\sqrt{1-\rho^2}} \end{pmatrix}
\tag{B.4.22}
$$

$$
(\nu_1, \nu_2)' = -\mathbf{C}(\mu_1, \mu_2)^T
$$

then U_1 and U_2 are independent and identically distributed standard normal random variables. Therefore, starting with any nondegenerate bivariate normal distribution we may by an affine transformation of the vector obtain any other bivariate normal distribution.

Another very important property of bivariate normal distributions is that normality is preserved under conditioning. That is,

Theorem B.4.2 *If (X, Y) has a nondegenerate $\mathcal{N}(\mu_1, \mu_2, \sigma_1^2, \sigma_2^2, \rho)$ distribution, then the conditional distribution of Y given $X = x$ is*

$$
\mathcal{N}\left(\mu_2 + \rho\frac{\sigma_2}{\sigma_1}(x - \mu_1),\ \sigma_2^2(1 - \rho^2)\right).
$$

Proof. Because X has a $\mathcal{N}(\mu_1, \sigma_1^2)$ distribution and (Y, X) is nondegenerate, we need only calculate

$$p(y \mid x) = \frac{p_{(\mathbf{Y},\mathbf{X})}(y, x)}{p_{\mathbf{X}}(x)}$$

$$= \frac{1}{\sigma_2 \sqrt{2\pi(1 - \rho^2)}} \exp \left\{ -\frac{1}{2(1 - \rho^2)} \left[\frac{(y - \mu_2)^2}{\sigma_2^2} - \frac{2\rho}{\sigma_2 \sigma_1}(y - \mu_2)(x - \mu_1) \right. \right.$$
$$\left. \left. + [1 - (1 - \rho^2)] \frac{(x - \mu_1)^2}{\sigma_1^2} \right] \right\}$$

$$= \frac{1}{\sigma_2 \sqrt{2\pi(1 - \rho^2)}} \exp \left\{ -\frac{1}{2(1 - \rho^2)} \left[\frac{(y - \mu_2)}{\sigma_2} - \rho \frac{(x - \mu_1)}{\sigma_1} \right]^2 \right\}.$$

(B.4.23)

This is the density we want.   □

Theorem B.4.2 shows that the conditional mean of Y given $X = x$ falls on the line $y = \mu_2 + \rho(\sigma_2/\sigma_1)(x - \mu_1)$. This line is called the *regression line*. See Figure B.4.2, which also gives the *contour plot* $S_c = \{(x, y) : f(x, y) = c\}$ where c is selected so that $P((X, Y) \in S_c) = \gamma$, $\gamma = 0.25$, 0.50, and 0.75. Such a contour plot is also called a $100\gamma\%$ *probability level curve*. See also Problem B.4.6.

By interchanging the roles of Y and X, we see that the conditional distribution of X given $Y = y$ is $\mathcal{N}(\mu_1 + (\sigma_1/\sigma_2)\rho(y - \mu_2), \sigma_1^2(1 - \rho^2))$.

With the convention that $0/0 = 0$ the theorem holds in the degenerate case as well. More generally, the conditional distribution of any linear combination of X and Y given any other linear combination of X and Y is normal (Problem B.4.9).

As we indicated at the beginning of this section the bivariate normal family of distributions arises naturally in limit theorems for sums of independent random vectors. The main result of this type is the bivariate central limit theorem. We postpone its statement and proof to Section B.6 where we can state it for the k-variate normal.

B.5 MOMENTS OF RANDOM VECTORS AND MATRICES

We generalize univariate notions from Sections A.10 to A.12 in this section. Let $\mathbf{U}$, respectively $\mathbf{V}$, denote a random k, respectively l, vector or more generally $\mathbf{U} = \|U_{ij}\|_{k \times l}$, a matrix of random variables. Suppose $E|U_{ij}| < \infty$ for all i, j. Define the expectation of $\mathbf{U}$ by

$$E(\mathbf{U}) = \|E(U_{ij})\|_{k \times l}.$$

B.5.1 Basic Properties of Expectations

If $\mathbf{A}_{m \times k}$, $\mathbf{B}_{m \times l}$ are nonrandom and $E\mathbf{U}$, $E\mathbf{V}$ are defined, then

$$E(\mathbf{A}\mathbf{U} + \mathbf{B}\mathbf{V}) = \mathbf{A}E(\mathbf{U}) + \mathbf{B}E(\mathbf{V}).$$

(B.5.1)

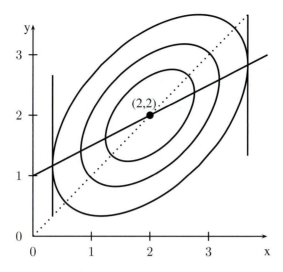

Figure B.4.2 25%, 50%, and 75% probability level-curves, the regression line (solid line), and major axis (dotted line) for the $\mathcal{N}(2, 2, 1, 1, 0.5)$ density.

This is an immediate consequence of the linearity of expectation for random variables and the definitions of matrix multiplication.

If $\mathbf{U} = \mathbf{c}$ with probability 1,

$$E(\mathbf{U}) = \mathbf{c}.$$

For a random vector $\mathbf{U}$, suppose $EU_i^2 < \infty$ for $i = 1, \ldots, k$ or equivalently $E(|\mathbf{U}|^2) < \infty$, where $|\cdot|$ denotes Euclidean distance. Define the *variance* of $\mathbf{U}$, often called the *variance-covariance* matrix, by

$$\begin{aligned} \text{Var}(\mathbf{U}) &= E(\mathbf{U} - E(\mathbf{U}))(\mathbf{U} - E(\mathbf{U}))^T \\ &= \|\text{Cov}(U_i, U_j)\|_{k \times k}, \end{aligned} \qquad (\text{B.5.2})$$

a symmetric matrix.

B.5.2   Properties of Variance

If $\mathbf{A}$ is $m \times k$ as before,

$$\text{Var}(\mathbf{AU}) = \mathbf{A}\,\text{Var}(\mathbf{U})\mathbf{A}^T. \qquad (\text{B.5.3})$$

Note that $\text{Var}(\mathbf{U})$ is $k \times k$, $\text{Var}(\mathbf{AU})$ is $m \times m$.

Let $\mathbf{c}_{k \times 1}$ denote a constant vector. Then

$$\mathrm{Var}(\mathbf{U} + \mathbf{c}) = \mathrm{Var}(\mathbf{U}). \tag{B.5.4}$$

$$\mathrm{Var}(\mathbf{c}) = \|0\|_{k \times k}. \tag{B.5.5}$$

If $\mathbf{a}_{k \times 1}$ is constant we can apply (B.5.3) to obtain

$$
\begin{aligned}
\mathrm{Var}(\mathbf{a}^T \mathbf{U}) &= \mathrm{Var}(\Sigma_{j=1}^{k} a_j U_j) \\
&= \mathbf{a}^T \mathrm{Var}(\mathbf{U}) \mathbf{a} = \sum_{i,j} a_i a_j \, \mathrm{Cov}(U_i, U_j).
\end{aligned}
\tag{B.5.6}
$$

Because the variance of any random variable is nonnegative and $\mathbf{a}$ is arbitrary, we conclude from (B.5.6) that $\mathrm{Var}(\mathbf{U})$ is a *nonnegative definite symmetric matrix*.

The following proposition is important.

Proposition B.5.1 *If $E|\mathbf{U}|^2 < \infty$, then $\mathrm{Var}(\mathbf{U})$ is positive definite if and only if, for every $\mathbf{a} \neq \mathbf{0}$, b,*

$$P[\mathbf{a}^T \mathbf{U} + b = 0] < 1. \tag{B.5.7}$$

Proof. By the definition of positive definite, (B.10.1), $\mathrm{Var}(\mathbf{U})$ is not positive definite iff $\mathbf{a}^T \mathrm{Var}(\mathbf{U}) \mathbf{a} = 0$ for some $\mathbf{a} \neq 0$. By (B.5.6) that is equivalent to $\mathrm{Var}(\mathbf{a}^T \mathbf{U}) = 0$, which is equivalent to (B.5.7) by (A.11.9). $\qquad \square$

If $\mathbf{U}_{k \times 1}$ and $\mathbf{W}_{k \times 1}$ are independent random vectors with $E|\mathbf{U}|^2 < \infty$, $E|\mathbf{W}|^2 < \infty$, then

$$\mathrm{Var}(\mathbf{U} + \mathbf{W}) = \mathrm{Var}(\mathbf{U}) + \mathrm{Var}(\mathbf{W}). \tag{B.5.8}$$

This follows by checking the identity element by element.

More generally, if $E|\mathbf{U}|^2 < \infty$, $E|\mathbf{V}|^2 < \infty$, define the covariance of $\mathbf{U}_{k \times 1}$, $\mathbf{V}_{l \times 1}$ by

$$
\begin{aligned}
\mathrm{Cov}(\mathbf{U}, \mathbf{V}) &= E(\mathbf{U} - E(\mathbf{U}))(\mathbf{V} - E(\mathbf{V}))^T \\
&= \|\mathrm{Cov}(U_i, V_j)\|_{k \times l}.
\end{aligned}
$$

Then, if $\mathbf{U}, \mathbf{V}$ are independent

$$\mathrm{Cov}(\mathbf{U}, \mathbf{V}) = \mathbf{0}. \tag{B.5.9}$$

In general

$$\mathrm{Cov}(\mathbf{AU} + \mathbf{a}, \ \mathbf{BV} + \mathbf{b}) = \mathbf{A}\mathrm{Cov}(\mathbf{U}, \mathbf{V})\mathbf{B}^T \tag{B.5.10}$$

for nonrandom $\mathbf{A}, \mathbf{a}, \mathbf{B}, \mathbf{b}$, and

$$\mathrm{Var}(\mathbf{U} + \mathbf{V}) = \mathrm{Var}(\mathbf{U}) + 2\,\mathrm{Cov}(\mathbf{U}, \mathbf{V}) + \mathrm{Var}(\mathbf{V}). \tag{B.5.11}$$

We leave (B.5.10) and (B.5.11) to the problems.

Define the *moment generating function* (m.g.f.) of $\mathbf{U}_{k \times 1}$ for $\mathbf{t} \in R^k$ by

$$M(\mathbf{t}) = M_{\mathbf{U}}(\mathbf{t}) = E(e^{\mathbf{t}^T \mathbf{U}}) = E(e^{\Sigma_{j=1}^{k} t_j U_j}).$$

Note that $M(\mathbf{t})$ can be ∞ for all $\mathbf{t} \neq \mathbf{0}$. In parallel[1] define the *characteristic function* (c.f.) of $\mathbf{U}$ by,

$$\varphi(\mathbf{t}) = E(e^{i\mathbf{t}^T \mathbf{U}}) = E(\cos(\mathbf{t}^T U)) + i E(\sin(\mathbf{t}^T \mathbf{U}))$$

where $i = \sqrt{-1}$. Note that φ is defined for all $\mathbf{t} \in R^k$, all $\mathbf{U}$. The proofs of the following theorems are beyond the scope of this book.

Theorem B.5.1 *Let $S = \{\mathbf{t} : M(\mathbf{t}) < \infty\}$. Then,*

(a) *S is convex. (See **B.9**).*

(b) *If S has a nonempty interior S^0, (contains a sphere $S(\mathbf{0}, \epsilon)$, $\epsilon > 0$), then M is analytic on S^0. In that case $E|\mathbf{U}|^p < \infty$ for all p. Then, if $i_1 + \cdots + i_k = p$,*

$$\frac{\partial^p M(\mathbf{0})}{\partial t_1^{i_1} \ldots \partial t_k^{i_k}} = E(U_1^{i_1} \ldots U_k^{i_k}). \tag{B.5.12}$$

In particular,

$$\left\| \frac{\partial M}{\partial t_j}(\mathbf{0}) \right\|_{k \times 1} = E(\mathbf{U}) \tag{B.5.13}$$

and

$$\left\| \frac{\partial^2 M(\mathbf{0})}{\partial t_i \partial t_j} \right\|_{k \times k} = E(\mathbf{U}\mathbf{U}^T). \tag{B.5.14}$$

(c) *If S^0 is nonempty, M determines the distribution of $\mathbf{U}$ uniquely.*

Expressions (B.5.12)–(B.5.14) are valid with φ replacing M if $E|\mathbf{U}|^{p/2} < \infty$, $E|\mathbf{U}| < \infty$, $E|\mathbf{U}|^2 < \infty$, respectively. The characteristic function always determines the distribution of $\mathbf{U}$ uniquely.

Proof. See Billingsley (1995).

The *cumulant generating function* of $\mathbf{U}$ is defined by $K(\mathbf{t}) = K_{\mathbf{U}}(\mathbf{t}) = \log M(\mathbf{t})$. If $S(t) = \{\mathbf{t} : M(\mathbf{t}) < \infty\}$ has a nonempty interior, then we define the cumulants as

$$c_{i_1 \ldots i_k} = c_{i_1 \ldots i_k}(\mathbf{U}) = \frac{\partial^p}{\partial t_1^{i_1} \ldots \partial t_k^{i_k}} K(\mathbf{t}) \Big|_{t=0} , \quad i_1 + \cdots + i_k = p.$$

An important consequence of the definitions and (A.9.3) is that if $\mathbf{U}_{k \times 1}$, $\mathbf{V}_{k \times 1}$ are independent then

$$M_{\mathbf{U}+\mathbf{V}}(\mathbf{t}) = M_{\mathbf{U}}(\mathbf{t}) M_{\mathbf{V}}(\mathbf{t}), \quad K_{\mathbf{U}+\mathbf{V}}(\mathbf{t}) = K_{\mathbf{U}}(\mathbf{t}) + K_{\mathbf{V}}(\mathbf{t}) \tag{B.5.15}$$

where we use subscripts to indicate the vector to which the m.g.f. belongs. The same type of identity holds for c.f.'s. Other properties of cumulants are explored in the problems. See also Barndorff–Nielsen and Cox (1989).

Example B.5.1 *The Bivariate Normal Distribution.*   If $(U_1, U_2)^T$ have a $\mathcal{N}(\mu_1, \mu_2, \sigma_1^2, \sigma_2^2, \rho)$ distribution then, it is easy to check that

$$E(\mathbf{U}) = \boldsymbol{\mu} \tag{B.5.16}$$

$$\text{Var}(\mathbf{U}) = \boldsymbol{\Sigma} \tag{B.5.17}$$

$$M_{\mathbf{U}}(\mathbf{t}) = \exp\left\{ \mathbf{t}^T \boldsymbol{\mu} + \frac{1}{2} \mathbf{t}^T \boldsymbol{\Sigma} \mathbf{t} \right\} \tag{B.5.18}$$

where $\mathbf{t} = (t_1, t_2)^T$, $\boldsymbol{\mu}$, and $\boldsymbol{\Sigma}$ are defined as in (B.4.7), (B.4.8). Similarly

$$\varphi_{\mathbf{U}}(\mathbf{t}) = \exp\left\{ i\mathbf{t}^T \boldsymbol{\mu} - \frac{1}{2} \mathbf{t}^T \boldsymbol{\Sigma} \mathbf{t} \right\} \tag{B.5.19}$$

obtained by substituting it_j for t_j, $1 \le j \le k$, in (B.5.18). The result follows directly from (A.13.20) because

$$E(\exp(\mathbf{t}^T \mathbf{U})) = E(\exp(t_1 U_1 + t_2 U_2)) \tag{B.5.20}$$

and by (B.4.20), $t_1 U_1 + t_2 U_2$ has a $\mathcal{N}(\mathbf{t}^T \boldsymbol{\mu}, \mathbf{t}^T \boldsymbol{\Sigma} \mathbf{t})$ distribution.

By taking the log in (B.5.18) and differentiating we find the first five cumulants

$$(c_{10}, c_{01}, c_{20}, c_{02}, c_{11}) = (\mu_1, \mu_2, \sigma_1^2, \sigma_2^2, \sigma_1 \sigma_2 \rho).$$

All other cumulants are zero.   □

B.6   THE MULTIVARIATE NORMAL DISTRIBUTION

B.6.1   Definition and Density

We define the multivariate normal distribution in two ways and show they are equivalent. From the equivalence we are able to derive the basic properties of this family of distributions rapidly.

Definition B.6.1 $\mathbf{U}_{k \times 1}$ has a *multivariate (k-variate) normal distribution* iff $\mathbf{U}$ can be written as

$$\mathbf{U} = \boldsymbol{\mu} + \mathbf{A} \mathbf{Z}$$

when $\boldsymbol{\mu}_{k \times 1}$, $\mathbf{A}_{k \times k}$ are constant and $\mathbf{Z} = (Z_1, \ldots, Z_k)^T$ where the Z_j are independent standard normal variables. This is the immediate generalization of our definition of the bivariate normal. We shall show that as in the bivariate case the distribution of $\mathbf{U}$ depends on $\boldsymbol{\mu} = E(\mathbf{U})$ and $\boldsymbol{\Sigma} \equiv \text{Var}(\mathbf{U})$, only.

Definition B.6.2 $\mathbf{U}_{k \times 1}$ has a multivariate normal distribution iff for every $\mathbf{a}_{k \times 1}$ nonrandom, $\mathbf{a}^T \mathbf{U} = \Sigma_{j=1}^k a_j U_j$ has a univariate normal distribution.

Theorem B.6.1 *Definitions* B.6.1 *and* B.6.2 *define the same family of distributions.*

Proof. If $\mathbf{U}$ is given by Definition B.6.1, then $\mathbf{a}^T\mathbf{U} = \mathbf{a}^T(\mathbf{AZ} + \boldsymbol{\mu}) = [\mathbf{A}^T\mathbf{a}]^T\mathbf{Z} + \mathbf{a}^T\boldsymbol{\mu}$, a linear combination of independent normal variables and, hence, normal. Conversely, if $X = \mathbf{a}^T\mathbf{U}$ has a univariate normal distribution, necessarily this is $\mathcal{N}(E(X), \mathrm{Var}(X))$. But

$$E(X) = \mathbf{a}^T E(\mathbf{U}) \tag{B.6.1}$$

$$\mathrm{Var}(X) = \mathbf{a}^T \, \mathrm{Var}(\mathbf{U})\mathbf{a} \tag{B.6.2}$$

from (B.5.1) and (B.5.4). Note that the finiteness of $E(|\mathbf{U}|)$ and $\mathrm{Var}(\mathbf{U})$ is guaranteed by applying Definition B.6.2 to $\mathbf{e}_j^T\mathbf{U}$, where $\mathbf{e}_j$ denotes the $k \times 1$ coordinate vector with 1 in the jth coordinate and 0 elsewhere. Now, by definition,

$$
\begin{aligned}
M_{\mathbf{U}}(\mathbf{a}) &= E(\exp(\mathbf{a}^T\mathbf{U})) = E(e^X) \\
&= \exp\left\{\mathbf{a}^T E(\mathbf{U}) + \tfrac{1}{2}\mathbf{a}^T \, \mathrm{Var}(\mathbf{U})\mathbf{a}\right\}
\end{aligned}
\tag{B.6.3}
$$

from (A.13.20), for all $\mathbf{a}$. Thus, by Theorem B.5.1 the distribution of $\mathbf{U}$ under Definition B.6.2 is completely determined by $E(\mathbf{U})$, $\mathrm{Var}(\mathbf{U})$. We now appeal to the principal axis theorem (B.10.1.1). If $\boldsymbol{\Sigma}$ is nonnegative definite symmetric, there exists $\mathbf{A}_{k \times k}$ such that

$$\boldsymbol{\Sigma} = \mathbf{AA}^T, \tag{B.6.4}$$

where $\mathbf{A}$ is nonsingular iff $\boldsymbol{\Sigma}$ is positive definite. Now, given $\mathbf{U}$ defined by B.6.2 with $E(\mathbf{U}) = \boldsymbol{\mu}$, $\mathrm{Var}(\mathbf{U}) = \boldsymbol{\Sigma}$, consider

$$\mathbf{V}_{k \times 1} = \boldsymbol{\mu} + \mathbf{AZ}$$

where $\mathbf{A}$ and $\mathbf{Z}$ are as in Definition B.6.1. Then

$$E(\mathbf{V}) = \boldsymbol{\mu}, \; \mathrm{Var}(\mathbf{V}) = \mathbf{A}\mathrm{Var}(\mathbf{Z})\mathbf{A}^T = \mathbf{AA}^T$$

because $\mathrm{Var}(\mathbf{Z}) = \mathbf{J}_{k \times k}$, the identity matrix.

Then, by definition, $\mathbf{V}$ satisfies Definition B.6.1 and, hence, B.6.2 and has the same first and second moments as $\mathbf{U}$. Since first and second moments determine the k-variate normal distribution uniquely, $\mathbf{U}$ and $\mathbf{V}$ have the same distribution and the theorem follows. $\square$

Notice that we have also proved:

Corollary B.6.1. *Given arbitrary $\boldsymbol{\mu}_{k \times 1}$ and $\boldsymbol{\Sigma}$ nonnegative definite symmetric, there is a unique k-variate normal distribution with mean vector $\boldsymbol{\mu}$ and variance-covariance matrix $\boldsymbol{\Sigma}$.*

We use $\mathcal{N}_k(\boldsymbol{\mu}, \boldsymbol{\Sigma})$ to denote the k-variate normal distribution of Corolary B.6.1. Arguing from Corollary B.2.1 we see the following.

Theorem B.6.2 *If $\boldsymbol{\Sigma}$ is positive definite or equivalently nonsingular, then if $\mathbf{U} \sim \mathcal{N}_k(\boldsymbol{\mu}, \boldsymbol{\Sigma})$, $\mathbf{U}$ has a density given by*

$$p_{\mathbf{U}}(\mathbf{x}) = \frac{1}{(2\pi)^{k/2}[\det(\boldsymbol{\Sigma})]^{1/2}} \times \exp\left\{-\frac{1}{2}(\mathbf{x} - \boldsymbol{\mu})^T\boldsymbol{\Sigma}^{-1}(\mathbf{x} - \boldsymbol{\mu})\right\}. \tag{B.6.5}$$

Proof. Apply Definition B.6.1 with $\mathbf{A}$ such that $\boldsymbol{\Sigma} = \mathbf{A}\mathbf{A}^T$.

The converse that $\mathbf{U}$ has a density only if $\boldsymbol{\Sigma}$ is positive definite and similar more refined results are left to Problem B.6.2.

There is another important result that follows from the spectral decomposition theorem (B.10.1.2).

Theorem B.6.3 *If $\mathbf{U}_{k \times 1}$ has $\mathcal{N}_k(\boldsymbol{\mu}, \boldsymbol{\Sigma})$ distribution, there exists an orthogonal matrix $\mathbf{P}_{k \times k}$ such that $\mathbf{P}^T\mathbf{U}$ has an $\mathcal{N}_k(\boldsymbol{\nu}, \mathbf{D}_{k \times k})$ distribution where $\boldsymbol{\nu} = \mathbf{P}^T\boldsymbol{\mu}$ and $\mathbf{D}_{k \times k}$ is the diagonal matrix whose diagonal entries are the necessarily nonnegative eigenvalues of $\boldsymbol{\Sigma}$. If $\boldsymbol{\Sigma}$ is of rank $l < k$, necessarily only l eigenvalues are positive and conversely.*

Proof. By the spectral decomposition theorem there exists $\mathbf{P}$ orthogonal such that

$$\boldsymbol{\Sigma} = \mathbf{P}\mathbf{D}\mathbf{P}^T.$$

Then $\mathbf{P}^T\mathbf{U}$ has a $\mathcal{N}_k(\boldsymbol{\nu}, \mathbf{D})$ distribution since $\mathrm{Var}(\mathbf{P}^T\mathbf{U}) = \mathbf{P}^T\boldsymbol{\Sigma}\mathbf{P} = \mathbf{D}$ by orthogonality of $\mathbf{P}$. $\qquad\square$

This result shows that an arbitrary normal random vector can be linearly transformed to a normal random vector with independent coordinates, some possibly degenerate. In the bivariate normal case, (B.4.19) and (B.4.22) transformed an arbitrary nondegenerate bivariate normal pair to an i.i.d. $\mathcal{N}(0, 1)$ pair.

Note that if rank $\boldsymbol{\Sigma} = k$ and we set

$$\boldsymbol{\Sigma}^{\frac{1}{2}} = \mathbf{P}\mathbf{D}^{\frac{1}{2}}\mathbf{P}^T, \ \boldsymbol{\Sigma}^{-\frac{1}{2}} = \left(\boldsymbol{\Sigma}^{\frac{1}{2}}\right)^{-1} = \mathbf{P}\left(\mathbf{D}^{\frac{1}{2}}\right)^{-1}\mathbf{P}^T$$

where $\mathbf{D}^{\frac{1}{2}}$ is the diagonal matrix with diagonal entries equal to the square root of the eigenvalues of $\boldsymbol{\Sigma}$, then

$$\mathbf{Z} = \boldsymbol{\Sigma}^{-\frac{1}{2}}(\mathbf{U} - \boldsymbol{\mu}) \tag{B.6.6}$$

has a $\mathcal{N}(\mathbf{0}, \mathbf{J})$ distribution, where $\mathbf{J}$ is the $k \times k$ identity matrix.

Corollary B.6.2 *If $\mathbf{U}$ has an $\mathcal{N}_k(\mathbf{0}, \boldsymbol{\Sigma})$ distribution and $\boldsymbol{\Sigma}$ is of rank k, then $\mathbf{U}^T\boldsymbol{\Sigma}^{-1}\mathbf{U}$ has a χ_k^2 distribution.*

Proof. By (B.6.6), $\mathbf{U}^T\boldsymbol{\Sigma}^{-1}\mathbf{U} = \mathbf{Z}^T\mathbf{Z}$, where $\mathbf{Z}$ is given by (B.6.6). But $\mathbf{Z}^T\mathbf{Z} = \sum_{i=1}^k Z_i^2$ where Z_i are i.i.d. $\mathcal{N}(0, 1)$. The result follows from (B.3.1). $\qquad\square$

B.6.2   Basic Properties. Conditional Distributions

If $\mathbf{U}$ is $\mathcal{N}_k(\boldsymbol{\mu}, \boldsymbol{\Sigma})$ and $\mathbf{A}_{l \times k}$, $\mathbf{b}_{l \times 1}$ are nonrandom, then $\mathbf{A}\mathbf{U} + \mathbf{b}$ is $\mathcal{N}_l(\mathbf{A}\boldsymbol{\mu} + \mathbf{b}, \ \mathbf{A}\boldsymbol{\Sigma}\mathbf{A}^T)$. This follows immediately from Definition B.6.2. In particular, marginal distributions of blocks of coordinates of $\mathbf{U}$ are normal. For the next statement we need the following block matrix notation. Given $\boldsymbol{\Sigma}_{k \times k}$ positive definite, write

$$\boldsymbol{\Sigma} = \begin{pmatrix} \boldsymbol{\Sigma}_{11} & \boldsymbol{\Sigma}_{12} \\ \boldsymbol{\Sigma}_{21} & \boldsymbol{\Sigma}_{22} \end{pmatrix} \tag{B.6.7}$$

where Σ_{11} is the $l \times l$ variance of $(U_1, \ldots, U_l)^T$, which we denote by $\mathbf{U}^{(1)}$, Σ_{22} the $(k-l) \times (k-l)$ variance of $(U_{l+1}, \ldots, U_k)^T$ denoted by $\mathbf{U}^{(2)}$, and $\Sigma_{12} = \mathrm{Cov}(\mathbf{U}^{(1)}, \mathbf{U}^{(2)})_{l \times k - l}$, $\Sigma_{21} = \Sigma_{12}^T$. Similarly write $\boldsymbol{\mu} = \begin{pmatrix} \boldsymbol{\mu}^{(1)} \\ \boldsymbol{\mu}^{(2)} \end{pmatrix}$, where $\boldsymbol{\mu}^{(1)}$ and $\boldsymbol{\mu}^{(2)}$ are the mean vectors of $\mathbf{U}^{(1)}$ and $\mathbf{U}^{(2)}$.

We next show that independence and uncorrelatedness are the same for the k variate normal. Specifically

Theorem B.6.4 *If* $\mathbf{U}_{(k+l) \times 1} = \begin{pmatrix} \mathbf{U}^{(1)} \\ \mathbf{U}^{(2)} \end{pmatrix}$, *where* $\mathbf{U}^{(1)}$ *and* $\mathbf{U}^{(2)}$ *are* k *and* l *vectors, respectively, has a* $k + l$ *variate normal distribution, then* $\mathbf{U}^{(1)}$ *and* $\mathbf{U}^{(2)}$ *are independent iff*

$$\mathrm{Cov}(\mathbf{U}^{(1)}, \mathbf{U}^{(2)})_{k \times l} = \|\mathbf{0}\|_{k \times l}. \tag{B.6.8}$$

Proof. If $\mathbf{U}^{(1)}$ and $\mathbf{U}^{(2)}$ are independent, (B.6.8) follows from (A.11.22). Let $\mathbf{U}^{(1)*}$ and $\mathbf{U}^{(2)*}$ be independent $\mathcal{N}_k(E(\mathbf{U}^{(1)}), \mathrm{Var}(\mathbf{U}^{(1)})), \mathcal{N}_l(E(\mathbf{U}^{(2)}), \mathrm{Var}(\mathbf{U}^{(2)}))$. Then we show below that $\mathbf{U}^* \equiv \begin{pmatrix} \mathbf{U}^{(1)*} \\ \mathbf{U}^{(2)*} \end{pmatrix}$ has the same distribution as $\mathbf{U}$ and, hence, $\mathbf{U}^{(1)}, \mathbf{U}^{(2)}$ are independent. To see that $\mathbf{U}^*$ has the same distribution as $\mathbf{U}$ note that

$$E\mathbf{U}^* = E\mathbf{U} \tag{B.6.9}$$

by definition, and because $\mathbf{U}^{(1)}$ and $\mathbf{U}^{(2)}$ have $\Sigma_{12} = \mathbf{0}$ and by construction $\mathrm{Var}(\mathbf{U}^{(j)*}) = \Sigma_{jj}, j = 1, 2$, then

$$\mathrm{Var}(\mathbf{U}) = \mathrm{Var}(\mathbf{U}^*) \tag{B.6.10}$$

by (B.6.7). Therefore, $\mathbf{U}$ and $\mathbf{U}^*$ must have the same distribution by the determination of the k-variate normal by first and second moments. $\square$

Theorem B.6.5 *If* $\mathbf{U}$ *is distributed as* $\mathcal{N}_k(\boldsymbol{\mu}, \Sigma)$, *with* Σ *positive definite as previously, then the conditional distribution of* $\mathbf{U}^{(1)}$ *given* $\mathbf{U}^{(2)} = \mathbf{u}^{(2)}$ *is* $\mathcal{N}_l(\boldsymbol{\mu}^{(1)} + \Sigma_{12}\Sigma_{22}^{-1}(\mathbf{u}^{(2)} - \boldsymbol{\mu}^{(2)}), \Sigma_{11} - \Sigma_{12}\Sigma_{22}^{-1}\Sigma_{12})$. *Moreover* $\Sigma_{11} - \Sigma_{12}\Sigma_{22}^{-1}\Sigma_{21}$ *is positive definite so there is a conditional density given by (B.6.5) with* $(\Sigma_{11} - \Sigma_{12}\Sigma_{22}^{-1}\Sigma_{21})$ *substituted for* Σ *and* $\boldsymbol{\mu}^{(1)} + \Sigma_{12}\Sigma_{22}^{-1}(\mathbf{u}^{(2)} - \boldsymbol{\mu}^{(2)})$ *for* $\boldsymbol{\mu}$.

Proof. Identification of the conditional density as normal can be done by direct computation from the formula

$$p(\mathbf{u}^{(1)} \mid \mathbf{u}^{(2)}) = p_{\mathbf{U}}(\mathbf{u})/p_{\mathbf{U}^{(1)}}(\mathbf{u}^{(2)}) \tag{B.6.11}$$

after noting that Σ_{11} is positive definite because the marginal density must exist.

To derive this and also obtain the required formula for conditional expectation and variance we proceed as follows. That Σ_{11}, Σ_{22} are positive definite follows by using $\mathbf{a}^T \Sigma \mathbf{a} > 0$ with $\mathbf{a}$ whose last $k - l$ or first l coordinates are 0. Next note that

$$\Sigma_{12}\Sigma_{22}^{-1}\Sigma_{21} = \mathrm{Var}(\Sigma_{12}\Sigma_{22}^{-1}\mathbf{U}^{(2)}) \tag{B.6.12}$$

because

$$\text{Var}(\boldsymbol{\Sigma}_{12}\boldsymbol{\Sigma}_{22}^{-1}\mathbf{U}^{(2)}) = \boldsymbol{\Sigma}_{12}\boldsymbol{\Sigma}_{22}^{-1}\text{Var}(\mathbf{U}^{(2)})\boldsymbol{\Sigma}_{22}^{-1}\boldsymbol{\Sigma}_{21}$$
$$= \boldsymbol{\Sigma}_{12}\boldsymbol{\Sigma}_{22}^{-1}\boldsymbol{\Sigma}_{22}\boldsymbol{\Sigma}_{22}^{-1}\boldsymbol{\Sigma}_{21} \tag{B.6.13}$$

by (B.5.3). Furthermore, we claim

$$\text{Cov}(\boldsymbol{\Sigma}_{12}\boldsymbol{\Sigma}_{22}^{-1}\mathbf{U}^{(2)}, \ \mathbf{U}^{(1)} - \boldsymbol{\Sigma}_{12}\boldsymbol{\Sigma}_{22}^{-1}\mathbf{U}^{(2)}) = 0 \tag{B.6.14}$$

(Problem B.6.4) and, hence, by Theorem B.6.4, $\mathbf{U}^{(1)} - \boldsymbol{\Sigma}_{12}\boldsymbol{\Sigma}_{22}^{-1}\mathbf{U}^{(2)}$ and $\mathbf{U}^{(2)}$ are independent. Thus, the conditional distribution of $\mathbf{U}^{(1)} - \boldsymbol{\Sigma}_{12}\boldsymbol{\Sigma}_{22}^{-1}\mathbf{U}^{(2)}$ given $\mathbf{U}^{(2)} = \mathbf{u}^{(2)}$ is the same as its marginal distribution. By the substitution property of the conditional distribution this is the same as the conditional distribution of $\mathbf{U}^{(1)} - \boldsymbol{\Sigma}_{12}\boldsymbol{\Sigma}_{22}^{-1}\mathbf{u}^{(2)}$ given $\mathbf{U}^{(2)} = \mathbf{u}^{(2)}$. The result now follows by adding $\boldsymbol{\Sigma}_{12}\boldsymbol{\Sigma}_{22}^{-1}\mathbf{U}^{(2)}$ and noting that

$$\text{Var}(\mathbf{U}^{(1)} - \boldsymbol{\Sigma}_{12}\boldsymbol{\Sigma}_{22}^{-1}\mathbf{U}^{(2)}) = \boldsymbol{\Sigma}_{11} - \boldsymbol{\Sigma}_{12}\boldsymbol{\Sigma}_{22}^{-1}\boldsymbol{\Sigma}_{21} \tag{B.6.15}$$

and

$$\begin{aligned} E(\mathbf{U}_1 \mid \mathbf{U}_2 = \mathbf{u}_2) &= E(\mathbf{U}_1 - \boldsymbol{\Sigma}_{12}\boldsymbol{\Sigma}_{22}^{-1}\mathbf{U}_2 \mid \mathbf{U}_2 = \mathbf{u}_2) + \boldsymbol{\Sigma}_{12}\boldsymbol{\Sigma}_{22}^{-1}\mathbf{u}_2 \\ &= E(\mathbf{U}_1 - \boldsymbol{\Sigma}_{12}\boldsymbol{\Sigma}_{22}^{-1}\mathbf{U}_2) + \boldsymbol{\Sigma}_{12}\boldsymbol{\Sigma}_{22}^{-1}\mathbf{u}_2 \\ &= \boldsymbol{\mu}_1 - \boldsymbol{\Sigma}_{12}\boldsymbol{\Sigma}_{22}^{-1}\boldsymbol{\mu}_2 + \boldsymbol{\Sigma}_{12}\boldsymbol{\Sigma}_{22}^{-1}\mathbf{u}_2 \\ &= \boldsymbol{\mu}_1 + \boldsymbol{\Sigma}_{12}\boldsymbol{\Sigma}_{22}^{-1}(\mathbf{u}_2 - \boldsymbol{\mu}_2). \end{aligned}$$

$\square$

Theorem B.6.6 The Multivariate Central Limit Theorem. *Let* $\mathbf{X}_1, \mathbf{X}_2, \ldots, \mathbf{X}_n$ *be independent and identically distributed random k vectors with* $E|\mathbf{X}_1|^2 < \infty$. *Let* $E(\mathbf{X}_1) = \boldsymbol{\mu}$, $\text{Var}(\mathbf{X}_1) = \boldsymbol{\Sigma}$, *and let* $\mathbf{S}_n = \Sigma_{i=1}^n \mathbf{X}_i$. *Then, for every continuous function* $g : R^k \to R$,

$$g\left(\frac{\mathbf{S}_n - n\boldsymbol{\mu}}{\sqrt{n}}\right) \xrightarrow{\mathcal{L}} g(\mathbf{Z}) \tag{B.6.16}$$

where $\mathbf{Z} \sim \mathcal{N}_k(\mathbf{0}, \boldsymbol{\Sigma})$.

As a consequence, if $\boldsymbol{\Sigma}$ is positive definite, we can use Theorem B.7.1 to conclude that

$$P\left[\frac{\mathbf{S}_n - n\boldsymbol{\mu}}{\sqrt{n}} \leq \mathbf{z}\right] \to P[\mathbf{Z} \leq \mathbf{z}] \tag{B.6.17}$$

for all $\mathbf{z} \in R^k$. Here $\{\mathbf{x} : \mathbf{x} \leq \mathbf{z}\} = \{\mathbf{x} : x_i \leq z_i, \ i = 1, \ldots, k\}$ where as usual subscripts indicate coordinate labels.

A proof of this result may be found in more advanced texts in probability, for instance, Billingsley (1995) and in Chung (1974).

An important corollary follows.

Corollary B.6.7 *If the* $\mathbf{X}_i$ *are as in the statement of Theorem B.6.6, if* $\boldsymbol{\Sigma}$ *is positive definite and if* $\bar{\mathbf{X}} = \frac{1}{n}\Sigma_{i=1}^n \mathbf{X}_i$, *then*

$$n(\bar{\mathbf{X}} - \boldsymbol{\mu})^T \boldsymbol{\Sigma}^{-1}(\bar{\mathbf{X}} - \boldsymbol{\mu}) \xrightarrow{\mathcal{L}} \chi_k^2. \tag{B.6.18}$$

Proof. $\sqrt{n}(\bar{\mathbf{X}} - \boldsymbol{\mu}) = (\mathbf{S}_n - n\boldsymbol{\mu})/\sqrt{n}$. Thus, we need only note that the function $g(\mathbf{x}) \equiv \mathbf{x}^T \boldsymbol{\Sigma}^{-1} \mathbf{x}$ from R^k to R is continuous and that if $\mathbf{Z} \sim N(\mathbf{0}, \boldsymbol{\Sigma})$, then $\mathbf{Z}^T \boldsymbol{\Sigma}^{-1} \mathbf{Z} \sim \chi_k^2$ (Corollary B.6.2). $\qquad\square$

B.7 CONVERGENCE FOR RANDOM VECTORS: O_P AND o_P NOTATION

The notion of convergence in probability and convergence in law for random variables discussed in section A.1.5 generalizes to random vectors and even abstract valued random elements taking their values in metric spaces. We give the required generalizations for random vectors and, hence, random matrices here. We shall also introduce a unique notation that makes many computations easier. In the following, $|\cdot|$ denotes Euclidean distance.

B.7.1 A sequence of random vectors $\mathbf{Z}_n \equiv (Z_{n1}, \dots, Z_{nd})^T$ converges *in probability* to $\mathbf{Z} \equiv (Z_1, \dots, Z_d)^T$ iff

$$|\mathbf{Z}_n - \mathbf{Z}| \overset{P}{\to} 0$$

or equivalently $Z_{nj} \overset{P}{\to} Z_j$ for $1 \leq j \leq d$.

Note that this definition also makes sense if the $\mathbf{Z}_n$ are considered under probabilities P_n that depend on n. Thus, $\mathbf{Z}_n \overset{P_n}{\to} \mathbf{Z}$ iff

$$P_n[|\mathbf{Z}_n - \mathbf{Z}| \geq \epsilon] \to 0 \qquad \text{for every } \epsilon > 0.$$

WLLN (the weak law of large numbers). *Let* $\mathbf{Z}_1, \dots, \mathbf{Z}_n$ *be i.i.d. as* $\mathbf{Z}$ *and let* $\bar{\mathbf{Z}}_n = n^{-1} \sum_{i=1}^n \mathbf{Z}_i$. *If* $E|\mathbf{Z}| < \infty$, *then* $\bar{\mathbf{Z}}_n \overset{P}{\to} \boldsymbol{\mu} = E\mathbf{Z}$.

When $E|\mathbf{Z}|^2 < \infty$, the result follows from Chebychev's inequality as in Appendix A. For a proof in the $E|\mathbf{Z}| < \infty$ case, see Billingsley (1995).

The following definition is subtler.

B.7.2 A sequence $\{\mathbf{Z}_n\}$ of random vectors *converges in law* to $\mathbf{Z}$, written $\mathbf{Z}_n \overset{\mathcal{L}}{\to} \mathbf{Z}$ or $\mathcal{L}(\mathbf{Z}_n) \to \mathcal{L}(\mathbf{Z})$, iff

$$h(\mathbf{Z}_n) \overset{\mathcal{L}}{\to} h(\mathbf{Z})$$

for all functions $h : R^d \to R$, h continuous.

We saw this type of convergence in the central limit theorem (B.6.6).

Note that in the definition of convergence in law, the random vectors $\mathbf{Z}_n, \mathbf{Z}$ only play the role of defining marginal distributions. No requirement is put on joint distributions of $\{\mathbf{Z}_n\}, \mathbf{Z}$. Thus, if $\mathbf{Z}_1, \dots, \mathbf{Z}_n$ are i.i.d., $\mathbf{Z}_n \overset{\mathcal{L}}{\to} \mathbf{Z}_1$, but $\mathbf{Z}_n \overset{P}{\nrightarrow} \mathbf{Z}_1$.

An equivalent statement to (B.7.2) is

$$Eg(\mathbf{Z}_n) \to Eg(\mathbf{Z}) \tag{B.7.3}$$

for all $\mathbf{g} : R^d \to R$ continuous and bounded. Note that (B.7.3) implies (A.14.6). The following stronger statement can be established.

Theorem B.7.1 $\mathbf{Z}_n \overset{\mathcal{L}}{\to} \mathbf{Z}$ *iff (B.7.3) holds for every* $\mathbf{g} : R^d \to R^p$ *such that* $\mathbf{g}$ *is bounded and if* $A_{\mathbf{g}} \equiv \{\mathbf{z} : \mathbf{g}$ *is continuous at* $\mathbf{z}\}$ *then* $P[\mathbf{Z} \in A_{\mathbf{g}}] = 1$.

Here are some further properties.

Proposition B.7.1

(a) *If* $\mathbf{Z}_n \overset{P}{\to} \mathbf{Z}$ *and* $\mathbf{g}$ *is continuous from* R^d *to* R^p, *then* $\mathbf{g}(\mathbf{Z}_n) \overset{P}{\to} \mathbf{g}(\mathbf{Z})$.

(b) *The implication in* (a) *continues to hold if "P" is replaced by "$\mathcal{L}$" in premise and conclusion above.*

(c) *The conclusion of* (a) *and* (b) *continues to hold if continuity of* $\mathbf{g}$ *is replaced by* $P[\mathbf{Z} \in A_{\mathbf{g}}] = 1$ *where* $A_{\mathbf{g}} \equiv \{\mathbf{z} : \mathbf{g}$ *is continuous at* $\mathbf{z}\}$.

B.7.4 If $\mathbf{Z}_n \overset{P}{\to} \mathbf{Z}$ then $\mathbf{Z}_n \overset{\mathcal{L}}{\to} \mathbf{Z}$.

A partial converse follows.

B.7.5 If $\mathbf{Z}_n \overset{\mathcal{L}}{\to} \mathbf{z}_0$ (a constant), then $\mathbf{Z}_n \overset{P}{\to} \mathbf{z}_0$.

Note that (B.7.4) and (B.7.5) generalize (A.14.3), (A.14.4).

Theorem B.7.2 Slutsky's Theorem. *Suppose* $\mathbf{Z}^T{}_n = (\mathbf{U}^T{}_n, \mathbf{V}^T{}_n)$ *where* $\mathbf{Z}_n$ *is a* d *vector,* $\mathbf{U}_n$ *is* b-*dimensional,* $\mathbf{V}_n$ *is* $c = d - b$-*dimensional and*

(a) $\mathbf{U}_n \overset{\mathcal{L}}{\to} \mathbf{U}$

(b) $\mathbf{V}_n \overset{\mathcal{L}}{\to} \mathbf{v}$ *where* $\mathbf{v}$ *is a constant vector.*

(c) $\mathbf{g}$ *is a continuous function from* R^d *to* R^b.

Then

$$\mathbf{g}(\mathbf{U}_n^T, \mathbf{V}_n^T) \overset{\mathcal{L}}{\to} \mathbf{g}(\mathbf{U}^T, \mathbf{v}^T).$$

Again continuity of $\mathbf{g}$ can be weakened to $P[(\mathbf{U}^T, \mathbf{v}^T)^T \in A_{\mathbf{g}}] = 1$.
We next give important special cases of Slutsky's theorem:

Example B.7.1

(a) $d = 2, b = c = 1, g(u, v) = \alpha u + \beta v, g(u, v) = uv$ or $g(u, v) = \frac{u}{v}$ and $v \neq 0$. This covers (A.14.9)

(b) $\mathbf{V}_n = \|V_{nij}\|_{b \times b}, c = b^2, g(\mathbf{u}^T, \mathbf{v}^T) = \mathbf{vu}$ where $\mathbf{v}$ is a $b \times b$ matrix. To apply Theorem B.7.2, rearrange $\mathbf{V}_n$ and $\mathbf{v}$ as $c \times 1$ vectors with $c = b^2$.

Combining this with $b = c = d/2$, $g(\boldsymbol{u}^T, \boldsymbol{v}^T) = \boldsymbol{u} + \boldsymbol{v}$, we obtain, that if the $b \times b$ matrix $\|\boldsymbol{V}_n\| \xrightarrow{P} \|\boldsymbol{v}\|$ and $\boldsymbol{W}_n$, $b \times 1$, tends in probability to $\boldsymbol{w}$, a constant vector, and $\boldsymbol{U}_n \xrightarrow{\mathcal{L}} \boldsymbol{U}$, then

$$\boldsymbol{V}_n \boldsymbol{U}_n + \boldsymbol{W}_n \xrightarrow{\mathcal{L}} \boldsymbol{v}\boldsymbol{U} + \boldsymbol{w}. \qquad (\text{B.7.6})$$

The proof of Theorem B.7.1 and other preceding results comes from the following theorem due to Hammersley (1952), which relates the two modes of convergence. Skorokhod (1956) extended the result to function spaces.

Theorem B.7.3 Hammersley. *Suppose vectors* $\boldsymbol{Z}_n \xrightarrow{\mathcal{L}} \boldsymbol{Z}$ *in the sense of Definition B.7.2. There exists (on a suitable probability space) a sequence of random vectors* $\{\boldsymbol{Z}_n^*\}$ *and a vector* $\boldsymbol{Z}^*$ *such that*

(i) $\mathcal{L}(\boldsymbol{Z}_n^*) = \mathcal{L}(\boldsymbol{Z}_n)$ *for all* n, $\mathcal{L}(\boldsymbol{Z}^*) = \mathcal{L}(\boldsymbol{Z})$

(ii) $\boldsymbol{Z}_n^* \xrightarrow{P} \boldsymbol{Z}^*$.

A somewhat stronger statement can also be made, namely, that

$$\boldsymbol{Z}_n^* \xrightarrow{a.s.} \boldsymbol{Z}^*$$

where $\xrightarrow{a.s.}$ refers to almost sure convergence defined by

$$\boldsymbol{Z}_n \xrightarrow{a.s.} \boldsymbol{Z} \text{ if } P\left(\lim_{n \to \infty} \boldsymbol{Z}_n = \boldsymbol{Z} \right) = 1.$$

This type of convergence also appears in the following famous law.

SLLN (the strong law of large numbers). *Let* $\boldsymbol{Z}_1, \dots, \boldsymbol{Z}_n$ *be i.i.d. as* $\boldsymbol{Z}$ *and let* $\bar{\boldsymbol{Z}}_n = n^{-1} \sum_{i=1}^n \boldsymbol{Z}_i$, *then* $\bar{\boldsymbol{Z}}_n \xrightarrow{a.s.} \mu = E\boldsymbol{Z}$ *iff* $E|\boldsymbol{Z}| < \infty$.

For a proof, see Billingsley (1995).

The proof of Theorem B.7.3 is easy for $d = 1$. (Problem B.7.1) For the general case refer to Skorokhod (1956). Here are the proofs of some of the preceding assertions using Hammersley's theorem and the following.

Theorem B.7.4 *If the vector* $\boldsymbol{U}_n$ *converges in probability to* $\boldsymbol{U}$ *and* g *is bounded and* $P[\boldsymbol{U} \in A_g] = 1$, *then*

$$Eg(\boldsymbol{U}_n) \to Eg(\boldsymbol{U}).$$

For a proof see Billingsley (1995, p. 209). Evidently, Theorem B.7.4 gives the equivalence between (B.7.2) and (B.7.3) and establishes Theorem B.7.1.

The proof of Proposition B.7.1(a) is easy if g is uniformly continuous; that is, for every $\epsilon > 0$ there exists $\delta > 0$ such that

$$\{(\boldsymbol{z}_1, \boldsymbol{z}_2) : |g(\boldsymbol{z}_1) - g(\boldsymbol{z}_2)| \geq \epsilon\} \subset \{(\boldsymbol{z}_1, \boldsymbol{z}_2) : |\boldsymbol{z}_1 - \boldsymbol{z}_2| \geq \delta\}.$$

A stronger result (in view of Theorem B.7.4) is as follows.

Theorem B.7.5 Dominated Convergence Theorem. *If* $\{W_n\}, W$ *and* V *are random variables with* $W_n \overset{P}{\to} W$, $P[|W_n| \leq |V|] = 1$, *and* $E|V| < \infty$, *then* $EW_n \to EW$.

Proposition B.7.1(b) and (c) follow from the (a) part and Hammersley's theorem. Then (B.7.3) follows from the dominated convergence because if g is bounded by M and uniformly continuous, then for $\delta > 0$

$$|E_P g(\mathbf{Z}_n) - E_P g(\mathbf{Z})| \leq \qquad\qquad (B.7.7)$$
$$\sup\{|g(\mathbf{z}) - g(\mathbf{z}')| : |\mathbf{z} - \mathbf{z}'| \leq \delta\} + MP[|\mathbf{Z}_n - \mathbf{Z}| \geq \delta]$$

Let $n \to \infty$ to obtain that

$$\limsup_n |E_P g(\mathbf{Z}_n) - E_P g(\mathbf{Z})| \leq \sup\{|g(\mathbf{z}) - g(\mathbf{z}')| : |\mathbf{z} - \mathbf{z}'| \leq \delta\} \qquad (B.7.8)$$

and let $\delta \to 0$. The general argument is sketched in Problem B.7.3.

For B.7.5 let $h_\epsilon(\mathbf{z}) = 1(|\mathbf{z} - \mathbf{z}_0| \geq \epsilon)$. Note that $A_{h_\epsilon} = \{\mathbf{z} : |\mathbf{z} - \mathbf{z}_0| \neq \epsilon\}$. Evidently if $P[\mathbf{Z} = \mathbf{z}_0] = 1$, $P[\mathbf{Z} \in A_{h_\epsilon}] = 1$ for all $\epsilon > 0$. Therefore, by Problem B.7.4, $P[|\mathbf{Z}_n - \mathbf{z}_0| \geq \epsilon] \to P[|\mathbf{Z} - \mathbf{z}_0| \geq \epsilon] = 0$ because $P[\mathbf{Z} = \mathbf{z}_0] = 1$ and the result follows.

Finally Slutsky's theorem is easy because by Hammersley's theorem there exist $\mathbf{V}^*_n$, $\mathbf{U}^*_n$ with the same marginal distributions as $\mathbf{V}_n, \mathbf{U}_n$ and $\mathbf{U}^*_n \overset{P}{\to} \mathbf{U}^*, \mathbf{V}^*_n \overset{P}{\to} \mathbf{v}$. Then $(\mathbf{U}^*_n, \mathbf{V}^*_n) \overset{P}{\to} (\mathbf{U}^*, \mathbf{v})$, which by Proposition B.7.1 implies that $(\mathbf{U}_n, \mathbf{V}_n) \overset{\mathcal{L}}{\to} (\mathbf{U}, \mathbf{v})$, which by Theorem B.7.1 implies Slutsky's theorem.

In deriving asymptotic properties of some statistical methods, it will be convenient to use convergence of densities. We will use the following.

Theorem B.7.6 Scheffé's Theorem. *Suppose* $p_n(\mathbf{z})$ *and* $p(\mathbf{z})$ *are densities or frequency functions on* R^d *such that* $p_n(\mathbf{z}) \to p(\mathbf{z})$ *as* $n \to \infty$ *for all* $\mathbf{z} \in R^d$. *Then*

$$\int |p_n(\mathbf{z}) - p(\mathbf{z})| d\mathbf{z} \to 0 \text{ as } n \to \infty$$

in the continuous case with a sum replacing the integral in the discrete case.

Proof. We give the proof in the continuous case. Note that

$$|p_n(\mathbf{z}) - p(\mathbf{z})| = p_n(\mathbf{z}) - p(\mathbf{z}) + 2[p(\mathbf{z}) - p_n(\mathbf{z})]^+$$

where $x^+ = \max\{0, x\}$. Thus,

$$\int |p_n(\mathbf{z}) - p(\mathbf{z})| dx = \int [p_n(\mathbf{z}) - p(\mathbf{z})] d\mathbf{z} + 2\int [p(\mathbf{z}) - p_n(\mathbf{z})]^+ d\mathbf{z}.$$

The first term on the right is zero. The second term tends to zero by applying the dominated convergence theorem to $U_n = [p(\mathbf{Z}) - p_n(\mathbf{Z})]^+/p(\mathbf{Z})$ and $g(u) = u, u \in [0, 1]$, because $[p(\mathbf{z}) - p_n(\mathbf{z})]^+ \leq p(\mathbf{z})$. $\qquad\square$

Proposition B.7.2 *If* $\mathbf{Z}_n$ *and* $\mathbf{Z}$ *have densities or frequency functions* $p_n(\mathbf{z})$ *and* $p(\mathbf{z})$ *with* $p_n(\mathbf{z}) \to p(\mathbf{z})$ *as* $n \to \infty$ *for all* $\mathbf{z} \in R^d$, *then* $\mathbf{Z}_n \overset{\mathcal{L}}{\to} \mathbf{Z}$.

Proof. We give the proof in the continuous case. Let $g : R^d \to R$ be continuous and bounded, say $|g| \le M < \infty$. Then

$$|Eg(\mathbf{Z}_n) - Eg(\mathbf{Z})| = \left| \int g(\mathbf{z})[p_n(\mathbf{z}) - p(\mathbf{z})]d\mathbf{z} \right| \le M \int |p_n(\mathbf{z}) - p(\mathbf{z})|d\mathbf{z}$$

and the result follows from (B.7.3) and Theorem B.7.5. $\square$

Remark B.7.1 Theorem B.7.5 can be strengthened considerably with a suitable background in measure theory. Specifically, suppose μ is a sigma finite measure on $\mathcal{X}$. If g_n and g are measurable functions from $\mathcal{X}$ to R such that

(1) $g_n \to g$ in measure, i.e., $\mu\{x : |g_n(x) - g(x)| \ge \epsilon\} \to 0$ as $n \to \infty$ for all $\epsilon > 0$

and

(2) $\int |g_n|^r d\mu \to \int |g|^r d\mu$ as $n \to \infty$ for some $r \ge 1$, then $\int |g_n - g| d\mu \to 0$ as $n \to \infty$. A proof of this result can be found in Billingsley (1979, p. 184). $\square$

Theorem B.7.7 Polya's Theorem. *Suppose real-valued $X_n \overset{\mathcal{L}}{\to} X$. Let F_n, F be the distribution functions of X_n, X, respectively. Suppose F is continuous. Then*

$$\sup_x |F_n(x) - F(x)| \to 0.$$

Outline of Proof. By Proposition B.7.1, $F_n(x) \to F(x)$ and $F_n(x - 0) \to F(x)$ for all x. Given $\epsilon > 0$, choose $\underline{x}, \bar{x}$ such that $F(\underline{x}) \le \epsilon$, $1 - F(\bar{x}) \le \epsilon$. Because F is uniformly continuous on $[\underline{x}, \bar{x}]$, there exists $\delta(\epsilon) > 0$ such that for all $\underline{x} \le x_1, x_2 \le \bar{x}$, $|x_1 - x_2| \le \delta(\epsilon) \Rightarrow |F(x_1) - F(x_2)| \le \epsilon$. Let $\underline{x} = x_0 < x_1 \cdots < x_K = \bar{x}$ be such that $|x_j - x_{j-1}| \le \delta(\epsilon)$ for all j.

Then

$$\sup_{x_j \le x \le x_{j+1}} |F_n(x) - F(x)| \le \max\{|F_n(x_j) - F(x_j)|, |F_n(x_{j+1}) - F(x_{j+1})|\}$$

$$+ \sup_{x_j \le x \le x_{j+1}} \{\max\{(F_n(x) - F_n(x_j)), F_n(x_{j+1}) - F_n(x)\}$$

$$+\max\{(F(x) - F(x_j)), F(x_{j+1}) - F(x)\}\}.$$

The second term equals $(F_n(x_{j+1}) - F_n(x_j)) + (F(x_{j+1}) - F(x_j))$. Similarly,

$$\sup_{x \le \underline{x}} |F_n(x) - F(x)| \le F_n(\underline{x}) + F(\underline{x})$$

$$\sup_{x \ge \bar{x}} |F_n(x) - F(x)| \le (1 - F_n(\bar{x})) + (1 - F(\bar{x})).$$

Conclude that, $\overline{\lim}_n \sup_x |F_n(x) - F(x)| \le 3\epsilon$ and the theorem follows. $\square$

We end this section with some useful notation.

The $O_P, \asymp_P,$ and o_P Notation

The following asymptotic order in probability notation is useful.

$$
\begin{aligned}
\boldsymbol{U}_n = o_P(1) \qquad &\text{iff} \qquad \boldsymbol{U}_n \xrightarrow{P} 0 \\
\boldsymbol{U}_n = O_P(1) \qquad &\text{iff} \qquad \forall \epsilon > 0, \exists M < \infty \text{ such that } \forall n \quad P[\|\boldsymbol{U}_n\| \geq M] \leq \epsilon \\
\boldsymbol{U}_n = o_P(\boldsymbol{V}_n) \qquad &\text{iff} \qquad \frac{|\boldsymbol{U}_n|}{|\boldsymbol{V}_n|} = o_P(1) \\
\boldsymbol{U}_n = O_P(\boldsymbol{V}_n) \qquad &\text{iff} \qquad \frac{|\boldsymbol{U}_n|}{|\boldsymbol{V}_n|} = O_P(1) \\
\boldsymbol{U}_n \asymp_P \boldsymbol{V}_n \qquad &\text{iff} \qquad \boldsymbol{U}_n = O_P(\boldsymbol{V}_n) \quad \text{and} \quad \boldsymbol{V}_n = O_P(\boldsymbol{U}_n).
\end{aligned}
$$

Note that

$$O_P(1)o_P(1) = o_P(1), \; O_P(1) + o_P(1) = O_P(1), \tag{B.7.9}$$

and $\boldsymbol{U}_n \xrightarrow{\mathcal{L}} \boldsymbol{U} \Rightarrow \boldsymbol{U}_n = O_P(1)$.

Suppose $\mathbf{Z}_1, \ldots, \mathbf{Z}_n$ are i.i.d. as $\mathbf{Z}$ with $E|\mathbf{Z}| < \infty$. Set $\boldsymbol{\mu} = E(\mathbf{Z})$, then $\bar{\mathbf{Z}}_n = \boldsymbol{\mu} + o_P(1)$ by the WLLN. If $E|\mathbf{Z}|^2 < \infty$, then $\bar{\mathbf{Z}}_n = \boldsymbol{\mu} + O_P(n^{-\frac{1}{2}})$ by the central limit theorem.

B.8   MULTIVARIATE CALCULUS

B.8.1 A function $T : R^d \rightarrow R$ is *linear* iff

$$T(\alpha \mathbf{x}_1 + \beta \mathbf{x}_2) = \alpha T(\mathbf{x}_1) + \beta T(\mathbf{x}_2)$$

for all $\alpha, \beta \in R$, $\mathbf{x}_1, \mathbf{x}_2 \in R^d$. More generally, $T : \underbrace{R^d \times \cdots \times R^d}_{k} \rightarrow R$ is k *linear* iff $T(\mathbf{x}_1, \mathbf{x}_2, \ldots, \mathbf{x}_k)$ is linear in each coordinate separately when the others are held fixed.

B.8.2 $\mathbf{T} \equiv (T_1, \ldots, T_p)$ mapping $\underbrace{R^d \times \cdots \times R^d}_{k} \rightarrow R^p$ is said to be k linear iff $T_1, \ldots, T_p$ are k linear as in B.8.1.

B.8.3 T is k linear as in B.8.1 iff there exists an array $\{a_{i_1, \ldots, i_k} : 1 \leq i_j \leq d, \ 1 \leq j \leq k\}$ such that if $\mathbf{x}_t \equiv (x_{t1}, \ldots, x_{td}), 1 \leq t \leq k$, then

$$T(\mathbf{x}_1, \ldots, \mathbf{x}_k) = \sum_{i_k=1}^{d} \cdots \sum_{i_1=1}^{d} a_{i_1, \ldots, i_k} \prod_{j=1}^{k} x_{j i_j} \tag{B.8.4}$$

B.8.5 If $\mathbf{h} : \mathcal{O} \rightarrow R^p$, $\mathcal{O}$ open $\subset R^d$, $\mathbf{h} \equiv (h_1, \ldots, h_p)$, then $\mathbf{h}$ is *Fréchet differentiable* at $\mathbf{x} \in \mathcal{O}$ iff there exists a (necessarily unique) linear map $\mathbf{Dh}(\mathbf{x}) : R^d \rightarrow R^p$ such that

$$|\mathbf{h}(\mathbf{y}) - \mathbf{h}(\mathbf{x}) - \mathbf{Dh}(\mathbf{x})(\mathbf{y} - \mathbf{x})| = o(|\mathbf{y} - \mathbf{x}|) \tag{B.8.6}$$

where $|\cdot|$ is the Euclidean norm. If $p = 1$, $\mathbf{Dh}$ is the *total differential*.

More generally, $\mathbf{h}$ is m times *Fréchet differentiable* iff there exist l linear operators $\mathbf{D}^l \mathbf{h}(\mathbf{x}) : \underbrace{R^d \times \cdots \times R^d}_{l} \to R^p, 1 \le l \le m$ such that

$$\left| \mathbf{h}(\mathbf{y}) - \mathbf{h}(\mathbf{x}) - \sum_{l=1}^{m} \frac{\mathbf{D}^l \mathbf{h}}{l!}(\mathbf{x})(\mathbf{y} - \mathbf{x}, \ldots, \mathbf{y} - \mathbf{x}) \right| = o(|\mathbf{y} - \mathbf{x}|^m). \qquad \text{(B.8.7)}$$

B.8.8 If $\mathbf{h}$ is m times Fréchet differentiable, then for $1 \le j \le p$, h_j has partial derivatives of order $\le m$ at $\mathbf{x}$ and the jth component of $\mathbf{D}^l \mathbf{h}(\mathbf{x})$ is defined by the array

$$\left\{ \frac{\partial^l h_j(\mathbf{x})}{\partial x_1^{\epsilon_1} \ldots \partial x_d^{\epsilon_d}} : 1 \le i_j \le d,\ \epsilon_1 + \cdots + \epsilon_d = l,\ 0 \le \epsilon_i \le l,\ 1 \le i \le d \right\}.$$

B.8.9 $\mathbf{h}$ is m times Fréchet differentiable at $\mathbf{x}$ if h_j has partial derivatives of order up to m on $\mathcal{O}$ that are continuous at $\mathbf{x}$.

B.8.10 Taylor's Formula

If $h_j, 1 \le j \le p$ has continuous partial derivatives of order up to $m+1$ on $\mathcal{O}$, then, for all $\mathbf{x}, \mathbf{y} \in \mathcal{O}$,

$$\mathbf{h}(\mathbf{y}) = \mathbf{h}(\mathbf{x}) + \sum_{l=1}^{m} \frac{\mathbf{D}^l \mathbf{h}(\mathbf{x})}{l!}(\mathbf{y} - \mathbf{x}, \ldots, \mathbf{y} - \mathbf{x}) + \frac{\mathbf{D}^{m+1} \mathbf{h}(\mathbf{x}^*)}{(m+1)!}(\mathbf{y} - \mathbf{x}, \ldots, \mathbf{y} - \mathbf{x})$$

$$\text{(B.8.11)}$$

for some $\mathbf{x}^* = \mathbf{x} + \alpha^*(\mathbf{y} - \mathbf{x}), 0 \le \alpha^* \le 1$. These classical results may be found, for instance, in Dieudonné (1960) and Rudin (1991). As a consequence, we obtain the following.

B.8.12 Under the conditions of B.8.10,

$$\left| \mathbf{h}(\mathbf{y}) - \mathbf{h}(\mathbf{x}) - \sum_{l=1}^{m} \frac{\mathbf{D}^l \mathbf{h}(\mathbf{x})}{l!}(\mathbf{y} - \mathbf{x}, \ldots, \mathbf{y} - \mathbf{x}) \right|$$

$$\le ((m+1)!)^{-1} \sup\{|\mathbf{D}^{m+1} \mathbf{h}(\mathbf{x}')| : |\mathbf{x}' - \mathbf{x}| \le |\mathbf{y} - \mathbf{x}|\} |\mathbf{y} - \mathbf{x}|^{m+1}$$

for all $\mathbf{x}, \mathbf{y} \in \mathcal{O}$.

B.8.13 *Chain Rule.* Suppose $\mathbf{h} : \mathcal{O} \to R^p$ with derivative $\mathbf{Dh}$ and $\mathbf{g} : \mathbf{h}(\mathcal{O}) \to R^q$ with derivative $\mathbf{Dg}$. Then the composition $\mathbf{g} \circ \mathbf{h} : \mathcal{O} \to R^q$ is differentiable and

$$\mathbf{D}(\mathbf{g} \circ \mathbf{h})(\mathbf{x}) = \mathbf{Dg}(\mathbf{Dh}(\mathbf{x})).$$

As a consequence, we obtain the following.

B.8.14 Let $d = p$, $\mathbf{h}$ be $1-1$ and continuously Fréchet differentiable on a neighborhood of $\mathbf{x} \in \mathcal{O}$, and $\mathbf{Dh}(\mathbf{x}) = \left\| \frac{\partial h_i}{\partial x_j}(\mathbf{x}) \right\|_{p \times p}$ be nonsingular. Then $\mathbf{h}^{-1} : \mathbf{h}(\mathcal{O}) \to \mathcal{O}$ is Fréchet differentiable at $\mathbf{y} = \mathbf{h}(\mathbf{x})$ and

$$\mathbf{Dh}^{-1}(\mathbf{h}(\mathbf{x})) = [\mathbf{Dh}(\mathbf{x})]^{-1}.$$

B.9   CONVEXITY AND INEQUALITIES

Convexity

A subset S of R^k is said to be *convex* if for every $\mathbf{x}, \mathbf{y} \in S$, and every $\alpha \in [0, 1]$, $\alpha\mathbf{x} + (1 - \alpha)\mathbf{y} \in S$. When $k = 1$, convex sets are finite and infinite intervals. When $k > 1$, spheres, rectangles, and hyperplanes are convex. The point $\mathbf{x}_0$ belongs to the interior S^0 of the convex set S iff for every $\mathbf{d} \neq 0$,

$$\{\mathbf{x} : \mathbf{d}^T\mathbf{x} > \mathbf{d}^T\mathbf{x}_0\} \cap S^0 \neq \emptyset \text{ and } \{\mathbf{x} : \mathbf{d}^T\mathbf{x} < \mathbf{d}^T\mathbf{x}_0\} \cap S^0 \neq \emptyset \qquad (\text{B.9.1})$$

where $\emptyset$ denotes the empty set.

A function g from a convex set S to R is said to be *convex* if

$$g(\alpha\mathbf{x} + (1 - \alpha)\mathbf{y}) \leq \alpha g(\mathbf{x}) + (1 - \alpha)g(\mathbf{y}), \text{ all } \mathbf{x}, \mathbf{y} \in S, \ \alpha \in [0, 1]. \qquad (\text{B.9.2})$$

g is said to be *strictly convex* if (B.9.2) holds with $\leq$ replaced by $<$ for all $\mathbf{x} \neq \mathbf{y}$, $\alpha \notin \{0, 1\}$. Convex functions are continuous on S^0. When $k = 1$, if g'' exists, convexity is equivalent to $g''(\mathbf{x}) \geq 0$, $\mathbf{x} \in S$; strict convexity holds if $g''(\mathbf{x}) > 0$, $\mathbf{x} \in S$. For g convex and fixed $\mathbf{x}, \mathbf{y} \in S$, $h(\alpha) = g(\alpha\mathbf{x} + (1 - \alpha)\mathbf{y})$ is convex in α, $\alpha \in [0, 1]$. When $k > 1$, if $\partial g^2(\mathbf{x})/\partial x_i \partial x_j$ exists, convexity is equivalent to

$$\sum_{i,j} u_i u_j \partial^2 g(\mathbf{x})/\partial x_i \partial x_j \geq 0, \text{ all } \mathbf{u} \in R^k \text{ and } \mathbf{x} \in S.$$

A function h from a convex set S to R is said to be *(strictly) concave* if $g = -h$ is (strictly) convex.

Jensen's Inequality. If $S \subset R^k$ is convex and closed, g is convex on S, $P[\mathbf{U} \in S] = 1$, and $E\mathbf{U}$ is finite, then $E\mathbf{U} \in S$, $Eg(\mathbf{U})$ exists and

$$Eg(\mathbf{U}) \geq g(E\mathbf{U}) \qquad (\text{B.9.3})$$

with equality if and only if there are a and $\mathbf{b}_{k \times 1}$ such that

$$P[g(\mathbf{U}) = a + \mathbf{b}^T\mathbf{U}] = 1.$$

In particular, if g is strictly convex, equality holds in (B.9.3) if and only if $P[\mathbf{U} = \mathbf{c}] = 1$ for some $\mathbf{c}_{k \times 1}$.

For a proof see Rockafellar (1970). We next give a useful inequality relating product moments to marginal moments:

Hölder's Inequality. Let r and s be numbers with $r, s > 1$, $r^{-1} + s^{-1} = 1$. Then

$$E|XY| \leq \{E|X|^r\}^{\frac{1}{r}}\{E|Y|^s\}^{\frac{1}{s}}. \qquad (\text{B.9.4})$$

When $r = s = 2$, Hölder's inequality becomes the Cauchy–Schwartz inequality (A.11.17). For a proof of (B.9.4), see Billingsley (1995, p. 80) or Problem B.9.3.

We conclude with bounds for tails of distributions.

Bernstein Inequality for the Binomial Case. Let $S_n \sim \mathcal{B}(n, p)$, then

$$P(|S_n - np| \geq n\epsilon) \leq 2 \exp\{-n\epsilon^2/2\} \text{ for } \epsilon > 0. \tag{B.9.5}$$

That is, the probability that S_n exceeds its expected value np by more than a multiple $n\epsilon$ of n tends to zero exponentially fast as $n \to \infty$. For a proof, see Problem B.9.1.

Hoeffding's Inequality. The exponential convergence rate (B.9.5) for the sum of independent Bernoulli variables extends to the sum $S_n = \sum_{i=1}^{n} X_i$ of i.i.d. bounded variables X_i, $|X_i - \mu| \leq c_i$, where $\mu = E(X_1)$

$$P[|S_n - n\mu| \geq x] \leq 2 \exp\left\{ -\tfrac{1}{2}x^2 / \sum_{i=1}^{n} c_i^2 \right\}. \tag{B.9.6}$$

For a proof, see Grimmett and Stirzaker (1992, p. 449) or Hoeffding (1963).

B.10   TOPICS IN MATRIX THEORY AND ELEMENTARY HILBERT SPACE THEORY

B.10.1   Symmetric Matrices

We establish some of the results on symmetric nonnegative definite matrices used in the text and B.6. Recall $A_{p \times p}$ is *symmetric* iff $A = A^T$. A is *nonnegative definite* (nd) iff $\mathbf{x}^T A \mathbf{x} \geq 0$ for all $\mathbf{x}$, *positive definite* (pd) if the inequality is strict unless $\mathbf{x} = \mathbf{0}$.

B.10.1.1. The Principal Axis Theorem

(a) A is symmetric nonnegative definite (snd) iff there exist $C_{p \times p}$ such that

$$A = CC^T. \tag{B.10.1}$$

(b) A is symmetric positive definite (spd) iff C above is nonsingular.

The "if" part in (a) is trivial because then $\mathbf{x}^T A \mathbf{x} = \mathbf{x}^T CC^T \mathbf{x} = |C\mathbf{x}|^2$. The "only if" part in (b) follows because $|C\mathbf{x}|^2 > 0$ unless $\mathbf{x} = \mathbf{0}$ is equivalent to $C\mathbf{x} \neq \mathbf{0}$ unless $\mathbf{x} = \mathbf{0}$, which is nonsingularity. The "if" part in (b) follows by noting that C nonsingular iff $\det(C) \neq 0$ and $\det(CC^T) = \det^2(C)$. Parenthetically we note that if A is positive definite, A is nonsingular (Problem B.10.1). The "only if" part of (a) is deeper and follows from the spectral theorem.

B.10.1.2 Spectral Theorem

(a) $A_{p \times p}$ is symmetric iff there exists P orthogonal and $D = \text{diag}(\lambda_1, \ldots, \lambda_p)$ such that

$$A = PDP^T. \tag{B.10.2}$$

(b) The λ_j are real, unique up to labeling, and are the eigenvalues of A. That is, there exist vectors $\mathbf{e}_j$, $|\mathbf{e}_j| = 1$ such that

$$A\mathbf{e}_j = \lambda_j \mathbf{e}_j. \tag{B.10.3}$$

(c) If A is also snd, all the λ_j are nonnegative. The rank of A is the number of nonzero eigenvalues. Thus, A is positive definite iff all its eigenvalues are positive.

(d) In any case the vectors $\mathbf{e}_i$ can be chosen orthonormal and are then unique up to label.

Thus, Theorem B.10.1.2 may equivalently be written

$$A = \sum_{i=1}^{p} \mathbf{e}_i \mathbf{e}_i^T \lambda_i \tag{B.10.4}$$

where $\mathbf{e}_i \mathbf{e}_i^T$ can be interpreted as projection on the one-dimensional space spanned by $\mathbf{e}_i$ (Problem B.10.2).

(B.10.1) follows easily from B.10.3 by taking $C = P\, \mathrm{diag}(\lambda_1^{\frac{1}{2}}, \ldots, \lambda_p^{\frac{1}{2}})$ in (B.10.1).

The proof of the spectral theorem is somewhat beyond our scope—see Birkhoff and MacLane (1953, pp. 275–277, 314), for instance.

B.10.1.3 If A is spd, so is A^{-1}.

Proof. $A = P\, \mathrm{diag}(\lambda_1, \ldots, \lambda_p)P^T \Rightarrow A^{-1} = P\, \mathrm{diag}(\lambda_1^{-1}, \ldots, \lambda_p^{-1})P^T$.

B.10.1.4 If A is spd, then $\max\{\mathbf{x}^T A\mathbf{x} : \mathbf{x}^T\mathbf{x} \leq 1\} = \max_j \lambda_j$.

B.10.2   Order on Symmetric Matrices

As we defined in the text for A, B symmetric $A \leq B$ iff $B - A$ is nonnegative definite. This is easily seen to be an ordering.

B.10.2.1 If A and B are symmetric and $A \leq B$, then for any C

$$CAC^T \leq CBC^T. \tag{B.10.5}$$

This follows from definition of snd or the principal axis theorem because $B - A$ snd means $B - A = EE^T$ and then $CBC^T - CAC^T = C(B-A)C^T = CEE^TC^T = (CE)(CE)^T$.

Furthermore, if A and B are spd and $A \leq B$, then

$$A^{-1} \geq B^{-1}. \tag{B.10.6}$$

Proof. After Bellman (1960, p. 92, Problems 13, 14). Note first that, if A is symmetric,

$$\mathbf{x}^T A^{-1}\mathbf{x} = \max\{\mathbf{y} : 2\mathbf{x}^T\mathbf{y} - \mathbf{y}^T A\mathbf{y}\} \tag{B.10.7}$$

because $\mathbf{y} = A^{-1}\mathbf{x}$ maximizes the quadratic form. Then, if $A \leq B$,

$$2\mathbf{x}^T\mathbf{y} - \mathbf{y}^T A\mathbf{y} \geq 2\mathbf{x}^T\mathbf{y} - \mathbf{y}^T B\mathbf{y}$$

for all $\mathbf{x}, \mathbf{y}$. By (B.10.7) we obtain $\mathbf{x}^T A^{-1}\mathbf{x} \geq \mathbf{x}^T B^{-1}\mathbf{x}$ for all $\mathbf{x}$ and the result follows. □

B.10.2.2 The Generalized Cauchy–Schwarz Inequality

Let $\Sigma = \begin{pmatrix} \Sigma_{11} & \Sigma_{12} \\ \Sigma_{21} & \Sigma_{22} \end{pmatrix}$ be spd, $(p+q) \times (p+q)$, with Σ_{11}, $p \times p$, Σ_{22}, $q \times q$. Then Σ_{11}, Σ_{22} are spd. Furthermore,

$$\Sigma_{11} \geq \Sigma_{12}\Sigma_{22}^{-1}\Sigma_{21}. \tag{B.10.8}$$

Proof. From Section B.6 we have noted that there exist (Gaussian) random vectors $\mathbf{U}_{p\times 1}$, $\mathbf{V}_{q\times 1}$ such that $\Sigma = \mathrm{Var}(\mathbf{U}^T, \mathbf{V}^T)^T$, $\Sigma_{11} = \mathrm{Var}(\mathbf{U})$, $\Sigma_{22} = \mathrm{Var}(\mathbf{V})$, $\Sigma_{12} = \mathrm{cov}(\mathbf{U}, \mathbf{V})$. The argument given in B.6 establishes that

$$\Sigma_{11} - \Sigma_{12}\Sigma_{22}^{-1}\Sigma_{21} = \mathrm{Var}(\mathbf{U} - \Sigma_{12}\Sigma_{22}^{-1}\mathbf{V}) \tag{B.10.9}$$

and the result follows. □

B.10.2.3 We note also, although this is not strictly part of this section, that if $\mathbf{U}, \mathbf{V}$ are random vectors as previously (not necessarily Gaussian), then equality holds in (B.10.8) iff for some $\mathbf{b}$

$$\mathbf{U} = \mathbf{b} + \Sigma_{12}\Sigma_{22}^{-1}\mathbf{V} \tag{B.10.10}$$

with probability 1. This follows from (B.10.9) since $\mathbf{a}^T \mathrm{Var}(\mathbf{U} - \Sigma_{12}\Sigma_{22}^{-1}\mathbf{V})\mathbf{a} = 0$ for all $\mathbf{a}$ iff

$$\mathbf{a}^T(\mathbf{U} - \Sigma_{12}\Sigma_{22}^{-1}\mathbf{V} - \mathbf{b}) = 0 \tag{B.10.11}$$

for all $\mathbf{a}$ where $\mathbf{b}$ is $E(\mathbf{U} - \Sigma_{12}\Sigma_{22}^{-1}\mathbf{V})$. But (B.10.11) for all $\mathbf{a}$ is equivalent to (B.10.10). □

B.10.3   Elementary Hilbert Space Theory

A linear space $\mathcal{H}$ over the reals is a Hilbert space iff

(i) It is endowed with an inner product $(\cdot, \cdot) : \mathcal{H} \times \mathcal{H} \to R$ such that $(\cdot, \cdot)$ is *bilinear*,

$$(ah_1 + bh_2, ch_3 + dh_4) = ab(h_1, h_2) + ac(h_1, h_3) + bc(h_2, h_3) + bd(h_2, h_4),$$

symmetric, $(h_1, h_2) = (h_2, h_1)$, and

$$(h, h) \geq 0$$

with equality iff $h = 0$.

It follows that if $\|h\|^2 \equiv (h,h)$, then $\|\cdot\|$ is a norm. That is,

(a) $\|h\| = 0$ iff $h = 0$
(b) $\|ah\| = |a|\|h\|$ for any scalar a
(c) $\|h_1 + h_2\| \le \|h_1\| + \|h_2\|$. *Triangle inequality*

(ii) $\mathcal{H}$ is complete. That is, if $\{h_m\}_{m \ge 1}$ is such that $\|h_m - h_n\| \to 0$ as $m, n \to \infty$ then there exists $h \in \mathcal{H}$ such that $\|h_n - h\| \to 0$.

The prototypical example of a Hilbert space is Euclidean space R^p from which the abstraction is drawn. In this case if $\mathbf{x} = (x_1, \ldots, x_p)^T$, $\mathbf{y} = (y_1, \ldots, y_p)^T \in R^p$, $(\mathbf{x}, \mathbf{y}) = \mathbf{x}^T \mathbf{y} = \sum_{j=1}^p x_j y_j$, $\|\mathbf{x}\|^2 = \sum_{j=1}^p x_j^2$ is the squared length, and so on.

B.10.3.1 Orthogonality and Pythagoras's Theorem

h_1 is *orthogonal* to h_2 iff $(h_1, h_2) = 0$. This is written $h_1 \perp h_2$. This is the usual notion of orthogonality in Euclidean space. We then have

Pythagoras's Theorem. *If $h_1 \perp h_2$, then*

$$\|h_1 + h_2\|^2 = \|h_1\|^2 + \|h_2\|^2. \tag{B.10.12}$$

An interesting consequence is the inequality valid for all h_1, h_2,

$$|(h_1, h_2)| \le \|h_1\|\|h_2\|. \tag{B.10.13}$$

In R^2 (B.10.12) is the familiar "square on the hypotenuse" theorem whereas (B.10.13) says that the cosine between $\mathbf{x}_1$ and $\mathbf{x}_2$ is ≤ 1 in absolute value.

B.10.3.2 Projections on Linear Spaces

We naturally define that a sequence $h_n \in \mathcal{H}$ converges to h iff $\|h_n - h\| \to 0$. A linear subspace $\mathcal{L}$ of $\mathcal{H}$ is *closed* iff $h_n \in \mathcal{L}$ for all n, $h_n \to h \Rightarrow h \in \mathcal{L}$. Given a closed linear subspace $\mathcal{L}$ of $\mathcal{H}$ we define the projection operator $\Pi(\cdot \mid \mathcal{L}) : \mathcal{H} \to \mathcal{L}$ by: $\Pi(h \mid \mathcal{L})$ is that $h' \in \mathcal{L}$ that achieves $\min\{\|h - h'\| : h' \in \mathcal{L}\}$. It may be shown that Π is characterized by the property

$$h - \Pi(h \mid \mathcal{L}) \perp h' \text{ for all } h' \in \mathcal{L}. \tag{B.10.14}$$

Furthermore,

(i) $\Pi(h \mid \mathcal{L})$ exists and is uniquely defined.

(ii) $\Pi(\cdot \mid \mathcal{L})$ is a linear operator

$$\Pi(\alpha h_1 + \beta h_2 \mid \mathcal{L}) = \alpha \Pi(h_1 \mid \mathcal{L}) + \beta \Pi(h_2 \mid \mathcal{L}).$$

(iii) Π is *idempotent*, $\Pi^2 = \Pi$.

(iv) Π is *norm reducing*

$$\|\Pi(h \mid \mathcal{L})\| \le \|h\|. \tag{B.10.15}$$

In fact, and this follows from (B.10.12),

$$\|h\|^2 = \|\Pi(h \mid \mathcal{L})\|^2 + \|h - \Pi(h \mid \mathcal{L})\|^2. \tag{B.10.16}$$

Here $h - \Pi(h \mid \mathcal{L})$ may be interpreted as a projection on $\mathcal{L}^\perp \equiv \{h : (h, h') = 0 \text{ for all } h' \in \mathcal{L}\}$. Properties (i)–(iii) of Π above are immediate.

All of these correspond to geometric results in Euclidean space. If $\mathbf{x}$ is a vector in R^p, $\Pi(\mathbf{x} \mid \mathcal{L})$ is the point of $\mathcal{L}$ at which the perpendicular to $\mathcal{L}$ from $\mathbf{x}$ meets $\mathcal{L}$. (B.10.16) is Pythagoras's theorem again. If $\mathcal{L}$ is the column space of a matrix $A_{n \times p}$ of rank $p < n$, then

$$\Pi(\mathbf{x} \mid \mathcal{L}) = A[A^T A]^{-1} A^T \mathbf{x}. \tag{B.10.17}$$

This is the formula for obtaining the fitted value vector $\widehat{\mathbf{Y}} = (\widehat{Y}_1, \dots, \widehat{Y}_n)^T$ by least squares in a linear regression $\mathbf{Y} = A\beta + \epsilon$ and (B.10.16) is the ANOVA identity.

The most important Hilbert space other than R^p is $L_2(P) \equiv \{$All random variables X on a (separable) probability space such that $EX^2 < \infty\}$. In this case we define the inner product by

$$(X, Y) \equiv E(XY) \tag{B.10.18}$$

so that

$$\|X\| = E^{\frac{1}{2}}(X^2). \tag{B.10.19}$$

All properties needed for this to be a Hilbert space are immediate save for completeness, which is a theorem of F. Riesz. Maintaining our geometric intuition we see that, if $E(X) = E(Y) = 0$, orthogonality simply corresponds to uncorrelatedness and Pythagoras's theorem is just the familiar

$$\mathrm{Var}(X + Y) = \mathrm{Var}(X) + \mathrm{Var}(Y)$$

if X and Y are uncorrelated.

The projection formulation now reveals that what we obtained in Section 1.4 are formulas for projection operators in two situations,

(a) $\mathcal{L}$ is the linear span of $1, Z_1, \dots, Z_d$. Here

$$\Pi(Y \mid \mathcal{L}) = E(Y) + (\Sigma_{\mathbf{ZZ}}^{-1} \Sigma_{\mathbf{ZY}})^T (\mathbf{Z} - E(\mathbf{Z})). \tag{B.10.20}$$

This is just (1.4.14).

(b) $\mathcal{L}$ is the space of all $X = g(\mathbf{Z})$ for some g (measurable). This is evidently a linear space that can be shown to be closed. Here,

$$\Pi(Y \mid \mathcal{L}) = E(Y \mid \mathbf{Z}). \tag{B.10.21}$$

That is what (1.4.4) tells us.

The identities and inequalities of Section 1.4 can readily be seen to be special cases of (B.10.16) and (B.10.15).

For a fuller treatment of these introductory aspects of Hilbert space theory, see Halmos (1951), Royden (1968), Rudin (1991), or more extensive works on functional analysis such as Dunford and Schwartz (1964).

B.11   PROBLEMS AND COMPLEMENTS

Problems for Section B.1

1. An urn contains four red and four black balls. Four balls are drawn at random without replacement. Let Z be the number of red balls obtained in the first two draws and Y the total number of red balls drawn.

(a) Find the joint distribution of Z and Y and the conditional distribution of Y given Z and Z given Y.

(b) Find $E(Y \mid Z = z)$ for $z = 0, 1, 2$.

2. Suppose Y and Z have the joint density $p(z, y) = k(k-1)(z-y)^{k-2}$ for $0 < y \leq z < 1$, where $k \geq 2$ is an integer.

(a) Find $E(Y \mid Z)$.

(b) Compute $EY = E(E(Y \mid Z))$ using (a).

3. Suppose Z_1 and Z_2 are independent with exponential $\mathcal{E}(\lambda)$ distributions. Find $E(X \mid Y)$ when $X = Z_1$ and $Y = Z_1 + Z_2$.
Hint: $E(Z_1 + Z_2 \mid Y) = Y$.

4. Suppose Y and Z have joint density function $p(z, y) = z + y$ for $0 < z < 1, 0 < y < 1$.

(a) Find $E(Y \mid Z = z)$.

(b) Find $E(Y e^{[Z+(1/Z)]} \mid Z = z)$.

5. Let $(X_1, \ldots, X_n)$ be a sample from a Poisson $\mathcal{P}(\lambda)$ distribution and let $S_m = \sum_{i=1}^{m} X_i$, $m \leq n$.

(a) Show that the conditional distribution of $\mathbf{X}$ given $S_n = k$ is multinomial $\mathcal{M}(k, 1/n, \ldots, 1/n)$.

(b) Show that $E(S_m \mid S_n) = (m/n)S_n$.

6. A random variable X has a $\mathcal{P}(\lambda)$ distribution. Given $X = k$, Y has a binomial $\mathcal{B}(k, p)$ distribution.

(a) Using the relation $E(e^{tY}) = E(E(e^{tY} \mid X))$ and the uniqueness of moment generating functions show that Y has a $\mathcal{P}(\lambda p)$ distribution.

(b) Show that Y and $X - Y$ are independent and find the conditional distribution of X given $Y = y$.

7. Suppose that X has a normal $\mathcal{N}(\mu, \sigma^2)$ distribution and that $Y = X + Z$, where Z is independent of X and has a $\mathcal{N}(\gamma, \tau^2)$ distribution.

(a) What is the conditional distribution of Y given $X = x$?

(b) Using Bayes Theorem find the conditional distribution of X given $Y = y$.

8. In each of the following examples:

(a) State whether the conditional distribution of Y given $Z = z$ is discrete, continuous, or of neither type.

(b) Give the conditional frequency, density, or distribution function in each case.

(c) Check the identity $E[E(Y \mid Z)] = E(Y)$

(i)

$$p_{(Z,Y)}(z, y) \;=\; \frac{1}{\pi}, z^2 + y^2 < 1$$
$$\;=\; 0 \text{ otherwise.}$$

(ii)

$$p_{(Z,Y)}(z, y) \;=\; 4zy, \ 0 < z < 1, \ 0 < y < 1$$
$$\;=\; 0 \text{ otherwise.}$$

(iii) Z has a uniform $\mathcal{U}(0, 1)$ distribution, $Y = Z^2$.

(iv) Y has a $\mathcal{U}(-1, 1)$ distribution, $Z = Y^2$.

(v) Y has a $\mathcal{U}(-1, 1)$ distribution, $Z = Y^2$ if $Y^2 < \frac{1}{4}$ and $Z = \frac{1}{4}$ if $Y^2 \geq \frac{1}{4}$.

9. (a) Show that if $E(X^2)$ and $E(Y^2)$ are finite then

$$\text{Cov}(X, Y) = \text{Cov}(X, E(Y \mid X)).$$

(b) Deduce that the random variables Z and Y in Problem B.1.8(i) have correlation 0 although they are not independent.

10. (a) If $X_1, \ldots, X_n$ is a sample from *any* population and $S_m = \sum_{i=1}^{m} X_i$, $m \leq n$, show that the joint distribution of (X_i, S_m) does not depend on i, $i \leq m$.
Hint: Show that the joint distribution of $(X_1, \ldots, X_n)$ is the same as that of $(X_{i_1}, \ldots, X_{i_n})$ where $(i_1, \ldots, i_n)$ is any permutation of $(1, \ldots, n)$.

(b) Assume that if X and Y are any two random variables, then the family of conditional distributions of X given Y depends only on the joint distribution of (X, Y). Deduce from (a) that $E(X_1 \mid S_n) = \cdots = E(X_n \mid S_n)$ and, hence, that $E(S_m \mid S_n) = (m/n)S_n$.

11. Suppose that Z has a binomial, $\mathcal{B}(N, \theta)$, distribution and that given $Z = z$, Y has a hypergeometric, $\mathcal{H}(z, N, n)$, distribution. Show that

$$P[Z = z \mid Y = y] = \binom{N - n}{z - y} \theta^{z-y}(1 - \theta)^{N-n-(z-y)}$$

(i.e., the binomial probability of $z - y$ successes in $N - n$ trials).
    Hint: Divide the numerator and denominator of (B.1.5) by

$$\binom{N}{n}\binom{n}{y} \theta^y (1 - \theta)^{n-y}.$$

This gives the required binomial probability in the numerator. Because the binomial probabilities add to one, the denominator must be one.

Problems for Section B.2

1. If θ is uniformly distributed on $(-\pi/2, \pi/2)$ show that $Y = \tan \theta$ has a *Cauchy* distribution whose density is given by $p(y) = 1/[\pi(1 + y^2)]$, $-\infty < y < \infty$. Note that this density coincides with the Student t density with one degree of freedom obtainable from (B.3.10).

2. Suppose X_1 and X_2 are independent exponential $\mathcal{E}(\lambda)$ random variables. Let $Y_1 = X_1 - X_2$ and $Y_2 = X_2$.

    (a) Find the joint density of Y_1 and Y_2.

    (b) Show that Y_1 has density $p(y) = \frac{1}{2}\lambda e^{-\lambda|y|}$, $-\infty < y < \infty$. This is known as the *double exponential* or *Laplace density*.

3. Let X_1 and X_2 be independent with $\beta(r_1, s_1)$ and $\beta(r_2, s_2)$ distributions, respectively. Find the joint density of $Y_1 = X_1$ and $Y_2 = X_2(1 - X_1)$.

4. Show that if X has a gamma $\Gamma(p, \lambda)$ distribution, then

    (a) $M_X(t) = E(e^{tX}) = \left(\frac{\lambda}{\lambda-t}\right)^p$, $t < \lambda$.

    (b) $E(X^r) = \frac{\Gamma(r+p)}{\lambda^r \Gamma(p)}$, $r > -p$.

    (c) $E(X) = p/\lambda$, $\mathrm{Var}(X) = p/\lambda^2$.

5. Show that if X has a beta $\beta(r, s)$ distribution, then

    (a) $E(X^k) = \frac{r...(r+(k-1))}{(r+s)...(r+s+(k-1))}$, $k = 1, 2, \ldots$.

    (b) $\mathrm{Var}\, X = \frac{rs}{(r+s)^2(r+s+1)}$.

6. Let $V_1, \ldots, V_{n+1}$ be a sample from a population with an exponential $\mathcal{E}(1)$ distribution (see (A.13.24)) and let $S_m = \sum_{i=1}^{m} V_i$, $m \le n + 1$.

(a) Show that $\mathbf{T} = \left(\frac{V_1}{S_{n+1}}, \ldots, \frac{V_n}{S_{n+1}} \right)^T$ has a density given by

$$
\begin{aligned}
p_{\mathbf{T}}(t_1, \ldots, t_n) &= n!, \ t_i > 0, \ 1 \le i \le n, \ \textstyle\sum_{i=1}^{n} t_i < 1, \\
&= 0 \text{ otherwise.}
\end{aligned}
$$

Hint: Derive first the joint distribution of $\left(\frac{V_1}{S_{n+1}}, \ldots, \frac{V_n}{S_{n+1}}, S_{n+1} \right)^T$.

(b) Show that $\mathbf{U} = \left(\frac{S_1}{S_{n+1}}, \ldots, \frac{S_n}{S_{n+1}} \right)^T$ has a density given by

$$
\begin{aligned}
p_{\mathbf{U}}(u_1, \ldots, u_n) &= n!, \ 0 < u_1 < u_2 < \cdots < u_n < 1, \\
&= 0 \text{ otherwise.}
\end{aligned}
$$

7. Let $S_1, \ldots, S_r$ be r disjoint open subsets of R^n such that $P[\mathbf{X} \in \cup_{i=1}^{r} S_i] = 1$. Suppose that $\mathbf{g}$ is a transformation from $\cup_{i=1}^{r} S_i$ to R^n such that

(i) $\mathbf{g}$ has continuous first partial derivatives in S_i for each i.

(ii) $\mathbf{g}$ is one to one on each S_i.

(iii) The Jacobian of $\mathbf{g}$ does not vanish on each S_i.

Show that if $\mathbf{X}$ has density $p_{\mathbf{X}}$, $\mathbf{Y} = \mathbf{g}(\mathbf{X})$ has density given by

$$
p_{\mathbf{Y}}(\mathbf{y}) = \sum_{i=1}^{r} p_{\mathbf{X}}(\mathbf{g}_i^{-1}(\mathbf{y})) |J_{\mathbf{g}_i}(\mathbf{g}_i^{-1}(\mathbf{y}))|^{-1} I_i(\mathbf{y}) \text{ for } \mathbf{y} \in \mathbf{g}(\cup_{i=1}^{r} S_i)
$$

where $\mathbf{g}_i$ is the restriction of $\mathbf{g}$ to S_i and $I_i(\mathbf{y})$ is 1 if $\mathbf{y} \in \mathbf{g}(S_i)$ and 0 otherwise. (If $I_i(\mathbf{y}) = 0$, the whole summand is taken to be 0 even though $\mathbf{g}_i^{-1}$ is in fact undefined.)
Hint: $P[\mathbf{g}(\mathbf{X}) \in B] = \sum_{i=1}^{r} P[\mathbf{g}(\mathbf{X}) \in B, \mathbf{X} \in S_i]$.

8. Suppose that $X_1, \ldots, X_n$ is a sample from a population with density f. The X_i arranged in order from smallest to largest are called the *order statistics* and are denoted by $X_{(1)}, \ldots, X_{(n)}$. Show that $\mathbf{Y} = \mathbf{g}(\mathbf{X}) = (X_{(1)}, \ldots, X_{(n)})^T$ has density

$$
p_{\mathbf{Y}}(\mathbf{y}) = n! \prod_{i=1}^{n} f(y_i) \text{ for } y_1 < y_2 < \cdots < y_n
$$

Hint: Let

$$
\begin{aligned}
S_1 &= \{(x_1, \ldots, x_n) : x_1 < \cdots < x_n\}, \\
S_2 &= \{(x_1, \ldots, x_n) : x_2 < x_1 < \cdots < x_n\}
\end{aligned}
$$

and so on up to $S_{n!}$. Apply the previous problem.

9. Let $X_1, \ldots, X_n$ be a sample from a uniform $\mathcal{U}(0,1)$ distribution (cf. (A.13.29)).

(a) Show that the order statistics of $\mathbf{X} = (X_1, \ldots, X_n)$ have the distribution whose density is given in Problem B.2.6(b).

(b) Deduce that $X_{(k)}$ has a $\beta(k, n-k+1)$ distribution and that $X_{(l)} - X_{(k)} \sim \beta(l - k, n - l + k + 1)$.

(c) Show that $EX_{(k)} = k/(n+1)$ and $\text{Var } X_{(k)} = k(n-k+1)/(n+1)^2(n+2)$.
Hint: Use Problem B.2.5.

10. Let $X_1, \ldots, X_n$ be a sample from a population with density f and d.f. F.

(a)   Show   that   the   conditional   density   of $(X_{(1)}, \ldots, X_{(r)})^T$ given $(X_{(r+1)}, \ldots, X_{(n)})^T$ is

$$p(x_{(1)}, \ldots, x_{(r)} \mid x_{(r+1)}, \ldots, x_{(n)}) = \frac{r! \prod_{i=1}^{r} f(x_{(i)})}{F^r(x_{(r+1)})}$$

if $x_{(1)} < \cdots < x_{(r)} < x_{(r+1)}$.

(b) Interpret this result.

11. (a) Show that if the population in Problem B.2.10 is $\mathcal{U}(0,1)$, then

$$\left(\frac{X_{(1)}}{X_{(r+1)}}, \ldots, \frac{X_{(r)}}{X_{(r+1)}} \right)^T \text{ and } (X_{(r+1)}, \ldots, X_{(n)})^T \text{ are independent.}$$

(b) Deduce that $X_{(n)}, \ldots, \frac{X_{(n)}}{X_{(n-1)}}, \frac{X_{(n-1)}}{X_{(n-2)}}, \ldots, \frac{X_{(2)}}{X_{(1)}}$ are independent in this case.

12. Let the d.f. F have a density f that is continuous and positive on an interval (a, b) such that $F(b) - F(a) = 1$, $-\infty \le a < b \le \infty$. (The results are in fact valid if we only suppose that F is continuous.)

(a) Show that if X has density f, then $Y = F(X)$ is uniformly distributed on $(0, 1)$.

(b) Show that if $U \sim \mathcal{U}(0,1)$, then $F^{-1}(U)$ has density f.

(c) Let $U_{(1)} < \cdots < U_{(n)}$ be the order statistics of a sample of size n from a $\mathcal{U}(0,1)$ population. Show that then $F^{-1}(U_{(1)}) < \cdots < F^{-1}(U_{(n)})$ are distributed as the order statistics of a sample of size n from a population with density f.

13. Using Problems B.2.9(b) and B.2.12 show that if $X_{(k)}$ is the kth order statistic of a sample of size n from a population with density f, then

$$p_{X_{(k)}}(t) = \frac{n!}{(k-1)!(n-k)!} F^{k-1}(t)(1 - F(t))^{n-k} f(t).$$

14. Let $X_{(1)}, \ldots, X_{(n)}$ be the order statistics of a sample of size n from an $\mathcal{E}(1)$ population. Show that $nX_{(1)}, (n-1)(X_{(2)} - X_{(1)}), (n-2)(X_{(3)} - X_{(2)}), \ldots, (X_{(n)} - X_{(n-1)})$ are independent and identically distributed according to $\mathcal{E}(1)$.
Hint: Apply Theorem B.2.2 directly to the density given by Problem B.2.8.

15. Let T_k be the time of the kth occurrence of an event in a Poisson process as in (A.16.4).

(a) Show that T_k has a $\Gamma(k, \lambda)$ distribution.

(b) From the identity of the events, $[N(1) \leq k - 1] = [T_k > 1]$, deduce the identity

$$\int_\lambda^\infty g_{k,1}(s)ds = \sum_{j=0}^{k-1} \frac{\lambda^j}{j!}e^{-\lambda}.$$

Problems for Section B.3

1. Let X and Y be independent and identically distributed $\mathcal{N}(0, \sigma^2)$ random variables.

(a) Show that $X^2 + Y^2$ and $\frac{X}{\sqrt{X^2+Y^2}}$ are independent.

(b) Let $\theta = \sin^{-1}\frac{X}{\sqrt{X^2+Y^2}}$. Show that θ is uniformly distributed on $\left(-\frac{\pi}{2}, \frac{\pi}{2}\right)$.

(c) Show that X/Y has a Cauchy distribution.
Hint: Use Problem B.2.1.

2. Suppose that $Z \sim \Gamma\left(\frac{1}{2}k, \frac{1}{2}k\right)$, $k > 0$, and that given $Z = z$, the conditional distribution of Y is $\mathcal{N}(0, z^{-1})$. Show that Y has a $\mathcal{T}_k$ distribution. When $k = 1$, this is an example where $E(E(Y \mid Z)) = 0$, while $E(Y)$ does not exist.

3. Show that if $Z_1, \ldots, Z_n$ are as in the statement of Theorem B.3.3, then

$$\sqrt{n}(\bar{Z} - \mu)/\sqrt{\sum_{i=1}^n (Z_i - \bar{Z})^2/(n - 1)}$$

has a $\mathcal{T}_{n-1}$ distribution.

4. Show that if $X_1, \ldots, X_n$ are independent $\mathcal{E}(\lambda)$ random variables, then $T = 2\lambda \sum_{i=1}^n X_i$ has a χ_{2n}^2 distribution.
Hint: First show that $2\lambda X_i$ has a $\Gamma\left(1, \frac{1}{2}\right) = \chi_2^2$ distribution.

5. Show that if $X_1, \ldots, X_m; Y_1, \ldots, Y_n$ are independent $\mathcal{E}(\lambda)$ random variables, then $S = (n/m)\left(\sum_{i=1}^m X_i\right)/\left(\sum_{j=1}^n Y_j\right)$ has a $\mathcal{F}_{2m,2n}$ distribution.

6. Suppose that X_1 and X_2 are independent with $\Gamma(p, 1)$ and $\Gamma\left(p + \frac{1}{2}, 1\right)$ distributions. Show that $Y = 2\sqrt{X_1 X_2}$ has a $\Gamma(2p, 1)$ distribution.

7. Suppose X has density p that is symmetric about 0; that is, $p(x) = p(-x)$ for all x. Show that $E(X^k) = 0$ if k is odd and the kth moment is finite.

8. Let $X \sim \mathcal{N}(\mu, \sigma^2)$.

(a) Show that the rth central moment of X is

$$E(X - \mu)^r = \frac{r!\sigma^r}{2^{\frac{1}{2}r}(r/2)!}, \quad r \text{ even}$$

$$= 0, \quad r \text{ odd}.$$

(b) Show the rth cumulant c_r is xero for $r \geq 3$.

Hint: Use Problem B.3.7 for r odd. For r even set $m = r/2$ and note that because $Y = [(X - \mu)/\sigma]^2$ has a χ_1^2 distribution, we can find $E(Y^m)$ from Problem B.2.4. Now use $E(X - \mu)^r = \sigma^r E(Y^m)$.

9. Show that if $X \sim \mathcal{T}_k$, then

$$E(X^r) = \frac{k^{\frac{1}{2}r} \Gamma\left(\frac{1}{2}(1+r)\right) \Gamma\left(\frac{1}{2}(k-r)\right)}{\Gamma\left(\frac{1}{2}\right) \Gamma\left(\frac{1}{2}k\right)}$$

for r even and $r < k$. The moments do not exist for $r \geq k$, the odd moments are zero when $r < k$. The mean of X is 0, for $k > 1$, and Var $X = k/(k-2)$ for $k > 2$.

Hint: Using the notation of Section B.3, for r even $E(X^r) = E(Q^r) = k^{\frac{1}{2}r} E(Z^r) E(V^{-\frac{1}{2}r})$, where $Z \sim \mathcal{N}(0,1)$ and $V \sim \chi_k^2$. Now use Problems B.2.4 and B.3.7.

10. Let $X \sim \mathcal{F}_{k,m}$, then

$$E(X^r) = \frac{m^r \Gamma\left(\frac{1}{2}k + r\right) \Gamma\left(\frac{1}{2}m - r\right)}{k^r \Gamma\left(\frac{1}{2}k\right) \Gamma\left(\frac{1}{2}m\right)}$$

provided $-\frac{1}{2}k < r < \frac{1}{2}m$. For other r, $E(X^r)$ does not exist. When $m > 2$, $E(X) = m/(m-2)$, and when $m > 4$,

$$\text{Var } X = \frac{2m^2(k+m-2)}{k(m-2)^2(m-4)}.$$

Hint: Using the notation of Section B.3, $E(X^r) = E(Q^r) = (m/k)^r E(V^r) E(W^{-r})$, where $V \sim \chi_k^2$ and $W \sim \chi_m^2$. Now use Problem B.2.4.

11. Let X have a $\mathcal{N}(\theta, 1)$ distribution.

(a) Show that $Y = X^2$ has density

$$p_Y(y) = \frac{1}{2\sqrt{2\pi y}} e^{-\frac{1}{2}(y+\theta^2)} (e^{\theta\sqrt{y}} + e^{-\theta\sqrt{y}}), \; y > 0.$$

This density corresponds to the distribution known as *noncentral χ^2 with 1 degree of freedom and noncentrality parameter θ^2*.

(b) Show that we can write

$$p_Y(y) = \sum_{i=0}^{\infty} P(R = i) f_{2i+1}(y)$$

where $R \sim \mathcal{P}\left(\frac{1}{2}\theta^2\right)$ and f_m is the χ_m^2 density. Give a probabilistic interpretation of this formula.

Hint: Use the Taylor expansions for $e^{\theta\sqrt{y}}$ and $e^{-\theta\sqrt{y}}$ in powers of $\sqrt{y}$.

12. Let $X_1, \ldots, X_n$ be independent normal random variables each having variance 1 and $E(X_i) = \theta_i$, $i = 1, \ldots, n$, and let $\theta^2 = \sum_{i=1}^{n} \theta_i^2$. Show that the density of $V = \sum_{i=1}^{n} X_i^2$ is given by

$$p_V(v) = \sum_{i=0}^{\infty} P(R = i) f_{2i+n}(v), \ v > 0$$

where $R \sim \mathcal{P}\left(\frac{1}{2}\theta^2\right)$ and f_m is the χ_m^2 density. The distribution of V is known as the *noncentral χ^2 with n degrees of freedom and (noncentrality) parameter θ^2*.

   Hint: Use an orthogonal transformation $\mathbf{Y} = \mathbf{A}\mathbf{X}$ such that $Y_1 = \sum_{i=1}^{n} (\theta_i X_i / \theta)$. Now V has the same distribution as $\sum_{i=1}^{n} Y_i^2$ where $Y_1, \ldots, Y_n$ are independent with variances 1 and $E(Y_1) = \theta$, $E(Y_i) = 0$, $i = 2, \ldots, n$. Next use Problem B.3.11 and

$$p_V(v) = \int_0^{\infty} \left[\sum_{i=0}^{\infty} P(R = i) f_{2i+1}(v - s) \right] f_{n-1}(s) ds.$$

13. Let $X_1, \ldots, X_n$ be independent $\mathcal{N}(0, 1)$ random variables and let $V = (X_1 + \theta)^2 + \sum_{i=2}^{n} X_i^2$. Show that for fixed v and n, $P(V \geq v)$ is a strictly increasing function of θ^2. Note that V has a noncentral χ_n^2 distribution with parameter θ^2.

14. Let V and W be independent with $W \sim \chi_m^2$ and V having a noncentral χ_k^2 distribution with noncentrality parameter θ^2. Show that $S = (V/k)/(W/m)$ has density

$$p_S(s) = \sum_{i=0}^{\infty} P(R = i) f_{k+2i,m}(s)$$

where $R \sim \mathcal{P}\left(\frac{1}{2}\theta^2\right)$ and $f_{j,m}$ is the density of $\mathcal{F}_{j,m}$. The distribution of S is known as the *noncentral $\mathcal{F}_{k,m}$ distribution with (noncentrality) parameter θ^2*.

15. Let $X_1, \ldots, X_n$ be independent normal random variables with common mean and variance. Define $\bar{X}_{(m)} = (1/m) \sum_{i=1}^{m} X_i$, and $S_m^2 = \sum_{i=1}^{m} (X_i - \bar{X}_{(m)})^2$.

   (a) Show that

$$S_m^2 = S_{m-1}^2 + \frac{(m-1)}{m} (X_m - \bar{X}_{(m-1)})^2.$$

   (b) Let

$$Y_1 = \sqrt{n} \bar{X}_{(n)}, Y_2 = (X_2 - \bar{X}_{(1)}) \sqrt{\frac{1}{2}}, Y_3 = (X_3 - \bar{X}_{(2)}) \sqrt{\frac{2}{3}}, \ldots,$$

$$Y_n = (X_n - \bar{X}_{(n-1)}) \sqrt{\frac{n-1}{n}}.$$

Show that the matrix $\mathbf{A}$ defined by $\mathbf{Y} = \mathbf{A}\mathbf{X}$ is orthogonal and, thus, satisfies the requirements of Theorem B.3.2.

   (c) Give the joint density of $(\bar{X}_{(n)}, S_2^2, \ldots, S_n^2)^T$.

16. Show that under the assumptions of Theorem B.3.3, $\bar{Z}$ and $(Z_1 - \bar{Z}, \ldots, Z_n - \bar{Z})$ are independent.

Hint: It suffices to show that $\bar{Z}$ is independent of $(Z_2 - \bar{Z}, \ldots, Z_n - \bar{Z})$. This provides another proof that $\bar{Z}$ and $\sum_{i=1}^{n}(Z_i - \bar{Z})^2$ are independent.

Problems for Section B.4

1. Let $(X, Y) \sim \mathcal{N}\left(1, 1, 4, 1, \frac{1}{2}\right)$. Find

(a) $P(X + 2Y \leq 4)$.

(b) $P(X \leq 2 \mid Y = 1)$.

(c) The joint distribution of $X + 2Y$ and $3Y - 2X$.

Let (X, Y) have a $\mathcal{N}(\mu_1, \mu_2, \sigma_1^2, \sigma_2^2, \rho)$ distribution in the problems 2–6, 9 that follow.

2. Let $F(\cdot, \cdot, \mu_1, \mu_2, \sigma_1^2, \sigma_2^2, \rho)$ denote the d.f. of (X, Y). Show that

$$\left(\frac{X - \mu_1}{\sigma_1}, \frac{Y - \mu_2}{\sigma_2} \right)$$

has a $\mathcal{N}(0, 0, 1, 1, \rho)$ distribution and, hence, express $F(\cdot, \cdot, \mu_1, \mu_2, \sigma_1^2, \sigma_2^2, \rho)$ in terms of $F(\cdot, \cdot, 0, 0, 1, 1, \rho)$.

3. Show that $X + Y$ and $X - Y$ are independent, if and only if, $\sigma_1^2 = \sigma_2^2$.

4. Show that if $\sigma_1 \sigma_2 > 0$, $|\rho| < 1$, then the following expression has a χ_2^2 distribution.

$$\frac{1}{(1 - \rho^2)} \left\{ \frac{(X - \mu_1)^2}{\sigma_1^2} - 2\rho \frac{(X - \mu_1)(Y - \mu_2)}{\sigma_1 \sigma_2} + \frac{(Y - \mu_2)^2}{\sigma_2^2} \right\}$$

Hint: Consider (U_1, U_2) defined by (B.4.19) and (B.4.22).

5. Establish the following relation due to Sheppard.

$$F(0, 0, 0, 0, 1, 1, \rho) = \frac{1}{4} + (1/2\pi) \sin^{-1} \rho.$$

Hint: Let U_1 and U_2 be as defined by (B.4.19) and B.4.22, then

$$\begin{aligned} P[X < 0, Y < 0] &= P[U_1 < 0, \rho U_1 + \sqrt{1 - \rho^2} U_2 < 0] \\ &= P\left[U_1 < 0, \frac{U_2}{U_1} > \frac{-\rho}{\sqrt{1 - \rho^2}} \right]. \end{aligned}$$

6. *The geometry of the bivariate normal surface.*

(a) Let $S_c = \{(x, y) : p_{(X,Y)}(x, y) = c\}$. Suppose that $\sigma_1^2 = \sigma_2^2$. Show that $\{S_c; c > 0\}$ is a family of ellipses centered at (μ_1, μ_2) with common major axis given by $(y - \mu_2) =$

$(x - \mu_1)$ if $\rho > 0$, $(y - \mu_2) = -(x - \mu_1)$ if $\rho < 0$. If $\rho = 0$, $\{S_c\}$ is a family of concentric circles.

(b) If $x = c$, $p_{\mathbf{X}}(c, y)$ is proportional to a normal density as a function of y. That is, sections of the surface $z = p_{\mathbf{X}}(x, y)$ by planes parallel to the (y, z) plane are proportional to Gaussian (normal) densities. This is in fact true for sections by any plane perpendicular to the (x, y) plane.

(c) Show that the tangents to S_c at the two points where the line $y = \mu_2 + \rho(\sigma_2/\sigma_1)(x - \mu_1)$ intersects S_c are vertical. See Figure B.4.2.

7. Let $(X_1, Y_1), \ldots, (X_n, Y_n)$ be a sample from a $\mathcal{N}(\mu_1, \mu_2, \sigma_1^2, \sigma_2^2, \rho) = \mathcal{N}(\boldsymbol{\mu}, \boldsymbol{\Sigma})$ distribution. Let $\bar{X} = (1/n)\sum_{i=1}^{n} X_i$, $\bar{Y} = (1/n)\sum_{i=1}^{n} Y_i$, $S_1^2 = \sum_{i=1}^{n}(X_i - \bar{X})^2$, $S_2^2 = \sum_{i=1}^{n}(Y_i - \bar{Y})^2$, $S_{12} = \sum_{i=1}^{n}(X_i - \bar{X})(Y_i - \bar{Y})$.

(a) Show that $n(\bar{X} - \mu_1, \bar{Y} - \mu_2)^T \boldsymbol{\Sigma}^{-1}(\bar{X} - \mu_1, \bar{Y} - \mu_2)$ has a χ_2^2 distribution.

(b) Show that $(\bar{X}, \bar{Y})$ and (S_1^2, S_2^2, S_{12}) are independent.
Hint: (a): See Problem B.4.4.

(b): Let $\mathbf{A}$ be an orthogonal matrix whose first row is $\left(n^{-\frac{1}{2}}, \ldots, n^{-\frac{1}{2}}\right)$. Let $\mathbf{U} = \mathbf{AX}$ and $\mathbf{V} = \mathbf{AY}$, where $\mathbf{X} = (X_1, \ldots, X_n)^T$ and $\mathbf{Y} = (Y_1, \ldots, Y_n)^T$. Show that $(U_2, V_2), \ldots, (U_n, V_n)$ form a sample from a $\mathcal{N}(0, 0, \sigma_1^2, \sigma_2^2, \rho)$ population. Note that $S_1^2 = \sum_{i=2}^{n} U_i^2$, $S_2^2 = \sum_{i=2}^{n} V_i^2$, $S_{12} = \sum_{i=2}^{n} U_i V_i$, while $\bar{X} = U_1/\sqrt{n}$, $\bar{Y} = V_1/\sqrt{n}$.

8. In the model of Problem B.4.7 let $R = S_{12}/S_1 S_2$ and

$$T = \frac{\sqrt{(n-2)}R}{\sqrt{1 - R^2}}.$$

(a) Show that when $\rho = 0$, T has a $\mathcal{T}_{n-2}$ distribution.

(b) Find the density of R if $\rho = 0$.
Hint: Without loss of generality, take $\sigma_1^2 = \sigma_2^2 = 1$. Let $\mathbf{C}$ be an $(n - 1) \times (n - 1)$ orthogonal matrix whose first row is $(U_2, \ldots, U_n)/S_1$. Define $(W_2, \ldots, W_n)^T = \mathbf{C}(V_2, \ldots, V_n)^T$ and show that T can be written in the form $T = L/M$ where $L = S_{12}/S_1 = W_2$ and $M^2 = (S_1^2 S_2^2 - S_{12}^2)/(n - 2)S_1^2 = \sum_{i=3}^{n} W_i^2/(n - 2)$. Argue that given $U_2 = u_2, \ldots, U_n = u_n$, no matter what $u_2, \ldots, u_n$ are, T has a $\mathcal{T}_{n-2}$ distribution. Now use the continuous version of (B.1.24).

9. Show that the conditional distribution of $aX + bY$ given $cX + dY = t$ is normal.
Hint: Without loss of generality take $a = d = 1$, $b = c = 0$ because $(aX + bY, cX + dY)$ also has a bivariate normal distribution. Deal directly with the cases $\sigma_1 \sigma_2 = 0$ and $|\rho| = 1$.

10. Let p_1 denote the $\mathcal{N}(0, 0, 1, 1, 0)$ density and let p_2 be the $\mathcal{N}(0, 0, 1, 1, \rho)$ density. Suppose that (X, Y) have the joint density

$$p(x, y) = \frac{1}{2}p_1(x, y) + \frac{1}{2}p_2(x, y).$$

Show that X and Y have normal marginal densities, but that the joint density is normal, if and only if, $\rho = 0$.

11. Use a construction similar to that of Problem B.4.10 to obtain a pair of random variables (X, Y) that

    (i) have marginal normal distributions.

    (ii) are uncorrelated.

    (iii) are *not* independent.

Do these variables have a bivariate normal distribution?

Problems for Section B.5

1. Establish (B.5.10) and (B.5.11).

2. Let $\mathbf{a}_{k \times 1}$ and $\mathbf{B}_{k \times k}$ be nonrandom. Show that

$$M_{\mathbf{a} + \mathbf{B}\mathbf{U}}(\mathbf{t}) = \exp\{\mathbf{a}^T \mathbf{t}\} M_{\mathbf{U}}(\mathbf{B}^T \mathbf{t})$$

and

$$K_{\mathbf{a} + \mathbf{B}\mathbf{U}}(t) = \mathbf{a}^T \mathbf{t} + K_{\mathbf{U}}(\mathbf{B}^T \mathbf{t}).$$

3. Show that if $M_{\mathbf{U}}(\mathbf{t})$ is well defined in a neighborhood of zero then

$$M_{\mathbf{U}}(\mathbf{t}) = 1 + \sum_{p=1}^{\infty} \frac{1}{p!} \mu_{i_1 \ldots i_k} t_1^{i_1} \cdots t_k^{i_k}$$

where $\mu_{i_1 \cdots i_k} = E(U_1^{i_1} \cdots U_k^{i_k})$ and the sum is over all $(i_1, \ldots, i_k)$ with $i_j \geq 0, \sum_{j=1}^{k} i_j = p, p = 1, 2, \ldots$. Moreover,

$$K_{\mathbf{U}}(\mathbf{t}) = \sum_{p=1}^{\infty} \frac{1}{p!} c_{i_1 \cdots i_k} t_1^{i_1} \cdots t_k^{i_k}.$$

That is, the Taylor series for K_U converges in a neighborhood of zero.

4. Show that the second- and higher-degree cumulants (where $p = \sum_{j=1}^{k} i_j \geq 2$) are invariant under shift; thus, they depend only on the moments about the mean.

5. Establish (B.5.16)–(B.5.19).

6. In the bivariate case write $\boldsymbol{\mu} = E(\mathbf{U})$, $\sigma_{ij} = E(U_1 - \mu_1)^i (U_2 - \mu_2)^j$, $\sigma_1^2 = \sigma_{20}$, $\sigma_2^2 = \sigma_{02}$. Show that

$$(c_{10}, c_{01}, c_{20}, c_{02}, c_{11}, c_{30}, c_{03}, c_{21}, c_{12}) = (\mu_1, \mu_2, \sigma_1^2, \sigma_2^2, \sigma_{11}, \sigma_{30}, \sigma_{03}, \sigma_{21}, \sigma_{12})$$

and

$$(c_{40}, c_{04}, c_{22}, c_{31}, c_{13})$$
$$= (\sigma_{40} - 3\sigma_1^2, \sigma_{04} - 3\sigma_2^2, \sigma_{22} - \sigma_1^2\sigma_2^2 - 2\sigma_{11}^2, \sigma_{31} - 3\sigma_1^2\sigma_{11}, \sigma_{13} - 3\sigma_2^2\sigma_{11}).$$

7. Suppose V, W, and Z are independent and that $U_1 = Z + V$ and $U_2 = Z + W$. Show that

$$M_{\mathbf{U}}(\mathbf{t}) = M_V(t_1)M_W(t_2)M_Z(t_1 + t_2)$$
$$K_{\mathbf{U}}(\mathbf{t}) = K_V(t_1) + K_W(t_2) + K_Z(t_1 + t_2)$$

and show that $c_{ij}(\mathbf{U}) = c_{i+j}(Z)$ for $i \neq j$; $i, j > 0$.

8. (The bivariate log normal distribution). Suppose $\mathbf{U} = (U_1, U_2)^T$ has a bivariate $\mathcal{N}(\mu_1, \mu_2, \sigma_1^2, \sigma_2^2, \rho)$ distribution. Then $\mathbf{Y} = (Y_1, Y_2)^T = (e^{U_1}, e^{U_2})^T$ is said to have a *bivariate log normal distribution*. Show that

$$E(Y_1^i Y_2^j) = \exp\left\{ i\mu_1 + j\mu_2 + \frac{1}{2}i^2\sigma_1^2 + ij\sigma_{11} + \frac{1}{2}j^2\sigma_2^2 \right\}$$

where $\sigma_{11} = \sigma_1\sigma_2\rho$.

9. (a) Suppose $\mathbf{Z}$ is $\mathcal{N}(\boldsymbol{\mu}, \boldsymbol{\Sigma})$. Show that all cumulants of degree higher than 2 (where $p = \sum_{j=1}^{k} i_j > 2$) are zero.

    (b) Suppose $\mathbf{U}_1, \ldots, \mathbf{U}_n$ are i.i.d. as $\mathbf{U}$. Let $\mathbf{Z}_n = n^{-\frac{1}{2}} \sum_{i=1}^{n}(\mathbf{U}_i - \boldsymbol{\mu})$. Show that $K_{\mathbf{Z}_n}(\mathbf{t}) = nK_{\mathbf{U}}(n^{-\frac{1}{2}}\mathbf{t}) - n^{\frac{1}{2}}\boldsymbol{\mu}^T\mathbf{t}$ and that all cumulants of degree higher than 2 tend to zero as $n \to \infty$.

10. Suppose $\mathbf{U}_{k\times1}$ and $\mathbf{V}_{m\times1}$ are independent and $\mathbf{Z}_{(k+m)\times1} = (\mathbf{U}^T, \mathbf{V}^T)^T$. Let $C_{I,J}$ where $I = \{i_1, \ldots, i_k\}$ and $J = \{i_{k+1}, \ldots, i_{k+m}\}$ be a cumulant of $\mathbf{Z}$. Show that $C_{I,J} \neq 0$ unless either $I = \{0, \ldots, 0\}$ or $J = \{0, \ldots, 0\}$.

Problems for Section B.6

1. (a) Suppose $U_i = \mu + \alpha Z_i + \beta Z_{i-1}$, $i = 1, \ldots, k$, where $Z_0, \ldots, Z_k$ are independent $\mathcal{N}(0, \sigma^2)$ random variables. Compute the expectation and covariance matrix of $\mathbf{U} = (U_1, \ldots, U_k)$. Is $\mathbf{U}$ k-variate normal?

    (b) Perform the same operation and answer the same question for $\bar{U}_i$ defined as follows:

$$\bar{U}_1 = Z_1, \bar{U}_2 = Z_2 + \alpha\bar{U}_1, \bar{U}_3 = Z_3 + \alpha\bar{U}_2, \ldots, \bar{U}_k = Z_k + \alpha\bar{U}_{k-1}.$$

2. Let $\mathbf{U}$ be as in Definition B.6.1. Show that if $\boldsymbol{\Sigma}$ is not positive definite, then $\mathbf{U}$ does not have a density.

3. Suppose $\mathbf{U}_{k\times1}$ has positive definite variance $\boldsymbol{\Sigma}$. Let $U_{l\times1}^{(1)}$ and $U_{(k-l)\times1}^{(2)}$ be a partition of $\mathbf{U}$ with variances $\boldsymbol{\Sigma}_{11}$, $\boldsymbol{\Sigma}_{22}$ and covariance $\boldsymbol{\Sigma}_{12} = \text{Cov}(\mathbf{U}^{(1)}, \mathbf{U}^{(2)})_{l\times(k-l)}$. Show that

$$\text{Cov}(\boldsymbol{\Sigma}_{12}\boldsymbol{\Sigma}_{22}^{-1}\mathbf{U}^{(2)}, \mathbf{U}^{(1)} - \boldsymbol{\Sigma}_{12}\boldsymbol{\Sigma}_{22}^{-1}\mathbf{U}^{(2)}) = 0.$$

Problems for Section B.7

1. Prove Theorem B.7.3 for $d = 1$ when Z an Z_n have continuous distribution functions F and F_n.

Hint: Let U denote a uniform, $\mathcal{U}(0, 1)$, random variable. For any d.f. G define the left inverse by $G^{-1}(u) = \inf\{t : G(t) \geq u\}$. Now define $Z_n^* = F_n^{-1}(U)$ and $Z^* = F^{-1}(U)$.

2. Prove Proposition B.7.1(a).

3. Establish (B.7.8).

4. Show that if $\mathbf{Z}_n \xrightarrow{\mathcal{L}} \mathbf{z}_0$, then $P(|\mathbf{Z}_n - \mathbf{z}_0| \geq \epsilon) \to P(|\mathbf{Z} - \mathbf{z}_0| \geq \epsilon)$.
    Hint: Extend (A.14.5).

5. The L_p *norm* of a random vector $\mathbf{X}$ is defined by $|\mathbf{X}|_p = \{E|\mathbf{X}|^p\}^{\frac{1}{p}}$, $p \geq 1$. The sequence of random variables $\{\mathbf{Z}_n\}$ is said to *converge to* $\mathbf{Z}$ *in* L_p *norm* if $|\mathbf{Z}_n - \mathbf{Z}|_p \to 0$ as $n \to \infty$. We write $\mathbf{Z}_n \xrightarrow{L_p} \mathbf{Z}$. Show that

(a) if $p < q$, then $\mathbf{Z}_n \xrightarrow{L_q} \mathbf{Z} \Rightarrow \mathbf{Z}_n \xrightarrow{L_p} \mathbf{Z}$.
Hint: Use Jensen's inequality B.9.3.

(b) if $\mathbf{Z}_n \xrightarrow{L_p} \mathbf{Z}$, then $\mathbf{Z}_n \xrightarrow{P} \mathbf{Z}$.
Hint:

$$E|\mathbf{Z}_n - \mathbf{Z}|^p \geq E[|\mathbf{Z}_n - \mathbf{Z}|^p 1\{|\mathbf{Z}_n - \mathbf{Z}| \geq \epsilon\}] \geq \epsilon^p P(|\mathbf{Z}_n - \mathbf{Z}| \geq \epsilon).$$

6. Show that $|\mathbf{Z}_n - \mathbf{Z}| \xrightarrow{P} 0$ is equivalent to $Z_{nj} \xrightarrow{P} Z_j$ for $1 \leq j \leq d$.
    Hint: Use (B.7.3) and note that $|Z_{nj} - Z_j|^2 \leq |\mathbf{Z}_n - \mathbf{Z}|^2$.

7. Let $U \sim \mathcal{U}(0, 1)$ and let $U_1 = 1$, $U_2 = 1\{U \in [0, \frac{1}{2})\}$, $U_3 = 1\{U \in [\frac{1}{2}, 1)\}$, $U_4 = 1\{U \in [0, \frac{1}{4})\}$, $U_5 = 1\{U \in [\frac{1}{4}, \frac{1}{2})\}, \ldots, U_n = 1\{U \in [m2^{-k}, (m + 1)2^{-k})\}$, where $n = m + 2^k$, $0 \leq m \leq 2^k$ and $k \geq 0$. Show that $U_n \xrightarrow{P} 0$ but $U_n \xrightarrow{a.s.} 0$.

8. Let $U \sim \mathcal{U}(0, 1)$ and set $U_n = 2^n 1\{U \in [0, \frac{1}{n})\}$. Show that $U_n \xrightarrow{a.s.} 0$, $U_n \xrightarrow{P} 0$, but $U_n \xrightarrow{L_p} 0$, $p \geq 1$, where L_p is defined in Problem B.7.5.

9. Establish B.7.9.

10. Show that Theorem B.7.5 implies Theorem B.7.4.

11. Suppose that as in Theorem B.7.7, $F_n(x) \to F(x)$ for all x, F is continuous, and strictly increasing so that $F^{-1}(\alpha)$ is unique for all $0 < \alpha < 1$. Show that

$$\sup\{|F_n^{-1}(\alpha) - F^{-1}(\alpha)| : \epsilon \leq \alpha \leq 1 - \epsilon\} \to 0$$

for all $\epsilon > 0$. Here $F_n^{-1}(\alpha) = \inf\{x : F_n(x) \geq \alpha\}$.
    Hint: Argue by contradiction.

Problems for Section B.8

1. If $h : R^d \to R^p$ and $\dot{h}(x) = Dh(x)$ is continuous in a sphere $\{x : |x - x_0| < \delta\}$, then for $|z| < \delta$

$$h(x_0 + z) = h(x_0) + \left(\int_0^1 \dot{h}(x_0 + uz)zdu \right) z^T.$$

Here the integral is a $p \times d$ matrix of integrals.

Hint: Let $g(u) = h(x_0 + uz)$. Then by the chain rule, $\dot{g}(u) = \dot{h}(x_0 + uz)z$ and

$$\int_0^1 \dot{h}(x_0 + uz)zdu = \int_0^1 \dot{g}(u)du = g(1) - g(0) = h(x_0 + z) - h(x_0).$$

2. If $h : R^d \to R$ and $\ddot{h}(x) = D^2 h(x)$ is continuous in a sphere $\{x : |x - x_0| < \delta\}$, then for $|z| < \delta$,

$$h(x_0 + z) = h(x_0) + \dot{h}(x_0)z + z^T \left[\int_0^1 \int_0^1 \ddot{h}(x_0 + uvz)vdudv \right] z.$$

Hint: Apply Problem B.8.1 to $h(x_0 + z) - h(x_0) - \dot{h}(x_0)z$.

3. Apply Problems B.8.1 and B.8.2 to obtain special cases of Taylor's Theorem B.8.12.

Problems for Section B.9

1. State and prove Jensen's inequality for conditional expectations.

2. Use Hoeffding's inequality (B.9.6) to establish Bernstein's inequality (B.9.5). Show that if $p = \frac{1}{2}$, the bound can be improved to $2 \exp\{-2n/\epsilon^2\}$.

3. Derive Hölder's inequality from Jensen's inequality with $k = 2$.
   Hint: For $(x, y) \in R^2$, consider $g(x, y) = \frac{|x|^r}{r} + \frac{|y|^s}{s}$, $\frac{1}{r} + \frac{1}{s} = 1$.

4. Show that if $k = 1$ and $g''(x)$ exists, then $g''(x) \geq 0$, all $x \in S$, and convexity are equivalent.

5. Show that convexity is equivalent to the convexity of $g(\alpha x + (1 - \alpha)y))$ as function of $\alpha \in [0, 1]$ for all x and y in S.

6. Use Problem 5 above to generalize Problem 4 above to the case $k > 1$.

7. Show that if $\frac{\partial}{\partial x_i} \frac{\partial}{\partial x_j} g^2(x)$ exists and the matrix $\left| \frac{\partial}{\partial x_i} \frac{\partial}{\partial x_j} g^2(x) \right|$ is positive definite, then g is strictly convex.

8. Show that

$$P(X \geq a) \leq \inf\{e^{-ta} E e^{tX} : t \geq 0\}.$$

   Hint: Use inequality (A.15.4).

9. Use Problem 8 above to prove Bernstein's inequality.

10. Show that the sum of (strictly) convex functions is (strictly) convex.

Problems for Section B.10

1. Verify that if A is snd, then A is ppd iff A is nonsingular.

2. Show that if S is the one-dimensional space $S = \{a\mathbf{e} : a \in R\}$ for $\mathbf{e}$ orthonormal, then the projection matrix onto S (B.10.17) is just $\mathbf{e}\mathbf{e}^T$.

3. Establish (B.10.15) and (B.10.16).

4. Show that $h - \Pi(h \mid \mathcal{L}) = \Pi(h \mid \mathcal{L}^\perp)$ using (B.10.14).

5. Establish (B.10.17).

B.12   NOTES

Notes for Section B.1.2

(1) We shall follow the convention of also calling $E(Y \mid \mathbf{Z})$ any variable that is equal to $g(\mathbf{Z})$ with probability 1.

Notes for Section B.1.3

(1) The definition of the conditional density (B.1.25) can be motivated as follows: Suppose that $A(\mathbf{x})$, $A(\mathbf{y})$ are small "cubes" with centers $\mathbf{x}$ and $\mathbf{y}$ and volumes $d\mathbf{x}$, $d\mathbf{y}$ and $p(\mathbf{x}, \mathbf{y})$ is continuous. Then $P[\mathbf{X} \in A(\mathbf{x}) \mid \mathbf{Y} \in A(\mathbf{y})] = P[\mathbf{X} \in A(\mathbf{x}), \mathbf{Y} \in A(\mathbf{y})]/P[\mathbf{Y} \in A(\mathbf{y})]$. But $P[\mathbf{X} \in A(\mathbf{x}), \mathbf{Y} \in A(\mathbf{y})] \approx p(\mathbf{x}, \mathbf{y})d\mathbf{x}\, d\mathbf{y}$, $P[\mathbf{Y} \in A(\mathbf{y})] \approx p_Y(\mathbf{y})d\mathbf{y}$, and it is reasonable that we should have $p(\mathbf{x} \mid \mathbf{y}) \approx P[\mathbf{X} \in A(\mathbf{x}) \mid \mathbf{Y} \in A(\mathbf{y})]/d\mathbf{x} \approx p(\mathbf{x}, \mathbf{y})/p_Y(\mathbf{y})$.

Notes for Section B.2

(1) We do not dwell on the stated conditions of the transformation Theorem B.2.1 because the conditions are too restrictive. It may, however, be shown that (B.2.1) continues to hold even if f is assumed only to be absolutely integrable in the sense of Lebesgue and K is any member of $\mathcal{B}^k$, the Borel σ-field on R^k. Thus, f can be any density function and K any set in R^k that one commonly encounters.

Notes for Section B.3.2

(1) In deriving (B.3.15) and (B.3.17) we are using the standard relations, $[\mathbf{AB}]^T = \mathbf{B}^T\mathbf{A}^T$, $\det[\mathbf{AB}] = \det \mathbf{A} \det \mathbf{B}$, and $\det \mathbf{A} = \det \mathbf{A}^T$.

Notes for Section B.5

(1) Both m.g.f.'s and c.f.'s are special cases of the Laplace transform ψ of the distribution of $\mathbf{U}$ defined by

$$\psi(\mathbf{z}) = E(e^{\mathbf{z}^T\mathbf{U}}),$$

where $\mathbf{z}$ is in the set of k tuples of complex numbers.

B.13   REFERENCES

ANDERSON, T. W., *An Introduction to Multivariate Statistical Analysis* New York: J. Wiley & Sons, 1958.

APOSTOL, T., *Mathematical Analysis*, 2nd ed. Reading, MA: Addison–Wesley,1974.

BARNDORFF–NIELSEN, O. E., AND D. P. COX, *Asymptotic Techniques for Use in Statistics* New York: Chapman and Hall, 1989.

BILLINGSLEY, P., *Probability and Measure*, 3rd ed. New York: J. Wiley & Sons, 1979, 1995.

BIRKHOFF, G., AND S. MacLANE, *A Survey of Modern Algebra*, rev. ed. New York: Macmillan, 1953.

BIRKHOFF, G., AND S. MacLANE, *A Survey of Modern Algebra*, 3rd ed. New York: MacMillan, 1965.

BREIMAN, L., *Probability* Reading, MA: Addison–Wesley, 1968.

CHUNG, K. L., *A Course in Probability Theory* New York: Academic Press, 1974.

DEMPSTER, A. P., *Elements of Continuous Multivariate Analysis* Reading, MA: Addison–Wesley, 1969.

DIEUDONNÉ, J., *Foundation of Modern Analysis*, v. 1, Pure and Applied Math. Series, Volume 10 New York: Academic Press, 1960.

DUNFORD, N., AND J. T. SCHWARTZ, *Linear Operators*, Volume 1, *Interscience* New York: J. Wiley & Sons, 1964.

FELLER, W., *An Introduction to Probability Theory and Its Applications*, Vol. II, 2nd ed. New York: J. Wiley & Sons, 1971.

GRIMMETT, G. R., AND D. R. STIRSAKER, *Probability and Random Processes* Oxford: Clarendon Press, 1992.

HALMOS, P. R., *An Introduction to Hilbert Space and the Theory of Spectral Multiplicity*, 2nd ed. New York: Chelsea, 1951.

HAMMERSLEY, J., "An extension of the Slutsky–Fréchet theorem," *Acta Mathematica, 87,* 243–247 (1952).

HOEFFDING, W., "Probability inequalities for sums of bounded random variables," *J. Amer. Statist. Assoc., 58,* 13–30 (1963).

LOÈVE, M., *Probability Theory*, Vol. 1, 4th ed. Berlin: Springer, 1977.

RAO, C. R., *Linear Statistical Inference and Its Applications*, 2nd ed. New York: J. Wiley & Sons, 1973.

ROCKAFELLAR, R. T., *Convex Analysis* Princeton, NJ: Princeton University Press, 1970.

ROYDEN, H. L., *Real Analysis*, 2nd ed. New York: MacMillan, 1968.

RUDIN, W., *Functional Analysis*, 2nd ed. New York: McGraw–Hill, 1991.

SKOROKHOD, A. V., "Limit theorems for stochastic processes," *Th. Prob. Applic., 1,* 261–290 (1956).

Appendix C

TABLES

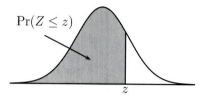

$$\Pr(Z \le z)$$

Table I The standard normal distribution

z	0.00	0.01	0.02	0.03	0.04	0.05	0.06	0.07	0.08	0.09
0.0	0.5000	0.5040	0.5080	0.5120	0.5160	0.5199	0.5239	0.5279	0.5319	0.5359
0.1	0.5398	0.5438	0.5478	0.5517	0.5557	0.5596	0.5636	0.5675	0.5714	0.5753
0.2	0.5793	0.5832	0.5871	0.5910	0.5948	0.5987	0.6026	0.6064	0.6103	0.6141
0.3	0.6179	0.6217	0.6255	0.6293	0.6331	0.6368	0.6406	0.6443	0.6480	0.6517
0.4	0.6554	0.6591	0.6628	0.6664	0.6700	0.6736	0.6772	0.6808	0.6844	0.6879
0.5	0.6915	0.6950	0.6985	0.7019	0.7054	0.7088	0.7123	0.7157	0.7190	0.7224
0.6	0.7257	0.7291	0.7324	0.7357	0.7389	0.7422	0.7454	0.7486	0.7517	0.7549
0.7	0.7580	0.7611	0.7642	0.7673	0.7704	0.7734	0.7764	0.7794	0.7823	0.7852
0.8	0.7881	0.7910	0.7939	0.7967	0.7995	0.8023	0.8051	0.8078	0.8106	0.8133
0.9	0.8159	0.8186	0.8212	0.8238	0.8264	0.8289	0.8315	0.8340	0.8365	0.8389
1.0	0.8413	0.8438	0.8461	0.8485	0.8508	0.8531	0.8554	0.8577	0.8599	0.8621
1.1	0.8643	0.8665	0.8686	0.8708	0.8729	0.8749	0.8770	0.8790	0.8810	0.8830
1.2	0.8849	0.8869	0.8888	0.8907	0.8925	0.8944	0.8962	0.8980	0.8997	0.9015
1.3	0.9032	0.9049	0.9066	0.9082	0.9099	0.9115	0.9131	0.9147	0.9162	0.9177
1.4	0.9192	0.9207	0.9222	0.9236	0.9251	0.9265	0.9279	0.9292	0.9306	0.9319
1.5	0.9332	0.9345	0.9357	0.9370	0.9382	0.9394	0.9406	0.9418	0.9429	0.9441
1.6	0.9452	0.9463	0.9474	0.9484	0.9495	0.9505	0.9515	0.9525	0.9535	0.9545
1.7	0.9554	0.9564	0.9573	0.9582	0.9591	0.9599	0.9608	0.9616	0.9625	0.9633
1.8	0.9641	0.9649	0.9656	0.9664	0.9671	0.9678	0.9686	0.9693	0.9699	0.9706
1.9	0.9713	0.9719	0.9726	0.9732	0.9738	0.9744	0.9750	0.9756	0.9761	0.9767
2.0	0.9772	0.9778	0.9783	0.9788	0.9793	0.9798	0.9803	0.9808	0.9812	0.9817
2.1	0.9821	0.9826	0.9830	0.9834	0.9838	0.9842	0.9846	0.9850	0.9854	0.9857
2.2	0.9861	0.9864	0.9868	0.9871	0.9875	0.9878	0.9881	0.9884	0.9887	0.9890
2.3	0.9893	0.9896	0.9898	0.9901	0.9904	0.9906	0.9909	0.9911	0.9913	0.9916
2.4	0.9918	0.9920	0.9922	0.9925	0.9927	0.9929	0.9931	0.9932	0.9934	0.9936
2.5	0.9938	0.9940	0.9941	0.9943	0.9945	0.9946	0.9948	0.9949	0.9951	0.9952
2.6	0.9953	0.9955	0.9956	0.9957	0.9959	0.9960	0.9961	0.9962	0.9963	0.9964
2.7	0.9965	0.9966	0.9967	0.9968	0.9969	0.9970	0.9971	0.9972	0.9973	0.9974
2.8	0.9974	0.9975	0.9976	0.9977	0.9977	0.9978	0.9979	0.9979	0.9980	0.9981
2.9	0.9981	0.9982	0.9982	0.9983	0.9984	0.9984	0.9985	0.9985	0.9986	0.9986
3.0	0.9987	0.9987	0.9987	0.9988	0.9988	0.9989	0.9989	0.9989	0.9990	0.9990
3.1	0.9990	0.9991	0.9991	0.9991	0.9992	0.9992	0.9992	0.9992	0.9993	0.9993
3.2	0.9993	0.9993	0.9994	0.9994	0.9994	0.9994	0.9994	0.9995	0.9995	0.9995
3.3	0.9995	0.9995	0.9995	0.9996	0.9996	0.9996	0.9996	0.9996	0.9996	0.9997
3.4	0.9997	0.9997	0.9997	0.9997	0.9997	0.9997	0.9997	0.9997	0.9997	0.9998

Table entry is probability at or below z.

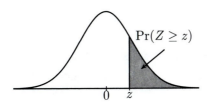

Table I′ Auxilliary table of the standard normal distribution

$\Pr(Z \geq z)$	.50	.45	.40	.35	.30	.25	.20	.15	.10
z	0	.126	.253	.385	.524	.674	.842	1.036	1.282

$\Pr(Z \geq z)$	.09	.08	.07	.06	.05	.04	.03	.025
z	1.341	1.405	1.476	1.555	1.645	1.751	1.881	1.960

$\Pr(Z \geq z)$	.02	.01	.005	.001	.0005	.0001	.00005	.00001
z	2.054	2.326	2.576	3.090	3.291	3.719	3.891	4.265

Entries in the top row are areas to the right of values in the second row.

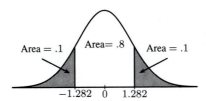

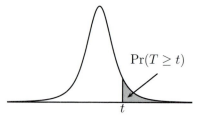

$$Pr(T \geq t)$$

t

Table II t distribution critical values

df	Right tail probability p									
	.25	.10	.05	.025	.02	.01	.005	.0025	.001	.0005
1	1.000	3.078	6.314	12.71	15.89	31.82	63.66	127.3	318.3	636.6
2	0.816	1.886	2.920	4.303	4.849	6.965	9.925	14.09	22.33	31.60
3	0.765	1.638	2.353	3.182	3.482	4.541	5.841	7.453	10.21	12.92
4	0.741	1.533	2.132	2.776	2.999	3.747	4.604	5.598	7.173	8.610
5	0.727	1.476	2.015	2.571	2.757	3.365	4.032	4.773	5.893	6.869
6	0.718	1.440	1.943	2.447	2.612	3.143	3.707	4.317	5.208	5.959
7	0.711	1.415	1.895	2.365	2.517	2.998	3.499	4.029	4.785	5.408
8	0.706	1.397	1.860	2.306	2.449	2.896	3.355	3.833	4.501	5.041
9	0.703	1.383	1.833	2.262	2.398	2.821	3.250	3.690	4.297	4.781
10	0.700	1.372	1.812	2.228	2.359	2.764	3.169	3.581	4.144	4.587
11	0.697	1.363	1.796	2.201	2.328	2.718	3.106	3.497	4.025	4.437
12	0.695	1.356	1.782	2.179	2.303	2.681	3.055	3.428	3.930	4.318
13	0.694	1.350	1.771	2.160	2.282	2.650	3.012	3.372	3.852	4.221
14	0.692	1.345	1.761	2.145	2.264	2.624	2.977	3.326	3.787	4.140
15	0.691	1.341	1.753	2.131	2.249	2.602	2.947	3.286	3.733	4.073
16	0.690	1.337	1.746	2.120	2.235	2.583	2.921	3.252	3.686	4.015
17	0.689	1.333	1.740	2.110	2.224	2.567	2.898	3.222	3.646	3.965
18	0.688	1.330	1.734	2.101	2.214	2.552	2.878	3.197	3.610	3.922
19	0.688	1.328	1.729	2.093	2.205	2.539	2.861	3.174	3.579	3.883
20	0.687	1.325	1.725	2.086	2.197	2.528	2.845	3.153	3.552	3.850
21	0.686	1.323	1.721	2.080	2.189	2.518	2.831	3.135	3.527	3.819
22	0.686	1.321	1.717	2.074	2.183	2.508	2.819	3.119	3.505	3.792
23	0.685	1.319	1.714	2.069	2.177	2.500	2.807	3.104	3.485	3.768
24	0.685	1.318	1.711	2.064	2.172	2.492	2.797	3.091	3.467	3.745
25	0.684	1.316	1.708	2.060	2.167	2.485	2.787	3.078	3.450	3.725
30	0.683	1.310	1.697	2.042	2.147	2.457	2.750	3.030	3.385	3.646
40	0.681	1.303	1.684	2.021	2.123	2.423	2.704	2.971	3.307	3.551
50	0.679	1.299	1.676	2.009	2.109	2.403	2.678	2.937	3.261	3.496
60	0.679	1.296	1.671	2.000	2.099	2.390	2.660	2.915	3.232	3.460
100	0.677	1.290	1.660	1.984	2.081	2.364	2.626	2.871	3.174	3.390
1000	0.675	1.282	1.646	1.962	2.056	2.330	2.581	2.813	3.098	3.300
∞	0.674	1.282	1.645	1.960	2.054	2.326	2.576	2.807	3.090	3.291
	50%	80%	90%	95%	96%	98%	99%	99.5%	99.8%	99.9%
					Confidence level C					

The entries in the top row are the probabilities of exceeding the tabled values. The left column gives the degrees of freedom.

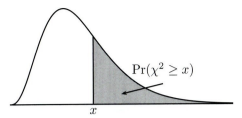

$$\Pr(\chi^2 \geq x)$$

x

Table III χ^2 distribution critical values

df	.25	.10	.05	.025	.02	.01	.005	.0025	.001	.0005
					Right tail probability p					
1	1.32	2.71	3.84	5.02	5.41	6.63	7.88	9.14	10.83	12.12
2	2.77	4.61	5.99	7.38	7.82	9.21	10.60	11.98	13.82	15.20
3	4.11	6.25	7.81	9.35	9.84	11.34	12.84	14.32	16.27	17.73
4	5.39	7.78	9.49	11.14	11.67	13.28	14.86	16.42	18.47	20.00
5	6.63	9.24	11.07	12.83	13.39	15.09	16.75	18.39	20.52	22.11
6	7.84	10.64	12.59	14.45	15.03	16.81	18.55	20.25	22.46	24.10
7	9.04	12.02	14.07	16.01	16.62	18.48	20.28	22.04	24.32	26.02
8	10.22	13.36	15.51	17.53	18.17	20.09	21.95	23.77	26.12	27.87
9	11.39	14.68	16.92	19.02	19.68	21.67	23.59	25.46	27.88	29.67
10	12.55	15.99	18.31	20.48	21.16	23.21	25.19	27.11	29.59	31.42
11	13.70	17.28	19.68	21.92	22.62	24.72	26.76	28.73	31.26	33.14
12	14.85	18.55	21.03	23.34	24.05	26.22	28.30	30.32	32.91	34.82
13	15.98	19.81	22.36	24.74	25.47	27.69	29.82	31.88	34.53	36.48
14	17.12	21.06	23.68	26.12	26.87	29.14	31.32	33.43	36.12	38.11
15	18.25	22.31	25.00	27.49	28.26	30.58	32.80	34.95	37.70	39.72
16	19.37	23.54	26.30	28.85	29.63	32.00	34.27	36.46	39.25	41.31
17	20.49	24.77	27.59	30.19	31.00	33.41	35.72	37.95	40.79	42.88
18	21.60	25.99	28.87	31.53	32.35	34.81	37.16	39.42	42.31	44.43
19	22.72	27.20	30.14	32.85	33.69	36.19	38.58	40.88	43.82	45.97
20	23.83	28.41	31.41	34.17	35.02	37.57	40.00	42.34	45.31	47.50
21	24.93	29.62	32.67	35.48	36.34	38.93	41.40	43.78	46.80	49.01
22	26.04	30.81	33.92	36.78	37.66	40.29	42.80	45.20	48.27	50.51
23	27.14	32.01	35.17	38.08	38.97	41.64	44.18	46.62	49.73	52.00
24	28.24	33.20	36.42	39.36	40.27	42.98	45.56	48.03	51.18	53.48
25	29.34	34.38	37.65	40.65	41.57	44.31	46.93	49.44	52.62	54.95
26	30.43	35.56	38.89	41.92	42.86	45.64	48.29	50.83	54.05	56.41
27	31.53	36.74	40.11	43.19	44.14	46.96	49.64	52.22	55.48	57.86
28	32.62	37.92	41.34	44.46	45.42	48.28	50.99	53.59	56.89	59.30
29	33.71	39.09	42.56	45.72	46.69	49.59	52.34	54.97	58.30	60.73
30	34.80	40.26	43.77	46.98	47.96	50.89	53.67	56.33	59.70	62.16
40	45.62	51.81	55.76	59.34	60.44	63.69	66.77	69.70	73.40	76.09
50	56.33	63.17	67.50	71.42	72.61	76.15	79.49	82.66	86.66	89.56
60	66.98	74.40	79.08	83.30	84.58	88.38	91.95	95.34	99.61	102.69
80	88.13	96.58	101.88	106.63	108.07	112.33	116.32	120.10	124.84	128.26
100	109.14	118.50	124.34	129.56	131.14	135.81	140.17	144.29	149.45	153.17

The entries in the top row are the probabilities of exceeding the tabled values. $p = \Pr(\chi^2 \geq x)$ where x is in the body of the table and p is in the top row (margin). df denotes degrees of freedom and is given in the left column (margin).

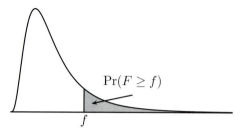

$\Pr(F \geq f)$

f

Table IV F distribution critical values

$\Pr(F \geq f)$	r_2	1	2	3	4	5	6	7	8	10	15
						r_1					
0.05	1	161	199	216	225	230	234	237	239	242	246
0.025		648	799	864	900	922	937	948	957	969	985
0.01		4052	4999	5403	5625	5764	5859	5928	5981	6056	6157
0.05	2	18.51	19.00	19.16	19.25	19.30	19.33	19.35	19.37	19.40	19.43
0.025		38.51	39.00	39.17	39.25	39.30	39.33	39.36	39.37	39.40	39.43
0.01		98.50	99.00	99.17	99.25	99.30	99.33	99.36	99.37	99.40	99.43
0.05	3	10.13	9.55	9.28	9.12	9.01	8.94	8.89	8.85	8.79	8.70
0.025		17.44	16.04	15.44	15.10	14.88	14.73	14.62	14.54	14.42	14.25
0.01		34.12	30.82	29.46	28.71	28.24	27.91	27.67	27.49	27.23	26.87
0.05	4	7.71	6.94	6.59	6.39	6.26	6.16	6.09	6.04	5.96	5.86
0.025		12.22	10.65	9.98	9.60	9.36	9.20	9.07	8.98	8.84	8.66
0.01		21.20	18.00	16.69	15.98	15.52	15.21	14.98	14.80	14.55	14.20
0.05	5	6.61	5.79	5.41	5.19	5.05	4.95	4.88	4.82	4.74	4.62
0.025		10.01	8.43	7.76	7.39	7.15	6.98	6.85	6.76	6.62	6.43
0.01		16.26	13.27	12.06	11.39	10.97	10.67	10.46	10.29	10.05	9.72
0.05	6	5.99	5.14	4.76	4.53	4.39	4.28	4.21	4.15	4.06	3.94
0.025		8.81	7.26	6.60	6.23	5.99	5.82	5.70	5.60	5.46	5.27
0.01		13.75	10.92	9.78	9.15	8.75	8.47	8.26	8.10	7.87	7.56
0.05	7	5.59	4.74	4.35	4.12	3.97	3.87	3.79	3.73	3.64	3.51
0.025		8.07	6.54	5.89	5.52	5.29	5.12	4.99	4.90	4.76	4.57
0.01		12.25	9.55	8.45	7.85	7.46	7.19	6.99	6.84	6.62	6.31
0.05	8	5.32	4.46	4.07	3.84	3.69	3.58	3.50	3.44	3.35	3.22
0.025		7.57	6.06	5.42	5.05	4.82	4.65	4.53	4.43	4.30	4.10
0.01		11.26	8.65	7.59	7.01	6.63	6.37	6.18	6.03	5.81	5.52
0.05	9	5.12	4.26	3.86	3.63	3.48	3.37	3.29	3.23	3.14	3.01
0.025		7.21	5.71	5.08	4.72	4.48	4.32	4.20	4.10	3.96	3.77
0.01		10.56	8.02	6.99	6.42	6.06	5.80	5.61	5.47	5.26	4.96
0.05	10	4.96	4.10	3.71	3.48	3.33	3.22	3.14	3.07	2.98	2.85
0.025		6.94	5.46	4.83	4.47	4.24	4.07	3.95	3.85	3.72	3.52
0.01		10.04	7.56	6.55	5.99	5.64	5.39	5.20	5.06	4.85	4.56
0.05	12	4.75	3.89	3.49	3.26	3.11	3.00	2.91	2.85	2.75	2.62
0.025		6.55	5.10	4.47	4.12	3.89	3.73	3.61	3.51	3.37	3.18
0.01		9.33	6.93	5.95	5.41	5.06	4.82	4.64	4.50	4.30	4.01
0.05	15	4.54	3.68	3.29	3.06	2.90	2.79	2.71	2.64	2.54	2.40
0.025		6.20	4.77	4.15	3.80	3.58	3.41	3.29	3.20	3.06	2.86
0.01		8.68	6.36	5.42	4.89	4.56	4.32	4.14	4.00	3.80	3.52

r_1 = numerator degrees of freedom, r_2 = denominator degrees of freedom.

INDEX